SUBSEA ENGINEERING HANDBOOK

SUBSEA ENGINEERING HANDBOOK

YONG BAI
QIANG BAI

AMSTERDAM • BOSTON • HEIDELBERG • LONDON
NEW YORK • OXFORD • PARIS • SAN DIEGO
SAN FRANCISCO • SINGAPORE • SYDNEY • TOKYO

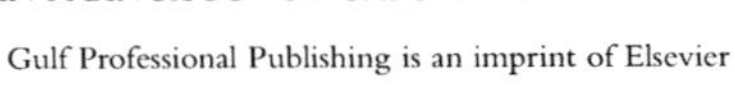

Gulf Professional Publishing is an imprint of Elsevier
225 Wyman Street, Waltham, MA 02451, USA
The Boulevard, Langford Lane, Kidlington, Oxford, OX5 1GB, UK

Notices

Knowledge and best practice in this field are constantly changing. As new research and experience broaden our understanding, changes in research methods, professional practices, or medical treatment may become necessary.

Practitioners and researchers must always rely on their own experience and knowledge in evaluating and using any information, methods, compounds, or experiments described herein. In using such information or methods they should be mindful of their own safety and the safety of others, including parties for whom they have a professional responsibility.

To the fullest extent of the law, neither the Publisher nor the authors, contributors, or editors, assume any liability for any injury and/or damage to persons or property as a matter of products liability, negligence or otherwise, or from any use or operation of any methods, products, instructions, or ideas contained in the material herein.

Library of Congress Cataloging-in-Publication Data

Bai, Yong.
Subsea engineering handbook / Yong Bai, Qiang Bai.
p. cm.
Includes bibliographical references and index.
ISBN 978-0-12-397804-2 (alk. paper)
1. Offshore structures–Handbooks, manuals, etc. 2. Offshore oil well drilling–Handbooks, manuals, etc. I. Bai, Qiang. II. Title.
TC1665.B355 2012
627'.98–dc23

2011052758

British Library Cataloguing-in-Publication Data

A catalogue record for this book is available from the British Library.

ISBN: 978-0-12-397804-2

For information on all Gulf Professional Publishing publications visit our Web site at www.elsevierdirect.com

Transferred to Digital Printing in 2012

Printed in USA

CONTENTS

PREFACE

MAY 2010

Subsea engineering is now a big discipline for the design, analysis, construction, installation and integrity management of subsea wellheads, trees, manifolds, jumpers, PLETS and PLEMs, etc. However, there is no book available that helps engineers understand the principles of subsea engineering.

This book is written for those who wish to become subsea engineers.

With the continuous encouragement of Mr. Ken McCombs of Elsevier, the authors spent a couple of years writing this book. The authors would like to thank those individuals who provided editing assistance (Ms. Lihua Bai & Ms. Shuhua Bai), initial technical writing for Chapters 1-4 (Mr. Youxiang Cheng), Chapters 6-8 (Mr. Xiaohai Song), Chapter 11 (Mr. Shiliang He), Chapter 5 (Mr. HongDong Qiao), Chapter 23 (Mr. Liangbiao Xu) and Chapter 27 (Mr. Mike Bian). They are employees of Offshore Pipelines & Risers (OPR) Inc. (bai@opr-inc.com, www.opr-inc.com, www.baiyongoe.com).

Thanks to all persons involved in reviewing the book, particularly Ms. Mohanambal Natarajan of Elsevier, who provided editory assistance.

We thank our families and friends for their support.

The first author would like to thank Zhejiang University for their support for publishing this book.

Prof. Yong Bai & Dr. Qiang Bai
Houston, USA

ABOUT THE AUTHORS

Professor Yong Bai is the president of Offshore Pipelines & Risers Inc. in Houston, and also the director of the Offshore Engineering Research Center at Zhejiang University. He has previously taught at Stavanger University in Norway where he was a professor of offshore structures. He has also worked with ABS as manager of the Offshore Technology Department and DNV as the JIP project manager.

Professor Yong Bai has also worked for Shell International E & P as a staff engineer. Through working at JP Kenny as manager of advanced engineering and at MCS as vice president of engineering, he has contributed to the advancement of methods and tools for the design and analysis of subsea pipelines and risers. Professor Bai is the author of the books *Marine Structural Design* and *Subsea Pipelines and Risers* and more than 100 papers on the design and installation of subsea pipelines and risers.

OPR has offices in Houston, Texas, USA; Kuala Lumpur, Malaysia; and Harbin, Beijing, and Shanghai, China. OPR is engaged in the design, analysis, installation, engineering, and integrity management of pipelines, risers, and subsea systems such as subsea wellheads, trees, manifolds, and PLET/PLEMs.

Dr. Qiang Bai has more than 20 years of experience in subsea/offshore engineering including research and engineering execution. He has worked at Kyushu University in Japan, UCLA, OPE, JP Kenny, and Technip. His experience includes various aspects of flow assurance and the design and installation of subsea structures, pipelines, and riser systems. Dr. Bai is coauthor of *Subsea Pipelines and Risers*.

LIST OF ABBREVIATIONS

A&R - Abandonment and recovery
AA - Anti agglomerate
AACE - Advancement of cost engineering
AAV - Annulus access valve
ACFM - Alternating current field measurement
AHC - Active heave compensation
AHV - Anchor handling vessel
AMV - Annulus master valve
APDU - Asphaltene precipitation detection unit
APV - Air pressure vessel
ASD - Allowable stress design
ASV - Annulus swab valve
AUV - Autonomous underwater vehicle
AWV - Annulus wing valve
B&C - Burial and coating
BM - Bending moment
BOPD - Barrels of oil per day
BR - Bend restrictor
C/WO - Completion and workover
CAPEX - Capital expenditures
CAPEX - Capital expenditures
CAT - Connector actuation tool
CCD - Charge-coupled device
CCO - Component change-out tool
CDTM - Control depth towing method
CFP – Cold flow pipeline
CG - Center of gravity
CI - Corrosion inhibitor
CII - Colloidal instability index
CIU - Chemical injection unit
CMC - Crown-mounted compensator
CoB - Cost of blowout
CoG - Center of gravity
CP - Cathodic protection
CPT - Compliant piled tower

CPT - Cone penetration test
CRA - Corrosion-resistant alloy
CV - Coefficient value
CVC - Pipeline end connector
CVI - Close visual inspection
DA - Diver assist
DCU - Dry completion unit
DDF - Deepdraft semi-submersible
DEG - Diethylene glycol
DFT - Dry film thickness
DGPS - Differential global positioning system
DH - Direct hydraulic
DHSV - Downhole safety valve
DOP - Dilution of position
DP - Dynamic positioning
DSS - Direct simple shear
DSV - Diving support vessel
EC - External corrosion
EDM - Electrical distribution module
EDP - Emergency disconnect package
EDU - Electrical distribution unit
EFAT - Extended factory acceptance test
EFL - Electric flying lead
EGL - Energy grade line
EH - Electrical heating
EI - External impact
EOS - Equation of state
EPCI - Engineering, procurement, construction and installation
EPU - Electrical power unit
EQD - Emergency quick disconnect
ESD - Emergency shutdown
ESP - Electrical submersible pump
FAR - Flexural anchor reaction
FAT - Factory acceptance test
FBE - Fusion bonded epoxy
FDM - Finite difference method
FE - Finite element
FEA - Finite element analysis
FEED - Front-end engineering design

FEM - Finite element Method
FMECA - Failure mode, effects, and criticality analysis
FOS - Factor of safety
FPDU -Floating production and drilling unit
FPS - Floating production system
FPSO - Floating production, storage and offloading
FPU - Floating production unit
FSHR - Free standing hybrid riser
FSO - Floating storage and offloading
FSV - Field support vessel
FTA - Fault tree analysis
GL - Guideline
GLL - Guideline-less
GoM - Gulf of Mexico
GOR - Gas/oil ratio
GPS - Global positioning system
GSPU - Polyurethane-glass syntactic
GVI - General visual inspection
HAZID - Hazard identification
HCLS - Heave compensated landing system
HCM - HIPPS control module
HCR - High collapse resistance
HDM -Hydraulic distribution module
HDPE - High density polyethylene
HFL - Hydraulic flying lead
HGL - Hydraulic grade line
HIPPS - High integrity pressure protection system
HISC - Hydrogen-induced stress cracking
HLV - Heavy lift vessel
HMI- Human machine interface
HP/HT - High pressure high temperature
HPU - Hydraulic power unit
HR - Hybrid riser
HSE - Health, safety, and environmental
HSP - Hydraulic submersible pump
HT - Horizontal tree
HTGC - High temperature gas chromatography
HXT - Horizontal tree
HXU - Heat exchanger unit

IA - Inhibitor availability
IBWM - International bureau of weights and measures
IC - Internal corrosion
ICCP - Impressed current cathodic protection
IE - Internal erosion
IMR - Inspection, maintenance, and repair
IPU - Integrated production umbilical
IRP - Inspection reference plan
IRR - Internal rate of return
ISA - Instrument society of America
ISO - International Organization for Standards
IWOCS – Installation and workover control system
JIC - Joint industry conference
JT - Joule Thompson
KI - Kinetic inhibitor
L/D - Length/diameter
LARS - Launch and recovery system
LBL - Long baseline
LC - Life cycle cost
LCWR - Lost capacity while waiting on rig
LDHI - Low dosage hydrate inhibitor
LFJ - Lower flexjoint
LOT - Linear override tool
LP - Low pressure
LPMV - Lower production master valve
LRP - Lower riser package
LWRP - Lower workover riser package
MAOP - Maximum allowable operating pressure
MASP - Maximum allowable surge pressure
MBR - Minimum bend radius
MCS - Master control station
MEG - Mono ethylene glycol
MF - Medium frequency
MIC - Microbiological induced corrosion
MMBOE - Million barrels of oil equivalent
MOPU - Mobile offshore drilling unit
MPI - Magnetic particle inspection
MPP - Multiphase pump
MQC - Multiple quick connector

MRP - Maintenance reference plan
MTO - Material take-off
NAS - National aerospace standard
NDE - None destructive examination
NDT - Nondestructive testing
NGS - nitrogen generating system
NPV - Net present value
NS - North sea
NTNU - The Norwegian university of science and technology
O&M - Operations and maintenance
OCR - Over consolidation ratio
OCS - Operational Control System
OHTC - Overall heat transfer coefficient
OPEX - Operation expenditures
OREDA - Offshore reliability data
OSI - Oil States Industries
OTC - Offshore Technology conference
PAN - Programmable acoustic navigator
PCP - Piezocone penetration
PGB - Production guide base
PHC - Passive heave compensator
PhS - Phenolic syntactic
PIP - Pipe in pipe
PLC - Programmable logic controller
PLEM - Pipeline end manifold
PLET - Pipeline end termination
PLL - Potential loss of life
PMV - Production master valve
PMV - Production master valve
PoB - Probability of blowout
POD - Point of disconnect
PP - Polypropylene
PPF - Polypropylene foam
PSCM - Procurement and supply chain management
PSV - Production swab valve
PT - Pressure transmitter
PTT - Pressure/Temperature Transducer
PU - Polyurethane
PWV - Production wing valve

QC - Quality control
QE - Quality engineer
QP - Quality program
QRA - Quantitative risk assessment
RAO -Response amplitude operator
RBD - Reliability block diagram
RBI - Risk-based inspection
RCDA - Reliability-centered design analysis
RCMM - Reliability capability maturity model
REB - Reverse end bearing
ROT - Remote operated tool
ROV - Remote operated vehicle
RPPF - Polypropylene-reinforced foam combination
RSV - ROV support vessel
SAM - Subsea accumulator module
SAMMB - Subsea accumulator module mating block
SBP - Sub-bottom profiler
SCF - Stress concentration factor
SCM - Subsea control module
SCMMB - Subsea control module mounting base
SCR - Steel catenary riser
SCSSV - Surface controlled subsurface safety valve
SDA - Subsea distribution assembly
SDS - Subsea distribution system
SDU - Subsea distribution unit
SEM - Subsea electronics module
SEP - Epoxy syntactic
SEPLA - Suction embedded plate anchor
SIS - Safety instrumented system
SIT - Silicon intensified target
SIT - System integration test
SLEM - Simple linear elastic model
SPCS - Subsea production control system
SPCU - Subsea production communication unit
SPS - Subsea production system
SPU - Polyurethane-syntactic
SSC - Sulfide stress cracking
SSCC - Stress corrosion cracking
SSP - Subsea processing

SSS - Side-scan sonar
SSTT - Subsea test tree
SU - Separator unit
SUTA - Subsea umbilical termination assembly
SV - Satellite vehicle
TDP - Touchdown point
TDS - Total dissolved solid
TDU - Tool deployment unit
TDZ - Touchdown zone
TEG - Triethylene glycol
TFL - Through-flowline
TGB - Temporary guide base
THI - Thermodynamic inhibitor
TLP - Tension leg platform
TMGB - Template-mounted guide base
TMS - Tether management system
TPPL - Total plant peak load
TPRL - Total plant running load
TRT - Tree running tool
TT - Temperature transmitter
TTF - Time to failure
TTR - Top tensioned riser
TUFFP - Tulsa university fluid flow project
TUTA - Topside umbilical termination assembly
TVD - True vertical depth
TWI - Thermodynamic wax inhibitor
UFJ - Upper flexjoint
UHF - Ultra high frequency
UPC - Pull out capacity
UPMV - Upper production master valve
UPS - Uninterruptible power supply
USBL - Ultra short baseline
USV - Underwater safety valve
UTA - Umbilical termination assembly
UTH - Umbilical termination head
UU - Unconsolidated, undrained
VG - Vetco Gray
VIM - Vortex Induced Motion
VIT - Vacuum insulated tubing

VIV - Vortex induced vibration
VRU - Vertical reference unit
VT - Vertical tree
VXT - Vertical Xmas tree
WA - West africa
WAT - Wax appearance temperature
WBS - Work breakdown structure
WD - Water depth
WHI - Wellhead growth index
WHP - Wellhead platform
WHU - Wellhead unit
WS - Winter storm
WSD - Working stress design
XLPE - Cross linked polyethylene
XOV- Crossover valve

PART One

Subsea Production Systems

Overview of Subsea Engineering

Contents

1.1. INTRODUCTION

The world's energy consumption has increased steadily since the 1950s. As shown in Figure 1-1, the fossil fuels (oil, natural gas, and coal) still amount to 80% of the world's energy consumption even though a considerable number of initiatives and inventions in the area of renewable energy resources have

Subsea Engineering Handbook
ISBN 978-0-12-397804-2, doi:10.1016/B978-0-12-397804-2.00001-1

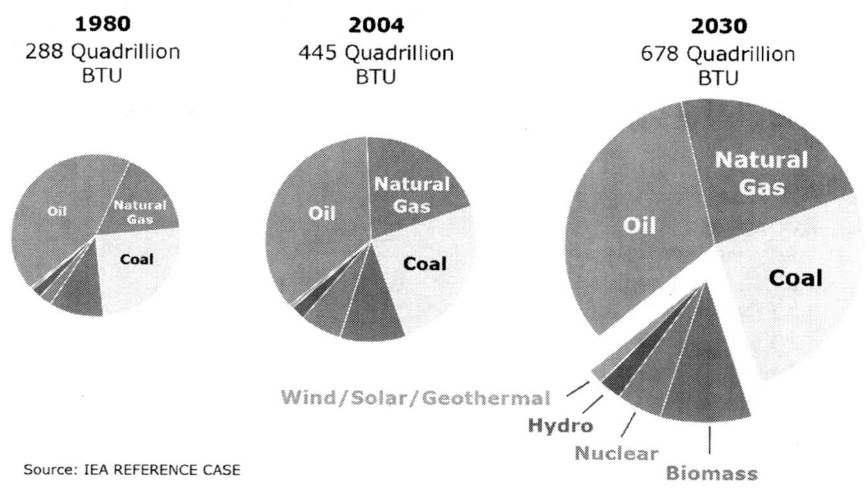

Figure 1-1 Coal, Oil, and Natural Gas Consumption [1]

decreased their use. The rapid rises in crude oil prices during the late 2000s is a response to increasing demand for oil and gas. Of the fossil fuels consumed, almost 80% are oil and gas; therefore, the production of oil and gas is of major importance to the stability of the world's energy supply.

The offshore oil and gas industry started in 1947 when Kerr-McGee completed the first successful offshore well in the Gulf of Mexico (GoM) off Louisiana in 15 ft (4.6 m) of water [2]. The concept of subsea field development was suggested in the early 1970s by placing wellhead and production equipment on the seabed with some or all components encapsulated in a sealed chamber [3]. The hydrocarbon produced would then flow from the well to a nearby processing facility, either on land or on an existing offshore platform. This concept was the start of subsea engineering, and systems that have a well and associated equipment below the water surface are referred as *subsea production systems*. Figure 1-2 shows the number of shallow and deepwater subsea completions in the GoM from 1955 to 2005. Subsea completions in less than 1,000 ft (305 m) water depths are considered to be shallow-water completions, whereas those at depths greater than 1,000 ft (305 m) are considered to be deepwater completions. In the past 40 years, subsea systems have advanced from shallow-water, manually operated systems into systems capable of operating via remote control at water depths of up to 3,000 meters (10,000 ft).

With the depletion of onshore and offshore shallow-water reserves, the exploration and production of oil in deep water has become a challenge to

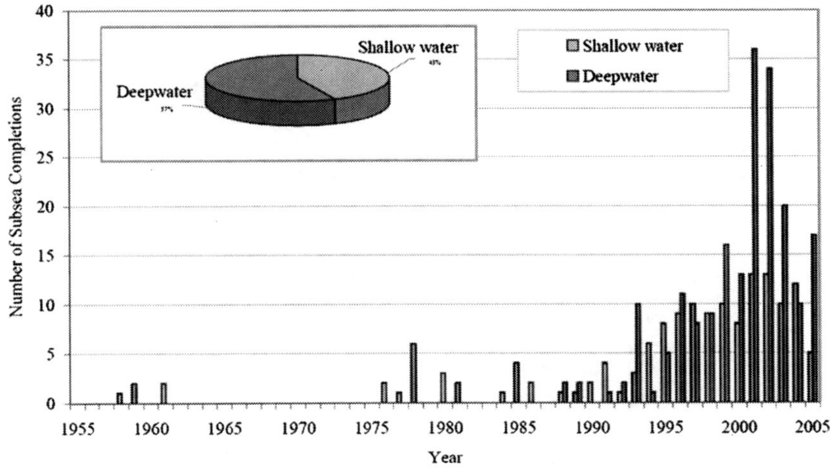

Figure 1-2 Number of Shallow and Deepwater Subsea Completions Each Year from 1955 to 2005 [4]

the offshore industry. Offshore exploration and production of oil and gas are advancing into deeper waters at an increasing pace. Figure 1-3 shows the maximum water depth of subsea completions installed each year in the GoM. Figure 1-4 illustrates offshore oil production trends in the GoM from shallow and deep water. Offshore oil production from deep water has

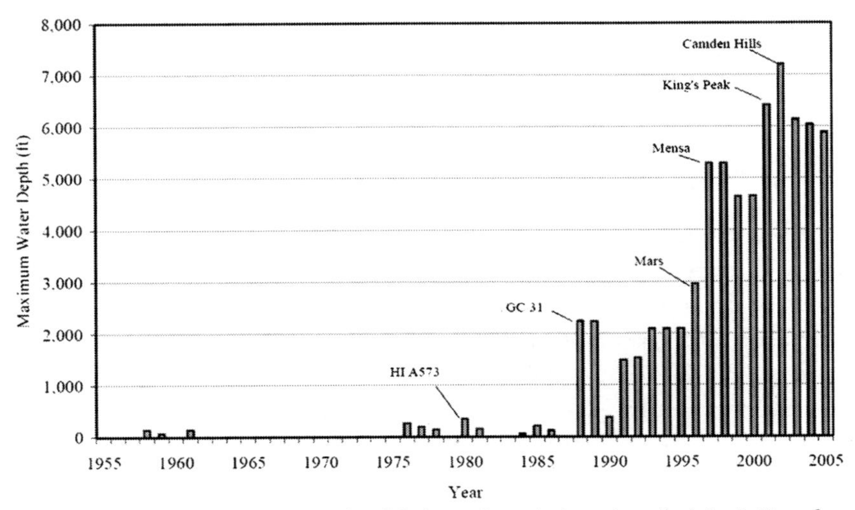

Figure 1-3 Maximum Water Depth of Subsea Completions Installed Each Year from 1955 to 2005 [4]

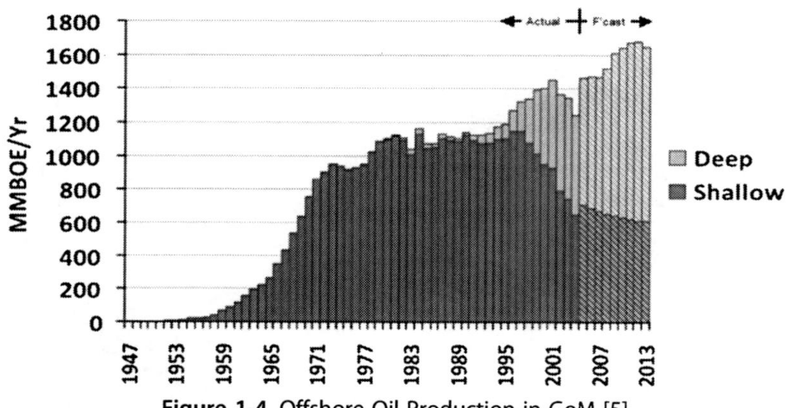

Figure 1-4 Offshore Oil Production in GoM [5]

increased sharply since 1995, starting at approximately 20 million barrels of oil equivalent (MMBOE) per year from deep water.

The subsea technology used for offshore oil and gas production is a highly specialized field of application that places particular demands on engineering. The subsea production system carries some unique aspects related to the inaccessibility of the installation and its operation and servicing. These special aspects make subsea production a specific engineering discipline. This book will discuss the topics of subsea engineering in four parts:

Part 1: Subsea Production Systems

Part 2: Flow Assurance and System Engineering

Part 3: Subsea Structures and Equipment

Part 4: Subsea Umbilicals, Risers, and Pipelines.

1.2. SUBSEA PRODUCTION SYSTEMS

A subsea production system consists of a subsea completed well, seabed wellhead, subsea production tree, subsea tie-in to flowline system, and subsea equipment and control facilities to operate the well. It can range in complexity from a single satellite well with a flowline linked to a fixed platform, FPSO (Floating Production, Storage and Offloading), or onshore facilities, to several wells on a template or clustered around a manifold that transfer to a fixed or floating facility or directly to onshore facilities.

As the oil and gas fields move further offshore into deeper water and deeper geological formations in the quest for reserves, the technology of drilling and production has advanced dramatically. Conventional techniques

Figure 1-5 All Segments of a Subsea Production System [6]

restrict the reservoir characteristics and reserves that can be economically exploited in the deep waters now being explored. The latest subsea technologies have been proven and formed into an engineering system, namely, the subsea production system, which is associated with the overall process and all the equipment involved in drilling, field development, and field operation, as shown in Figure 1-5. The subsea production system consists of the following components:

- Subsea drilling systems;
- Subsea Christmas trees and wellhead systems;
- Umbilical and riser systems;
- Subsea manifolds and jumper systems;
- Tie-in and flowline systems;
- Control systems;
- Subsea installation.

Figure 1-6 illustrates the detailed relationship among the major components of a subsea production system.

Most components of the subsea production system will be described in the chapters in Part 1, while the components of subsea structures and subsea equipment will be the focus of other parts of this book.

1.2.1. Field Architecture

Subsea production systems are generally arranged as shown in Figure 1-7. Some subsea production systems are used to extend existing platforms. For example, the geometry and depth of a reservoir may be such that a small

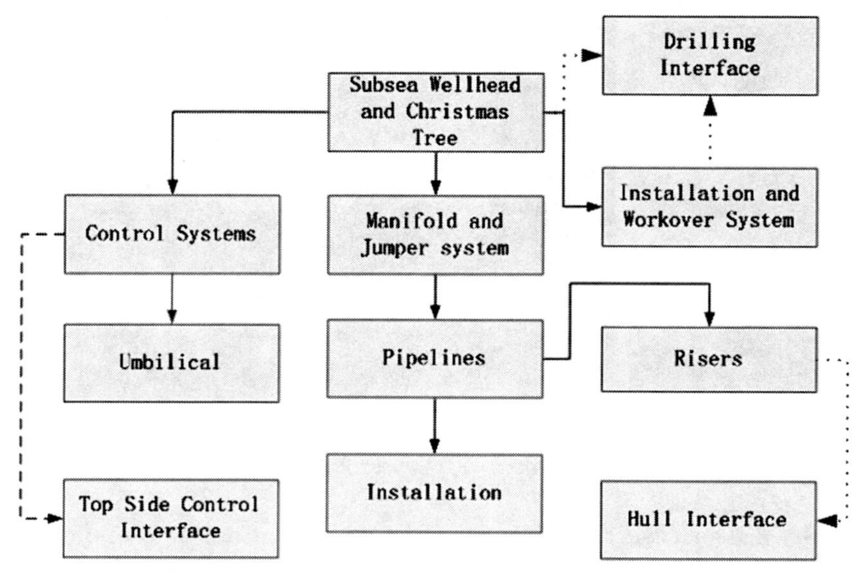

Figure 1-6 Relationship among the Major Components of a Subsea Production System

Figure 1-7 Typical Subsea Production System with Wet Tree [7]

section cannot be reached easily from the platform using conventional directional drilling techniques or horizontal wells. Based on the location of the tree installation, a subsea system can be categorized as a dry tree production system or a wet tree production system. Water depth can also impact subsea field development. For the shallower water depths, limitations on subsea development can result from the height of the subsea structures. Christmas trees and other structures cannot be installed in water depths of less than 30 m (100 ft). For subsea development in water depths less than 30 m (100 ft), jacket platforms consisting of dry trees can be used.

The goal of subsea field development is to safely maximize economic gain using the most reliable, safe, and cost-effective solution available at the time. Even though wet well systems are still relatively expensive, their attraction in reducing overall capital expenditures has already been made clear. Subsea tie-backs are becoming popular in the development of new oil and gas reserves in the 21st century. With larger oil and gas discoveries becoming less common, attention has turned to previously untapped, less economically viable discoveries.

In subsea field development, the following issues should be considered:

- Deepwater or shallow-water development;
- Dry tree or wet tree;
- Stand alone or tie-back development;
- Hydraulic and chemical units;
- Subsea processing;
- Artificial lift methods;
- Facility configurations (i.e., template, well cluster, satellite wells, manifolds).

The advantages, disadvantages, and limitations of the above issues will be described in the relevant sections of Chapter 2, which covers field architecture.

1.2.2. Distribution Systems

The subsea system is associated with the overall process and all equipment involved in the arrangement. It is designed in such a way that safety, environment protection, and flow assurance and reliability are taken into consideration for all subsea oil and gas exploitation. Subsea distribution systems consist of a group of products that provide communication between subsea controls and topside controls for all equipment via an umbilical system.

Subsea distribution systems may include, but not be limited to, the following major components [8]:

- Topside umbilical termination assembly (TUTA);
- Subsea accumulator module (SAM);
- Subsea umbilical termination assembly (SUTA), which includes:
  - Umbilical termination head (UTH);
  - Hydraulic distribution manifold/module (HDM);
  - Electric distribution manifold/module (EDM);
  - Flying leads.
- Subsea distribution assembly (SDA);
- Hydraulic flying leads (HFLs);
- Electric flying leads (EFLs);
- Multiple quick connector (MQC);
- Hydraulic coupler;
- Electrical connector;
- Logic caps.

The advantages, disadvantages, and limitations of the above components will be described in the relevant sections of the chapter on subsea distribution systems, Chapter 3.

1.2.3. Subsea Surveys

The subsea survey for positioning and soil investigation is one of the main activities for subsea field development. As part of the planned field development, a detailed geophysical and geotechnical field development survey together with soil investigation is performed. The purpose of the survey is to identify the potential man-made hazards, natural hazards, and engineering constraints of a proposed subsea field area and pipeline construction; to assess the potential impact on biological communities; and to determine the seabed and sub-bottom conditions. In Chapter 4, the following issues related to subsea surveys are discussed:

- Establishing vertical route profiles, a contour plan, and the seabed's features, particularly any rock outcrops or reefs;
- Obtaining accurate bathymetry, locating all obstructions, and identifying other seabed factors that may affect the development of the selected subsea field area including laying, spanning, and stability of the pipeline;
- Carrying out a geophysical survey of the selected subsea field and route to define the shallow sub-seabed geology;

- Carrying out geotechnical sampling and laboratory testing in order to evaluate precisely the nature and mechanical properties of soils at the selected subsea field area and along the onshore and offshore pipelines and platform locations;
- Locating existing subsea equipment (e.g., manifold, jumper, and subsea tree), pipelines, and cables, both operational and redundant, within the survey corridor;
- Determining the type of subsea foundation design that is normally used for subsea field development.

1.2.4. Installation and Vessels

The development of subsea production systems requires specialized subsea equipment. The deployment of such equipment requires specialized and expensive vessels, which need to be equipped with diving equipment for relatively shallow equipment work, and robotic equipment for deeper water depths. Subsea installation refers to the installation of subsea equipment and structures in an offshore environment for the subsea production system. Installation in an offshore environment is a dangerous activity, and heavy lifting is avoided as much as possible. This is achieved fully by subsea equipment and structures that are transmitted to the installation site by installation vessels.

Subsea installation can be divided into two parts: installation of subsea equipment and installation of subsea pipelines and subsea risers. Installation of subsea equipment such as trees and templates can be done by a conventional floating drilling rig, whereas subsea pipelines and subsea risers are installed by an installation barge using S-lay, J-lay, or reel lay. The objective of Chapter 5 is to review existing vessels used for the installation of subsea equipment such as trees, manifolds, flowlines, and umbilicals. This includes special vessels that can run the trees and rigless installation. Subsea equipment to be installed is categorized based on weight, shapes (volume versus line type), dimensions, and water depth (deep versus shallow).

1.2.5. Cost Estimation

When considering a subsea system as a development option for a specific reservoir and number of wells required, the subsea cost is relatively flat with increasing water depth. For the rigid platform case, however, costs increase rapidly with water depth. Therefore, deeper water tends to favor the use of subsea systems. Conversely, for a given water depth and location, platform costs are less sensitive to an increasing number of wells; well drilling from a platform is relatively inexpensive, and the platform structure cost is governed more by water depth, process requirements, and environment. The use of mobile drilling units for subsea wells increases drilling costs. Therefore, situations where a relatively small number of wells are needed favor the use of a subsea system.

Subsea costs refer to the cost of the whole subsea project and generally include capital expenditures (CAPEX) and operating expenditures (OPEX). CAPEX is the total amount of investment necessary to put a project into operation and includes the cost of initial design, engineering, construction, and installation. OPEX is the expenses incurred during the normal operation of a facility, or component after the installation, including labor, material, utilities, and other related expenses. OPEX contains operational costs, maintenance costs, testing costs, and other related costs. Chapter 6 covers cost estimates in detail.

1.2.6. Subsea Control

The subsea production control system is defined as the control system operating a subsea production system during production operations according to ISO 13628-6 [9]. The subsea control system is the heart of any subsea production system, and it is a relatively low-cost item compared to the cost of drilling, line pipe, installation, etc. Therefore, control systems are usually low on the list of initial project priorities. However, ignoring the complexity, the number of components and interfaces can lead to problems with installation and commissioning and to long-term reliability issues.

In the chapter on subsea control, Chapter 7, the principles and characteristics of subsea production control systems are explained and the advantages, disadvantages, and limitations are compared. The government regulations, industry codes, recommended practices, and environmental specifications that apply to subsea control systems are detailed.

1.2.7. Subsea Power Supply

Power supply is a key factor in subsea processing. The subsea power supply is an important component in the systems necessary for processing the well stream at the seabed close to the wells. Not having the power supply system in place can stop the development of subsea processing.

Chapter 8 focuses on the following three main areas:

- Electrical power unit (EPU);
- Uninterruptible power supply (UPS);
- Hydraulic power unit (HPU).

The power supply system's components and technologies are also described in Chapter 8.

1.2.8. Project Execution and Interfaces

The success of any project depends significantly on project execution. Project execution will allow for timely corrective action or redirection of the project. Once the project execution plan has been defined, a formal process of regular reports and reviews is required. Project execution is relevant at all stages of a project but the issues become more intense and complex as activities increase in number, diversity, and geographical spread. The project manager must set the expectation that the project management team understand its project execution system and the quality of data available. Project execution does not have its own momentum and it is, therefore, critical that it be proactively driven by the project manager.

The subject of Chapter 9 is to provide a clear understanding of the requirements for subsea project execution and interfaces with the accumulated knowledge and experience of project managers to guide all involved parties in managing project activities. Additionally, this chapter addresses the challenges of creating seamless interfaces and establishes the methodology and tools for identifying interface issues between the various functional groups.

1.3. FLOW ASSURANCE AND SYSTEM ENGINEERING

System engineering discipline activities are broadly classified into three primary service areas:

- Production system design;
- System integration;
- Equipment application and development.

Figure 1-8 shows the details of these three service areas.

1.3.1. Subsea Operations

After the production system has been installed, numerous operations are in place to ensure safe and pollution-free operations and support the continued

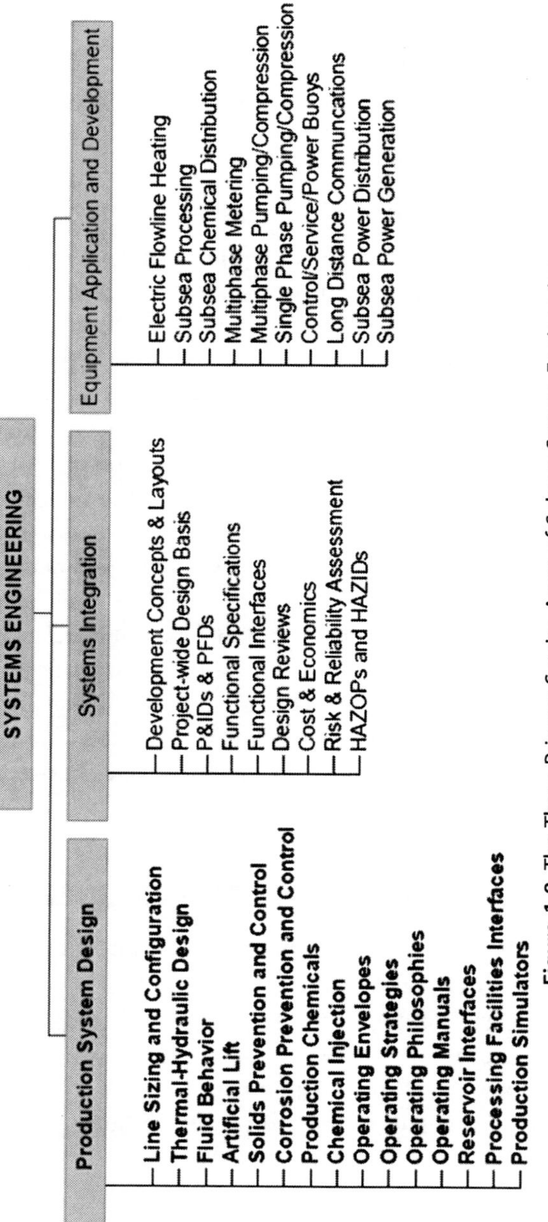

Figure 1-8 The Three Primary Service Areas of Subsea System Engineering

flow of hydrocarbons. The following are typical of postinstallation operations:

- Commissioning and start-up (start-up could be "cold" or "hot");
- Normal operations;
- Production processing;
- Chemical injection;
- Routine testing;
- Maintenance and repairs (remotely operated vehicle [ROV], routine surface);
- Emergency shutdown;
- Securing facilities (e.g., from extreme weather events);
- Intervention.

In many cases the technology and techniques applied to support production activities in deepwater are similar in scope to developments in shallow water. Deep water does add a level of complexity to the project, particularly subsea developments, since the facilities may be located remote from the control (host) facility and not readily accessible. For example, a workover may require a dedicated riser and control system, as well as a deepwater capable rig and all of the support that comes with the drilling unit or even a specialized intervention vessel. A significant amount of work is necessary for proper planning, simulations (steady-state and transient), design, testing, and system integration before the deepwater development moves forward.

1.3.2. Commissioning and Start-Up

Operations typically begin with systems integration testing (SIT) at a shore base or vendor or manufacturer's facility. Particularly for subsea projects, the remote tools installation that will be used to make connections (for example, ROV) will perform tests that simulate the actual installation. Mobilization to and the start-up of installation at the offshore location can involve a large number of vessels, including the drilling unit, support boats, derrick barge, transport barges and tugs, pipelay vessels, ROVs, and divers.

Production from the well at this point will include completion fluids and reservoir fluids. These may be flared/burned, treated, and discharged overboard, or transported to shore for disposal at an approved location. The cleanup phase of bringing a well/field on line typically last 2 to 5 days.

1.3.3. Production Processing

Production processing equipment is generally the same for the both shallow and deepwater developments. The production system may involve several separators, a series of safety valves, treaters, compressors, pumps, and associated piping. For deepwater facilities, the production system may be designed to process higher rates of flow. These could include production from multiple developments commingled at a common host facility.

The main surface production processing system components might involve crude oil separation, water injection equipment, gas compression, chemical injection, control systems for subsea production equipment, and associated piping. The processing system varies little from other development concepts (for example, a fixed platform serving as a host for subsea development). One area that does differ is the need to account for vessel motion that can be induced by environmental forces on these floating production facilities. In these conditions, production separators require specialized designs.

1.3.4. Chemicals Injection

Fluid problems in deepwater are critical issues (such as colder seabed temperatures, produced water, condensates, paraffin, and asphaltene contents in the oil) that can compromise the viability of a development project. To remedy that concern, chemicals are being increasingly relied on for production assurance. The use of chemicals in offshore oil production processes is not a new approach. Some of the chemicals used are corrosion inhibitors, workover/packer fluids (weighted clear fluids, bromides, chlorides, etc.), hydrate and paraffin inhibitors, defoamers, solvents (soaps, acids), glycol, and diesel. These chemicals are typically used for batch treatments, small-volume continuous injections, and remedial treatments such as workover operations. Material safety data sheets are required for all chemicals used offshore.

Corrosion inhibitors are used to protect carbon-steel components of the production systems that are wetted by the produced fluids. Material selection is a critical factor in the proper design of a production system, requiring information about the composition of the produced fluids.

1.3.4.1. Hydrate Inhibition

Hydrate inhibition is normally associated with batch treatments for the processes of start-up and shut-down (planned or unplanned). Continuous injection also occurs when there is induced cooling likely due to chokes and the natural cooling of pipelines by the cold ambient temperatures of the seabed. Methanol is one of the most common hydrate inhibitors used, particularly for subsea wells and in arctic regions where rapid cooling of the produced fluid flow (gas and water) can cause hydrate formation. Methanol is injected into the tree and sometimes downhole just above the subsurface safety valve while the fluids are hot. Some subsea developments in the deepwater GoM area inject methanol at rates of 20% to 40% of the water production rate. In the chapter of Hydrates, Chapter 15, the characteristics and formation of hydrates in subsea production systems are detailed. The solution methodology and hydrate control designs are summarized.

1.3.4.2. Paraffin Inhibitors

Paraffin inhibitors are used to protect the wellbore, production tree, and subsea pipelines/flowlines from plugging. The injection of these chemical inhibitors is dependent on the composition of the produced fluids. Injection can occur continuously at the tree, pipeline, manifold, and other critical areas while the production flow is hot, and to batch treatments at production start-up and shut-down processes. The wax content, pour point, and other factors are determined prior to beginning production to determine the chemical(s) needed, if any, and the best method for treatment. For a 10,000-BOPD (Barrels of Oil Per Day) well, the paraffin inhibitor could be injected at a rate of 30,000 gal per year (enough to ensure a 200-ppm concentration in the produced fluid flow).

1.3.4.3. Asphaltene Inhibitors

Asphaltene inhibitors are injected in the same manner as other inhibitors, but on a continuous basis. Asphaltenes can form in the production system as the pressure declines to near the bubble point.

Most development projects require one or all of these chemical inhibitors to avoid produced fluid problems. Efforts are under way to improve the performance of the inhibiting chemicals and to reduce the toxicity of the chemicals. In the chapter of Was and Asphaltenes, Chapter 16, the characteristics and formation of wax and asphaltenes in subsea production systems are detailed. The solution methodology and control designs are summarized.

1.3.5. Well Testing

Flow testing is done to confirm the producibility of the reservoir and to locate any boundary effects that could limit long-term production. In some instances, an extended well test may be necessary to confirm the development potential. A well test could last for several days to a month. For an extended test, the actual production time (well flowing) is typically less than one-half of the total test time. Significant data are gathered about the system from the pressure build-up stage of a well test. Oil recovered as part of the well test will be stored and reinjected, burned, or transported to shore for sales or disposal; gas is normally flared during the test.

1.3.6. Inspection and Maintenance

Facilities and pipelines require periodic inspections to ensure that no external damage or hazards are present that will affect the system's integrity. Unlike the shallow-water platform and subsea completions where diver access is possible, a deepwater system requires the use of ROVs for surveys and some repairs. For floating systems such as the TLP (Tension Leg Platform), the survey would examine the tendons as well as the hull and production riser.

Inspections of other systems would investigate the mooring system components as well as the production components (trees if subsea, pipelines, risers, umbilical, manifold, etc.). Many of the components of subsea equipment are modular, with built-in redundancy to expedite retrievals in the event of a failure. Mobilization of a drilling rig or specialized intervention vessel would be required for intervention into any of the subsea systems. If the production equipment is surface based, the maintenance, retrieval, and repair would be similar in scope to the conventional fixed platforms.

1.4. SUBSEA STRUCTURES AND EQUIPMENT

1.4.1. Subsea Manifolds

Subsea manifolds have been used in the development of oil and gas fields to simplify the subsea system, minimize the use of subsea pipelines and risers, and optimize the flow of fluid in the system. The manifold, as shown in Figure 1-9, is an arrangement of piping and/or valves designed to combine, distribute, control, and often monitor fluid flow. Subsea manifolds are installed on the seabed within an array of wells to gather production or to inject water or gas into wells. The numerous types of manifolds range from a simple pipeline end manifold (PLEM/PLET) to large structures such as a subsea process system. The manifold may be anchored to the seabed with piles or skirts that penetrate the mudline. Size is dictated by the number of the wells and throughput, as well as how the subsea wells are integrated into the system.

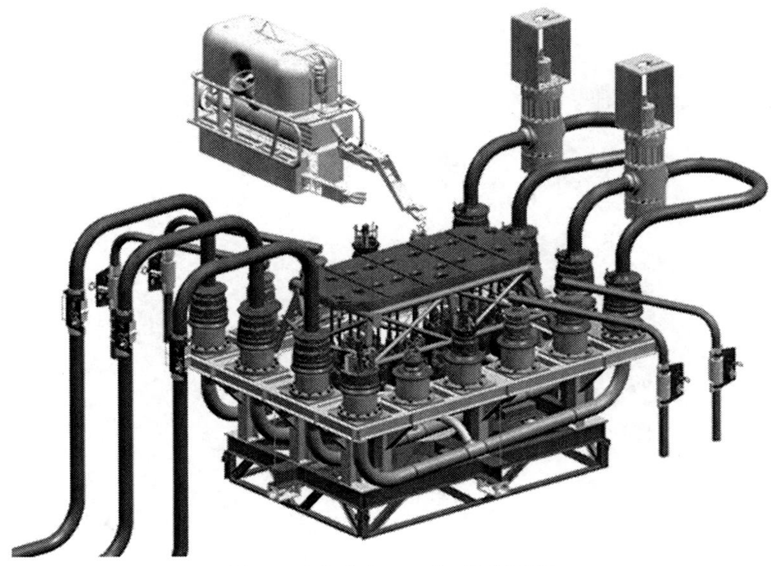

Figure 1-9 Subsea Manifold [10]

1.4.2. Pipeline Ends and In-line Structures

Pipeline end terminations (PLET)/pipeline end manifold (PLEM), and in-line structure (ILS) are subsea structures designed to attach the pipeline end and then lowered to the seabed in the desired orientation. The PLET/PLEM is located at the end of a subsea pipeline, while the inline structure is located in the middle of the pipeline. The design and installation of PLET/ILS include first-end, middle, and second-end options. The components of them may include from a single hub with manual isolation valve, to two or three hubs with ROV actuated valves, chemical injection, pig launching capabilities and more. The foundation of PLET/ILS may be a mudmat, or a single suction pile. A rigid or flexible jumper is utilized to tie-in the PLET/ILS to the other subsea structures e.g. tree, manifold, or other PLET/PLEM. Figure 1-10 shows a subsea PLET is ready for installation on the J-lay tower of a pipe-lay vessel.

1.4.3. Jumpers

In subsea oil/gas production systems, a subsea jumper, as shown in Figure 1-11, is a short pipe connector that is used to transport production fluid between two subsea components, for example, a tree and a manifold,

Figure 1-10 Subsea PLET *(Courtesy Shell)*

a manifold and another manifold, or a manifold and an export sled. It may also connect other subsea structures such as PLEM/PLETs and riser bases. In addition to being used to transport production fluid, a jumper can also be used to inject water into a well. The offset distance between the components (such as trees, flowlines, and manifolds) dictates the jumper length and characteristics. Flexible jumper systems provide versatility, unlike rigid jumper systems, which limit space and handling capability.

1.4.4. Subsea Wellheads

Wellhead is a general term used to describe the pressure-containing component at the surface of an oil well that provides the interface for drilling, completion, and testing of all subsea operation phases. It can be located on the offshore platform or onshore, in which case it is called a *surface wellhead*; it can also be settled down on the mudline, in which case it is called a *subsea wellhead* or *mudline wellhead* as shown in Figure 1-12.

Subsea wells can be classified as either satellite wells or clustered wells. Satellite wells are individual and share a minimum number of facilities with

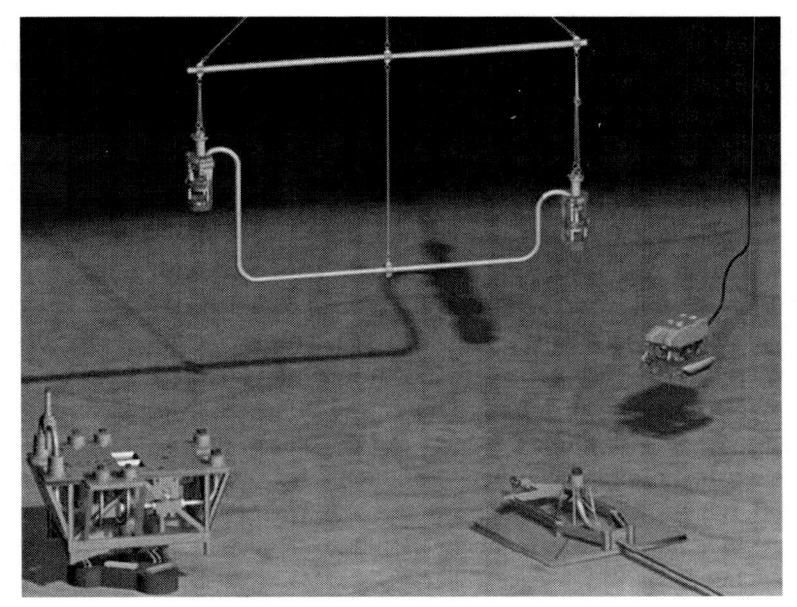

Figure 1-11 Subsea Rigid Jumper [11]

Figure 1-12 Subsea Wellhead

other wells. They are usually drilled vertically. Satellites wells can produce directly to a surface facility (the platform of a floating vessel) or through a subsea manifold that commingles the production of several satellite wells. The primary advantage of satellite wells is the flexibility of individual well location, installation, control, and service. Each well is handled separately, so that its production and treatment can be optimized. Exploration or delineation wells in a field can also be reused by completing them as satellite wells, thereby eliminating the drilling costs associated with a new well.

When several subsea wellheads are located on a central subsea structure, the system is referred to as a *clustered system*. This arrangement provides the possibility of sharing common functions among several wells, such as manifolded service or injection lines and common control equipment, which then require fewer flowlines and umbilicals, thus reducing costs. In addition, because maintainable components are centralized on a clustered system, it is possible to service more than one well with a single deployment of a service vessel, thereby saving mobilization costs. On the other hand, shared functions can reduce the capability to treat each well separately. Clustered systems, however, introduce the need for subsea chokes to allow individual well control. Other disadvantages of clustered systems are that drilling or workover operations on one well of the cluster may interrupt production from others and special simultaneous drilling and production procedures need to be implemented.

1.4.5. Subsea Trees

The subsea production tree is an arrangement of valves, pipes, fittings, and connections placed on top of a wellbore. Orientation of the valves can be in the vertical bore or the horizontal outlet of the tree, as shown in Figure 1-13. The valves can be operated by electrical or hydraulic signals or manually by a diver or ROV.

1.4.6. Umbilical Systems

An umbilical, as shown in Figure 1-14, is a bundled arrangement of tubing, piping, and/or electrical conductors in an armored sheath that is installed from the host facility to the subsea production system equipment. An umbilical is used to transmit the control fluid and/or electrical current necessary to control the functions of the subsea production and safety equipment (tree, valves, manifold, etc.). Dedicated tubes in an umbilical are used to monitor pressures and inject fluids (chemicals such as methanol) from

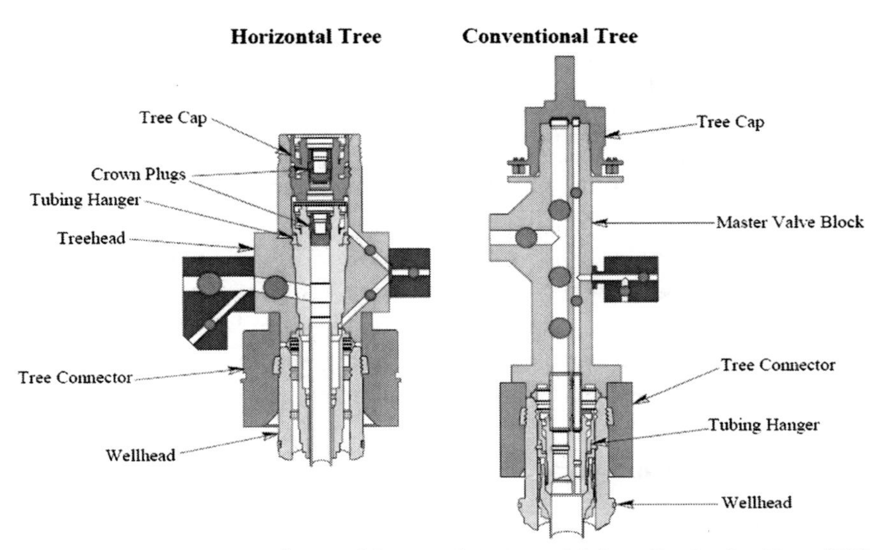

Figure 1-13 Comparison of Vertical-Bore and Horizontal Subsea Production Trees (ABB Vetco Gray Inc.) [12]

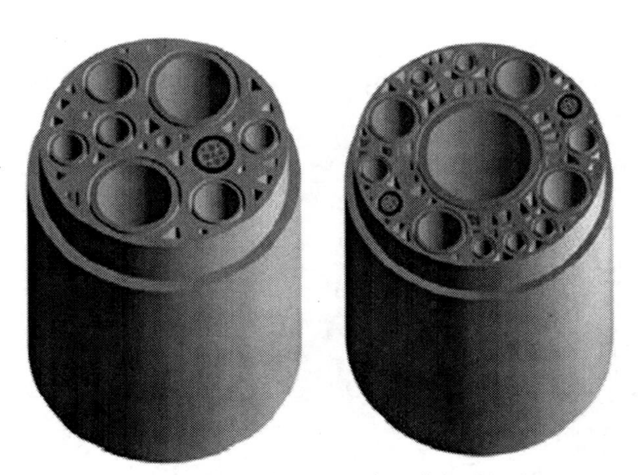

Figure 1-14 Subsea Steel Umbilical [13]

the host facility to critical areas within the subsea production equipment. Electrical conductors transmit power to operate subsea electronic devices.

Umbilical dimensions typically range up to 10 in. (25.4 cm) in diameter. The umbilical will include multiple tubings normally ranging in size up to 2 in. (5.08 cm); the number of tubes is dependent on the complexity of the production system. The length of an umbilical is defined by the spacing of the subsea components and the distance these components are located from the host facility.

1.4.7. Production Risers

The production riser is the portion of the flowline that resides between the host facility and the seabed adjacent to a host facility. Riser dimensions range from 3 to 12 in (76.2 to 304.8 mm). in diameter. Riser length is defined by the water depth and riser configuration, which can be vertical or a variety of wave forms. Risers can be flexible or rigid. They can be contained within the area of a fixed platform or floating facility, run in the water column. Figure 1-15 illustrates part of a subsea drilling riser.

1.5. SUBSEA PIPELINES

Subsea flowlines are the subsea pipelines used to connect a subsea wellhead with a manifold or the surface facility. The flowlines may be made of flexible pipe or rigid pipe and they may transport petrochemicals, lift gas, injection water, and chemicals. For situations where pigging is required, flowlines are connected by crossover spools and valves configured to allow pigs to be circulated. Flowlines may be single pipe, or multiple lines bundled inside

Figure 1-15 Subsea Drilling Riser

a carrier pipe. Both single and bundled lines may need to be insulated to avoid problems associated with the cooling of the produced fluid as it travels along the seabed.

Subsea flowlines are increasingly being required to operate at high pressures and temperatures. The higher pressure condition results in the technical challenge of providing a higher material grade of pipe for high-pressure, high-temperature (HP/HT) flowline projects, which will cause sour service if the product includes H_2S and saltwater. In addition, the higher temperature operating condition will cause the challenges of corrosion, down-rated yield strength, and insulation coating. Flowlines subjected to HP/HT will create a high effective axial compressive force due to the high fluid temperature and internal pressure that rises when the flowline is restrained.

REFERENCES

[1] C. Haver, Industry and Government Model for Ultra-Deepwater Technology Development, OTC 2008, Topical Luncheon Speech, Houston, 2008.

[2] C.W. Burleson, Deep Challenge: The True Epic Story or Our Quest for Energy Beneath the Sea, Gulf Publishing Company, Houston, Texas, 1999.

[3] M. Golan, S. Sangesland, Subsea Production Technology, vol. 1, NTNU (The Norwegian University of Science and Technology), 1992.

[4] Minerals Management Service, Deepwater Gulf of Mexico 2006: America's Expanding Frontier, OCS Report, MMS 2006-022, 2006.

[5] J. Westwood, Deepwater Markets and Game-Changer Technologies, presented at U.S. Department of Transportation 2003, Conference, 2003.

[6] FMC Corporation, Subsea System, http://www.fmctechnologies.com/en/SubseaSystems.aspx, 2010.

[7] H.J. Bjerke, Subsea Challenges in Ice-Infested Waters, USA-Norway Arctic Petroleum Technology Workshop, 2009.

[8] International Standards Organization, Petroleum and Natural Gas Industries-Design and Operation of the Subsea Production Systems, Part 1: General Requirements and Recommendations, ISO, 2005, 13628-1.

[9] International Standards Organization, Petroleum and Natural Gas Industries-Design and Operation of the Subsea Production Systems, Part 6: Subsea Production Control Systems, ISO, 2000, 13628-6.

[10] M. Faulk, FMC ManTIS (Manifolds & Tie-in Systems), SUT Subsea Awareness Course, Houston, 2008.

[11] C. Horn, Flowline Tie-in Presentation, SUT Subsea Seminar, 2008.

[12] S. Fenton, Subsea Production System Overview, Vetco Gray, Clarion Technical Conferences, Houston, 2008.

[13] P. Collins, Subsea Production Control and Umbilicals, SUT, Subsea Awareness Course, Houston, 2008.

CHAPTER 2

Subsea Field Development

Contents

2.1. SUBSEA FIELD DEVELOPMENT OVERVIEW

Subsea field development is a long and complicated procedure that begins with the primary survey and ends with the last reservoir recovery.

Figure 2-1 illustrates the life cycle of the development of an oil or gas subsea field. Initially, mapping and reconnaissance are conducted by

Subsea Engineering Handbook
ISBN 978-0-12-397804-2, doi:10.1016/B978-0-12-397804-2.00002-3

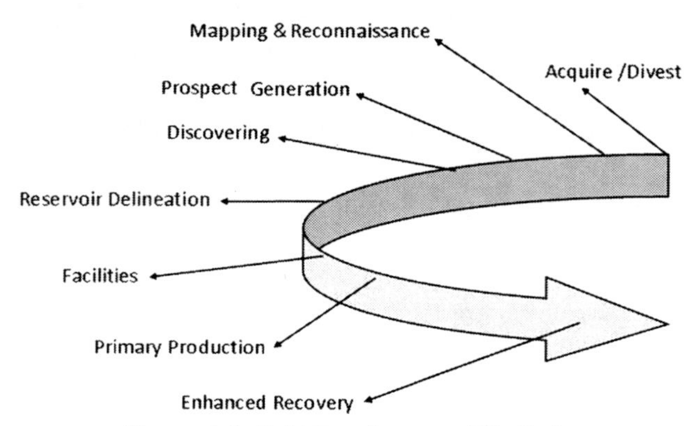

Figure 2-1 Field Development Life Cycle

exploration geologists and geophysicists. They ultimately delineate the development area's geology based on the data gathered from old wells, seismic analysis, and any other information that is available. The initial issues at this stage concern the following aspects:

- Structure of the basin and the subregional features (i.e., fault and/or fold traps for hydrocarbons);
- The stratigraphy (i.e., whether the reservoir rocks exhibit porosity and permeability);
- The burial history of the basin (i.e., whether the source rocks have been buried sufficiently for hydrocarbon generation).

By addressing these concerns, investigators may identify and select parts of the larger area for further study and may ultimately generate a prospect evaluation.

After the initial investigations, the reservoir description phase begins, which involves drilling delineation wells and perhaps conducting 3D seismic analyses. This new information allows reservoir engineers and geologists to calculate the volume of oil and/or gas that is present in the reservoir. Then it is time to ascertain the optimum subsea field layout and pipeline route; the production facilities will also be selected based on field layout and installation considerations. After all well and equipment testing, the field begins to produce oil and gas. However, as more and more oil and gas are transported to the host structure from the reservoir, the reservoir pressure will decrease, and need to recovery to keep the production being transported from the reservoir.

This chapter provides guidelines for the main disciplines associated with the development of a field architecture without topside facilities and also for system integration and interfacing, which are the most important parts of a field development project.

When defining a field architecture, the following issues should be considered:

- Deepwater or shallow-water development;
- Dry tree or wet tree;
- Stand-alone or tie-back development;
- Subsea processing;
- Artificial lift methods;
- Facility configurations (i.e., template, well cluster, satellite wells, manifolds).

2.2. DEEPWATER OR SHALLOW-WATER DEVELOPMENT

Subsea field development can be categorized according to the water depth:

- A field is considered a shallow-water subsea development if the water depth at the location is less than 200 m (656 ft). In practice, shallow water is the water depth within a diver's reach.
- A field is considered a deepwater subsea development if the water depth ranges between 200 and 1500 m (656 and 5000 ft);
- Ultra-deepwater subsea developments are those in which the water depths are greater than 1500 m (5000 ft).

The difference between shallow-water and deepwater field developments in terms of design considerations are listed in Table 2-1.

2.3. WET TREE AND DRY TREE SYSTEMS

Two kinds of subsea production systems are used in deepwater fields: dry tree systems and wet tree systems, as shown in Figure 2-2.

For the dry tree system, trees are located on or close to the platform, whereas wet trees can be anywhere in a field in terms of cluster, template, or tie-back methods. Dry tree platforms have a central well bay for the surface trees, providing direct access to the wells for workover and recovery. Tension leg platforms (TLPs) and spars are normally utilized in a dry tree system.

The size of the central well bay on dry tree platforms is dictated by well count and spacing. Topside equipment has to be arranged around the well bay. The surface trees are designed for full reservoir shut-down pressures. A large production manifold is required on deck, and a skiddable rig is required for individual well interventions [1].

Table 2-1 Difference between Shallow-Water and Deepwater Field Developments

Items	**Shallow-Water Development**	**Deepwater Development**
Hardware design	Because diver-assisted intervention is possible, an ROV–related structure is not necessary. Mudline trees usually will be used.	Because an ROV assists with all interventions, an ROV-related structure is needed. Insulation is needed for pipes due to high pressures and temperatures. A horizontal or vertical tree will be used.
Installation requirement	Limited by the size of vessel.	More difficult than in shallow water due to higher tension, especially horizontal load.
Umbilical design	Smaller umbilicals can be utilized due to the short distance of power transportation.	Umbilicals are bigger and more expensive.
Intervention, maintenance, and repair	Diver-assisted intervention is feasible.	Deepwater maintenance and repairs require the use of ROVs for surveys and some repairs. Deepwater subsea developments are high cost and high risk.

For wet tree systems, the Christmas tree and its associated components are exposed to the ambient seabed conditions. In deepwater fields, the wet tree system normally utilizes a remotely controlled subsea tree installation tool for well completions, but for shallow water the diver can assist with

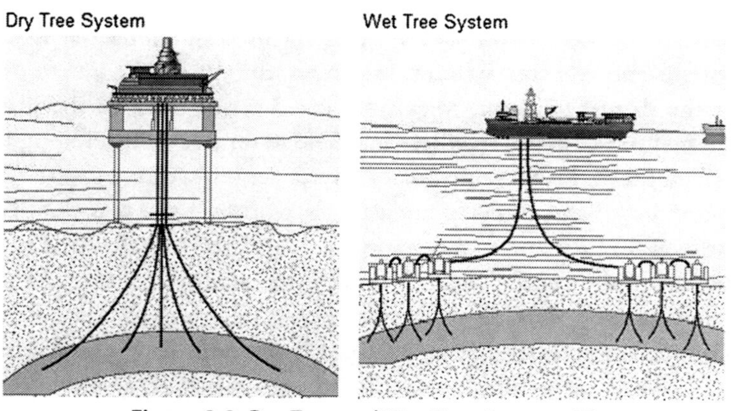

Figure 2-2 Dry Tree and Wet Tree Systems [1]

installation and operation. Wet tree platforms have a central moon-pool for running marine risers and trees, which might also be preferred for installing other equipment, such as the manifold and blow-out preventer (BOP), if the sizes are suitable. Wet tree systems are suitable for widespread reservoir structures. They provide a degree of vessel and field expansion flexibility with simplified riser interfaces, but at the expense of high drilling and workover costs.

In recent years, operators have been compelled to reappraise their strategy of rapid development in ultra-deepwater areas due to the commercial competitiveness and technical issues related to dry tree versus wet tree systems. Globally, more than 70% of the wells in deepwater developments that are either in service or committed are wet tree systems. These data demonstrate the industry's confidence in wet tree systems.

Compared to wet tree systems, dry tree systems require motion-optimized hulls to accommodate the riser systems; they are considered to be limited with respect to water depth and development flexibility. Globally, most subsea field developments that are either in service or committed to are wet tree systems. Although widely used for developments in shallow to medium water depths, the dry tree units are still not considered optimum for deepwater and ultra-deepwater situations.

2.3.1. Wet Tree Systems

For wet tree system, the subsea field layout usually comprises two types: subsea wells clusters and direct access wells.

Direct access is only applied in marginal field development. All such developments are usually based on semi-submersible floating production and drilling units (FPDUs) with oil export either via pipeline or to a nearby floating storage and offloading (FSO) unit, providing direct and cost-effective access from the surface to the wells to allow for workover or drilling activities directly from the production support, especially for deepwater interventions.

A subsea cluster of wells gathers the production in the most efficient and cost-effective way from nearby subsea wells, or (when possible) from a remote /distant subsea tie-back to an already existing infrastructure based on either a floating production, storage and offloading (FPSO) or a floating production unit (FPU), depending on the region considered.

Figures 2-3, 2-4, and 2-5 illustrate subsea well clusters and subsea direct-access wells. The subsea tie-back system normally can be regarded as a supplement for subsea cluster well developments, as shown in Figure 2-3.

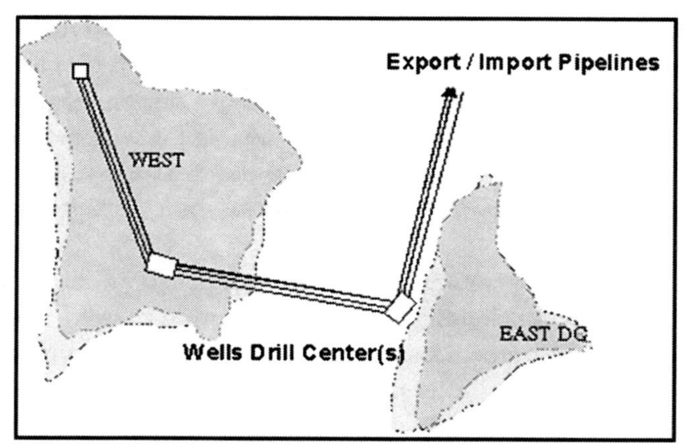

Figure 2-3 Tie-Back Field Architecture [1]

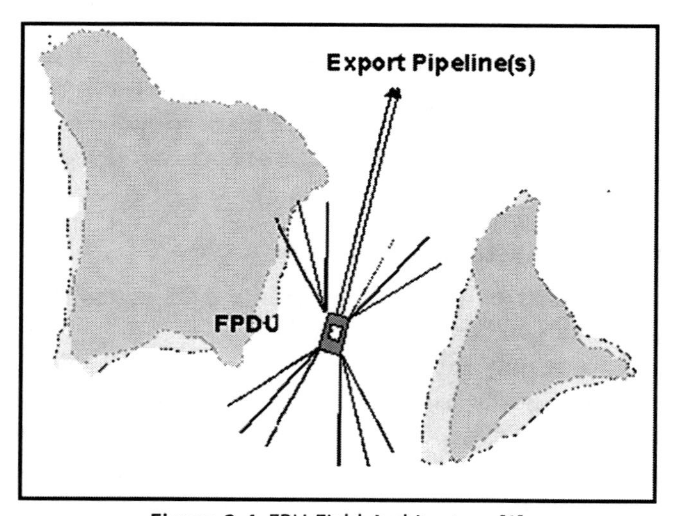

Figure 2-4 FPU Field Architecture [1]

The FPU has various ways to export the oil or gas, as shown in Figure 2-6. Typically a barge, semi-submersible, or even a mini-TLP type vessel is used. Figure 2-4 illustrates an FPU field architecture.

Figure 2-5 shows a field with an FPSO field architecture. The FPSO usually utilizes a ship- shaped or barge-type vessel as the host structure that is moored either via a turret and weathervaning system to allow for tandem offloading or spread moored with offloading via a distant buoy or still in tandem mode [1].

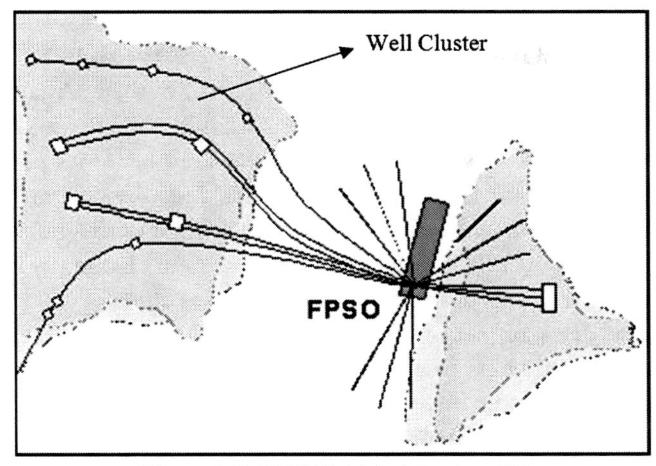

Figure 2-5 FPSO Field Architecture [1]

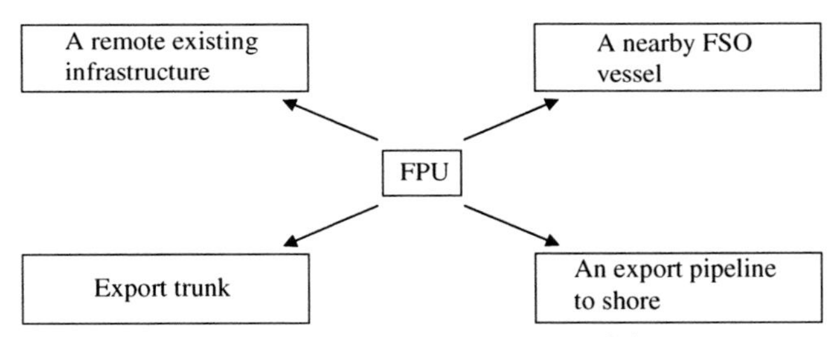

Figure 2-6 FPU Connection Module

For both subsea well clusters or subsea direct-access wells, the three main riser options would be vertical top tensioned risers, steel catenary risers, and flexible risers. Pending use of the above field architectures and risers in conjunction with particular flow assurance design criteria, these wet tree system would be the most effective.

2.3.2. Dry Tree Systems

Dry tree system production systems are the main alternative to the subsea well cluster architecture. Their surface well architectures provide direct access to the wells.

Current dry tree system architectures consist of an FPDU hub based either on a TLP, on a Spar, or even (in some cases) on a compliant piled tower (CPT) concept. Alternative concepts in the form of barge or deep-draft semi-submersible (DDF) floaters could also be considered as possible options in some cases but have not yet materialized.

Deepwater surface well architectures in the form of a wellhead platform (WHP or FDU) associated with either an FPSO or an FPU are starting to emerge in other parts of the world (West Africa, Southeast Asia). This type of association between a WHP and an FPSO could also eventually become reality in the Gulf of Mexico as the FPSO concept becomes progressively authorized. Existing (or actually planned) WHP hubs are currently all based on a TLP concept with either full drilling capacities or with tender assistance (mini-TLP). However, other concepts such as Spar, barge, or deep-draft semi-submersible units could also be considered as alternatives.

Risers for dry completion units (DCUs) could be either single casing, dual casing, combo risers (used also as drilling risers), or tubing risers and could include a split tree in some cases.

The riser tensioning system also offers several options such as active hydropneumatic tensioners, air cans (integral or nonintegral), locked-off risers, or king-post tensioning mechanism.

2.3.3. Systems Selection

A thorough and objective assessment of wet and dry tree options should be conducted during the selection process and include costs, risks, and flexibility considerations to ensure that the development concept that best matches the reservoir characteristics is selected.

Operators are often faced with the problem that commercial metrics, such as capital cost or net present value (NPV), are inconclusive. In the base case of developments examined, the commercial metrics are shown to generally favor the wet tree concept over the life cycle of the system, but the benefit over the dry tree concept is not overwhelming. When this is the case, the operator must thoroughly assess other important differentiators before making the final selection.

The best tree system matchup with the reservoir characteristic can be selected by experience and technical analysis. The following are the basic selection points:

- *Economic factors:* Estimated NPV, internal rate of return (IRR), project cash flow, project schedule, and possibly enhanced proliferation control initiative (EPCI) proposals (if any available at the time of the selection) will most certainly be the key drivers of this choice.
- *Technical factors:* These factors are driven primarily by reservoir depletion plans and means, field worldwide location, operating philosophy, concept maturity and reliability, feasibility, and industry readiness.

- *External factors:* These factors are in the form of project risks, project management, innovative thinking, operator preferences, and people (the evaluation method may vary between each individual).

2.4. SUBSEA TIE-BACK DEVELOPMENT

Subsea developments only made sense for big reservoirs before, because the CAPEX and OPEX costs are high and it is difficult to justify the return versus the risk. So, in some cases, the small marginal oil fields are normally ignored. In recent years, subsea tie-backs were set to become popular in the development of marginal oil and gas reserves in an effective and economical way. Operators began to realize that the overall capital expenditure can be decreased by utilizing the processing capacity on existing platform infrastructures, rather than by continuing to build new structures for every field. Thus much smaller accumulations can be developed economically.

Generally subsea tie-backs require significantly lower initial investments, compared with developments using FPSOs or other fixed installations. The economics of having a long tie-back are governed, however, by a number of factors specific to that field:

- Distance from existing installation;
- Water depth;
- Recoverable volumes, reservoir size, and complexity;
- Tariffs for processing the produced fluids on an existing installation;
- The potentially lower recovery rates from subsea tie-backs versus standalone development, due to limitations in the receiving facility's processing systems;
- The potentially higher recovery rates from platform wells, due to easier access to well intervention and workovers.

The host of subsea tie-backs can be categorized as follows:

- Tie-back to floating production unit (see Figure 2-7);
- Tie-back to fixed platform (see Figure 2-8);
- Tie-back to onshore facility (see Figure 2-9).

2.4.1. Tie-Back Field Design

A subsea tie back system generally includes a subsea wellhead and a flowline to an existing production platform for example. Some serious limitations of flow assurance are expected with a longer subsea tie-back, such as hydrate formation induced plugging of the flowline due to the heat loss to the environment and therefore a decrease of temperature along the flowline.

Figure 2-7 Subsea Tie-Back to FPSO [2]

Figure 2-8 Subsea Tie-Back to TLP [2]

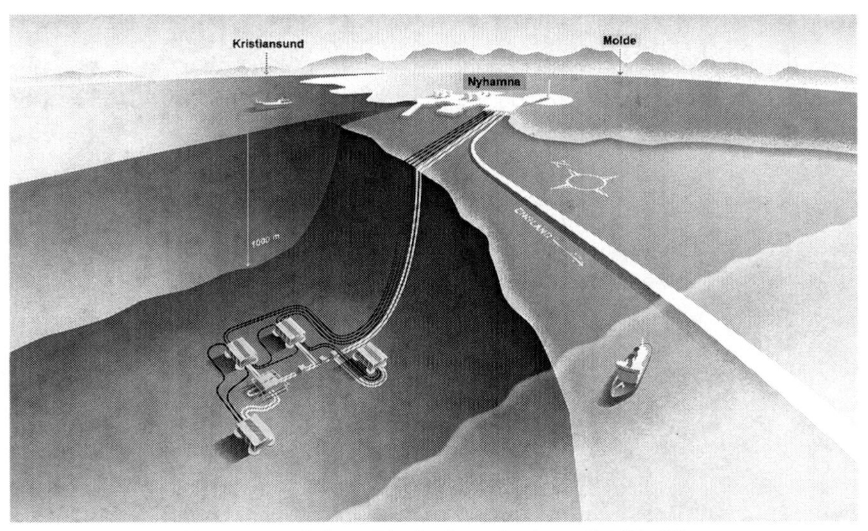

Figure 2-9 Subsea Tie-Back to Onshore Facility [3]

The conventional remedial methods include thermal insulation of flowlines or injection of chemical inhibitors to prevent the formation of hydrates (refer to Part II, flow assurance part of this book). Such chemicals can be transported from the host platform to the subsea wellhead with an umbilical, and can be injected into the flowline at the wellhead. The umbilical can also be used to control the subsea wellhead. The cost of such umbilical is typically very high, and the economics of a subsea tie-back is often threatened by the excessive umbilical cost for tie-back distances greater than 30 km.

An alternative development scenario of flow assurance consists of providing a small offshore platform near the wellhead with remote control from the host platform and injection of chemical stored on the small offshore platform via a short umbilical connected to the subsea system. The presence of a platform directly above the wellhead enables pig launcher capabilities and well logging tools support.

When multiphase hydrocarbon flow is expected in the flowline, the tie-back distance is limited because of flow assurance problems. Current technological developments are aimed at providing subsea separation facilities on seafloor to separate the fluid into hydrocarbon liquid, gas and water, allowing hydrocarbons to flow over a longer distance.

Subsea equipment such as subsea pumps may be required to increase the pressure of fluid and assist flow assurance over the tie-back length. Such pump system also requires power which can be provided by a surface facility.

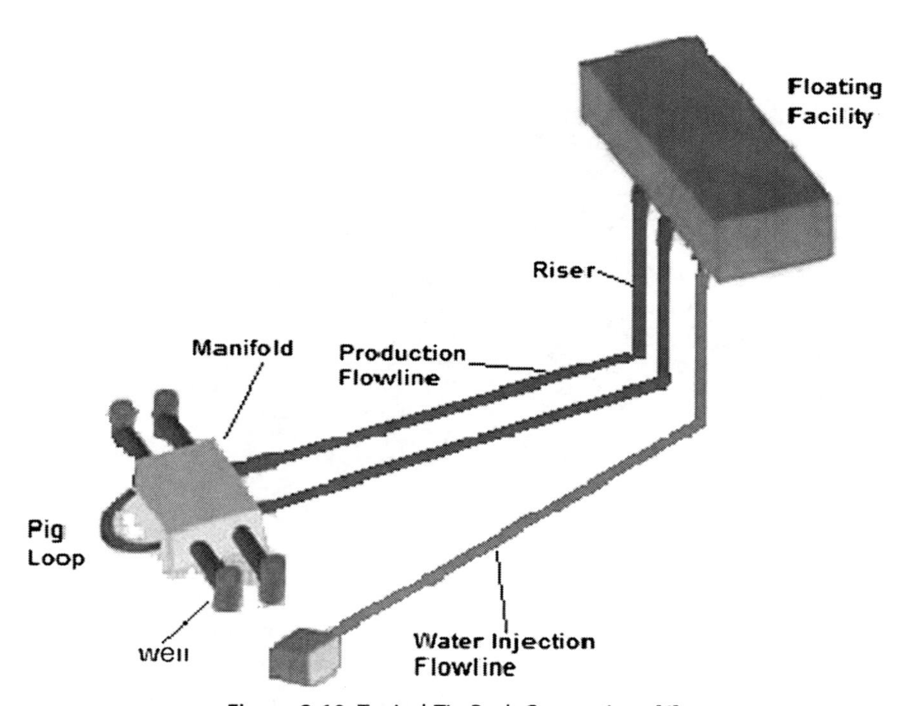

Figure 2-10 Typical Tie-Back Connections [4]

Dual-flowlines as shown in Figure 2-10 are used widely in the subsea tie-back system to provide a circuit for pig to start from the production platform, go through the flowlines to remove the solids in the flowlines, such as wax, asphaltene, sand, and return to the production platform.

2.4.2. Tie-Back Selection and Challenges

Advances in flow assurance and multiphase transport now allow the use of tie-backs over much longer distances, while the introduction of subsea processing will strengthen the business case for subsea tie-backs in future field developments. However, we can try to choose the best development plan, based on an overall consideration of the following factors:

- *Cost:* Lowest life-cycle cost (i.e. lower CAPEX and OPEX);
- *Safety:* Safety of personnel and other stakeholders in construction and operation;
- *Environment:* Impact of development on the environment;
- *Technology innovation or transfer:* Trial of new technology or transfer of existing technology and know-how;

- *Capacity utilization:* Use of existing infrastructure, facilities, and elongation of useful life;
- Recoverable volumes, reservoir size, and complexity;
- Tariffs for processing the produced fluids on an existing installation;
- The potentially lower recovery rates from subsea tie-backs versus stand-alone development, due to limitations in the receiving facility's processing systems;
- The potentially higher recovery rates from platform wells, due to easier access to well intervention and workover.

Many marginal fields are developed with subsea completions and with the subsea tie-back flowlines to existing production facilities some distance away. Subsea tie-backs are an ideal way to make use of existing infrastructure. Long tie-back distances impose limitations and technical considerations:

- Reservoir pressure must be sufficient to provide a high enough production rate over a long enough period to make the development commercially viable. Gas wells offer more opportunity for long tie-backs than oil wells. Hydraulic studies must be conducted to find the optimum line size.
- Because of the long distance travelled, it may be difficult to conserve the heat of the production fluids and they may be expected to approach ambient seabed temperatures. Flow assurance issues of hydrate formation, asphaltene formation, paraffin formation, and high viscosity must be addressed. Insulating the flowline and tree might not be enough. Other solutions can involve chemical treatment and heating.
- The gel strength of the cold production fluids might be too great to be overcome by the natural pressure of the well after a prolonged shutdown. It may be necessary to make provisions to circulate out the well fluids in the pipeline upon shutdown, or to push them back down the well with a high-pressure pump on the production platform, using water or diesel fuel to displace the production fluids.

Subsea long tie-back developments will be utilized widely in the future with the advent of new technology such as subsea processing and subsea electrical power supply and distribution.

2.5. STAND-ALONE DEVELOPMENT

Stand alone field development needs to construct a new host platform. Installation of new infrastructure in deep water is exceedingly expensive.

Using the existing infrastructure is the first consideration for starting a new development. This includes existing production platforms, pipelines and wells. Figure 2-11 illustrates a typical stand-alone field development.

Following issues are the main considerations for a stand-alone field development:

- Well groupings. Clustering wells or installing well templates;
- Optimizing flowline configuration;
- Pigging requirements;
- Possible needs for subsea production boosting or pumping as part of the initial development or future needs.

Well grouping scenarios and location should be determined according to the reservoir data and drilling engineering. Types of wells and their locations can be determined once the reservoir is mapped and the number of wells is created according to the reservoir model. Wells are typically grouped as follows:

- Satellite wells: typically used for small filed development requiring few wells, for example, concept of tie-back to the existing structures;

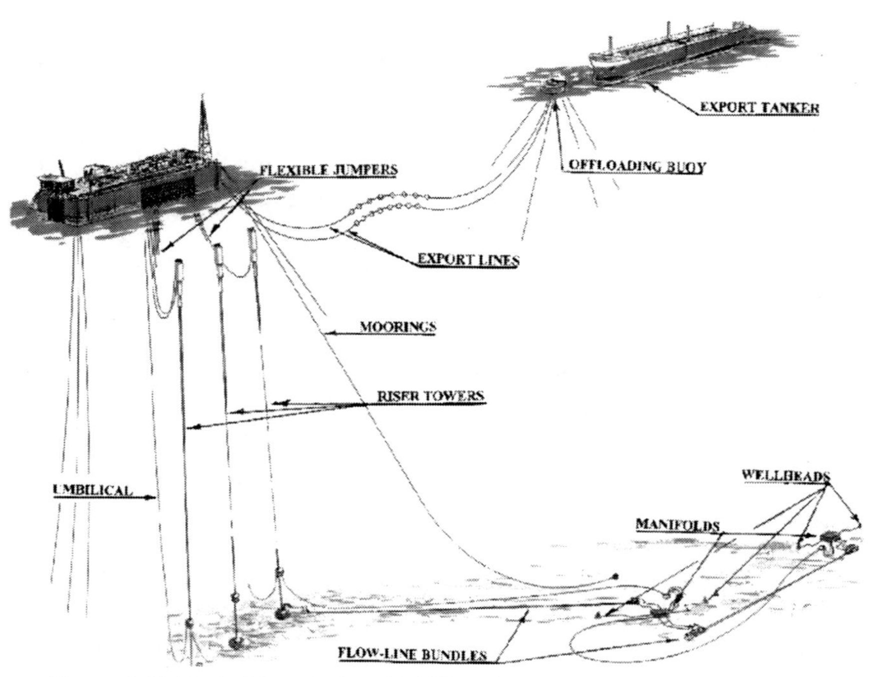

Figure 2-11 Typical Stand–alone Field Development *(Courtesy SapuraAcergy)*

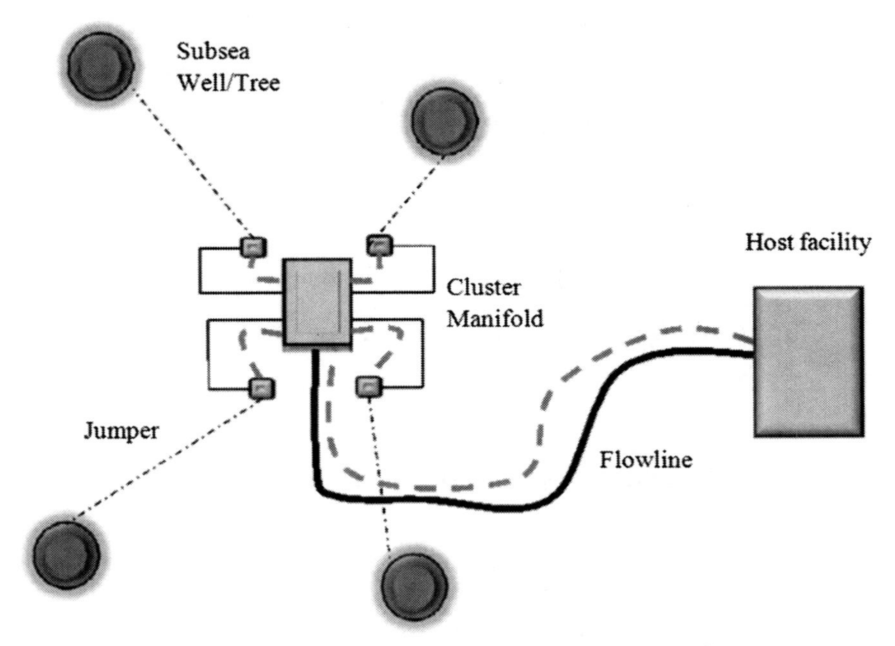

Figure 2-12 Typical Cluster Wells Concept *(Courtesy Technip)*

- Cluster wells; common concept for a stand-alone field development. Normally there are 3 to 8 subsea Xmas trees located in the surrounding of a central production manifold, as shown in Figure 2-12;
- Template wells; the subsea wells are grouped closely together. This concept usually utilizes a template in which the well guide bases and the manifold are integrated. Subsea Xmas trees are landed and locked on each slots of the template;
- Combination of the above.

2.5.1. Comparison between the Stand-Alone and Tie-Back Developments

For the stand-alone concept, the main subsea structure is normally constructed as a drilling platform and will also be used as a production platform. The structure acts as a stabilizer and conditioning for the subsea well production. Subsea risers, helicopter landing pads, and mooring facilities for boats are necessities and supported by the structure. Table 2-2 summarizes the differences between the tie-back development and the stand-alone development.

Table 2-2 Comparison between the Stand-alone and Tie-back Developments

Concept	Features
Subsea tie-back development	• Investment is reduced using the spare capacity in the existing platform; • Very suitable option for small developments e.g. marginal fields;
Stand-alone development	• No existing subsea structures and the recoverable reserves are large; • Or the distance to shore is too far; • New stand-alone development can be considered using a platform that is the most suitable for the environmental conditions.

2.5.2. Classification of Stand-Alone Facilities

The stand-alone facility is a host facility that receives the production from a field. Figure 2-13 illustrates four typical host facilities, which vary from the wellhead and process platform receiving production from surface trees to subsea trees of subsea tiebacks. The host facility could

Figure 2-13 Typical Host Faculties

even be a land based facility receiving production from a subsea tieback to beach.

The stand-alone facilities used in subsea field development can be divided into two categories: fixed platforms and floating systems. Bottom of the fixed platform is located on the seabed to support the decks to be fixed above the water surface. The floater systems have to be moored in place with tendons or wire ropes to keep connection with the subsea systems below. Floating systems can be used from 300 m to more than 1500 m water depth. Following are the definitions and main features of the host facilities.

- Fixed platforms: fixed platforms are built on concrete or steel jackets which are directly anchored on the seabed. Various types of platforms, e.g. concrete caisson, and steel jacket are used for this concept. A jack up platform is also used for the production of oil and gas. Fixed platforms are usually installed in the water depth within 500 m;
- TLP: tension leg platform (TLP) is a vertically moored floating structure. A group of tethers called "leg" is used to moor the each corners of the floating structure. The lateral movement is allowed, but vertical movement is prevented by the legs. The TLP can be used in the water depth from 300 m to 1500 m. Both dry trees and wet trees can be used for the TLP production system;
- SPAR: the Spar platform utilize a large-diameter, single vertical cylinder buoy floating on the sea water surface to support the decks. Spars can be used in the water depth around 1500 m;
- Semi-submersible platforms: Semi-submersible platform can be operated for drilling, subsea equipments installation and oil/gas production.
- FPSO: a floating vessel for the processing and storage of oil and gas. The FPSO is designed to receive the production oil and gas from nearby platform or subsea production systems. The crude oil can be offloaded onto a tanker or transported through pipelines.

Comparison of the wave responses of the floaters are as following:

- FPSO is very responsive because of large water plane, large surface area (beam seas), and softer restoring force;
- Semi-submersible has a slow natural period due to softer restoring force, vortex-shedding, shallow drafts increase pitch/roll response;
- SPAR has a slow natural period due to softer restoring force, vortex shedding, deep draft reduces pitch/roll response;
- TLP has a short natural period due to highly tensioned tendons and limited water plane;

2.6. ARTIFICIAL LIFT METHODS AND CONSTRAINTS

2.6.1. General

Artificial lift is widely used in the shallow water of the Gulf of Mexico, but its use in deepwater (>1,000 ft) is limited. However, most deepwater oil fields ultimately require the artificial lift to maintain the production flow and achieve economic objectives. Planning for an artificial lift in deepwater is critical, as the environment is operationally more difficult and economically more challenging.

The following are the main artificial lift methods:

- Gas Lift (GL);
- Subsea Boosting;
- Electrical submersible pumping (ESP).

2.6.2. Gas Lift

Gas lift (GL) is a predominant artificial lift method used in the offshore environment to date; however, as operators progressively move into deepwater, GL applications become more limited due to higher operating pressures and ESPs applications become more suitable. Figure 2-14 shows a typical subsea gas lift system.

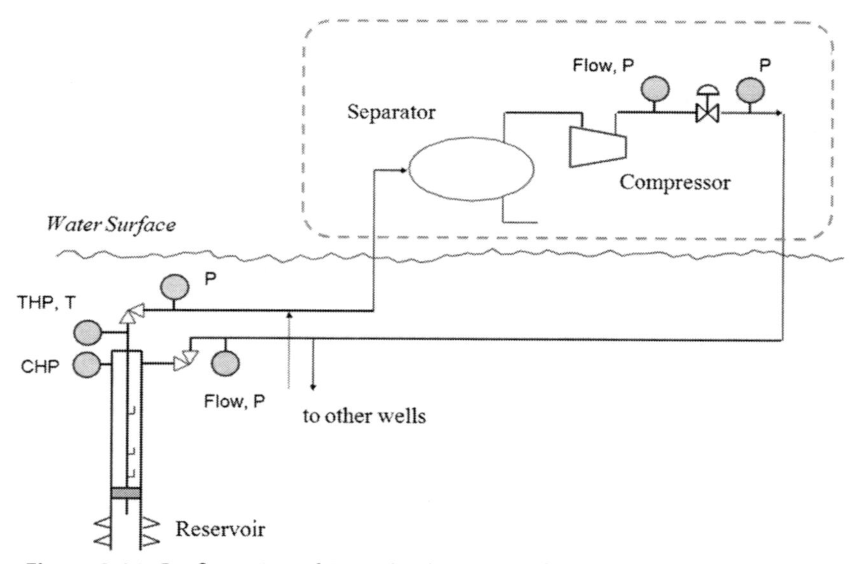

Figure 2-14 Configuration of Typical Subsea Gas Lift System *(Courtesy Shell E&P)*

The selection and determination of the gas lift system depends on:

- High water cut of the reservoir;
- Low GOR of the reservoir;
- Long offset flowline;
- etc.

The designs of gas lift for subsea wells have several requirements that are not normally encountered in the designs of traditional gas lift. First, the cost of intervention in a subsea well is considerably higher than for a traditional completion, the subsurface gas lift equipment must be designed with special attention in reliability and longevity. Secondly, the sizing of the port in the operating valve must anticipate production conditions for the life of the well. Figure 2-15 shows the details of a subsea gas lift in downhole.

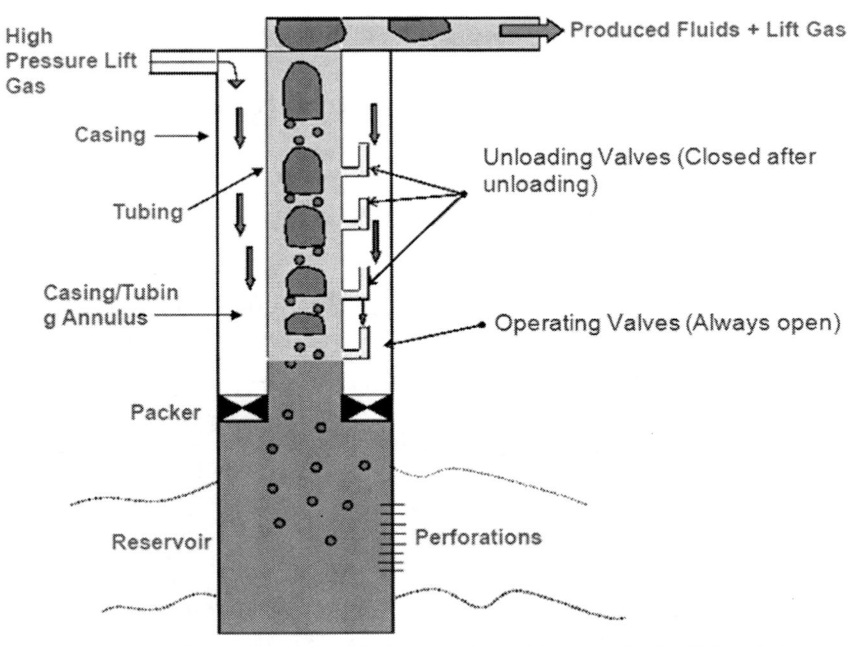

Figure 2-15 Details of Gas Lift in Downhole *(Courtesy Curtin University)*

The design of gas lift system involves two key parameters: gas lift volume and gas lift pressure. Gas lift volume is the total requirement for the field determined by individual well requirements. Production will increase as a function of lift gas volume until a point of maximum production is reached. Gas lift pressure should be determined carefully since this parameter will influence the system operating pressure, material and equipment specification of the well system.

2.6.3. Subsea Pressure Boosting

Some reservoirs have sufficient pressure to push the production fluid from the reservoir to the platform without the use of downhole artificial lift. However, due to the reduction of the reservoir pressure after a long time production, or because of ultra-deepwater light-oil and deepwater heavy-oil reservoirs having pressure that is near hydrostatic pressure, it is quite difficult to produce the fluids to the sea surface.

Subsea boosting as shown in Figure 2-16 reduces or eliminates the backpressure on the wells resulting from the riser hydrostatic head and the riser and flowline pressure drop caused by high viscosity. The pressure increase between the output of the boosting and the backpressure on the well will increase the flow from the well. Components of a subsea boosting station may include:

- Subsea gas compressor, normally used for gas re-injection into the reservoir for pressure maintenance;
- Subsea multiphase pump, used for reduce the back-pressure on well-heads thus increase the transport distance;
- Subsea wet gas compressors, used for gas transportation to remote offshore host facilities or onshore factory;

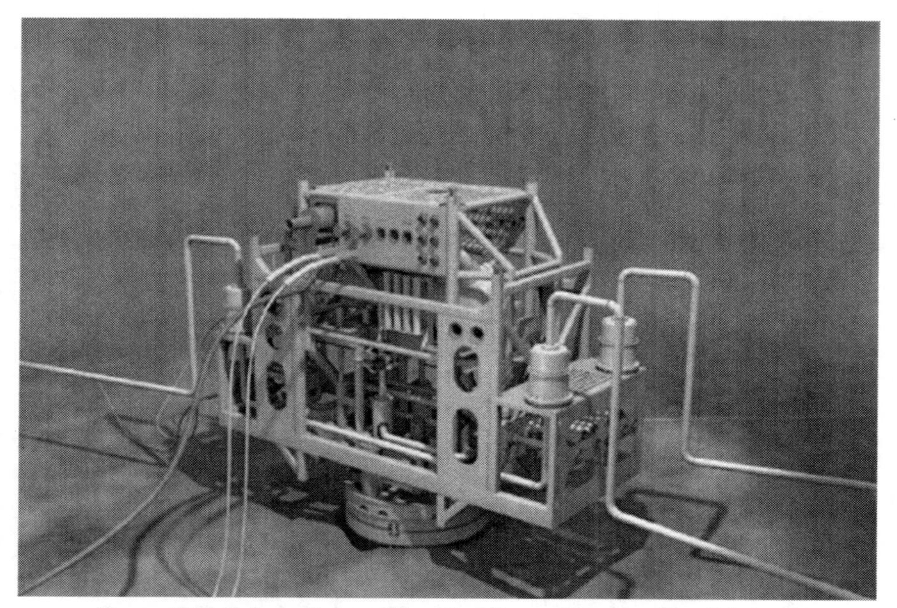

Figure 2-16 Subsea Pressure Boosting Pump *(Courtesy Aker Solutions)*

Subsea boosting may require high consumption of the electric power during the production thus the power sources should be considered.

Subsea pressure boosting system enables longer subsea tiebacks, which potentially could enable the economics of exploiting small, remote, marginal fields. Subsea separation could provide an economic alternative for de-bottlenecking existing surface process facilities, allowing better utilization of these installations by adding new subsea tiebacks which currently would not be economic to develop. Subsea gas separation may allow oil and gas to be separated at the seabed and be transported to different production facilities, which may solve flow assurance problem due to low temperature at seabed.

2.6.4. Electric Submersible Pump (ESP)

Due to the reduction of the driving force which lifts the reservoir from downhole naturally, pumps were commonly used to increase the back-pressure for production. The electric submersible pump (ESP) is an effective and economical method of lifting large volume of the fluids from downhole under different well conditions. ESP system requires a large electricity supply, but it is less complex and more efficient than delivering gas to gas lift systems.

An ESP system may include following components:

- Three phase electric motor;
- Seal assembly;
- Rotary gas separator;
- Multi-stage centrifugal pump;
- Electrical power cable;
- Motor controller;
- Transformers.

Different from the surface pump system, the ESP systems are particularly designed to be immersed in fluid. It can be either located in a well or on the seabed. The ESP motors are pressure balanced with the environment, whether that is downhole pressure or water pressure in subsea conditions.

Optional components of the ESP system may include tubing joints, check valve, drain valve, downhole pressure and temperature transmitters, etc. Figure 2-17 shows a typical ESP configuration in downhole.

The selection of ESP types mainly depends on the well fluid properties. Following are the three major types of ESP applications:

- High water-cut wells producing fresh water or brine;
- Multi-phase flow well with high GOR;
- Highly viscous fluid Well.

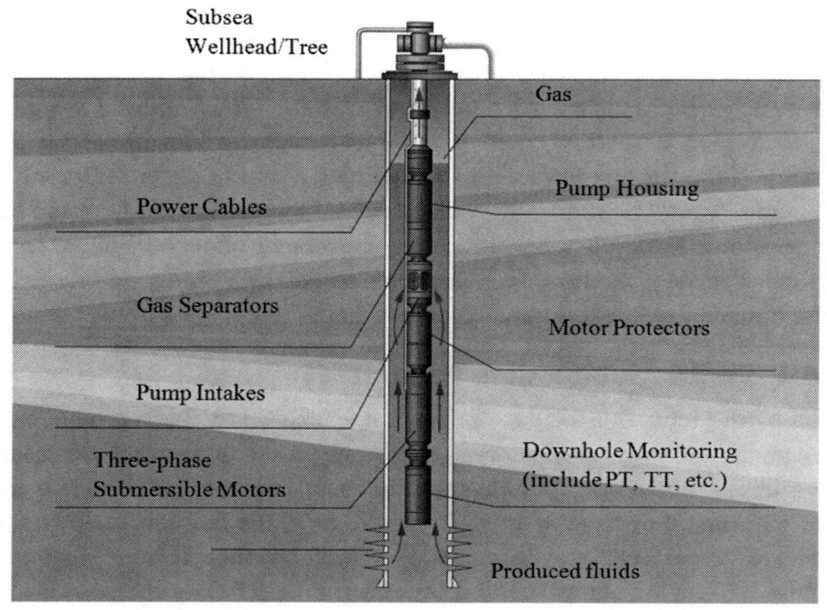

Figure 2-17 Typical Downhole ESP Configuration *(Courtesy Schlumberger)*

The pump rate is a function of the rotational speed, the number of stages, the dynamic head acting against the ESP and the pumped fluid viscosity. These factors dictate the differential pressure across a pump system, and therefore the flow rate. However, for a given pump, there is an optimal design flow rate that maximizes pump efficiency and run life. Figure 2-18 shows the operating range recommended by ESP manufacturers.

Sizing of ESP is based on predicted completion performance, or flow rate. This usually involves examination of the well inflow performance relationship (IPR), which describes the production response to changes in bottomhole pressure (BHP).

Data required for calculation and sizing of ESP includes well data, production data, well fluid conditions, power sources and possible problems etc. Calculations for designing an ESP system include:

- Determination of Pump Intake Pressure;
- Calculation of total dynamic head ;
- Selection of pump type;
- Check of load limits;
- Selection of accessory and optional equipments.

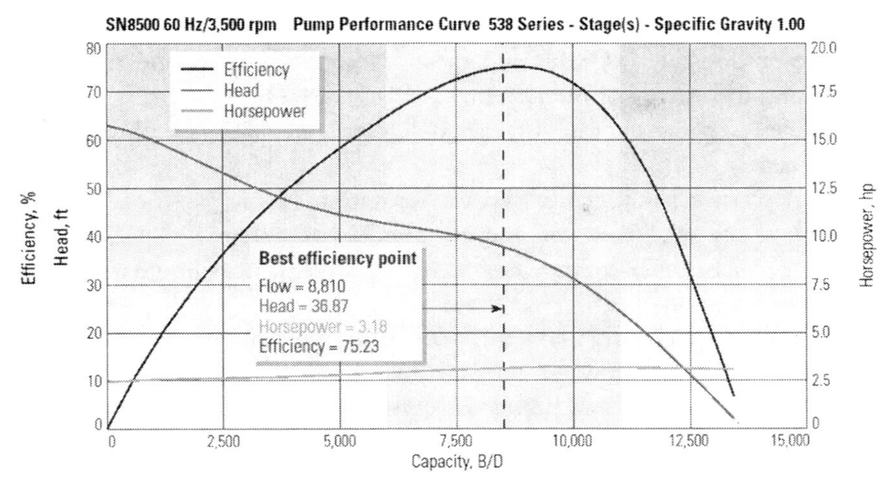

Figure 2-18 Pump Performance Curve *(Courtesy Schlumberger)*

2.7. SUBSEA PROCESSING

Subsea processing (SSP) can be defined as any handling and treatment of the produced fluids for mitigating flow assurance issues prior to reaching the platform or onshore. This includes:

- Boosting;
- Separation;
- Solids management;
- Heat exchanging;
- Gas treatment;
- Chemical injection.

The benefits of introducing subsea processing in a field development could be [4]:

- Reduced total CAPEX, by reducing the topside processing and/or pipeline CAPEX;
- Accelerated and/or increased production and/or recovery;
- Enabling marginal field developments, especially fields at deepwater/ultra-deepwater depths and with long tie-backs;
- Extended production from existing fields;
- Enabling tie-in of satellite developments into existing infrastructure by removing fluid;
- Handling constraints;
- Improved flow management;

- Reduced impact on the environment.

Subsea boosting, as explained in an earlier section, is one means of increasing the energy of the system.

Subsea separation can be based either on two- or three-phase separation:

- Two-phase separators are used for separation of any gas–liquid system such as gas–oil, gas–water, and gas–condensate systems.
- Three-phase separators are used to separate the gas from liquid phase and water from oil.

As can be seen from Figure 2-19, the technology and products required for subsea boosting and water removal are available and in operation, while three-phase separation and subsea gas compression still require some qualification before being put into operation.

Thus, a three-phase separator is useful for the crude consisting of all three phases, namely, oil, water, and gas, whereas a two-phase separator is used for the system consisting of two phases such as gas–oil, gas–water, or gas condensate. Further, subsea separation could have a positive effect on flow assurance, including the risk related to hydrate formation and internal corrosion protection derived from the presence of the produced water in combination with gas.

As opposed to the traditional methods of processing reservoir fluids at a process station, subsea processing holds great promise in that all of the processing to the point where the product is final salable crude is done at the seabed itself. This offers cost benefits and also improves recovery factors from the reservoir. Other advantages include a lesser susceptibility to hydrate formation and lower operating expenditures.

Table 2-3 summarizes the classification of subsea processing systems. Characteristics, equipment, water disposal, and sand disposal are compared for four types of subsea processing systems.

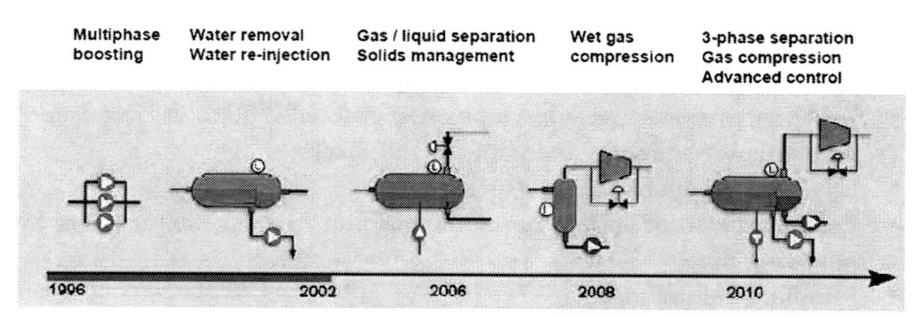

Figure 2-19 Subsea Separation Development [9]

Table 2-3 Classification of Subsea Processing Systems [10]

Classification	Characteristic	Equipment	Water Disposal	Sand Disposal
Type 1	Multiphase mixture is handled directly	Multiphase pump	None; pumped with other produced fluids	None; pumped with other produced fluids
Type 2	Partial separation of the production stream	Separator and multiphase pump; possible use of wet-gas compressor	Possible reinjection of partial water stream, i.e., "free" water	None; pumped with liquid stream
Type 3	Complete separation of the production stream at subsea conditions	Separator and scrubber stages with single or multiphase pump; possible use of gas compressor	Reinjection/ disposal of majority of water stream	Must be addressed
Type 4	Export pipeline quality oil and gas	Multistage separator and fluid treatment; single-phase pumps and compressors	Re-injection/ Disposal of entire of water stream	Must be addressed

2.8. TEMPLATE, CLUSTERED WELL SYSTEM, AND DAISY CHAIN

Subsea production equipment can be configured in multiple ways based on the field specifications and operator's approach to operation.

Field development planners need to work closely with the reservoir and drilling engineers early in the planning stages to establish a good well location plan. Once the reservoir is mapped and reservoir models created, the number of wells, types of wells, and their locations can be optimized. Well layout is usually an exercise of balancing the need to space the wells out for good recovery of the reservoir fluids against the cost savings of grouping the wells in clusters. Add to this the consideration of using extended reach wells, and the number of possible variables to consider becomes great. A further consideration, reservoir conditions permitting, is the use of fewer, high production rate wells through horizontal well completions or other well technology. Here again, there are cost trade-off considerations.

2.8.1. Satellite Well System

A satellite well is an individual subsea well. Figure 2-20 illustrates a typical satellite tie-back system. Satellite wells are typically used for small developments requiring few wells. Often the wells are widely separated and the production is delivered by a single flowline from each well to a centrally located subsea manifold or production platform. Various field layouts must be examined. This evaluation must involve hydraulic calculations and cost sensitivity analyses taking into consideration flowline cost, umbilical cost, and installation cost and flow assurance issues.

2.8.2. Template and Clustered Well System

If subsea wells can be grouped closely together, the development cost will usually be less than that for an equivalent number of widely dispersed wells. Well groupings may consist of satellite wells grouped in a cluster, or a well template, in which the well spacing is closely controlled by the template structure. Figure 2-21 shows a typical manifold cluster layout.

2.8.2.1. Clustered Satellite Wells

Clustered satellite subsea well developments are less expensive than widely spaced satellite wells mainly because of flowline and control umbilical savings. If several satellite wells are in proximity to one another, a separate production manifold may be placed near the wells to collect the production from all of the wells and deliver it in a single production flowline that is

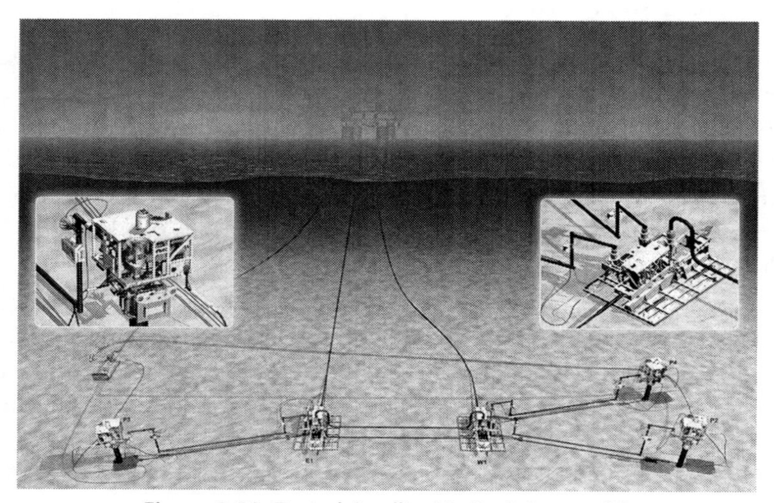

Figure 2-20 Typical Satellite Tie-Back System [2]

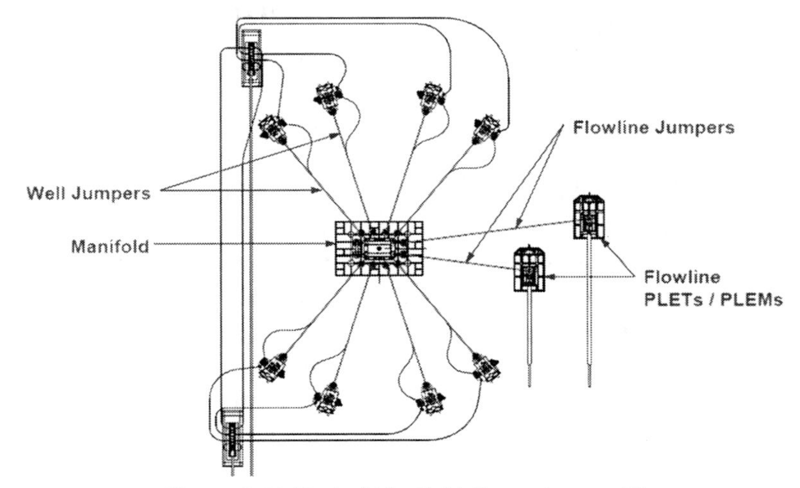

Figure 2-21 Typical Manifold Cluster Layout [2]

connected to the production facility. In addition, a single umbilical and umbilical terminal assembly (UTA) can be used between the well cluster and the production platform. Figure 2-22 shows a field with eight clustered satellite wells, two subsea production manifolds, and a production umbilical and UTA.

In the case of clustered satellite wells, wells may be placed from several meters to tens of meters from one another. The wider well spacing is often

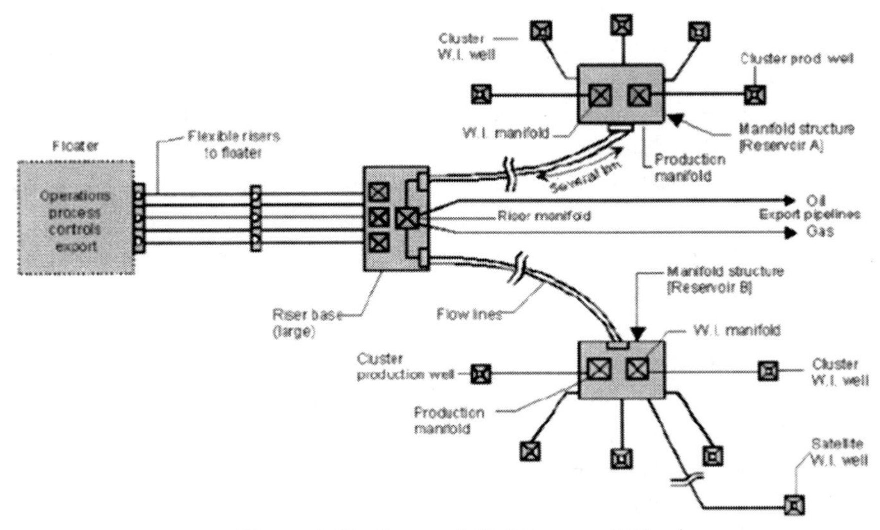

Figure 2-22 Clustered Well Systems [11]

dictated by a desire to be able to position the drilling rig over one well without imposing dropped object risk on adjacent wells. It is hard to precisely control the spacing of individual satellite wells, so crossover piping and control umbilicals must be able to accommodate the variations in spacing.

2.8.2.2. Production Well Templates

A template is a seabed structure that provides guidance for drilling and/or other equipment. It also works as the structural framework supporting other equipment, such as manifolds, risers, wellheads, drilling and completion equipment, and pipeline pull-in and connection equipment. The structure should be designed to withstand any load from thermal expansion of the wellheads and snag loads on the pipelines. Production from the templates may flow to floating production systems, platforms, shore or other remote facilities.

The production well template is used to support a manifold for produced fluids. Wells would not be drilled through such a template, but may be located near it or in the vicinity of the template.

The clustering wells can also be arranged by means of a well template. Well templates are structural weldments that are designed to closely position group of well conductors. The well template is typically used to group several subsea wells at a single seabed location. Apart from reservoir considerations, the number of wells in a well template is limited by the size of the well template that can be handled by the installation vessel. Small templates are usually deployed from the drilling rig. Larger ones may require a special installation vessel with heavier lift capacity or better handling characteristics.

The advantages of the production well template compared to the Clustered satellite Wells are summarized as,

- Well are precisely located and manifold piping and valves are incorporated in the templates.
- Piping and Jumpers may be prefabricated and tested before deployment offshore, therefore, installation time is reduced and less expensive.

2.8.3. Daisy Chain

A daisy chain configuration connects wells in serious, one after the other by flowlines. The flowline can be either connected to the wells by subsea jumpers, or directly connected to the flowbase of the wells if applicable. Figure 2-23 shows a flowline loop connected to the individual subsea wells along its route, which forms the daisy chain layout. The daisy chain field

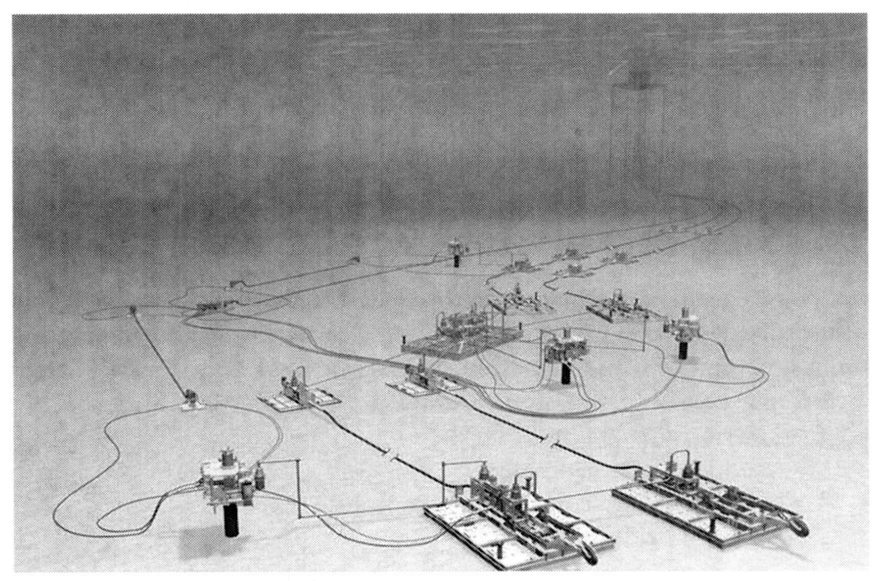

Figure 2-23 Typical Manifold Daisy Chain Layout [2]

layout is considered as an economic solution comparing to the cluster manifold layout in case there are several satellite wells e.g. from different marginal fields.

Typical example of daisy chain field architecture is Canyon Express field which is located in GoM. This project involvs 10 subsea wells in three different deepwater fields. Staring from the Canyon Station platform, one flowline connects 2 wells in the first field, 2 in the second, and is ended by a sled, which is then connected to the third field by a subsea jumper. Another flowline connects 2 wells in the third field, 2 in the second and the remained 2 wells in the first and finally tied back to the platform. These dual flowlines formed a daisy chain piggy loop.

The daisy chain approach of Canyon Express not only makes use of flowline and equipment already used and paid for, but also saves on the capital expenses comparing to a cluster manifold solution according to a conceptual study made in the early phase of the project .

Typical features of daisy chain field architecture include:

- Inline sleds may be installed at each well location thus Xmas trees can be connected to the flowline by means of jumpers;
- Subsea multiphase flow meters may be required to ensure accurate flow allocation among different wells;

- Flow assurance analysis is key in formulating the production envelop for the daisy chain flowlines;
- Subsea chokes may be necessary on each well/Xmas Tree;
- Round-trip pigging is possible to ensure timely removal of wax build-up in the flowline

2.9. SUBSEA FIELD DEVELOPMENT ASSESSMENT

For subsea field development, field layout designs are selected based on historical data obtained from other projects, and various brainstorming and discussion sessions with teams from all other subdisciplines. Critical design factors may consist of, but are not limited to, the following [4]:

- Engineering and design;
- Cost and schedule involved;
- Well placement and completion complexity;
- Flexibility of field expansion;
- Ease of construction and fabrication for the subsea hardware;
- Intervention capability (easy, moderate, or difficult to intervene);
- Rig movement (offset from mean under extreme environmental conditions);
- Installation and commissioning, such as ease of installation and commissioning, flexibility of installation sequence;
- Reliability and risk of the field architecture;
- ROV accessibility.

The inputs from the entire project team are used to assign percentage weights to each of these design factors. Normally, all of the various types of field layout designs, such as the subsea tie-back, subsea stand-alone, or subsea daisy chain, can be applied in the field. Reliability, risk assessment, and economic balance are the dominant factors when deciding what kinds of field layout will be chosen.

The following material presents an overview of the available means of adding energy to the production fluid and assessing the best location for the artificial lift to be introduced in a generic deepwater field development. The objectives are:

- To evaluate the artificial lift options including lift gas, electrical motor-driven pumps (ESPs or multiphase pumps) and hydraulic submersible pumps (HSPs);
- To determine the best location (riser base, subsea manifold, or down-hole) for artificial lifts using lift gas;

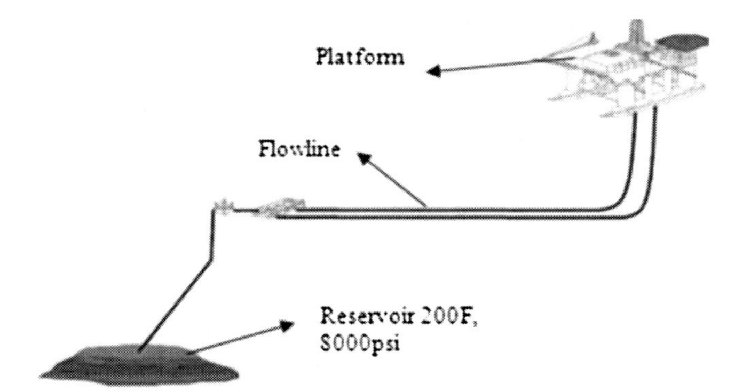

Figure 2-24 Overview of Deepwater Field Development [12]

- To determine the best location for installing electrical motor-driven pumps for artificial lifts (ESPs downhole versus multiphase pumps at subsea manifold or riser base);
- To determine the best configuration (open loop versus closed loop) for installing an HSP in a riser in conjunction with coiled tubing for artificial lift.

Figure 2-24 shows an overview of deepwater field development. An example is given in the following sections to show how to calculate and analyze a field development.

2.9.1. Basic Data

Reservoir

- Pressure = 8000 psia;
- Temperature = 200°F;
- Well PI = 20 bpd/psi;
- Reservoir pressure maintenance by water injection.
- True vertical depth (TVD) = 11,000 ft from mudline;
- Tubing size = 5.5-in. OD (0.36-in. WT);
- Kick-off point = 2,000 ft below mudline;
- Kick-off angle = 45°;
- Roughness of tubing = 0.0018";
- U value for wellbore = 2.0 Btu/(ft^2 hr °F).

Subsea Development

- Water depth of 8000 ft;
- Subsea well completion, dual flowlines, three identical wells per flowline;

- Tie-back distance from subsea manifold to topsides (flowline + riser) of 25,000 ft;
- Steel catenary riser, pipe-in-pipe configuration considered for both riser and flowline with an assumed U value of 0.2 Btu/(ft^2 hr °F);
- Seabed terrain assumed to be flat;
- Typical environmental conditions including seawater temperature profile and air temperature were used;
- Riser/flowline roughness = 0.0018 in.;
- 10- and 12-in. nominal riser/flowline sizes were evaluated.

A reservoir's inflow performance relation is achieved by using Fetkovick's equation:

$$Q(P_{wf} \cdot P_R) = \text{AOFP}(P_R) \cdot \left[1 - \left(\frac{P_{wf}}{P_R}\right)^2\right]^n$$

where

$$\text{AOFP}(PR) = \left(\frac{P_R^2}{\text{bar}^2 \cdot e^C}\right)^n \cdot \frac{\text{Sm}^3}{\text{day}}$$

The following slope factor and intercept factor has been selected: $n = 0.82$, $C = 0.35$.

The productivity index for the selected inflow performance relation with a wellbore flowing pressure of 325 bar and a reservoir pressure of 350 bar is as follows:

$$PI(P_{wf} \cdot P_R) = \left|\frac{d}{dp_{wf}} Q(P_{wf} \cdot P_R)\right|$$

$$PI(325 \cdot \text{bar}, 350 \cdot \text{bar}) = 69.271 \cdot \frac{\text{Sm}^3}{\text{day} \times \text{bar}}$$

Correction to the inflow performance with regard to increased water saturation of the reservoir has not been made.

2.9.2. Water-Cut Profile

The water cut of the well fluid from the reservoir is assumed as described in the Figure 2-25. An increase in water production occurs much more slowly in this depletion case than would be the case if water injection for pressure maintenance was included. The water-cut profile is described as a function of accumulated liquid from the reservoir.

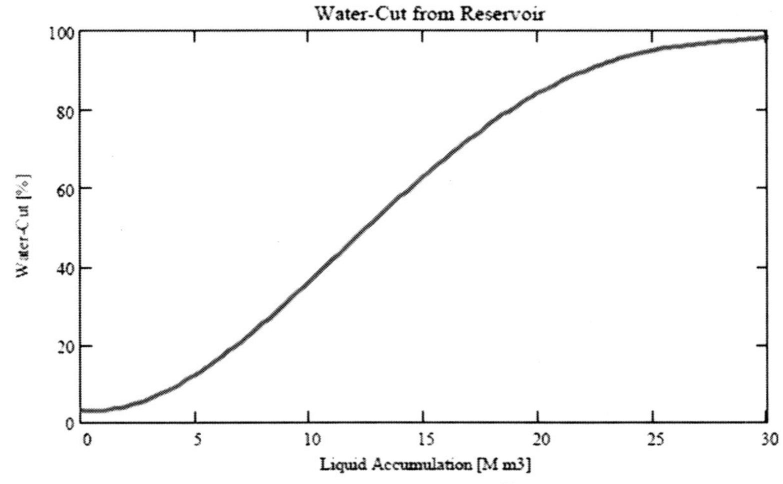

Figure 2-25 Water-Cut Profile [12]

When a volume of liquid of 30 million cubic meters (Mm^3) has been extracted from the reservoir, the water-cut has reached about 99%. The total volume of oil produced to the topsides specification is shown next when 30 Mm^3 liquid has been extracted:

$$\frac{1}{B_{oil}}\int_{0\times m^3}^{30\times Mm^3}(1 - \text{Water-Cut}(N_p)dN_p) = 11.69\cdot Mm^3$$

2.9.3. Process Simulations

Process simulations have been carried out with HYSYS process simulator. The OLGAS correlation has been applied for pressure drop calculations in pipe segments. The model is illustrated in Figure 2-26, showing a simulation of a multiphase pump at the riser base in a situation when 5 Mm^3 liquid has been produced from the reservoir.

A series of simulations are performed for various reservoir accumulations to establish the liquid production from the reservoir as a function of the reservoir condition. These simulations are then repeated for the various production cases that are investigated.

The time it takes to produce a certain volume from the reservoir is calculated by integrating the liquid rate function with regard to volume accumulation. Accumulated production versus time is illustrated in Figure 2-26 for a multiphase pump at the wellhead in the 1000-m case.

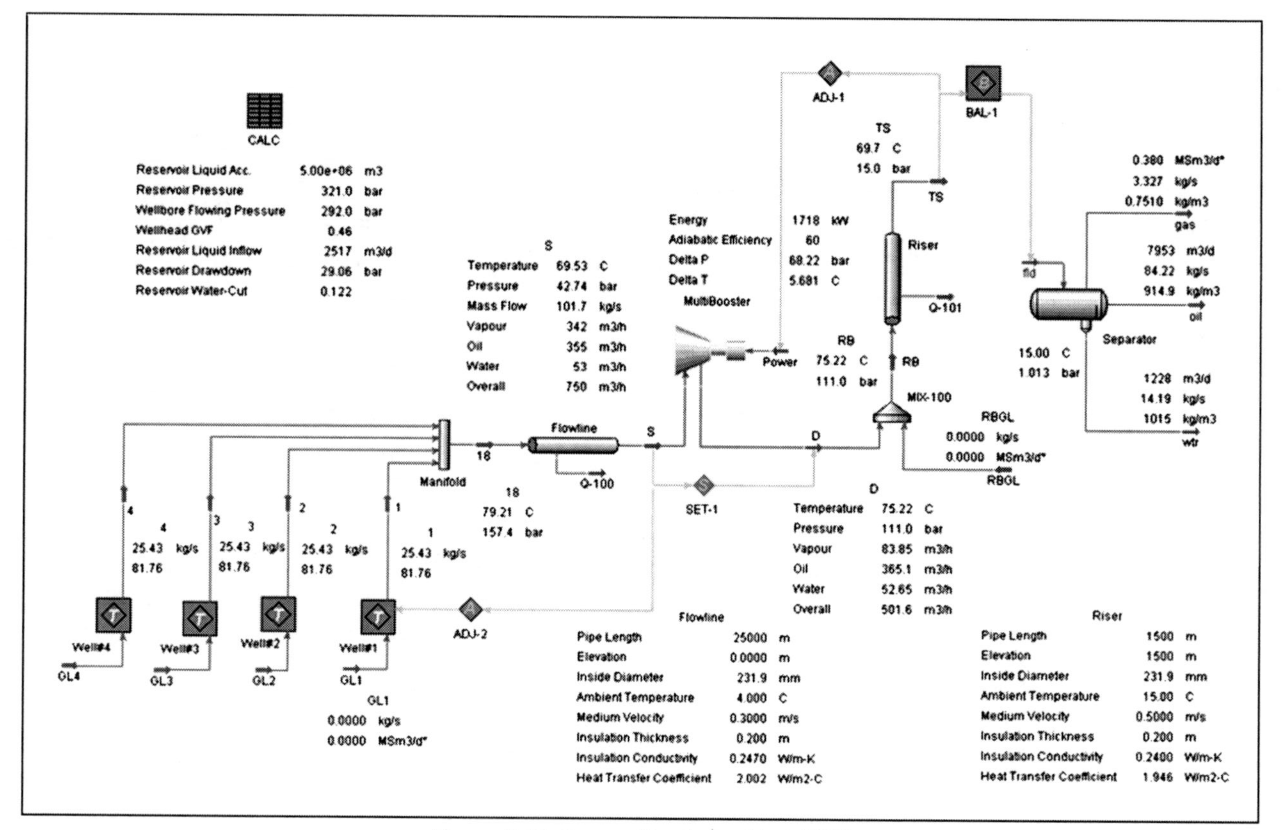

Figure 2-26 Process Simulation Model [12]

REFERENCES

[1] C. Claire, L. Frank, Design Challenges of Deepwater Dry Tree Riser Systems for Different Vessel Types, ISOPE Conference, Cupertino, 2003.
[2] M. Faulk, FMC ManTIS (Manifolds & Tie-in Systems), SUT Subsea Awareness Course, Houston, 2008.
[3] R. Eriksen, et al., Performance Evaluation of Ormen Lange Subsea Compression Concepts, Offshore, May 2006.
[4] CITEPH, Long Tie-Back Development, Saipem, 2008.
[5] R. Sturgis, Floating Production System Review, SUT Subsea Awareness Course, Houston, 2008.
[6] Y. Tang, R. Blais, Z. Schmidt, Transient Dynamic Characteristics of Gas-lift unloading Process, SPE 38814, 1997.
[7] DEEPSTAR, The State of Art of Subsea Processing, Part A, Stress Engineering Services (2003).
[8] P. Lawson, I. Martinez, K. Shirley, Improving Deepwater Production through Subsea ESP Booster Systems, inDepth, The Baker Hughes Technology Magazine, vol. 13 (No 1) (2004).
[9] G. Mogseth, M. Stinessen, Subsea Processing as Field Development Enabler, FMC, Kongsberg Subsea, Deep Offshore Technology Conference and Exhibition, New Orleans, 2004.
[10] S.L. Scott, D. Devegowda, A.M. Martin, Assessment of Subsea Production & Well Systems, Department of Petroleum Engineering, Texas A&M University, Project 424 of MMS, 2004.
[11] International Standards Organization, Petroleum and Natural Gas Industries-Design and Operation of the Subsea Production Systems, Part 1: General Requirements and Recommendations, ISO 13628-1, 2005.
[12] O. Jahnsen, G. Homstvedt, G.I. Olsen, Deepwater Multiphase Pumping System, DOT International Conference & Exhibition, Parc Chanot, France, 2003.

CHAPTER 3

Subsea Distribution System

Contents

Subsea Engineering Handbook
ISBN 978-0-12-397804-2, doi:10.1016/B978-0-12-397804-2.00003-5

3.1. INTRODUCTION

A subsea distribution system (SDS) consists of a group of products such as umbilical and other in-line structures that provide communication from subsea controls to topside. This chapter describes the main components of the SDS currently used in subsea oil/gas production, and defines its design and the functional requirements of the system.

The type of system to be discussed in this chapter should be designed to perform the following functions:

- Hydraulic power distribution;
- Chemical injection distribution;
- Electrical power distribution;
- Communication distribution.

3.1.1. System Architecture

The SDS normally includes, but is not limited to, the following major components:

- Topside umbilical termination assembly;
- Subsea accumulator module;
- Subsea umbilical termination assembly, which includes:
  - Umbilical termination head (UTH);
  - Hydraulic distribution module;
  - Electrical distribution module;
  - Flying leads.
- Subsea distribution assembly;
- Hydraulic flying leads;
- Electrical flying leads;
- Multiple quick connects;
- Hydraulic coupler;
- Electrical connector;
- Logic caps.

Figure 3-1 illustrates the relationships of the main structures.

The subsea umbilical termination assembly mainly consists of inboard multiple quick connect (MQC) plates, mounting steel structures, a lifting device, mudmat, logic cap, long-term cover, field assembled cable termination, and electrical connectors.

The subsea distribution assembly mainly consists of a hydraulic distribution module (HDM) and electrical distribution module (EDM). The HDM consists of inboard MQC plates, mounting steel structures, lifting

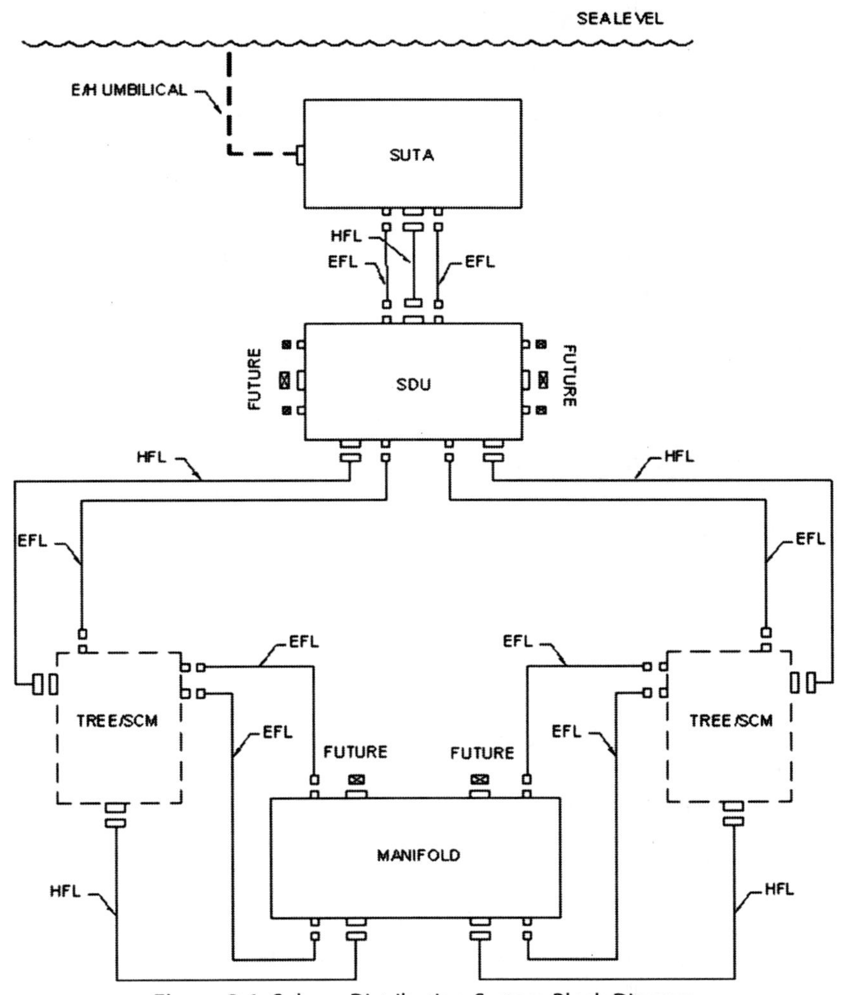

Figure 3-1 Subsea Distribution System Block Diagram

padeyes, a mudmat, logic cap, and long-term cover. The EDM consists of bulkhead electrical connectors and cables and, in some cases, an electrical transformer module.

Hydraulic flying leads (HFLs) mainly consist of two outboard MQC plates with holding structures and steel tubes. Electrical flying leads (EFLs) is mainly consist of two electrical connectors and a number of cables.

3.2. DESIGN PARAMETERS

3.2.1. Hydraulic System

The following main parameters need to be determined for the hydraulic system:

- Reservoir sizing;
- The time to prime the hydraulic system from a depressurized state;
- Opening and closing response times of the process valves under conditions of minimum and maximum process pressure;
- The time for the pressure to recover following a process valve opening;
- The time to carry out a sequence of valve openings, such as the opening of a tree (neglecting choke valve operation);
- The stability of opened control and process valves to pressure transients, caused by operation of the other control and process valves (sympathetic control valve de-latching, process valve partial closing, etc.);
- Response time to close process valves in the event of a common close command, such as an emergency shutdown at the surface, venting off hydraulic control valves via supply lines;
- The time to vent the umbilical hydraulic supplies;
- The impact that failure of subsea accumulation has on the safe operation and closure of the process valves;
- The extent of control fluid leakage rate that can be accommodated by the system;
- System response times for simultaneously opening and closing multiple choke valves.

3.2.2. Electrical Power System and Communication

The following main parameters need to be determined by means of a system power demand analysis:

- Voltage at subsea electronic module (SEM) for maximum and minimum SEM power loads;
- Voltages at each SEM at maximum and minimum numbers of subsea control modules (SCMs) on the subsea electrical distribution line;
- Voltages at SEM at minimum and maximum designed umbilical lengths;
- Voltages at SEM at cable parameters for dry and wet umbilical insulations;
- Minimum and maximum subsea power requirements;
- Maximum current load;
- Topsides electrical power unit power factor versus SCM voltage.

A communication analysis is conducted to determine the minimum specifications of SCM and master control station (MCS) modems:

- Modem transmit level;
- Modem receive sensitivity;
- Modem source/load impedance.

3.3. SDS COMPONENT DESIGN REQUIREMENTS

3.3.1. Topside Umbilical Termination Assembly (TUTA)

The TUTA as shown in Figure 3-2 provides the interface between the topside control equipment and the main umbilical system. This fully enclosed unit incorporates electrical junction boxes for the electrical power and communication cables, as well as tube work, gauges, and block and bleed valves for the appropriate hydraulic and chemical supplies.

The TUTA will typically be located near the umbilical J-tube on the host. Basically, it includes an electrical enclosure in a lockable stainless steel cabinet certified for its area of classification with ingress protection to fit its

Figure 3-2 Topside Umbilical Termination Assembly (TUTA) [1]

location. Additionally the valves in the TUTA comply with requirements for valves in flammable services as stated on fire testing standards.

The termination unit at the topside end of the umbilical is designed for hang-off and includes a bull-nose suitable for pulling the umbilical up through the host guide tube onto a termination support. For a free-flooded umbilical, it is sufficient to seal off the individual ends of the conductors and tubes and use an open bull-nose.

Tubes must be individually sealed to prevent hydraulic oil loss and water ingress during the pulling operations. Electrical conductors must be sealed to prevent water ingress along the insulation.

If the umbilical is intended for temporary laydown, pressure relief during retrieval is considered.

Basically, the J-tube seals can withstand a 100-year maximum wave and maintain its differential pressure capability during service life. Free span corrections between the J-tube bell mouth and seabed are also be designed according to the project's specific design basis.

3.3.2. Subsea Umbilical Termination Assembly (SUTA)

The SUTA shown in Figure 3-3 is the subsea interface for the umbilical and may serve as the distribution center for the hydraulic and chemical services at the seabed. The SUTA is connected to the subsea trees via HFLs.

SUTA is typically composed of the following:

- UTH;
- Flying leads to connect the UTH and HDM;

Figure 3-3 Subsea Umbilical Termination Assembly *(Courtesy OCEANEERING)*

- HDM if available;
- Mudmat foundation assembly with stab and hinge-over mechanism;
- MQC plates to connect to HFLs.

The type of SUTA to be used is determined by field architecture considerations, and is further defined during detailed design.

The SUTA is designed with the following flexibilities:

- Provision for possible links to additional umbilicals;
- Provision for spare header included in case of internal piping failure;
- Demonstration of design that allows flexible installation;
- Retrievability and reconfiguration options;
- Redirection of any umbilical line to any tree service via flying leads.

SUTAs are installable with or without drilling rig assistance. The mudmat includes attachment points for adjusting the subsea position, with application of external horizontal forces, when direct vertical access for installation is not possible.

A SUTA's dimensions allow for ground transportation from the fabrication facility to the final destination.

The umbilical is permanently terminated in the UTH.

The infield first end SUTA is a smaller unit, its purpose is to link the electrical of hydraulic/chemical lines.

HFLs are used to connect hydraulic/chemical lines between SUTAs and subsea trees/manifolds.

EFLs are used to connect electrical power/communication from the umbilical termination assembly (UTA) to the manifold and tree-mounted SCM.

The UTH, HDM, and EDMs are each independently retrievable from the UTA's mudmat.

Each of the electrical quads (umbilical cables with four conductors) is terminated in electrical connectors at the UTH.

The EFL interconnects between these UTH connectors and the EDM connectors, routing power and communication from the UTH to the EDM. Subsea electrical distribution is done from the EDM to the subsea trees and production manifolds.

Two flying leads provide the hydraulic/chemical interconnections between the UTH and the HDM.

3.3.3. Umbilical Termination Head (UTH)

The UTH, shown in Figure 3-4, consists of a structural frame, hinging stab, MQC plates, super-duplex tubing, and bulkhead style ROV electrical

Figure 3-4 Umbilical Termination Head (UTH) [2]

connectors. Umbilical services are routed to the MQC plates and electrical connectors for distribution to subsea production equipment.

The SUTA terminates the hydraulic subsea umbilical and provides a flange connection to attach the umbilical termination. The umbilical is composed of super-duplex tubes.

All tubing in the UTH and HDM is welded to the hydraulic couplers located within the MQC plates. The number of welded connections between the couplers and the tubing is kept at a minimum. Butt-weld joints are preferred over socket-welded connections. The design allows for full opening through the tubing and the welded area. The design also avoids any area where crevice corrosion may occur.

The termination flange connection between the SUTA and the umbilical is designed to accomplish the following condition:

- Because the SUTA is installed with its stab and hinge-over stinger, the flange is capable of supporting the weight of the SUTA as well as all installation loads.
- After installation, the SUTA may need to be lifted from its mudmat and stab and hinge-over funnel. As the SUTA is lifted and brought back to the surface, the flange is capable of supporting the "unsupported" umbilical weight with a 50% safety factor.

The UTH at minimum complies with the following requirements:

- At minimum, the UTH is designed to allow termination of a minimum of nine umbilical steel tube lines and two electrical quads.

- The structural frame is designed to securely attach to and support the umbilical and end termination as well as provide mounting locations for the MQC plates and bulkhead electrical connectors.
- The hinging stab is attached to the frame and is the interface between the UTH and mudmat structure.
- The UTH frame size is minimized in order to simplify handling and overboarding.
- The UTH are kept small to fit inside most umbilical overboarding chutes, and can be maneuvered through most vessel umbilical handling systems.
- The recovery padeyes are designed to take the full recovery load of the UTH and the umbilical.
- The UTH, combined with the umbilical split barrel (supplied by the umbilical manufacturer), is designed to sit on the umbilical reel of an installation vessel at the end and/or at the beginning of the umbilical.
- Tubing connections and other possible points of failure are reduced as much as possible in the UTH to avoid having to retrieve the umbilical to repair a failed component.

3.3.4. Subsea Distribution Assembly (SDA)

The SDA, as shown in Figure 3-5, distributes hydraulic supplies, electrical power supplies, signals, and injection chemicals to the subsea facilities. The facilities can be a subsea template, a satellite well cluster, or a distribution to satellite wells. The SDA connects to the subsea umbilical through the SUTA.

3.3.4.1. Construction

The SDA frame is fabricated from carbon steel coated in accordance with a subsea paint specification. The frame is designed for lifting and lowering onto a location on a subsea production structure. Alternatively, the SDA can be located on a mudmat, simple protective frame, or monopile.

3.3.4.2. Interface with the Umbilical

The SUTU can connect to the SDA with a vertical/horizontal stab and hinge-over/clamp connection. Alternatively, it can connect via electrical and hydraulic jumpers at a seabed level pull-in location or manifold structure pull-in location using an ROV or diver connectors. If the field layout demands, the jumpers can route through a weak link breakaway connector.

3.3.4.3. Interface with SCM

The jumpers from the SDA to the subsea accumulator module mounting base (SCMMB) are connected using an ROV.

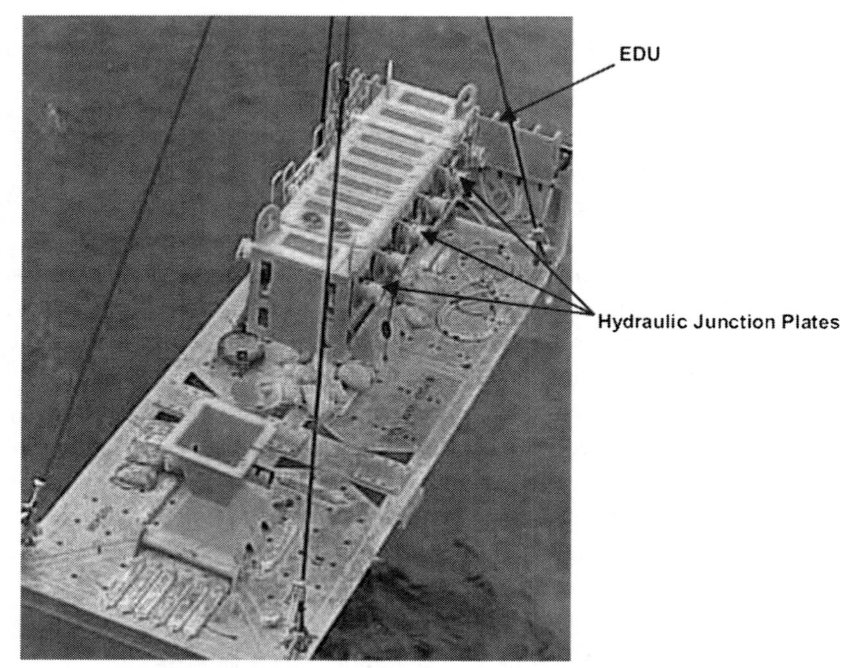

Figure 3-5 Subsea Distribution Assembly *(Courtesy TotalFinaElf)*

3.3.4.4. Electrical Distribution

The electrical distribution is usually contained in an oil-filled, pressure-balanced, fabricated and coated carbon steel housing called an electrical distribution unit (EDU). (Non–pressure-balanced, resin-filled junction boxes are sometimes used, but these do not allow future maintenance and require the encapsulated components to be suitable for use at depth; designs requiring current-limiting devices may be housed in a one-atmosphere enclosure.) Entry and exit of the EDU is by flange-mounted electrical controlled environment-type connectors.

The connectors are configured so that any connections that may be accidentally disconnected live have the live conductors protected from the seawater.

Cable tails from the back of the electrical connectors within the oil-filled housing are connected for distribution as required by the control system architecture and the system redundancy capability.

The requirement for fault protection is dependent on the system design and the number of wells that could potentially be disabled by a subsea cable fault. Three types of electrical protection are used: fuses, circuit breakers, and thermal resetting trip devices.

Fuses are not effective, as slow-blow fuses are necessary in order to cater to the inrushing current while charging up the umbilical. This makes fuses ineffective in isolating a fault in the distribution system without overloading the remainder of distribution outlets and, generally, a fuse would not blow before the line insulation fault trip in the EPU is activated.

Circuit breakers have been used subsea in EDUs, but are not commonly used because the circuit breaker reset mechanism has to penetrate through the EDU housing using O-rings, which introduces a potential fault path.

The thermal resetting devices are semiconductor devices and due to the technology required, they are not available from all suppliers.

3.3.4.5. Hydraulic and Chemical Distribution

The hydraulic distribution is by tubing from the incoming interface connection routed around the structure to the distribution outlets. The stab connections and the tubing are generally made of type 316 stainless steel. The tubing terminations are all welded for integrity. The tubing, which is usually installed at a fabrication site, has to be flushed and cleaned to the integrity required by the subsea control system.

Chemical injection systems generally require larger volume flows during normal operation, and are also subject to increased viscosity at lower seabed temperatures. Therefore, larger bore tubing or piping is generally used, again welded to maintain integrity.

Multiple stab plate hydraulic connections must have some movement in order to allow for alignment during makeup. Also tubing is often installed on structures using clamps with plastic inserts. This can leave the tubing and end connections floating without cathodic protection. It is essential that these items be electrically bonded to the main structure cathodic protection system to avoid rapid corrosion of the system.

Other material that may be considered for the distribution piping or tubing is carbon steel for the chemical injection system, or more exotic materials such as Duplex or Super Duplex stainless steel.

To ensure correct mating of the respective parts, guide pins are used on stab plates, and single connections may have different size quick-connect couplings or may be keyed for proper orientation.

3.3.4.6. ROV Connection

The access of an ROV to the SDU has to be carefully considered. It is not necessary to have a docking station for ROV makeup, but docking may make certain tasks easier. If the field survey shows strong currents at the seabed and changeable directions, then ROV docking is necessary.

With multiwell applications where the ROV must remove connectors from parking positions and hook up at positions on the SDU, it is essential that the ROV does not get entangled in any of the other flying leads. This can cause damage to the flying leads or may entangle the ROV where it would need to cut flying leads in order to free itself.

Clear marking of the connection point is essential to ensure that the ROV pilot can orient the ROV at the desired location and to ensure that the correct hookup in low-visibility subsea.

3.3.5. Hydraulic Distribution Manifold/Module (HDM)

The purpose of the HDM is to distribute hydraulic fluid (LP and HP) and chemicals to each of the subsea trees, production manifolds, and to future expansions. The HDM as shown in Figure 3-6 consists of multiple inboard MQC plates on a frame that is mounted on an umbilical termination assembly mudmat (UTA-MM).

The hydraulic distribution depends on:

- The number of valve functions;
- The actuator sizes;
- The frequency of valve operation while operating the system;
- The location of the valves, trees, and manifolds.

Stored hydraulic energy in accumulator bottles (SAM) is required to allow fast sequential opening of subsea valves without affecting the operation

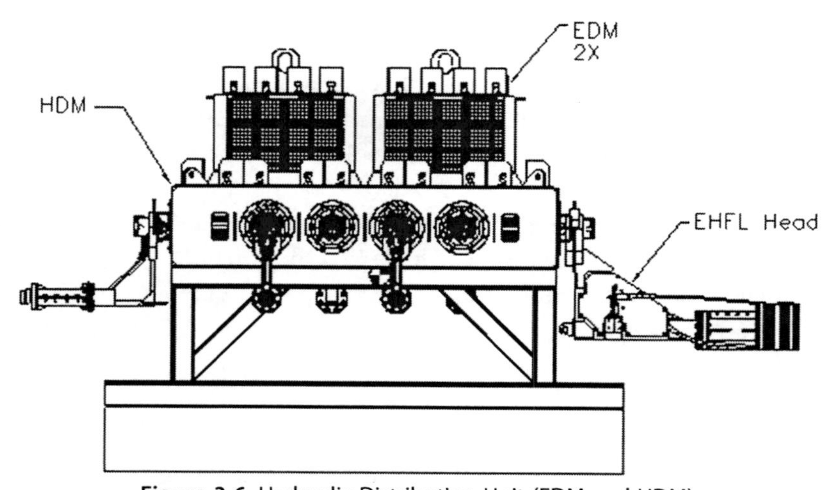

Figure 3-6 Hydraulic Distribution Unit (EDM and HDM)

of other valves in the system. The accumulators are charged up by the hydraulic supply in the umbilical.

The HDM allows for the reconfiguration of services by changing logic caps during the service life.

Each HDM has a spare line going to all inboard MQC plates to allow the umbilical services to be reconfigured in case of an umbilical line failure or a change in the operability and/or chemical requirements of the existing field or an expansion location.

ROV valves are being used where desired functionality or isolation cannot be achieved by a logic cap. A spare manifold is included in the HDM to provide flexibility for reconfigurations.

The minimum component test pressure is 50% above design pressure.

Basically, all parts and components of the subsea distribution system are clean before and during assembly.

The subsea hydraulic system is robust and designed to tolerate accidental contamination by seawater and particles. It should be designed to avoid accumulation of such contamination. All components are qualified for operation in fluid contaminated with solid particles. Vulnerable parts with very low fluid consumption (e.g., directional control valve pilot stages) should be protected by filters. Otherwise strainers should be used. A logic/crossover cap is used to reconfigure spare umbilical line to replace failed line.

3.3.6. Electrical Distribution Manifold/Module (EDM)

The number of electrical connectors in series is kept to a minimum. Redundant routing to the module connectors should follow different paths if possible.

The cables should be installed into self-pressure compensated fluid-filled hoses. The fluid is a dielectric type. Dual barriers are provided between water and the conductor. Both barriers should each be designed for operation in seawater.

Manifold electrical distribution cabling and jumper cables from umbilical termination to the subsea control module (SCM) can be replaced by an ROV. Removal of the faulty cable is not necessary.

In the case of a failure in the distribution system, it should be possible to run several SEMs on one pair after using an ROV for reconfiguration.

If one electrical line is supplying more than two SEMs, the distribution and isolation units can be located in a separate, retrievable distribution module.

Connection of electrical distribution cabling and electrical jumpers is made by an ROV using simple tools, with minimum implications for rig/vessel time.

The cable assemblies should be designed and installed such that any seawater entering the oil will move away from the end terminations by gravity.

The EDM is configured with the following in mind:

- Each UTA is configured with a single EDM.
- The EDM is capable of distributing the electrical and communication services of two umbilical quads from a UTH to the multiple subsea assemblies.
- There are four electrical parking positions for parking of the flying lead end of the distribution harness during installation and retrieval, each protected long term by an ROV universal protection cap.
- No other electrical parking positions are provided on the EDM. Additional parking provisions will be provided by the deployment of a separate parking frame assembly if required.
- There are eight electrical output connectors (receptacles), each protected long term with an ROV universal protection cap.
- The EDM consists of oil-filled electrical splitter harnesses mounted within a retrievable structure. The splitter harness multiple outputs are terminated in bulkhead-mounted ROV electrical plug connectors with socket contacts.
- At the input end, which connects to the UTH, the single leg of the harness is terminated with an ROV-mateable plug connector with socket contacts.
- The EDM interfaces with the tree/manifold supply EFLs and step-out EFLs to the static umbilicals, as shown earlier in Figure 3-6.
- Power and communication are provided from a female electrical connector to a male connector.

3.3.7. Multiple Quick Connects (MQCs)

The MQC plates are outfitted with the appropriate number of couplers to match the number of tubes in the hydraulic flying leads. Any unused positions in the MQC plate are outfitted with blanked-off couplers. All tubes and coupling assignments need to match the tree assignments. Figure 3-7 shows an MQC assembly.

All couplers are energized, and the MQC is capable of withstanding full coupler pressures (design pressure and test pressure). The design of couplers is considered to have the potential for external leakage due to external pressure being greater than internal pressure.

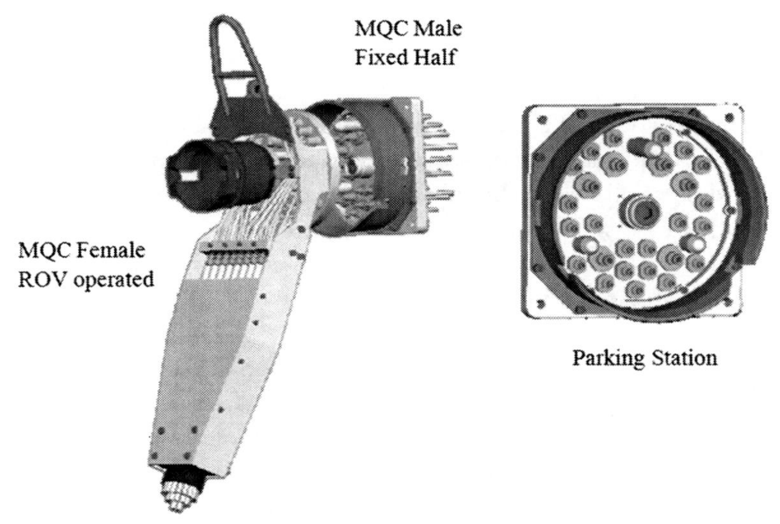

Figure 3-7 MQC Assembly *(Courtesy UNITECH)*

MQC plates are located within the SUTA envelope/tree/manifold side, for example, and are protected from any damage during installation. The location and arrangement of the MQC plates is ROV friendly and allow for connection and disconnection of any MQC connection without disturbing the others.

MQC plates are supplied with pressure-retaining protective covers. These may be required to remain subsea for extended periods of time and to be removable by an ROV, prior to installation of flying leads.

MQC plates have ROV-operable attachment means in order to connect to the flying leads. All standard and special ROV tools required for this connection must be specified. Besides it is also required to specify the following ROV makeup and disengage torque requirements:

- Starting torque;
- Running torque;
- Maximum torque required to engage fully energized couplers;
- Minimum torque required to disengage the couplers or break-out torque with and without internal pressure at the subsea ambient condition;
- Maximum torque required to engage the override mechanism.

The MQC designs provide means for the ROV to verify the orientation and alignment of the outboard plate, prior to insertion into the inboard MQC plate.

The necessary tools to release the outboard plate from the inboard plate in the event the plates cannot be separated by the ROV should be well equipped. MQC plates are outfitted with an emergency release device. The design also prevents the MQC plates from engaging if they are not aligned properly. The couplers are not engaged prior to proper inboard plate alignment. For interface information purposes, all data for the acceptable angular, axial, and radial misalignment tolerance to ROVs must be recorded.

The designs allow for the full guidance of the attachment mechanism and avoid any possibility of thread engagement. The design also allows for the verification of the alignment and engagement of the threads.

MQC plates have identification numbers that are visible for ROV inspection.

MQC plates have windows for ROV inspection. These windows will be used for inspecting full engagement of the MQC plates as a means to verify that installation was properly made, and to inspect for leaks during operation.

3.3.8. Hydraulic Flying Leads and Couplers

The hydraulic flying leads deliver hydraulic and chemical services from the SUTA to the tree or from the SDA to the tree, as shown in Figure 3-8.

The HFLs are made up of three main components: the tubing bundle, the steel bracket assembly heads, and the MQCs. The bundle is terminated at both ends with an MQC plate, which is used to connect the HFL to the trees, or the UTA. These assemblies have padeyes for handling and deployment purposes.

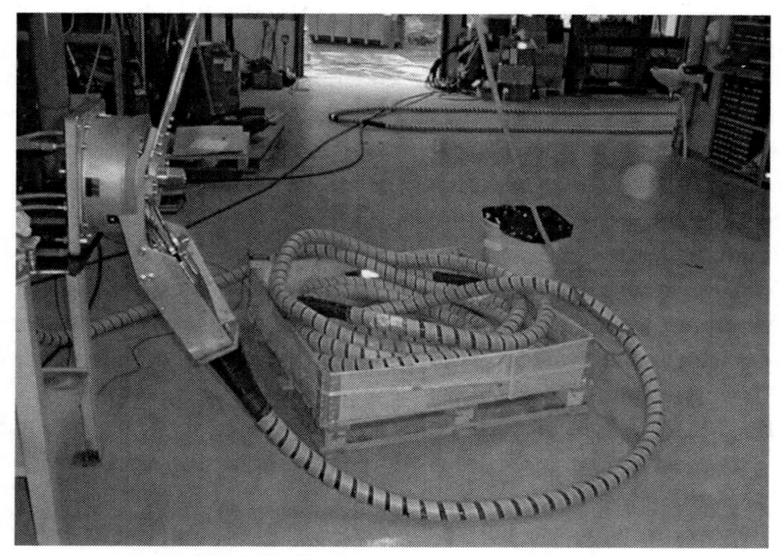

Figure 3-8 HFL Assembly *(Courtesy UNITECH)*

3.3.8.1. Construction

Tubes are bundled and jacketed for protection and to prevent kinks in the tubes.

End terminations from the tubes to the MQC assemblies and to the couplers are welded and NDE (None Destructive Examination) examined. Termination assemblies are structurally sound and capable of withstanding all transportation, installation, and operation loads.

3.3.8.2. Connection Plates

The hydraulic flying leads are supplied with MQC plates designed for a number of couplers. All tubes and coupling assignments need to match UTA and tree assignments.

All couplers and the HFL are capable of withstanding full design pressure and test pressure. The design takes into consideration the fact that all couplers will be energized to a full test pressure of 1.5× design pressure during a FAT (Fabrication Assemble Testing) proof test.

Design of all MQCs and couplers also considers the potential for leakage due to the external pressure being higher than the internal pressure.

All seals are compatible with the chemicals, methanol, and hydraulic control fluid used.

Hydraulic couplers are leak free in the unmated condition with high pressure of full working pressure or low pressure of 1 psi.

MQC plates on the HFL have ROV-operable attachment means in order to connect to the inboard MQC plates at the UTAs, trees, and manifold. Makeup at depth must be achieved with all positions pressurized or only one side pressurized. Breakout at depth must be achieved with no positions pressurized. No special tooling is required for installation or removal of the terminations.

Hydraulic couplings must be qualified for the duty and be capable of mating and unmating at worst case angles after coarse alignment, without failure. The MQC must be capable of mating and unmating 50 times on land and 30 times at design water depth without seal replacement and without leakage when made up and subjected to internal pressures between 0 psi and the system working pressure, and maximum hydrostatic pressure.

The following potential misalignments should be considered in MQC design:

- Linear misalignments: 1.5 in. in each direction;
- Axial misalignment between male and female connector limited to 3 degrees;
- Rotational misalignment limited to 5 degrees.

MQC plates are outfitted with an emergency release device. The design also prevents the MQC plates from engaging if they are not aligned properly. The couplers are not engaged prior to proper alignment. The emergency release mechanism makes provision for the separating forces necessary at depth.

MQC plates are visible for ROV inspection. This visibility must be capable of indicating full engagement of the MQC plates and means to verify that installation was properly made. It must be possible to inspect for leaks during operation.

3.3.8.3. Installation

All circuit paths in the HFL are proof tested to 1.5× maximum design pressure.

HFLs are visible to the ROV at the design water depth and in installed configurations.

HFL assembly design incorporates the means for offshore deployment. The design also incorporates the capability for HFLs to be lifted and installed by ROVs subsea. Maximum weights in air and water (for empty and filled tubing) and any buoyancy/flotation requirements must be defined.

Methods for handling, installation, and retrieval and for providing any necessary permanent flotation must be specified. Offshore test/verification procedures and long-term storage procedures must be provided. All assembly drawings, list of materials, and interface drawings to ROV operation and installation must be provided.

HFLs are outfitted with ROV API-17F Class 4 buckets or as referenced in ISO 13628-8 [3] for handling purposes.

The type of fluid that is required inside the HFL during shipping and deployment must be defined. Storage fluids must be compatible with chemicals and hydraulic fluid. HFLs are outfitted with MQC protection caps during shipping.

The maximum allowable pull on a termination and the minimum bend radius for assembled HFLs must be specified.

HFLs have clear permanent markings visible to an ROV during design life at service subsea.

3.3.8.4. Hydraulic Couplers

The hydraulic couplers for deepwater applications need a spring strong enough to seal against the external pressure head in order to prevent seawater from contaminating the hydraulic fluid, and need to be designed such that only a low-pressure force is required for makeup. Figure 3-9 and Figure 3-10 illustrate female and male hydraulic coupler structures, respectively.

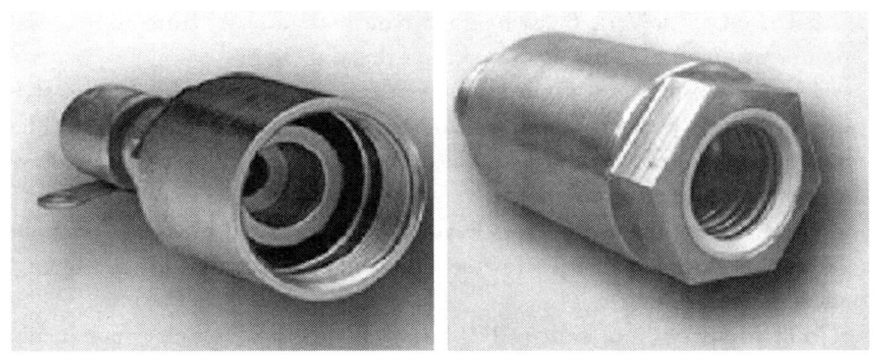

Figure 3-9 Female Hydraulic Couplers *(Courtesy SUBSEA COMPONENTS)*

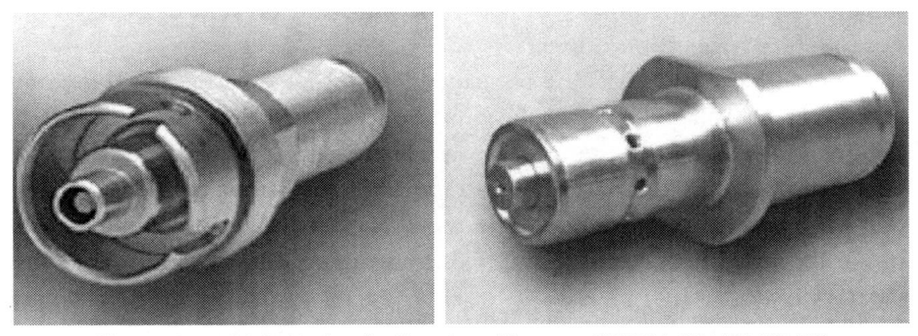

Figure 3-10 Male Hydraulic Couplers *(Courtesy SUBSEA COMPONENTS)*

Couplings are available that require a low force for makeup, and these are available with dual redundant resilient seals and with a combination of resilient and metal seals.

For deepwater applications a fully pressure-balanced coupling is preferable. These are fairly new concepts, but are available in single coupler pair design and also the four coupler hydraulic circuits design. The single pair is a resilient seal with a porting design that allows for inherent pressure balancing across the poppets, which provides pressure assistance to the spring closure force when the coupling is disconnected. The design uses a combined metal and resilient seal arrangement acting in a shear seal arrangement as the coupler mates and disconnects.

Both designs use the principle that the flow path is radial and hence produces no resultant separation force.

It is important for hydraulic couplers to have adequate flow paths to ensure that adequate hydraulic response times are achievable. The poppets are balanced to prevent them being driven hydraulically from the central open position and sealing against one of the seal faces.

The female couplers are usually assembled into a hydraulic jumper stab plate so that they can be retrieved and the seals replaced if necessary. The female couplers are assembled so that they are floating on the plate to allow for any manufacturing tolerances.

The backs of the connectors have to be terminated by screw-on hose termination couplers or by screw seal or welded assemblies for the termination of steel tubing.

Joint Industry Conference (JIC) hose terminations, which swage inside the central core tube of a standard thermoplastic hose, can only be used when the hose can be maintained full of fluid of a specific gravity equal to or similar to the specific gravity of seawater. This is to prevent collapse of the hose in deepwater applications.

Alternatively, high collapse resistance (HCR) hose with a spiral flexible metal former under the core tube must be used. The flexible inner core is designed to withstand the external seawater pressure and to prevent the hose core from collapsing. The HCR hose requires a different type of coupling that has a welded construction. The metal former inside the coupler slides inside the spiral hose support and seals by swaging onto the outside of the thermoplastic liner.

When stab plates are densely populated, it can be difficult to turn and orient all of the hoses through 90 degrees and into the hose/cable restraint. Right-angled connectors are used to orient the hoses into the clamp. It may also be necessary to have these connectors of stepped heights in order to allow hose makeup and to avoid tight bends or kinking of the hoses.

The basic design requirements for hydraulic couplers should follow ISO 13628-6 [3] and ISO 13628-4 [4]; However, some specific design requirements may vary according to its applications, such as subsea production control modules, junction plate assemblies, flying lead connectors, and etc. The following requirements cover different types of hydraulic couplers, but not limited to, (1) a coupler with poppet, (2) a coupler without poppet, (3) a male blank coupler, and (4) a female blank coupler.

- The couplers include inboard and outboard MQC plates that provide the mechanism for mating, demating, and locking multiple coupler connections within a single assembly.
- The hydraulic coupler system is configured to ensure that the replaceable seals are located in the hydraulic flying leads.

- Design of the hydraulic system should consider water hammer, high-pressure pulses, and vibration on couplers. This includes external sources, for example, chokes.
- Where high cyclic loads are identified, the design and manufacturing should be reviewed to mitigate associated risks, for example, the use of butt-weld hydraulic connections.
- The designs minimize ingress of external fluid during running and makeup operation.
- The couplers are designed for reliable and repeatable subsea wet mating under turbid environmental conditions.
- The couplers have a minimum of two seal barriers to the environment unless the barrier is seal welded.
- All chemical and hydraulic circuits within the same component are rated to the same design pressure.
- All hydraulic and chemical tubing circuits are NAS 1638-64 Class 6 [5] (or equivalent ISO 4406 Code 15/12) [6].
- The couplers are designed for operation and sealing under the maximum torque and bending moment applied to mated couplers through MQC engagement and misalignment.
- The couplers have metal-to-metal seals with elastomeric backup seals. Elastomeric seals must be compatible with the operating fluid.
- Couplers are furnished with necessary protection equipment in order to protect the equipment when being unmated and in-service and to prevent calcareous buildup and marine growth.
- Poppet couplers are to be used on all hydraulic and low-flow chemical services.
- Poppetless couplers are to be used on full-bore and high-flow chemical injection lines to reduce pressure losses and eliminate trash buildup through the poppet area (i.e., methanol supply and annulus vent lines).
- Spare umbilical tubes have poppet couplers.
- Consideration must be given to the ability to bleed trapped pressure in poppet circuits when recovered to surface (i.e., residual operation pressure or head pressure once disconnected from system).
- Special consideration is given to scale buildup prior to connection.

3.3.9. Electrical Flying Leads and Connectors

The EFL connects the EDU to the SCM on the tree. Each SCM utilizes two independent EFLs from the EDU for the redundant power on communication circuits, as shown in Figure 3-11.

Figure 3-11 EFL Assembly *(Courtesy FMC Technologies)*

3.3.9.1. Manufacturing

The EFL assembly is composed of one pair of electrical wires enclosed in a thermoplastic hose, fitted at both ends with soldered electrical connectors. The assembly constitutes an oil–filled, pressure-compensated enclosure for all wires and their connections to ROV-mateable connectors.

3.3.9.2. Construction

Wires are continuous and are, at a minimum, of 16 AWG. A twisted-pair configuration is recommended.

Voltage and current ratings for the wires are sized to not significantly degrade overall circuit performance based on the results from the electrical analysis.

Wires are soldered to the connector pins and protected by boot seals of compatible material. Pin assignment matches the system requirements.

A hose with low collapse resistance, specifically selected for subsea use, with titanium or equivalent end fittings, connects both electrical connectors of the flying lead, to ensure compatibility of materials used.

The length of the wire within the hose is sized to allow for any stretching of the hose up to failure of the hose or the end fittings. Hose stretching does not allow for any pull load on the soldered connections.

Hoses are a continuous length with no splices or fittings for lengths under 300 ft (91m). Any use of splices or fittings is brought forward for approval on a case-by-case basis.

Hoses are filled with Dow Corning –200 dielectric fluids.

The compatibility of the hoses, boot seals, and wire insulation with the compensating fluid and seawater is confirmed.

The wire insulation is a single-pass extrusion and suitable for direct exposure to seawater.

All wires are100 % tested for voids and pinholes by immersion in water hipot.

Connectors are marked appropriately to simplify ROV operations. An alignment key or other device is incorporated to ensure correct orientation.

The electrical connectors must be qualified for the duty and be capable of making and breaking at worst case angles, before and after course alignment, without failure. They must be capable of making and breaking 100 times under power on the female pin half without any sign of damage to pins or sockets and still remain capable of excluding seawater.

3.3.9.3. Installation

Connectors are provided with shipping protection covers.

All assemblies are identified with tags on both ends of the EFL. Tags must be designed such that they do not come off under severe handling onshore, offshore, and during installation and must be visible to ROVs at working water depth.

The color of the hose is visible to ROVs when in subsea use. Such colors include yellow or orange.

In order to be easily visualized and identified by ROV in underwater situations, the ROV handles and the bottom plastics sleeves of flying leads should be painted with colorful marks according to standards and codes. These marks should be identified throughout the life time of the system.

All EFL assemblies are filled with compensating fluid to a slight positive pressure (10 psi) prior to deployment.

Flying leads installed subsea are protected with mating connectors when not in use. These EFLs are temporarily located on parking positions on the E-UTA assembly, at the tree, or on a parking stand installed for that purpose.

3.3.9.4. Electrical Connectors

Electrical connectors have the following basic requirements:

- An electrical connector is a termination for electrical cables used to transmit electrical power of low voltage and communication signals between subsea production control system components.
- Electrical connectors at the very minimum meet all requirements as stated in the latest revision of ISO 13628-4 [4] and ISO 13628-6 [3].
- The number of electrical connectors in series is kept to a minimum. Redundant routing follows different paths. Consideration should be given to keeping voltage levels as low as practical in order to minimize electrical stresses on conductive connectors.
- Connectors are either Tronic or ODI.
- The electrical connector is capable of making wet mateable electrical connections utilizing an ROV. They are designed and constructed for normal and incidental loads imparted by ROVs during make-or-break operations.
- It is important to confirm the type of connector halves – whether it is "cable end" type or "bulkhead connector" type.
- The Christmas tree side has male (pin) connectors and the flying leads have female (socket) connectors.
- Connectors are configured to ensure that no male pins are powered up while exposed. Electrical distribution systems should be designed such that "live disconnect" is not required during normal maintenance or if possible during failure mode operation or recovery periods.
- Connectors are furnished with the necessary equipment to protect it from being unmated while in-service and to prevent calcareous buildup and marine growth.
- Optical connectors for any fiber-optic lines are fitted with long-term protective caps.

3.3.10. Logic Caps

Hydraulic and chemical distribution equipment includes dedicated MQC plates, known as logic caps (see Figure 3-12), which provide the ability to redirect services by replacing an outboard MQC plate with an ROV. Logic caps provide the flexibility to modify distribution of hydraulic or chemical services due to circuit failures or changes in system requirements.

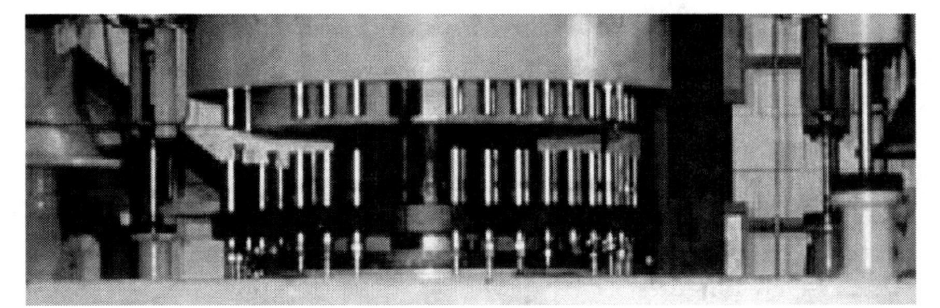

Figure 3-12 Typical Remote Make-up Logic Cap Interface *(Courtesy TotalFinaElf)*

The logic cap consists of stab plate mounted hydraulic couplers connected to HFL tubing and plumbed accordingly to suit the application.

Logic caps meet the following minimum requirements:

- They are always used for direct or manifold services in lieu of valves when possible.
- Designed for 15 years.
- Rated to maximum 5000 psi.
- Redundant LP supplies (e.g., LP1 and LP2) are in separate logic caps on separate sides of the frame, where possible.
- High-pressure (HP) supplies must be routed through a logic cap.
- Redundant HP supplies (e.g., HP1 and HP2) also must be in separate logic caps on separate sides of the frame, and in the same logic cap with a corresponding low-pressure (LP) circuit (i.e., HP1 and LP1 will be routed to same logic cap where possible).
- All chemical and annulus vent lines are distributed through a logic cap.
- Chemical tubes dedicated for step-out connections are not required to pass through logic caps.
- The logic cap configuration does not allow cross flow between trees.
- Logic caps are ROV installable and retrievable. Grab rails are provided to assist in ROV operations.
- All features conform to field wide standard designs in order to facilitate interchangeability and to minimize tooling.
- Logic cap utilization is based on the following priority list:
  - Hydraulic services;
  - Methanol (or other hydrate mitigation chemical) distribution;
  - Corrosion inhibitor (CI);
  - Asphaltene dispersant/inhibitor;
  - Other chemicals as defined by project team.

3.3.11. Subsea Accumulator Module (SAM)

The SAM, illustrated in Figure 3-13, is the subsea unit that stores hydraulic fluid such that adequate pressure is always available to the subsea system even when other valves are being operated.

SAM is used to improve the hydraulic performance of the subsea control system in trees and manifolds. Basically, it will improve hydraulic valve actuation response time and system hydraulic recovery time, at minimum system supply pressures.

3.3.11.1. Description

When a subsea system is required to operate a number of trees located a long distance away from the host, the hydraulic fluid from the topsides HPU will take a considerable time to reach the subsea equipment, particularly where small hoses are used in the umbilical. This can result in a drop in pressure at the subsea tree when a valve is opened, as the pressure cannot then be restored immediately via the umbilical.

If the pressure drops, other open tree valves may begin to close, before the pressure can be restored.

Figure 3-13 Subsea Accumulator Module [7]

If the pressure drops too much, the pilot valves in the SCM will "drop out," that is, close, causing one or more tree valves to close irrespective of whether pressure is then restored via the umbilical.

To maintain an adequate level of pressure at the subsea location, some degree of local accumulation may be required. This can be provided by individual accumulators on the SCM itself, but more usually, a self-contained skid containing several accumulator bottles is often provided, this being termed a *subsea accumulator module* or SAM.

The SAM will house sufficient LP accumulation to maintain pressure during valve operations. In addition, it is also sized to hold sufficient fluid to perform a number of subsea control operations, even if the supply from the surface no longer functions, thus giving a degree of reserve power. A trade-off must be made against the size of skid required and the amount of accumulation and this is done via a hydraulic analysis performed by the manufacturer against the specifications. Sometimes the analysis will demonstrate that HP accumulation is also required.

The requirement for an increased nitrogen precharge pressure in deepwater applications will decrease the efficiency of subsea accumulation, which will increase the number of accumulator bottles required.

3.3.11.2. Components

A SAM is a simple skid, primarily housing accumulators. Nevertheless, it must be designed, manufactured, and tested as part of the overall system, and may be designed to be retrievable for maintenance, because the accumulator precharges may periodically need replenishing.

The design may also incorporate a filter and block/bleed valves to allow flushing and testing.

Figure 3-14 shows a block diagram for the subsea accumulator module. The SAM is usually a stand-alone skid, connected via a mounting base, which is connected to the SDU or the hydraulic tubing supplying the subsea control modules. The SAM is run and retrieved in a manner similar to that for the subsea control module. To allow the SAM to be retrieved for maintenance, ROV-operable block, vent, and bypass valves are sometimes incorporated into the manifold/template tubing (where the hydraulic distribution is hard piped). This valve will allow production to continue while the unit is replaced, but tree valves should not be operated while the bypass is in operation, due to the risk of pressure drop as outlined earlier.

When block and vent valves are not used, the SAM can be retrieved by pulling off its mounting base and relying on the self-sealing, quick-connect

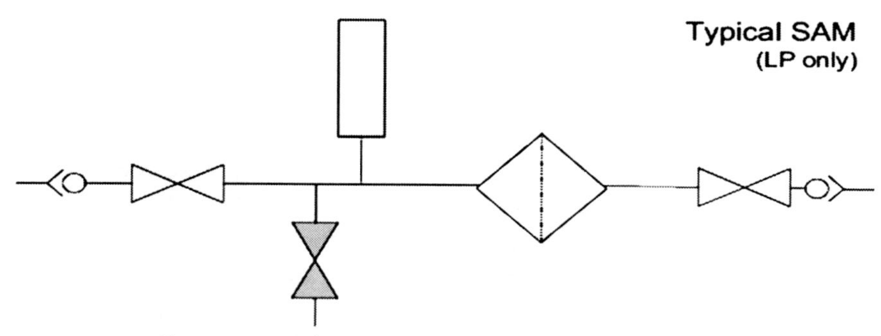

Figure 3-14 Subsea Accumulator Module Block Diagram

hydraulic couplings sealing to maintain the pressure in the hydraulic system.

Installation of the SAM is difficult, because a large force is required to mate the SAM and SAMMB (Subsea Accumulator Module Mating Block) against the closed force exerted behind the closed hydraulic poppets. It is essential that the hydraulic quick-connect couplings used in this application be fully pressure balanced to counteract the coupling mating forces, particularly in deepwater applications.

A running tool is required to run and set the SAM onto the SAMMB in deepwater applications.

REFERENCES

[1] P. Collins, Subsea Production Control Umbilicals, SUT Subsea Awareness Course, Houston, 2008.

[2] T. Horn, G. Eriksen, W. Bakke, Troll Pilot - Definition, Implementation and Experience, OTC 14004, Houston, 2002.

[3] International Standards Organization, Petroleum and Natural Gas Industries – Design and Operation of Subsea Production Systems – Part 8: Remotely Operated Vehicle (ROV) Interfaces on Subsea Production Systems, ISO 13628-8/API 17F, 2002.

[4] International Standards Organization, Petroleum and Natural Gas Industries – Design and Operation of Subsea Production Systems – Part 4: Subsea Wellhead and Tree Equipment, ISO 13628-4, 1999

[5] National Aerospace Standard, Cleanliness Requirements of Parts Used in Hydraulic Systems, NAS 1638-64, Class 6, 2001.

[6] International Standards Organization, Hydraulic Fluid Power – Fluids – Method for Coding the Level of Contamination by Solid Particles, ISO 4406, 1999.

[7] Deep Down Company, Subsea Accumulator Module, <http://www.deepdowncorp.com/deepdown/products/sams>.

CHAPTER 4

Subsea Surveying, Positioning, and Foundation

Contents

Subsea Engineering Handbook
ISBN 978-0-12-397804-2, doi:10.1016/B978-0-12-397804-2.00004-7

4.1. INTRODUCTION

The study of the subsea soil, including the subsea survey, positioning, soil investigation, and foundation, is one of the main activities for the subsea field development. This chapter provides minimal functional and technical requirements for the subsea soil issue, but these guidelines can be used as a general reference to help subsea engineers make decisions.

As part of the planned field development, a detailed geophysical and geotechnical field development survey and a soil investigation based on the survey results are to be performed. The purpose of the survey is to identify potential man-made hazards, natural hazards, and engineering constraints when selecting a subsea field area and also flowline construction; to assess the potential impact on biological communities; and to determine the seabed and sub-bottom conditions.

This chapter briefly explains these topics:

- Establishing vertical route profiles, a contour plan, and the seabed features, particularly any rock outcrops or reefs;
- Obtaining accurate bathymetry, locating all obstructions, and identifying other seabed factors that can affect the development of the selected subsea field area including laying, spanning, and stability of the pipeline;
- Carrying out a geophysical survey of the selected subsea field and route to define the shallow sub-seabed geology;
- Carrying out geotechnical sampling and laboratory testing in order to evaluate precisely the nature and mechanical properties of soils at the selected subsea field area and along the onshore and offshore pipelines and platform locations;
- Locating existing subsea equipment (examples: manifold, jumper, and subsea tree), pipelines and cables, both operational and redundant, within the survey corridors;
- Determining the type of subsea foundation design that is normally used for subsea field development.

4.2. SUBSEA SURVEY

The subsea survey is described as a technique that uses science to accurately determine the terrestrial or 3D space position of points and the distances and angles between them in the seabed area for subsea field development.

4.2.1. Subsea Survey Requirements

Geophysical and geotechnical surveys are conducted to evaluate seabed and subsurface conditions in order to identify potential geological constraints for a particular project.

4.2.1.1. Survey Pattern for Selected Subsea Field and Each Pipeline Route

The base survey covers the whole subsea field development, which includes the infield pipelines, mobile offshore drilling unit (MOPU) footprints, pipeline end manifolds (PLEMs), manifolds, Christmas tree, umbilicals, etc.

The nominal width of the pipeline route survey corridor is generally 1640ft (500 m) with maximum line spacing of 328ft (100 m). Different scenarios can be proposed provided full route coverage is achieved.

4.2.1.2. Geotechnical Study

A geotechnical study is necessary to establish data that allows an appropriate trenching design and equipment to be selected. It is also important to identify any possibility of hard grounds, reefs, shallows, and man-made debris.

The electromagnetic properties of the soil are also of interest, and the potential effect that the ferrous content may have on the sacrificial anodes of certain subsea equipment such as manifolds and PLEMs needs to be assessed. A grab sample/cone penetration test (CPT) is conducted at locations determined from review of the geophysical survey. Based on the test, the characteristics of the seabed soil around the subsea field development area can be determined. If there is a drastic change in one of the core samples, additional samples will be taken to determine the changes in condition. A piezocone penetration test (PCPT) should be obtained at the MODU and PLEM locations and FSO anchor locations.

Geotechnical gravity core, piston core, or vibracore samples are obtained 5 to 10 m from the seabed of the subsea equipment locations such as PLEMs, umbilicals, PLETs, etc. Samples should be suitable for a laboratory test program geared toward the determination of strength and index properties of the collected specimens. On board, segments (layers) of all samples at 1-m intervals will be classified by hand and described. Samples for density measurements are also taken. At least one sample from each layer shall be adequately packed and sent to the laboratory for index testing and/or sieve analysis and unconsolidated, undrained (UU) triaxial testing. Cohesion will be measured on clayey parts of the core by torvane and a pocket penetrometer on board and by unconfined compression tests in the laboratory. The minimum internal diameter of the samples is generally 2.75 in. (70 mm).

4.2.1.3. Survey Vessel

The vessel proposed for survey should be compliant with all applicable codes and standards (see Figure 4-1). The vessel must follow high safety standards and comply with all national and international regulations, and the marine support must be compatible for survey and coring operations.

The survey vessels provided are collectively capable of the following:

- Minimum offshore endurance of 2 to 3 weeks.
- Operating in a maximum sea state of 2.5 to 3.5 m.
- Survey at speed of 3 to 10 knots.
- Supplying the necessary communication and navigation equipment.
- Supplying minimum required survey equipment: multibeam echo sounder, precision depth sounder, side-scan sonar, sub-bottom profiler, grab sampler/CPT, piston/vibracore coring equipment, and differential GPS (dual system with independent differential corrections).
- Supplying lifting equipment capable of safely deploying, recovering, and handling coring and geophysical equipment.

Figure 4-1 Survey Vessel [1]

- Supplying adequate AC power to operate all geophysical systems simultaneously without interference.
- Accommodating all personnel required to carry out the proposed survey operations.
- Accommodating a minimum of two representative personnel
- Providing office space/work area. The area is fitted with a table/desk large enough to review drawings produced on board and to allow the installation of notebook computers and printers.
- The vessel should have radio, mobile telephone, and fax equipment. This equipment should be capable of accepting a modem hookup.
- The vessel should have a satellite or cellular phone link to report progress of the work on a daily basis. Communications with this system should not cause interference with the navigation or geophysical systems.
- At the time of mobilization, a safety audit is carried out of the nominated vessel to ensure compliance with standards typical for the area of operations, or as agreed. A current load certificate is supplied for all lifting equipment to be used for the survey operations (i.e., deep tow-fish, coring). Safety equipment, including hard hats, safety boots, and safety glasses, should be worn during survey operations.

4.2.1.4. Survey Aids

The survey vessel is normally equipped with an A-frame and heave-compensated offshore cranes that are capable of operating the required survey equipment. Winches are used for handling of sampling and testing equipment in required water depths. The winches have a free-fall option if required, such as for hammer sampling and chiseling. However, the winch speed should be fully controllable in order to achieve safe deployment. Geotechnical sampling and testing equipment are remotely operated. The tools are guided remotely and in a safe manner off and onto the deck. The vessels have laboratory facilities and equipment that can perform routine laboratory work.

A vessel in the soil drilling mode should include the following:

- Sampling and testing equipment that is fully remotely operated;
- A pipe centralizer in the moon-pool is used;
- A pipe stab guide that is used when pipe stringing.

4.2.1.5. Gyrocompass

A gyrocompass is similar to a gyroscope, as shown in Figure 4-2. It is a compass that finds true north by using an electrically powered, fast-spinning

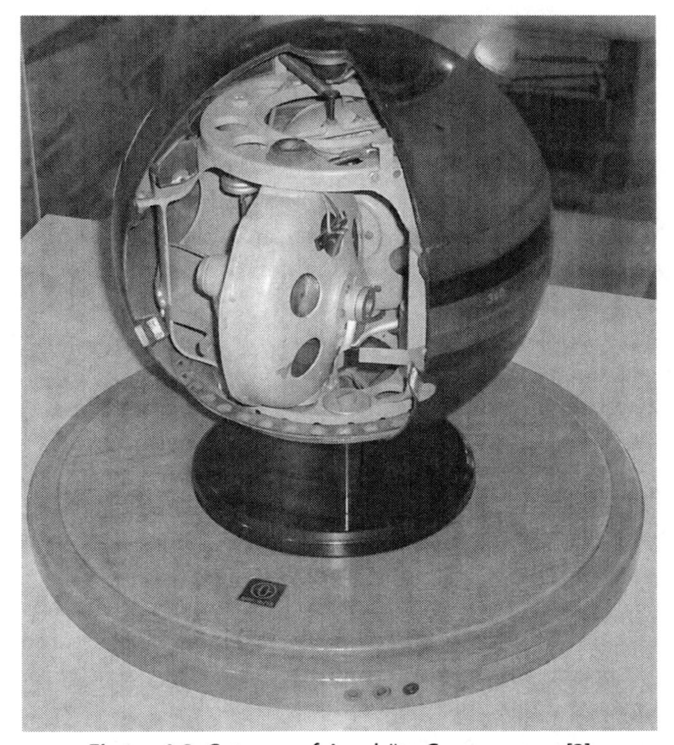

Figure 4-2 Cutaway of Anschütz Gyrocompass [2]

wheel and friction forces in order to exploit the rotation of the Earth. Gyrocompasses are widely used on vessels. They have two main advantages over magnetic compasses:

- They find true north, that is, the direction of Earth's rotational axis, as opposed to magnetic north.
- They are far less susceptible to external magnetic fields, for example, those created by ferrous metal in a vessel's hull.

The gyrocompass can be subject to certain errors. These include steaming errors, in which rapid changes in course speed and latitude cause deviations before the gyro can adjust itself [3]. On most modern ships the GPS or other navigational aid feeds into the gyrocompass allowing a small computer to apply a correction. Alternatively, a design based on an orthogonal triad of fiber-optic or ring laser gyroscopes will eliminate these errors as they depend on no mechanical parts, instead using the principles of optical path difference to determine rate of rotation [4].

A dedicated survey gyro is installed on the vessel and interfaced to the navigation computer. During mobilization, calibration of the gyros is carried out while the vessel is at the dock.

4.2.1.6. Navigation Computer and Software

The navigation computer and software is capable of:

- Simultaneous acquisition of all navigation and sensor data as interfaced;
- Generation of closures to all geophysical acquisition equipment recorders simultaneously;
- Helmsman display showing vessel and fish position, proposed pipeline route, and intended survey line.
- Producing a header sheet and fix printout containing all relevant survey constants with bathymetry and position fix information.

4.2.1.7. Personnel

Besides the full complement of vessel operations personnel normally aboard a survey vessel, additional qualified personnel may be utilized to safely and efficiently carry out the survey and geotechnical operations. The number of personnel should be adequate to properly interpret and document all data for the time period required to complete the survey work, without operational shutdown due to operator fatigue.

Qualified personnel interpret data during the survey and make route recommendations or changes based on the information gathered. This geophysical data interpretation should be performed by a qualified marine engineering geologist or geophysicist experienced in submarine pipeline route analysis.

4.2.2. Subsea Survey Equipment Requirements

For the main survey vessel, survey equipment is used to meet this specification. Vessels must follow high safety standards and comply with all national and international regulations, and marine support will be compatible for survey and coring operations. All survey systems are able to operate simultaneously with minimal interference.

4.2.2.1. Multibeam Echo Sounder (MBES)

The MBES, or swath echo sounder, is a high-precision method for conducting bathymetric surveys obtained at water depths and seabed gradients over the corridor along the proposed pipeline routes, as shown in Figure 4-3. During data acquisition, the data density should be sufficient to

Figure 4-3 MBES Working Model [5]

ensure that 95% of the processed bins contain a minimum of four valid depth points.

The following issues are required for subsea survey equipment:

- Equipment specifications;
- Method of integrating the system with the vessel's;
- Surface positioning system;
- Method of calibration;
- Method of postprocessing of data;
- On- and off-line quality control, with particular reference to overlap swaths.

The swath bathymetric system provides coherent data across the full width of the swath. Alternatively, the portion of the swath that does not provide coherent data should be clearly identified and the data from that portion are not be used.

A 50% overlap of adjacent swaths is arranged to provide overlap of acceptable data for verifying accuracy. In areas where swath bathymetry overlaps occur, the resulting differences between data after tidal reduction are less than $\pm 0.5\%$ of water depth. Line spacing is adjusted according to the water depth to provide sufficient overlap (50%) between adjacent swaths to facilitate correlation of the data of the adjacent swaths.

Consideration should be given to installing a tidal gauge(s) or acquiring actual tidal data from an existing tide gauge in the area. If no nearby benchmarks of known height are available for reference, the tide gauge must be deployed for at least one lunar cycle.

4.2.2.2. Side-Scan Sonar

Side-scan sonar is a category of sonar system that is used to efficiently create an image of large areas of the seafloor. This tool is used for mapping the seabed for a wide variety of purposes, including creation of nautical charts and detection and identification of underwater objects and bathymetric features. Side-scan sonar imagery is also a commonly used tool to detect debris and other obstructions on the seafloor that may be hazardous to shipping or to seafloor installations for subsea field development. In addition, the status of pipelines and cables on the seafloor can be investigated using side-scan sonar. Side-scan data are frequently acquired along with bathymetric soundings and sub-bottom profiler data, thus providing a glimpse of the shallow structure of the seabed.

A high-precision, dual-frequency side-scan sonar system can obtain seabed information along the routes for example, anchor/trawl board scours, large boulders, debris, bottom sediment changes, and any item on the seabed having a horizontal dimension in excess of 1.64 ft (0.5 m). Side-scan sonar systems consist of a dual-channel tow-fish capable of operating in the water depths for the survey and contain a tracking system. The equipment is use to obtain complete coverage of the specified areas and operates at scales commensurate with line spacing, optimum resolution, and 100% data overlap.

The height of the tow-fish above the seabed and the speed of the vessel are adjusted to ensure full coverage of the survey area. The maximum tow-fish height is 15% of the range setting. Recorder settings are continuously monitored to ensure optimum data quality. Onboard interpretation of all contacts identified during the survey is undertaken by a geophysicist suitably experienced in side-scan sonar interpretation.

4.2.3. Sub-Bottom Profilers

The sub-bottom profilers are tested under tow at each transmitting frequency available using maximum power and repetition rates for a period of half an hour.

During mobilization, the outgoing pulse from the transducer /seismic source is monitored to ensure a sharp, repeatable signature utilizing a suitably calibrated hydrophone. The monitored pulse shall be displayed on board on an oscilloscope with storage facility and a copy generated for approval. The pulse should conform to the manufacturer's specifications and a printout of the signature should be included in the final report.

A static or dynamic pulse test may be used to demonstrate a stable and repeatable seismic source signal producing a far-field signature at a tow depth of 3.28 ft (1 m):

- Pulse in excess of 1 bar meter peak-to-peak;
- Pulse length not exceeding 3 ms;
- Bandwidth of at least 60 to 750 Hz (–6 dB);
- Primary to secondary bubble ratio > 10:1.

4.2.3.1. High-Resolution Sub-Bottom Profiler

A high-precision sub-bottom profiler system is provided and operated to obtain high-resolution data in the first 10 m of sediment. The operating frequency and other parameters are adjusted to optimize data within the first 5 m from the seafloor. Vertical resolution of less than 1 m is required.

A dual-channel "chirp" sub-bottom profiler may be capable of operating within the 3.5 to 10kHz range with a pulse width selectable between 0.15 and 0.5 m and transmitting power selectable between 2 and 10 kW. The system is capable of transmitting repetition rates up to 10 Hz. Transmitting frequency, pulse length, power output, receiving frequency, bandwidth, and TVG (Time Varying Gain) are adjustable. A heave compensator is required when using a sub-bottom profiler.

The system may either be collocated with the side-scan sonar system utilizing the same tracking system or hull mounted/over the side and reference the ship's navigation antenna. Onboard interpretation of sub-bottom profiler records is carried out by a geophysicist suitably experienced in interpretation of such records.

4.2.3.2. Low-Resolution Sub-Bottom Profiler

The generic term *mini air gun* covers a range of available hardware that uses explosive release of high-pressure air to create discrete acoustic pulses within the water column, of sufficient bandwidth and high-frequency component to provide medium-resolution data for engineering and geohazard assessment.

The system is capable of delivering a stable, short-duration acoustic pulse at a cycle repetition rate of 1 sec. The hydrophone comprises a minimum of 20 elements, linearly separated with an active length not exceeding 32.8 ft(10 m), with a flat frequency response across the 100 to 2000Hz bandwidth.

4.2.4. Magnetometer

A magnetometer is a scientific instrument used to measure the strength and direction of the magnetic field in the vicinity of the instrument. Magnetism varies from place to place because of differences in Earth's magnetic field caused by the differing nature of rocks and the interaction between charged particles from the sun and the magnetosphere of a planet.

Magnetometers are used in geophysical surveys to find deposits of iron because they can measure the magnetic field variations caused by the deposits. Magnetometers are also used to detect shipwrecks and other buried or submerged objects.

A towed magnetometer (cesium, overhauser, or technical equivalent) has a sensor head capable of being towed in a stable position above the seabed. The sensor head comprises a three-component marine gradiometer platform synchronized to within less than 0.1 m and able to measure 3D gradient vectors. The tow position should be far enough behind the vessel to minimize magnetic interference from the vessel.

In normal operation, the sensor is towed above the seabed at a height not exceeding 5 m. In the case of any significant contacts, further profiles across such contacts may be required. In these circumstances, the magnetometer should be drifted slowly across the contact position to permit maximum definition of the anomaly's shape and amplitude.

The magnetometer has field strength coverage on the order of 24,000 to 72,000 gammas with a sensitivity of 0.01 nanotesla and is capable of a sampling rate of 0.1 sec. The equipment incorporates depth and motion sensors and operates in conjunction with a tow-fish tracking system. Onboard interpretation of all contacts made during the course of the survey should be carried out by a geophysicist suitably experienced in magnetometer interpretation.

4.2.5. Core and Bottom Sampler

The gravity corer, piston corer, or vibracore can be deployed over the side or through the A-frame of a vessel or operated from a crane configured with a 70 mm-ID core barrel and clear plastic liners. The barrel length has 5 to 10 m barrel options.

The grab sampler can be of the Ponar or Van Veen type, which can be handled manually. Both systems are capable of operating in a water depth 135% deeper than the maximum anticipated water depth. At all core locations, up to three attempts are made to acquire samples to the target depth of 5 to 10 m. After three unsuccessful attempts, the site is abandoned.

4.2.6. Positioning Systems

4.2.6.1. Offshore Surface Positioning

A differential global positioning system (DGPS) is utilized for surface positioning. A DGPS is capable of operating continually on a 24-hour basis. Differential corrections are supplied by communications satellites and terrestrial radio links. In either case, multiple reference stations are required. A dual-frequency DGPS is required to avoid problems associated with ionospheric activity.

A secondary system operates continuously with comparison between the two systems recorded along with the bathymetry data (geophysical vessel only). The two systems utilize separate correction stations, receivers, and processors. The system is capable of a positioning accuracy of less than ±3.0 m with an update rate of better than 5 sec.

During geophysical operations, the receiver can display quality control (QC) parameters to the operator via an integral display or remote monitor. The QC parameters to be displayed include the following items:

- Fix solution;
- Pseudo-range residuals;
- Error ellipse;
- Azimuth, altitude of satellite vehicles (SVs) tracked;
- Dilution of position (DOP) error figures for fix solution;
- Identity of SVs and constellation diagram;
- Differential correction stations, position comparisons.

A transit fix at a platform, dock, or trestle near the survey area is carried out to ensure the DGPS is set up and operating correctly. The transit fix is carried out in both clockwise and counterclockwise directions.

4.2.6.2. Underwater Positioning

An ultra-short baseline (USBL) tracking system is provided on board the offshore survey vessel for tracking the positions and deployments of the towed, remote/autonomous vehicles or the position determination of geotechnical sampling locations.

The positioning systems interface with the online navigation computer. All position tracking systems provide 100% redundancy (a ship-fit USBL may be backed up by a suitably high-precision portable system), encompassing a fully backed up autonomous system. In addition, a complete set of the manufacturer's spares are kept for each piece of positioning instrumentation such that continuous operations may be guaranteed.

The system, including the motion compensator, is installed close to the center of rotation of the vessel. It incorporates both fixed and tracking head transducers to allow the selection of the most optimum mode of performance for the known range of water depths and tow/offset positions.

The hull-mounted transducer is located so as to minimize disturbances from thrusters, machinery noise, air bubbles in the transmission channel, and other acoustic transmissions. In addition, the transponder/responder mounted on the ROV or AUV will require suitable positioning and insulation to reduce the effects of ambient noise.

A sufficient number of transponders/responders, with different codes and frequencies, are used to allow the survey operation to be conducted without mutual interference. The system should perform to an accuracy of better than 1% of slant range.

4.3. SUBSEA METROLOGY AND POSITIONING

Metrology is defined by the International Bureau of Weights and Measures (IBWM) as "the science of measurement, embracing both experimental and theoretical determinations at any level of uncertainty in any field of science and technology" [6]. This section describes the subsea positioning systems, which are integrated with the main survey computer in order to provide accurate and reliable absolute positioning of the surface and subsurface equipment.

4.3.1. Transducers

A transducer is a device for transforming one type of wave, motion, signal, excitation, or oscillation into another. Transducers are installed on board the vessels accordingly. A plan is created for the locations of all acoustic transducers and their coordinates that refers to a fixed reference point. A high-quality motion sensor (motion reference unit) is used to compensate for transducer movement.

4.3.2. Calibration

Calibration is the process of comparing a measuring instrument with a measurement standard to establish the relationship between the values indicated by the instrument and those of the standard. Calibration of the positioning systems, including all spare equipment, is carried out to ensure that each piece of individual equipment is working properly. Field calibration is performed during both prefield and postfield work. Depending on the length of the field work, additional field calibrations may be required during the course of the work. The following general procedures and requirements are adopted during the calibration process:

- No uncalibrated equipment, including cables and printed circuit boards, is used during any part of the position fixing.
- Each calibration setup should last for at least 20 min and the resulting data is logged for processing and reporting.
- The results of the calibrations, including relevant information on equipment settings, are presented for review and acceptance. The report includes measured minimum, maximum, mean, and standard deviations for each measurement and recommendations for operating figures. Any odd figures, anomalies, or apparent erroneous measurements are highlighted and explained in the report. If, after further examination of results, doubt still exists as to the integrity of any equipment, the faulty equipment is replaced with similarly calibrated equipment prior to the start of the survey.
- In the event that positioning equipment must be repaired or circuit boards changed, and if such actions alter the position information, recalibration is performed.

4.3.3. Water Column Parameter

A water column is a conceptual column of water from surface to bottom sediments. The application of the correct speed of sound through seawater is critical to the accuracy of the acoustic positioning. Sound velocity is a function of temperature, salinity, and density. All three properties change randomly and periodically; therefore, regular measurements for velocity changes are required.

A salinity, temperature, and depth profiler is used to determine the propagation velocity of sound through seawater. The computed sound velocity value or profile is then entered into the appropriate acoustic system. All procedures need to be properly followed and the results applied correctly.

4.3.3.1. Field Procedure

A velocity value or profile is obtained at the beginning of the survey and thereafter. Observations are made at suitable depth or intervals during descent and ascent through the water column. The velocity profile is determined with the value at common depth agreed to within ±3 m/sec; otherwise, the observation has to be repeated. The sound velocity at the sea bottom level is determined within ±1.5 m/sec. Having observed and recorded these values, a computation of the speed of sound is made.

4.3.3.2. Calibration

The temperature/salinity/depth probe has a calibration certificate verifying that it has been checked against an industrial standard thermometer in addition to testing against a calibrated saline solution. A strain gauge pressure sensor certificate is supplied.

4.3.4. Acoustic Long Baseline

Acoustic long baseline (LBL), also called range/range acoustic, navigation provides accurate position fixing over a wide area by ranging from a vessel, towed sensor, or mobile target to three or more transponders deployed at known positions on the seabed or on a structure. The line joining a pair of transponders is called a baseline. Baseline length varies with the water depth, seabed topography, and acoustic frequency band being used, from more than 5000 m to less than 100 m. The LBL method provides accurate local control and high position repeatability, independent of water depth. With the range redundancy that results from three or more range measurements, it is also possible to make an estimation of the accuracy of each position fix. These factors are the principal reasons for a major increase in the utilization of this method, particularly for installation position monitoring.

LBL calibration and performance can be improved significantly by using “intelligent” transponders. These devices calibrate arrays by making direct measurements of the baselines and acoustically telemetering the data to the surface equipment for computation and display. They also reduce errors inherent in the conventional LBL due to ray-bending effects, as measurements are made close to the seabed where propagation changes are generally slight. In addition, they can be supplied with environmental sensors to monitor the propagation conditions.

4.3.4.1. Field Procedure

The system is operated by personnel who have documented experience with LBL operations, to the highest professional standards and manufacturers’ recommendations. Local seabed acoustic arrays consist of networks with at least six LBL transponders. Ultra-high-frequency (UHF) arrays are used for the installations with the highest requirements for accurate installations. The system includes:

- A programmable acoustic navigator (PAN) unit for interrogation of LBL;
- Transponders;
- A transducer for vessel installation;
- All necessary cables and spare parts.

All equipment is interfaced to the online computer. The online computer system is able to handle the LBL readings without degrading other computation tasks. The software routines allow for the efficient and accurate use of all LBL observations and in particular deal with the problems encountered in surveying with a LBL system. The vessel's LBL transducer is rigidly mounted. The operating frequency for the required medium-frequency (MF) system is typically 19 to 36 kHz. The operating frequency for the required UHF system is typically 50 to 110 kHz. A minimum of five lines of position for each fix are available at all times.

4.3.4.2. MF/UHF LBL Transponder

The latest generation LBL transponder has the following minimum requirements (data presented as MF/UHF):

- Transducer beam shape: hemispherical/hemispherical;
- Frequency range: 19 to 36 kHz/50 to 110 kHz;
- Acoustic sensitivity: 90 dB re 1 μPa/90 to 125 dB;
- Acoustic output: 192 dB re 1 μPa at 1 m/190 dB;
- Pulse Length: 4 ms/1 ms;
- Timing resolution: 1.6 μsec/8.14 μsec;
- Depth rating: according to project requirements.

At least two of the transponders in each array include depth, temperature, and conductivity options. The MF transponder includes anchor weights (minimum of 80 kg) attached to the release mechanism on the base of the unit by a strop (preferably nylon to avoid corrosion) 1.5 to 2 m in length. A synthetic foam collar is used for buoyancy.

The UHF array setup includes frames with transponders rigidly installed in, for example, baskets 2.0 to 2.5 m above the seabed. The deployment is performed following a special procedure in accordance with the sea bottom depth. The setup is visually inspected by an ROV after installation. Concrete reference blocks or other transponder stands for MF arrays may be required.

4.3.5. Acoustic Short Baseline and Ultra-Short Baseline

4.3.5.1. Acoustic Short Baseline

A short baseline (SBL) acoustic positioning system [7] is one of the three broad classes of underwater acoustic positioning systems that are used to track underwater vehicles and divers. The other two classes are USBL and LBL systems. Like USBL systems, SBL systems do not require any seafloor-mounted transponders or equipment and are thus suitable for tracking underwater targets from boats or ships that are either anchored or under way. However, unlike USBL systems, which offer a fixed accuracy, SBL positioning accuracy improves with transducer spacing [8]. Thus, where space permits, such as when operating from larger vessels or a dock, the SBL system can achieve a precision and position robustness that is similar to that of seafloor-mounted LBL systems, making the system suitable for high-accuracy survey work. When operating from a smaller vessel where transducer spacing is limited (i.e., when the baseline is short), the SBL system will exhibit reduced precision.

4.3.5.2. Ultra-Short Baseline

A complete USBL system consists of a transceiver, which is mounted on a pole under a ship, and a transponder/responder on the seafloor or on a tow-fish or ROV. A computer, or "topside unit," is used to calculate a position from the ranges and bearings measured by the transceiver.

An acoustic pulse is transmitted by the transceiver and detected by the subsea transponder, which replies with its own acoustic pulse. This return pulse is detected by the shipboard transceiver. The time from the transmission of the initial acoustic pulse until the reply is detected is measured by the USBL system and converted into a range.

To calculate a subsea position, the USBL calculates both a range and an angle from the transceiver to the subsea beacon. Angles are measured by the transceiver, which contains an array of transducers. The transceiver head normally contains three or more transducers separated by a baseline of 10 cm or less. A method called *phase differencing* within this transducer array is used to calculate the angle to the subsea transponder.

4.3.5.3. Description

SBL systems conventionally replace the large baselines formed between transponders deployed on the seabed with baselines formed between reference points on the hull of a surface vessel. The three or four reference points are marked by hydrophones, which are typically separated by distances of 10 to 50 m and connected to a central control unit.

Seabed locations or mobile targets are marked by acoustic beacons whose transmissions are received by the SBL hydrophones. It is more convenient than the LBL method because multiple transponder arrays and their calibration are not required; however, position accuracy is lower than the LBL method and decreases in deeper water or as the horizontal offset to a beacon increases. Additional factors such as vessel heading errors and roll and pitch errors are significant in the accuracy measurements.

In the USBL, the multiple separate SBL hull hydrophones are replaced by a single complex hydrophone that uses phase comparison techniques to measure the angle of arrival of an acoustic signal in both the horizontal and vertical planes. Thus, a single beacon may be fixed by measuring its range and bearing relative to the vessel. Although more convenient to install, the USBL transducer requires careful adjustment and calibration.

4.3.5.4. Field Procedure

A USBL system, with tracking and the latest generation fixed narrow transducer, can be used. A high-precision acoustic positioning (HIPAP) or similar system can also be used. This subsurface positioning system is integrated with the online computer system to provide an accurate and reliable absolute position for the transponders and responders.

All necessary equipment is supplied so that a fully operational USBL system can be interfaced to an online computer for integration with the surface positioning systems. It must also meet the operational requirements set forth in this section. The installation of equipment should comply with manufacturers' requirements, and special attention should be given to the following requirements:

- A system check is performed within the last 12 months prior to fieldwork. Documentation must be submitted for review.
- The installation and calibration of the acoustic positioning system should provide accuracy better than 1% of the slant range.
- The hull-mounted USBL transducer should be located so as to minimize disturbances from thrusters and machinery noise and/or air bubbles in the transmission channel or other acoustic transmitters.
- The USBL equipment is supplied with its own computer and display unit, capable of operating as a stand-alone system.

- The vertical reference unit (VRU) is fabricated based on recommendations by the USBL manufacturer and installed as recommended.
- The system is capable of positioning at least nine transponders and/or responders.

4.3.5.5. Calibration of the USBL System

Calibration and testing of the USBL and VRU should be performed according to the latest revision of the manufacturers' procedures. If any main component in the USBL system has to be replaced, a complete installation survey/calibration of the system must be performed.

In the USBL, the multiple separate SBL hull hydrophones are replaced by a single complex hydrophone that uses phase comparison techniques to measure the angle of arrival of an acoustic signal in both the horizontal and vertical planes. Thus, a single beacon may be fixed by measuring its range and bearing relative to the vessel.

Although more convenient to install, the USBL transducer requires careful adjustment and calibration. A compass reference is required and the bearing measurements must be compensated for the roll and pitch of the vessel. Unlike the LBL method, there is no redundant information from which to estimate position accuracy.

4.4. SUBSEA SOIL INVESTIGATION

Subsea soil investigations are performed by geotechnical engineers or engineering geologists to obtain information on the physical properties of soil and rock around the subsea field development for use in the design of subsea foundations for the proposed subsea structures. A soil investigation normally includes surface exploration and subsurface exploration of the field development. Sometimes, geophysical methods are used to obtain data about the field development. Subsurface exploration usually involves soil sampling and laboratory tests of the soil samples retrieved. Surface exploration can include geological mapping, geophysical methods, and photogrammetry, or it can be as simple as a professional diver diving around to observe the physical conditions at the site.

To obtain information about the soil conditions below the surface, some form of subsurface exploration is required. Methods of observing the soils below the surface, obtaining samples, and determining physical properties of the soils and rocks include test pits, trenching (particularly for locating faults and slide planes), boring, and *in situ* tests.

4.4.1. Offshore Soil Investigation Equipment Requirements

The general requirements for soil investigations are as follows:

- Drill, sample, and downhole test to a minimum of 120 m below seabed.
- Carry out relevant seabed *in situ* testing, for example, a cone penetration test (CPT) to a maximum of 10 m depending on soil conditions.
- The actual sampling and subsequent handling are carried out with minimum disturbance to the sediments. The choice of sampler and sampling tubes reflects the actual sediment conditions and the requirements for the use of the sediment data. Therefore, different types of equipment are required.
- All equipment capable of electronic transmissions is designed to sustain the water pressure expected in the field.
- Records of experience with the use of the equipment and routines and procedures for interpretation of measurements for assessment of sediment parameters are documented and made available.

A detailed description of the sampling is provided as is testing equipment, which includes the following:

- Geometry and weight in air and water of all sampling and testing equipment;
- Handling of the seabed equipment over the side, over the stern, or through the moon-pool as applicable;
- Required crane and/or A-frame lifting force and arm length;
- Any limitations as to crane and A-frame capacity, water depth, sediment type, penetration depth, etc.;
- Zeroing of the PCPT before deployment;
- During testing, recording of the zero readings of all sensors before and after each test.

Calibration certificates for all cones are presented on commencement of operations. Sufficient spare calibrated cone tips should be provided to ensure work can be completed.

4.4.1.1. Seabed Corer Equipment

The coring equipment used should be of well-proven types and have a documented history of satisfactory operation for similar types of work. The seabed corers have a nonreturn valve at the top of the tube to avoid water ingress and sample washing out when pulling the sampler back to the surface. Both penetration and recovery are measured and recorded.

The main operational requirements for the corers are as follows:

- The corer is capable of operating at seabed.
- The corer is monitored continuously in the water column using a transponder.

4.4.1.2. Piezocone Penetration Test

The main operational requirements for the PCPT are as follows:

- PCP equipment is capable of operating at seabed.
- All cones are of the electric type, and cone end point resistance, sleeve friction, and pore water pressure are continuously recorded with depth during penetration.
- The PCP rig is monitored continuously in the water column using a transponder.
- Typical penetration below the seabed is up to 5 m pending soil conditions.
- During PCPT operations, prior to the start of the penetration of the push rods into the soil, the following data are recorded: water head, the resistance at the penetrating probe, the lateral friction, and the pore pressure starting from an elevation of 1 m above the seabed.
- The penetrometer is positioned in such a way as to provide the perfect verticality of push roads.

A typical scheme for a PCPT is shown in Figure 4-4.

4.4.1.3. Drilling Rig

Figure 4-5 illustrates a typical jack-up drilling rig. The drilling rig should be provided with all drill string components: the drill pipe, drill bits, insert bits, subs, crossovers, etc. The capability for the drill string on the drilling rig to be heave compensated such that the drill bit has a minimum of movement while drilling and performing downhole sampling and testing is very important.

Borings are drilled, using rotary techniques with a prepared drilling mud, from the seabed to the target depth. The objective of the borings is to obtain high-quality samples and perform *in situ* testing.

4.4.1.4. Downhole Equipment

Equipment for performing sampling and testing in downhole operation mode through a drill string is relevant to the investigation:

- PCPT;
- Push sampling;
- Piston sampling;
- Hammer sampling.

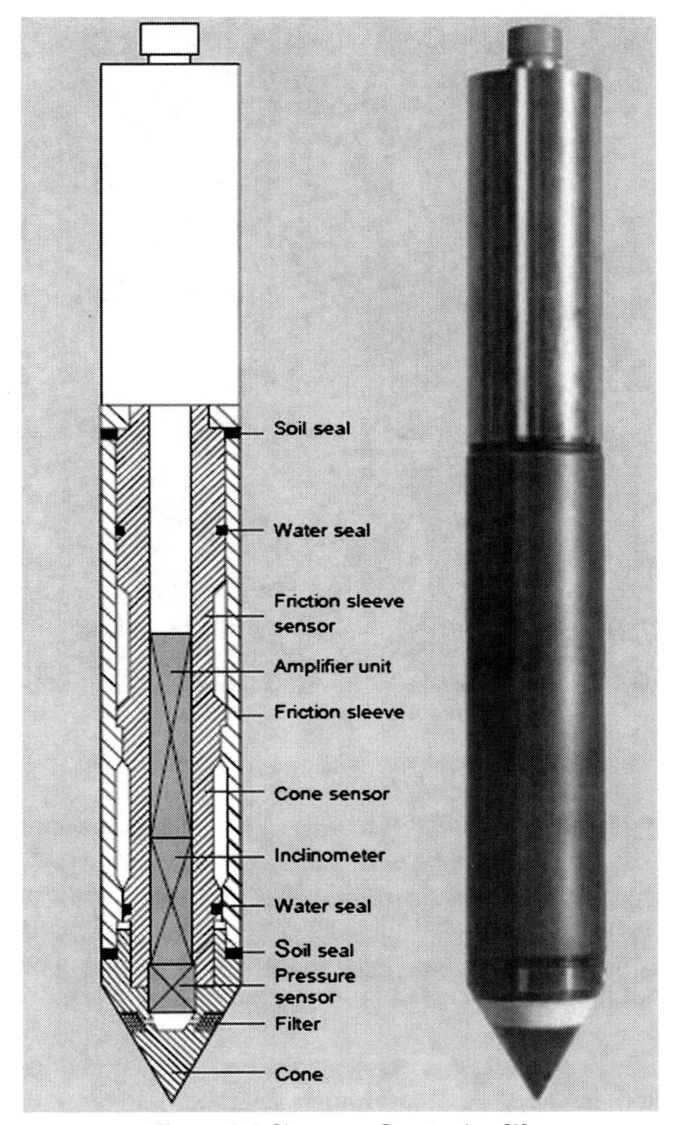

Figure 4-4 Piezocone Penetration [9]

An ample number of cones and sample tubes should be available. Push sampling is performed with thin-wall or thick-wall sample tubes, depending on the soil conditions. The main operational requirement for downhole equipment is that the equipment be used in the maximum relevant water and drilling depths.

Figure 4-5 Jack-Up Drilling Rig

4.4.1.5. Laboratory Equipment

The vessel is provided with either a room or a container to act as an offshore soil testing laboratory with sufficient equipment and personnel for 24 hour per day operation. All necessary supplies and equipment for cutting liners and sealing and waxing samples, including transportation boxes for shipping of samples to the onshore laboratory, have to be carefully provided for.

The offshore laboratory varies depending on the nature of the project. Equipment is required for performing the following types of standard laboratory tests:

- Extrusion of samples;
- Description of samples;
- Bulk density;
- Specific gravity;
- Water content;
- Shear strength of cohesive sediment.

4.4.2. Subsea Survey Equipment Interfaces

4.4.2.1. Sound Velocity Measurement

Velocity profiles are recorded whenever necessary to ensure that the correct speed of sound in seawater is utilized for the calibrations of the geophysical and bathymetric instruments. The velocity of sound in seawater can be calculated with a recognized formula.

All equipment is operated in accordance with manufacturers' published instructions and conforms to manufacturers' specifications. The velocity probe and winch system are capable of operating efficiently in survey water depths. Data are to be recorded on the descent to the seabed and recovery to the surface.

The instrumentation is calibrated to the standards set by the National Bureau of Standards within the 12 months prior to the mobilization date. Calibration certificates should be included with the survey procedures.

4.4.2.2. Sediment Handling and Storage Requirements

Sediment samples are carefully marked, handled, and transported. Samples from the corer are cut in 1-m sections. The core samples are then stored in a cool place, but not frozen where shaking and shock are limited to a minimum. The sealed cylinders or waxed samples are clearly labeled with:

- Top (nearest seabed);
- Bottom;
- A "Top Up" indicator (arrow pointing upward);
- Core location, attempt number, date, and company project number;
- Section number and depths at top/bottom;
- Length of core in meters.

An identification label is placed inside the top cap.

The sealed and marked sample cylinders and waxed samples are placed in boxes suitable for transportation. If possible, the core barrels should be stored vertically. Rooms adjacent to heavy engines or generators, which generate excessive vibrations, are avoided.

The boxes with sealed sediment material are transported to the onshore laboratory with caution and handled with care. Special precautions are made to prevent shock and impact loads to the sediment material during handling of the boxes.

The sediment must not be exposed to temperatures below 0°C. Whether samples are air freighted or trucked to the onshore laboratory must be decided in each case.

Each cylinder and waxed sample is registered and stored for convenient retrieval.

On completion of the fieldwork, a sample log for each sample is prepared. The sample log includes the following information:

- Project number;
- Site area;
- Borehole or core number;
- Sample number;
- Water depth;
- Date of sampling;
- Type of sampler;
- Diameter of sampling tube;
- Length of core material;
- Length of sediment penetration;
- Core catcher material;

Whether core material is extruded on board or sealed in a tube or liner;

A short description of sediment type should be prepared based on contents in the core catcher and in each end of the liners.

4.4.2.3. Onboard Laboratory Test

The cores are cut into sections no more than 1 m in length. Disturbance of the cores is avoided during cutting and at other times. The following tests are conducted at each end of the 1-m samples:

- Pocket penetrometer;
- Torvane;
- Motorized miniature vane.

Sediment samples obtained by the Ponar/Van Veen grab sampler are described, bagged, and sealed for transportation with the cores. A motorized miniature vane measurement is conducted within the box core sample near the center of the core where the soils are undisturbed.

4.4.2.4. Core Preparation

Prior to sealing, a visual classification of the sediment types is performed. Pocket penetrometer and shear vane tests are undertaken at the top and bottom of each core section. All cores are then labeled and sample tubes are cut to minimize air space, sealed to prevent moisture loss, and then stored vertically. Minimum labeling includes this information:

- Company;
- Project name;

- Core location reference number;
- Date;
- Water depth;
- Clear indications of the top and bottom of the core (e.g., use different color caps or mark the cores "Top" and "Bottom");
- An "UP" mark indicate proper storage orientation.

4.4.2.5. Onshore Laboratory Tests

The following tests, as applicable depending on soil types and locations, are carried out in a geotechnical laboratory on core samples sealed and undisturbed in the field as soon as possible after recovering the samples:

- Sample description;
- Sieve analysis;
- UU (Un-consolidated, undrained) and triaxial (cohesive soil);
- Miniature vane (cohesive soils);
- Classification tests (Atterberg limits, water content, submerged unit weight);
- Carbonate content;
- Ferrous content;
- Thermal properties;
- Organic matter content;
- Hydrometer.

The onshore laboratory program is approved prior to commencement of testing.

4.4.2.6. Near-Shore Geotechnical Investigations

To carry out geotechnical investigations in near-shore areas, a self-elevating jack-up is fully utilized or, as an alternative, an anchored barge for drilling operations in up to 20 m of water depth (WD) and as shallow as 2 m of WD.

The general requirements for certification, integrity, and safe/efficient working described in preceding sections are applied. In addition, the acceptable sanitary conditions and messing conditions are guaranteed which can reduce the impaction of environment in near-shore areas.

For support of the geotechnical drilling unit, any small boat operations should comply with the following guidelines:

- Small boats will be equipped with spare fuel, basic tool kit, essential engine spares, radar reflector, portable radio, mobile telephone, potable water, first aid kit, and distress signals/flares (secure in a water proof container).

- Small boats will only be driven by members of the crew or other personnel who have undergone a specialized small boat handling course.

4.5. SUBSEA FOUNDATION

A foundation is a structure that transfers loads to the earth. Foundations are generally broken into two categories: shallow foundations and deep foundations. A subsea production structure may be supported by piles, by nudmats with skirt or directly by the seabed. It could also be supported by a combination of these three structures. Table 4-1 shows the typical foundation selection matrix.

4.5.1. Pile- or Skirt-Supported Structures

The foundation piles of a pile-supported structure should be designed for compression, tension, lateral loads, and also shear stress as applicable.

The structure is properly connected to the pile/skirt (see Figure 4-6). This can be accomplished with a mechanical device or by grouting the annulus between the pile and sleeve.

4.5.2. Seabed-Supported Structures

The foundation of a seabed-supported structure is designed to have sufficient vertical and horizontal bearing capacity for the loads in question.

Depending on seabed conditions, high contact stresses may develop. This should be considered in the design. Underbase grouting may have to be used to achieve the required stability and load distribution on the seabed.

4.5.3. Pile and Plate Anchor Design and Installation

4.5.3.1. Basic Considerations

The technology for the evaluation of the geotechnical capacity of a suction pile and plate anchor is still under development; therefore, specific and detailed recommendations cannot be given at this point. Instead, general

Table 4-1 Foundation Selection Matrix

Geography\Vessel		Type of Installation Vessel for Different Foundations: Drilling Rig	Pipe Lay Barge/Construction Vessel
Soil Conditions	Hard	—	Mudmat
	Medium	Suction pile	Mudmat or suction pile
	Soft	Suction pile	Suction pile

Figure 4-6 Suction Piles *(http://www.offshore-technology.com/)*

statements are used to indicate that consideration should be given to some particular points and references are given. Designers are encouraged to utilize all research advances available to them. It is hoped that more specific recommendations can be issued after completion of the research in this area.

4.5.4. Geotechnical Capacity of Suction Piles

4.5.4.1. Basic Considerations

Suction pile anchors resist vertical uplift loads by various mechanisms depending on the load type:

- Storm loading:
  - External skin friction;
  - Reverse end bearing (REB) at the tip of the piles;
  - Submerged weight of the anchor.
- Long-term loop current loading:
  - External skin friction appropriately reduced for creep and cyclic effects due to long-term duration of the event;
  - Reduced value of reverse end bearing;
  - Submerged weight of the anchor.
- Pretension loading:
  - External skin friction;
  - Smaller of internal skin friction or soil plug weight;
  - Submerged weight of anchor.
- Suction piles resist horizontal loads by the following mechanisms:
  - Passive and active resistance of the soil;
  - External skin friction on the pile wall sides (as appropriate);
  - Pile tip shear.

The geotechnical load capacity of the anchor should be based on the lower bound soil strength properties. This is derived from the site-specific soil investigation and interpretation. Anchor adequacy with respect to installation should be based on the upper bound soil strength properties.

Because this geometry may change the relationship between the horizontal and vertical anchor loads, the impact of the mooring line geometry in the soil on the anchor loads should be considered. For example, the inverse catenary of the mooring line in the soil may make the mooring line angle steeper at the anchor. This steeper angle could result in a reduced horizontal anchor load, but an increased vertical anchor load. Both an upper and lower bound inverse catenary should be checked to ensure that the worst case anchor loading is established.

Axial safety factors take into consideration the fact that piles are primarily loaded in tension and are therefore higher than for piles loaded in compression. As with other piled foundation systems, the calculated ultimate axial soil resistance should be reduced if soil setup, which is a function of time after pile installation, will not be complete before significant loads are imposed on the anchor pile.

Because the lateral failure mode for piles is considered to be less catastrophic than vertical failure, lower factors of safety have been recommended for lateral pile capacity. Use of different safety factors for vertical and lateral pile capacities may be straightforward for simple beam-column analysis of, for example, mobile moorings, but more complex methodologies do not differentiate between vertical and lateral pile resistance. The following formula is proposed to provide a combined factor of safety for the latter situation:

$$\mathrm{FOS}_{\mathrm{combined}} = \sqrt{(\mathrm{FOS}_{\mathrm{lateral}} \cos\theta)^2 + (\mathrm{FOS}_{axial} \sin\theta)^2} \tag{4-1}$$

where

$\mathrm{FOS}_{\mathrm{combined}}$: combined factor of safety (FOS);

$\mathrm{FOS}_{\mathrm{lateral}}$: lateral FOS;

$\mathrm{FOS}_{\mathrm{axial}}$: axial FOS;

θ: angle of mooring line from horizontal at pile attachment point.

4.5.4.2. Analysis Method

It is recommended that suction pile design for permanent moorings use advanced analysis techniques such as limit equilibrium analysis or finite element analysis of the pile and adjacent soil. For example, the resultant capacity may be calculated with limit equilibrium methods, where the circular area is transformed into a rectangle of the same area and width equal to the diameter, and with 3D effects accounted for by side shear factors.

For mobile moorings, simpler beam column analysis using load transfer displacement curves (i.e., *P-y*, *T-z*, *Q-z*) described in API RP 2A [11] are considered adequate if suitably modified.

In areas where tropical cyclonic storms may exceed the capacity of the mobile mooring or anchoring system, such as the Gulf of Mexico, the design of suction piles should consider an anchor failure mode that reduces the chance of anchor pullout. For example, the mooring line anchor point can be located on the suction anchor such that the anchor does not tilt during soil failure, but "plows" horizontally in a vertical orientation.

4.5.5. Geotechnical Capacity of Plate Anchors

4.5.5.1. Basic Considerations

The ultimate holding capacity is usually defined as the ultimate pull-out capacity (UPC), which is the load for the soil around the anchor reaching failure mode for plate anchors. For anchors that dive with horizontal movement, the ultimate holding capacity is reached when excessive lateral drag distance has occurred. At UPC, the plate anchor starts moving through the soil in the general direction of the applied anchor load with no further increase in resistance or the resistance starts to decline. The ultimate pull-out capacity of a plate anchor is a function of the soil's undrained shear strength at the anchor fluke, the projected area of the fluke, the fluke shape, the bearing capacity factor, and the depth of penetration. The disturbance of the soil due to the soil failure mode should be considered when analyzing the plate anchor's ultimate pull-out capacity. This mode is generally accounted for in the form of a disturbance factor or capacity reduction factor. The bearing capacity factor and disturbance factor should be based on reliable test data, studies, and references for such anchors. The plate anchor's penetration is usually in a range of two to five times the fluke width, depending on the undrained shear strength of the soil, in order to generate a deep failure mode. If the final depth does not generate a deep failure mode, a suitable reduction in bearing capacity factor should be used.

Plate anchors get their high holding capacity from being embedded into competent soil. Therefore, it is important for the anchor's penetration depth to be established during the installation process. Furthermore, a plate anchor gets its high ultimate pull-out capacity by having its fluke oriented nearly perpendicular to the applied load. To ensure that the fluke will rotate to

achieve a maximum projected bearing area, the plate anchor design and installation procedure should:
- Facilitate rotation of the fluke when loaded by environmental loads or during installation or both;
- Ensure that no significant or unpredicted penetration is lost during anchor rotation, which may move the fluke into weaker soil;
- Have the structural integrity to allow such fluke rotation to take place during installation and keying operations or while subject to the ultimate pull-out capacity load. Depending on the type of plate anchor and its installation orientation, this item may also apply to fluke rotation about both horizontal and vertical axes.

As appropriate, the anchor capacities should be reduced to account for anchor creep under long-term static loading and cyclic degradation.

4.5.5.2. Prediction Method for a Drag Embedded Plate Anchor

Three aspects of drag embedded plate anchor performance require prediction methods:
- Anchor line mechanics;
- Installation performance;
- Holding capacity performance.

All three mechanisms are closely linked and influence one another, as explained next.

4.5.5.3. Prediction Method for Direct Embedded Plate Anchor

Anchor capacity determination for direct embedded plate anchors is identical to that shown for drag embedded anchors with the following exceptions:
- Final penetration depth is accurately known.
- Nominal penetration loss during keying should be included (usually taken as 0.25 to 1.0 times the fluke's vertical dimension, or B in Figure 4-7, depending on the shank and keying flap configuration).
- Calculation of the effective fluke area should use an appropriate shape factor and projected area of the fluke with a keying flap in its set position.

Safety factors for drag embedded plate anchors are higher because overloading of the anchor normally results in the anchor pulling out, whereas a drag anchor may drag horizontally or dig in deeper, developing constant or higher holding capacity under a similar situation. For plate anchors that exhibit overloading behaviors similar to those of drag anchors, consideration may be given to using drag anchor safety factors, assuming the behavior can be verified by significant field tests and experience.

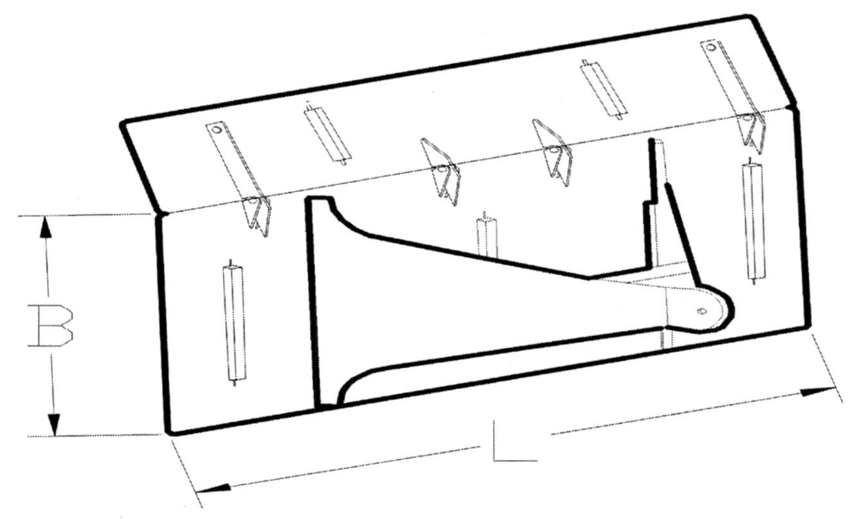

Figure 4-7 Special Consideration of Safety Factors for Drag Embedded Plate Anchors

4.5.6. Structural Design of Suction Piles

4.5.6.1. Basic Considerations

The purpose of this section is to provide guidance and criteria for the structural design of suction piles. Some of the guidance and criteria are also applicable to driven piles.

4.5.6.2. Design Conditions

The suction pile structure should be designed to withstand the maximum loads applied by the mooring line, the maximum negative pressure required for anchor embedment, the maximum internal pressure required for anchor extraction, and the maximum loads imposed on the anchor during lifting, handling, launching, lowering, and recovery. The fatigue lives of critical components and highly stressed areas of the anchor should be determined and checked against the required minimum fatigue life.

Mooring Loads on Global Anchor Structure

The load case that provides the maximum horizontal and vertical loads at the mooring padeye should be used for the global structural design of the anchor. The soil reactions generated by the geotechnical analysis will be used in these calculations. Sensitivity checks should be performed to ensure that a load case with less than the maximum load, but applied at a more onerous angle at the padeye, does not control the design.

Mooring Loads on Anchor Attachment

The mooring line attachment padeye or lug is a critical structural component. To meet fatigue resistance criteria, the padeye is often an integral cast lug and base structure. This avoids the use of heavy weldments, which can result in a lower fatigue life. The attachment padeye should be designed to satisfy both strength and fatigue requirements. The padeye should be designed for the controlling design load with an appropriate factor of safety. Designing the padeye for a maximum load equal to a factor times the break strength of the mooring line may lead to a significantly overdesigned padeye, which may not integrate well with the anchor shell and backup structures.

The mooring line padeye should be designed for the controlling load case, and sensitivity checks should be performed to ensure that a load case with less than the maximum load but applied at a more onerous angle does not control the design. The orientation of the applied load at the padeye will be affected by the inverse catenary of the mooring line, vertical misalignment due to anchor tilt, and rotational misalignment due to deviation from the target orientation. These factors should be properly accounted for.

Embedment Loads

For anchor embedment, the estimated upper bound suction pressure required to embed the anchor to its design penetration should be used for the design of the anchor wall and anchor cap structure. However, the maximum suction pressure used should not be higher than the suction at which internal plug uplift occurs.

Extraction Loads

With respect to anchor extraction, two conditions require evaluation:

- *Temporary condition:* Extraction of a suction pile may be required for permanent moorings. For example, after all suction piles have been preinstalled along with the mooring lines, one of the mooring lines is accidentally dropped to the seabed and damaged during the hookup operation with the vessel. At this time, a decision to extract the suction pile and recover the mooring leg may be made. Typically, such situations may occur 30 to 60 days after the first suction pile has been installed. For mobile moorings, the suction piles are often extracted at the end of the current drilling or testing operation and reused in other locations.

- *Terminal condition:* The suction piles for a permanent mooring may be extracted at the end of their service life. The estimated maximum internal pressure required to extract the anchor for these two situations should be used for the design of the anchor wall and anchor cap structure. However, the maximum extraction pressure used should not be higher than the pressure causing overload of soil-bearing capacity at the anchor tip. The vessel removing the anchor is often capable of applying a lifting force on the anchor with the recovery line. This assistance can significantly reduce the required extraction pressure and therefore should be included in the removal analysis.

Transportation and Handling of Loads

The suction pile structure and its installation appurtenances should be designed for the maximum loads generated during suction pile handling, transportation, lifting, upending, lowering, and recovery. The suction pile designer should interface closely with the installation contractor when determining these load cases. Design of appurtenances for these load cases is typically performed using the installation contractor's in-house design guidelines or other recognized codes. Nevertheless, all lifting appurtenances and their supporting structures should meet the minimum requirements of API RP 2A [11].

4.5.6.3. Structural Analysis Method

Pile analysis in accordance with Section 3 of API RP 2A [11] is appropriate for piles with diameter-to-thickness ratios (D/t) of less than 120. For cylindrical piles with D/t ratios exceeding 120, it is recommended that a detailed structural finite element model be developed for the global structural anchor analysis to ensure that the anchor wall structure and appurtenances have adequate strength in highly loaded areas. Supplementary manual calculations may be appropriate for members or appurtenances subjected to local loading.

4.5.6.4. Space Frame Model

A space frame model generally consists of beam elements plus other elements needed to model specific structural characteristics. This is appropriate for piles with D/t ratios of less than 120 and for preliminary design of the top cap or padeye backup structures on large-diameter piles (i.e., $D/t > 120$).

Finite Element Model

Finite element analysis is recommended for the global shell structure, top cap plate and supporting members, and the padeye backup structure for piles with $D/t > 120$. Complex shapes such as the padeye casting or welding should also be analyzed by finite element method.

Manual Calculations

Manual calculations using empirical formulas and basic engineering principles can be performed when detailed finite element analysis is not needed.

Stress Concentration Factors

Stress concentration factors can be determined by detailed finite element analysis, physical models, and other rational methods or published formulas.

Stability Analysis

Formulas for the calculation of the buckling strength of structural elements are presented in API Recommended Practice 2A [11]; API Bulletin 2U, "Stability Design of Cylindrical Shells" [12]; and API Bulletin 2V, "Design of Flat Plate Structures" [13]. As an alternative, buckling and postbuckling analysis or model tests of specific shell or plate structures may be performed to determine buckling and ultimate strength.

Dynamic Response

Significant dynamic response is not expected for the anchor in its in-place condition; therefore, anchor structures are often analyzed statically. Transportation analysis, however, will typically include dynamic loads generated by harmonic motions of a simple single-degree-of-freedom model.

4.5.6.5. Structural Design Criteria

Design Codes

The method for structure design is the working stress design method, where stresses in all components of the structure are kept within specified values. In general, cylindrical shell elements should be designed in accordance with API RP 2A [11] for D/t ratios of less than 300 or API Bulletin 2U [8] when D/t exceeds 300, flat plate elements in accordance with API Bulletin 2V [9], and all other structural elements in accordance with API RP 2A, as applicable. In cases where the structure's configurations or loading conditions are not specifically addressed by these codes, other accepted codes of practice can be used. In this case, the designer must ensure that the safety levels and design philosophy implied in the API Recommended Practice 2SK [14] are adequately met.

In API RP 2A, allowable stress values are expressed, in most cases, as a fraction of the yield or buckling stress. In API Bulletin 2U, allowable stress values are expressed in terms of critical buckling stresses. In API Bulletin 2V, the allowable stresses are classified in two basic limit states: ultimate limit states and serviceability limit states. Ultimate limit states are associated with the failure of the structure, whereas serviceability limit states are associated with adequacy of the design to meet its functional requirements. For the purpose of suction anchor design, only the ultimate limit state is considered.

Safety Categories

There are two safety categories: Category A safety criteria are intended for normal design conditions, and Category B safety criteria are intended for rarely occurring design conditions. The criteria listed in Table 4-2 are recommended.

Allowable Stresses

For structural elements designed in accordance with API RP 2A [11], the allowable stresses recommended in these codes should be used for normal design conditions associated with Category A safety criteria. For extreme design conditions associated with Category B safety criteria, the allowable stresses may be increased by one-third if the working stress design method is utilized (e.g., API RP 2A-WSD [15]).

For shell structures designed in accordance with API Bulletin 2U [12], a factor of safety equal to 1.67Ψ is recommended for buckling modes for Category A safety criteria. For Category B safety criteria, the corresponding factor of safety is equal to 1.25Ψ. The parameter Ψ varies with buckling stress and is defined in API Bulletin 2U. It is equal to 1.2 for elastic buckling

Table 4-2 Suction Pile Safety Criteria [13]

Load Condition	Safety Criteria
Maximum intact	A
Maximum one-line damaged	B
Anchor embedment	A
Anchor extraction (temporary)	A
Anchor extraction (terminal)	B
Handling/lifting/lowering/recovery	A
Transportation	B

stresses at the proportional limit and reduces linearly for inelastic buckling to 1.0 when the buckling stress is equal to the yield stress.

For flat plate structures designed in accordance with API Bulletin 2V [13], the allowable stress is obtained by dividing the ultimate limit state stress by an appropriate factor of safety, which is 2.0 for Category A safety criteria and 1.5 for Category B safety criteria.

For cylindrical elements with D/t ratios exceeding 120, it is recommended that global strength be analyzed using finite element techniques. Local buckling formulations for axial compression, bending, and hydrostatic pressure given in API RP 2A [11] for D/t <300 and API Bulletin 2U [12] D/t $\geq$ 300 are considered valid if due consideration is made for variable wall thicknesses (when it occurs) and buckling length (which may extend below the mudline when performing suction embedment analysis).

The nominal Von Mises (equivalent) stress at the element's extreme fiber should not exceed the maximum permissible stress as calculated below:

$$\sigma_A = \eta_i \sigma_y \tag{4-2}$$

where

σ_A: Allowable Von Mises stress;

η_i: Design factor for specified load condition;

σ_y: Specified minimum yield stress of anchor material.

Design factors for the listed load conditions are given in Table 4-3.

4.5.7. Installation of Suction Piles, Suction Caissons, and Plate Anchors

4.5.7.1. Suction Piles and Suction Caissons

Installation Procedure, Analysis, and Monitoring

Installation procedures should be developed and installation analyses should be performed for suction pile and suction caisson anchors to verify that the

Table 4-3 Design Factors for Finite Element Analysis

Load Condition	Design Factor ηi
Maximum intact	0.67
Maximum one-line damaged	0.90
Anchor embedment	0.67
Anchor extraction (temporary)	0.67
Anchor extraction (terminal)	0.90
Handling/lifting/lowering/recovery	0.67
Transportation	0.90

anchors can penetrate to the design depth. The installation analysis should also consider anchor retrieval for the following cases:

- Mobile moorings where anchor removal is needed for reuse of the anchor or to clear the seabed. The suction pile retrieval procedures and analysis should account for the estimated maximum setup time.
- Permanent moorings where it is required by authorities that the anchors be removed after the system service life. The suction pile retrieval procedures and analysis should be based on full soil consolidation.

For suction pile embedment analysis, the risk of causing uplift of the soil plug inside the anchor should be considered. The allowable underpressure to avoid uplift should exceed the required embedment pressure by a factor of 1.5.

Anchor installation tolerances should be established and considered in the suction pile anchor geotechnical, structural, and installation design process. The following typical tolerances should be considered:

- Allowable anchor tilt angle in degrees;
- Allowable deviation from target orientation of the mooring line attachment to limit padeye side loads and rotational moments on the anchor;
- Minimum penetration required to achieve the required holding capacity.

Suction pile installation analysis should provide the relevant data needed for the suction pile design and installation procedures. The following typical information is required:

- Anchor self-weight penetration for applicable soil properties or range of properties;
- Embedment pressure versus penetration depth for applicable soil properties;
- Allowable embedment pressure to avoid plug uplift;
- Penetration rate;
- Estimated internal plug heave.

To verify that the suction pile installation is successful and in agreement with the design assumptions, the following data should be monitored and recorded during the installation of suction piles:

- Self-weight penetration;
- Embedment pressure versus penetration depth;
- Penetration rate;
- Internal plug heave (by direct or indirect means);
- Anchor heading and anchor tilt in degrees;

- Final penetration depth.

For the installation of temporary mooring suction pile anchors, measurement of the internal plug heave is not required if the anchor reaches its design embedment depth.

Skirt Penetration of Suction Caissons

The following points should be given attention when designing the skirt penetration capability of suction caissons:

- The skirt penetration resistance should account for the reduced shear strength along the skirt wall due to the disturbance during penetration. Normally, the remolded shear strength is applied.
- Stiffeners (outside and inside) may influence the penetration resistance, because additional force may be required to penetrate them, and the failure mode around internal stiffeners should be given attention. On the other hand, a gap may form above outside and inside stiffeners. This may reduce the penetration resistance and form potential flow paths. In the case of several ring stiffeners, clay from the upper part of the profile may be trapped between the stiffeners and give low resistance at larger depth. In stiff clays the soil plug may stand open, giving essentially no skin friction along the inside wall.
- The allowable underpressure for penetration should be calculated as the sum of the inverse bearing capacity at skirt tip and the internal skirt wall friction. There is some controversy as to whether the conventional bearing capacity factor can be used to calculate the end bearing capacity below the skirt tip, but most designers tend to assume conventional bearing capacity factors.
- If the outside or inside suction caisson skirt wall is given surface treatment (e.g., painting), this may cause reduction in the skirt wall friction, which should be taken into account in the calculations.

4.5.7.2. Plate Anchors

Direct Embedded Plate Anchors

Direct embedment of plate anchors can be achieved by suction, impact or vibratory hammer, propellant, or hydraulic ram. The suction embedded plate anchor (SEPLA) has been used for major offshore mooring operations. As an example, the SEPLA uses a so-called suction follower, which is essentially a reusable suction anchor with its tip slotted for insertion of a plate anchor. The suction follower is immediately retracted by reversing the pumping action once the plate anchor is brought to the design depth, and can be used to install additional plate anchors. In the SEPLA concept, the plate anchor's fluke is embedded in the vertical position and necessary fluke rotation is achieved during a keying process by pulling on the mooring line.

Installation procedures should be developed and installation analyses should be performed for direct embedded plate anchors to verify that the anchors can penetrate to the design depth. The installation analysis should also consider plate anchor retrieval if applicable.

For the embedment analysis, the risk of causing uplift of the soil plug inside the suction embedment tool should be considered. The allowable underpressure to avoid uplift should exceed the required embedment pressure by a factor of 1.5.

Plate anchor installation tolerances should be established and considered in the anchor's geotechnical, structural, and installation design process. The following typical tolerances should be considered:

- Allowable deviation from target heading of the mooring line attachment to limit padeye side loads and rotational moments on the anchor padeye;
- Minimum penetration required before keying or test loading to achieve the required holding capacity;
- Allowable loss of anchor penetration during plate anchor keying or test loading.

Suction embedment analysis should provide relevant data needed for the design and installation procedures, which should allow verification of the assumptions used in the anchor design. The following typical information is required:

- Suction embedment tool self-weight penetration for applicable soil properties or range of properties;
- Embedment pressure versus penetration depth for applicable soil properties;
- Allowable embedment pressure to avoid plug uplift;
- Penetration rate;
- Estimated internal plug heave.

To verify that the plate installation is successful and in agreement with the assumptions in design, the following data should be monitored and recorded during the installation of the suction embedment tool to verify the assumption used in design:

- Self-weight penetration;
- Embedment pressure versus penetration depth;
- Penetration rate;

- Internal plug heave, if it is expected that plug heave could be a concern;
- Anchor orientation;
- Final penetration.

Drag Embedded Plate Anchors

For drag embedded plate anchors used in permanent moorings, the installation process should provide adequate information to ensure that the anchor reaches the target penetration, and that the drag embedment loads are within the expected load range for the design soil conditions. The following typical information should be monitored and verified:

- Line load in drag embedment line;
- Catenary shape of embedment line based on line tension and line length to verify that uplift at the seabed during embedment is within allowable ranges and to verify anchor position;
- Direction of anchor embedment;
- Anchor penetration.

4.5.7.3. Test Loading of Anchors

For suction piles, suction caissons, and plate anchors, the installation records should demonstrate that the anchor penetration is within the range of upper and lower bound penetration predictions developed during the anchor geotechnical design. In addition, the installation records should confirm the installation behavior, that is, self-weight penetration, embedment pressures, and drag embedment loads and that the anchor orientation is consistent with the anchor design analysis. Under these conditions, test loading of the anchor to a full intact storm load should not be required.

Plate anchors should be subjected to adequate keying loads to ensure that sufficient anchor fluke rotation will take place without further loss of anchor penetration. The keying load required and amount of estimated fluke rotation should be based on reliable geotechnical analysis and verified by prototype or scale model testing. The keying analysis used to establish the keying load should also include analysis of the anchor's rotation when subjected to the maximum intact and one-line damage survival loads. If the calculated anchor rotation during keying differs from the anchor rotation in survival conditions, then the anchor's structure should be designed for any resulting out-of-line loading to ensure that the anchor's structural integrity is not compromised.

In cases where the installation records show significant deviation from the predicted values and these deviations indicate that the anchor holding capacity may be compromised, test loading of the anchor to the maximum dynamic intact load may be required and may be an acceptable option to prove holding capacity for temporary moorings. However, testing anchors to the maximum intact load does not necessarily prove that required anchor holding safety factors have been met, which is of special concern for permanent mooring systems. Consequently, if the installation records show that the anchor holding capacity is significantly smaller than calculated and factors of safety are not met, then other measures to ensure adequate factors of safety should be considered:

- Additional soil investigation at the anchor location to establish and/or confirm soil properties at the anchor site;
- Retrieval of the anchor and reinstallation at a new undisturbed location;
- Retrieval of the anchor, redesign and reconstruction of the anchor to meet design requirements, and reinstallation at a new undisturbed location;
- Delay of vessel hookup to provide additional soil consolidation.

4.5.8. Driven Pile Anchor

4.5.8.1. Basic Considerations

Driven pile anchors provide a large vertical load capacity for taut catenary mooring systems. The design of driven pile anchors builds on a strong industry background in the evaluation of geotechnical properties and the axial and lateral capacity prediction for driven piles. The calculation of driven pile capacities, as developed for fixed offshore structures, is well documented in API RP 2A. The recommended criteria in API RP 2A should be applied for the design of driven anchor piles, but with some modifications to reflect the differences between mooring anchor piles and fixed platform piles.

The design of a driven pile anchor should consider four potential failure modes:

1. Pull-out due to axial load;
2. Overstress of the pile and padeye due to lateral bending;
3. Lateral rotation and/or translation;
4. Fatigue due to environmental and installation loads.

The safety factors of holding capacity is defined as the calculated soil resistance divided by the maximum anchor load from dynamic analysis.

4.5.8.2. Geotechnical and Structural Strength Design

In most anchor pile designs, the mooring line is attached to the pile below the seafloor, to transfer lateral load to stronger soil layers. As a result, the design should consider the mooring line angle at the padeye connection resulting from the "reverse catenary" through the upper soil layers. Calculation of the soil resistance above the padeye location should also consider remolding effects due to this trenching of the mooring line through the upper soil layers.

Driven pile anchors in soft clay typically have aspect ratios (penetration/diameter) of 25 to 30. Piles having such an aspect ratio would be fixed in position about the pile tip and consequently would deflect laterally and fail in bending before translating laterally as a unit. Driven pile anchors are typically analyzed using a beam-column method with a lateral load-deflection model (p-y curves) for the soil. These computations should include the axial loading in the pile, as well as the mooring line attachment point, which will influence the deflection, shear, and bending moment profiles along the pile. Pile stresses should be limited to the basic allowable values in API RP 2A [11] under intact conditions. Basic allowable stresses may be increased by one-third for rarely occurring design conditions such as a one-line damaged condition.

Because an anchor pile near its ultimate holding capacity experiencing the largest deflection always engages new soil, "static" p-y curves may be considered for calculation of lateral soil resistance. "Cyclic" p-y curves may be more appropriate for fatigue calculations, because the soil close to the pile will be more continually disturbed due to smaller, cyclic deflections. The p-y curve modifications developed by Stevens and Audibert [16] are recommended in place of the API RP 2A p-y curves, to obtain more realistic deflections. Consideration should be given to degrading the p-y curves for deflections greater than 10% of the pile diameter. In addition, when lateral deflections associated with cyclic loads at or near the mudline are relatively large (e.g., exceeding y_c as defined in API RP 2A for soft clay), consideration should be given to reducing or neglecting the soil-pile adhesion (skin friction) through this zone.

The design of driven anchor piles should consider typical installation tolerances, which may affect the calculated soil resistance and the pile structure. Pile verticality affects the angle of the mooring line at the padeye, which changes the components of the horizontal and vertical mooring line loads that the pile must resist. Underdrive will affect the axial pile capacity and may result in higher bending stresses in the pile. Padeye orientation (azimuth) may affect the local stresses in the padeye and connecting shackle. Horizontal positioning may affect the mooring scope and/or angle at the vessel fair lead, and should be considered when balancing mooring line pretensions.

4.5.8.3. Fatigue Design

Basic Considerations

Anchor piles should be checked for fatigue caused by in-place mooring line loads. Fatigue damage due to pile driving stresses should also be calculated and combined with in-place fatigue damage. For typical mooring systems, fatigue damage due to pile driving is much higher than that caused by in-place mooring line loads.

In-Place Loading

A global pile response analysis accounting for the pile–soil interaction should be carried out for the mooring line reactions due to the fatigue sea states acting on the system. The local stresses that accumulate fatigue damage in the pile should be obtained by calculating a stress concentration factor (SCF), relative to the nominal stresses generated by the global analysis, at the fatigue critical locations. These locations are typically at the padeye, at the girth welds between the padeye and the pile, and between subsequent pile cans.

The evaluation of SCFs for girth welds needs to account for the local thickness misalignment at the weld. Note that the calculated SCF needs to be corrected by the ratio of the nominal thickness used in the pile response analysis to the lesser of the pile wall thicknesses joining at the weld. The SCF is to be applied to the nominal pile stress range obtained at the weld location due to in-place loads, from which damage is to be calculated.

Installation Loading

Dynamic loads due to hammer impact during pile installation will induce fatigue damage on both padeye and pile girth welds. The evaluation of the cyclic loads involves the dynamic response of the pile–soil system due to the hammer impact. This requires a wave equation analysis per blow for a given hammer type and efficiency, pile penetration, and soil resistance. Various such analyses are to be conducted for judiciously selected pile penetrations. For each analysis, traces of stress versus time at the critical locations along the pile are to be developed, as well as the number of blows associated with the assumed penetration.

For either welds or padeye, fatigue load calculations should be carried out at various pile locations using local stress range, derived from the wave equation analysis at the selected pile penetrations. The location of the girth weld should be determined by the pile makeup schedule. The local response should include the corresponding SCF effect. The number of cycles of the stress history per blow is obtained using a variable amplitude counting method, such as the reservoir methods.

Fatigue Resistance

Applicable SN curves (stress-number of cycles to failure) depend on the manufacturing processes and defect acceptance criteria. Typically, pile sections are welded by a two-sided SAW (Submerged Arc Welding) process and are left in the as-welded condition. For this case, the D curve may be used. Use of a higher SN curve for this application, without additional treatment of the weld, should be demonstrated by relevant data. Use of weld treatment methods, such as grinding, may support the upgrading of the SN curve, provided that (1) the grinding process is properly implemented, (2) weld inspection methods and defect acceptance criteria are implemented, and (3) pertinent fatigue data are generated to qualify the weld to a performance level higher than that implied by the D curve.

Total Fatigue Damage and Factor of Safety

Once the fatigue loading and resistance have been determined, fatigue damage due to in-place and installation loads can be evaluated. The total fatigue damage should satisfy the following equation for the critical structural elements:

$$D = D_1 + D_2 < 1/F \tag{4-3}$$

where D_1 is calculated fatigue damage for phase 1, that is, the installation (pile driving) phase and transportation phase, if significant. The D_2 term is the calculated fatigue damage for phase 2, that is, the in-service phase, during the service life (e.g., 20 years). The factor of safety is F and is equal to 3.0.

4.5.8.4. Test Loading of Driven Pile Anchors

The driven pile installation records should demonstrate that the pile self-weight penetration, pile orientation, driving records, and final penetration are within the ranges established during pile design and pile driving analysis. Under these circumstances, test loading of the anchor to full intact storm load should not be required. However, the mooring and anchor design should define a minimum acceptable level of test loading. This test loading should ensure that the mooring line's inverse catenary is sufficiently formed to prevent unacceptable mooring line slacking due to additional inverse catenary cut-in during storm conditions. Another function of the test loading is to detect severe damage to the mooring components during installation.

REFERENCES

[1] GEMS, Vessel Specification of MV Kommandor Jack, <www.gems-group.com>.
[2] K.F. Anschütz, Cutaway of Anschütz Gyrocompass, <http://en.wikipedia.org/wiki/Gyrocompass>.
[3] Navis.gr, Gyrocompass - Steaming Error http://www.navis.gr/navaids/gyro.htm.
[4] D.J. House, Seamanship Techniques: Shipboard and Marine Operations, Butterworth-Heinemann, 2004.
[5] L. Mayer, Y. Li, G. Melvin, 3D Visualization for Pelagic Fisheries Research and Assessment, ICES, Journal of Marine Science vol. 59 (2002).
[6] B.M. Isaev, Measurement Techniques, vol. 18, No 4, Plenum Publishing Co, 2007.
[7] P.H. Milne, Underwater Acoustic Positioning Systems, Gulf Publishing, Houston, 1983.
[8] R.D. Christ, R.L. Wernli, The ROV Manual, Advantages and Disadvantages of Positioning Systems, Butterworth-Heinemann 2007.
[9] Fugro Engineers B.V., Specification of Piezo-Cone Penetrometer, <http://www.fugro-singapore.com.sg>.
[10] M. Faulk, FMC ManTIS (Manifolds & Tie-in Systems), SUT Subsea Awareness Course, Houston, 2008.
[11] American Petroleum Institute, Recommended Practice for Planning, Designing and Constructing Fixed Offshore Platforms - Load and Resistance Factor Design, API RP 2A-LRFD (1993).
[12] American Petroleum Institute, Bulletin on Stability Design of Cylindrical Shells, API Bulletin 2U (2003).
[13] American Petroleum Institute, Design of Flat Plate Structures, API Bulletin 2V (2003).
[14] American Petroleum Institute, Design and Analysis of Station Keeping Systems for Floating Structures, API RP 2SK (2005).
[15] American Petroleum Institute, Recommended Practice for Planning, Designing and Constructing Fixed Offshore Platforms - Working Stress Design - Includes Supplement 2, API 2A WSD, 2000.
[16] J.B. Stevens, J.M.E. Audibert, Re-Examination of P-Y Curve Formulations, OTC 3402, Houston, 1979.

Installation and Vessels

Contents

5.1. INTRODUCTION

Most subsea structures are built onshore and transported to the offshore installation site. The process of moving subsea hardware to the installation site involves three operations: load-out, transportation, and installation operations. A typical subsea installation also includes three phases: lowering, landing, and locking (e.g., subsea tree, manifold, and jumper).

Subsea Engineering Handbook
ISBN 978-0-12-397804-2, doi:10.1016/B978-0-12-397804-2.00005-9

The objective of this chapter is to provide a basic understanding of the installation concepts and vessel requirements for subsea engineering. The main topics of this chapter are as follows:

- Typical installation vessels;
- Vessel requirements and selection;
- Installation positioning;
- Installation analysis.

For typical installation methods for each subsea structure, please refer to the specific topics and chapters.

5.2. TYPICAL INSTALLATION VESSELS

Installation of subsea structures may involve the following vessels:

- Transportation barges and tug boats
- Drilling vessels including jack-up rigs, semi-submersibles, and drill ships;
- Pipe-laying and umbilical-laying vessels;
- Heavy lift vessels;
- Offshore support vessels, such as ROV support vessels, diving support vessels, and field support vessels.

5.2.1. Transportation Barges and Tug Boats

Subsea structures are normally transported from onshore to the offshore installation site by a transportation barge. Once at the offshore installation site, the subsea structure is transferred from the transportation barge to the

Figure 5-1 Transportation Barge [1]

Figure 5-2 Tug Boat [2]

drilling rig or construction vessels. Figure 5-1 shows a subsea Christmas tree that was transported to its offshore location by a barge and prepared for lowering by a rig winch.

Selection of the proper transportation barge and the arrangement of the structures on the deck depends primarily on the following features:

- Dimensions, weight, and center of gravity (CoG) height of the structure;
- Distance and transportation route;
- Schedule constraints;
- Cost;
- Ability to avoid bad weather.

Transportation barges are normally towed by tug boats from one location to another. A typical tug boat is shown in Figure 5-2.

5.2.2. Drilling Vessels

Drilling vessels are mainly designed for drilling activities, but they are also used for the Subsea Production System (SPS) installation because of their installation capacities such as water range, lifting capacity, and positioning capacity. Drilling vessels normally include jack-up rigs, semi-submersibles, and drilling ships.

5.2.2.1. Jack-Up Rigs

The jack-up, or self-elevating, rig was first built in 1954 and rapidly became one of the most popular designs in mobile offshore drilling units. Jack-ups were popular for drilling activities in shallow-water areas-up to 360 ft (110 m). They provide a very stable drilling platform because part of the structure is in firm contact with the seafloor, and they are fairly easy to move from one location to another. Figure 5-3 shows a typical jack-up platform.

Figure 5-3 Jack-Up Rig [3]

5.2.2.2. Semi-Submersibles

Semi-submersibles are the choice for drilling from a floating position, due to the deep water depths. Semi-submersible ("semi") rigs float like a ship while being towed into position, where their pontoons can then be flooded, partially submerging the rig. Figure 5-4 illustrates a typical semi-submersible. Because the main part of the structure is beneath the sea surface, semis are not as susceptible to wave actions as drilling ships. Semis may be stationed using dynamic positioning or anchoring. This type of platform has the advantages of a smaller waterline plane, less susceptibility to wave effects, good stability, long self-sustaining period, and a great working depth. Thus, they are suitable for installation of subsea structures ranging from 400 to 4000 ft (120-1200 m), and maybe deeper. For example, a spread-moored semi in 6152 ft (1875 m) water depth (WD) was set offshore in Malaysia in 2002, and a semi at a WD of 9,472 ft (2890 M) with a DP system was set in Brazil in 2003. A semi for a "taut line" mooring system was set in a WD of 8,950 ft (2730 m) in the GoM in 2003, which also set a record for subsea completions in 7,571 ft (2300 m).

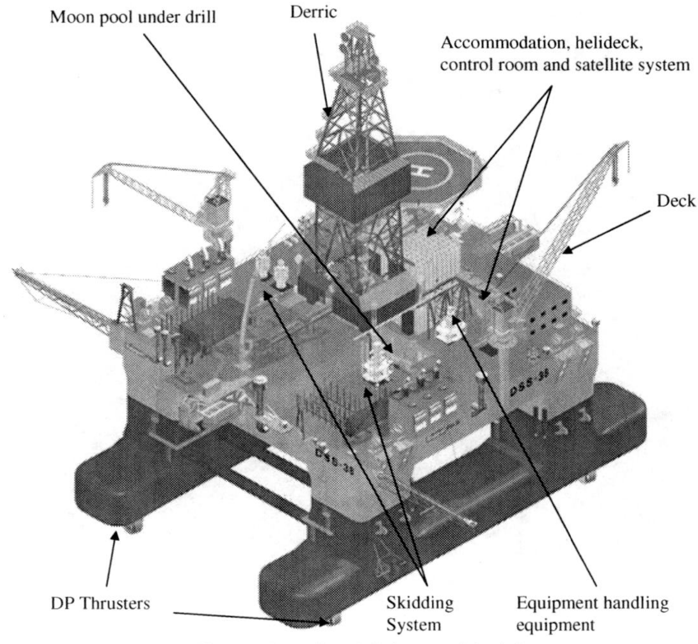

Figure 5-4 Semi-Submersible [4]

5.2.2.3. Drill Ships

A drill ship is a vessel with a drilling rig and station-keeping equipment for drilling oil fields in deepwater. Drill ships have a larger payload capacity than semi-submersible drilling vessels, but face shortcomings in terms of their motion characteristics. Figure 5-5 illustrates a drill ship.

5.2.3. Pipe-Laying Vessels

Pipe-laying vessels can be categorized according to their pipe-laying methods based on site characteristics, such as water depth and weather. The three most common types are the S-lay, J-lay, and reel-lay types.

5.2.3.1. S-Lay Vessels

The S-lay method has been used for many years to lay offshore pipelines. An S-lay vessel has a broad deck operating area. The pipe joint assembly line is deployed in the middle or at the side of the main deck, which includes pipe conveyors units, welding, NDT inspection and coating stations etc. The stern of the vessel is constructed into a sloping slideway with a stinger, which is used to modify and control the configuration (stress/stress distributions along

Figure 5-5 Drill Ship [5]

Figure 5-6 S-Lay Barge [6]

the pipeline) of the pipeline in an "S" shape down to the seabed. Figure 5-6 shows a picture of a conventional S-lay installation vessel.

5.2.3.2. J-Lay Vessels

The J-lay vessel uses a J-lay tower/ramp to install the pipeline instead of the stingers used in the S-lay vessel. Figure 5-7 shows a sketch of a J-lay vessel with a J-lay ramp. The J-lay method differs from the S-lay method in that the pipe departs from the lay vessel at a near-vertical angle (e.g., 60° to 87°). There is no overbend to maintain as in S-lay, thus no stinger is required. This type of operation is developed to cater to deepwater pipeline installations.

5.2.3.3. Reel-Lay Vessels

A typical reel-lay vessel usually provides an economical tool for installing long, small-diameter pipelines (typically smaller than 16 in. (0.406m)). The

Figure 5-7 J-Lay Vessel with a J-lay Ramp [7]

pipeline is made up onshore and is reeled onto a large drum (e.g., approximate diameter × width = 20 m × 6 m)) on the middle deck of a purpose-built vessel in a spool base 66 ft × 20 ft. During the reeling process in the spool base, the pipe undergoes plastic deformation on the drum. During the offshore installation, the pipe is unreeled and straightened using a special straight ramp. The pipeline on drum will be unreeled accompanying with vessel speed at e.g. 12km/day (usually 3-15 km depending on pipe diameter) [14]. The pipe is then placed on the seabed in a configuration similar to that used by S-lay vessels, although in most cases a steeper ramp can be used and overbend curvature is eliminated as with a J-lay. This kind of pipe-laying vessel is easily manipulated with simple pipe-laying devices; in addition, it has a good pipe-laying speed. Figure 5-8 shows a typical reel-lay vessel.

5.2.4. Umbilical-Laying Vessels

Umbilicals can be installed either by a reel-lay vessel or a carousel-lay vessel. Reels provide easy load-out. The maximum installed umbilical length is 1.9–9.3 mile (3 to 15 km), depending on the diameter of the reels. The critical gross weight of reels with umbilicals is about 250 ton (226 metric ton). However, a carousel allows for a longer length of umbilical to be installed (>62 mile (100 km)), and it can avoid field splices. It does, however, require a dedicated vessel and longer load-out times of, for example, 22-33 ft (6 to 10 m)/min. Figure 5-9 illustrates a typical the carousel-lay vessel.

Figure 5-8 Reel-Lay Vessel [8]

Figure 5-9 Carousel-lay Vessel [9]

5.2.5. Heavy Lift Vessels

A heavy lift vessel (HLV) is a vessel with a specific crane that has a large lifting capacity of up to thousands of tonnes. Large and heavy subsea structures such as templates may require HLVs to perform lifting. The lifting capacities of most HLVs range between 500 and 1000 ton (454 to 907 metric ton), whereas crane capacity on normal construction vessels is below 250 ton (226 metric ton). An HLV's stability and sea-keeping abilities are its most important characteristics. Figure 5-10 illustrates a typical heavy lift vessel.

5.2.6. Offshore Support Vessels

Offshore support vessels are special vessels that provide support for field drilling, construction, decommissioning, and abandonment. The support vessels

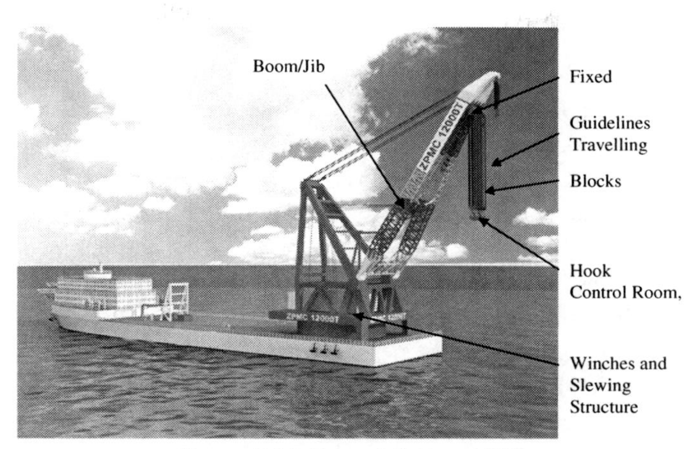

Figure 5-10 Heavy Lift Vessel [10]

normally include survey, standby, inspection, and installation assistance (e.g., monitoring). The following types of offshore support vessels may be utilized:

- ROV support vessel (RSV);
- Diving support vessel (DSV);
- Survey ship;
- Offshore supply ship or field support vessel (FSV).

An ROV support vessel is a platform with specialized equipment and spaces to store, deploy, and support ROVs for their subsea intervention.

A diving support vessel is a platform with specialized diving equipment, such as diver-to-surface communication system, submersible, on-site diving hyperbaric chamber, compression chamber, and so on for subsea interventions by professional divers.

A survey ship is a platform with specialized instruments and laboratories for the study of the ocean physics, chemistry, geology, topography, aerography, and hydrology required for SPS installation.

A field support vessel is a multipurpose vessel that can provide transportation, supplies, and rescue and diving support.

5.3. VESSEL REQUIREMENTS AND SELECTION

A proven subsea arrangement is the use of pipelines and umbilicals connecting subsea wells to the moored host facility (semi-submersible, FPSO, etc.). This subsea development scenario may require the following installation tasks:

- Lifting and installation of subsea Christmas trees, manifolds, PLET/PLEMs, UTA, SDU, etc.;

- Laying of umbilicals and pipelines;
- Subsea tie-ins.

The following factors need to be taken into account for selecting vessels for subsea development:

- Installation task: What is to be installed?
- Environmental conditions: How deep is the water in the installation site?
- Vessel features: What requirements should the vessel meet?
- Costs: What is the estimated budget?

This section emphasizes vessel features. The requirements of vessel features are divided into two categories in this chapter: basic requirements and functional requirements. The basic requirements describe what is needed to ensure vessel basic performance and safety. The functional requirements describe the critical actions needed to perform installation activities.

5.3.1. Basic Requirements for Vessels and Barges

A vessel's performance and structure are basic requirements to be considered when selecting vessels and barges.

5.3.1.1. Vessel Performance

Vessel performance issues for and installation vessel include the following:

- *Buoyancy:* The vessel should be buoyant at a certain loaded condition. There may be four buoyancy conditions for the vessel: upright position, trim, heel, and a combination of those conditions.
- *Stability:* Stability means that the vessel can get back to its initial balanced position once the forces or moments applied on it have disappeared. Stability can be divided into "initial stability at small angle of inclination" ($\leq 10°$) and "stability at a large angle of inclination." Initial stability has a-linear relationship between inclination angle and uprighting moment.
- *Insubmersibility:* Insubmersibility means that the vessel can remain buoyant and stable once one space or multiple spaces of the vessel are flooded. Sufficient stability and reserve buoyancy are ensured for all floating vessels in all stages of marine operations.
- *Sea-keeping:* Sea-keeping means that the vessel can remain safe while navigating or operating at sea, even though it may be exposed to the severe forces and moments created by wind, waves, and current.
- *Maneuverability, speed, and resistance:* Maneuverability of the vessel refers to the capability of the vessel to keep a constant navigation direction or change direction only according to the pilot's desire. Speed and

resistance refers to the speed capability of the vessel at the rated power of the main engine.

5.3.1.2. Vessel Strength

Vessel structures can experience three types of failures: strength failures, stability failures, and fatigue failures. A strength failure normally means that the stress of the vessel structure is larger than its Specified Minimum Yield Strength (SMYS). A stability failure means that the compressive stress of the structure is larger than the critical stability stress (e.g., Euler stress) and results in large displacements. Fatigue failure refers to the cracking or fracturing of the vessel's structure due to continuous stress circulations.

The strength of the vessel normally includes longitudinal strength and local strength. The strength of the vessel should not be exceeded during all vessel activities such as load-out, lifting, and transportation.

5.3.2. Functional Requirements

Generally speaking, a vessel is just a stable platform that sustains the specific equipment performing the required activities. The specific equipment is mobilized on board before carrying out the installation activities, and demobilized to the shore base once the required activities have been completed. The mobilization and demobilization consumes time and money. Consequently, specialized vessels with specific equipment for a certain type of job, such as pipe-laying vessels or diving support vessels, are often used. The main differences between the function requirements of a subsea hardware installation vessel and that of a pipe-laying/umbilical-laying vessel are the different requirements for specific equipment. This section discusses common vessel features that need to be considered during vessel selection.

5.3.2.1. Vessel for Subsea Hardware Installation

Generally, subsea hardware can be installed by a discretional vessel with sufficient winch rope length and crane capacity with or without a heave compensator. The vessel may be a drill vessel, pipe-laying vessel, umbilical-laying vessel, or offshore support vessel. The critical vessel features for subsea installation normally include the following:

- Deck space for the project equipment;
- Deck load capacity;
- Crane capacity and coverage;
- Vessel sea behaviors (RAOs);

- ROV requirements, such as two work-class ROVs for subsea structure installation, one work-class ROV, and one survey ROV for pipeline/umbilical laying, monitoring, etc.
- Accommodation capacity;
- Transit speed (high transit speed has become important due to long transit distances as field developments are placed in deeper and deeper water);
- Positioning requirements.

Lifting and installation of subsea structures must be performed in a timely and safe manner, taking into account all site limitations:

- Weather conditions;
- Seabed soil conditions and visibility;
- Other site constraints (mooring lines, other subsea structures, etc.).

The deck handling system and the lifting devices on the vessel must be designed to control and prevent the pendulum movements of the subsea structures as they leave the deck storage area, pass through the splash zone, and land safely on the seabed.

This dedicated installation equipment (cranes, winches, etc.) must be designed, built, and operated to suit the conditions in which they are to perform. More specifically, they also must take into account the structure's installation criteria and constraints, as specified by the designers/manufacturers and national codes/certifying authority rules (DNV, API, ABS, etc.).

5.3.2.2. Vessel for Pipe and Umbilical Laying

Compared with the subsea hardware installation vessel, the selection of a pipe-laying vessel should include the following additional requirements for some specific equipment:

- *Capacity of tensioner:* This is required based on the water depth and pipeline unit weight and buoyancy.
- *Abandon and recovery winch:* This is required at the end of the pipe-laying process and if emergency conditions arise.
- *Davits capacity:* This is required once davit activity is necessary for the offshore connection of pipeline or umbilicals.
- *Product storage capacity:* This is required if the pipe joints or umbilical reels cannot be transferred from the storage barge to the laying vessel due to bad weather.

5.4. INSTALLATION POSITIONING

Installation of subsea hardware requires the vessel to keep its position well so that the subsea hardware can be placed onto the target foundation with the required accuracy. The positioning operation includes surface positioning and subsea positioning. Surface positioning or vessel positioning refers to positioning that maintains the vessel at the correct position at all times during installation. It is the first step of subsea production system installation because it brings the hardware to be installed near its ultimate aiming position. Subsea positioning refers to positioning to monitor and control the equipment underwater relative to the installation vessel and seabed target area during the process of lowering equipment through the water and landing or locking.

5.4.1. Surface Positioning

A surface positioning system can be divided into these components:

- Power system, supplying power to all of the following systems;
- Position reference system, normally using DGPS a hydroacoustic measuring system such as USBL, SBL, LBL, and a taut wire system;
- Controlling system;
- Station-keeping system such as a mooring system including anchor gear, anchor lines, and anchors for positioning (see Figure 5-11) or a thruster system for dynamic positioning (see Figure 5-12).

5.4.2. Subsea Positioning

Once the surface vessel has been positioned, subsea hardware will be deployed from the vessel through water to the target location on the seabed. During the lowering and landing process, the hardware will be tracked with a hydroacoustic unit (e.g., transponder) for position measuring and

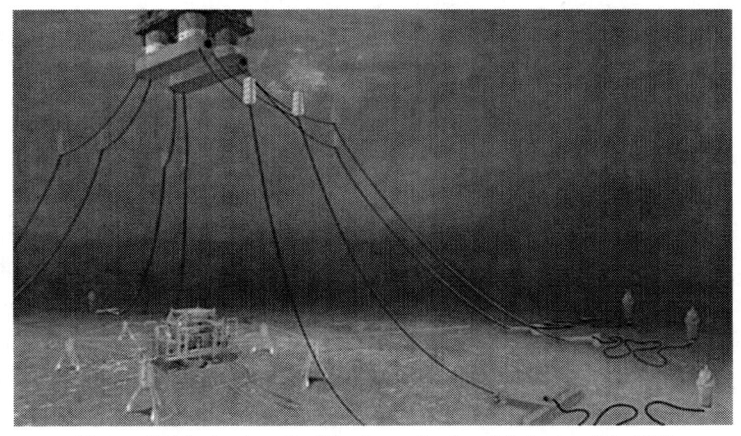

Figure 5-11 Anchor Positioning of a Semi-Submersible [11]

Figure 5-12 Dynamic Positioning System [11]

a gyrocompass (and ROV) for inclination detection (roll, pitch, and heading), which were tied to the hardware on board before lowering.

Two methods are widely used for subsea hardware installation: the guideline (GL) method and the guideline-less (GLL) method. The GL method uses guidelines (normally four tensioned wires) to deploy subsea hardware to the seabed. The subsea positioning for this method is quite convenient, but has been limited by the guidelines. It also consumes time and costs money for deepwater installations. The GLL method deploys subsea hardware in deepwater without guidelines. The subsea positioning for this method is relatively complex for hardware landing and requires limited installed tolerance such as inclination and position bias in the x, y direction. It also requires heavy structures on all equipment to be in the proper orientation.

The hardware heave motion should be strictly limited during its landing process, so active or passive heave compensator systems have been introduced into the lifting system of installation vessels used for subsea deployment.

5.5. INSTALLATION ANALYSIS

Installation analysis is a type of calculation for the validation of lifting capacity, strength capacity for lifting objects and vessel structures, etc. Installation analysis can be divided into two categories based on different phases of installation engineering: (1) preliminary installation analysis of the front-end engineering design (FEED) for determining the installation method, installation vessels and equipment, as well as the relevant installation duration and cost estimations, and (2) detailed installation analysis of the engineering design with the goal of developing and installation procedure and relevant installation drawings. The preliminary installation analysis is just used for some critical installation activities such as SPS splash zone lowering and pipeline installation at maximum water depth to prove the feasibility/ capability of the vessel and equipment, whereas the detailed installation analysis is designed to provide a step-by-step analysis of the SPS installation.

The subsea installation can be categorized into three types of installation events: (1) subsea equipment deployment installation, (2) pipeline/riser installation, and (3) umbilical installation. The minimum required data for an installation analysis normally include:

- Environment and geotechnical data;
- Vessel motion characteristics;
- Subsea production system data;
- Rigging system data.

5.5.1. Subsea Structure Installation Analysis

The subsea structure installation procedure normally includes the following:

- Load-out and sea-fastening;
- Transportation;
- Site survey;
- Deployment, typically including overboarding, splash zone lowering, midwater lowering, landing, and positioning and setting (see Figure 5-13);
- As-built survey.

This section describe only the structure deployment analysis, which may be the most critical analysis for subsea installation, while other analyses such as barge strength verification for transportation and lifting analysis for load-out, are not included in this section.

Analysis of subsea structure deployment provides the maximum allowable sea states and maximum expected cable tensions and the motions of installed equipment during installation. Finite element software (e.g., Orcaflex) is used for the installation analysis. The objective of the detailed installation analysis is to provide a step-by-step analysis to aid in the generation of the equipment deployment procedure.

The installation analysis can be divided into two stages: static analysis without any environmental loading, and dynamic analysis with environmental loading such as current and wave. The static analysis determines the relationship between vessel position, wire payouts, and tensions for the system in a static state. The dynamic analysis is performed for the system under environmental loads in order to determine the maximum allowable

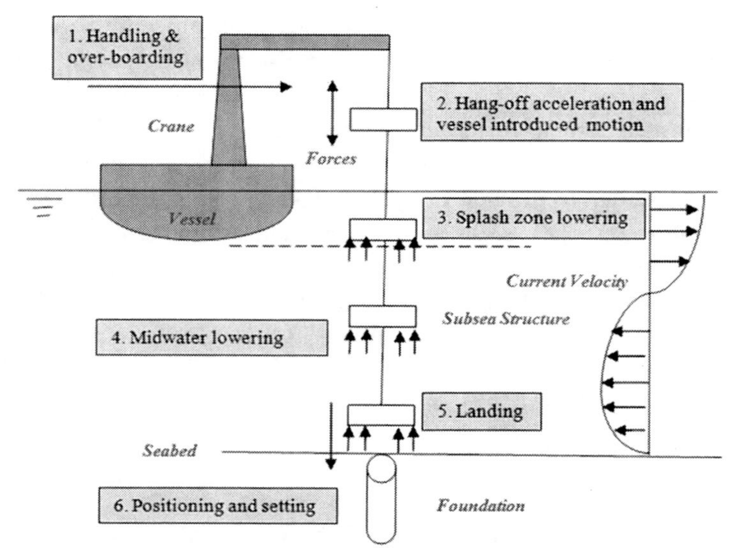

Figure 5-13 Sketch of Analysis Steps for Typical Subsea Structure

installation sea states and the maximum tension required. The analyses are carried out for a range of wave heights and wave periods.

The analysis model comprises the installation vessel associated with its RAOs, drilling pipe with running tools or a crane with its winch wires, and rigging systems and equipment.

Figure 5-13 illustrates the installation procedures for a typical subsea structure with the forces modeled during each step. For detailed installation analysis of the manifold, refer to Section 19.5.

5.5.2. Pipeline/Riser Installation Analysis

The four normal installation methods for pipelines/risers are as follows:

- S-laying method, in which the pipe is laid from a near-horizontal position on a lay barge using a combination of horizontal tension and a stinger (bend-limiting support).
- J-laying method, in which the pipe is laid from an elevated tower on a lay barge using longitudinal tension without an overbend configuration at the sea surface. Load-out and transportation of pipe joints will be performed by the transportation barge at the same time pipe is being laid by the pipe-laying vessel for the S-laying and J-laying method.
- Reel-laying method, in which the pipe is made up at some remote location onshore, spooled onto a large radius reel aboard a reel-lay vessel,

and then unreeled, straightened, and laid down to the seabed at the offshore installation location.

- Towing lay method, in which the pipe is made up at some remote location onshore, transported to the offshore installation site by towing, and laid down. The towing is either on the water surface (surface towing), at a controlled depth below the surface (control depth towing method [CDTM]), or on the sea bottom (bottom towing), where the different water depths are used mainly to reduce the fatigue damage due to wave action.

Figure 5-14 shows the pipeline configurations for the S-laying, J-laying, reel-laying, and towing lay methods.

A subsea riser or pipeline is exposed to different loads during installation from a laying vessel depending on the installation methods. The loads include hydrostatic pressure, axial tension, and bending. The normal failure modes may be local buckling and buckling propagation due to external pressure and bending moments for pipeline/riser installation.

The installation analysis includes two parts: static analysis and dynamic analysis. The stress or strain criteria used for the installation calculation are different based on the project. In most cases, the strain criteria are used for the analysis. However, in some projects, stress criteria are also used. For example, the stress criteria are 72% SMYS and 96% SMYS for pipeline sag-bend area and pipeline overbend area (as shown in Figure 5-15), respectively, according to DNV OS F101 2007 [15]. The stress analysis during the pipeline installation procedure should be carried out in detail to check the stress criteria.

The analysis provides the maximum allowable sea states and maximum expected tension loads and stress/strain distribution during installation procedure. The pipeline/riser installation analysis is normally carried out by using OFFPIPE or Orcaflex software.

5.5.3. Umbilical Installation Analysis

Umbilicals are laid using one of the following typical methods:

- The umbilical is initiated at the manifold with a stab and hinge-over connection or a pull-in/connection method and terminated near the subsea well with a second end lay down sled (i.e., infield umbilical connection from manifold to satellite well). The connection between the umbilical and the subsea well is later made using a combination of the following tie-in methods: (1) rigid or a flexible jumper, (2) junction plates, and (3) flying leads.

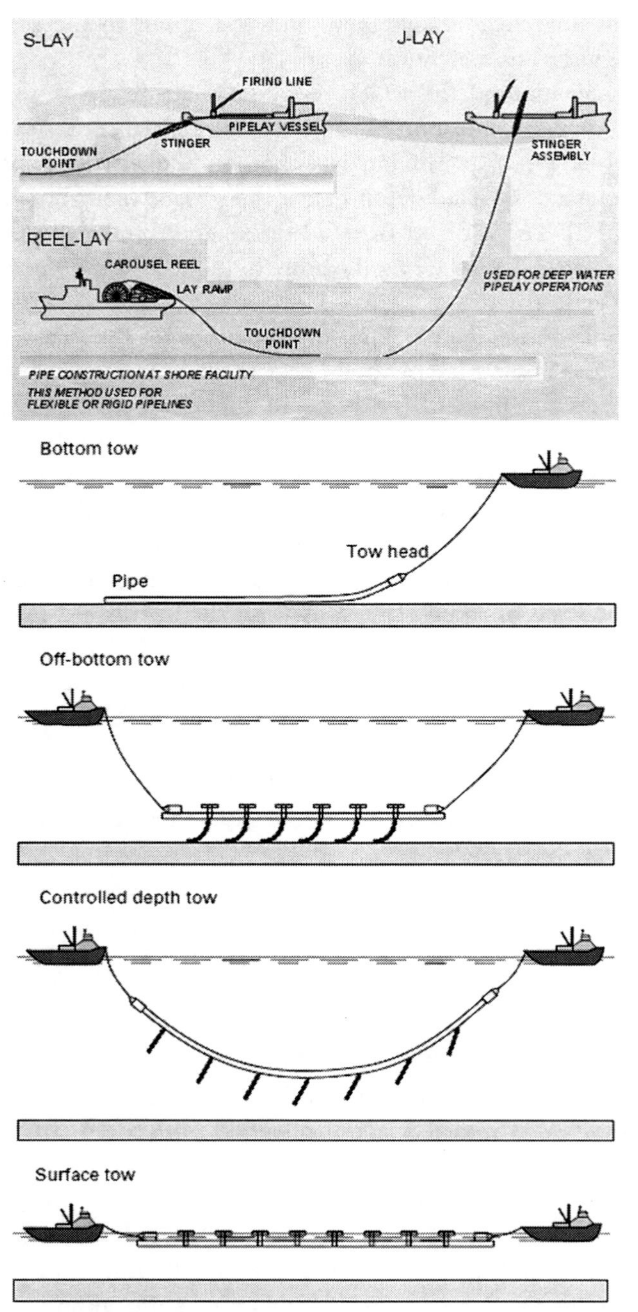

Figure 5-14 Pipeline Installation Configuration for the S-Laying, J-Laying, Reel–Laying, and Towing-Laying Methods [12,13]

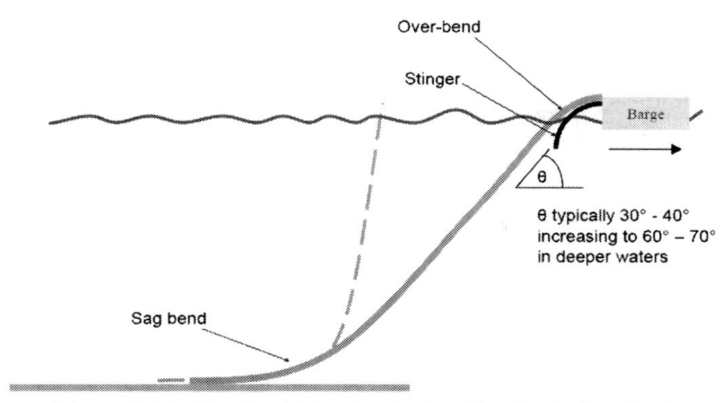

Figure 5-15 Sketch of Pipeline Installation Analysis (S-lay)

- The umbilical is initiated at the manifold with a stab and hinge-over connection or a pull-in/connection method. It is laid in the direction to the fixed or floating production system and pulled through an I/J tube or cross-hauled from the laying vessel to the floating production vessel.
- The umbilical can also be initiated at the fixed or floating production system and terminated near the subsea structure with a second end umbilical termination assembly (i.e., termination head, lay down sled, umbilical termination unit). A pull-in/connection tool operated by an ROV may be used to connect the umbilical to the subsea structure.

The objective of the umbilical installation analysis is to provide a step-by-step analysis that can aid in the generation of the umbilical installation procedure. The analysis also provides the maximum allowable sea states, maximum expected loads, and guidelines for vessel offsets.

The analysis is performed in two stages: static analysis without any environmental loading on the model and dynamic analysis with environmental loading such as current and wave. Dynamic analysis is performed by selecting the worst cases (based on minimum bending radius and tension) from the static analysis and then applying environmental loading such as current action, wave action, peak period, and directionality to the model. The directions of wave and current may be conservatively assumed to be the same.

The design criteria for an umbilical installation are as follows:

- Minimum bending radius (MBR) of umbilical;
- Maximum allowable tension and compression loads;
- Maximum allowable crushing loads, which can be translated to check the maximum allowable top tension;
- Lateral stability on the seabed, which can be translated to check the maximum allowable tension force of the touchdown point.

The analysis model for a normal umbilical lay includes the installation vessel with its RAOs and umbilical. The umbilical for lay initiation analysis is modeled additionally with the umbilical termination head (UTH). The umbilical for laydown analysis is modeled in Orcaflex additionally with a wire attached to the end of the umbilical, together with the end termination/bend stiffener/pulling head assembly.

REFERENCES

[1] J. Pappas, J.P. Maxwell, R. Guillory, Tree Types and Installation Method, Northside Study Group, SPE, 2005.
[2] Dredge Brokers, Offshore Tug Boat, http://www.dredgebrokers.com, 2007.
[3] Energy Endeavour, Jack-Up Rig, http://www.northernoffshorelimited.com/rig_fleet.html, Northern Offshore Ltd, 2008.
[4] Maersk Drilling, DSS 21 deepwater rigs, www.maersk-drilling.com.
[5] Saipem S.P.A, Saipem 12000, Ultra deepwater drillship, http://www.saipem.it.
[6] Allseas Group, Solitaire, the Largest Pipelay Vessel in the World, http://www.allseas.com/uk.
[7] Heerema Group, DCV "Balder", Deepwater Crane Vessel, http://www.heerema.com.
[8] Subsea 7, Vessel Specification of Seven Navica, http://www.subsea7.com/v_specs.php.
[9] Solstad Offshore ASA, CSV: Vessel Specification of Normand Cutter, http://www.solstad.no.
[10] People Heavy Industry, 12,000 Ton Full Revolving Self-propelled Heavy Lift Vessel, http://www.peoplehi.com.
[11] Abyssus Marine Service, Anchor and Dynamic Positioning Systems, http://www.abyssus.no.
[12] International Marine Contractors Association, Pipelay Operations, http://www.imca-int.com
[13] M.W. Braestrup, et al., Design and Installation of Marine Pipelines, Blackwell Science Ltd, Oxford, UK, 2005.
[14] Y. Bai, Q. Bai, Subsea Pipelines and Risers, Elsevier, Oxford, UK, 2005.
[15] DNV, Submarine Pipeline Systems, DNV-OS-F101, 2007.

Subsea Cost Estimation

Contents

6.1. INTRODUCTION

Subsea cost refers to the cost of the whole project, which generally includes the capital expenditures (CAPEX) and operation expenditures (OPEX) of the subsea field development, as shown in Figure 6-1.

From Figure 6-1 we can see that expenditures are incurred during each period of the whole subsea field development project. Figure 6-2 illustrates

Subsea Engineering Handbook
ISBN 978-0-12-397804-2, doi:10.1016/B978-0-12-397804-2.00006-0

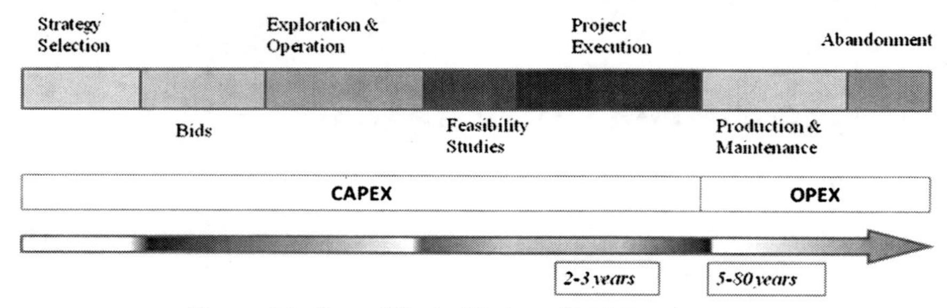

Figure 6-1 Cost of Typical Subsea Field Development

the feasibility studies in different phases of a subsea field development project. The feasibility studies are performed before execution of the project, which may include three phases as shown in the figure:

- Prefield development;
- Conceptual/feasibility study;
- Front-end engineering design (FEED).

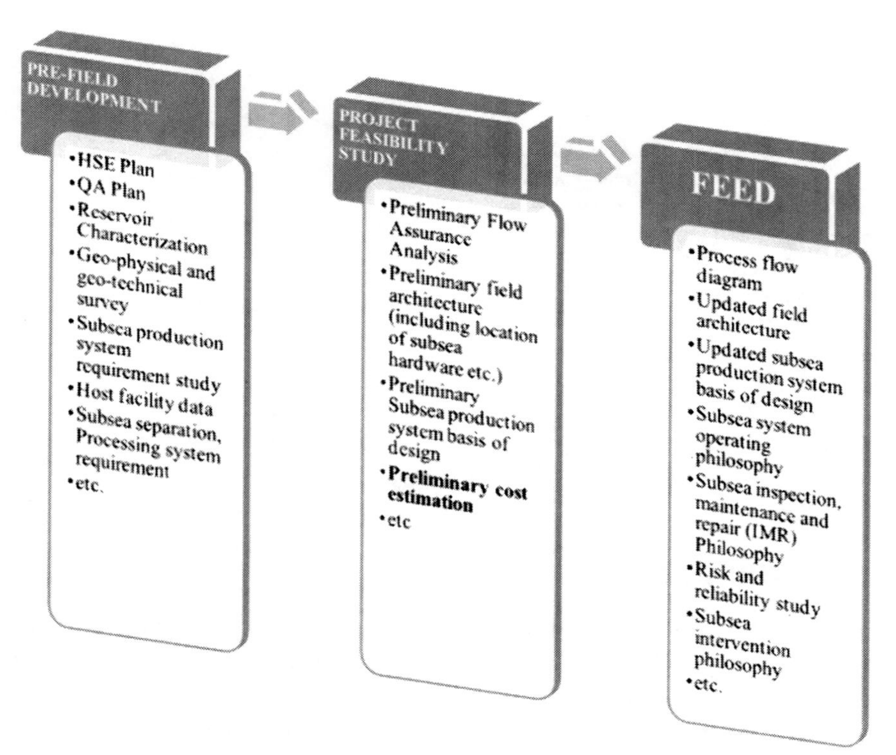

Figure 6-2 Feasibility Studies in Subsea Field Development Project

Table 6-1 Cost Estimation Classification Matrix (AACE) [1]

Estimate Class	Level of Project Definition	End Usage	Methodology	Expected Accuracy Range	Preparation Effort
Class 5	0 to 2%	Screening or feasibility	Stochastic or judgment	4 to 20	1
Class 4	1% to 15%	Concept study or feasibility	Primarily stochastic	3 to 12	2 to 4
Class 3	10% to 40%	Budget, authorization, or control	Mixed, but primary stochastic	2 to 6	3 to 10
Class 2	30% to 70%	Control or bid/ tender deterministic	Primarily 1 to 3	5 to 20	
Class 1	50% to 100%	Check estimate or bid/tender	Deterministic	1	10 to 100

Cost estimations are made for several purposes, and the methods used for the estimations as well as the desired amount of accuracy will be different. Note that for a "preliminary cost estimation" for a "project feasibility study," the accuracy will normally be ±30%. Table 6-1 shows cost estimation classifications according to Association for the Advancement of Cost Engineering (AACE):

- Level of project definition: expressed as percentage of complete definition;
- End usage: typical purpose of estimation;
- Methodology: typical estimating method;
- Expected accuracy range: typical ± range relative to best index of 1 (if the range index value of "1" represents +10/−5%, then an index value of 10 represents +100/−50%);
- Preparation effort: typical degree of effort relative to least cost index of 1 (if the cost index of "1" represents 0.005% of project costs, then an index value of 100 represents 0.5%).

This chapter provides guidelines for cost estimation during a project feasibility study, where the accuracy range is between ±30% for subsea field development projects.

6.2. SUBSEA CAPITAL EXPENDITURES (CAPEX)

Based on Douglas-Westwood's "The Global Offshore Report," [2] the global subsea CAPEX and OPEX in 2009 was about $250 billion and will

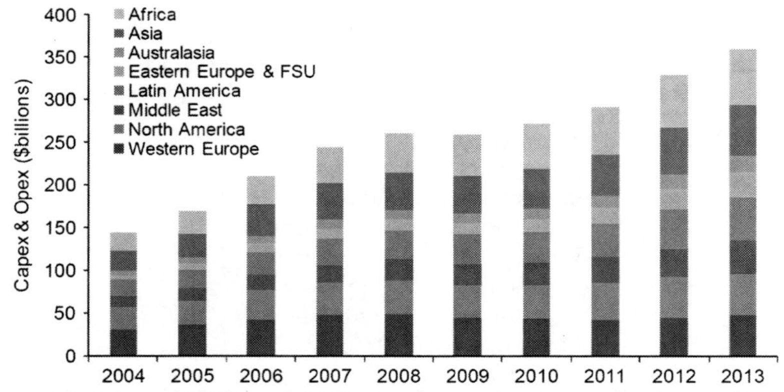

Figure 6-3 Subsea CAPEX and OPEX Distribution by Geographical Area (Douglas-Westwood [2])

be $350 billion in 2013. Figure 6-3 shows the distribution of subsea costs by geographical areas.

Figure 6-4 and Figure 6-5 show examples of the breakdown of deepwater subsea CAPEX and shallow-water subsea CAPEX, respectively. The major cost components of subsea CAPEX are equipment, testing, installation, and commissioning. The key cost drivers for subsea CAPEX are number of wells, water depth, pressure rating, temperature rating, materials requirement, and availability of an installation vessel.

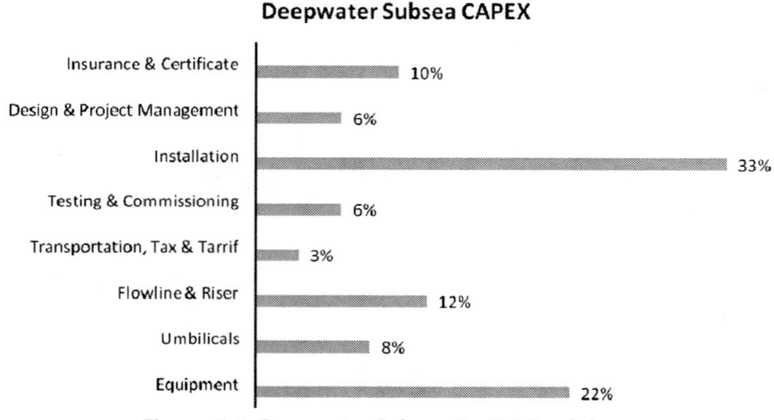

Figure 6-4 Deepwater Subsea CAPEX Breakdown

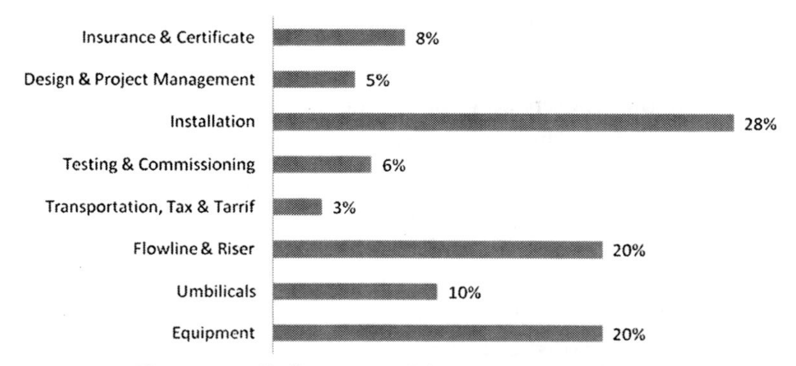

Figure 6-5 Shallow-Water Subsea CAPEX Breakdown

6.3. COST ESTIMATION METHODOLOGIES

Different cost estimating methods are used according to different phases of the project and how much data/resources have been obtained. Three methods are introduced in this book:

- Cost–capacity estimation;
- Factored estimation;
- Work breakdown structure.

The cost–capacity estimation method is an order-of-magnitude estimation method, based on similar previous cost data. This method's accuracy range is within ±30%.

The factored estimation method is based on several cost-driving factors. Each of the factors is considered as a "weight" on basic cost data. The basic cost is normally the price of a standard product based on the proven technology at that time. Accuracy can be within ±30%. This book gives upper and lower bounds for the estimated costs in Section 6.4. Note that the suggested cost-driving factors as well as recommended values for the methodology should be used within the given range and be updated based on the actual data for the time and location in question.

Another estimation method is the working breakdown structure (WBS) methodology, which is commonly used in budget estimation. This method is based on more data and details than are the two methods just mentioned. By describing the project in detail, the costs can be listed item by item, and at the end, the total costs are calculated.

6.3.1. Cost–Capacity Estimation

Cost–capacity factors can be used to estimate a cost for a new size or capacity from the known cost for a different size or capacity. The relationship has a simple exponential form:

$$C_2 = C_1\left(\frac{Q_2}{Q_1}\right)^x \tag{6-1}$$

where

C_2: estimated cost of capacity;

C_1: the known cost of capacity;

x: cost–capacity factor.

The capacities are the main cost drivers of the equipment, such as the pressure ratings, weight, volume, and so on. The exponent x usually varies from 0.5 to 0.9, depending on the specific type of facility. A value of $x = 0.6$ is often used for oil and gas processing facilities. Various values of x for new projects should be calculated based on historical project data. Let's look at the procedures for calculating x values.

Revise Equation (6-1) to:

$$x = \ln\left(\frac{C_2}{C_1}\right)\Big/\ln\left(\frac{Q_2}{Q_1}\right) \tag{6-2}$$

The resulting cost–capacity curve is shown in Figure 6-6. The dots in the figure are the calculation results based on a database built by the user. The slope of the line is the value of exponent x.

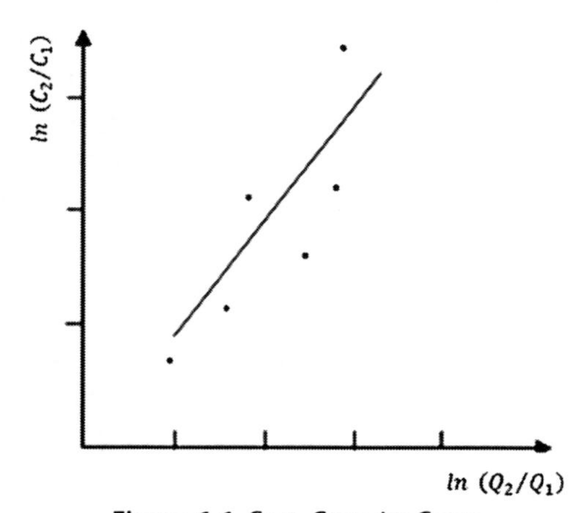

Figure 6-6 Cost–Capacity Curve

6.3.2. Factored Estimation

6.3.2.1. Cost Estimation Model

Costs are a function of many influencing factors and are expressed as:

$$C = F(f_1, f_2, f_3, \text{K} f_i) \quad (i = 1, 2, 3, \text{K}) \tag{6-3}$$

where

C: cost of the subsea Christmas tree;

F: calculation function;

f_i: cost factors.

Assume that:

$$C = f_1 \bullet f_2 \bullet f_3 \bullet \text{K} f_n \bullet C_0 + C_{misc} \quad (i = 1, 2, 3, \text{K}) \tag{6-4}$$

where

C: cost of the subsea equipment;

f_i: cost-driving factors;

C_0: basic cost;

C_{misc}: miscellaneous cost.

Equation (6-4) also shows that the cost is the product of the factors based on a fixed cost C_0. It is clear that the cost C can be estimated by multiplying the cost drivers by the basic cost, which is generally the cost of a standard product. The C_{misc} term refers to other miscellaneous costs that are related to the equipment but not typical to all types.

The basic cost C_0 is the cost of the typical/standard product among the various types of a product; for example, for a subsea Christmas tree, there are mudline trees, vertical trees, and horizontal trees. Currently, the standard product in the industry is a 10 ksi vertical tree, so the cost of a 10 ksi vertical tree will be considered the basic cost while calculating the other trees.

6.3.2.2. Cost-Driving Factors

The following general factors are applied to all subsea cost estimation activities.

Inflation Rate

Inflation is a rise in the general level of prices of goods and services in an economy over a period of time. When the price level rises, each unit of currency buys fewer goods and services. A chief measure of price inflation is the inflation rate, the percentage change in a general price index (normally the Consumer Price Index) over time. The cost data provided in this book are in U.S. dollars unless otherwise indicated, and all the cost data are based on the year 2009, unless a specific year is provided.

To calculate the cost for a later/target year, the following formula should be used:

$$C_t = C_b \bullet (1 + r_1) \bullet (1 + r_2) \bullet \mathrm{K}(1 + r_i)(i = 1, 2, 3, \ldots) \qquad (6\text{-}5)$$

where

C_t: cost for target year;

C_b: cost for basic year;

r_i: inflation rate for the years between base year and target year.

For example, if the cost of an item is 100 for the year 2007, and the inflation rates for 2007 and 2008 are 3% and 4%, respectively, then the cost of that item in year 2009 will be $C_{2009} = C_{2007} * (1 + r_{2007}) * (1 + r_{2008}) = 100 * 1.03 * 1.04 = 1.0712$.

Raw Materials Price

The price of raw material is one of the major factors affecting equipment costs. Figure 6-7 shows the trends for steel and oil prices over time between 2001 and 2006.

Market Condition

Supply and demand is one of the most fundamental concepts of economics and it is the backbone of a market economy. Demand refers to how much (quantity) of a product or service is desired by buyers. Supply represents how much the market can offer. Price, therefore, is a reflection of supply

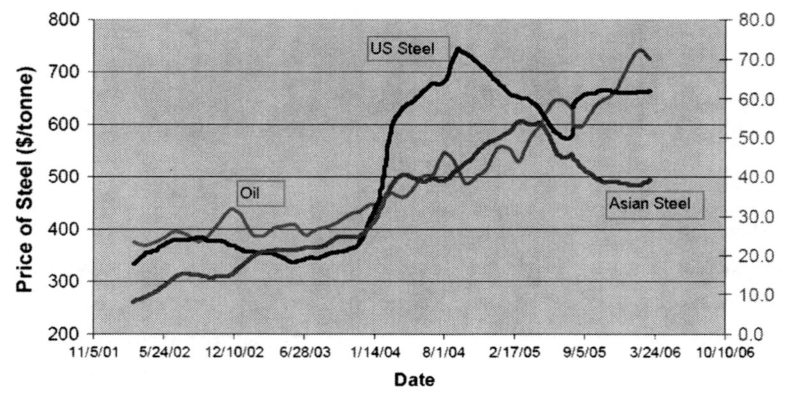

Figure 6-7 Steel and Oil Prices over Time [3]

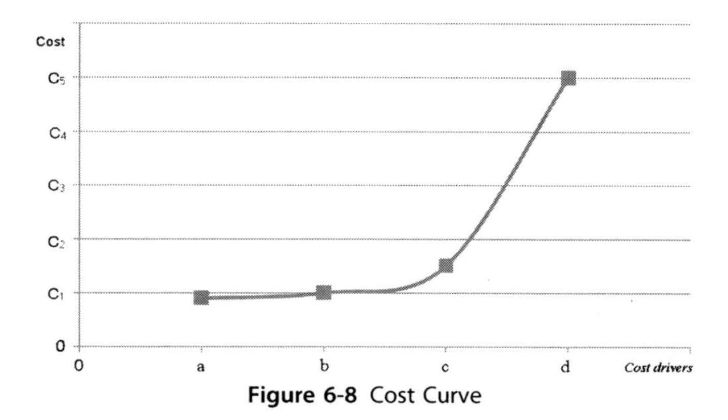

Figure 6-8 Cost Curve

and demand. The law of demand states that if all other factors remain equal, higher demand for a good will raise the price of the good.

For subsea development, the availability/supply of fabrication capacity and installation vessels is one of the major cost drivers. Tight supply will increase costs sharply, as shown in Figure 6-8.

Figure 6-8 also explains cost trends resulting from technology conditions, which influence market supply. Costs change very slowly within normal capacity, but they increase smartly after point *c* as technology's limits are reached. For example, a 10-ksi subsea Christmas tree is now the standard product in the market, so the cost of a 5-si subsea Christmas tree will not change too much. However, the 15-ksi subsea Christmas tree is still a new technology, so its costs will increase a project's cost by a large amount.

Subsea-Specific Factors

Besides the general factors just introduced, cost estimations for subsea field developments have their own specific factors:

- *Development region:* Affects the availability of a suitable installation vessel, mob/demob (mobilization/demobilization) costs, delivery/transportation cost, etc.
- *Distance to existing infrastructure:* Affects the pipeline/umbilical length and design.
- *Reservoir characteristics:* These characteristics, such as pressure rating and temperature rating, affect equipment design.
- *Water depth,* metocean *(normally refers to wind, wave, and current data) and soil condition:* Affects equipment design, installation downtime, and installation design.

For more details on these factors, see specific topics in Section 6.4.

6.3.3. Work Breakdown Structure

The work breakdown structure (WBS) is a graphic family tree that captures all of the work of a project in an organized way. A subsea field development project can be organized and comprehended in a visual manner by breaking the tasks into progressively smaller pieces until they are a collection of defined tasks, as shown in Figure 6-9. The costs of these tasks are clear and easy to estimate for the project after it has been broken down into the tasks.

The level 1 breakdown structure is based primarily on the main costs of subsea equipment, system engineering, installation, and testing and commissioning. The subsea elements are further broken down into the units of subsea Christmas trees, wellheads, manifolds, pipelines, and so on. At level 3 the breakdown reflects the equipment, materials, and fabrication required.

The cost estimation uses the elements of the WBS described. The cost of the project is estimated based on the costs established for each element, and is the sum of all elements. The material costs and fabrication costs are obtained by requesting budget prices from a selection of preapproved bidders for each type of material or work. The scope of the fabrication work provides details about the work relevant to that material. It contains the project description, a list of free issue materials,

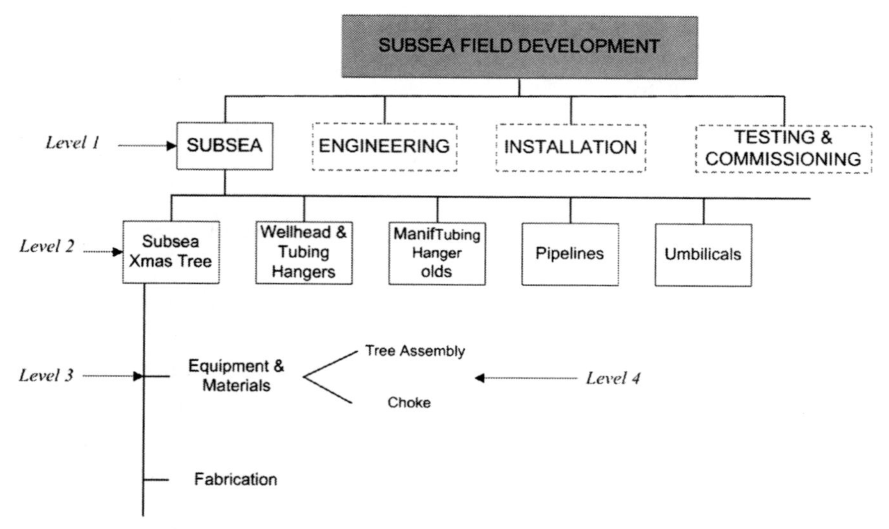

Figure 6-9 Typical WBS for a Subsea Field Development

a detailed list of work scope, and a fabrication, construction, and delivery schedule.

In some cases, the scope of work document is replaced with a drawing. The costs for the engineering elements are based on experience and knowledge of the project requirements. The man-hours required for the individual engineering activities are estimated and the cost derived by application of the appropriate man-hour rates.

6.3.4. Cost Estimation Process

Cost estimates for subsea equipment can also be obtained by combining the above two methods (factored estimation and WBS estimation). As shown in the flowchart of Figure 6-10, we first select the basic cost, which is decided based on the WBS of a standard product, and then choose the cost-driving factors. By listing the data in several tables, choosing the appropriate data, and putting them in Equation (6-4), we can arrive at the final estimation cost.

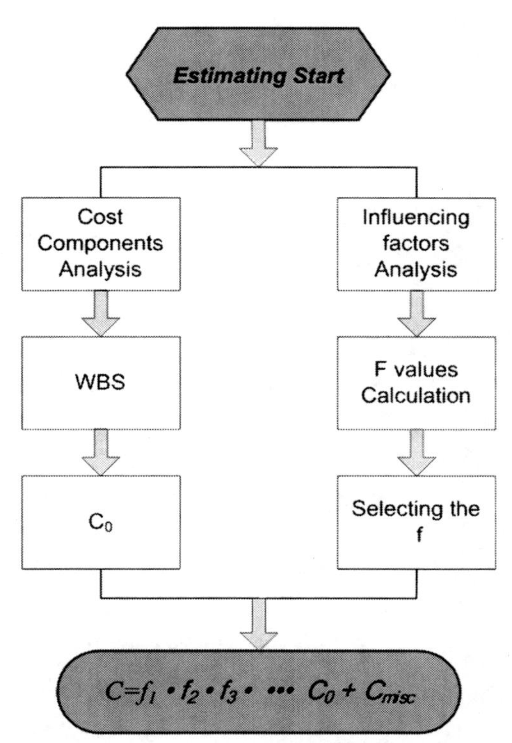

Figure 6-10 Cost-Estimating Process Flowchart

6.4. SUBSEA EQUIPMENT COSTS

6.4.1. Overview of Subsea Production System

Subsea production system includes varies of subsea structures or equipment such as wellheads, trees, jumpers, manifolds, etc., which depends on the field architecture types and the topside equipment. Typical filed developments may utilize several wellheads and a cluster manifold located in the center of them. For marginal fields, it is more flexible and economic to use a satellite well tie-back. Typical components and equipment in subsea production system as shown in Figure 6-11 are:

- Subsea wellhead: a structure used for supporting the casing strings in the well. It usually includes a guide base thus the wellhead is also used for guiding while install the tree;
- Subsea Xmas tree: an assembly of piping and valves and associated controls, instrumentations that landing and locking on top of the subsea wellhead for controlling production fluid from the well;
- Jumper: a connector or tie-in between the subsea structures, e.g. tree and manifold, manifold and PLET, PLET and PLET. Jumpers include flexible jumpers and rigid jumpers;
- Manifold: equipment used for gathering the production fluid from trees/ wellheads, and then transporting the production fluid to the floaters

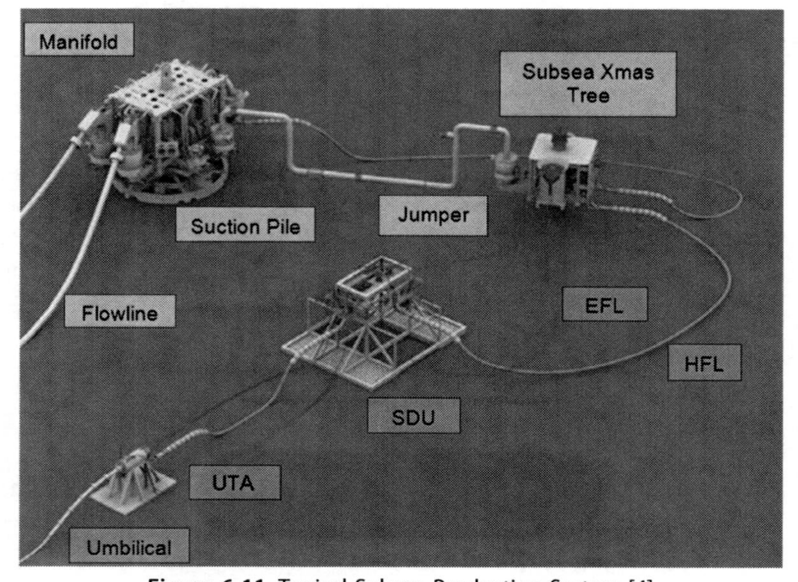

Figure 6-11 Typical Subsea Production System [4]

through subsea pipelines. A cluster manifold with 4, 6, 8 or 10 slots is the typical manifold;
- Template: a subsea structure to support the subsea wells or the manifolds;
- PLET: a subsea structure set at the end of pipeline to connect the pipeline with other subsea structures, such as manifolds or trees;
- Subsea foundation: a component to support subsea structures on seabed. Mudmat, suction pile, and drilling pile are typical subsea foundations;
- Subsea production control system (SPCS): components such as master control station (MCS), electrical power unit (EPU), hydraulic power unit (HPU) and etc.;
- Umbilical termination assembly (UTA): the termination that mates with the umbilical flange for installation and pull-in of the umbilical to the required subsea structure;
- Flying lead: a connector between UTA and other subsea equipment, it includes hydraulic flying leads (HFL), and electrical flying leads (EFL);
- Subsea distribution unit (SDU): a connector with the subsea umbilical through the UTA, distributing hydraulic supplies, electrical power supplies, signals, and injection chemicals to the subsea facilities;
- HIPPS: equipment designed to protect low-rated equipment against overpressure or abject flow accompanying the upset condition by either isolating or diverting the upset away from the low-rated equipment;
- Umbilical: a component that contains two or more functional elements, e.g. thermoplastic hoses and/or metal tubes, electrical cables and optical fibres. Umbilical is the main medium for power and signal transmission between topside and subsea;
- Chemical injection unit (CIU): equipment located on the topside platform to provide the chemical injection (e.g. the corrosion inhibitor) into subsea equipments;
- Subsea control module (SCM), subsea control equipment normally located on subsea trees for transferring the data and signal from the topside to operate the valves or other mechanisms.

This book will focus on some typical and common equipment and introduce the cost estimation processes for them, instead of covering all subsea equipment costs.

6.4.2. Subsea Trees

Figure 6-12 shows a subsea tree being deployed. The cost of subsea trees in a subsea field development can simply be estimated by multiplying the unit price by the number of trees. Tree type and number are selected and

Figure 6-12 A Subsea Tree Being Deployed [5]

estimated according to the field conditions such as water depth, reservoir characteristic, and production fluid type. The unit price can be provided by proven contractors and manufacturers.

6.4.2.1. Cost-Driving Factors

The following components are typically included while estimating the cost of trees:

- Tubing hanger assembly;
- ROV tree cap;
- EFAT testing.

The tree parameters should be selected and specified during the FEED, based on the information of oil/gas production, reservoir characteristics, production rate, water depth and etc.,

- Main components of tree: tree body, tree valves, tree piping, protection frame, SCM, production choke, tree connector, ROV panel, etc.;
- Typical bore sizes: 5" and 7";
- Standard pressure ratings: 5 ksi, 10 ksi, and 15 ksi;

The types of tree can be summarized into two categories: horizontal tree (HT) and vertical tree (VT). The main differences between HTs and VTs are the configuration, size, and weight. In a HT, the tubing hanger is in the tree body, whereas in a VT, the tubing hanger is in the wellhead. In addition, HT is usually smaller in the size than VT.

The bore size is standardized to 5", so the prices of the trees will not change too much for which bore size is 5" below. However, the trees with 7" or ever bigger bore sizes are still new technologies and the cost will changes largely.

The pressure ratings for subsea Christmas trees are 5, 10, and 15 ksi, in accordance with API 17D [6] and API 6A [7]. Different pressure ratings are used at different water depths. The technology of 5- and 10-ksi trees is commonly used in water depths greater than 1000 m. The main difference in the cost is determined by the weight and size. Few companies can design and fabricate the 15-ksi tree, so costs are high because of market factors.

The temperature ratings of subsea Christmas trees influence the sealing system, such as the sealing method and sealing equipment. API 6A [7] temperature ratings are K, L, P, R, S, T, U, and V. Typical subsea Christmas tree ratings are LV, PU, U, and V. Many manufacturers supply the equipment with a wide range of temperature ratings so that they work in various types of conditions. The temperature ratings do not have too great influence on the total cost of a subsea Christmas tree.

6.4.2.2. Cost Estimation Model

The cost estimation of subsea Xmas tree is produced by taking the individual cost driving factors of the equipment, multiplying with the basic cost which is usually the normal cost range of a standard product on the market at that time. A correction must be made if some cases are different from those considered.

The cost model of the subsea Xmas tree can be expressed as:

$$C_1 = C_0 \bullet f_1 \bullet f_2 \bullet f_3 \bullet \cdots + C_{corr} \tag{6-6}$$

where

$f_1, f_2, f_3 \ldots$: the cost driving factors for subsea Xmas trees, such as tree type, pressure rating and bore size;

C_1 : the cost of new subsea Xmas tree to be estimated;

C_0 : the basic cost of a subsea Xmas tree;

C_{corr} : the correction cost.

Tree types, pressure ratings, and bore sizes are the main cost driving factors of the cost estimation for a subsea Xmas tree. Following is an example of the cost estimation for a 5 inch 10 ksi vertical Xmas tree.

Base Cost C_0 (x10^6 USD)		**Min. = 2.20 | Average = 2.45 | Max. = 2.70**			
Pressure	10 ksi	**Bore Size**	5 in	**Tree Type**	VT
Material	Carbon Steel				

Tree Type		**Mudline Tree**	**VT**	**HT**
	Min.	0.25		1.20
Cost Factor, f_1	Average	0.35	1.00	1.25
	Max.	0.45		1.30

Pressure Rating		**5 ksi**	**10 ksi**	**15 ksi**
	Min.			1.10
Cost Factor, f_2	Average	0.90	1.00	1.15
	Max.			1.25

Bore Size		**3 inch**	**5 inch**	**7 inch**
	Min.			1.10
Cost Factor, f_3	Average	0.85	1.00	1.15
	Max.			1.25

Correction Cost, C_{corr} (×10^3 USD)	Min. = 150 | Average = 250 | Max. = 350

6.4.3. Subsea Manifolds

Several concepts are applied to manifolds and associated equipment in a subsea field development. Figure 6-13 shows the installation of a subsea manifold from the moon-pool of the installation vessel. Some fields use templates instead of manifolds. Actually the templates have the functions of a manifold. PLET/PLEMs are subsea structures (simple manifolds) set at the end of a pipeline that are used to connect rigid pipelines with other subsea structures, such as a manifold or tree, through a jumper. This equipment is used to gather and distribute the production fluids between wells and flowlines.

The costs of this type of equipment are mainly driven by the cost of the manifold, because it generally makes up about 30% to 70% of the total equipment cost, depending on the type and size of the field.

Figure 6-13 Subsea Manifold [8]

6.4.3.1. Cost-Driving Factors

The following components and issues should be included while estimating the cost of a subsea manifold or other subsea structures:

- Foundation of manifold;
- Controls of manifolds (e.g. SCM)
- Pressure and temperature;
- Pigging loop;
- EFAT testing.

Typical manifolds are cluster type with 4, 6, 8, 10 slots. Cost of the manifold mainly depends on the slots, as each slot needs one set of valves and pipe works. This will increase not only the purchase cost but also the material cost. Besides, size and weight of the structures also influence the installation cost.

Subsea structures such as manifolds and trees shall follow the standard pressure ratings specified as discussed in the sections before. Pressure ratings mainly influence the piping design in the manifold, e.g. the wall thickness.

6.4.3.2. Cost Estimation Model

Cost estimation for subsea structures follows the method used in the cost model of subsea Xmas tree. The cost model of the subsea manifold can be expressed as below:

$$C_1 = C_0 \cdot f_1 \cdot f_2 \cdot f_3 \cdot \cdots + C_{corr} \qquad (6\text{-}7)$$

where

f_1, f_2, f_3 ...: the cost driving factors for subsea manifolds, such as slot number and pipe size;

C_1 : the cost of subsea manifold to be estimated;

C_0 : the basic cost of a subsea manifold;

C_{corr} : the correction cost.

PLEM is a type of manifold, which normally has 1 to 3 hubs. Structure of PLEM is similar to that of manifold. However, if there are only 2 wells, for example, a PLEM is much more flexible and economic to be installed and connected to the wells, comparing to a cluster manifold. Following is an example of the cost estimation for a subsea manifold.

Base Cost C_0 (x10^6 USD)			**Min. = 1.8 | Avg. = 3.0 | Max. = 4.2**		
Number of Slots	4	Type	Cluster	Pipe OD	10"
Materials	Carbon Steel		Pressure		10 ksi

Number of Slots		**(PLEM)**	**4**	**6**	**8**	**10**
	Min.	0.50		1.05	1.25	1.60
Cost Factor, f_1	Avg.	0.65	1.00	1.25	1.45	1.90
	Max.	0.80		1.45	1.65	2.20

Pipe Size (OD)		**8"**	**10"**	**12"**	**16"**	**20"**
	Min.	0.85		1.04	1.14	1.18
Cost factor, f_2	Avg.	0.90	1.00	1.07	1.18	1.29
	Max.	0.95		1.10	1.22	1.40

Correction Cost, C_{corr} ($\times 10^3$ USD)	Min. = 150 | Average = 250 | Max. = 350

6.4.4. Flowlines

Flowlines are used to connect the wellbore to the surface facility and allow for any required service functions. They may transport oil or gas products, lift gas, injection water, or chemicals and can provide for well testing. Flowlines may be simple steel lines, individual flexible lines, or multiple lines bundled in a carrier pipe. All may need to be insulated to avoid problems associated with cooling of the production fluid as it travels along the seafloor.

The cost of flowlines is usually calculated separately from the costs for other subsea equipment. The estimation can be simply arrived at by multiplying the length of the line and the unit cost. Installation costs are discussed in Section 6.5.

6.4.4.1. Cost-Driving Factors

The main cost drivers for flowline procurement are:

- Type (flexible, rigid);
- Size (diameter and wall thickness, based on pressure rating and temperature rating);
- Material class;
- Coating;
- Length.

The steels applied in the offshore oil and gas industry vary from carbon steels (API standards Grade B to Grade X70 and higher) to exotic steels such as duplex. The higher grade steel obviously commands a higher price. However, as the costs of producing high-grade steels have been reduced, the general trend in the industry has been to use the higher grade steels, typically subsea flowline grades X70 and X80 for nonsour service and grades X65 and X70 with a wall thickness of up to 40 mm for sour service.

Flexible flowlines make the laying and connection operations relatively easy and fast. Material costs for flexible lines are considerably higher than that of conventional steel flowlines, but this may be offset by typically lower installation costs.

High-pressure ratings require high-grade pipe materials, thus the cost of steel increases for high-pressure projects. However, the increase in grade may permit a reduction of pipeline wall thickness. This results in an overall reduction of fabrication costs when using a high-grade steel compared with a low-grade steel. The pressure rating–cost curve is shown in Figure 6-14.

The factor of pressure rating is combined into pipe size factor.

6.4.4.2. Cost Estimation Model

Subsea flowline cost is usually estimated by multiplying the unit price by the total length. The cost model of the flowline can be expressed as below:

$$C_1 = C_u \cdot L \cdot f_1 \cdot f_2 \cdot f_3 \cdot \cdots + C_{corr} \tag{6-8}$$

where

f_1, f_2, f_3 ...: the cost driving factors for subsea flowlines, such as pipeline OD and pipe wall thickness;

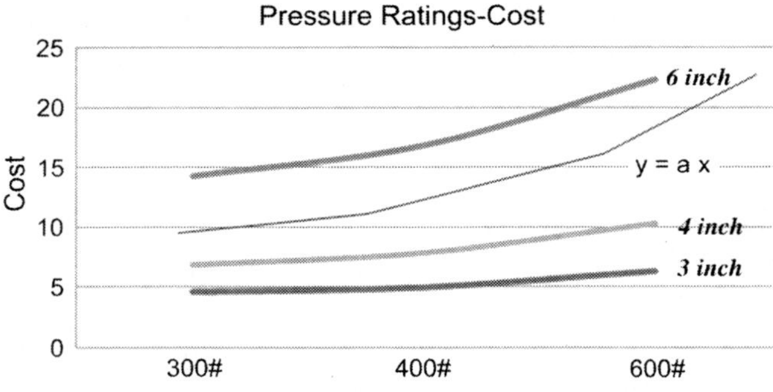

Figure 6-14 Pressure Rating–Cost Curve

C_1: the cost of subsea flowline to be estimated;
C_u: the unit cost of the flowline (per meter);
L: the total length of the flowline;
C_{corr}: the correction costs for joints and coatings and etc.

Outside diameter (OD), wall thickness of flowline, pressure rating, and water depth are the main cost driving factors of the cost estimation for flowlines. Following is an example of the cost estimation for a subsea flowline.

Base Cost C_u (USD/meter)	**Min. = 150 | Avg. = 215 | Max. = 280**
	OD: 10" Material grade : API 5L X65 Rigid Pipe

Pipe OD		**4"**	**10"**	**12"**	**16"**	**20"**
Cost Factor, f_1	Min.	0.15		1.20	1.60	2.20
	Avg.	0.25	1.00	1.30	1.80	2.60
	Max.	0.35		1.40	2.00	3.00

Coatings (USD/meter)		**4"**	**10"**	**12"**	**16"**	**20"**
Correction Cost, C_{corr}	Min.	140	350	380	460	580
	Avg.	170	450	485	580	710
	Max.	200	550	590	700	840

Flexible pipe or composite pipe is different from rigid pipeline. A typical flexible pipeline contains several layers. However, if the base cost of the flexible pipe is available, the cost model for flexible flowline is still applicable for the cost estimation of the flowlines with different sizes.

6.5. TESTING AND INSTALLATION COSTS

6.5.1. Testing Costs

The Factory Acceptance Test (FAT) is a test of subsea equipment before installation, which is always performed on newly manufactured equipment, to check whether the equipment satisfy the performance and function requirements or not. Extended Factory Acceptance Test (EFAT) may be only applicable for several equipment or subsea structures, e.g. subsea trees with sub-assemblies. The System Integrity Test (SIT) is performed to verify the whole system no matter from one supplier only or different suppliers, which shall interface with each other acceptably.

Typical FAT items of subsea equipment are:

- Flushing, functioning and pressure testing of hydraulic control circuits;
- Assembly hydrostatic pressure testing;
- Function testing of ROV interfaces;
- Gas testing;
- Testing for installation and function SCM;
- Thermal insulation test;
- Continuity check for cathodic protection (CP);

The following are typical items for SIT:

- To confirm the interchangeability of individual assembly, e.g. tubing hanger and tree system, pigging loop and manifold;
- To confirm the physical interface of individual equipments, e.g. the tree on the wellhead stack , and the jumper landing on the hub of manifold;
- To identify any interface issues between the individual equipment;
- To confirm the physical interface of ROV tools to the control panel of the equipment;
- To determine the maximum angular offset that will allow for proper installation of well jumper kit.

The testing may need special tools or equipment, which should be considered for the cost estimation:

- Water filled test pool;
- Cranes;
- ROV hot stab assembly;
- Handling wire rope sling;
- Running tool & tool kits;
- Test stumps;
- Inspection stands;

Table 6-2 SIT Cost Estimation for Subsea Production System

Item	Unit Cost	Quantity	Unit
Test facilities preparation	\$3,000 ~ 3,500	3 ~ 5	Days
Site Mobilization	\$2,500 ~ 3,500	1 ~ 2	Each
Site Demobilization	\$2,500 ~ 3,500	1 ~ 2	Each
Crane Rental	\$1,700 ~ 2,300	3 ~ 5	Days
Onshore Service Engineers (at least 2)	\$1,000 ~ 1,500	3 ~ 5	Days

Normally the FAT and EFAT cost are included in the procurement of the individual equipment. The SIT cost estimation model for subsea production system is listed in Table 6-2.

6.5.2. Installation Costs

Installation costs for a subsea field development project are a key part of the whole CAPEX, especially for deepwater and remote areas. Planning for the installation needs to be performed at a very early stage of the project in order to determine the availability of an installation contractor and/or installation vessel, as well as a suitable weather window. Also, the selection of installation vessel/method and weather criteria affects the subsea equipment design.

The following main aspects of installation need to be considered at the scope selection and scope definition stages of subsea field development projects:

- Weather window;
- Vessel availability and capability;
- Weight and size of the equipment;
- Installation method;
- Special tooling.

Different types of subsea equipment have different weights and sizes and require different installation methods and vessels. Generally the installation costs for a subsea development are about 15% to 30% of the whole subsea development CAPEX. The costs of subsea equipment installation include four major components:

- Vessel mob/demob cost;
- Vessel day rate and installation spread;
- Special tooling rent cost;
- Cost associated with vessel downtime or standby waiting time.

The mob/demob costs range from a few hundred thousand dollars to several million dollars depending on travel distance and vessel type.

The normal pipe-laying vessel laying speed is about 3 to 6 km (1.8 to 3.5 miles) per day. Welding time is about 3 to 10 minutes per joint depending on diameter, wall thickness, and welding procedure. Winch lowering speeds range from 10 to 30 m/s (30 to 100 ft/s) for deployment (pay-out) and 6 to 20 m/s (20 to 60 ft/s) for recovery (pay-in).

For subsea tree installation, special tooling is required. For a horizontal tree, the tooling rent cost is about USD $7000 to $11,000 per day. For a vertical tree, the tooling rent cost is about USD $3000 to $6000 per day. In addition, for a horizontal tree, an additional subsea test tree (SSTT) is required, which costs USD $4000 to $6000 per day. Tree installation (lowering, positioning, and connecting) normally takes 2 to 4 days.

Table 6-3 shows the typical day rates for various vessels, and Table 6-4 lists some typical subsea equipment installation duration times.

Table 6-3 Day rates for Different Vessel Types

Vessel Types	Average Day Rate
Drill Ship < 1500 m WD	$240,000
Semisub < 500 m WD	$250,000
Semisub > 500 m WD	$290,000
Semisub > 1500 m WD	$430,000
Jack-up 100 m WD	$90,000
Jack-up >100 m WD	$140,000
Anchor Handling Vessel	$50,000
Pipe Laying Vessel, <200 m WD	$300,000
Pipe Laying Vessel, >200 m WD	$900,000

Table 6-4 Typical Subsea Structures Installation Duration (WD: 4000ft/1200 m)

Installation Activities	Parameters	Installation Duration (Days)
Flowline, PLET and Riser (SCR) Installation	8 inch, 8 km flowline 12 inch, 8 km flowline 8 inch 4 km riser; and 3 PLETs	27 ~ 30
Umbilical and UTA Installation	3 umbilicals with lengths of 1500 m, 2600 m, and 5500 m	17 ~ 20
Installation Tree	5 inch horizontal tree	1 ~ 2
Manifold Installation	<8 slots	1 ~ 3
Jumper Installation	4 rigid Jumpers	11 ~ 13
Flying Leads Hook-up	6 flying leads	5 ~ 8

6.6. PROJECT MANAGEMENT AND ENGINEERING COSTS

The project management of a subsea field development mainly includes planning, execution, processes monitoring and controlling, resources managing, and etc. The cost should be estimated for each stage of the project.

The costs for each stage mainly depend on the personnel, e.g. the senior consultant fee, the engineers fee, etc. On average, the costs of different engineers are shown below:

- Project Engineer: USD 24 K/month;
- Tree Engineer: USD 20 K/month;
- Controls Engineer: USD 20 K/month;
- Flowline Engineer: USD 20 K/month;
- Riser Engineer: USD 20 K/month;
- Flow Assurance Engineer: USD 25 K/month.

6.7. SUBSEA OPERATION EXPENDITURES (OPEX)

An offshore well's life includes five stages: planning, drilling, completion, production, and abandonment. The production stage is the most important stage because when oil and gas are being produced, revenues are being generated. Normally a well's production life is about 5 to 20 years.

During these years, both the planned operations and maintenance (O&M) expenditures and the unplanned O&M expenditures are needed to calculate life cycle costs. OPEX includes the operating costs to perform "planned" recompletions. OPEX for these planned recompletions is the intervention rig spread cost multiplied by the estimated recompletion time for each recompletion. The number and timing of planned recompletions are uniquely dependent on the site-specific reservoir characteristics and the operator's field development plan.

Each of the identified intervention procedures is broken into steps. The duration of each step is estimated from the historical data. The non-discounted OPEX associated with a recompletion is estimated as:

$$\text{OPEX} = (\text{Intervention duration}) \times (\text{Rig spread cost})$$

Figure 6-15 shows a distribution for the typical cost components of OPEX for a deepwater development. The percentage of each cost component of the total OPEX varies from company to company and location to location. Cost distributions among OPEX components for shallow-water

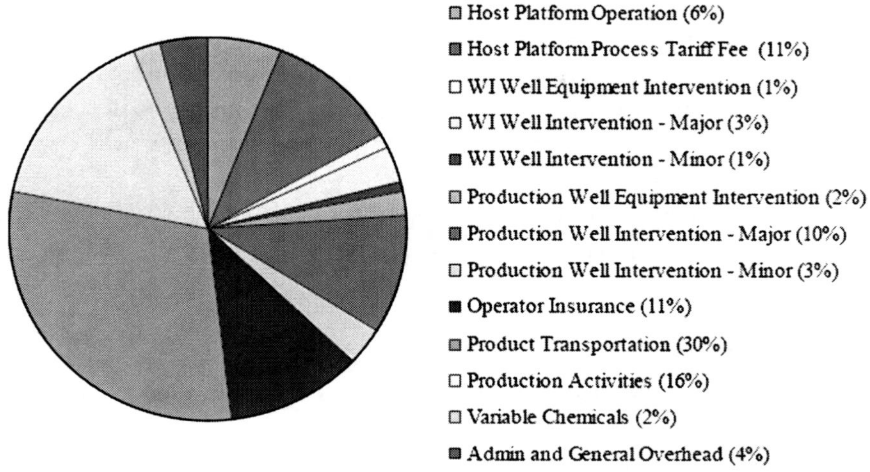

Figure 6-15 Typical Cost Distribution of Deepwater OPEX [9]

development are similar to those for deepwater developments, except that the cost of product transportation is significantly lower.

6.8. LIFE CYCLE COST OF SUBSEA SYSTEM

Many cost components/aspects must be considered to determine the most cost-effective subsea system for a particular site. The risks associated with blowouts are often an important factor during drilling/installation. Another often overlooked important factor is the cost of subsea system component failures. As oil exploration and production moves into deeper and deeper water, the costs to repair subsea system component failures escalate dramatically.

Therefore, besides CAPEX and OPEX, two other cost components are introduced for determining the total life cycle cost of a subsea system [9]:

- *RISEX:* risk costs associated with loss of well control (blowouts) during installation, normal production operations, and during recompletions;
- *RAMEX:* the reliability, availability, and maintainability costs associated with subsea component failures.

Let's also revisit the definitions of CAPEX and OPEX at this time:

- *CAPEX:* capital costs of materials and installation of the subsea system. Materials include subsea tress, pipelines, PLEMs, jumpers, umbilicals, and controls systems. Installation costs include vessel spread costs

multiplied by the estimated installation time and for rental or purchase of installation tools and equipment.
- *OPEX:* operating costs to perform well intervention/workovers. The number and timing of these activities are uniquely dependent on the site-specific reservoir characteristics and operator's field development plan.

The life cycle cost (LC) of a subsea system is calculated by:

$$\text{LC} = \text{CAPEX} + \text{OPEX} + \text{RISEX} + \text{RAMEX} \tag{6-9}$$

6.8.1. RISEX

RISEX costs are calculated as the probability of uncontrolled leaks times assumed consequences of the uncontrolled leaks:

$$\text{RI} = \text{PoB} \bullet \text{CoB} \tag{6-10}$$

where

RI: RISEX costs;

PoB: probability of blowout during lifetime;

CoB: cost of blowout.

Blowout of a well can happen during each mode of the subsea system: drilling, completion, production, workovers, and recompletions. Thus, the probability of a blowout during a well's lifetime is the sum of each single probability during each mode:

$$\text{PoB} = P(\text{dri}) + P(\text{cpl}) + P(\text{prod}) + \sum P(\text{wo}) + \sum P(\text{re} - \text{cpl}) \tag{6-11}$$

The cost of a subsea well control system failure (blowout) is made up of several elements. Considering the pollution response, it is likely to be different among different areas of the world. Table 6-5 shows this kind of costs in the industry from last decades.

6.8.2. RAMEX

RAMEX costs are related to subsea component failures during a well's lifetime. A component failure requires the well to be shutdown, the workover vessel to be deployed, and the failed component to be repaired. Thus, the main costs will fall into two categories:

- The cost to repair the component, including the vessel spread cost;
- The lost production associated with one or more wells being down.

Table 6-5 Cost of Blowouts in Different Geographic Areas [9]

Area	Type of Incident	Date	Cost* ($ MM)	Type of Damage
North Sea	Surface blowout	09/1980	16.1	Cost of cleanup
			13.9	Redrilling costs
France	Underground blowout on producing well	02/1990	9.0	Redrilling costs
			12.0	Cost of cleanup
GoM	Underground blowout	07/1990	1.5	Cost of cleanup
Middle East	Underground blowout when drilling	11/1990	40.0	
Mexico	Exploration and blowout	08/1991	16.6	Operator's extra expenditure
North Sea	Blowout of high-pressure well during exploration drilling	09/1991	12.25	Operator's extra expenditure
GoM	Blowout	02/1992	6.4	Operator's extra cost
North Sea	Underground blowout during exploration drilling	04/1992	17.0	Operator's extra expenditure
India	Blowout during drilling	09/1992	5.5	Operator's extra expenditure
Vietnam	Surface gas blowout followed by underground flow	02/1993	6.0	Redrilling costs
			54.0	Cost of the well
GoM	Blowout	01/1994	7.5	Operator's extra expenditure
Philippines	Blowout of exploration well	08/1995	6.0	Cost of the well
GoM	Surface blowout of producing well (11 wells lost)	11/1995	20.0	Cost of wells and physical damage costs

**Note:* The costs are based on the specific years. If considering the inflation rate, see Section 6.3.2.

Actually, the repair cost of a failed component is also a workover cost, which should be an item of OPEX. Normally, however, only the "planned" intervention/workover activities are defined and the cost estimated. With "unplanned" repairs, RAMEX costs are calculated by multiplying the probability of a failure of the component (severe enough to warrant a workover) by the average consequence cost associated with the failure. The total RAMEX cost is the sum of all of the components' RAMEXs:

$$\mathrm{RA} = C_r + C_p \tag{6-12}$$

where

RA: RAMEX cost;

C_r: cost of repair (vessel spread cost and the component repair/change cost);

C_p: lost production cost.

The procedures for calculating this cost are illustrated in the Figure 6-16.

The vessel spread costs are similar to the installation vessel costs; see Section 6.5.2. For more information about failures of subsea equipment, see Chapter 11.

Figure 6-17 illustrates the costs that arise as a result of lost production time where TTF is time to failure, LCWR is lost capacity while waiting on rig, T_{RA} is the resource's availability time (vessel), and T_{AR} is the active repair time.

The mean time to repair is dependent on the operation used to repair the system. A repair operation is required for each component failure. Each operation will have a corresponding vessel, depending on the scenario (subsea system type or field layout; see Chapter 2).

From Figure 6-17, we can clearly see the production lost cost: the loss of oil or gas production. Note that this cost is the sum of all of the individual subsea wells.

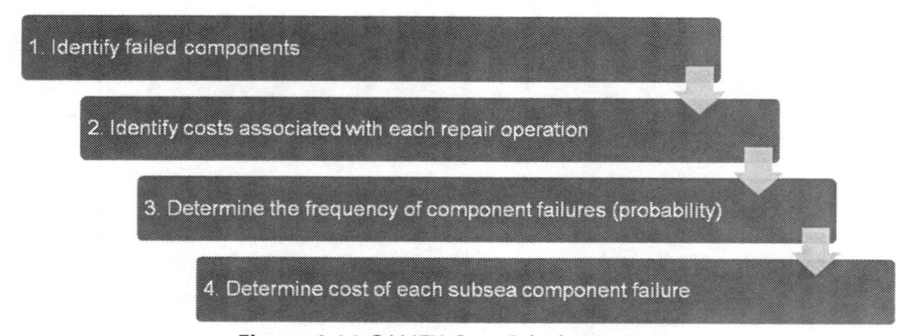

Figure 6-16 RAMEX Cost Calculation Steps

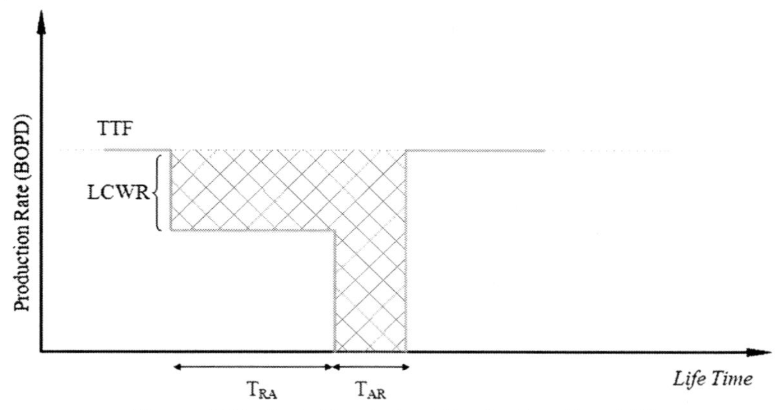

Figure 6-17 Costs Due to Lost Production Time [10]

6.9. CASE STUDY: SUBSEA SYSTEM CAPEX ESTIMATION

Too often, only CAPEX is estimated in detail on a sheet listing items one by one (the WBS method). OPEX, RISEX, and RAMEX, in contrast, depend largely on reservoir characteristics, specific subsea system designs, and operating procedures. The flowchart shown in Figure 6-18 details the CAPEX estimation steps in a feasibility study for a subsea field development. Note that the data provided in this flowchart should be used carefully and modified if necessary for each specific project.

We look now at an example of CAPEX estimation using the WBS method and the steps illustrated in Figure 6-18.

Field Description

- Region: Gulf of Mexico;
- Water depth: 4500ft;
- Number of trees: 3;
- Subsea tie-back to a SPAR.

Main Equipment

- Three 5-in. × 2-in.10-ksi vertical tree systems;
- One manifold;
- Two production PLETs;
- One SUTA;
- 25,000-ft umbilical;
- 52,026-ft flowline.

Calculation Steps

- See Table 6-6.

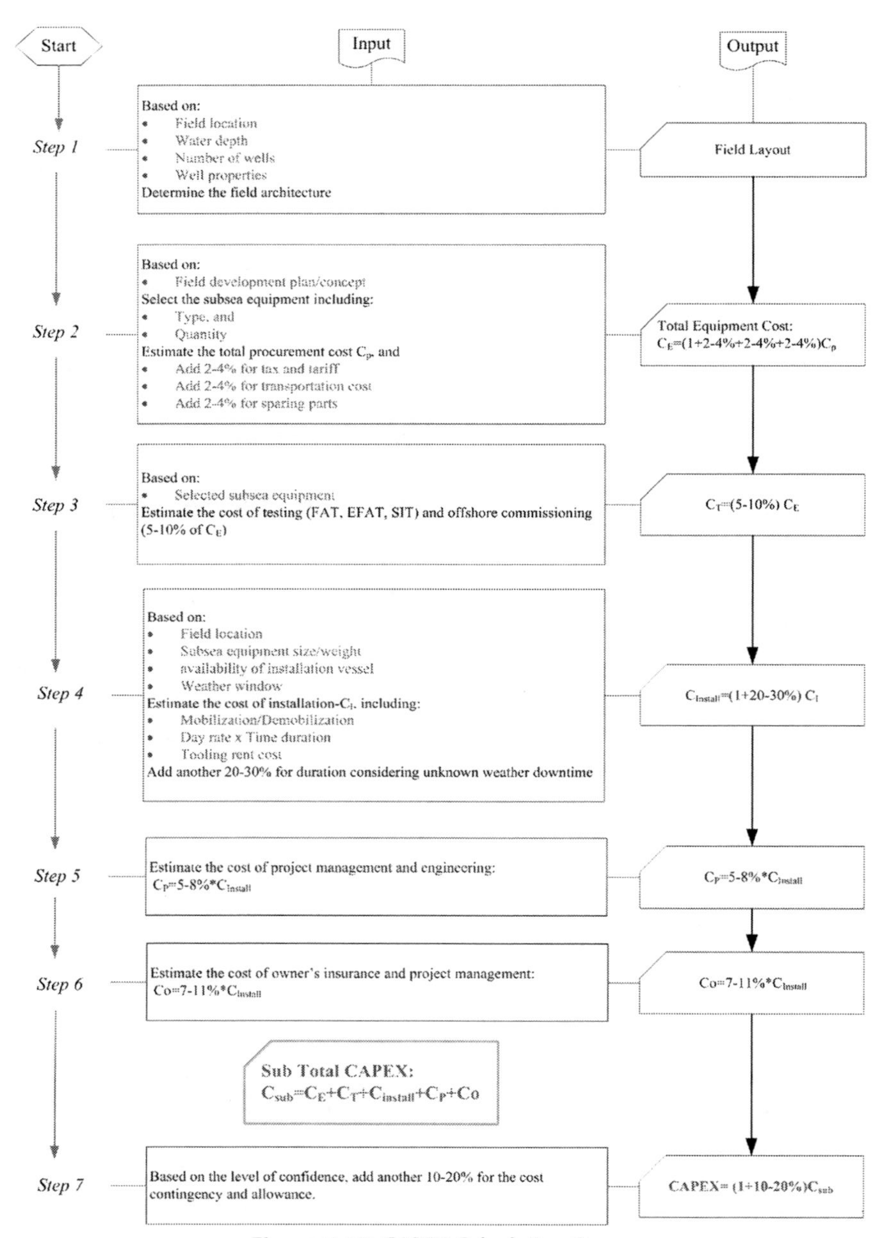

Figure 6-18 CAPEX Calculation Steps

Table 6-6 CAPEX Estimation Example (2007 Data)

1. Subsea Equipment Cost

Subsea Trees		Unit	Cost
Subsea Tree Assembly		**3**	**$4,518,302**
(each)	5-inch × 2-inch 10-ksi vertical tree assembly	1	included
	Retrievable choke assembly	1	included
	Tubing hanger 5-in. 10 ksi	1	included
	High-pressure tree cap	1	included
	5-in. tubing head spool assembly	1	included
	Insulation	1	included
Subsea Hardware			
Subsea Manifold			
	(EE trim)	1	$5,760,826
Suction Pile			
	Suction pile for manifold	1	$1,000,000
Production PLET		2	$3,468,368
Production Tree Jumpers		3	$975,174
Pigging Loop		1	$431,555
Production PLET Jumpers		2	$1,796,872
Flying Leads			$1,247,031
	Hydraulic flying lead SUTA to tree		
	Electrical flying lead SUTA to tree		
	Hydraulic flying lead SCM to manifold		
	Electrical flying lead SUTA to manifold		
Other Subsea Hardware			
Multiphase Flow Meter		1	$924,250
Controls			
Topsides Equipment		1	$2,037,000
	Hydraulic power unit (include gas lift outputs)	1	$569,948
	Master control station (with serial links to OCS)	1	$204,007
	Topside umbilical termination assembly (TUTA) (split)	1	$156,749
	Electrical power unit (incl. UPS) (*Note:* Check capacity of existing UPS.)	1	included above
Tree-Mounted Controls		3	$5,108,940
Manifold Equipments		1	$1,104,163
SUTA		1	$2,764,804

(*Continued*)

Table 6-6 CAPEX Estimation Example (2007 Data)—cont'd

1. Subsea Equipment Cost			
Subsea Trees		**Unit**	**Cost**
Umbilicals			
Umbilical			$11,606,659
	25,000 ft Length		
Risers			
Riser			$6,987,752
	Prod. 8.625-in. × 0.906-in. × 65 SCR, 2 × 7500 ft		
Flowlines			
Flowline			$4,743,849
	Dual 10-in. SMLS API 5L X-65, flowline, 52,026 ft		
		Total Procurement Cost	**$54,264,324**
2. Testing Cost			
Subsea Hardware FAT,EFAT			$27,132,162
Tree SIT & Commissioning			$875,000
Manifold & PLET SIT			$565,499
Control System SIT			$237,786
		Total Testing Cost	**$28,810,447**
3. Installation Cost			
Tree	3 days × $1000 k per day		$3,000,000
Manifold & Other hardware			$48,153
Jumpers	(1 day per jumper + downtime)		$32,102
ROV Vessel Support			$1,518,000
Other Installation Cost			$862,000
Pipe-lay	52,0260 ft		$43,139,000
		Total Installation Cost	**$63,179,032**
4. Engineering & Project Management Cost			
		Total Engineering Cost	**$4,738,427**
5. Insurance			
		Total Insurance Cost	**$6,002,008**
		Sub-Total of CAPEX	**$156,994,238**
6. Contigency and Allowance			
		Total Allowance Cost	**$12,559,539**
Total Estimated CAPEX: $169,553,778			

REFERENCES

[1] AACE International Recommended Practice, Cost Estimation Classification System, AACE, 1997, NO. 17R-97.
[2] Douglas-Westwood, The Global Offshore Report, 2008.
[3] C. Scott, Investment Cost and Oil Price Dynamics, IHS, Strategic Track, 2006.
[4] U.K. Subsea, Kikeh – Malaysia's First Deepwater Development, Subsea Asia, 2008.
[5] Deep Trend Incorporation, Projects, http://www.deeptrend.com/projects-harrier.htm, 2010.
[6] American Petroleum Institute, Specification for Subsea Wellhead and Christmas Tree Equipment, first ed., API Specification 17D, 1992.
[7] American Petroleum Institute, Specification for Wellhead and Christmas Tree Equipment, nineteenth ed., API Specification 6A, 2004.
[8] FMC Technologies, Manifolds & Sleds, FMC Brochures, 2010.
[9] Mineral Management Service, Life Time Cost of Subsea Production Cost JIP, MMS Subsea JIP Report, 2000.
[10] R. Goldsmith, R. Eriksen, M. Childs, B. Saucier, F.J. Deegan, Life Cycle Cost of Deepwater Production Systems, OTC 12941, Offshore Technology Conference, Houston, 2001.

CHAPTER 7

Subsea Control

Contents

7.1. INTRODUCTION

The subsea control system operates the valves and chokes on subsea trees, manifold/templates, and pipelines. It also receives and transmits the data between the surface and subsea, which helps engineers monitoring the status of production by indicating temperatures, pressures, sand detection, etc. The location of control devices is extremely important. Careful consideration to the location of controls can result in a reduction in the amount of piping and cabling and the number of connections required, which in turn influences the subsea installation and retrieval tasks.

Subsea Engineering Handbook
ISBN 978-0-12-397804-2, doi:10.1016/B978-0-12-397804-2.00007-2

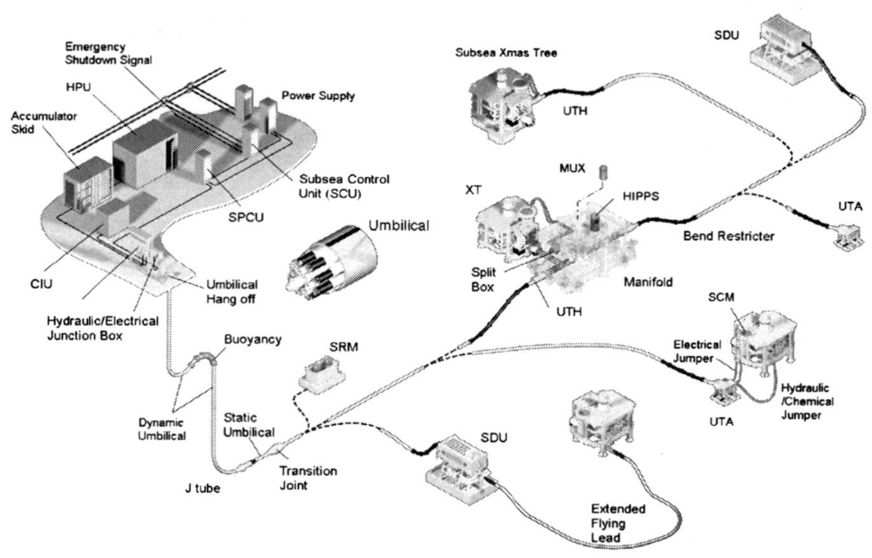

Figure 7-1 Subsea Control System Overview *(Courtesy of FMC)*

Figure 7-1 gives an overview of a subsea control system, and Figure 7-2 shows the main building blocks in a typical multiplex electrohydraulic control system. The typical control elements include the following:

- *Topside:* Electrical power unit, hydraulic power unit, master control station, topside umbilical termination assembly, etc.;

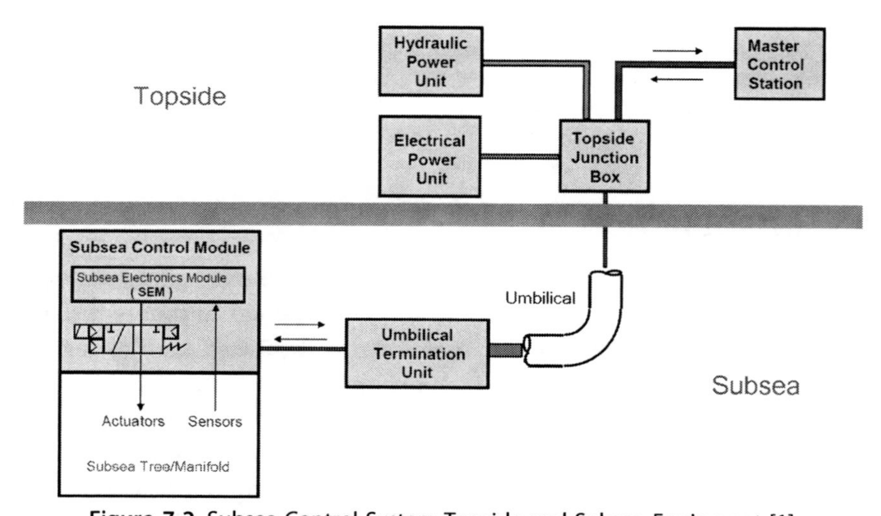

Figure 7-2 Subsea Control System Topside and Subsea Equipment [1]

- *Subsea:* umbilicals, subsea umbilical termination assembly, electrical and hydraulic flying leads, subsea control module, etc.

7.2. TYPES OF CONTROL SYSTEMS

The fundamental purpose of a control system is to open and close valves. However, other properties, such as instrumentation, provide chock control and important diagnostics.

The five types of fundamental control systems are:

- Direct hydraulic;
- Piloted hydraulic;
- Sequenced hydraulic;
- Multiplex electrohydraulic;
- All-electric.

Since the 1960s, the evolution of control system technology has proceeded from direct hydraulic to piloted and sequenced systems to provide improved response time and allow for long-distance tie-backs. Today, most subsea developments use the multiplex electrohydraulic control system. This is essentially a subsea computer/communication system consisting of hydraulic directional control valves. These electrically actuated valves allow stored pressure within subsea accumulators to be routed to individual hydraulic lines and onward to actuated gate valves and chokes on subsea production equipment. All-electric control systems are an attractive addition and an alternative to existing electrohydraulic systems. The all-electric subsea electric controls will reduce the cost of topside power generation and subsea umbilicals [2].

7.2.1. Direct Hydraulic Control System

The simplest remotely operated system for control and monitoring of a subsea system is the direct hydraulic control system. In this system each valve actuator is controlled through its own hydraulic line. This system is typically used for workover applications and small systems, and is especially common in single-satellite oil/gas fields of distances less than 15 km (9.3 miles). The principle of a direct hydraulic control system is shown in Figure 7-3.

When the operator sets the control valve to open, the direct hydraulic pressure control fluid flows to the actuator. To close the valve, the operator sets the wellhead control panel valve to the closed position, venting hydraulic fluid from the actuator back to the reservoir.

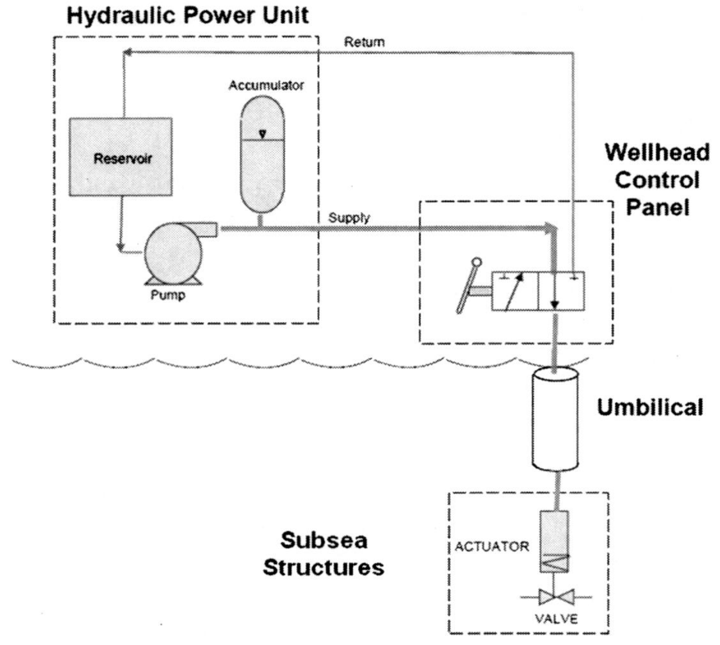

Figure 7-3 Direct Hydraulic Control System [1]

The major components of the system are as follows:

- *Hydraulic power unit (HPU):* The HPU supplies filtered and regulated hydraulic fluid to the subsea installations; it also applies the pressure that drives motors, cylinders, and other complementary parts of a hydraulic system. Unlike standard pumps, these power units use multistage pressurization networks to move fluid, and they often incorporate temperature control devices.
- *Wellhead control panel:* The control panel is arranged for initiating predetermined commands to the wellheads. A logic circuit is connected to provide an output control signal designated by the control panel.
- *Control umbilical:* A control umbilical connects topside equipment to subsea equipment. The control umbilical can transfer high- and low-pressure fluid supplies, chemical injection fluids, annulus fluids, and electrical power/signals.
- *Subsea tree:* The subsea tree is the primary well control module. It provides a mechanism for flow control and well entry.

The prominent features, advantages, and disadvantages of direct hydraulic control systems are summarized in Table 7-1.

Table 7-1 Prominent Features, Advantages, and Disadvantages of Direct Hydraulic Control Systems

Advantages	Disadvantages	Prominent Features
• Minimum subsea equipment • Low cost • Reliability is high because the critical components are on the surface • Maintenance access is very good because all critical components are on the surface • Large umbilical	• Very slow • Large number of hoses • Limited monitoring capabilities and distance limitations due to long response time and umbilical costs • Limited operational flexibility	• Few system components • Simple operating theory • Minimum subsea equipment • Inherent reliability

7.2.2. Piloted Hydraulic Control System

The piloted hydraulic control system has a dedicated hydraulic pilot supply (hose) for each subsea function and a hydraulic supply line to a simple subsea control module (SCM). At the SCM, a hydraulic accumulator provides a reserve of hydraulic energy to speed up the tree valve opening response time. This system is typically used in a single satellite of short to medium distances (4–25 km). Figure 7-4 shows the principle of a piloted hydraulic control system.

The difference of the system from the direct hydraulic system is that the umbilical does not include large bore hoses to achieve the performance requirements. The pilot system uses a small bore hose for the pilot line and a larger bore hose for the supply line. The hydraulic pilot volume to actuate the pilot valve is very small and there is very little volume flow required to energize the pilot valve and to open the subsea valve. As a result, the valve actuation time is improved. The features, advantages and limitations of the pilot control system are summarized in Table 7-2.

7.2.3. Sequenced Hydraulic Control System

The sequenced hydraulic control system consists of several sequence valves and accumulators. Various complicated program actions are designed by series-parallel for sequence valves. The principle of a sequenced hydraulic control system is shown in Figure 7-5.

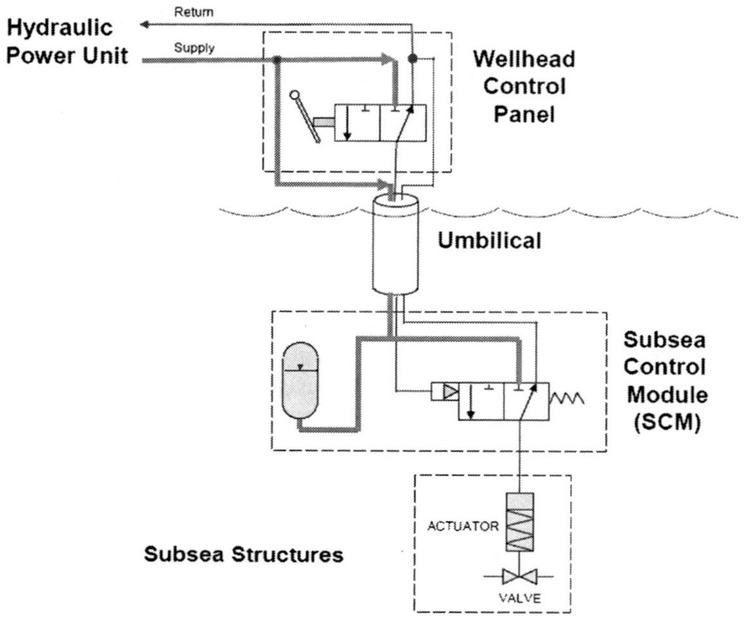

Figure 7-4 Piloted Hydraulic Control System [1]

Table 7-2 Prominent Features, Advantages, and Disadvantages of Piloted Hydraulic Control Systems

Advantages	Disadvantages	Prominent Features
• Low cost • Reliability is high because the critical components are on the surface • Maintenance access is good because the majority of the components are on the surface • Proven and simple subsea equipment	• Still slow • Large number of hoses • Limitation in distance because response is slow • No subsea monitoring because there are no electrical signals	• Improved response time • Slight reduction in umbilical size • Excellent backup configuration

Valves in this system are opened in pre-determined sequences, depending on the magnitude of the signal from the topside. The system works by adjusting the regulator up to the opening pilot pressure of the first valve to open it. Figure 7-5 shows a typical system, with the first valve in the sequence

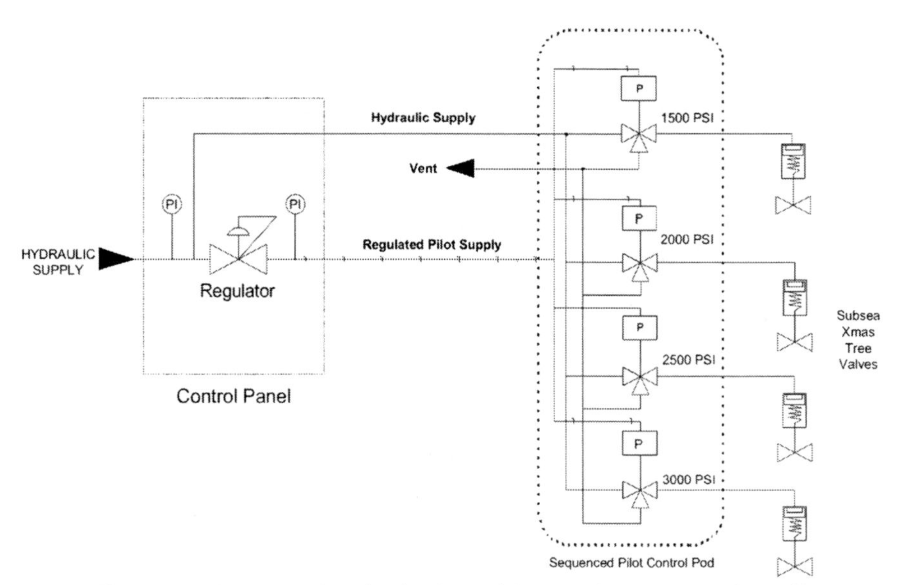

Figure 7-5 Sequenced Hydraulic Control System *(Courtesy TotalFinaElf)*

opening at 1,500 psi (103.5 bar) and each subsequent valve opening at 500 psi (34.5 bar) pressure increments. In order to keep sufficient differential between valve actuations to prevent spurious opening, and to keep the system hydraulic working pressure within the normal system design working pressures, the number of valve actuations in a sequence is limited. Table 7-3 summarizes the prominent features, advantages, and disadvantages of sequenced hydraulic control systems.

7.2.4. Multiplexed Electrohydraulic Control System

The master control station (MCS) is implemented by a computer, and communicates with the microprocessor in the subsea electronics module (SEM), which is the communication link with the MCS and performs MCS commanded functions. The multiplexed electrohydraulic system allows many SCMs to be connected to the same communications, electrical, and hydraulic supply lines. The result is that many wells can be controlled via one simple umbilical, which is terminated at a subsea distribution unit (SDU). From the SDU, the connections to the individual wells and SCMs are made with jumper assemblies.

The cost of a multiplexed electrohydraulic system is high due to the electronics within the SEM, the addition of the computer topside, and the

Table 7-3 Prominent Features, Advantages, and Disadvantages of Sequenced Hydraulic Control Systems

Advantages	Disadvantages	Prominent Features
• Improved system response compared to direct hydraulic control and piloted hydraulic control systems • Reduced umbilical compared to direct hydraulic control and piloted hydraulic control systems • Small number of hydraulic hoses	• Slow operation • The sequence of valve opening and closing is fixed • Distances are limited because response is slow • No subsea monitoring because there are no electrical signals • Increase in number of surface components • Increase in number of subsea components	• Improved response time • Capable of complex control operations • Substantial reduction in umbilical size • Excellent backup configuration • Excellent simple-sequence potential

required computer software. These costs, however, are balanced against smaller and less complex umbilicals and advancing technology, reducing the cost of the electronics. Figure 7-6 shows the principle of a multiplexed electrohydraulic control system. The system is typically used in complex fields of long distances (more than 5 km).

When a digital signal is sent to the SEM, it excites the selected solenoid valve, thereby directing hydraulic fluid from the supply umbilical to the associated actuator. The multiplex electrohydraulic control system is capable of monitoring pressure, temperature, and valve positions by means of electrical signals, without further complicating the electrical connections through the umbilical.

The prominent features, advantages and limitations of the multiplexed electrohydraulic control system are shown in Table 7-4.

7.2.5. All Electrical Control System

The all-electric control system is an all-electric–based system without the conventional hydraulic control of subsea components. The elimination of hydraulics means that any control system commands are sent in rapid succession without the usual retardation time required for accumulators to charge. This system is typically used in complex fields and marginal fields of long distances (usually greater than 5 km) and for high-pressure and

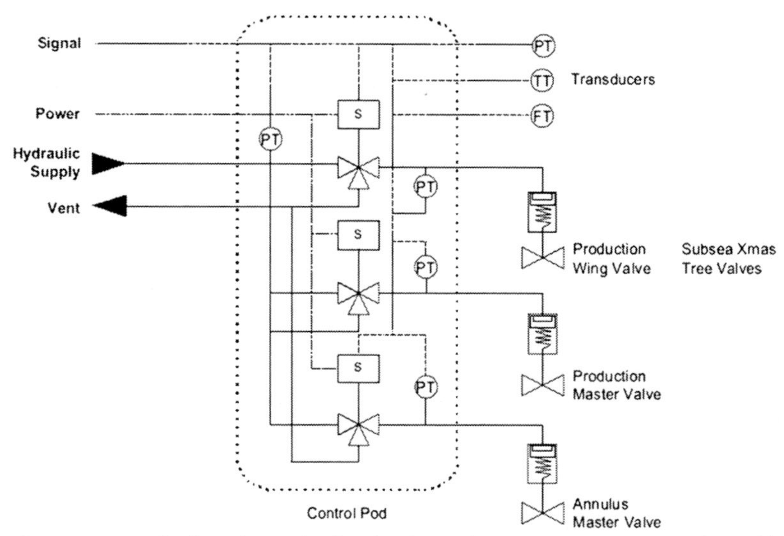

Figure 7-6 Multiplex Electro-hydraulic Control System *(Courtesy TotalFinaElf)*

high-temperature wells. The principle of an all-electric control system is shown in Figure 7-7.

For the subsea Xmas tree, the gate valves and the insert-retrievable choke are fitted with electric actuators. The tree contains dual, all-electric subsea control modules (SCM), which supply power and signal to individual actuators. The SCMs are normally subsea-retrievable.

The main feature of this system is that the operation of the electric motors in valve actuators is performed by locally stored power from rechargeable Li-Ion batteries. The total power consumption of the system is quite low since only electronic supply power and the battery charging power is transferred from topside to subsea.

The benefits of using all-electric control systems are very clear. An all-electric control system is simpler compared to a conventional electrohydraulic control system. It is favorable to use when developing marginal fields at long distances from a processing facility because of the lower cost of umbilicals, and it also provides solutions to the problems associated with high-pressure and high-temperature wells because there is no need for hydraulic fluid. In addition, it provides a higher degree of flexibility when expanding an existing system and when introducing new equipment into the system. Finally, the removal of the hydraulic system erases environmental and economic problems related to the leakage of hydraulic control fluids and the complexity of working with hydraulics.

Table 7-4 Prominent Features, Advantages, and Disadvantages of Multiplex Electrohydraulic Control Systems

Advantages	Disadvantages	Prominent Features
• Give good response times over long distances • Smaller umbilical diameter • Allows control of many valves/wells via a single communication line • Redundancy is easily built in • Enhanced monitoring of operation and system diagnostics • Ideal for unmanned platform or complex reservoirs • Able to supply high volume of data feedback • No operational limitations	• High level of system complexity • Requires subsea electrical connectors • Increase in surface components • Increase in subsea components • Recharging of the hydraulic supply over such a long distance • Hydraulic fluid cleanliness • Materials compatibility • Limitations over long distance tie-backs	• Real-time system response • Virtually no distance limitations • Maximum reduction in umbilical size • Subsea status information available • High level of operational flexibility

7.3. TOPSIDE EQUIPMENT

The following subsea control system components should be located on the host facility (topside):

- Master control station, including the human/machine interface;
- Electrical power unit;
- Hydraulic power unit;
- Topside umbilical termination assembly.

7.3.1. Master Control Station (MCS)

The MCS is necessary for any electrohydraulic multiplexed subsea control system. It provides the interface between the operator and the subsea equipment. Because the system operates using electronic messages between the surface and subsea equipment (SCM, sensors, etc.), some form of communication subsystem is required in i t. The MCS interfaces with the topside

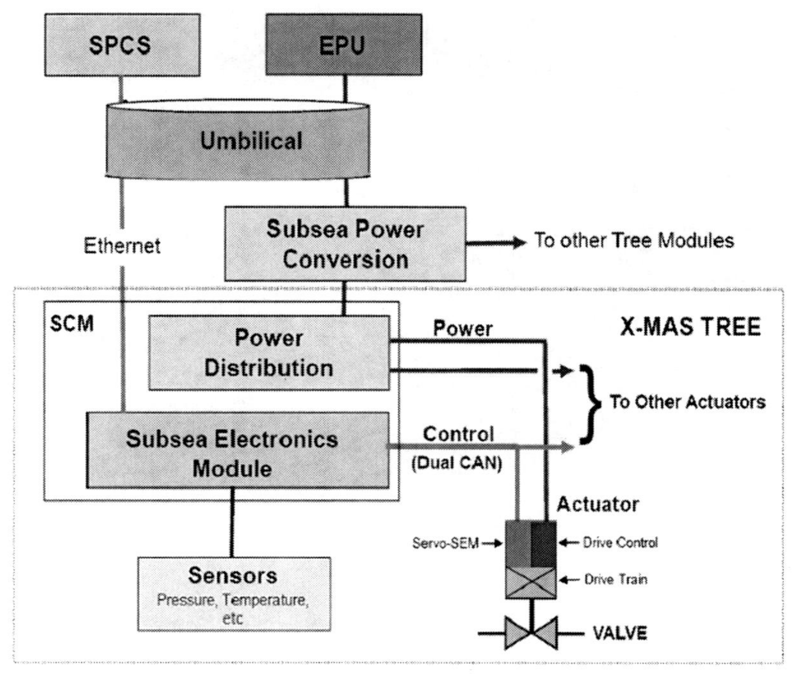

Figure 7-7 All Electric Control System

equipment via a supervisory control network. Functions include executing safety shutdown, subsea well and manifold control, and data acquisition. The operator interfaces with the MCS through a human/machine interface (HMI), which comprises a keyboard and a visual display. A separate HMI may be provided for downhole sensors and their associated electronic equipment. Figure 7-8 shows a single MCS; sometimes dual MCSs are used.

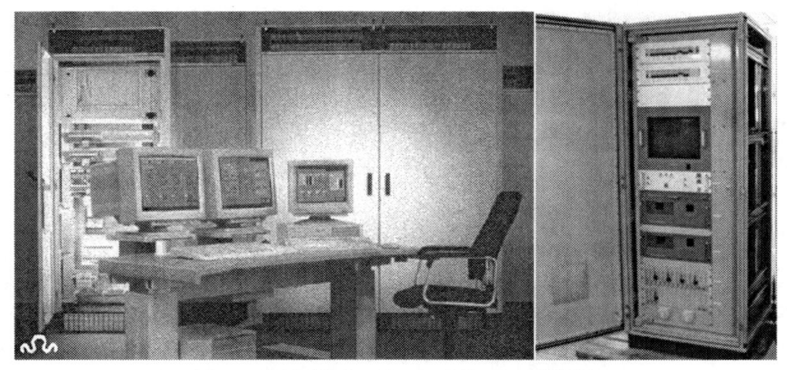

Figure 7-8 Mater Control Station (MCS)

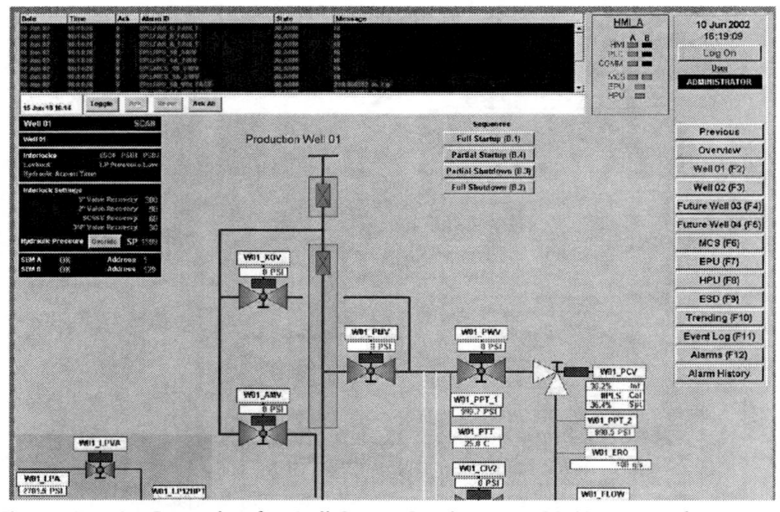

Figure 7-9 An Example of a Well Status Display on MCS *(Courtesy of Cameron)*

The functions of an MCS usually include the following operating modes:

- Tree valve control;
- Choke control;
- Subsea sensor monitoring/fast scan;
- Interlocks;
- Alarm announcements;
- Workover;
- Emergency shutdown;
- Well test management;
- Hydrate region warning;
- Trends/historical data reporting and data logging.

The MCS is typically made up of a number of units, including computers, displays, and controls. Figure 7-9 illustrates an example of well status being displayed on an MCS computer.

7.3.2. Electrical Power Unit (EPU)

Any electrohydraulic multiplexed subsea control system requires a topside unit to control and provide the necessary power to the subsea equipment. The EPU shown in Figure 7-10 provides conditioned electrical power

Figure 7-10 Electrical Power Unit (EPU)

to the topside and subsea system components. The EPU supplies dual, isolated, single-phase power for the subsea system through the composite service umbilical, together with power supply modules for the MCS and HPU.

The EPU supplies electrical power at the desired voltage and frequency to subsea users. Power transmission is performed via the electrical umbilical and the subsea electrical distribution system.

For further information on EPUs, see Chapter 8.

7.3.3. Hydraulic Power Unit (HPU)

The HPU shown in Figure 7-11 provides a stable and clean supply of hydraulic fluid to the remotely operated subsea valves. The fluid is supplied via the umbilical to the subsea hydraulic distribution system, and to the subsea control module (SCM) to operate subsea valve actuators.

The HPU supplies the necessary hydraulic fluid and comprises, among others:

- Hydraulic pump aggregate;
- Hydraulic tank;

Figure 7-11 Hydraulic Power Unit (HPU) *(Courtesy of Oceaneering)*

- Eventual pressure compensation system if the HPU is running submerged;
- Main filters and system pressure valves.

For more information about HPUs, see Chapter 8.

7.4. SUBSEA CONTROL MODULE MOUNTING BASE (SCMMB)

The SCMMB shown in Figure 7-12 is the interface between the SCM and the subsea Christmas tree/manifold valves and remote sensors. It is usually a welded structure, bolted and earth bonded to the frame of a subsea tree or manifold.

Figure 7-12 Typical SCMMB *(Courtesy of FMC)*

The vertical face on the side of the SCMMB is the interface with the incoming connections: electrical couplers for power and signals as well as the hydraulic couplers for low-pressure (LP) and high-pressure (HP) supplies.

The SCMMB will have course alignment and fine alignment adjustments for locating the SCM. For fine location via a guide pin, the SCM will not land if the orientation is incorrect. For self-aligning SCMs, a central helix will automatically rotate the SCM to the correct orientation before it is able to land.

7.5. SUBSEA CONTROL MODULE (SCM)

The SCM is an independently retrievable unit. SCMs are commonly used to provide well control functions during the production phase of subsea oil and gas production. Figure 7-13 shows a typical SCM.

Typical well control functions and monitoring provided by the SCM are as follows:

- Actuation of fail-safe return production tree actuators and downhole safety valves;
- Actuation of flow control choke valves, shutoff valves, etc.;
- Actuation of manifold diverter valves, shutoff valves, etc.;
- Actuation of chemical injection valves;

Figure 7-13 Typical SCM *(Courtesy of FMC)*

- Actuation and monitoring of surface-controlled reservoir analysis and monitoring systems, sliding sleeve, choke valves;
- Monitoring of downhole pressure, temperature, and flow rates;
- Monitoring of sand probes and production tree and manifold pressures, temperatures, and choke positions.

7.5.1. SCM Components

The typical SCM receives electrical power, communication signals, and hydraulic power supplies from surface control equipment. Figure 7-14 shows typical SCM components. The subsea control module and production tree are generally located in a remote location relative to the surface control equipment. Redundant supplies of communication signals and electrical and hydraulic power are transmitted through umbilical hoses and cables ranging from 1000 ft to several miles in length, linking surface equipment to subsea equipment. Electronics equipment located inside the SCM conditions the electrical power, processes communications signals, transmits status, and distributes power to solenoid piloting valves, pressure transducers, and temperature transducers.

The subsea electrical module (SEM) is fed with AC power directly from the EPU. It normally incorporates two AC-DC converters and output to DC busbars (typically 24 and 5 V). Actuation of the electrohydraulic valves

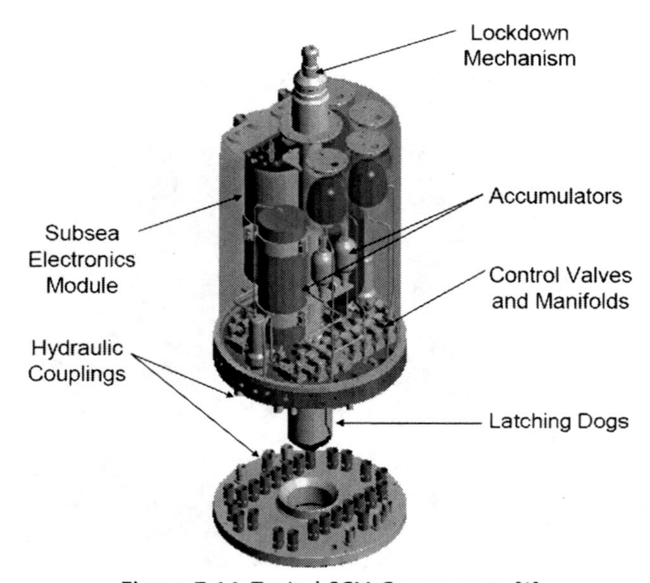

Figure 7-14 Typical SCM Components [1]

Figure 7-15 Subsea Electrical Module

is tapped into a 24- busbar, while the sensors are tapped into a 5-V busbar. An example of an SEM is shown in Figure 7-15.

Low-flow-rate solenoid piloting valves are typically used to pilot high-flow-rate control valves. These control valves transmit hydraulic power to end devices such as subsea production tree valve actuators, choke valves, and downhole safety valves. The status condition of control valves and their end devices is read by pressure transducers located on the output circuit of the control valves.

Auxiliary equipment inside the typical SCM consists of hydraulic accumulators for hydraulic power storage, hydraulic filters for the reduction of fluid particulates, electronics vessels, and a pressure/temperature compensation system. Previous devices have used an oil-filled chamber to compensate for hydrostatic pressure increases outside of the device during use to keep seawater away from cable assemblies.

The SCM is typically provided with a latching mechanism that extends through the body of the SCM and that has retractable and extendable dogs or cams thereon to engage a mating receptacle in a base plate [3].

7.5.2. SCM Control Mode Description

The SCM utilizes single LP and HP hydraulics, as shown in Figure 7-16, supplied from the umbilical system. Each supply has a "last chance" filter

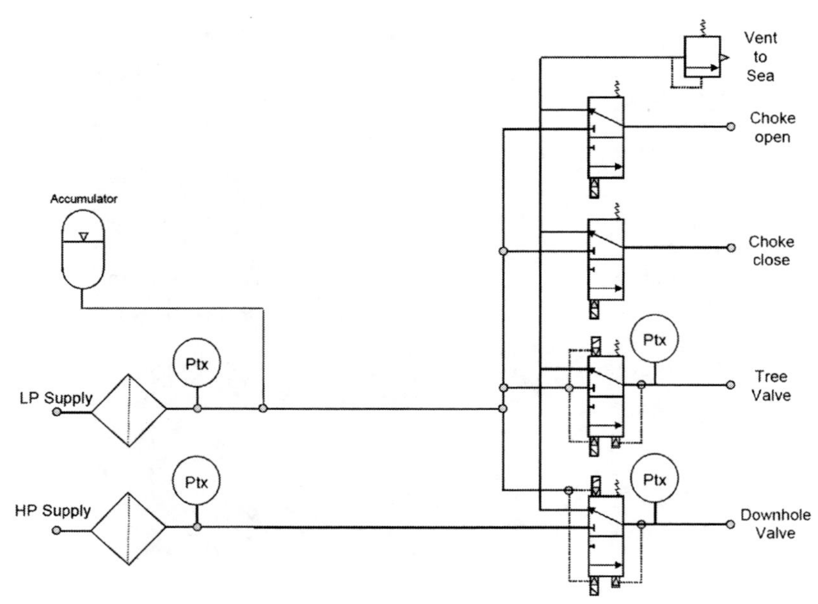

Figure 7-16 SCM Hydraulics [1]

before supply is manifolded to the main supply and pilot supply ports of the solenoid-operated pilot valves.

The LP supply has an accumulator that is mounted internally to the SCM. The HP supply may also have an accumulator. A pressure transducer measures the HP and LP supply pressures for display at the MCS. As shown in Figure 7-16, there are four pilot valves: three on the LP for tree valves and choke, and one on the HP for the downhole safety valve (DHSV), usually referred to the surface-controlled subsurface safety valve (SCSSV). The hydraulic return line has a nonreturn valve to vent used fluid to the sea and prevent seawater ingress.

7.5.2.1. Valve Actuation

Most of the pilot valves have two solenoids to operate them, one to open and one to close (see Figure 7-17). The solenoids are driven by the solenoid drivers in the SEM.

To open a tree valve, the appropriate solenoid is commanded from the MCS, and the microprocessor in the SEM activates the solenoid driver, which energizes the open solenoid for 2 sec. (typically; this may be adjusted at the MCS). This allows hydraulic fluid to flow into the function line to the tree valve actuator. The pressure in this line will rise very quickly to a value

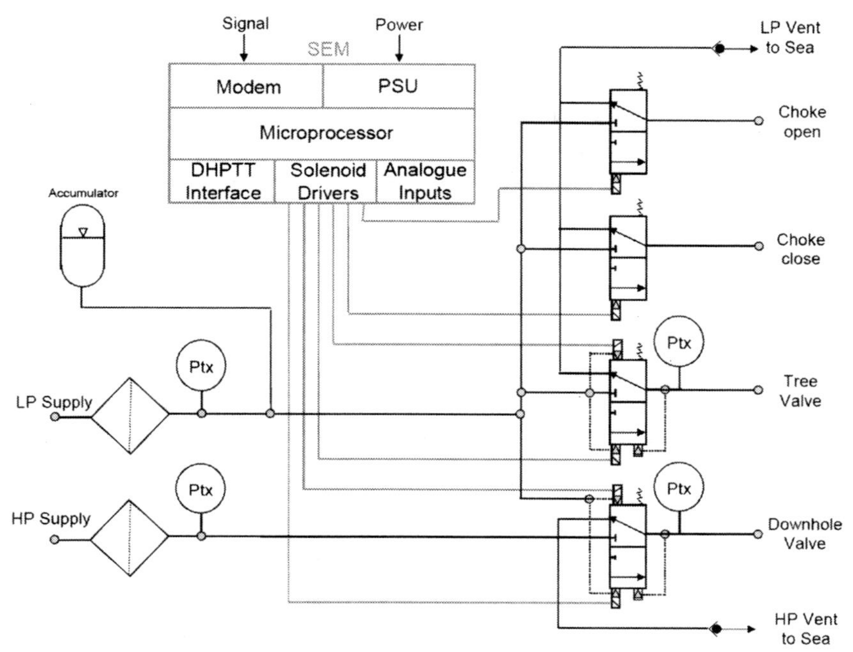

Figure 7-17 SCM Mode-Valve Actuation [1]

that allows the valve to latch open hydraulically. Thereafter, the valve will remain open as long as the hydraulic supply pressure remains above about 70 bar.

To close a tree valve, the close solenoid is energized in a similar manner to the open solenoid. This causes the spool in the pilot valve to move, venting the hydraulic fluid from the tree valve actuator. The used fluid is vented to sea via a nonreturn valve.

7.5.2.2. Choke Operation

The choke has two hydraulic actuators, one to open and one to close, that move the choke via a pawl and ratchet mechanism.

The choke is moved by applying a series of hydraulic pressure pulses to the appropriate actuator. On each pressurize/vent cycle, the choke will move by one step. To provide these pressure pulses, the SCM has two pilot valves, one for opening the choke, and one for closing the choke. These valves are not hydraulically latched and will only pass hydraulic fluid when the solenoid is energized. Figure 7-18 shows an SCM mode-choke operation chart.

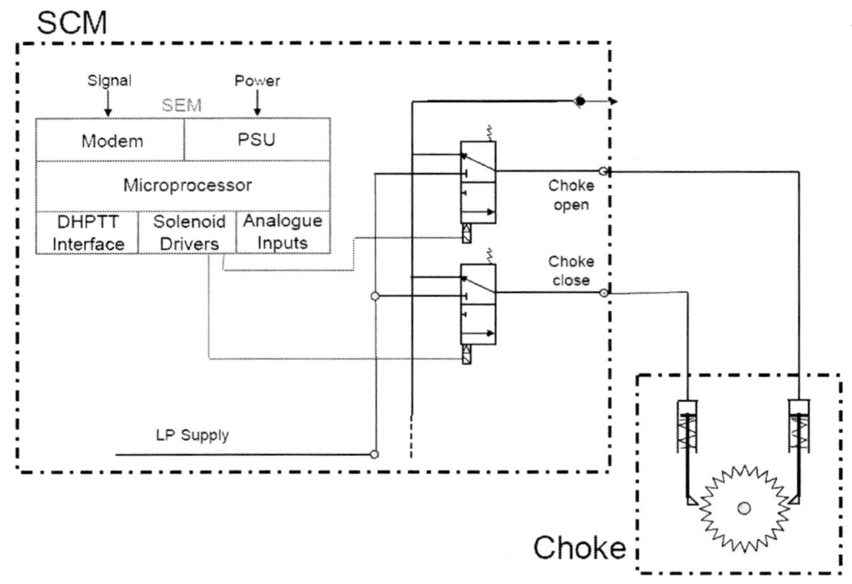

Figure 7-18 SCM Mode-Choke Operation [1]

To operate the choke, the MCS sends a series of commands to the SCM. For example, if the operator wishes to move the choke by 20 steps, the MCS will send 20 appropriately timed valve commands to the SCM to energize the appropriate solenoid.

7.6. SUBSEA TRANSDUCERS/SENSORS

Subsea sensors are at multiple locations on the trees, manifold, and flowlines. Figure 7-19 shows the subsea transducers/sensors' locations on a subsea tree. Tree-mounted pressure sensors and temperature sensors measure upstream and downstream of the chokes. Software and electronics in the SCMs compile sensor data and system status information with unique addresses and time-stamp validations to transmit to the topside MCS as requested.

Various transducers and sensors are used in a subsea production system:

- Pressure transducer;
- Temperature transducer;
- Combined pressure and temperature transducer;
- Choke position indicator;
- Downhole pressure temperature transducer;
- Sand detector;

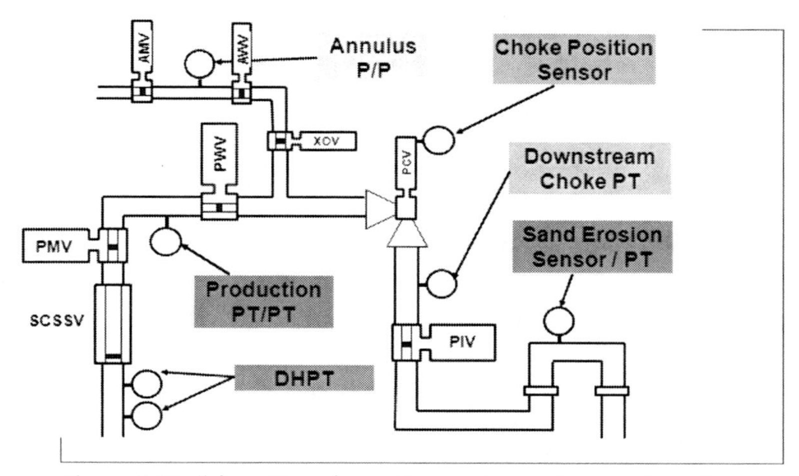

Figure 7-19 Subsea Transducers/Sensors Located on a Subsea Tree

- Erosion detection;
- Pig detector.

7.6.1. Pressure Transducer (PT)

Pressure transducers, as shown in Figure 7-20, are typical subsea sensors located on a subsea tree, manifold, etc.. A subsea PT typically operates using the force-balance technique, in which the current required by a coil to resist the movement of the detecting diaphragm gives a measure of the applied pressure. The diaphragm therefore does not actually deform; hence, devices can be built to withstand high pressures. Such devices can achieve an

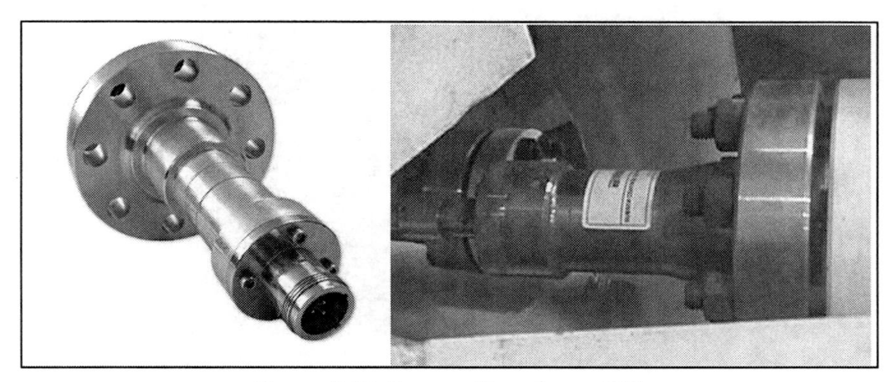

Figure 7-20 Pressure Transducers (PT)

accuracy of ±0.15%, although the output current circuitry may reduce this to an overall error band of around ±0.5% of full scale.

Because pressure transducers are flange mounted, they cannot be removed subsea if they fail. It is possible, however, to specify a transducer with dual gauges in one housing, providing a dual redundant sensor.

7.6.2. Temperature Transducer (TT)

Some TTs operate by measuring the output of a thermocouple, which is a simple device whose output is proportional to the difference in temperature between a hot and a cold junction. The hot junction is the one measuring the process, and the cold junction is at the head itself.

Temperature transducers, as shown in Figure 7-21, present somewhat of a dilemma with regard to their installation. In theory, the sensing element should be as close to the process fluid as possible. However, if the sensor were simply fitted into the production bore, there would be no other physical barrier between the process and the environment, other than the sensor itself. The usual approach therefore is to install a thermowell or use a pocket drilled into the tree block.

7.6.3. Pressure/Temperature Transducer (PTT)

A sensor design is available in which pressure and temperature elements are combined into one package. In this design, the temperature sensor is located in a probe, which is designed to be flush mounted into the process

Figure 7-21 Temperature Transducers (TT)

pipework. This also helps reduce errors due to hydrate formation. For sensors upstream of the production master valve, regulations may require that there be one or two block valves before the sensor, so, although the pressure measurement is unaffected, the temperature measurement may be subject to thermal inertia depending on its location.

7.6.4. Sand Detector

Detecting sand in the produced fluids can be an important part of a proactive strategy for:

- Managing short-term damage of vulnerable equipment (e.g., chokes);
- Managing long-term damage of pipe/flowline;
- Warning of reservoir collapse;
- Preventing separator level control problems.

The detector works either by monitoring the noise generated by sand impacts on a solid surface or by measuring the erosion damage of a target inserted into the flow. This leads to two types: acoustic and electrical detectors. Figure 7-22 shows acoustic sand detectors. Acoustic signals generated by the particles impinging on the walls of piping are monitored by the detector, which is usually installed immediately past a bend of the piping. However, the accuracy is affected by the noise of flowlines or pumps. The electrical detector is based on measuring the change of electrical resistance of thin sensing elements, which can be eroded by sand.

Figure 7-22 Acoustic Sand Detectors

The system reads the erosive effect directly with high accuracy. Software tools can be used to predict the worst case erosion on critical pipe elements such as chokes.

7.7. HIGH-INTEGRITY PRESSURE PROTECTION SYSTEM (HIPPS)

The HIPPS is a specific application of a safety instrumented system (SIS) designed in accordance with IEC 61508 [4]. The function of a HIPPS is to protect the downstream equipment against overpressure by closing the source. Usually this is done by timely closing of one or more dedicated safety shutoff valves to prevent further pressure rise of the piping downstream from those valves. A HIPPS is also commonly referred to as a "high-integrity pipeline protection system," because many HIPPS design/cost studies are associated with field layout pipeline/flowline designs [5]. Figure 7-23 shows how a HIPPS is applied in a subsea field development.

Application of HIPPS allows for the use of a lower pressure rated flowline compared to the wellhead and tree equipment. Because length, wall thickness, and pressure ratings are the major cost-driving factors for flowlines, the flowline can be an expensive item in subsea field development. Because a subsea HIPPS is installed upstream of the flowline, the design pressure of it can be lower than the well shutdown pressure. Thus the wall thickness can be reduced, which directly cuts its cost.

Figure 7-24 shows a subsea HIPPS philosophy chart. The HIPPS is designed as a system that detects rising pressure in a flowline and quickly shuts one or more isolation barrier valves before the pressure can rise too high. This requires a very reliable and highly available system as well as a fast-acting system. Therefore, the main components of a HIPPS are as follows:

- Pressure (and other) transmitters;
- Logic;
- Redundant barrier valves.

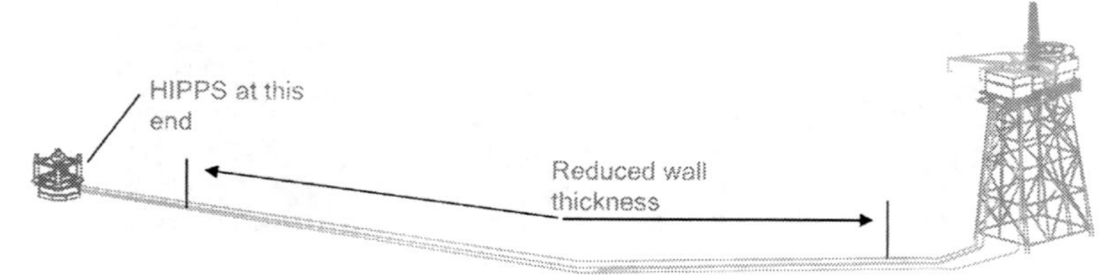

Figure 7-23 HIPPS Layout in Subsea Field *(Courtesy of Vetco Gray)*

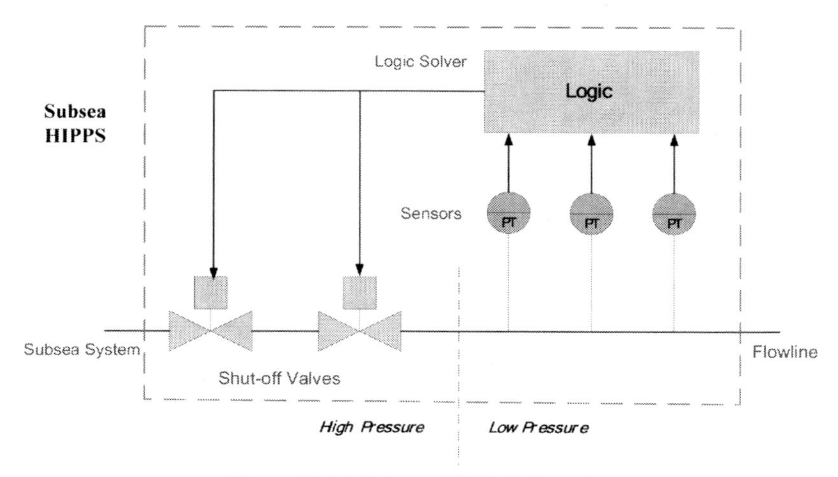

Figure 7-24 Subsea HIPPS Philosophy

Topside equipment will be needed to supply the necessary electrical and hydraulic power, which will be transmitted to the subsea equipment via an umbilical. Subsea equipment located adjacent to the barrier valves will provide the actual detection and actuation mechanisms. A second HIPPS barrier valve is also monitored on either side by pressure transmitters, which provides redundancy in case of failure or leakage of the first valve.

The topsides equipment comprises:

- Master control system;
- Electrical power unit;
- Uninterruptible power supply (UPS);
- Hydraulic power unit.

The subsea equipment comprises:

- Umbilical and termination/distribution unit;
- HIPPS subsea control module and mounting base;
- SAM and mounting base;
- Hydraulic jumpers;
- Process barrier valves and position indicators;
- Maintenance, venting, and test valves.

To ensure availability and prevent erroneous shutdowns, usually more than one transmitter is employed in a "two out of three" (2oo3) voting (triplicate) configuration, and different makes of sensor can be used to avoid common fault modes. Either the transmitters themselves and/or the detection logic will act in such a way that a failed transmitter is seen as a "high" pressure and therefore trips the system. Figure 7-25 shows a 2oo3 voting loop.

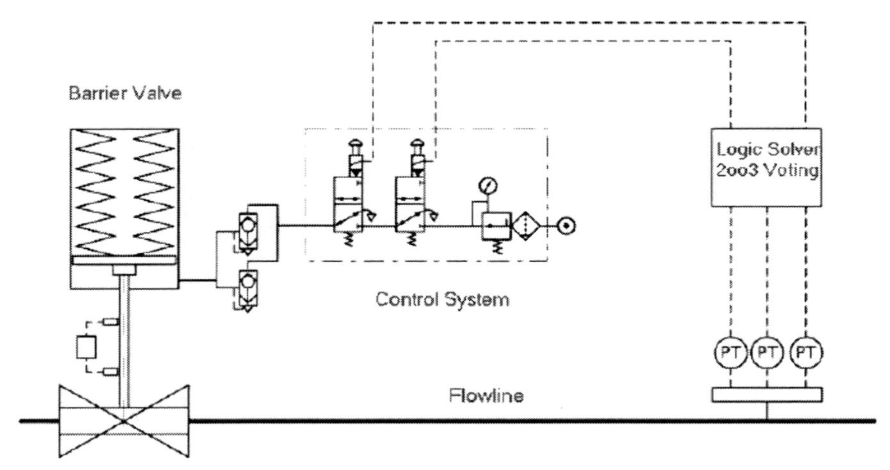

Figure 7-25 2oo3 Voting Loop

The control hardware monitors the transmitters and closes the barrier valves when high pressure is detected. The detection and activation system is inevitably an electronic system, and it should be independent of any other platform or subsea system. This usually results in use of a dedicated subsea controller that is located in an SCM adjacent to the barrier valves themselves. The HIPPS control module (HCM) monitors the transmitters and closes the valves in the event excessive pressure is detected. This operates via a certified fixed-logic electronic system and is not dependent on microprocessors or links to the surface. Thus, each trip is preset and cannot be changed without physical access to the controller. This requires careful attention to hydraulic design to ensure pulses do not inadvertently trip the system.

Figures 7-26 and 7-27 show two typical subsea HIPPS layouts: single-well tie-back and manifold/template tie-back. Note that the HIPPS isolation barrier valves are controlled by the HCM, which is located in the SCM.

7.8. SUBSEA PRODUCTION CONTROL SYSTEM (SPCS)

As illustrated in Figures 7-28 and 7-29, the SPCS generally can be divided into three main sections:

- The equipment installed on the surface;
- The umbilical and its terminations;
- The equipment installed on the subsea Christmas tree.

Because the components of a subsea production system are inaccessible for operator intervention and must work together in a coordinated manner, it is necessary to design an integrated control system that works remotely.

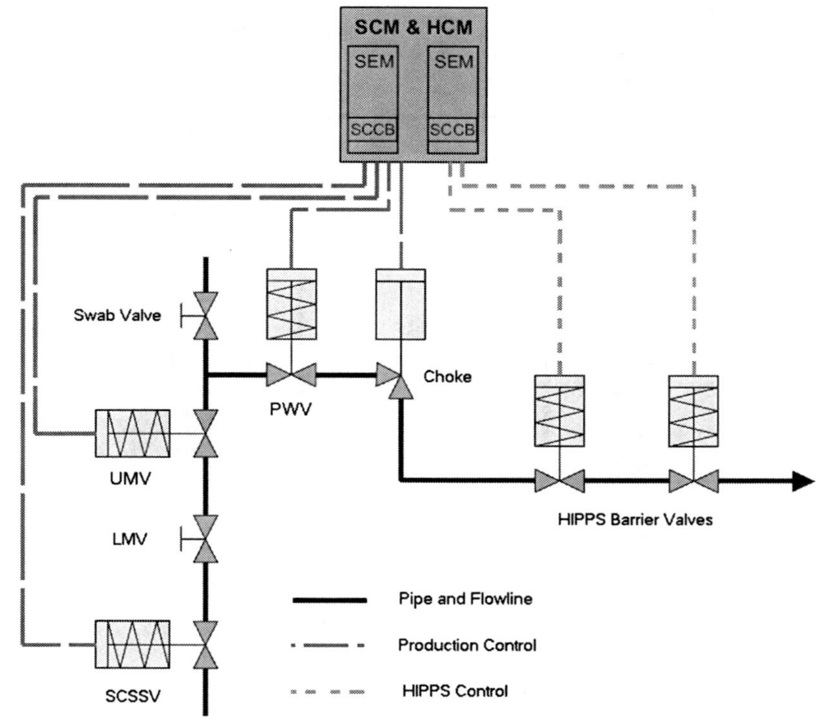

Figure 7-26 Single-Well Tie-Back HIPPS Control System *(Courtesy of Vetco Gray)*

That is the subsea production control system. The functions of an SPCS include the following:

- Opening and closing subsea Christmas tree production, annulus, and crossover valves;
- Opening and closing SCSSVs;
- Opening and closing subsea production manifold flowline valves and pigging valves;
- Opening and closing chemical injection valves;
- Adjusting subsea production chokes;
- Monitoring temperature, pressure, flow rate, and some other data from tree-mounted, manifold-mounted, or downhole instrumentation.

Generally, there are four types of subsea production control systems:

- Direct hydraulic (DH) control system;
- Piloted hydraulic (PH) control system;
- Electrohydraulic (EH) piloted control system;
- Electrohydraulic multiplexed (EH-MUX) control system.

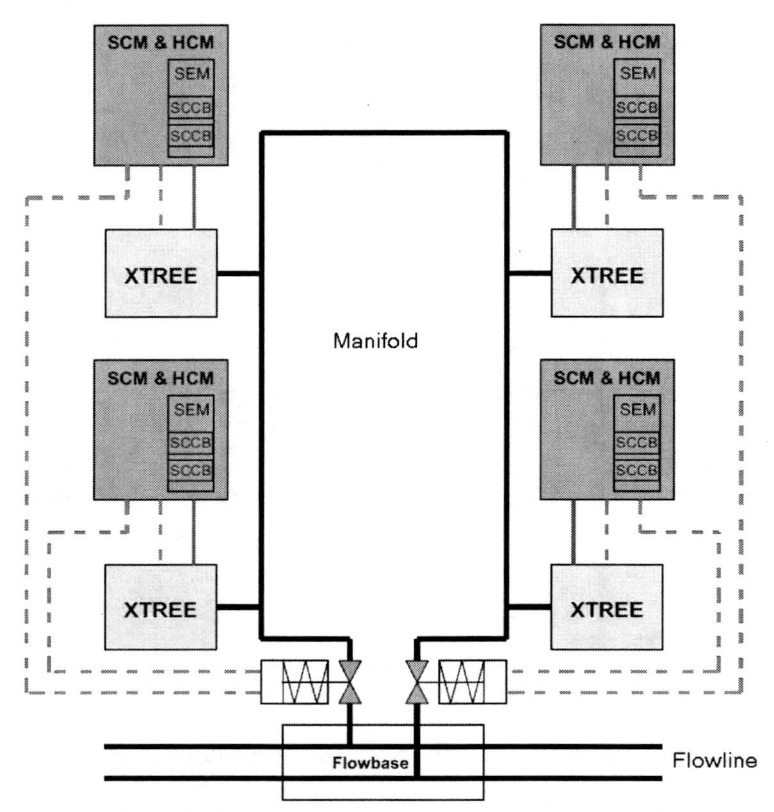

Figure 7-27 Manifold Tie-Back HIPPS Control System *(Courtesy of Vetco Gray)*

A DH control system is the simplest and cheapest production control system. It consists of a topside HPU with one dedicated control line for each remotely actuate valve on the subsea Christmas tree. This type of control system is recommended for a subsea production system with one or two well tie-backs that is located within 3 miles of the host platform.

A PH control system is similar to a DH control system except that the valves that require fast closing times will have a pilot valve and will vent to the sea on closing.

An EH piloted control system is used for medium-offset subsea tie-backs. It consists of a topside electrical and hydraulic control system tied to one or more service umbilicals to the field. Each subsea Christmas tree and manifold has an SCM that takes LP and HP supplies and directs them to local valves when commanded to do so by the topside equipment.

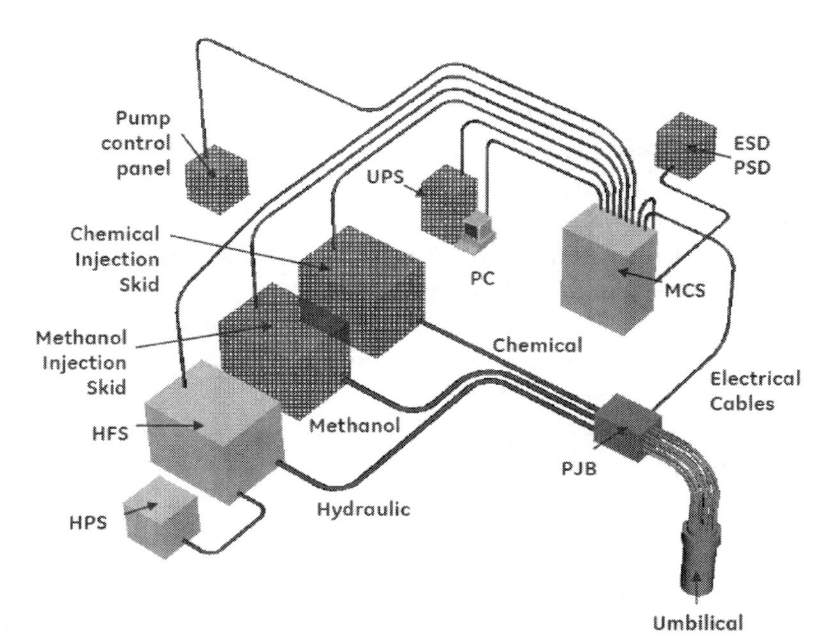

Figure 7-28 Subsea Production Control System—Topsides *(Courtesy of Vetco Gray)*

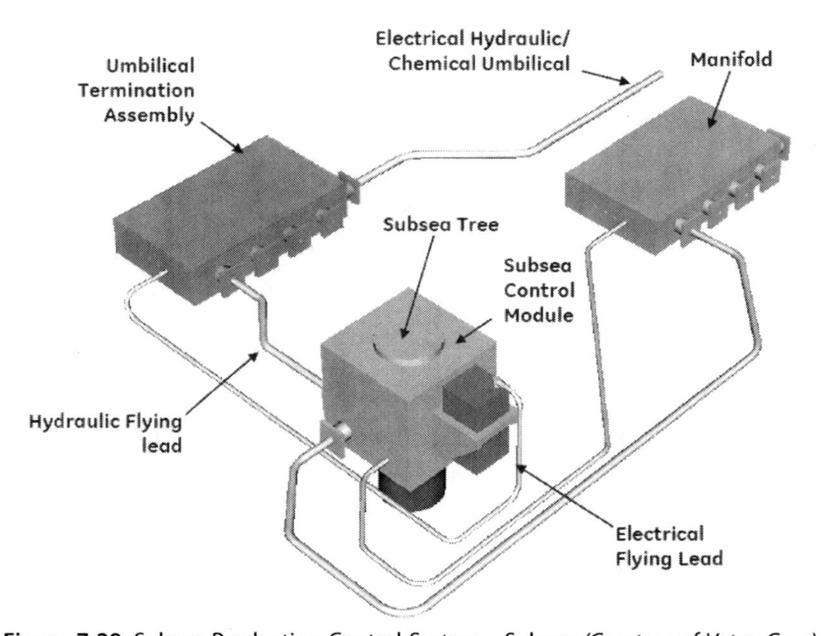

Figure 7-29 Subsea Production Control System—Subsea *(Courtesy of Vetco Gray)*

An EH-MUX control system is used for medium- to long-offset subsea tie-backs. It consists of a topside electrical and hydraulic control system tied through a service umbilical to one or more subsea Christmas trees. Each tree and manifold has an SCM, which receives the multiplexed electrical control signals and LP and HP hydraulic supplies, which are directed to tree- or manifold-mounted valves or other equipment. This system is common in large multiwell deepwater developments. The main advantage is the use of a multiplexed electrical control signal over a single pair of conductors, resulting in a smaller control umbilical, which accommodates future expansion easily and reduces umbilical costs significantly.

7.9. INSTALLATION AND WORKOVER CONTROL SYSTEM (IWOCS)

As illustrated in Figure 7-30, the main functions of the IWOCS are as follows:

- Control and monitor subsea equipment during installation, retrieval, and workover.

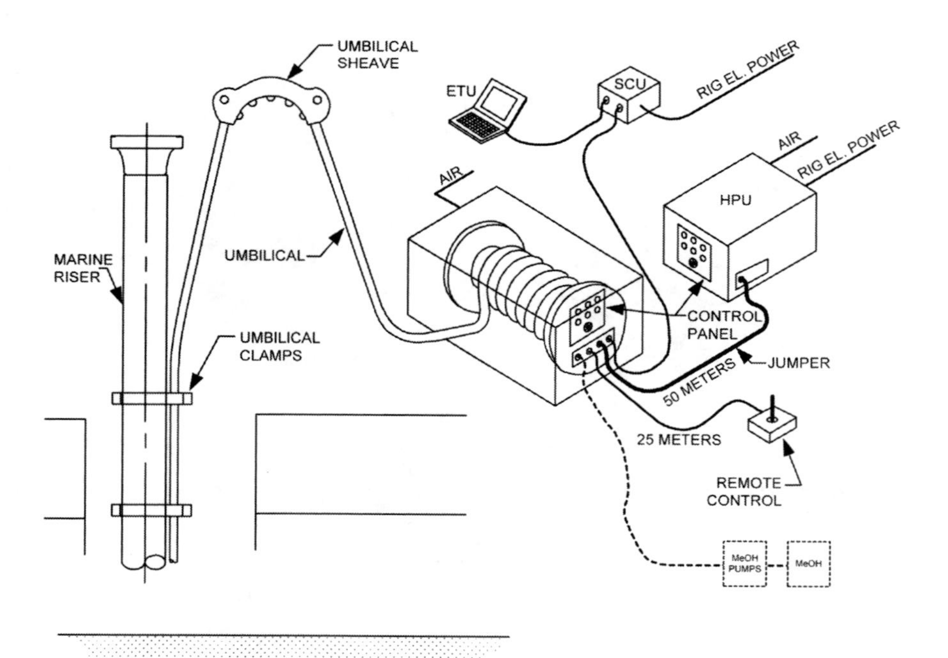

Figure 7-30 IWOCS

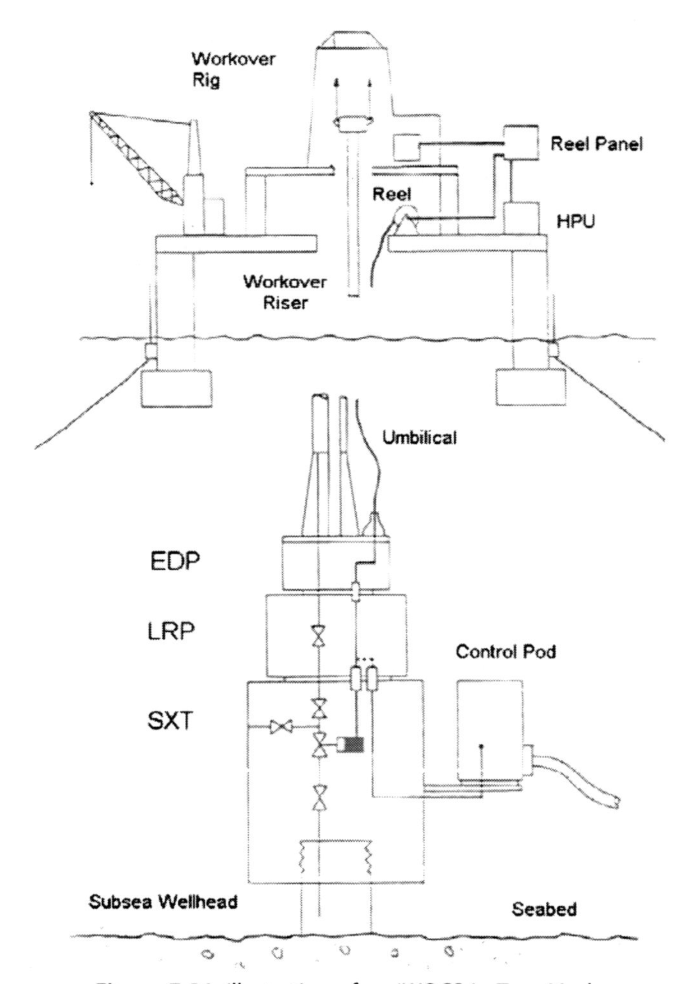

Figure 7-31 Illustration of an IWOCS in Tree Mode

- Control and monitor downhole equipment during completion, flow testing, and workover.
- Log data during completion, flow testing, and workover operations.
- Test manifolds and/or PLEMs during installation, if applicable.

Because the IWOCS is used for intervention of subsea wells, certain safety features are required:

- *Emergency shutdown (ESD)*: It must be possible to initiate emergency shutdown. The shutdown shall be automatically sequenced such that the system is shutdown safely and release of hydrocarbons is prevented. ESD panels are provided on different panels located on board the rig to enable ESD even if there is a fire or release of hydrocarbons on the drill floor.

- *Emergency Quick Disconnect (EQD)*: It must be possible to disconnect the lower riser package (LRP) within a certain time delay from initiation. An ESD must automatically be executed prior to disconnection of the LRP.

The IWOCS usually consists of the following components:

- HPU;
- Main control panel (on HPU);
- Remote control panel (on the drill floor);
- Emergency shutdown panels (at main escape routes);
- Umbilicals on reels;

Primary tools controlled by IWOCS are the tubing hanger running tool and the tree running tool/tree cap running tool.

To control these tools, a riser system is utilized that may consists of various riser joints, LRP, etc. (The riser system is discussed in Part 4). The whole system is illustrated in Figure 7-31.

As with the SPCS, the IWOCS also has four types: direct, piloted, electrohydraulic, and electrohydraulic multiplexed; see Section 7.2 for more information.

REFERENCES

[1] Society for Underwater Technology, Subsea Production Control, SUT, Subsea Awareness Course, 2008.
[2] B. Laurent, P.S. Jean, L. Robert, First Application of the All-Electric Subsea Production System Implementation of a New Technology, OTC 18819, Offshore Technology Conference, Houston, 2006.
[3] C.P. William, Subsea Control Module, U.S. Patent 6,161,618, 2000.
[4] International Electro-technical Commission, Functional safety of electrical/electronic/programmable electronic safety-related systems, IEC 61508, 2010.
[5] J. Davalath, H.B. Skeels, S. Corneliussen, Current State of the Art in the Design of Subsea HIPPS Systems, OTC 14183, Offshore Technology Conference, Houston, 2002.
[6] International Organization of Standards, Petroleum and natural gas industries - Design and operation of subsea production Systems - Part 6: Subsea production control systems, ISO 13628, 2000.

CHAPTER 8

Subsea Power Supply

Contents

8.1. INTRODUCTION

The power supply for a subsea production system is designed according to the subsea control system (Chapter 7). Different control system types (direct hydraulic, electrohydraulic, all-electric, etc.) require different power system designs. However, basically two types of power systems are used: an electrical power system or a hydraulic power system.

The power system supplies either electrical or hydraulic power to the subsea equipment: valves and actuators on subsea trees/manifolds, transducers and sensors, SCM, SEM, pumps, motors, etc.. The power sources can come from either an onshore factory (in a subsea-to-beach field layout), as shown in Figure 8-1, or from the site (platform or subsea generators).

Figure 8-2 illustrates the subsea power distribution when the power sources are coming from a surface vessel.

Subsea Engineering Handbook
ISBN 978-0-12-397804-2, doi:10.1016/B978-0-12-397804-2.00008-4

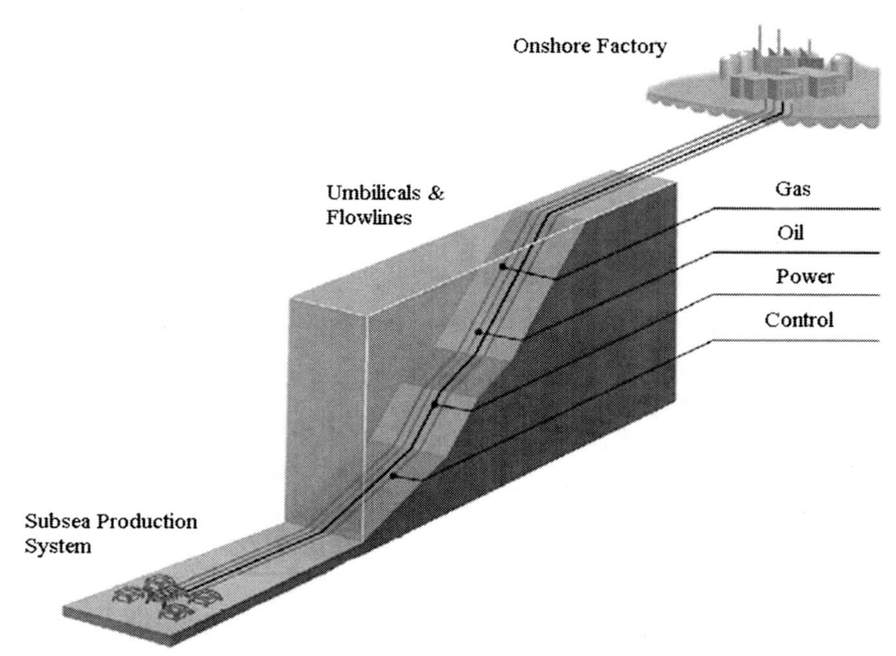

Figure 8-1 Power Supply from Onshore to Subsea *(Courtesy of Vetco Gray)*

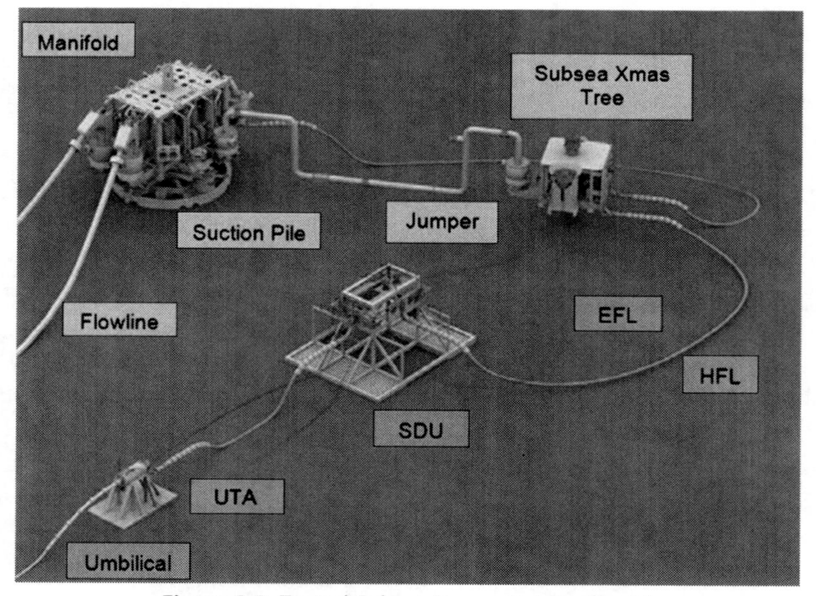

Figure 8-2 Typical Subsea Power Distribution [1]

8.2. ELECTRICAL POWER SYSTEM

The electrical power system in a typical subsea production system that provides power generation, power distribution, power transmission, and electricity from electric motors. The power is either generated on site (from a platform) or onshore (in a subsea-to-beach filed layout). To ensure continuous production from a subsea field, it is of utmost importance that the subsea system's associated electrical power system be designed adequately. Figure 8-3 shows the design process for an electrical power system.

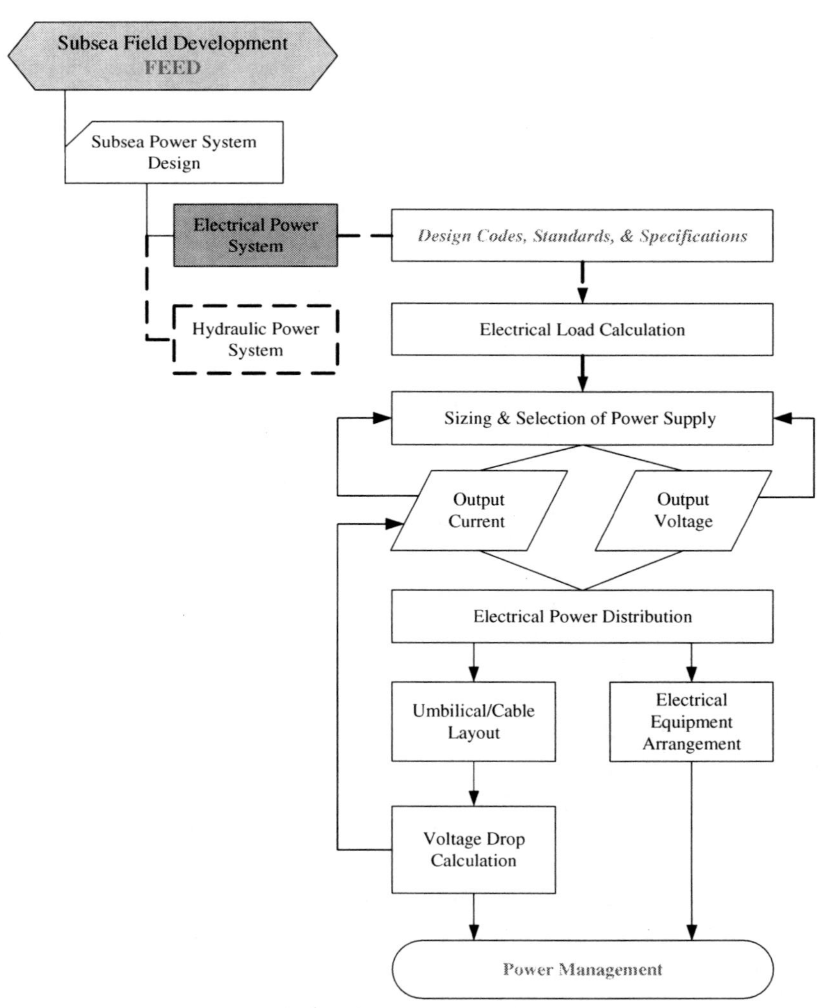

Figure 8-3 Electrical Power System Design Process

8.2.1. Design Codes, Standards, and Specifications

Various organizations have developed many electrical codes and standards that are accepted by industries throughout the world. These codes and standards specify the rules and guidelines for the design and installation of electrical systems. Tables 8-1 to 8-4 list some of the major international codes and standards used for subsea field development.

8.2.2. Electrical Load Calculation

Electrical load calculation is one of the earliest tasks during electrical power system design. Engineers should estimate the required electrical load of all of the subsea elements that will consume the electricity so that they can select an adequate power supply.

Each local load may be classified into several different categories, for example, vital, essential, and nonessential. Individual oil companies often

Table 8-1 American Petroleum Institute

API RP 14F	Recommended Practice for Design and Installation of Electrical Systems for Fixed and Floating Offshore Petroleum Facilities for Unclassified and Class I, Division 1 and Division 2 Locations
API RP 17A	Recommended Practice for Design and Operation of Subsea Production Systems
API RP 17H Draft	ROV Interfaces with Subsea Equipment
API RP 500	Recommended Practice for Classification of Locations for Electrical Installations at Petroleum Facilities Classified as Class I, Division 1 and Division 2
API SPEC 17D	Specification for Subsea Wellhead and Christmas Tree Equipment
API SPEC 17E	Specification for Subsea Production Control Umbilicals

Table 8-2 International Electrotechnical Commission

IEC 50 (426)	International Electrotechnical Vocabulary (IEV)-Chapter 426-Electrical Apparatus for Explosive Atmosphere

Table 8-3 Institute of Electrical and Electronics Engineers

Std. 100	Standard Dictionary of Electrical and Electronics Terms
Std. 141	Electrical Power Distribution or Industry Plants
Std. 399	Recommended Practice for Power Systems Analysis

Table 8-4 International Standards Organization

ISO 13628-5	Petroleum and Natural Gas Industries—Design and Operation of Subsea Production Systems—Part 5: Subsea Control Umbilicals
ISO 13628-6	Petroleum and Natural Gas Industries—Design and Operation of Subsea Production Systems—Part 6: Subsea Production Control Systems

Table 8-5 Typical Electrical Load Categories [2]

Load Categories	**Classification Questions**
Vital	Will the loss of power jeopardize safety of personnel or cause serious damage within the platform/vessel? (YES)
Essential	Will the loss of power cause a degradation or loss of the oil/gas production? (YES)
Nonessential	Does the loss have no effect on safety or production? (YES)

use their own terminology and terms such as "emergency" and "normal" are frequently encountered. In general terms, there are three ways of considering a load or group of loads and these may be cast in the form of questions as shown in Table 8-5.

All of the vital, essential, and nonessential loads can typically be divided into three duty categories [2]:

- Continuous duty;
- Intermittent duty;
- Standby duty (those that are not out of service).

Hence, each particular switchboard (e.g., from the EPU) will usually cover all three of these categories. We will call these C for continuous duty, I for intermittent duty, and S for standby duty. Let the total amount of each at this particular switchboard be C_{sum}, I_{sum}, and S_{sum}. Each of these totals will consist of the active power and the corresponding reactive power.

To estimate the total consumption for this particular switchboard, it is necessary to assign a diversity factor to each total amount. Let these factors be D. The total load can be considered in two forms, the total plant running load (TPRL) and the total plant peak load (TPPL), thus:

$$\text{TPRL} = \sum^{n}(D_c \times C_{sum} + D_i \times I_{sum}) \text{ kW} \tag{8-1}$$

$$\text{TPPL} = \sum^{n}(D_c \times C_{sum} + D_i \times I_{sum} + D_s \times S_{sum})\ \text{kW} \tag{8-2}$$

where

n: number of switchboards;

D_c: diversity factor for sum of continuous duty (C_{sum});

D_i: diversity factor for sum of intermittent duty (I_{sum});

D_s: diversity factor for sum of standby duty (S_{sum}).

Oil companies that use this approach have different values for their diversity factors, largely based on experience gained over many years of designing plants. Besides, different types of host facilities may warrant different diversity factors [2]. Typically,

$D_c = 1.0–1.1$;

$D_s = 0.3–0.5$;

$D_i = 0.0–0.2$.

The continuous loads are associated with power consumption that remains constant during the lifetime of the system regardless of the operation taking place at any one time. Such consumers would include the subsea production communication unit (SPCU, located on the platform) and the monitoring sensors.

Intermittent loads are considered the loads that depend on the operational state of the system. A typical example would be a load due to valve actuation or HPU system activation. For the duration of each operation, the power requirement for the system increases to accommodate the operation. For the definition of the momentary loads, apart from the corresponding power requirement, it is essential to identify the duration and frequency of operations as well as a statistical description of operating occurrences in a specified time period.

Note that at no point during its lifetime should the subsea power system run idle (without load), except for the case of a temporary production shutdown. Tables 8-6 and 8-7 present typical values for continuous and intermittent loads during the operation of electrohydraulic and all-electric production systems, respectively. The data are presented in terms of electrical loads. Note that the use of a choke valve can be either continuous or intermittent, depending on field requirements.

8.2.3. Power Supply Selection

After the load has been carefully estimated, the ratings for the power supply sources must be selected. For the electrical system applied in offshore oil/gas

Table 8-6 Load Schedule for an Electrohydraulic Control System [3]

Operation	Type	Power Requirement	Frequency (per day)	Duration
HPU	Intermittent	11 kW/pump	2	2 min
Single valve actuation	Intermittent	10 W	1–3	2 sec
Choke valve actuation	Intermittent or continuous	10 W (int)	N/A	2 sec (int)
SEM	Continuous	Max of 80 W	—	—
Sensors	Continuous	Max of 50 W	—	—

Table 8-7 Load Schedule for an All-Electric Control System [3]

Operation	Type	Power Requirement	Frequency (/day)	Duration
Single valve actuation	Intermittent	3–5 kW	1~3	45–60 sec.
Single valve normal operation	Continuous	20–50 W	—	—
Choke valve actuation	Intermittent or continuous	1–2 kW (int) 60 W (cont)	N/A	2 sec (int)
SEM	Continuous	Max of 80 W	—	—
Sensors	Continuous	Max of 50 W	—	—

fields, the power transmission can be from onshore or offshore. Offshore power transmission can occur on the surface or subsea.

8.2.3.1. Power Supply from Topside UPS

Typically, the electrical power supply for a subsea production system is from the UPS, which has its own rechargeable batteries.

Figure 8-4 shows that for an electrohydraulic control system type, the UPS supplies electrical power to the MCS, EPU, and HPU, which then combines the power and other data to the TUTU.

Figure 8-5 shows a picture of a UPS and battery rack. The UPS protects the system from electrical power surges and blackouts. Electric power should be supplied from the host platform main supply. The UPS typically operates by rectifying and smoothing the incoming supply, converting it to DC, which can then be used to charge associated batteries. The output from the batteries is then converted back to AC and is ready for use to power the

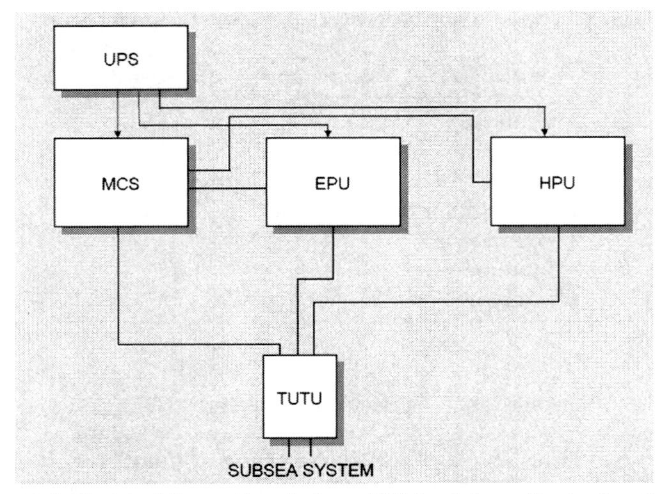

Figure 8-4 Electrical Power Supply from UPS for Subsea Production System

Figure 8-5 UPS and Battery Rack

subsea system. In the case of failure of the main incoming supply, the output from the batteries is quickly switched to power the DC-to-AC converter, thus ensuring a constant supply.

UPS systems are well known in industry, offices, and today even at home in situations where a power-consuming device must not lose its power supply. UPSs are available in small versions which are able to provide power from about 100 W up to several hundred kilowatts for from a few minutes up to many hours. During this time span the critical equipment supplied with power has either to be transferred into a power-off tolerable state or external power has to be resupplied; that is, grid power has to return or alternative power has to be provided. Usually the UPS is purchased from a specialist manufacture of such devices and is not built by a subsea control system supplier.

8.2.3.2. Power Supply from a Subsea UPS

A UPS is always located as close as possible to the power-consuming device to avoid as many fault sources as possible and is usually under control of the responsible operator of the power-consuming device.

By installing a subsea UPS system, costs may be reduced because fewer cables are required compared to having the UPS located topside because the UPS can be fed from the subsea main power supply. The short-circuit level of a UPS is low and the challenge of having enough short-circuit power available in a subsea installation to achieve the correct relay protection and discrimination philosophy can be solved by having the UPS subsea close to the power consumers.

In general, a subsea UPS can be used in all applications where distribution of low voltage (typically 400 V) is required subsea. The following are typical consumers of low-voltage subsea power supplied by a UPS:

- Several control systems located in a geographically small area;
- Electric actuators for valves;
- Magnetic bearings;
- Switchgear monitoring and control;
- Measuring devices for current and voltage in switchgear, transformers, motors, and other electrical installations.

A conventional UPS comprises an energy storage means and two power converters. A control and monitoring system is also a part of a UPS. Because power conversion involves losses resulting in heat, UPS systems may need cooling systems to transfer the heat to a heat sink [4].

The UPS should be designed to operate safely in the sited environment. The UPS is designed to enable the system to ride through short (seconds to minutes) power losses and to permit sufficient time for a graceful shutdown if necessary.

8.2.3.3. Power Supply from Subsea Generators

Electrical power can also come from subsea generators. Several types of subsea generators have been used in subsea field developments.

Autonomous systems consist of an electrical power source, which is typically seawater batteries or thermoelectric couplers. The power source utilizes the difference in temperature between the well stream and ambient seawater. The seawater battery solution requires a DC-to-AC converter to transform the voltage from, for example, 1 to 24 V (e.g., an SEM requires a 24-V electrical power supply). A seawater battery should have the capacity to operate the system for 5 years or more. The thermocoupler solution requires an accumulator to be able to operate the system when the well is not producing.

8.2.4. Electrical Power Unit (EPU)

The EPU shown in Figure 8-6 supplies dual, isolated, single-phase power for the subsea system through the composite service umbilical, together with power supply modules for the MCS and HPU.

Figure 8-6 Electrical Power Unit (EPU)

The EPU supplies electrical power at the desired voltage and frequency to subsea users. Power transmission is performed via the electrical umbilical and the subsea electrical distribution system.

The EPU should be designed to operate safely in the sited environment and allow for individual pair connection/disconnection and easy access to individual power systems for maintenance and repair. The EPU should contain redundant communication modems and filters to allow user definition of system monitoring, operation, and reconfiguration unless those modems reside in the MCS. The EPU prevents the potential damage (to subsea control modules and MCS) caused by voltage spikes and fluctuations and receives input voltage from a UPS.

The EPU usually has two outputs: a DC busbar and an AC line. The energy storage units are tapped to the DC busbar, whereas the AC output is connected to the SEM.

The typical features of the EPU are as follows [5]:

- Fully enclosed, proprietary powder-coated steel enclosure, incorporated into the MCS suite, with front and rear access;
- Standard design suitable for safe area, that is, a nonhazardous gases area and air-conditioned environment;
- Dedicated dual-channel power supplies, including fault detection to the subsea electronics module;
- Modems and signal isolation to effect the "communications on power" transmission system;
- Control and monitoring to the master control station;
- Electrical power backup input terminal in the event of a power supply outage to both the MCS and EPU.

8.2.5. Electrical Power Distribution

The subsea electrical distribution system distributes electrical power and signals from the umbilical termination head to each well. As introduced in Chapter 3, electrical power (as well as hydraulic pressure, chemical supply, and communications) is provided to a subsea system through an electro-hydraulic umbilical.

The SUTA is the main distribution point for the electrical supplies (also hydraulic and chemical) to various components of a subsea production system. The SUTA is permanently attached to the umbilical. Hydraulic and chemical tubes from the umbilical can have dedicated destinations or may be shared between multiple subsea trees, manifolds, or flowline sleds.

Electrical cables from the umbilical can also have dedicated destinations to electrical components of a subsea production system or may be shared by multiple SCMs or other devices. The electrical connection is made through electrical connectors on electrical flying leads (EFLs). The number of electrical connectors in series should be kept to a minimum. Redundant routing should, if possible, follow different paths. To minimize electrical

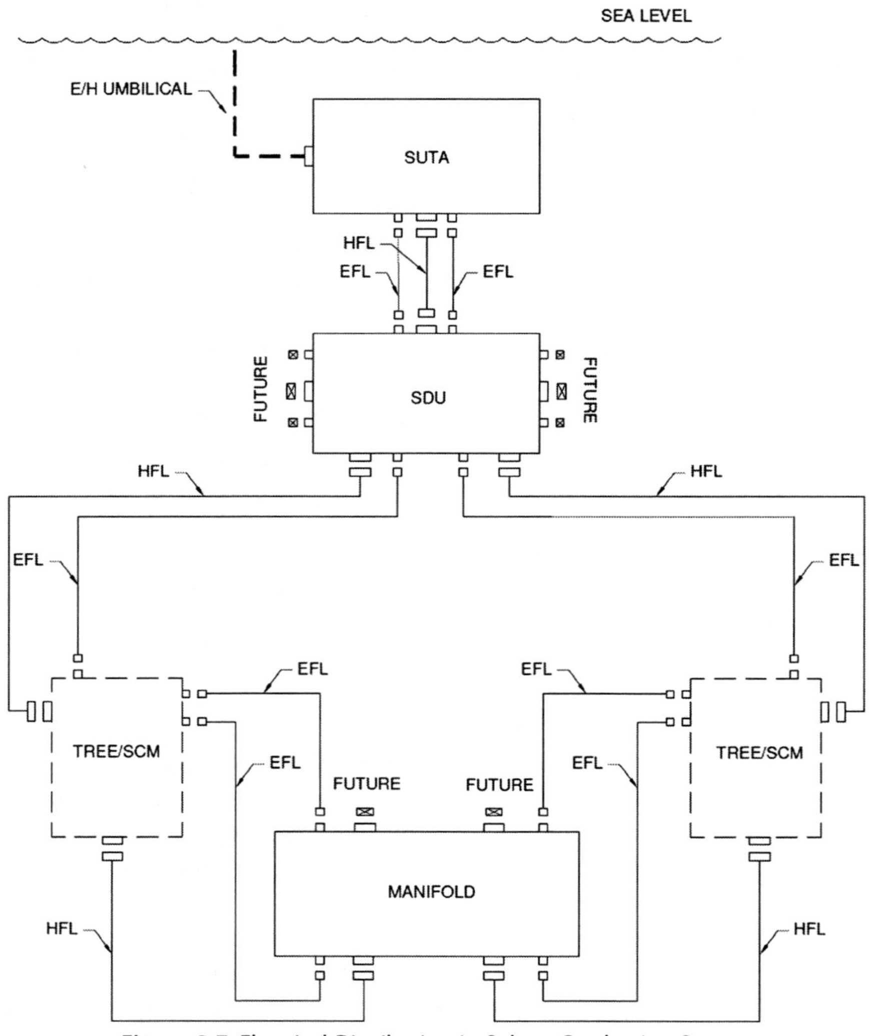

Figure 8-7 Electrical Distribution in Subsea Production System

stresses on conductive connectors, voltage levels should be kept as low as practical. Figure 8-7 shows the electrical as well as hydraulic power distribution in subsea production system.

Connection of electrical distribution cabling and electrical jumpers should be made by ROV or diver using simple tools, with minimum implications on rig/vessel time. Manifold electrical distribution cabling and jumper cables from the umbilical termination to the SCM should be repairable or reconfigurable by the ROV or diver.

The subsea electrical power distribution system differs from a topside system by being a point-to-point system with limited routing alternatives. The number of components shall be kept to a minimum, without losing required flexibility. Detailed electrical calculations and simulations are mandatory to ensure operation/transmission of the high-voltage distribution network under all load conditions (full load, no load, rapid change in load, short circuits).

8.3. HYDRAULIC POWER SYSTEM

The hydraulic power system for a subsea production system provides a stable and clean supply of hydraulic fluid to the remotely operated subsea valves. The fluid is supplied via the umbilical to the subsea hydraulic distribution system, and to the SCM to operate subsea valve actuators. Figure 8-8 illustrates a typical hydraulic power system.

Hydraulic systems for control of subsea production systems can be categorized in two groups:

- Open circuits where return fluid from the control module is exhausted to the sea;
- Closed circuits where return fluid is routed back to the HPU through a return line.

Open circuits utilize simple umbilicals, but do need equipment to prevent a vacuum in the return side of the system during operation. Without this equipment, a vacuum will occur due to a check valve in the exhaust line, mounted to prevent seawater ingress in the system. To avoid creating a vacuum, a bladder is included in the return line to pressure compensate the return line to the outside water.

The hydraulic system comprises two different supply circuits with different pressure levels. The LP supply will typically have a 21.0-MPa differential pressure. The HP supply will typically be in the range of 34.5 to

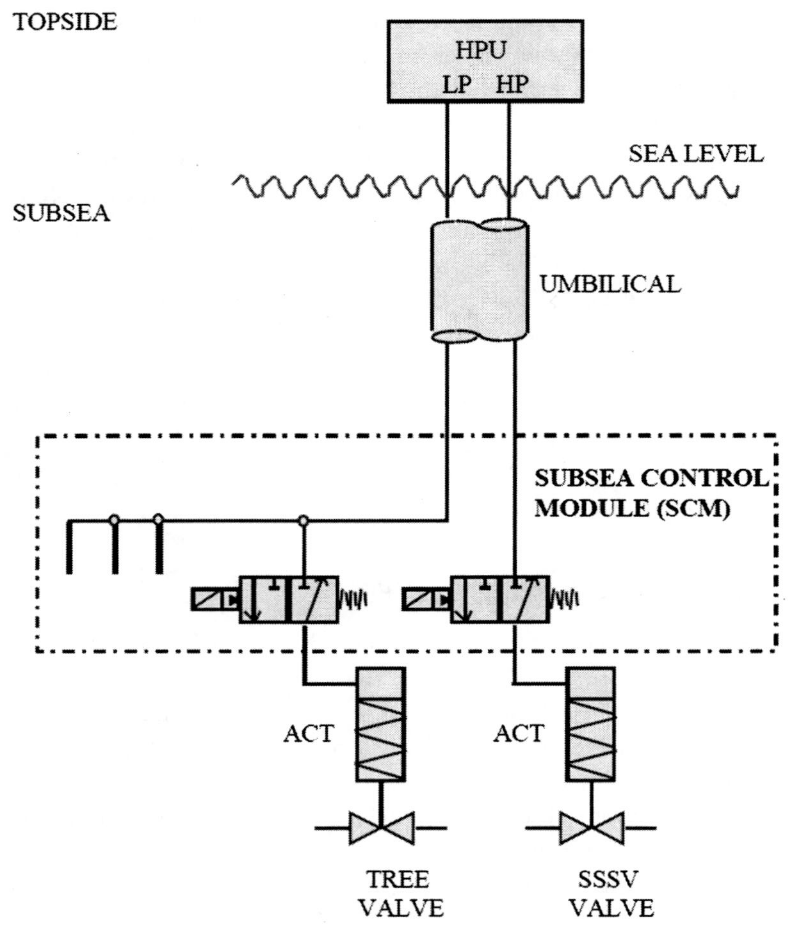

Figure 8-8 Typical Hydraulic Power System

69.0 Mpa (5000 to 10,000 psi) differential pressure. The LP circuits are used for subsea tree and manifold functions, whereas the HP circuit is for the surface-controlled subsurface safety valve (SCSSV).

The control valves used in a hydraulic control system will typically be three-way, two-position valves that reset to the closed position on loss of hydraulic supply pressure (fail-safe closed). The valves will typically be pilot operated with solenoid-operated pilot stages to actuate the main selector valve. To reduce power consumption and solenoid size, but increase reliability, it is common practice to operate the pilot stages for the HP valves on the lower pressure supply.

8.3.1. Hydraulic Power Unit (HPU)

The HPU is a skid-mounted unit designed to supply water-based biodegradable or mineral oil hydraulic fluid to control the subsea facilities that control the subsea valves. Figure 8-9 shows a typical HPU.

The HPU normally consists of the following components:

- Pressure-compensated reservoir;
- Electrical motors;
- Hydraulic pumps;
- Accumulators;
- Control valves;
- Electronics;
- Filters;
- Equipment to control start and stop of pumps.

Figure 8-10 shows a typical hydraulic power unit schematic. As introduced before, the hydraulic power unit includes two separate fluid reservoirs. One reservoir is used for filling of new fluid, return fluid from subsea (if implemented), and return fluid from depressurization of the system. The other reservoir is used for supplying clean fluid to the subsea system.

Figure 8-9 Hydraulic Power Unit (HPU) *(Courtesy of Oceaneering)*

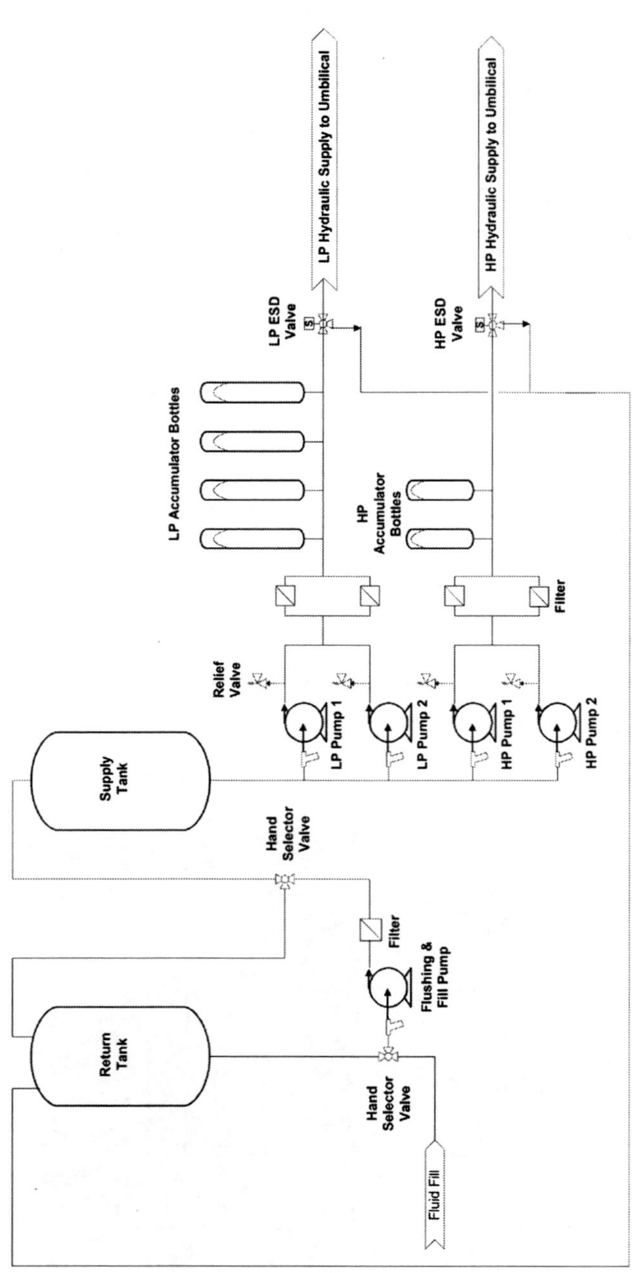

Figure 8-10 Typical Hydraulic Power Unit Schematic

The HPU also provides LP and HP hydraulic supplies to the subsea system. Self-contained and totally enclosed, the HPU includes duty and backup electrically driven hydraulic pumps, accumulators, dual redundant filters, and instrumentation for each LP and HP hydraulic circuit. The unit operates autonomously under the control of its dedicated programmable logic controller (PLC), which provides interlocks, pump motor control, and an interface with the MCS.

Dual hydraulic supplies are provided at both high pressure (SCSSV supply) and low pressure (all other functions). LP supplies are fed to the internal SCM headers via a directional control changeover valve. The changeover valve should be independently operated from the HMI, such that the header can be connected to either supply. Supply pressure measurement should be displayed. HP supplies should be controlled and monitored in a similar manner. The hydraulic discharge pressure of each function is monitored and displayed on the HMI.

8.3.1.1. Accumulators

Accumulators on the HPU should provide pump pressure damping capabilities. They should have sufficient capacity for the operation of all valves on one subsea tree with the HPU pumps disabled. Accumulators would also be of sufficient capacity to accommodate system cycle rate and recharging of the pumps. If all electric power to pumps was lost, the accumulator would have sufficient capacity to supply certain redundancies.

8.3.1.2. Pumps

All pumps should be operational when initially charging the accumulators or initially filling the system on start-up. The pump (and accumulator) sizes should be optimized to avoid excessive pump cycling and premature failure.

The quantity of pumps (and other components) per supply circuit should be determined through a reliability analysis. Pump sizing is determined by hydraulic analysis. Both analyses are performed prior to starting the detailed design for the HPU. Pulsation dampeners are provided immediately downstream of the pumps, if required, for proper operation of the HPU.

All pumps should be electrically driven, supplied from the platform electrical power system. Pumps should have the capacity to quickly regain operating pressures after a hydraulic depressurization of all systems.

There are different types of pumps, but the most common type uses accumulators that are charged by fixed pumps. These pumps, which start and stop at various preprogrammed pressures, are controlled by a PLC.

8.3.1.3. Reservoir

The HPU has a low-pressure fluid storage reservoir to store control fluid and a high-pressure (3000-psi) storage reservoir. One of the two separate fluid reservoirs is used for filling of new fluid, return fluid from subsea (if implemented), and return fluid from depressurization of the system. The other reservoir is used for supplying clean fluid to the subsea system.

Fluid reservoirs should be made from stainless steel and equipped with circulating pumps and filters. Sample points should be made at the lowest point of the reservoir and at pump outlets [6]. The hydraulic fluid reservoirs should also be equipped with visual level indicators. Calibration of level transmitters should be possible without draining of tanks.

The HPU reservoir should contain level transmitters, level gauges, drain ports, filters, air vents, and an opening suitable for cleanout. The supply and return reservoirs may share a common tank structure utilizing a baffle for separation of clean and dirty fluid. The baffle should not extend to the top of the reservoir so that fluid from overfilling or ESD venting can spill over into the opposite reservoir.

Level Sensor

The reservoir level-sensing system should meet the following requirements:

- The low-level switch should be at a level sufficient to provide a minimum of 5 min of pump operating time.
- The low-level switch should be located at a level above the drain port to prevent the pumps from ingesting air into the suctions.
- The high-level switch should be located at a level equal to 90% of the reservoir capacity.

Control Fluid

The control fluids are oil-based or water-based liquids that are used to convey control and/or hydraulic power from the surface HPU or local storage to the SCM and subsea valve actuators. Both water-based and oil-based fluids are used in hydraulic systems.

The use of synthetic hydrocarbon control fluids has been infrequent in recent years, and their use is usually confined to electrohydraulic control systems. Water-based hydraulic fluids are used most extensively. The characteristics of high water content–based control fluids depend on the ethylene glycol content (typically 10% to 40%), and viscosity varies with temperature (typically 2° to 10°C). Because government regulations do not allow venting of mineral-based oil into the sea, if the system uses this type of fluid, it must be a closed-loop system, which adds an extra conduit in the umbilical, making it more complex. Required fluid cleanliness for control systems is Class 6 of National Aerospace Standard (NAS) 1638 [7].

The water-based hydraulic fluid should be an aqueous solution. The oil-based hydraulic fluid should be a homogeneous miscible solution. The fluid should retain its properties and remain a homogeneous solution, within the temperature range, from manufacture through field-life operation.

The first synthetic hydrocarbon control fluid was utilized on Shell's Cormorant Underwater Manifold Centre in the early 1980s. This type of control fluid has low viscosity, great stability, and excellent materials compatibility, and is tolerant of seawater contamination. This fluid requires the control system to incorporate return lines and an oil purification system (filter, vacuum dehydration to remove water). The cost of synthetic hydrocarbon control fluids is approximately four times that of mineral hydraulic oils.

The first water-based control fluid was utilized on Statoil's Gullfaks development in the early 1980s. This type of control fluid has a very low viscosity and is discharged to the sea after use. This fluid requires the control system to incorporate higher specification metals, plastics, and elastomers. The cost of water-based control fluids is approximately twice that of mineral hydraulic oils.

Control fluid performance influences control system safety, reliability, and cost of ownership. Control fluids also affect the environment. The control fluid performances are as follows:

- The control fluid must be capable of tolerating all conditions and be compatible with all materials encountered throughout the control system.
- The control fluid is a primary interface between components and between subsystems. It is also an interface between different but connected systems.
- To maintain control system performance, system components must continue to function within their performance limits for the life of the system—and that includes the control fluid.

- Any reduction in control fluid performance can have an adverse effect throughout the control system. The factors that can reduce the performance of a control fluid in use are as follows:
  - Conditions exceeding the operating parameters of the control fluid;
  - Poor product stability, resulting in a reduction in control fluid performance over time;
  - Contaminants interfering with the ability of the control fluid to function.

8.3.1.4. Control and Monitoring

The HPU is typically supplied with an electronic control panel, including a small PLC with a digital display, control buttons, and status lamps. The electronics interface with other system modules, for remote monitoring and control.

The HPU parameters monitored from the safety automation system should typically be:

- Nonregulated supply pressure;
- Regulated supply pressure;
- Fluid levels;
- Pump status;
- Return flow (if applicable).

The control panel may be a stand-alone or an integral part of the HPU. It utilizes a series of valves to direct the hydraulic and/or electric signals or power to the appropriate functions.

Displays should be required to indicate hydraulic power connections from the HPU to the topside umbilical termination (or distribution) units, riser umbilicals, and subsea distribution to the individual hydraulic supplies to the SCMs. Links should be provided to individual hydraulic circuit displays.

REFERENCES

[1] U.K. Subsea, Kikeh – Malaysia's First Deepwater Development, Subsea Asia, 2008.
[2] A.L. Sheldrake, Handbook of Electrical Engineering: For Practitioners in the Oil, Gas and Petrochemical Industry, John Wiley & Sons Press, West Sussex, England, 2003.
[3] M. Stavropoulos, B. Shepheard, M. Dixon, D. Jackson, Subsea Electrical Power Generation for Localized Subsea Applications, OTC 15366, Offshore Technology Conference, Houston, 2003.
[4] G. Aalvik, Subsea Uninterruptible Power Supply System and Arrangement, International Application No. PCT/NO2006/000405, 2007.

[5] Subsea Electrical Power Unit (EPU). http://www.ep-solutions.com/solutions/CAC/Subsea_Production Control_System_SEM.htm., 2010.
[6] NORSOK Standards, Subsea Production Control Systems, NORSOK, U-CR-005, Rev. 1. (1995).
[7] National Aerospace Standard, Cleanliness Requirements of Parts Used in Hydraulic Systems, NAS, 1638, 2001.

Project Execution and Interfaces

Contents

Subsea Engineering Handbook
ISBN 978-0-12-397804-2, doi:10.1016/B978-0-12-397804-2.00009-6

9.1. INTRODUCTION

The success of any project depends significantly on project execution. Well-planned project execution allows for timely corrective action or redirection of a project. Once the project execution plan has been defined, a formal process of regular reports and reviews is required. Project execution is relevant at all stages of the project, but the issues become more intense and complex as the activities increase in number, diversify, and spread geographically. The project manager must set the expectation that members of the project management team understand their project execution system and the quality of data available. Project execution does not have its own momentum and it is, therefore, critical for it to be proactively driven by the project manager.

The goal of this chapter is to provide a clear understanding of the requirements for subsea project execution and of the interfaces among parties involved in managing project activities. Additionally, this chapter addresses the challenges of creating seamless interfaces and establishes the methodology and tools for identifying interface issues that can arise among various functional groups.

9.2. PROJECT EXECUTION

9.2.1. Project Execution Plan

To successfully complete any project, the scope of the project should reflect the vision of the goal it aims to achieve, and a plan is made to ensure that all aspects of the project are accounted for. Thorough comprehension of the requirements for project execution can enhance the chances for the successful implementation of the project.

The project execution plan will change as the project progresses, especially as the design changes. Likely design changes include the following:
- Resources and duration needs for the project;
- Methods for developing and producing the proposed project solution;
- Identification of significant work items;
- Contingency plans for specific elements of the project;
- Program optimization elements.

All necessary structures for a project are defined with a standard structure as the basis (top-down analysis). All project activities are then defined for the different elements, and a logical planning network is created. Following a time analysis, the project's schedule is defined (bottom-up analysis). Once the project is established, monitoring and reporting will start and continue throughout the life cycle of the project.

9.2.2. Schedule Versions and Baseline Updates

To be able to control the project, two different reference versions of the schedule are stored for comparison purposes. These reference versions are defined as the original and baseline version. A current version is updated regularly, at least monthly, with new progress and forecast dates. All additionally approved work that changes the scope of the project (e.g., variation orders) is included in the current version.

Baseline updating is carried out as required and each new version replaces the old baseline. An agreement is often in place between the client and the company for scheduling reviews of the project for this purpose. The bases for baseline updates are:
- The previous baseline;
- Variation orders issued up to the baseline cutoff that were not previously included;
- Implementation of contract options that were not previously included.

9.2.3. Project Organization

Most projects are defined at the outset by a purpose and a set of contract obligations. For the execution of the project, a project organization is established that is responsible for fulfilling the contract obligations. The responsibilities of individuals handling the various project functions are discussed next.

9.2.3.1. Project Manager

The project manager is responsible for performing the project within the agreed time and budget according to the scope of work. The detailed work includes these tasks:

- Organize the project in order to achieve the best possible cooperation between the involved sections.
- Establish an order plan and a budget assisted by cost controls and structured according to the project schedule.
- Make sure that the project plan is an integrated part of the total plan for the basis organization to ensure an on-time completion of milestones.
- Report to the company according to the contract and give information about the progress or deviations within, but not limited to, planning, technical solutions, procurement, manufacturing, economy, invoicing, legal conditions, and personnel conditions.

9.2.3.2. Deputy Project Manager

The deputy project manager assists the project manager with the tasks just described and executes the project manager's tasks when he or she is out of the office. The deputy project manager reports to the project manager and has responsibility for oversight of other tasks as delegated by the project manager.

9.2.3.3. Lead Engineer

The lead engineer (known as the systems engineer in some organizations) has responsibility for the technical integration of the project and identifying the necessary skills for the project. She or he will play a primary role in defining the project's deliverables with the project's customer. The lead engineer will also play a primary role in the technical development of the project once the project has started. In industry today, the function of the lead engineer is often combined with the function of project manager and handled by the same person.

9.2.3.4. Manufacturing Manager

The manufacturing manager is responsible for the following activities:

- Manage the production team and equipment team to fulfill the required targets.
- Plan for and control the stock of raw materials, work in progress, and finished goods.
- Manage the efficiency, quality, cost, and safety of the production to meet the company's target.

- Conduct a related incentive program; work out the training plan for the employees and implement it.
- Perform in-process inspections according to control instructions.
- Control production scrap procedures.
- Monitor the materials and logistics to ensure the project operates smoothly; control and improve the production scrap rate.
- Report deviations from the schedule.
- Monitor and implement a standardized management role for the production; make sure the manufacturing capacity meets the production requirements.
- Manage operational activities.
- Find and use advanced manufacturing technology, equipment, and fixtures in the project to upgrade the manufacturing ability and reduce manufacturing costs.

9.2.3.5. Cost and Finance Manager

The cost and finance manager is responsible for:

- Analysis of manufacturing contribution margin, budget and forecast differences, and cost reduction plans;
- Cost control of the project managing necessary foreign exchange
- Providing assistance to the project manager and all project personnel in economical evaluations and decisions;
- Development of standard costs;
- Producing reports, including deviations, for internal use in cooperation with the project manager;
- Monthly analysis of actual product cost versus standard costs.

9.2.3.6. Planning Manager

The planning manager cooperates closely with the project manager and is responsible for establishing and updating the main plan of the project and issuing reports in accordance with the requirements of the contract.

9.2.3.7. Quality Manager

The quality manager is responsible for:

- Formulating and managing the development and implementation of goals, objectives, policies and procedures, and systems pertaining to QA and regulatory functions;
- Verifying the functionality of the QA system according to specifications by performing reviews and audits;

- Developing, implementing, communicating, and maintaining a quality plan to bring the company's quality systems and policies into compliance with quality system requirements;
- Assisting the project manager and the project personnel in QA matters;
- Managing documentation related to quality system guidelines;
- Analyzing and reporting failure cost data;
- Providing leadership for developing and directing QA and quality improvement initiatives (cost-of-quality reductions, audit system, etc.) for all products, processes, and services;
- Coordinating health, safety, and environmental (HSE) matters according to the project's HSE program.

9.2.3.8. Procurement Manager

The procurement program should be an integrated part of the project plan, and the procurement manager should report progress according to this plan. The procurement manager will place orders for all procured items for the project.

9.2.3.9. Secretary

The secretary is responsible for document control, including archiving incoming and outgoing correspondence, tracking transmittals and documents throughout the review process, and assisting the project team with day-to-day tasks.

9.2.3.10. Contract Manager

The contract manager assists the project manager with contractual issues and establishing the project.

The project manager has to operate in a complex organization that includes people from different professions, different departments, different companies, and other organizations. Even in the best circumstances, managing a project can be a difficult task. The project manager has to plan, organize, coordinate, communicate, lead, and motivate all participants to achieve a successful project outcome.

In the typical project organization, all of these people must work together, but many will have no direct line responsibility to the project manager. The project manager is not responsible for their performance assessments, coaching and development, promotions, pay increases, welfare, and other line relationships. Project personnel have different loyalties and objectives, have probably never worked together before, and might never work together again. The one thing that binds them all together is the project organization structure. The project manager must, therefore, deal with the human problems of developing a project team out of the diverse groups working on the project. This involves complex relationships with managers in many departments and companies, not all of whom will be directly employed on the project.

The temporary nature of project organizations allows insufficient time for interpersonal relationships to reach the static state possible in routine operations management. Good group performance is necessary from the beginning of a project, because it is difficult or even impossible to recover from mistakes made and time lost at the start. Also, the total management group on a project changes frequently, with new members joining and the roles of others diminishing as the project proceeds through its life cycle. In spite of all these disruptions and difficulties, project managers must work under pressure to deliver their time, cost, and performance targets, which can require the application of pressure or other motivational measures on the people and groups involved [1].

9.2.4. Project Management

The project management personnel are responsible for effectively and efficiently implementing the project to agreed-on performance measures. Performance criteria are ultimately based on the requirements set out in the business case and project brief. The project management team is an agent of the development management activity zone, which is likely to consist of project management professionals. The project management team is ultimately responsible for preparing the project execution plan and ensuring that all relevant inputs from other activity zones are guided and integrated toward the successful implementation of the project.

Project management is the process of managing, allocating, and timing resources in order to achieve a given objective in an expedient manner. The objective may be stated in terms of time, monetary, or technical results. Project management is used to achieve objectives by utilizing the combined capabilities of available resources. It represents a systematic execution of tasks needed to achieve project objectives. Project management covers the following basic functions:

- Planning;
- Organizing;
- Scheduling;
- Control.

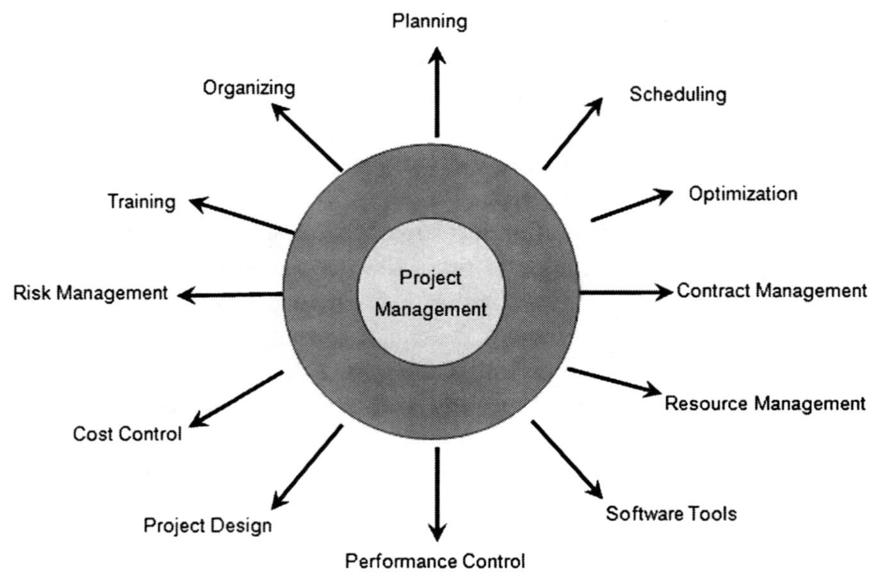

Figure 9-1 Multidimensionality of Project Management [2]

Figure 9-1 illustrates the multidimensionality of the project management process.

9.2.5. Contracting Strategy

The actual contracting strategy for a project is a result of multiple contractor approaches. The managing team will direct these contractors to execute major project segments. An umbrella agreement may tie these contracts together for interface management, alignment, and common systems and controls. Each contractor will be able to subcontract suitable portions of the work scope.

The project contract strategy is based on competitive bidding of both major contracts and purchase orders for free-issue equipment. Another key strategy is the decision of the project team to design certain elements, such as the manifold and the jumper systems, in house. To continue the example, key components of these systems can then be free-issued to a manifold fabricator. Fabrication of the manifold is negotiated with, say, the successful supplier of the valves and connection systems, thereby reducing interfaces and potential construction errors and delays. Two primary reasons to do this are (1) to allow the team to maintain better control over the multiple interfaces involving hardware suppliers and internal interfaces involving drilling and installation and (2) to take advantage of having the technical expertise to do the design work available in house [3].

9.2.6. Quality Assurance

The project has different phases representing different milestones toward completion; some phases are scheduled in parallel and some in series. Throughout these phases, the project team monitors progress and reports to the company according to the contract. Deviations are analyzed and treated according to a designated QA system. The project team then executes corrective actions.

The contracting party that will complete the work should have and maintain a QA system to ensure that products and services are delivered according to the agreed-on specifications as well as statutory requirements, thereby satisfying the needs of customers and internal users. The term *quality assurance system* refers to a formal management system you can use to strengthen your organization. It is intended to raise standards of work and to make sure everything is done consistently. A QA system sets out the expectations that a quality organization should meet. Typically, organizations implement a QA system for these purposes:

- To agree on standards, which are concerned with the performance that staff, trustees, and users expect from the organization.
- To carry out a self-assessment, which means that the project team's performance is compared to the expectations set by the standards.
- To draw up an action plan, which includes what needs to be done, who will do it, how it will be done, and when.
- To implement and do the work.
- To conduct a review, during which the project team determines what changes have been made and whether they have made the difference the team was hoping to achieve.

The practice applied to all activities leading to delivery of products or services should be: "Perform all activities according to established requirements, and when changes are applicable, a formal change approval procedure should be followed."

The project management team will actively support the quality manager and the quality department in its overall monitoring and auditing of the system, of both the contracted and subcontracted parties, to ensure compliance with requirements. Audit results should be reported to the top management as appropriate.

A number of challenges arise when planning and executing a quality program (QP) for a project. For a subsea systems group, the key challenge is often the extreme geographical spread of contractors and vendors.

The QA philosophy at the start of the project is to rely on project contractors' existing QPs, based on ISO-9001 and the development of specific quality plans, which include production, inspection, and test procedures. Initially the role of QA and quality control programs is to function as a monitoring overseer, but as the project progresses, increased levels of surveillance are required in congruence with the increased complexity of the project as it matures.

The quality team of the subsea group for the project consists of a QA coordinator and a quality engineer (QE), who report to the subsea manager. The responsibilities of the QA coordinator are to develop and oversee the implementation of the QP and associated procedures and to conduct internal and external quality audits. The quality engineer is responsible for the daily interfacing with third-party inspectors and the monitoring of contractor and vendor inspection and test plan implementations.

The third-party inspectors are selected based on the vendor inspection execution plan criteria jointly prepared by the discipline lead engineers and the QE. The basic inspection personnel contracting philosophy is to employ cognizant and capable inspectors by geographic area instead of using a blanket agreement with an inspection agency.

The QP is intended to cover all aspects of a subsea system group's involvement in the project and to complement the general quality management system established by the project services. It is also used to ensure that the change guidelines are followed by the project [3].

9.2.7. Systems Integration Manufacturing and Testing

Quality management of the manufacture and testing of all subsea equipment should rigorously follow the project's QA plan. The criticality of the equipment and the high cost of subsequent repairs require a level of inspection disproportionate to the cost of the equipment. A robust organization staffing plan is needed to ensure that resources are available to support the quality management activities and to ensure that the required inspection and test requirements are delivered. The project's team members should be assigned specific accountability for the manufacture and delivery of each subsea component and ensure that the required level of inspection and monitoring takes place. Subsea systems consist of a large number of unique components, which makes overall status monitoring difficult to achieve and the performance management system must be designed to take this into account. The project plan must recognize that many subsea items may be delivered close to the time period designated for installation and, hence, contingency plans should be identified. The contingency plans are prepared to satisfy customer requirements in the event of an emergency such as utility interruptions, labor shortages, key equipment failures, or field returns. The manufacturing schedule must allow sufficient time for stack up and integrated system testing and rectification of failures prior to installation. Planners must ensure that the necessary testing and dummy equipment will be available for testing.

The systems integration testing (SIT) program is designed to be thorough, with the objective of testing each installation and operational feature on land before going offshore. Better preparation on land and a high degree of testing will result in an efficient installation and smooth start-up once the system is moved offshore.

The SIT is conducted at the manifold fabricator's yard and is completed in two phases. The first phase consists of performing all of the mechanical interfaces between the pipeline end manifolds, suction piles, manifolds, and subsea trees. One size of each flowline and well jumper is fabricated and tested with the maximum or worst case tolerance stack-ups confirmed. The second phase of the SIT involves the integration of the production control system with the manifolds, trees, and flying leads.

The overall planning and execution of the SIT is the responsibility of the subsea systems group. The planning effort involves the development of a SIT plan that identifies the scope of work, detailed test procedures, equipment testing and handling requirements, personnel requirements, schedule and cost controls, and safety practices.

During the execution of the SIT, the fabricator essentially provides only labor, equipment, and facilities, with all detailed procedures, schedules, control budgets, and day-to-day activities provided and directed by the integrated project team and executed by the SIT coordinator. The contractor scope of work is executed under a time and materials contract. The SIT is staffed with a full-time subsea systems team and peaks at four individuals during execution of the SIT.

During the course of the SIT program, representatives from the installation and drilling groups, installation contractors, and operations personnel are invited to witness and participate in the actual SIT tests. This will help to familiarize the various parties with the equipment and allow for hands-on training to occur.

It is desirable to have a significant amount of expertise within the subsea group that helps plan and execute a successful SIT program. It is extremely important to have the right personnel involved during the execution of the SIT (subsea, operations, and installation personnel) and to get started as early as possible in the planning stages [3].

9.2.8. Installation

The installation activities are major project events and require detailed planning and management. The project manager must ensure that a planning and interface management process is in place among the engineering team, installation team, installation contractor, and facility fabrication team.

The management of installation activities is a specialized activity and requires specific knowledge and competence within the team. Preparation activities for the management of an installation include the following:

- Determine the load-out and installation method during FEED because this has a significant impact on facility design.
- Engage the installation procurement and supply chain management (PSCM) sector teams in leveraging the company's scope and scale.
- Ensure that the installation method is matched to available installation vessels and to the PSCM strategy with early lock-in of vessels to ensure availability when required.
- Ensure that, through early engagement of the installation contractor, the installation requirements are integrated with the facility's design. There is a tendency for installation contractors to review the facility late in the design or construction process. This can result in redesign and construction rework.
- Appoint an installation manager ahead of the development of the detailed installation strategy to manage the proposed installation activities and installation contract.

The following planning activities also must be considered:

- Ensure that timely detailed design reviews are undertaken between the engineering, installation and construction teams to ensure full alignment prior to start of activities.
- Initiate a peer review of the proposed installation design and methodology.
- Ensure detailed walk-throughs of the installation procedures have been carried out by the installation team prior to implementation.
- Implement a go/no go process to confirm that all requirements are in place prior to execution of load-out, sailaway, and installation.

Ensure that accountability for the approval to proceed is absolutely defined.

- Ensure that appropriate metocean and weather forecasting expertise is available to support installation planning and go/no go processes.

The installation of subsea equipment is the overall responsibility of the installation group that functions as a subgroup within the construction group.

The installation group's responsibilities consist primarily of negotiating, contracting, coordinating, scheduling, and budget controlling of the various contractors that will perform the subsea installation activities. This organization captures the synergies of optimizing vessel utilization, field safety, and overall construction management across the entire two-field installation campaigns. The technical responsibility still resides within the subsea group for providing technical input, review, and endorsement of contractor installation procedures and detailed testing instructions.

Overall, an organizational structure like that described can be successful in installing the subsea equipment with significant cost savings realized by the project. Identifying and selecting the installation contractors as early as possible allows for early input into subsea equipment designs as well as the development of early installation options and details. It also allows for any issues or risks associated with an installation to be identified. The installation risk assessments are also conducted for the primary components and have proven invaluable in the planning phase, where key issues and potential gaps are identified and resolution plans implemented prior to the offshore execution of the work [3].

9.2.9. Process Management

Process management is an agent of the development management activity zone that is responsible for planning and monitoring each phase. The responsibilities of process management include:

- Formulating the process execution plan in close collaboration with the project management team;
- Reviewing the phase review plans and reports;
- Determining and examining the inputs and outputs of the process in terms of the deliverables at each phase;
- Offering expert recommendations to the development management activity zone with regard to the satisfactory execution of the process for delivery of the product.

The process management activity zone should consist of professionals who are independent of the project.

9.2.10. HSE Management

9.2.10.1. HSE Philosophy

The protection of health and ensuring safety and conservation of the environment, apart from being subject to legislation, constitute assets and provide added value to any project. As a philosophy, awareness of them as essential and integral parts of products and services is necessary. The health, safety, and environmental (HSE) philosophy should function as the overall guideline for ensuring safety within a project.

The project team should aim at achieving technical and operational integrity as far as practically and economically feasible in connection with engineering, design, procurement, manufacturing, termination work, and testing [4].

9.2.10.2. HSE Management Plan

Achieving the goals of no accidents, no harm to people, and no damage to the environment requires planning that is commensurate with the scale and complexity of a project.

An HSE management plan is created to ensure safe and environmentally responsible operations throughout the design, construction, fabrication, installation, and operating phases. An HSE management plan provides the information needed to enable the entire project team, including contractors and others, to understand how HSE concerns and issues are identified, eliminated, mitigated, or managed, and how accountabilities are established at various levels of management. It is also used as an HSE training tool for members of the project team.

The objectives of the HSE management plan are as follows [5]:

- To clearly state the project HSE objectives and expectations and provide a tool for the project team to use in achieving them.
- To detail the project organization, responsibilities, and methods of management control as they relate to HSE.
- To identify the HSE studies and deliverables.
- To specify the HSE-related reviews and audits to be performed and their timing.

9.3. INTERFACES

9.3.1. General

An interface may relate to definitions of scope of responsibility between two functional groups, exchange of information between two functional groups, or a physical interface between equipment supplied by two different functional groups. A key component of project success is managing interface differences. Interface differences among the various manufacturers, contractors, and subcontractors involved in a project can lead to serious problems such as complications with coordination of all interface activities. All interfaces between different work elements are identified and specified together with proposals for mutually acceptable solutions. The work element means full contract scope, well-defined partial or subcontract scope, or facility/operation. These are discussed with the respective parties and adjusted as necessary. The company specifies the agreed-on interface boundary limits and the exact engineering solution within the scope of work and supply. Each interface is then detailed and progressively agreed to with the respective party. An interface register is established and maintained to reflect the development and solution of each interface.

The subsea facilities team should participate in the overall project management interface. The subsea facility interfaces should be managed by the team interface coordinator with a single point of contact who is responsible for maintaining an interface register and responding by assigning responsibility for follow-up action items. A management tool for this purpose is an interface responsibility matrix that allocates responsible key personnel to all activities through all project phases. Team members can input queries and/or answers into the interface database and then register and maintain the query until it is successfully closed out to the satisfaction and agreement of all parties. A register is a legally recognized entity, such as the Copyright Office, that makes a public record of ownership for the purposes of legal protection and recourse in the event of infringement. Registers normally charge a fee for the administrative work and issue a certificate and reference number for the ownership record. A database is merely a collection of records that is searchable. These records may be public records of ownership, books available for checkout from a library, or people looking for dates on the Internet. However, the presence of a record in a database does not confer on it the status of a public record of ownership.

9.3.2. Roles and Responsibilities

9.3.2.1. Project Manager

The project manager has the overall responsibility for the coordination of all interface activities and also delegates the day-to-day running of the interface system to a responsible member of the project team.

9.3.2.2. Interface Manager

The interface manager is responsible for managing and coordinating those project activities directly related to identification of interface issues, maintaining the interface register, organizing regular interface meetings, and facilitating the close-out of interface issues through the exchange of information among the various functional groups.

9.3.2.3. Interface Coordinator

An interface coordinator is designated within each functional group to be the point of contact with the interface manager for identifying interface issues, communicating those issues to other members within the functional group, and assisting in the close-out of interface issues through the exchange of information with the interface manager and other functional groups.

9.3.2.4. Lead Engineers

The lead engineers may be members of any of the functional groups engaged in the project. The lead engineers are responsible for identifying interface issues affecting the work for which they are responsible, and reporting these issues to the interface coordinator within their functional group. A lead engineer may also serve as an interface coordinator.

9.3.2.5. Project Controls Coordinator

The project controls coordinator is responsible for maintaining schedule and budget reports.

9.3.3. Interface Matrix

The subsea interface matrix defines the points of contact between the company and the main contractor engaged in the detailed engineering, construction, and installation of subsea facilities. The interface matrix is a list of all project interfaces, such that any member of the contractor's team may obtain a reference to the latest project information for use within the project activities. The interface matrix is a living document and is to be continuously updated as new interfaces are highlighted and the status of existing ones changes.

Each interface shown on the interface matrix is allocated a specific item number to remove the possibility of any ambiguity from interface queries, correspondence, and other references. Table 9-1 is an example of the interface matrix.

Table 9-1 Example of Interface Matrix

Interface No.	Interface Item	Detailed Description	Providing Company	Receiving Company	Required by Date	Status	Comments
			Company name	Company name		Open/ closed	

9.3.4. Interface Scheduling

Each interface is scheduled such that the company may receive information in sufficient time to be incorporated into the project activities. This date is called the *required-by date* on the interface register and within the interface matrix.

9.3.5. Interface Management Plan

The interface management plan addresses the management of the various functional and physical interfaces between organizations and equipment for the project. The goal of the plan is to ensure early communication between different parties regarding information on the systems and procedures and on activities or deliverables that are required to ensure timely completion and resolution of the interface. Identifying and declaring this information early also minimizes any potential disruption to the project schedule—effective handling of interface issues across the entire project will be to everyone's benefit.

Various functional groups are engaged in various aspects of the design, procurement, fabrication, installation, and commissioning of facilities for the project. The interface management plan establishes the methodology and tools for identifying interface issues among the various functional groups, recording and tracking the issues, and bringing the issues to closure [6].

9.3.6. Interface Management Procedure

All interface issues are entered into the interface database for tracking and reporting purposes. All of the interface managers and select others will be given access to the system.

9.3.6.1. Process

Interface management requires the early identification, prioritization, and quick resolution of interfaces to avoid any negative impact on a project's cost, schedule, and quality. The overall interface management process involves the planning, identification, assessment, monitoring and control, and closeout of all interfaces that also affect other company and contractor processes, such as document control, project control processes, and knowledge management systems.

Interface management influences all phases of the project from detailed engineering through start-up and final handover to operations. Interfaces exist in all phases of the project and need to be proactively managed and resolved to support the contractors' progress and execution of the work [7].

9.3.6.2. Communications

Open communications among all contractors and client companies are required to identify interface issues at the earliest possible time and to afford timely resolution of the issues. In fact, open communications should be encouraged between all interface personnel contractually involved in the project.

It is also advisable for the external interface teams to meet with each other on a regular basis during both the co-located and virtual team phases. Face-to-face meetings are important to facilitate and maintain communication among the contractors and a client company [7].

9.3.6.3. Meetings

Weekly meetings are to be held with all of the interface managers in attendance. The purpose of the meetings is to conduct direct discussion regarding existing interface issues, determination of "Need" and "Forecast" dates and the status of close-out items.

9.3.6.4. Database Tool

For staff working outside of the company network, access to the database will commonly be made through shared documentation and the Internet.

9.3.6.5. Interface Queries

Each interface issue should be raised as a specific interface query. The interface queries may be raised by any functional group, but must be submitted by the group's interface coordinator. The query numbers are assigned in sequential receiving order after the query is submitted by the interface manager, and a copy of the query is transmitted to all involved parties.

9.3.6.6. Action Plans

Complex or difficult interface issues may require a multistep solution that is best presented via an action plan. One of the functional groups should be responsible for developing the action plan and noted in the interface register.

9.3.6.7. Closeout of Interface Issues

When an interface issue is closed it should be noted in the interface register with a date in the "Date Closed" column, and a close-out reference should be cited. The close-out reference may be a correspondence number or document number reference.

9.3.6.8. Freezing Interfaces

An interface is considered "frozen" when the documents that define the interface have been issued as "Approved for Design" or "Approved for Construction." Unresolved interface areas should be identified with a "HOLD" notation, and a corresponding interface issue should be entered in the interface register for follow-up.

9.3.7. Interface Register

The interface register should comprise a list of all interface queries raised by the contractor, with a separate interface register for queries received from company, such that any missing queries may be easily identified.

9.3.8. Internal Interface Management

An internal interface is defined as an interface that occurs within a specific company and the design disciplines. The responsibility for internal interface management usually falls on the contractor, whether it is a full engineer, procure, install, and commission project or only a portion of a larger contract.

9.3.9. External Interface Management

An external interface is defined as an interface that occurs between a specific company and external organizations. External interface management can be applied as a policy tool by the production company or as a control tool operating within the project group.

9.3.10. Interface Resolution

Resolution of external interfaces should first be attempted between the parties involved in the respective interface, but subteams can be utilized to help resolve specific issues. If this team is unable to achieve a resolution, it should become the responsibility of the external interface team. If the interface conflict still cannot be resolved, then the client company should act as mediator and force resolution of the issue [7].

9.3.11. Interface Deliveries

For the project, the company should devise an interface deliverable database, which is to be maintained and administered. The database contains a list of all highlighted interfaces between the various contractors, with associated need and source dates.

REFERENCES

[1] H. Frederick, L. Dennis, Advanced Project Management: A Structured Approach, fourth ed., Gower Publishing, Aldershot, 2004.
[2] A.B. Badiru, Project Management in Manufacturing and High Technology Operations, John Wiley & Sons, New York, 1996.
[3] N.G. Gregory, Diana Subsea Production System: An Overview, OTC 13082, Offshore Technology Conference, Houston, 2001.
[4] International Association of Oil & Gas Producers, HSE Management-Guidelines for Working Together in a Contract Environment, OGP Report No. 423, 2010, June.
[5] M.G. Byrne, D.P. Johnson, E.C. Graham, K.L. Smith, BPTT Bombax Pipeline System: World-Class HSE Management, OTC 15218, Offshore Technology Conference, Houston, 2003.
[6] J.J. Kenney, K.H. Harris, S.B. Hodges, R.S. Mercie, B.A. Sarwono, Na Kika Hull Design Interface Management Challenges and Successes, OTC 16700, Offshore Technology Conference, Houston, 2004.
[7] U. Nooteboom, I. Whitby, Interface Management-the Key to Successful Project Completion, DOT 2004, Deep Offshore Technology Conference and Exhibition, New Orleans, Louisiana, 2004.

CHAPTER 10

Subsea Risk and Reliability

Contents

Subsea Engineering Handbook
ISBN 978-0-12-397804-2, doi:10.1016/B978-0-12-397804-2.00010-2

10.1. INTRODUCTION

The exploration and production of oil and gas resources entail a variety of risks, which, if not adequately managed, have the potential to result in a major incident.

All subsea field development procedures involved in designing, manufacturing, installing, and operating subsea equipment are vulnerable to a financial impact if poor reliability is related to the procedure. Equipment reliability during exploration and production is one of the control factors on safety, production availability, and maintenance costs. In the early design phases, the target levels of reliability and production availability can be controlled through application of a systematic and strict reliability management program.

This chapter presents a recommended systematic risk management program and further describes a methodology for analyzing field architectures to improve the reliability of system design and to reduce operating expenses by using reliable engineering tools.

10.1.1. Overview of Risk Management

Risk is defined as the potential for the realization of unwanted, negative consequences of an event. Risk can be measured by the product of the probability of the event occurring and the consequence of the event [1]. Risk management is the systematic process of identifying, analyzing, and responding to project risk. It involves minimizing the probability and consequences of adverse events and maximizing the probability and consequences of positive events based on project objectives.

The subsea business is a unique industry in many ways, but its uniqueness is not limited to the diversity inherent in any project or to the differences between all projects; its diversity also includes the people working within the industry. As a whole, these people are at the cutting edge of a high-risk and potentially highly rewarding industry. It would therefore seem beneficial for the industry, as a whole, to adopt an overall risk management methodology and philosophy in order to utilize joint efforts that can help maximize the success of its projects [2].

Risk management can be regarded as a continuous assessment process applied throughout the design, construction, operation, maintenance, and decommissioning of a project. The integrity risk management life cycle is illustrated in Figure 10-1.

10.1.2. Risk in Subsea Projects

A subsea project is complex and involves uncertainties from a wide range of sources. Managing these uncertainties in a systematic and efficient manner with a focus on the most critical uncertainties is the objective of a successful risk management plan. Some of the areas of uncertainty that need to be considered are as follows [3]:

- Technical;
- Financial;
- Organizational;
- Contract/procurement;
- Subcontractors;
- Political/cultural, etc.

Many cost issues need to be considered when assessing the economics of a field development. In the ranking of different investment

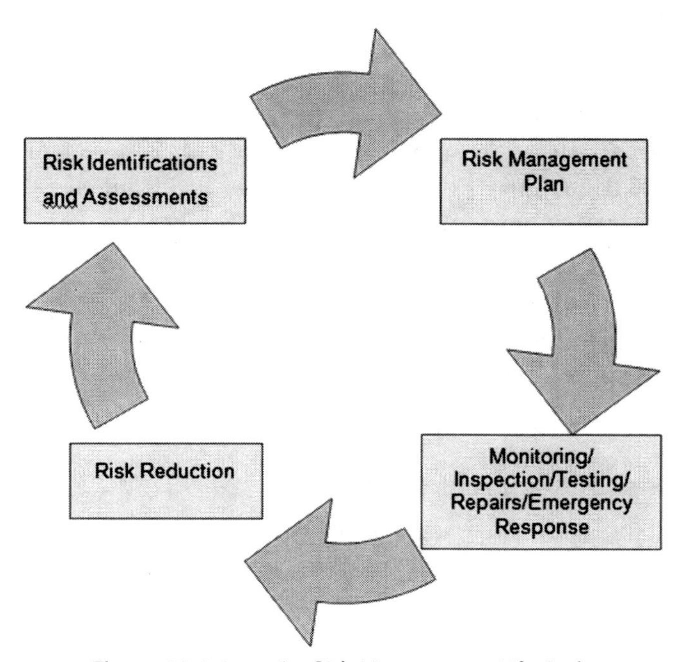

Figure 10-1 Integrity Risk Management Life Cycle

opportunities for a field development project, the following issues need to be considered:

- Costs of capital expenditure and operating expenditures;
- Schedule of project tasks and completion of milestones;
- Taxes and depreciation;
- Health, safety, and environmental concerns (meeting regulations and company requirements);
- Structural reliability (using a design that meets requirements);
- Flow assurance (analysis that meets availability requirements).

10.2. RISK ASSESSMENT

10.2.1. General

Risk assessment is the process of assessing risks and factors influencing the level of safety of a project. It involves researching how hazardous events or states develop and interact to cause an accident. The risk assessment effort should be tailored to the level and source of technical risk involved with the project and the project stage being considered. The assessment of technical risk will take different forms in different stages of the project; for example [1]:

- A simple high-level technical review may filter out equipment with technical uncertainty.
- Consequence/severity analyses can be used to identify equipment with the greatest impact on production or safety and environment.
- Potential failure modes or risk of failure can be identified.
- Technical risk reviews can be used to identify where equipment is being designed beyond current experience.

10.2.2. Assessment Parameters

When assessing risk exposure, the parameters listed in Table 10-1 should be evaluated.

10.2.3. Risk Assessment Methods

When assessing risk, the parameter of probability must be considered to obtain an overall assessment because not all risks will evolve into project certainties. During the assessment, risks are removed to get a global view. This method is based on functional expertise, and a fixed scoring value is used to achieve balanced results. For example, if a risk is assessed as having

Table 10-1 Assessment Parameters [4]

Assessment Parameter	Keywords for Assessment
Personnel exposure	- Qualification and experience of personnel - Organization - Required presence - Shift arrangements - Deputy/backup arrangements
Overall project particulars	- Delay - Replacement time/cost - Repair possibilities - Number of interfaces and contractors/subcontractors - Project development period
Existing field infrastructure	- Surface and subsea infrastructure
Handled object	- Value, structural strength, and robustness
Marine operation method	- Novelty/feasibility - Robustness - Type of operations - Previous experience - Installability
Equipment used	- Margins/robustness - Condition/maintenance - Previous experience - Suitability - Experience with operators/contractors (track record)
Operational aspects	- Cost of mobilized equipment and spread - Language barriers/hindrance - Season/environmental conditions - Local marine traffic - Proximity to shore

a probability of occurrence between 1% and 20%, then the mean of the range, 10%, will be used in the calculation. Table 10-2 illustrates the values utilized for different probabilities at various risk levels in a risk assessment.

Table 10-2 Probability in Risk Assessment [2]

Risk	Probability	Utilize
Improbable	<20%	10%
Not likely	20−40%	30%
Possible	40−60%	50%
Probable	60−80%	70%
Near certain	>80%	90%

A 100% probability does not appear in the table because 100% probability is a project certainty. The risk evaluation deals only with scenarios that *might* happen. Once having identified the probability and established the level of risk, it is necessary to prioritize the actions to be undertaken.

10.2.4. Risk Acceptance Criteria

The risk criteria define the level at which the risk can be considered acceptable /tolerable. During the process of making decisions, the criteria are used to determine if risks are acceptable, unacceptable, or need to be reduced to a reasonably practicable level. Numerical risk criteria are required for a quantitative risk assessment.

As described previously, risk assessment involves uncertainties. It may not be suitable to use the risk criteria in an inflexible way. The application of numerical risk criteria may not always be appropriate because of the uncertainties of certain inputs. The risk criteria may be different for different individuals and also vary in different societies and alter with time, accident experience, and changing expectations of life. Therefore, the risk criteria are only able to assist with informed judgments and should be used as guidelines for the decision-making process [5].

In risk analysis, the risk acceptance criteria should be discussed and defined first. Three potential risk categories are proposed in DNV-RP-H101 [4]:

- Low;
- Medium;
- High.

The categorization is based on an assessment of both consequence and probability, applying quantitative terms. The categories should be defined for the following aspects:

- Personnel safety;
- Environment;
- Assets;
- Reputation.

A risk matrix is recommended for defining the risk acceptance criteria, a sample of which is presented in Figure 10-2.

10.2.5. Risk Identification

Many tools and techniques are used when identifying risk. Some of them are introduced in this section.

Consequence					Probability (Increasing Probability →)			
Descriptive	Personnel	Environment	Assets	Reputation	Remote (A) Occurred- Unlikely	Unlikely (B) Could Occur	Likely (C) Easy to Postulate	Frequent (D) Occur Regularly
1. Extensive	Fatalities	Global or national effect. Restoration time > 10 yr.	Project Prod consequence costs > USD 10 mill.	International impact neg. exposure.	A1 = S	B1 = S	C1 = U	D1 = U
2. Severe	Major injury	Restoration time > 1 yr. Restoration cost > USD 1 mill.	Project Prod consequence costs > USD 1 mill.	Extensive National Impact.	A2 = A	B2 = S	C2 = S	D2 = U
3. Moderate	Minor Injury	Restoration time > 1md. Restoration cost > USD 1 K.	Project Prod consequence costs > USD 100K.	Limited National Impact.	A3 = A	B3 = A	C3 = S	D3 = S
4. Minor	Illness or slight injury	Restoration time < 1 md. Restoration cost < USD 1 K.	Project Prod consequence costs < USD 1K.	Local impact.	A4 = A	B4 = A	C4 = A	D4 = S

High Risk

If the undesired event after mitigating measures is evaluated to have *unacceptable risk (U)* the operation shall not be carried out. If the operation is still to be carried out, formal application for deviation shall be filed according to established procedures.

Medium Risk

Operation can be executed after cost efficient measures are implemented and the analyses team has found the risk satisfactory (S).

Low Risk

Acceptable risk subject to application of the principle of ALARP (as low as reasonably practicable).

Figure 10-2 Sample Risk Matrix [4]

10.2.5.1. Hazard Identification Analysis

The hazard identification (HAZID) technique is used to identify all hazards with the potential to cause a major accident. Hazard identification should be done in the early stage of the project and be conducted in the conceptual and front-end engineering stages. HAZID is a technique involving the use of trained and experienced personnel to determine the hazards associated with a project. Significant risks can be chosen through HAZID by screening all of the identified risks. The technique is also used to assess potential risks at an early stage of the project.

10.2.5.2. Design Review

The design review is used to evaluate the design based on expert opinions at various stages. It is also used to identify the weaknesses of a design for a particular system, structure, or component.

10.2.5.3. FMECA

A failure mode, effects, and criticality analysis (FMECA) is conducted to identify, address, and, if possible, design out potential failure modes. The use of a process FMECA to identify potential failures that could occur during each step of the procedure with a view toward finding better (risk-reducing) ways of completing the task (high risk operations only) should be considered. All procedure-related actions from any detailed design FMECA and peer reviews should be incorporated into the project [1].

The advantages of FMECA are as follows:

- Applicable at all project stages;
- Versatile—applicable to high-level systems, components, and processes;
- Can prioritize areas of design weakness;
- Systematic identification of all failure modes.

FMECA also has two weaknesses:

- Does not identify the real reason of the failure mode;
- Can be a time-consuming task.

10.2.6. Risk Management Plan

The risk management plan includes resources, roles and responsibilities, schedules and milestones, and so on. However, it should only involve items that can be achieved within the schedule and budget constraints. By applying the risk management plan to the total development project, risks will be reduced and decisions can be made with a better understanding of the total risks and possible results.

10.3. ENVIRONMENTAL IMPACT ASSESSMENT

When a subsea spill occurs, environmental consequences can be very severe. Assessing environmental damage is extremely difficult because of the many factors involved in cleanup effects and in estimating the costs for possible civil penalties or fines. Environmental damage is typically assessed based on a dollar-per-barrel estimate for the material and location of release.

The consequences of a release from process equipment or pipelines vary depending on such factors as physical properties of the material, its toxicity or flammability, weather conditions, release duration, and mitigation actions. The effects may impact plant personnel or equipment, population in the nearby residences, and the environment.

Environmental impact assessment is estimated in four phases:

- Discharge;
- Dispersion;
- Cleanup costs;
- Ecological effects.

10.3.1. Calculate the Volume Released

Sources of hazardous release include pipe and vessel leaks and ruptures, pump seal leaks, and relief valve venting. The mass of material, its release rate, and material and atmospheric conditions at the time of release are key factors in calculating their consequences.

Release can be instantaneous, as in the case of a catastrophic vessel rupture, or constant, as in a significant release of material over a limited period of time. The nature of release will also affect the outcome. With appropriate calculations, it is possible to model either of the two release conditions: instantaneous or constant.

For a subsea release case, the leak rate and detection times are major factors when determining the volume of the leak:

$$V_{rel} = V_{leak} \times t_{detect}$$

where

V_{rel}: Volume of liquid released from equipment;
V_{leak}: Leak rate;
t_{detect}: Detection time.

10.3.2. Estimate Final Liquid Volume

When a vapor or volatile liquid is released, it forms a vapor cloud that may or may not be visible. The vapor cloud is carried downwind as vapor and suspended liquid droplets. The cloud is dispersed through mixing with air until the concentration eventually reaches a safe level or is ignited.

Initially, a vapor cloud will expand rapidly because of the internal energy of the material. Expansion occurs until the material pressure reaches that of ambient conditions. For heavy gases, the material spreads along the ground and air is entrained in the vapor cloud, due to the momentum of the release. Turbulence in the cloud assists in mixing.

As the concentration drops, atmospheric turbulence becomes the dominant mixing mechanism, and a concentration profile develops across the vapor cloud. This concentration profile is an important feature in determining the effects of a vapor cloud.

Several factors determine the phenomena of dispersal:

- *Density:* The density of the cloud relative to air is a very important factor affecting cloud behavior. If denser than air, the cloud will slump and spread out under its own weight as soon as the initial momentum of the release starts to dissipate. A cloud of light gas does not slump, but rises above the point of release.
- *Release height and direction:* Releases from a high elevation, such as a stack, can result in lower ground-level concentrations for both light and heavy gases. Also, upward releases will disperse more quickly than those directed horizontally or downward, because air entrainment is unrestricted by the ground.
- *Discharge velocity:* For materials that are hazardous only at high concentrations, such as flammable materials, the initial discharge velocity is very important. A flammable high-velocity jet may disperse rapidly due to initial momentum mixing.
- *Weather:* The rate of atmospheric mixing is highly dependent on weather conditions at the time of release. Weather conditions are defined by three parameters: wind direction, speed, and stability.

When liquid escapes to the environment, the major influence is the environmental pollution, which is determined by the liquid remaining in the water.

Following a spill, a certain fraction of lighter hydrocarbons can evaporate, thus reducing the volume of liquid that needs to be cleaned up. The persistence factor F_{liquid} is used to quantify the amount of unevaporated liquid. In general, F_{liquid} is found as follows:

$$F_{\text{liquid}} = e^{-kt}$$

where

t: the time required to complete half the cleanup effort, in hours;

k: evaporation rate constant, in hours^{-1}.

The time, t, includes the time to initiate cleanup, which is estimated as the time required to begin the cleanup effort in earnest, including the time to plan a cleanup strategy and mobilize all necessary equipment.

Methods for calculating evaporation rate constants for pure components and mixtures are provided in Table 10-3.

10.3.3. Determine Cleanup Costs

In general, the cleanup costs for a leak to the environment are estimated using the following expression:

$$\text{Cost}_{\text{cleanup}} = V_{\text{env}} \times F_{\text{liquid}} \times C$$

Table 10-3 Evaporation Rate Constants for Pure Components [6]

	Evaporation Constant (hr^{-1})		
Hydrocarbon Component	**40°F**	**60°F**	**80°F**
n-C9	0.231	0.521	0.777
n-C10	0.0812	0.2	0.311
n-C12	0.0103	0.0302	0.0514
n-C14	0.0013	0.0046	0.0085
n-C16	5.65E-05	3.30E-04	7.60E-04
n-C18	2.46E-05	1.20E-04	2.70E-04

where

V_{env}: Volume released to the environment;

F_{liquid}: Fraction of liquid remaining;

C: Unit cost of cleanup.

The estimates of the cost to clean up spills of various liquids are the most uncertain variable in an environmental consequence analysis. Every attempt must been made to estimate cleanup costs in a reasonable manner. Based on historical data, cleanup costs for crude oil on open water can vary from $50 to $250 per gallon.

10.3.4. Ecological Impact Assessment

Oil spills on the sea surface can affect a number of marine species. The species most vulnerable to "oiling" are seabirds, marine mammals, and sea turtles that may come into direct contact with the hydrocarbons, although any interaction with the spilled hydrocarbons depends on the time these animals spend on the sea surface. The exact distribution and feeding areas of seabirds, marine mammals, and sea turtles in the offshore environment are unknown. Large swimming animals such as cetaceans and turtles are mobile and could move away from spilled oil and are less likely to be affected. Fishes living beneath the surface can detect and avoid oil in the water and are seldom affected.

Fish and Invertebrates

- Atlantic cod
- Capelin
- American lobster
- Atlantic herring
- Lumpfish

- Snow crab
- Redfish
- Yellowtail flounder
- Sea scallop

Sea-Associated Birds

- Northern gannet
- Greater shearwater
- Cormorants
- Common eider
- Black guillemot
- Harlequin duck
- Bald eagle
- Greater yellowlegs
- Purple sandpiper
- Piping plover

Marine Mammals

- River otter
- Fin whale
- Atlantic white-sided dolphin
- Humpback whale
- Harbor seal
- Blue whale
- White-beaked dolphin

Sea Turtle Each of these species is discussed next in terms of their habitats, life stages, and overall vulnerability to oil spills.

10.3.4.1. Fish and Invertebrates

Near-shore and shallow waters are important for the spawning and early stages of several important fish species, which could be vulnerable to spilled oil depending on the timing of the spill.

10.3.4.2. Sea-Associated Birds

Seabirds are the species most vulnerable to the effects of oiling; the species listed above all come into contact with the sea surface and, hence, potentially oil from a spill. Murres and black guillemots are particularly vulnerable because they fly infrequently, spending the majority of their time on the sea surface.

10.3.4.3. Marine Mammals

With the exception of seals, otters, and polar bears, marine mammals are not particularly susceptible to the harmful effects of oiling. Harbor seal pups may be susceptible to the effects of oiling. Although not classified as a marine mammal, river otters are included in the list because they spend a great deal of time in the marine environment.

10.3.4.4. Sea Turtles

The vulnerability of sea turtles to oiling is uncertain.

10.4. PROJECT RISK MANAGEMENT

One of the key uncertainties related to a subsea deepwater development is the capacity to deliver equipment reliably on time so that the project is on schedule and the time to first oil or gas production is not delayed. The number of interfaces, equipment suppliers, and subcontractors involved, however, can make this process complex and difficult to manage. Project risk management can provide an excellent tool for systematically managing these challenges.

Project risk management is a systematic approach for analyzing and managing threats and opportunities associated with a specific project and can, thereby, increase the likelihood of attaining the project's objectives in terms of cost, schedule, and operational availability. The use of a project risk management process will also enhance the understanding of major risk drivers and how these affect project objectives. Through this insight, decision makers can develop suitable risk strategies and action plans to manage and mitigate potential project threats and exploit potential project opportunities.

Project risk management in field development projects has the following set of goals:

- Identify, assess, and control risks that threaten the achievement of the defined project objectives, such as schedule, cost targets, and performance. These risk management activities should support the day-to-day management of the project as well as contribute to efficient decision making at important decision points.
- Develop and implement a framework, processes, and procedures that ensure the initiation and execution of risk management activities throughout the project.
- Adapt the framework, processes, and procedures so that the interaction with other project processes flows in a seamless and logical manner.

The project risk management process should be assisted by a set of tools that supports these processes and also allows a graphical representation of the project schedule and risks that potentially could affect these plans [3].

10.4.1. Risk Reduction

Risk reduction processes are focused on the generation of alternatives, cost effectiveness, and management involvement in the decision-making process. Use of these processes is designed to reduce the risks associated with significant hazards that deserve attention.

In safety analysis, safety-based design/operation decisions are expected to be made at the earliest stages in order to reduce unexpected costs and time delays. A risk reduction measure that is cost effective at the early design stage may not be as low as reasonably practicable at a later stage. Health, safety, and environmental (HSE) aim to ensure that risk reduction measures are identified and in place as early as possible when the cost of making any necessary changes is low. Traditionally, when making safety-based design/operation decisions, the cost of a risk reduction measure is compared with the benefit due to the reduced risks. If the benefit is larger than the cost, then it is cost effective; otherwise it is not. This kind of cost/benefit analysis based on simple comparisons has been widely used in safety analysis [5].

Figure 10-3 illustrates a route for improving safety.

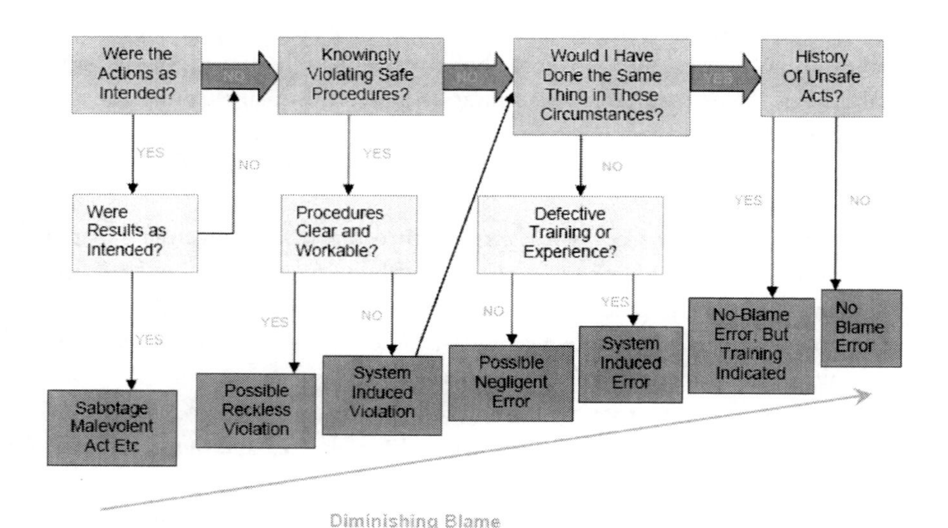

Figure 10-3 A Route for Improving Safety [7]

10.5. RELIABILITY

10.5.1. Reliability Requirements

Reliability refers to the ability of a device, system, or process to perform its required duty without failure for a given time when operated correctly during a specified time in a specified environment.

At present, it is obvious that numerical reliability requirements are rarely set at the invitation to tender stage of projects. A numerical analysis is generally performed later during the detailed design stage.

Reliability requirements are often imposed on new contracts following particular instances/experiences of failure. This is an understandable response; however, this is not a sound strategy for achieving reliability because the suppliers may feel they have met the reliability requirements if they have dealt with the listed issues.

If no reliability target is set, the underlying signal to the supplier is "supply at whatever reliability can be achieved at the lowest cost" [8].

10.5.2. Reliability Processes

API RP 17N lists twelve interlinked key processes that have been identified as important to a well-defined reliability engineering and risk management capability. These reliability key processes provide a supporting environment for reliability activities. When these are implemented across an organization, the reliability and technical risk management effort for each project is increased. Figure 10-4 illustrates the key processes for reliability management. The 12 reliability key processes are as follows [1]:

- **Definition of Availability Goals and Requirements** Ensures that the project goals are fully aligned with overall business performance objectives and provides the focus for design and manufacture for availability and reliability assurance. The trade-off between the purchase cost and the operational expenditure needs to be understood when considering the need for reliability improvement. This should then be considered when setting goals and requirements.
- **Organizing and Planning for Availability** Allocates leadership and resources to the required reliability activities such that they add value to the project overall and do not adversely affect the project schedule. The reliability activities identified should be considered an integral part of the engineering process and integrated with conventional engineering tasks in the project management system.

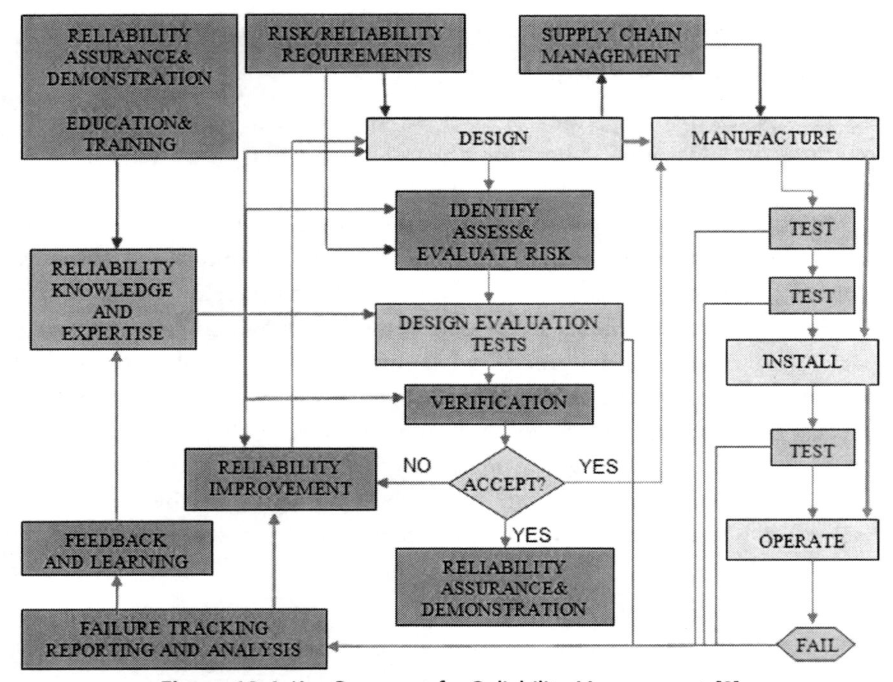

Figure 10-4 Key Processes for Reliability Management [8]

- **Design and Manufacture for Availability** Should be considered an extension of good engineering practice but requires increased focus on understanding how and why failures occur in operation. Information gathered during reliability analysis activities should be considered during the design process to drive the design's ability to achieve and deliver the specified availability requirements.
- **Reliability Assurance** This is the essential element of managing technical risk because it is the process of identifying, assessing, justifying, and, most importantly, communicating the information pertaining to risks to the technical effectiveness of the system.
- **Risk and Availability Analysis** Provides reliability management support by identifying potential faults and failure mechanisms and their effects on the system in advance of operation and to quantify risk and reliability. Analysis and models usually focus on function, hardware, or process.
- **Verification and Validation** Confirms that any given activity is the correct one and that it has been carried out correctly.

- **Project Risk Management** Addresses nontechnical risks throughout the project life cycle to enable all risks to be identified, quantified, managed, and preferably eliminated.
- **Reliability Qualification and Testing** This is the process by which systems are examined and evidence is provided to demonstrate that the technology employed meets the specified requirements for the intended use. The qualification process in some projects may start as early as the feasibility stage if it is known that a specific piece of unqualified hardware will be required to exploit a field.
- **Performance Tracking and Data Management** collects and organizes reliability performance data from all projects at all project stages to support the assessment of reliability, availability, and production efficiency. Historical reliability and availability data can be used to determine the availability goals and requirements for projects with field-proven equipment, and also understand any failures that have been experienced to provide inputs to reliability analysis and improved design.
- **Supply Chain Management** Ensures that reliability and technical risk management goals, requirements, achievement, and lessons learned are communicated among all organizations involved in the project. The ability to manage the various interfaces between the customers and suppliers down the supply chain is expected.
- **Management of Change** Ensures that any changes are consistent with the project reliability and technical risk management goals and that their impacts are fully assessed and managed.
- **Organizational Learning** Provides resources to ensure that information is fed back to the whole organization involved in design and system integration, and the whole organization understand the lessons to be learned from failure. The lessons learned usually cover the whole life of the project from strategic thinking and decision making through project execution and the delivery of benefits and should include both good and bad practices.

10.5.3. Proactive Reliability Techniques

In a proactive environment, the orientation toward reliability is changed. Reliability engineers become involved in product design at an early point to identify reliability issues and concerns and begin assessing reliability implications as the design concept emerges. Because of the continuous requirement for improved reliability and availability, an integrated systems engineering approach, with reliability as a focal point, is now the state of the art in product and systems design. Frequently, it is desirable to understand, and be able to predict, overall system failure characteristics.

Reliability and availability are typically increased [9]:

- Through improved hardware and software fault tolerance design;
- By implementing more efficient screening tests during manufacturing, which reduces the quantity of induced failures;
- By reducing the number of incidents where an apparent failure cannot be verified;
- By increasing the time between preventive maintenance actions.

10.5.4. Reliability Modeling

To predict system reliability, the reliabilities of the subsystems should be assessed and combined to generate a mathematical description of a system and its operating environment by reliability modeling. Once the system reliability has been calculated, other measures can be evaluated as inputs for decision making. In this circumstance, a mathematical description of a system constitutes a model of the system failure definitions; that is, the model expresses the various ways a system can actually fail. The complexity of a reliability model relies on the complexity of the system and its use but also to a large extent on the questions at hand, which the analysis attempts to answer. The addition of failure consequences and their costs, maintainability, downtime, and other considerations to the complexity of a model not only influences future economic realities of the system design and use, but also becomes the metrics of interest for decision makers.

A reliability model attempts to represent a system and its usage in such a way that it mirrors reality as closely as possible. To produce useful results in a timely fashion, models often have to balance the effort of reflecting a close-to-reality and using practical simplifications. This implies that oftentimes models contain approximations based on engineering judgments and reasoned arguments to reduce the complexity of the model. This can easily lead to a "false" impression of accuracy and a potentially incorrect interpretation of the results by using sophisticated mathematical models in situations in which the quality of the input data is not equally high. The main objective of reliability modeling is to identify weak points of a system as areas for improvement, although it is a quantitative tool. So there is a need to focus on the comparison of results for various system design solutions rather than on the often prompted question about the absolute results. In this respect a reliability model can be an extremely valuable tool for making design decisions.

The process of reliability modeling always begins with the question about the objective of the analysis. Only if the objective and the expected outcome are clearly defined can an appropriate modeling technique be chosen. The most common analysis techniques include, but are not limited to, reliability block diagrams, fault tree analysis, and Markov analysis, which is also known as state-time analysis. Oftentimes a combination of those tools is required to address the objective of an analysis adequately [10].

10.5.5. Reliability Block Diagrams (RBDs)

The general purpose of RBDs is to derive reliability predictions during the design phase of hardware developments and to provide a graphical representation of a system's reliability logic.

10.5.5.1. Concept

Rather than making absolute predictions of system reliability, RBDs establish a basis for design trade-off analyses. Like the other reliability engineering tools, RBDs allow product evaluations and identification of problem areas before a design is finalized. An RBD represents a clear picture of system functions and interfaces and shows the failure logic of a given system. It models the interdependencies of logical system components and allows the overall system reliability to be computed. The results can be used to verify compliance of subsystems within system-level requirements. Especially in cases of complex systems with redundancies, RBDs are helpful tools.

Using reliability engineering software allows the user to take into account various types of redundancies, such as active, standby, and others. Even the failure probability of imperfect switches can be included in a model. Once a model is established, sensitivity analyses can be performed by changing the configuration, adding redundancies, and modifying maintenance strategies–to name just a few of the options. The results help to prepare design decisions between various competing configurations. Reliability block diagrams can also be used to visualize the system structure and to aid in training and troubleshooting [11].

10.5.5.2. Timing of Application

A top-level RBD can be constructed as soon as a system layout has been defined. An RBD may be required whenever a design FMECA is performed. The best timing for use of a reliability block diagram is as early as possible in the design process.

10.5.5.3. Requirements

RBD users should understand the various types of redundancies as well as the system architecture and functionality. In addition, thorough knowledge of how the system operates and reliability parameters (e.g., hazard rate) for each block in the diagram are needed.

10.5.5.4. Strengths and Weaknesses

API RP 17N illustrates the strengths and weaknesses of RBDs [1]:

Strengths

- The best method of graphically representing complex system logic;
- Good visualization of redundant system logic;
- Good precursor to all other analysis methods.

Weaknesses

- Construction of RBDs can be difficult for complex systems;
- Numerical assessment of the RBD can be very time consuming (if performed manually or if multiple nested RBDs are apparent);
- Becomes very data intensive when there are more detailed levels;
- May require multiple RBDs for multiple functional modes.

10.6. FAULT TREE ANALYSIS (FTA)

FTA is another tool in the reliability engineering toolkit. The general purpose of FTA is to identify the technical reason for the specified unwanted events and to estimate or predict the system reliability performance. FTA logically represents all possible failure modes of a system or package.

10.6.1. Concept

FTA is a systematic and deductive method for defining a single undesirable event and determining all possible reasons that could cause that event to occur. The undesired event constitutes the top event of a fault tree diagram, and generally represents a complete or catastrophic failure of a product or process. As well as a FMECA, an FTA can also be used for identifying product safety concerns.

Contrary to a FMECA, which is a bottom-up analysis technique, a FTA takes a top-down approach to assess failure consequences. An FTA can be applied to analyze the combined effects of simultaneous, noncritical events on the top event, to evaluate system reliability, to identify potential design defects and safety hazards, to simplify maintenance and trouble-shooting, to identify root causes during a root cause failure analysis, to logically eliminate causes for an observed failure, etc. It can also be used to evaluate potential corrective actions or the impact of design changes [11].

10.6.2. Timing

FTA is best applied during the front-end engineering design (FEED) phase as an evaluation tool for driving preliminary design modifications. Once a product is already developed or even on the market, an FTA can help to identify system failure modes and mechanisms.

10.6.3. Input Data Requirements

Intimate product knowledge of the system logic is required for tree construction, and reliability data for each of the basic units/events are required by quantitative analyses.

10.6.4. Strengths and Weaknesses

API RP 17N illustrates the strengths and weaknesses of FTA [1]:

Strengths

- Can support common cause failure analysis;
- Can predict the probability of occurrence of a specific event;
- Can support root cause analysis;
- Compatible with event trees for cause/consequence analyses;
- Supports importance analysis.

Weaknesses

- Complex systems may become difficult to manage and resolve manually;
- Not suited to the consideration of sequential events.

10.6.5. Reliability Capability Maturity Model (RCMM) Levels

The reliability capability maturity model provides a means of assessing the level of maturity of the practices within organizations that contribute to reliability, safety, and effective risk management.

Reliability capability is categorized into five maturity levels as shown in Figure 10-5. Each level introduces additional or enhanced processes and

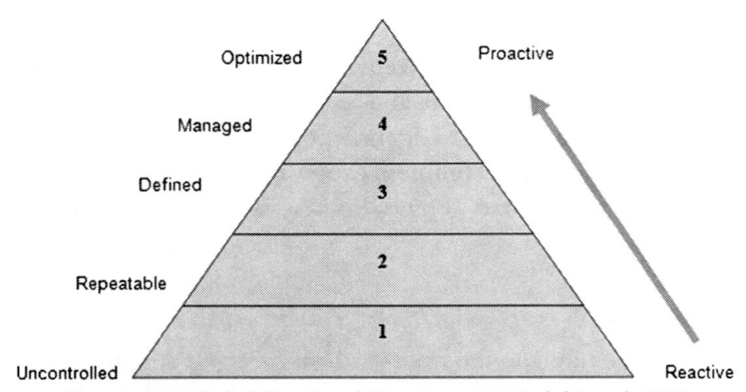

Figure 10-5 Reliability Capability Maturity Model Levels [12]

Table 10-4 Overview of Reliability Capability Maturity Levels [1]

Maturity Level	Description
1	No understanding of reliability concepts.
2	Prescriptive procedures that are repeatable but do not directly relate to reliability.
3	Understanding of historical achievements in reliability but with limited capability to learn from lessons and improve reliability.
4	Understanding of design for availability and how to correct designs to improve reliability given the observation of failure.
5	Understanding of design for availability and implementation into a proactive continuous improvement program (both managerial and operational).

these will lead to higher levels of reliability capability. An overview of RCMM levels is given in Table 10-4.

10.6.6. Reliability-Centered Design Analysis (RCDA)

Reliability centered design analysis is a formalized methodology that follows a step-by-step process. RCDA lowers the probability and consequence of failure, resulting in the most reliable, safe, and environmentally compliant design.

The direct benefits of using RCDA in FEED are as follows [13]:

- Higher mechanical availability, which results in longer operating intervals between major outages for maintenance, significantly increasing revenue.
- Reduced risk. RCDA results in designs that lower the probability and consequence of failure.

- RCDA is a functional-based analysis. It focuses on maximizing the reliability of critical components required to sustain the primary functions for a process.
- Shorter maintenance outages. Reduced downtime results in fewer days of lost production, significantly increasing revenue.
- Safer, more reliable operations, better quality control, more stable operation with the ability to respond to transient process upsets.
- Lower operating expenses. RCDA results in designs that cost less to maintain over the operating life of the asset.
- Optimized preventive and predictive maintenance programs and practices. A comprehensive program is created during RCDA. Training to these practices is performed in advance, so assets are maintained from the minute the project is commissioned.
- Emphasis on condition-based maintenance practices. Equipment condition is continuously monitored, maximizing the full potential of the assets, and avoiding unnecessary inspections and costly overhauls.
- RCDA can be used as a training tool for operators and maintenance personnel. RCDA documents the primary modes of failure and their consequences and causes for failure well in advance of building the platform.
- Spare parts optimization. Because the dominant failure causes are identified for each piece of equipment, the spare parts requirements are also known. Because this analysis is performed on the entire platform, stock levels and reorder levels can also be established.

The RCDA process is integrated into project management stages, that is, FEED. As a process, it follows a uniform set of rules and principles. Figure 10-6 illustrates the RCDA process flow diagram.

10.7. QUALIFICATION TO REDUCE SUBSEA FAILURES

The methodology for reducing subsea failures is formalized in DNV-RP-A203 [14]. It provides a systematic risk-based approach for obtaining the goals of the qualification.

The qualification process comprises the main activities listed next [15]. At each step of the process there is a need for documentation making the process traceable.

- Establish an overall plan for the qualification. This is a continuous process and needs updating after each step using the available knowledge on the status of the qualification.

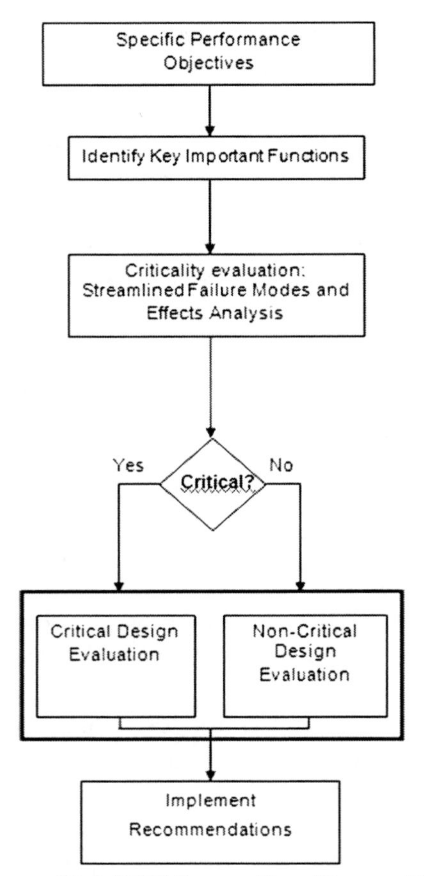

Figure 10-6 RCDA Process Flow Diagram [13]

- Establish a qualification basis comprising requirements, specification, and description. Define the functionality and limiting parameters.
- Screen the technology based on identification of failure mechanisms and their risk, and classification of the technology according to degree of newness to focus the effort where the related uncertainty is most significant.
- Assess maintenance, monitor conditions, and review possible modifications for their effect on the qualification basis.
- Plan and execute reliability data collection. The data are used to analyze the risk involved in not meeting the specifications through experience, numerical analysis, and tests.
- Analyze the reliability and, thereby, the risk of the failure modes related to the functional requirements.

REFERENCES

[1] American Petroleum Institute, Recommended Practice for Subsea Production System Reliability and Technical Risk Management, API RP 17N, 2009, March.
[2] R. Cook, Risk Management, England, 2004.
[3] H. Brandt, Reliability Management of Deepwater Subsea Field Developments, OTC 15343, Offshore Technology Conference, Houston, 2003.
[4] Det Norsk Veritas, Risk Management in Marine and Subsea Operations, DNV-RP-H101, 2003.
[5] J. Wang, Offshore Safety Case Approach and Formal Safety Assessment of Ships, Journal of Safety Research No. 33 (2002) 81–115.
[6] J. Aller, M. Conley, D. Dunlavy, Risk-Based Inspection, API Committee on Refinery Equipment BRD on Risk Based Inspection, 1996, October.
[7] International Association of Oil & Gas Producers, Managing Major Incident Risks Workshop Report, 2008, April.
[8] C. Duell, R. Fleming, J. Strutt, Implementing Deepwater Subsea Reliability Strategy, OTC 12998, Offshore Technology Conference, Houston, 2001.
[9] M. Carter, K. Powell, Increasing Reliability in Subsea Systems, E&P Magazine, Hart Energy Publishing, LP, Houston, 2006, February 1.
[10] H.B. Skeels, M. Taylor, F. Wabnitz, Subsea Field Architecture Selection Based on Reliability Considerations, Deep Offshore Technology (DOT), 2003.
[11] F. Wabnitz, Use of Reliability Engineering Tools to Enhance Subsea System Reliability, OTC 12944, Offshore Technology Conference, Houston, 2001.
[12] K. Parkes, Human and Organizational Factors in the Achievement of High Reliability, Engineers Australia/SPE, 2009.
[13] M. Morris, Incorporating Reliability Centered Maintenance Principles in Front End Engineering and Design of Deep Water Capital Projects, http://www.reliabilityweb.com/art07/rcm_design.htm, 2007.
[14] Det Norsk Veritas, Qualification Procedures for New Technology, DNV-RP-A203, 2001.
[15] M. Tore, A Qualification Approach to Reduce Subsea Equipment Failures, in: Proc. 13th Int. Offshore and Polar Engineering Conference, 2003.

CHAPTER 11

Subsea Equipment RBI

Contents

Subsea Engineering Handbook
ISBN 978-0-12-397804-2, doi:10.1016/B978-0-12-397804-2.00011-4

11.1. INTRODUCTION

Subsea equipment includes wellheads, wellhead connectors, trees, manifolds, jumpers, PLETs, pipeline connectors, pipelines and risers, and umbilicals and UTA. This chapter presents the concept of a risk-based inspection (RBI) study for subsea systems that is designed to ensure the integrity of subsea equipment.

As subsea operations move into ever deeper water, the costs and challenges associated with subsea systems are becoming more severe than ever. Therefore, it is important to develop a subsea equipment RBI that can deliver great benefits in terms of ensuring the integrity of a subsea system.

A subsea equipment RBI is a method that uses equipment criticality and failure modes as criteria for establishing the maintenance and inspection plan for each item of subsea equipment.

11.2. OBJECTIVE

The prime objective of a subsea equipment RBI is to reveal the major failure types and establish a mitigation and control plan to maximize the reliability and availability of subsea systems and ensure the integrity of those same systems by avoiding and reducing the frequency of poor outcomes. A subsea equipment RBI operates on many levels throughout the project. It is essential for information on major risks to flow upward into a central register where it can be aggregated, prioritized, and managed. The scope of the RBI system must be multidisciplinary and include technical, commercial, financial, and political risks, and must equally consider both threats and opportunities.

The objectives of a subsea equipment RBI are summarized as follows:
- Carefully define steps to improve facility performance.
- Make use of a systematically applied equipment database.
- Incorporate advances in technology promptly.
- Optimize inspection, test, and maintenance work.
- Eliminate unplanned equipment failures.
- Improve plant performance.
- Reduce costs.

With a subsea RBI module, a qualitative RBI assessment can be performed. All significant failure causes and modes will be identified and analyzed to develop a cost-effective inspection plan for subsea equipment.

11.3. SUBSEA EQUIPMENT RBI METHODOLOGY

11.3.1. General

The RBI approach prioritizes and optimizes inspection efforts by balancing the benefits (risk reduction) against the inspection costs. The results of RBI planning include a specification of future inspections in terms of:
- Which subsea systems to inspect;
- What degradation modes to inspect for;
- How to inspect;
- When to inspect;
- How to report inspection results;
- Directions for actions if defects are found or not found.

RBI planning is a "living process." It is essential for the analyses to utilize the most recent information about subsea systems in terms of design, construction, inspection, and maintenance of subsea systems in order to optimize future inspections. From the inspection results, new and better knowledge of operational conditions provides a basis for an updating of the structural reliability and further a correction of the time to next inspection.

A subsea equipment RBI can be analyzed in two ways. One way is to use a pipeline RBI by assuming that the subsea equipment, such as a manifold or jumper, is a pipeline. (PaRIS software, developed by OPR, is able to do

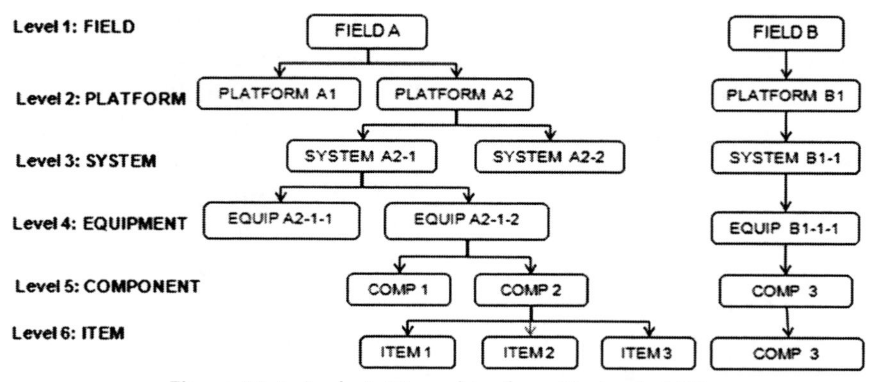

Figure 11-1 Analysis Hierarchies for a Mechanical RBI

a subsea equipment RBI.). The other way is to use the mechanical RBI involved in the failure of component parts.

In a mechanical RBI, the efforts focus mainly on the component degradation mechanisms of the subsea equipment, after which the equipment failure conditions can be determined based on these components. Figure 11-1 is an illustration of the analysis hierarchy for a mechanical RBI.

11.3.2. Subsea RBI Inspection Management

Subsea equipment RBI planning, execution, and evaluation should not be a one-time activity, but a continuous process in which information and data from the process and the inspection/maintenance/operation activities are fed back to the planning process, as indicated in Figure 11-2.

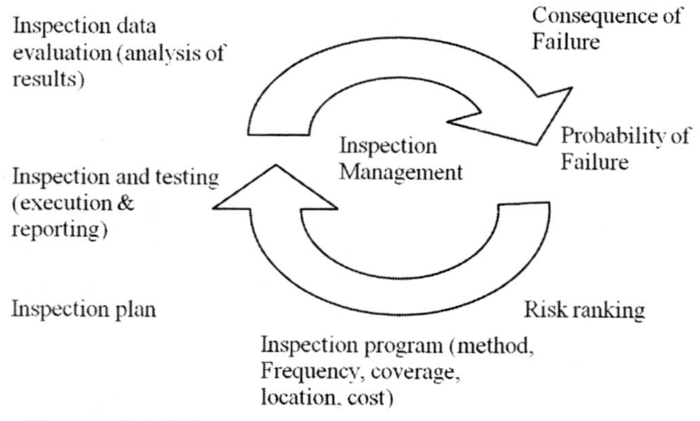

Figure 11-2 Subsea Equipment RBI Management Process [1]

11.3.3. Risk Acceptance Criteria

Risk acceptance criteria are the limits above which an operator will not tolerate risk on the installation. These criteria must be defined for each type of risk to be assessed. Similar to the traditional pipeline RBI, the subsea equipment RBI quantifies risk from the aspects of Safety, environment, and economy. Most importantly, the safety level depends on product, manned condition, and location class. If the product is toxic or the location is in a sensitive area, then the safety class should be considered to be high.

The risk acceptance criteria are used to derive the time of inspection, which is carried out prior to the acceptance limit being breached. This would allow either the reassessment of the risk level based on better information, a detailed evaluation of any damage, or the timely repair or replacement of the degraded component.

The acceptance criteria are defined for each of the different consequence categories. Acceptance criteria may be based on previous experience, design code requirements, national legislation, or risk analysis. The acceptance criteria for a function may be "broken down" into acceptance criteria for the performance of the individual items comprising the function.

Generally, due to the quick reaction ability of extensive valves and sensors, the main risk is the economic loss of the subsea tree or manifold. For a pipeline and riser, however, safety, environmental, and economic risks should be considered.

For each type of item and for each deterioration process, the acceptance criteria with regard to personnel risk, environmental impact, or economic risk may be represented by a risk matrix as illustrated in Figure 11-3.

11.3.4. Subsea RBI Workflow

Figure 11-4 illustrates the RBI workflow, which consists of the following items:

- Data gathering;
- Initial assessment;
- Detailed assessment;
- Inspection reference plan (IRP) and maintenance reference plan (MRP).

Collection of information is a fundamental task at the start of any RBI study. The information that is actually required depends on the level of detailed RBI assessment. For screening, only some basic information is required. With an increased level of assessment, more documentation is required.

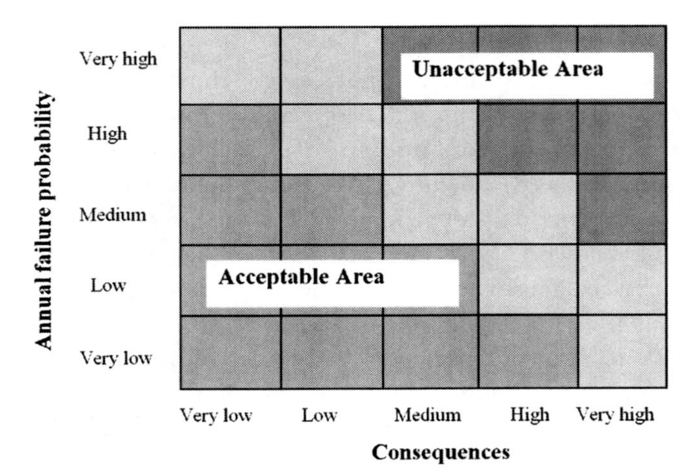

Figure 11-3 Illustration of the Principle of Acceptance Criteria and Risk Matrix [2]

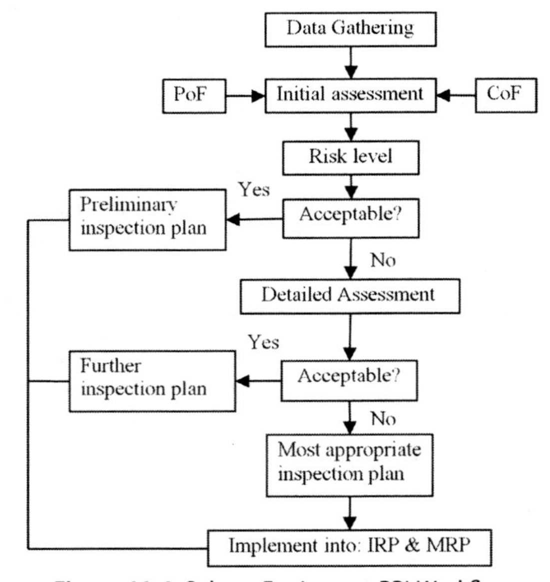

Figure 11-4 Subsea Equipment RBI Workflow

Typical data required for an RBI analysis may include, but are not limited to:

- Type of equipment;
- Materials of construction;
- Inspection, repair, and replacement records;
- Process fluid compositions;
- Inventory of fluids;

- Operating conditions;
- Safety systems;
- Detection systems;
- Deterioration mechanisms, rates, and severity;
- Personnel densities;
- Coating, cladding, and insulation data;
- Business interruption costs;
- Equipment replacement costs;
- Environmental remediation costs.

The initial assessment is intended to be an efficient initial qualitative assessment not requiring a detailed description of the system. In many cases this level can be the most appropriate approach for the inspection planning if detailed information or models are not available, or the benefit of a more costly assessment is marginal. The RBI initial assessment is mainly based on sound engineering judgment.

The detailed assessment is performed on a component level, defining the different sections of the subsea equipment and analyzing the reason for each individual degradation mechanism to provide results that can guide the development of an optimized inspection plan. This is different from the initial assessment, which considers the individual subsea equipment as one component. The detailed assessment is carried out for different detail levels with advanced and accurate prediction models. The detailed assessment incorporates both deterministic and probabilistic assessment of the probability of failure.

The MRP encompasses the action plan for "high-risk" items in the RBI detailed assessment. The MRP outlines the method of mitigation inclusive of remedial action to extend the remaining life of the components and to reduce the risk. The IRP is a document that describes how the RBI initial assessment will be implemented. Following the IRP will ensure that the failures occurring in the pipeline systems are managed in a cost-effective manner and kept within the acceptable limits of safety and economic risk.

11.3.5. Subsea Equipment Risk Determination

11.3.5.1. Subsea Equipment PoF Identification

The offshore reliability data (OREDA) database [2] can be used as the starting point to calculate the failure rate of subsea equipment and the process can be summarized as follows:

- Failure database from OREDA;
- Equipment operation condition evaluation;

- Failure database modification;
- Subsea equipment component probability of failure (PoF) identification;
- Subsea equipment PoF calculation.

The U.K. PARLOC database can also be used to predict pipeline failure rates based on the following two assumptions [3]:

- The development of failure rates is coherent with the historical statistical results in the PARLOC database;
- The PARLOC database from the North Sea is applicable to pipeline maritime space analyses.

It is reasonable to use the PARLOC database as a starting point for pipeline hazard identification, leaking hole size predictions, and PoF calculations.

Before calculating the failure rate of the subsea equipment, the operational conditions, including the corrosion rate, erosion rate, flow assurance problem and the properties and characteristics of the equipment, should be evaluated in order to modify the failure data recorded in the OREDA.

The CO_2 corrosion rate calculation module in the Norsok code M-506 is used to calculate the corrosion rate [4]:

- Water cut;
- Pressure;
- CO_2 concentration;
- pH;
- Temperature;
- Oxygen concentration;
- Inhibitor efficiency;
- Flow regime;
- Biological activity;
- Etc.

The erosion rate can also be determined by the equation below based on the following items and can be monitored continuously [5]:

- Fluid density and viscosity;
- Sand size and concentration;
- Sand shape and hardness;
- Flow rate of fluid and sand;
- Pipe diameter;
- Flow regime;
- Geometry;
- Etc.

Flow assurance analyzes the formation of wax, asphalting, hydrate, scale, etc. The prediction of the choke operation is based on the knowledge of flow assurance. In the analysis, the following items should be considered:

- Wax appearance temperature;
- Pour point temperature;
- Hydrate formation curve.

According to the monitored pressure and temperature, flow assurance software (such as PIPESIM or OLGA) can be used to determine the temperature at which wax or hydrate will appear in the monitoring pressure condition. Then this temperature can be compared with the monitored temperature to determine whether the wax or hydrate may occur.

The method for using statistical failure data to calculate the PoF of subsea equipment components is summarized as follows. The failure rate estimate is given by:

$$\lambda = n/t$$

A 90% confidence interval is given by:

$$(\text{low.high}) = \left(\frac{1}{2t}z_{0.95,zn}, \frac{1}{2t}z_{0.05,2(n+1)}\right)$$

Finally, the subsea equipment failure rate can be determined by the equation:

$$\text{PoF} = 1 - \sum_{i=1}^{n}(1 - P_1)(1 - p_2)\ldots(1 - p_i)$$

Also some subsea equipment can be simplified according to its structure and characteristics; for example, the manifold can be equal to the pipe and its failure rate calculation method can be used to determine the manifold PoF.

11.3.5.2. Subsea Equipment CoF Identification

The failure of subsea equipment is unlikely to cause a large quantity of leakage. Leakage is limited by the quick reaction of extensive valves and sensors, so the main consequence is the economic loss due to the delays caused by needed repairs and the repair costs. But sometimes subsea equipment failure will also cause significant leakage, for instance the 2010 GOM incident is environment disaster due to the failure of BOP system. The economic consequences of failure (CoF) of the subsea equipment can be analyzed on the base of:

- Product type;

Table 11-1 Qualitative Economic CoF Category

	Economy (%)				
Product	**<2**	**2–5**	**5–10**	**10–20**	**>20**
Gas, well fluid	A	B	C	D	E
Gas, semiprocessed	A	B	C	D	E
Gas, dry	A	B	C	D	E
Oil, well fluid	A	B	C	D	E
Oil, semiprocessed	A	B	C	D	E
Oil, dry	A	B	C	D	E

- Flow rate;
- Delay time of production.

Also the repair time of subsea equipment components can be gained from the OREDA database. Relative to overall field production throughput (%) can be used to determine the CoF. Table 11-1 is an example of qualitative CoF estimates for a tree failure.

11.3.5.3. Subsea Equipment Risk Identification

Risk is calculated by the equation risk = PoF × CoF. Risk criteria for subsea equipment should be much stricter than that for export pipelines, and the risk can be qualitative or quantitative.

11.3.6. Inspection Plan

The result from the subsea RBI assessment defines a proposed inspection plan for the subsea equipment system.

In this phase proposed inspection plans for all items are collected and grouped into suitable inspection intervals. Deliverables from this activity are handbooks giving recommendations for inspection scheduling. These handbooks describe the following for each of the considered pipelines systems:

- When to inspect (inspection time);
- Where to inspect (which items to be inspected);
- How to inspect (inspection methods, or level of inspection accuracy).

Figure 11-5 shows the setting of inspection intervals for both the time-based damage cause and the event-based damaged cause. Limited assessment models and data are utilized to predict the failure probability and consequences rating in an RBI assessment. Inspection planning from RBI must therefore apply a pragmatic view and sound engineering judgment. As

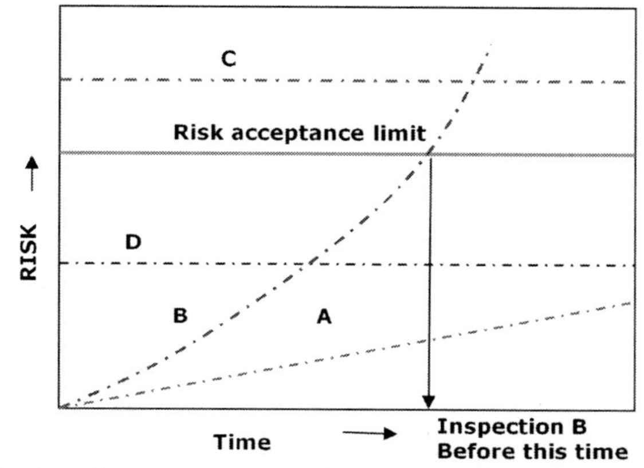

Figure 11-5 ***Setting of Inspection Intervals.*** *Note:* Line A and line B represent time-based damage causes. Lines C and D represented event-based damage causes.

described earlier, the calculated PoF will increase with time for some damage causes, and will maintain constant over time for others. This will have some impact on the inspection planning and two basic methodologies are applied:

- Time-based damage causes;
- Event-based damage causes.

The first methodology addresses cases for which PoF increases with time. The criterion for initiating an inspection is when a defined acceptable level of risk is exceeded. Over time, the PoF increases and hence the risk increases, and the time for inspection is determined as the point in time when the risk exceeds the acceptance limit.

The second methodology for inspection planning covers the case when PoF is constant with time, and hence the risk is also constant with time (assuming no variation in consequence). In this case, provided that the initial estimate of risk is acceptable, the risk will not exceed the acceptable risk limits and hence will not generate a date for inspection. However, for those cases where the risk is above a certain level, inspection should be performed at regular intervals, and more often for high-risk items than for low-risk items.

RBI competence group should take out inspection schedule and perform inspection during the time period when the PoF exceeds the limit PoF for the internal corrosion degradation mechanism.

11.3.7. Offshore Equipment Reliability Data

The OREDA project was established in 1981 in cooperation with the Norwegian Petroleum Directorate [2]. The initial objective of OREDA was to collect reliability data for safety equipment, and the main objective of OREDA now is to contribute to improved safety and cost effectiveness in the design and operation of oil and gas exploration and production facilities. The scope of OREDA was extended to cover reliability data from a wide range of equipment used in oil and gas exploration and production.

The OREDA database can be divided into four levels:

- *Field/installation:* This is an identifier for the subsea field and its installation(s). For each field several installations may be included.
- *Equipment unit:* This refers to an equipment unit on the highest equipment level used in OREDA, which typically includes a unit with one main function, for example, a Christmas tree or control system.
- *Subunit:* An equipment unit is subdivided in several subunits, each of which performs function(s) that are required by the equipment unit to perform its main function. Typical subunits include umbilicals and the HTP. The subunits may be redundant, for example, there may be two independent HPUs.
- *Components:* These are subsets of each subunit and will typically consist of the lowest level items that are being required or replaced as a whole (e.g., valve, sensor).

For each topside equipment unit and subsea equipment unit, the following information is presented:

- A drawing illustrating the boundary of the equipment unit, that is, a specification of subunits and components that are part of the equipment unit;
- A listing of all components;
- The observed number of failures for each component;
- The aggregated observed time in service for the equipment unit, classified as calendar time;
- An estimate of the failure rate for each component with associated uncertainty limits;
- A repair time estimate, that is, the elapsed time in number of hours required to repair the failure and restore functioning. This time is the active repair time, that is, the time when actual repair work was done;
- Supportive information, for example, the number of items and installations;
- A cross-tabulation of component versus failure mode, of subunit versus failure mode, of equipment unit versus failure mode, and of failure descriptor versus failure mode.

11.4. PIPELINE RBI

11.4.1. Pipeline Degradation Mechanisms

Damage is defined as an undesired physical deviation from a defined baseline condition that can be detected by an inspection technique or combination of techniques. The reasons for damage can be grouped into the following three categories:

- Event-based damage, for example, a dropped object, dragging trawl gear, landslide, or dropped anchor;
- Condition-based damage, for example, a change in pH, change in operating parameters, or a change in CP system;
- Time-based damage, for example, corrosion, erosion, or fatigue.

An RBI assessment considers the following reasons for damage to a pipeline:

- Internal corrosion;
- External corrosion;
- Internal erosion;
- External impact;
- Free-spans;
- On-bottom stability.

Each of these damage reasons is detailed in the next section for a pipeline system with respect to the likelihood of occurrence and the consequence of failure.

11.4.2. Assessment of PoF Value

In this section, the models for assessing the various reasons for damage are described. The user should keep in mind that these models are intended to be conservative, identifying an inspection plan based on a minimum level of provided data. As an alternative to an inspection plan, a more detailed assessment or remedial actions could also be proposed.

Proper actions and measures should be taken to ensure the integrity of the pipeline when the inspection outcome and risk assessment indicate that the risk is close to or exceeding the risk limit. These measures can include closer inspection, or mitigation actions as reduced pressure or fitness for service assessment

11.4.2.1. Internal Corrosion

Internal corrosion is the major factor of wall thinning during operation. Corrosion is a complex mechanism, depending on fluid composition, presence of water, operational changes, etc.

For an RBI qualitative assessment, the PoF for a pipeline is classified based on the following information:

- Outcome of last inspection, if any;
- Time since last inspection, if any;
- Corrosivity based on the product;
- Monitoring and maintenance level.

The category of PoF at the time of inspection is determined based on the inspection findings of an insignificant, moderate, or significant level of observed corrosion defects.

Also the PoF increases with time due to potential growth of corrosion defects. The corrosion rate is dependent on the corrosivity of the product. Except for the most corrosive products, credit is given for a good *condition monitoring level*, which is a measure of the follow-up of the operator.

In the corrosion quantitative assessment, internal corrosion damage may be caused by different mechanisms of corrosion degradation. In hydrocarbon pipeline systems, corrosion damage may be due to the following items:

- CO_2 corrosion;
- H_2S stress corrosion cracking (SSCC);
- Microbiological-induced corrosion (MIC).

The annual failure probability is calculated for both burst and leak failure modes, based on the PoF calculation procedure for metal loss defects. The most important input parameters are whether the corrosion takes place, and the corrosion rate. The factors affecting the corrosion rate can be summarized as follows:

- Material;
- Product type;
- Water content;
- Temperature;
- CO_2 partial pressure;
- Inhibition efficiency (if any);
- Flow regime.

The following items should be considered in order to calculate the PoF using the structural reliability method:

- Outer diameter;
- Internal pressure, maximum allowable operating pressure (MAOP);
- Nominal wall thickness;
- Material strength, SMTS;
- Commissioning year.

11.4.2.2. External Corrosion

Normally pipelines are protected against external corrosion by a corrosion coating that covers the complete external surface of the pipeline. An *impressed current cathodic protection* (ICCP) system in conjunction with sacrificial anodes is used in pipeline systems when the corrosion coating of a pipeline is damaged.

External coating damage could be caused by impacts from vessels, anchors, trawls, etc. In practice, external corrosion is normally not a big problem for the submerged section of a submarine pipeline.

In a qualitative assessment, the PoF categories for the offshore pipeline are based on the following issues and are summarized in Table 11-2:

- Outcome of last inspection, if any;
- Time since last inspection, if any;
- Abnormal anode depletion inspection, if any;
- IC "potential" readings, if any;
- Operating temperature.

The PoF increases with time due to potential growth of corrosion defects. Even though the corrosion occurs in the splash zone, the temperature of the riser surface is assumed to be equal to the operation temperature, which is used for the calculation of corrosion rate.

Table 11-3 shows the operating years for a one-unit increase in the PoF category for external corrosion, which is dependent on the operating temperature and the inspection results from the last inspection. There may be no corrosion for several years, but once corrosion is initiated, the defects can develop at a high rate.

Table 11-2 PoF Category for External Corrosion of Risers Depending on Inspection Results

Inspection Finding	PoF	Description
Insignificant	1	None of the inspection defect population challenges the 50% target level of the current pipeline MAOP.
Moderate	3	The most severe defect from the defect population that shall challenge the 50% target level of the current pipeline MAOP.
Significant	5	The most severe defect from the defect population that shall challenge the 80% target level of the current pipeline MAOP.
No inspection or blank	1	Time of "inspection" is set equal to commissioning/installation date.

Table 11-3 Years of Operating for a One-Unit Increase in PoF Category for Eternal Corrosion

Inspection Finding	Years of Operating for a One-Unit Increase in PoF When Operating Temperature Is	
	$<40°C$	$\geq 40°C$
Insignificant	4	3
Moderate	3	2
Significant	2	1
No inspection or blank	3	2

11.4.2.3. Internal Erosion

Erosion is not a common reason for pipeline failures. However, for high-velocity fluid containing sand particles, erosion could occur, especially at bends, at reduced diameters, and where pipeline connections or other geometrical details are present.

Erosion is usually not a problem if the velocity is less than about 3 to 4 m/s. The erosion rate is proportional to the mass of sand in the fluid, and large particles cause more severe erosion than smaller particles. The velocity is a very important parameter when considering erosion, because the erosion rate is proportional to the power of 2.5 to 3.0 for the velocity.

The defect characteristics of internal erosion can be similar to corrosion defects, and the same inspection categorization as for corrosion defects is used.

The PoF increases with time due to the potential growth of the erosion defects. The erosion rate is dependent on the velocity of the product. High velocities will result in a more rapid increase in the PoF, and the number of years for a one-unit increase of PoF is therefore dependent on sand (product) velocity.

Table 11-4 shows the operating years for a one-unit increase in PoF category for internal erosion. If sand is not present, the PoF is constant with time and equal to 1.

Table 11-4 Operating Years for a One-Unit Increase in PoF Category for Internal Erosion

Sand Velocity (*V*)	Years of Unit Increase in PoF (years)
$V < 3$m/s	10
3 m/s $\leq V <$ 8 m/s	3
$V \geq 8$ m/s	1
Unknown velocity	1

11.4.2.4. External Impact

Damage due to an external impact may arise from dropped objects, anchor impacts, anchor dragging, trawling, boat impacts on risers, fish-bombing, etc.

An external impact is an event-based damage reason, and if the annual probability of an impact is constant, the PoF is also close to constant. An inspection will have no or limited impact on the PoF, but it is still preferable to inspect the line at regular intervals.

The PoF category for an external impact for a pipeline is based on the following information:

- Outcome of last inspection, if any;
- Trawling activity;
- Pipeline diameter and concrete coating thickness;
- Buried;
- Marine operation activity.

Table 11-5 lists the PoF category for an external impact from trawling. Table 11-6 shows the PoF category for marine operation activities. The PoF category is independent of whether the line is buried or not.

11.4.2.5. Free-Spans

Free-spans occur in almost every pipeline, unless special conditions exist (e.g., burial). Seabed scouring is one of the major reasons for free-spans. Free-spans may be subject to in-line or cross-flow vibrations, which can eventually lead to fatigue failure.

The PoF category at time of inspection is determined in Table 11-7 based on inspection findings for offshore pipelines only.

To assign an inspection program for free-spans, a detailed assessment should be performed. The PoF in the initial assessment is assumed not to increase with time to initiate a detailed assessment. For the PoF evaluation, keep in mind that free-spans in soft soils tend to shift positions and change in length. For free-span records in soft soil, special consideration is made for a static span.

Table 11-5 PoF Categories for External Impact from Trawling

Description of Pipeline Condition	PoF
Diameter < 8 in.	4
8 in. ≤ *Diameter* < 16 in.	3
Diameter > 16 in. and coating < 40 mm	3
Diameter > 16 in. and coating ≥ 40 mm	1
Buried (all *Diameter*)	1

Table 11-6 PoF Categories for Marine Operations

Marine Operation Activity	PoF	Description
Low	1	Little or insignificant marine operation activity in the area compared to the overall spectrum of marine activity levels to which the total pipeline assets are exposed (as compared to historical marine activity for the pipeline location and risk assessment studies).
Medium	3	Moderate marine activity compared to the overall spectrum of marine activity levels to which the total pipeline assets are exposed (as compared to historical marine activity for the pipeline location and risk assessment studies).
High	4	High marine activity compared to the overall spectrum of marine activity levels to which the total pipeline assets are exposed.

Table 11-7 PoF Category Dependent on Inspection Outcome of Maximum Free-Span Length

Inspection Finding	PoF	Description
Insignificant	1	$L/Diameter < 20$
Moderate	3	$20 \leq Length/Diameter < 30$
Significant	5	$Length/Diameter \geq 30$, or dynamic spans
No inspection or blank	3	No inspection or blank
Dynamic free-span	5	The inspection history defines the span as dynamic. Span developed on soft soil.

11.4.2.6. On-Bottom Stability

Offshore pipelines may move under strong current conditions if the pipeline has insufficient capacity to ensure on-bottom stability. Except for small-diameter pipelines, a weight coating is often required to obtain sufficient stability. Concrete coating is normally used as a weight coating. Such a coating also protects the pipeline system against external impacts.

In an RBI qualitative assessment, the PoF category for the on-bottom stability of a pipeline is based on the following information:

- Outcome of last inspection, if any;
- Time since last inspection, if any;
- Buried or not;
- Location (onshore/offshore).

Table 11-8 PoF Category Dependent on Inspection Outcome for On-bottom Stability

Inspection Finding	PoF	Description
Insignificant	1	No, or only insignificant, movement of pipeline (displacement $\leq$ 10 m). Whole pipeline at deepwater, say, water deeper than 150 m.
Moderate	3	Moderate lateral movements; no sharp curves (10 m $<$ displacement $\leq$ 20 m).
Significant	5	Lateral movements resulting in a nonpreferable condition; sharp curves; signs of buckle or wrinkling; close to debris or outside route corridor (displacement $>$ 20 m).
No inspection or blank	1	Time of "inspection" is set equal to commissioning/installation date.

Table 11-8 shows the PoF category based on inspection findings when the inspection determines the lateral movement is insignificant, moderate, or significant.

11.4.3. Assessment of CoF Values

The CoF is measured in terms of safety, economic loss, and environmental pollution.

11.4.3.1. Safety Consequences

Safety consequences considers personnel injury or loss of life, which can be obtained from quantitative risk assessment (QRA) studies and presented in terms of *potential loss of life* (PLL) [0]. Table 11-9 lists the safety consequence rankings.

Table 11-9 Safety Consequence Rankings

CoF Factor	CoF Identification	Description
A	Very low	Personnel injury not likely.
B	Low	Potential minor injury; no lost time due to injury or loss of life.
C	Medium	Potential lost time due to injury limited to no more than one/a few persons. No potential for fatalities.
D	High	Potential multiple lost time due to injury. Potential for one fatality.
E	Very high	Potential for multiple fatalities.

11.4.3.2. Economic Consequences

The direct economic loss should be the spill volume of oil or gas and the repair costs. However, the deferred production time should be considered if there is a need to repair equipment. The repair can be divided into two parts, namely, consequences for a leak and consequences for a rupture. The repair consequence is also dependent on the location of the failure (e.g., above water, splash zone area, or underwater). Economic consequences due to a business interruption or deferred production relate to the costs due to the shutdown of the pipeline. An important factor to be considered is the redundancy in the system, whereby production is maintained by using by-pass lines. Table 11-10 lists economic consequence rankings.

11.4.3.3. Environmental Consequences

Environmental consequences are concerned with the impact on the environment of various types of product releases. The volume of oil dispersed in water can be modeled using software called Adios.

The severity of any environmental pollution is determined by the volume of oil dispersed in water and the local conditions, for example, the fishing resources. The environmental pollution ranking is determined by the recovery years of the natural resource, which is decided by the recovery of the local resource and local governmental efforts. Table 11-11 shows the environmental consequence rankings for different events.

11.4.4. Risk Identification and Criteria

Risk for a pipeline is calculated by this equation:

$$\text{Risk} = \text{PoF} \times \text{CoF}$$

Note that risk criteria can be qualitative or quantitative. Table 11-12 lists qualitative risk criteria.

Table 11-10 Economic Consequence Rankings

		Costs Related to Loss of Production	
CoF Factor	**CoF Identification**	**Relative to Overall Field Production Throughput (%)**	**Comment**
A	Very low	< 2	Minor flowlines
B	Low	2–5	Small flowlines
C	Medium	5–10	Medium flowlines
D	High	10–20	Important flowlines
E	Very high	> 20	Trunk lines

Table 11-11 Environmental Consequence Rankings

CoF Factor	CoF Identification	Description
A	Very low	None, small, or insignificant impact on the environment. There is either no release of internal media or only insignificant release of low toxic or nonpolluting media.
B	Low	Minor release of polluting or toxic media. The released media will be dispersed, decompose, or be neutralized rapidly by air or seawater.
C	Medium	Minor release of polluting or toxic media or large release of low-polluting or toxic media. The released media will take some time to disperse, decompose, or be neutralized by air or seawater, or can easily be removed.
D	High	Large releases of polluting and toxic media, which can be removed, or will after some time disperse, decompose, or be neutralized by air or seawater.
E	Very high	Large release of highly polluting and toxic media, which cannot be removed and will take a long time to disperse, decompose, or be neutralized by air or seawater.

Table 11-12 Qualitative Risk Criteria [6]

PoF Category	Annual Probability of Failure					
5	Expected failure	M	H	H	VH	VH
4	High	M	M	H	H	VH
3	Medium	L	M	M	H	H
2	Low	VL	L	M	M	H
1	Virtually nil	VL	VL	L	M	M
CoF of Failure Category		A	B	C	D	E

11.5. SUBSEA TREE RBI

A subsea tree is critical equipment in the subsea field. It consists primarily of a steel frame structure, connectors, valves, chokes, piping, tubing hanger, control systems, and a cap. The monitor data for a tree focus on the temperature, pressure, and sand conditions.

The RBI for a subsea tree can be thought an extension of a pipeline RBI in that the methodology and principles are similar to those of the pipeline RBI. The differences are that the RBI for a tree focuses on the degradation of equipment and the failure analysis of the components according to the inspection data.

11.5.1. Subsea Tree RBI Process

This section gives an example of a subsea tree risk determination based on many assumptions. The following steps are carried out for a subsea tree RBI analysis:

- Develop risk acceptance criteria.
- Collect information.
- Prepare a quantitative risk assessment.
- Generate an inspection plan.
- Input data into the inspection management system.

11.5.1.1. Collection of Information

The information collected for a subsea tree RBI includes design data, operation data, and a corrosion/erosion study report. These data are used to determine the PoF and CoF. These data are also used to predict the risk and establish an optimized inspection plan.

11.5.1.2. Risk Acceptance Criteria

Similar to the traditional pipeline RBI, the subsea equipment RBI risk bases its quantifications on the aspects of safety, environment, and economy. Most importantly, the safety class dependent on product, manned condition and location class. If the product is toxic or the location is in a sensitive area, then the safety class should be considered to be high.

Economic consequences are concerned with repair costs and production losses due to the delay time. The relative throughput has been used as the basis for quantification of the business consequences as shown earlier in Figure 11-3.

11.5.1.3. Degradation Mechanisms and Failure Mode

The degradation mechanisms include primarily internal/external corrosion, internal erosion, flow assurance problems, and mechanical damage. In the OREDA database, subsea tree equipment is considered and failure data have been recorded. A subsea tree is divided into the following parts:

- Chemical injection coupling;
- Connector;
- Debris cap;
- Flow spool;

- Hose (flexible piping);
- Hydraulic coupling;
- Piping (hard pipe);
- Tree cap;
- Tree guide frame;
- Valve, check;
- Valve, control;
- Valve, other;
- Valve, process isolation;
- Valve utility isolation.

Tubing Hanger

- Chemical injection coupling;
- Hydraulic coupling;
- Power/signal coupler;
- Tubing hanger body;
- Tubing hanger isolation plug.

In the OREDA, the failure mode of a tree consists of the following items:

- External leakage-process medium;
- External leakage-utility medium;
- Fail to close/lock;
- Fail to function on demand;
- Fail to open/unlock;
- Internal leakage-process medium;
- Internal leakage-utility medium;
- Leakage in closed position;
- Loss of barrier;
- Loss of redundancy;
- No immediate effect;
- Other;
- Plugged/choked;
- Spurious operation;
- Structural failure.

11.5.2. Subsea Tree Risk Assessment

In subsea tree RBI assessment, the PoF will be determined for different failure mechanisms and for different components as illustrated in Section 11.3.5. The consequences can be identified in terms of economic loss according to the repair times and repair costs.

11.5.2.1. Failure Database from 2002 OREDA Database

The 2002 OREDA database is used as our starting point in calculating PoF. In the OREDA database, the failure rate is presented in form of three categories: lower bound, best estimate, and upper bound.

For the process, the table titled "Failure Descriptor versus Failure Mode, Wellhead & Xmas Tree" on page 835 of OREDA [2] should be used as a reference for the failure data of the subsea tree components. In Table 11-13, the failure modes of every component are summarized.

11.5.2.2. Failure Database Modification

The failure data are statistical results deduced from the 2002 OREDA database. Note, however, that individual subsea equipment may have quite different histories, properties, characteristics, and functions. Therefore, these values require further modification based on the special conditions and properties of the equipment in question and the experience of the engineer. For example, if the material is high-strength steel, material failure can be neglected.

In the process of modification, the corrosion/erosion rate should be calculated. Flow assurance problems and operating conditions should also be analyzed.

We assume that the corrosion/erosion condition is not severe and that the flow assurance problem is serious according to the Section 11.3.5.1. Table 11-13 is a modified result from the OREDA table on page 835 [2], in which ELP means External Leakage-Process Medium, FTS means Failure to Start, FTQ means Failure to Open, PLU means Plugged, ELU means External Leakage-Utility Medium, OTH means Others.

Table 11-13 Failure Descriptor versus Failure Mode for Christmas Tree

Component	Descriptor	ELP	FTS	FTO	PLU	ELU	OTH
Connector	Leakage	1					
Flow spool	Blockage				1		
Hydraulic coupling	Leakage					1	
Piping	Blockage				1		
Tubing hanger body	Control failure	1					
	Looseness						2
Tree guide frame	Trawl board Impact						2
Valve, choke	Blockage				1		
	Control failure		4				
	Leakage		6				
Valve, process isolation	Blockage			1			
	Leakage						3

11.5.2.3. PoF Identification

The methodology for PoF determination was illustrated earlier in Section 11.3.5. The failure rate of every tree component described in Section 11.3.5.1 should be calculated, and then the subsea tree PoF can be identified. Table 11-14 lists the failure rate of subsea tree components.

Choosing the low mean or the upper value depends on the experience of the engineer, on operating conditions, and so on. In this example, we assume that the failure rate of all parts is high, so the subsea tree failure rate can be determined by the equation specified in Section 11.3.5. The result is 2.05×10^{-6} per hour or 1.8×10^{-2} per year.

11.5.2.4. CoF Calculation

The consequence of the failure is mainly the economic loss due to the delay time and the repair costs. The delay time can be determined according to the OREDA database or the experience of the project engineers. In our example, we assume that the delay time is 2 days. Then the total economic loss can be identified according to the daily production capacity and repair costs and in this case the CoF assigned is A.

The failure of tree components is unlikely to cause a large quantity of leakage. Leakage is limited by the quick reaction of extensive valves and sensors, so the main consequence is the economic loss due to the delay caused by the repair. The economic CoF of the tree can be analyzed on the base of:

- Product type;
- Flow rate;
- Delay time of production.

Table 11-1 shown in section 11.3.5.2 shows an example of a qualitative CoF for tree failure. The overall field production throughput (%) can be used to determine the CoF.

Table 11-14 Failure Rate of Subsea Tree Components

Component/Failure Rate (per hour)	Low	Mean	Upper
Connector	0.24×10^{-9}	0.24×10^{-8}	0.2×10^{-7}
Flow spool	0.65×10^{-9}	0.65×10^{-8}	0.7×10^{-7}
Hydraulic coupling	0.4×10^{-10}	0.4×10^{-9}	0.37×10^{-8}
Piping	0.1×10^{-8}	0.1×10^{-7}	0.96×10^{-6}
Tubing hanger body	0.11×10^{-7}	0.2×10^{-7}	0.1×10^{-6}
Tree guide frame	0.86×10^{-8}	0.24×10^{-7}	0.15×10^{-6}
Valve, choke	0.7×10^{-7}	0.73×10^{-7}	0.25×10^{-6}
Valve, process isolation	0.75×10^{-7}	0.11×10^{-6}	0.5×10^{-6}

11.5.2.5. Risk Determination

The risk can be qualitative or quantitative as illustrated in Section 11.3.5.3.The risk associated with the tree is acceptable according to the assessment results.

11.5.3. Inspection Plan

Once the risk exceeds the acceptance risk, the subsea tree should be inspected according to the risk results. According to the inspection methodology, the failure data in the OREDA database should be modified according to the different operating phase conditions, and a different type of failure rate should be determined for the different operating phases, after which the risk can be identified for the different phases.

11.6. SUBSEA MANIFOLD RBI

A manifold is an arrangement of piping and/or valves designed to combine, distribute, control, and often monitor fluid flow. Subsea manifolds are installed on the seabed within an array of wells to gather production fluids or to inject water or gas into wells.

11.6.1. Degradation Mechanism

In the subsea manifold RBI, the following degradation mechanisms are analyzed:

- Internal corrosion (IC);
- External corrosion (EC);
- Internal erosion (IE);
- External impact (EI).

11.6.2. Initial Assessment

11.6.2.1. CoF Identification

The CoF is divided into three categories with a qualitative ranking system:

- *Safety consequences:* considers personnel injury or PLL;
- *Economic consequences:* considered the repair costs and business loss due to interruptions in production;
- *Environmental consequences:* considers the impact of various types of production release to the environment and the cost of cleanup.

Figure 11-6 is a workflow for PoF identification during the initial assessment.

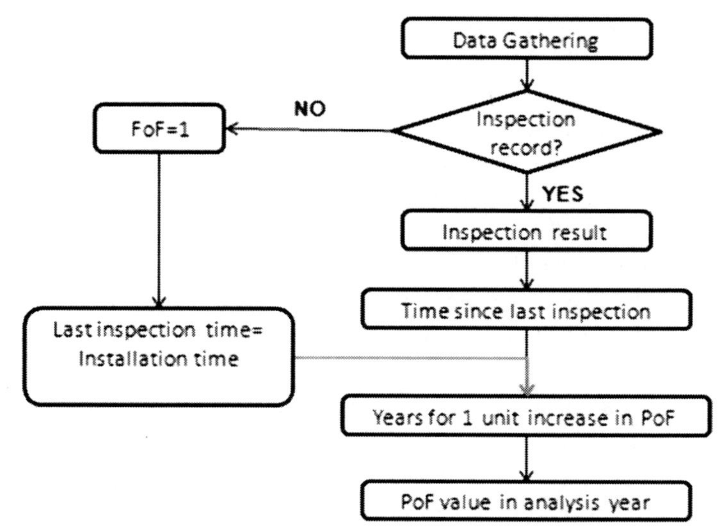

Figure 11-6 PoF Identification during Initial Assessment

11.6.2.2. Risk Criteria

During the initial assessment, the risk limit is settled as high. A risk level that exceeds "high" risk is regarded as not acceptable. Table 11-12 (earlier in the chapter) shows the criteria matrix.

11.6.2.3. Outcome

The objective of the initial assessment is to identify the risk level for each degradation mechanism using a qualitative method. According to the frame flowchart shown in Figure 11-6, if the risk level obtained in the initial assessment is acceptable (i.e., risk level is below high), then a primary inspection is carried out; if the risk level is not acceptable (i.e., risk level exceeds high), more data are required, and the assessment will proceed to a detailed assessment.

11.6.3. Detailed Assessment

11.6.3.1. Safety Class

The safety principles adopted are also risk principles, in which the target annual PoF is dependent on the consequences of failure. The safety classes for different manifolds are different, depending on the piping content (product) and the location. The safety classes include:

- Safety class–high;

- Safety class–normal;
- Safety class–low.

11.6.3.2. PoF Calculation

The PoF value for a detailed assessment is calculated at the quantitative level, using a probability statistic methodology. The best distribution that reflects actual conditions will be adopted as the basis of the theory. In the specific CoF calculation, the following items are essential:

- The distribution of component capacity;
- The distribution of component subjected loading.

11.6.3.3. Target Criteria

Acceptable criteria for the detailed assessment are defined according to the safety class level described in the Section 11.6.3.1.

- For safety class–high: target annual PoF is 10E-5;
- For safety class–normal: target annual PoF is 10E-4;
- For safety class–low: target annual PoF is 10E-3.

11.6.3.4. Outcome

The objective of the detailed assessment is to reassess the degradation mechanisms that were classified as not acceptable during the initial assessment using a quantitative method. According to the frame flowchart described for an RBI, if the assessment result in the detailed assessment is acceptable (the annual PoF is below the criteria target), then a further inspection will be carried out; if the results are not acceptable (the annual PoF exceeds the criteria target), then an immediate inspection will be performed according to IRP and MRP.

11.6.4. Example for a Manifold RBI

The example that follows describes the specific work flow for a manifold RBI assessment. In this example, we focus only on internal corrosion. Internal corrosion on the manifold piping system was detected by pigging via the MFL (Magnetic Flux Leakage) inspection method, assuming that CO_2 is the only reason for the internal defect in the carbon steel material. The defect depth and length are provided by a inspection report. As the defects grow, the capacity will decrease. The RBI methodology for a manifold will ensure that the system safety level is acceptable at any time. The analysis follows.

11.6.4.1. Initial Assessment

Input Data

The input data for the initial assessment are summarized in Table11-15.

Calculation

First the code B31G [7] is used to calculate the piping capacity subjected to internal pressure only under a corrosion defect:

$$A = 0.893 \frac{L}{\sqrt{D \cdot t}} = 0.845$$

Because the value of A is larger than 0.4, the failure pressure is:

$$P_{\text{corr}} = 2 \cdot \text{SMYS} \cdot t \cdot F/D = 21.735 \text{ MPa}$$

Because the value of P_{corr} is greater than two times the MAOP, the inspection finding in the inspection year is insignificant. So the PoF value for the inspection year is 1 according to Table 11-16. Then the years step for a one-unit increase in PoF is determined to be 4 years according to Table 11-17. So the PoF value is 3 after 9 years.

Second, we identify the CoF value for the corrosion defect. For input data, the manned condition, piping diameter, and product are known. Using Table 11-18, we can then identify the CoF value for safety, environmental, and economic factors separately.

- CoF value for safety factor is C;
- CoF value for environmental factor is D;
- CoF value for economic factors is D.

Table 11-15 Input Data for Initial Assessment

Parameters	Value
Nominal outside diameter (mm)	406
Nominal wall thickness (mm)	12.7
Defect depth (mm)	4.2
Defect length (mm)	68
Material Grade	X70 (API5L)
Product Type	Oil, semiprocessed
Maximum allowable operating pressure (MPa)	10
Design factor	0.72
Last inspection year:	2000
Corrosivity level:	Medium
Condition monitoring level	Good
Manned condition	Manned

Table 11-16 PoF Categories Dependent on Inspection Outcome for Internal Corrosion

Inspection finding	PoF	Description
Insignificant	1	None of the inspection defect population challenges the "2 times target" level of the current pipeline MAOP.
Moderate	3	The most severe defect from the defect population that challenges the "2 times target" level of the current pipeline MAOP.
Significant	5	The most severe defect from the defect population that challenges the "1.2 times target" level of the current pipeline MAOP.
No inspection or blank	1	Time of "inspection" is set equal to commissioning/installation date.

Table 11-17 Years of Operation for a One-Unit Increase in the PoF Category for Internal Corrosion

	Condition Monitoring Level		
Corrosivity	**Good**	**Normal**	**Poor**
High	1	1	1
Medium	4	3	2
Low	7	5	3

The final CoF value is chosen as the most severe value among the three conditions; therefore, the final CoF value is D.

In the initial assessment, the risk level is expressed by a 5 × 5 matrix, and the risk limit is high. In this example, the final risk in the initial assessment is still "high," which is obtained from the PoF multiplied by the CoF according to the matrix in Table 11-12 in section 11.4.4. This result implies that the risk level is not acceptable, so a more accurate detailed assessment is required.

11.6.4.2. Detailed Assessment

Input Data

The zone type is zone 2 and the inspection method for the pigging is relative. Table 11-19 summarizes the input data for a detailed assessment.

Calculation

In the detailed assessment, the defect failure pressure has to be recalculated using a more accurate method by taking the probability into consideration. To predict the defect growth tendency, the corrosion rate of the defect also needs to be assessed.

Table 11-18 General Consequence Model for RBI Initial Assessment for Loss of Containment

PRODUCT	SAFETY					ENVIRONMENT				ECONOMY	
	Manned	Occ	Unmanned	$D < 8$ in.	$D > 8$ in.	$D > 16$ in.	$D > 32$ in.	$D < 8$ in.	$D > 8$ in.	$D > 16$ in.	$D > 32$ in.
Gas, well fluid	E	D	B	B	B	B	C	B	C	D	E
Gas, semiprocessed	E	C	A	A	A	A	B	B	C	D	E
Gas, dry	E	C	A	A	A	A	B	B	C	D	E
Oil, well fluid	D	C	B	B	C	D	E	B	C	D	E
Oil, semiprocessed	C	B	A	B	C	D	E	B	C	D	E
Oil, dry	C	B	A	B	C	D	E	B	C	D	E
Condensate, well fluid	E	D	B	B	B	C	D	C	D	E	E
Condensate, semiprocessed	E	C	A	B	B	C	D	C	D	E	E
Condensate, dry	E	C	A	B	B	C	D	C	D	E	E
Treated seawater	B	A	A	A	A	A	A	A	B	C	D
Raw seawater	B	A	A	A	A	A	A	A	B	C	D
Produced water	B	A	A	B	B	B	C	A	B	C	D

Table 11-19 Input Data for Detailed Assessment

Parameters	Value
Confidence level of pigging (%)	80
Sizing accuracy of pigging (%)	10
Operating temperature (°C)	120
Operating pressure (MPa)	8
Water depth (m)	100
pH value of the product (kg/m^3)	6
Product density (kg/m^3)	780
Seawater density (kg/m^3)	1250
Mole percent of CO_2 in the gas phase (%)	5
Volume flow of liquid (m^3/day)	5184
Water cut (%)	25
Water cut at inversion point (%)	50
Specific gravity of gas relative to air	0.8
Gas compressibility (Z)	0.88
Pipe roughness (10E-6 m)	50
Viscosity of oil (Ns/m^2)	0.0088
Maximum relative viscosity (relative to oil)	7.06
Viscosity of gas ($Ns/m^{2)}$	0.00004
Corrosion inhibitor (*Fc*)	0.85
Weight -percent of glycol (%)	60

First, the CO_2 corrosion rate calculation module in the Norsok code M-506 [4] is used to calculate the corrosion rate. According to the input data, the middle variable values are shown below:

$$Kt = 7.77$$

$$f_{CO_2} = 5.282$$

$$f(\mathrm{pH})_t = 0.119$$

$$S = 0.87$$

$$Cr = Kt \cdot f_{CO_2} \cdot \left(\frac{S}{19}\right)^{0.146+0.0324\cdot\log(f_{CO_2})} \cdot f(\mathrm{pH})_t \cdot Fc = 0.18 \text{ Mm/year}$$

Second, we check whether the failure probability in the current year 2009 is not acceptable as identified in the initial assessment. In the detailed assessment, the probability of failure will be obtained via the quantitative method in which a normal distribution loading and capacity distribution are used.

Code DNV-RP F101 Part A [8] is used to calculate the failure pressure capacity, P_{corr}. Nine years later, a new defect depth will be:

$$d = 4.2 + Cr \cdot 9 = 5.82 \text{ mm}$$

The new defect length is:

$$L = 68 \cdot (1 + Cr \cdot 9/d) = 94.22 \text{ mm}$$

Then,

$$P_{corr} = \gamma_m \cdot \frac{2 \cdot t \cdot f_u}{(D - t)} \cdot \frac{(1 - \gamma d \cdot (d/t)^*)}{(1 - \gamma d \cdot (d/t)^*/Q)} = 16.6 \text{ MPa}$$

$$Q = \sqrt{1 + 0.31 (\frac{L}{\sqrt{D \cdot t}})^2} = 1.238$$

$$(d/t)^* = d/t + \varepsilon d \cdot StD = 0.539$$

To use the normal distribution to calculate the probability of failure, we have to know the loading distribution, which contains the operating pressure and the associated deviation. The deviation is the default value of 0.05 times the operating pressure. The capacity distribution contains P_{corr} and the deviation. The capacity deviation is expressed by $StD^* P_{corr}$.

$$\phi = \frac{\sqrt{2P_{corr}{}^2 - OP^2}}{\sqrt{P_{corr}{}^2 \cdot StD^2 + 0.05^2 \cdot OP^2}} = 6.33$$

$$StD = 0.078$$

3Then look for the normal distribution table and use the value of ϕ = 6.33. Before the analysis is performed, the probability of failure is identified to an acceptable limit, which is determined by the safety class, which is in turn determined by zone type, manned condition, and product. In this example, the safety class is "high" according to Table 11-20, and the limit PoF is 0.00001 according to the criteria described in the Section 11.6.3.3.

From the normal distribution, the PoF value is equal to 0.00001, which means ϕ_{limit} is equal to 4.5. Therefore, the PoF is acceptable because ϕ is larger than ϕ_{limit}.

In next job, we will predict when the PoF will exceed the limit 0.00001. To do so, we need to recalculate the value of ϕ year by year using the same

Table 11-20 Safety Class Identification Criteria

	Unmanned	Manned and Occasionally Manned	
Product	**Zone 1 and Zone 2**	**Zone1**	**Zone 2**
Gas, well fluid	Normal	Normal	High
Gas, semiprocessed	Normal	Normal	High
Gas, dry	Normal	Normal	High
Oil, well fluid	Normal	Normal	High
Oil, semiprocessed	Normal	Normal	High
Oil, dry	Normal	Normal	High
Condensate, well fluid	Normal	Normal	High
Condensate, semiprocessed	Normal	Normal	High
Condensate, dry	Normal	Normal	High
Treated seawater	Low	Low	Normal
Raw seawater	Low	Low	Normal
Produced water	Low	Low	Normal

step performed above until ϕ is larger than 4.5. In this example, the result is the year 2014.

New defect depth:

$$d = 4.2 + Cr \cdot 14 = 6.72 \text{ mm}$$

New defect length:

$$L = 68 \cdot (1 + Cr \cdot 9/d) = 108 \text{ mm}$$

Then

$$P_{\text{corr}} = \gamma_m \cdot \frac{2 \cdot t \cdot f_u}{(D - t)} \cdot \frac{(1 - \gamma d \cdot (d/t)^*)}{(1 - \gamma d \cdot (d/t)^*/Q)} = 12.72 \text{ MPa}$$

$$Q = \sqrt{1 + 0.31 \left(\frac{L}{\sqrt{D \cdot t}} \right)^2} = 1.304$$

$$(d/t)^* = d/t + \varepsilon d \cdot StD = 0.609$$

$$\phi = \frac{\sqrt{P_{\text{corr}}{}^2 - OP^2}}{\sqrt{P_{\text{corr}}{}^2 \cdot StD^2 + 0.05^2 \cdot OP^2}} = 4.4$$

So, the PoF value will exceed 0.00001 in 2014.

11.6.4.3. Inspection Planning

In RBI assessment, the inspection plan should be established based on both initial assessment and detailed assessment. However, the detailed assessment result is the major factor. So the RBI competence group should make an inspection schedule and perform the inspection in 2014 for the internal corrosion degradation mechanism.

11.7. RBI OUTPUT AND BENEFITS

The results of the RBI analysis are the inspection strategies with specification of components to inspect, locations to inspect, inspection methods to use, time intervals between the inspections, and the coverage assumed at the different inspections. The results of the inspections are used to update the modes for deterioration and thus to modify future inspection plans in accordance with the performance of the inspected components.

The results of the RBI analysis are inspection plans for subsea equipment that, on an overall basis, satisfy the acceptance criteria and at the same time minimize the economical risk for the operator of the facility.

RBI planning is a systematic approach to an installation-specific, cost–effective, and targeted inspection strategy for operation within acceptable safety, economic, and environmental criteria.

The RBI assessment may cause changes to the existing inspection plan. Any change will be based on balancing risk, that is, on balancing inspections to meet the acceptance criteria. By postponing costly inspections there is a potential of cost savings on inspection.

REFERENCES

[1] M. Humphreys, Subsea Reliability Study into Subsea Isolation System, HSE, London, United Kingdom, 1997.
[2] Det NorskeVeritas, OREDA Offshore Reliability Data Handbook, fourth ed., Det Norske Veritas Industri Norge as DNV Technica, Norway, 2002.
[3] Mott MacDonald Ltd, PARLOC 2001, The Update of the Loss of Containment Data for Offshore Pipelines, fifth ed., HSE, London, United Kingdom, 2003.
[4] Norwegian Technology Standards Institution, CO2 Corrosion Rate Calculation Model, NORSOK Standard No. M-506, (2005).
[5] M.H. Stein, A.A. Chitale, G. Asher, H. Vaziri, Y. Sun, J.R. Collbert, Integrated Sand and Erosion Alarming on NaKika, Deepwater Gulf of Mexico, SPE 95516, 2005, SPE Annual Technical Conference and Exhibition, Dallas, Texas, 2005.

[6] O.H. Bjornoy, C. Jahre-Nilsen, O. Eriksen, K. Mork, RBI Planning for Pipelines Description of Approach, OMAE2001/PIPE-4008, OMAE 2001, Rio de Janeiro, Brazil, 2001.
[7] American Society of Mechanical Engineers, Manual for Determining the Remaining Strength of Corroded Pipelines, ASME B31G-1991, New York, 1991.
[8] Det Norske Veritas, Corroded Pipelines, DNV-RP-F101, 2004.

PART Two

Flow Assurance and System Engineering

CHAPTER 12

Subsea System Engineering

Contents

12.1. INTRODUCTION

Flow assurance is an engineering analysis process that is used to ensure that hydrocarbon fluids are transmitted economically from the reservoir to the end user over the life of a project in any environment. With flow assurance, our knowledge of fluid properties and thermal-hydraulic analyses of a system are utilized to develop strategies for controlling solids such as hydrates, wax, asphaltenes, and scale from the system.

The term *flow assurance* was first used by Petrobras in the early 1990s; it originally referred to only the thermal hydraulics and production chemistry issues encountered during oil and gas production. Although the term is relatively new, the problems related to flow assurance have been a critical issue in the oil/gas industry from very early days. Hydrates were observed to cause blockages in gas pipelines as early as the 1930s and were solved with

Subsea Engineering Handbook
ISBN 978-0-12-397804-2, doi:10.1016/B978-0-12-397804-2.00012-6

chemical inhibition using methanol, as documented in the pioneering work of Hammerschmidt [1].

12.1.1. Flow Assurance Challenges

Flow assurance analysis is a recognized critical part of the design and operation of subsea oil/gas systems. Flow assurance challenges focus mainly on the prevention and control of solid deposits that could potentially block the flow of product. The solids of concern generally are hydrates, wax, and asphaltenes. Sometimes scale and sand are also included. For a given hydrocarbon fluid, these solids appear at certain combinations of pressure and temperature and deposit on the walls of the production equipment and flowlines. Figure 12-1 shows the hydrate and wax depositions formed in hydrocarbons flowlines, which ultimately may cause plugging and flow stoppage.

The solids control strategies used for hydrates, wax, and asphaltenes include the following:

- *Thermodynamic control:* Keep the pressure and temperature of the entire system out of the regions where the solids may form.
- *Kinetic control:* Control the conditions under which solids form so that deposits do not form.
- *Mechanical control:* Allow solids to deposit, but periodically removing them by pigging.

Flow assurance has become more challenging in recent years in subsea field developments involving long-distance tie-backs and deepwater. The challenges include a combination of low temperature, high hydrostatic

Hydrate

Wax

Figure 12-1 Solid Depositions Formed in Hydrocarbon Flowlines [2]

pressure for deepwater and economic reasons for long offsets. The solutions to solids deposition problems in subsea systems are different for gas versus oil systems.

For gas systems, the main concern of solids usually is hydrates [3]. Continuous inhibition with either methanol or mono-ethylene-glycol (MEG) is a common and robust solution, but low-dosage hydrate inhibitors (LDIs) are finding more applications in gas systems. The systems using methanol for inhibition are generally operated on a once-through basis. The methanol partitions into gas and water phases and is difficult to recover. Systems using MEG on the other hand normally involve the reclamation of MEG. If a hydrate plug forms, the remediation method may be a depressurization.

For oil systems, both hydrates and paraffins are critical issues. In the Gulf of Mexico (GoM), a *blowdown strategy* is commonly used [4]. The strategy relies on the insulation coating on the flowline to keep the fluids out of the hydrate and paraffin deposition regions during operation. During start-ups and shutdowns, a combination of inhibition, depressurization, and oil displacement is performed to prevent hydrate and paraffin deposition. Wax is removed by pigging. The strategy is effective, but depends on successful execution of relatively complex operational sequences. If a hydrate plug forms, it is necessary to depressurize the line to a pressure usually below 200 psi for a deepwater subsea system and wait for the plug to disassociate, which could take a very long time in a well-insulated oil system.

12.1.2. Flow Assurance Concerns

Flow assurance is only successful when the operations generate a reliable, manageable, and profitable flow of hydrocarbon fluids from the reservoir to the end user. Some flow assurance concerns are:

- *System deliverability:* Pressure drop versus production, pipeline size and pressure boosting, and slugging and emulsion. These topics are discussed in detail in Chapter 13 on hydraulics.
- *Thermal behavior:* Temperature distribution and temperature changes due to start-up and shutdown, and insulation options and heating requirements. These topics are discussed in Chapter 14 on heat transfer and thermal insulation.
- *Solids and chemistry inhibitors:* Hydrates, waxes, asphaltenes, and scaling. These topics are discussed in Chapters 15, 16, 17, and 18.

12.2. TYPICAL FLOW ASSURANCE PROCESS

As mentioned earlier, flow assurance is an engineering analysis process of developing a design and operating guidelines for the control of solids deposition in subsea systems. Depending on the characteristics of the hydrocarbons fluids to be produced, the processes corrosion, scale deposition, and erosion may also be considered in the flow assurance process. The main part of the flow assurance analysis should be done prior to or during the earlier front-end engineering and design (FEED) process. The requirements for each project are different and, therefore, project-specific strategies are required for flow assurance problems. However, during the past several decades, the flow assurance process itself has become standardized, and a typical procedure is shown in Figure 12-2. The main issues associated with the flow assurance process are as follows:

- Fluid characterization and flow property assessments;
- Steady-state hydraulic and thermal performance analyses;
- Transient flow hydraulic and thermal performance analyses;
- System design and operating philosophy for flow assurance issues.

Detailed explanations for each issue are given in the following sections. Some issues may occur in parallel, and there is considerable "looping back" to earlier steps when new information, such as a refined fluids analysis or a revised reservoir performance curve, becomes available.

12.2.1. Fluid Characterization and Property Assessments

The validity of the flow assurance process is dependent on careful analyses of samples from the wellbores. In the absence of samples, an analogous fluid, such as one from a nearby well in production, may be used. This always entails significant risks because fluid properties may vary widely, even within the same reservoir. The key fluid analyses for the sampled fluid are PVT properties, such as phase composition, GOR (gas/oil ratio), and bubble point; wax properties, such as cloud point, pour point, or WAT; and asphaltene stability.

Knowledge of the anticipated produced water salinity is also important, but water samples are seldom available and the salinity is typically calculated from resistivity logs. The composition of the brine is an important factor in the hydrate prediction and scaling tendency assessment. In cases where a brine sample is not available, predictions about composition can be made based on information in an extensive database of brine composition for deepwater locations.

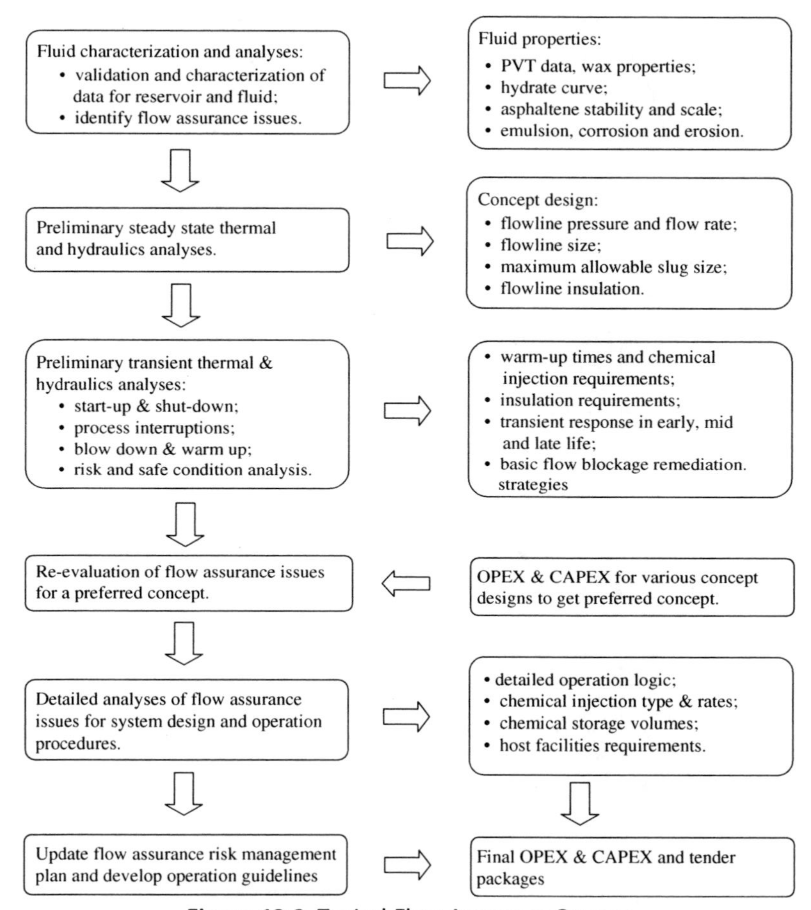

Figure 12-2 Typical Flow Assurance Process

The hydrate stability curves are developed based on PVT data and salinity estimates, and methanol dosing requirements are also obtained. A thermal-hydraulic model of the well(s) is developed to generate flowing wellhead temperatures and pressures for a range of production conditions. Then wellbore temperature transient analyses are carried out. Figure 12-3 demonstrates typical production profiles for oil, water, and gas for which the water content increases with time. Water content (water cut) is very important when choosing a flow assurance strategy to prevent hydrate formation.

Hydrate stability curves show the stability of natural gas hydrates as a function of pressure and temperature, which can be calculated based on

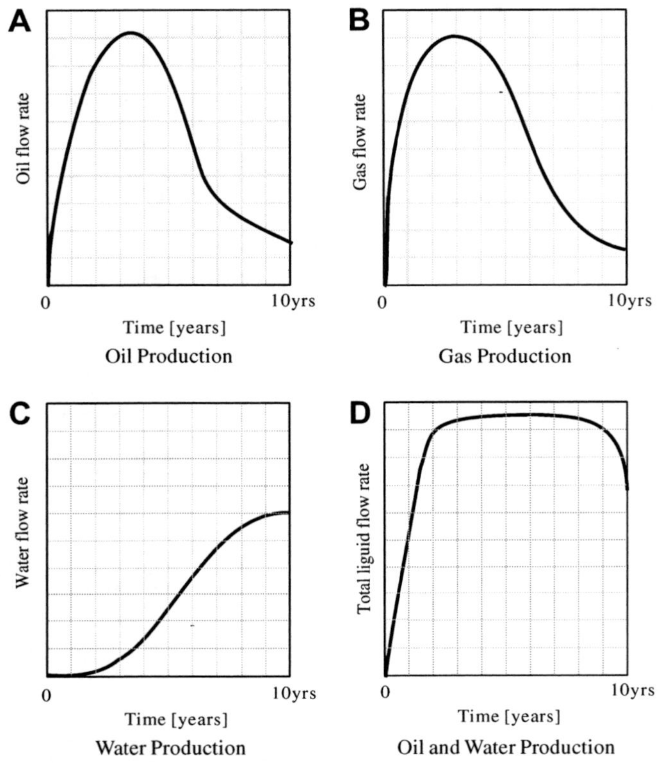

Figure 12-3 Typical Oil, Water, and Gas Production Profiles with Time

the hydrocarbon phase and aqueous phase compositions. A thermodynamic package such as Multiflash is used for the calculations. The hydrate curves define the temperature and pressure envelope. The dosing calculations of the hydrate inhibitor, such as methanol or MEG, indicate how much inhibitor must be added to the produced water at a given system pressure to ensure that hydrates will not form at the given temperature. Hydrate inhibitor dosing is used to control hydrate formation when system temperatures drop into the range in which hydrates are stable during the steady state or transient state of a subsea system during start-up, normal operations, and shutdown. The inhibitor dosing requirements are used to determine the requirements for the inhibitor storage, pumping capacities, and number and size of inhibitor flowlines in order to ensure that the inhibitor can be delivered at the required rates for treating a well and subsea system during start-up, normal operation, and shutdown.

12.2.2. Steady-State Hydraulic and Thermal Performance Analyses

The steady-state flowline model can be generated with software such as PIPESIM or HYSYS. Steady-state modeling has several objectives:

- To determine the relationship between flow rate and pressure drop along the flowline. The flowline size is decided based on the maximum allowable flow rate and the minimum allowable flow rate.
- To check temperature and pressure distributions along flowlines in a steady-state condition to ensure that the flowline never enters the hydrate-forming region during steady-state operation.
- To choose an insulation combination that prevents the temperature at the riser base of a tie-back subsea system from falling below the minimum value for cooldown at the maximum range of production rates. The riser base temperature is determined as a function of flow rate and the combined wellbore/flowline insulation system.
- To determine the maximum flow rate in the system to ensure that arrival temperatures do not exceed any upper limits set by the separation and dehydration processes or by the equipment design.

12.2.3. Transient Flow Hydraulic and Thermal Performances Analyses

Transient flowline system models can be constructed with software packages such as OLGA and ProFES. Transient flowline analyses generally include the following scenarios:

- Start-up and shutdown;
- Emergent interruptions;
- Blowdown and warm-up;
- Ramp up/down;
- Oil displacement;
- Pigging/slugging.

During these scenarios, fluid temperatures in the system must exceed the hydrate dissociation temperature corresponding to the pressure at every location; otherwise, a combination of an insulated pipeline and the injection of chemical inhibitors into the fluid must be simulated in the transient processes to prevent hydrate formation.

12.2.3.1. Start-Up

Hydrate inhibitor should generally be injected downhole and at the tree during start-up. When the start-up rate is high, inhibitor is not required downhole, but the hydrocarbon flow should be treated with inhibitors at the tree. Otherwise, the hydrocarbon flow is required to be treated with inhibitors downhole. Once the tree is outside the hydrate region, hydrate inhibitor can be injected at the tree and the flow rate increased to achieve system warm-up. The start-up scenario is different for the combination of a cold well with a cold flowline and a hot flowline.

12.2.3.2. Shutdown

Shutdown scenarios include planned shutdowns and unplanned shutdowns from a steady state and unplanned shutdowns during warm-up. In general, the planned and unplanned shutdowns from the steady state are the same with the exception that for a planned shutdown, hydrate inhibitor can be injected into the system prior to shutdown. Once the system is filled with inhibited product fluids, no further inhibitor injection or depressurization is needed prior to start-up.

After shutdown, the flowline temperature will decrease because of heat transfer from the system to surrounding water. The insulation system of the flowline is designed to keep the temperature of fluids above the hydrate dissociation temperature until the "no-touch time" has passed. When considering minimum cooldown times, the "no-touch time" is the one in which operators can try to correct problems without having to take any action to protect the subsea system from hydrates. Operators always want a longer "no-touch time," but it is a cost/benefit balancing problem and is decided on a project-by-project basis. Analyses of platform operation experience in West Africa indicate that many typical process and instrumentation interruptions can be analyzed and corrected in 6 to 8 hours.

Let's use a tie-back subsea system in West Africa as an example. If the system is shut down from a steady state, the first step is to see if the system can be restarted within 2 hours. If so, start-up should begin. If not, one option for hydrate control is for the riser to be bullheaded with MeOH (if MeOH is chosen as a hydrate inhibitor) to ensure that no hydrates can form in the base of the riser where fluids are collecting. Next the tree piping will be dosed with methanol. After that, the fluid in the flowline will begin to be fully treated with methanol. Once 8 hours have passed, operators must determine if the system can be started up or not. If it can be started, they will proceed to the start-up procedure outlined previously. If it cannot be started up, the flowlines will be depressurized. The intention of depressurization is to reduce the hydrate dissociation

Table 12-1 Unplanned Shutdown Sequence of Events

Time (hr)	Activity	
0	Shutdown	
3	No-touch time	
6	Blowdown with gas-lift assist	Jumper/tree, MeOH injection
12	Dead oil displacement	

temperature to below the ambient sea temperature. Once the flowlines have been depressurized, the flowlines, jumpers, and trees are in a safe state. If the wells have been shutdown for 2 days without a system restart, then the wellbores need to be bullheaded with MeOH to fill the volume of the wellbore down to the SCSSV. Once these steps have been taken, the entire system is safe.

Table 12-1 shows a typical sequence of events during an unplanned shutdown of a flowline system for the tie-back subsea system in West Africa. The shutdown event is followed by 3 hours of no-touch time. After these 3 hours, the wellbore and the jumpers are treated with methanol. Simultaneously, the flowline is depressurized. To ensure that this procedure can be finished in 3 hours, the blowdown is carried out with a gas-lift assist. Blowdown is followed by dead oil displacement.

12.2.3.3. Blowdown

To keep the flowline system out of the hydrate-forming region when the shutdown time is longer than the cooldown period, flowline blowdown or depressurization may be an option. The transient simulation of this scenario shows how long blowdown and liquid carryover during blowdown take. The simulation also indicates whether the target pressure to avoid hydrate formation can be reached. Figure 12-4 shows that shorter blowdown times are accompanied by greater liquid carryover. The blowdown rate may also need to be limited to reduce the amount of Joule-Thompson cooling downstream of the blowdown valve, to prevent the possibility of brittle fracture of the flowline. During blowdown, sufficient gas must evolve to ensure that the remaining volume of depressurized fluids exerts a hydrostatic pressure that is less than the hydrate dissociation pressure at ambient temperature. Until the blowdown criterion is met, the only way to protect the flowline from hydrate formation is to inject inhibitor.

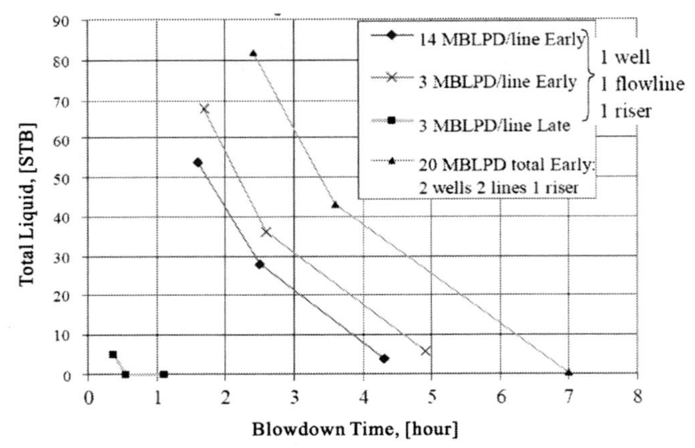

Figure 12-4 Liquid Carryover versus Blowdown Time [5]

12.2.3.4. Warm-Up

During the warm-up process, hydrate inhibitor must be injected until the flowline temperatures exceed the hydrate dissociation temperature at every location for a given pressure. Figure 12-5 shows the effects of insulation material on the warm-up time. Hot oiling has two beneficial effects. First, it reduces or eliminates the time required to reach the hydrate dissociation temperature in the flowline. Once the minimum void fraction has been reached, methanol injection can be safely stopped. Reduction of the methanol injection time is a tremendous advantage for projects with limited

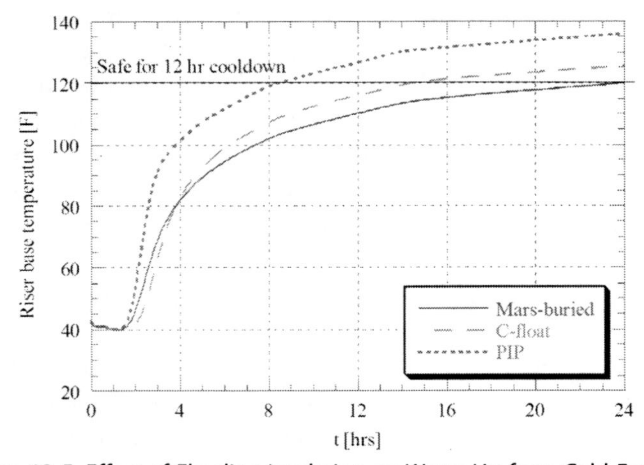

Figure 12-5 Effect of Flowline Insulation on Warm-Up from Cold Earth [5]

available methanol volumes. Hot oiling also warms up the pipeline and surrounding earth, resulting in a much longer cooldown time during the warm-up period than is accomplished by warming with product fluids. This gives more flexibility at those times when the system must be shutdown before it has reached steady state.

12.2.3.5. Riser Cooldown

The most vulnerable portion of the subsea system, in terms of hydrate formation, is typically the riser base. The steady-state temperatures at the riser base are near the lowest point in the whole system. The available riser insulation systems are not as effective as the pipe-in-pipe insulation that is used for some flowlines. The riser is subject to more convective heat transfer because it has a higher current velocity than a pipeline and, finally, it may be partially or completely gas filled during shutdown conditions, leading to much more rapid cooling.

The desired cooldown time before the temperature at the riser base reaches the hydrate temperature, is determined by the following formula:

$$t_{\text{min}} = t_{\text{no touch}} + t_{\text{treat}} + t_{\text{blowdown}} \tag{12-1}$$

where

t_{min}: minimum cooldown time, in hours;

$t_{\text{no touch}}$: no touch time, 2 to 3 hours;

t_{treat}: time to treat wellbores, trees, jumpers, and manifolds with inhibitors, in hours;

t_{blowdown}: time to blow down flowlines, in hours.

Cooldown times are typically on the order of 12to 24 hours. Figure 12-6 shows cooldown curves for an 8–in. oil riser with various insulation materials. The increase in the desired cooldown time requires better flowline insulation and or higher minimum production rates. The desired cooldown dictates a minimum riser base temperature for a given riser insulation system. This temperature becomes the target to be reached or exceeded during steady-state operation of the flowline system.

12.3. SYSTEM DESIGN AND OPERABILITY

In a system design, the entire system from the reservoir to the end user has to be considered to determine applicable operating parameters; flow diameters and flow rates; insulation for tubing, flowlines, and manifolds; chemical injection requirements; host facilities; operating strategies and procedures;

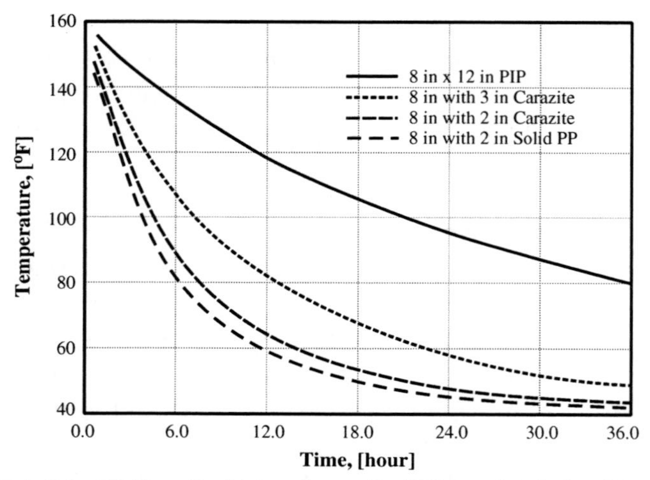

Figure 12-6 8-in. Oil Riser Cooldown Curves for Different Insulation Materials [5]

etc., to ensure that the entire system can be built and operated successfully and economically. All production modes, including start-up, steady state, flow rate change, and shutdown throughout the system life, must be considered.

Operating strategies and procedures for successful system designs are developed with system unknowns and uncertainties in mind and can be readily adapted to work with the existing system, even when that is different from that assumed during design. In deepwater projects, the objective of operating strategies is to avoid the formation of hydrate or wax plugs at any time, especially hydrates in the subsea system including wellbores, trees, well jumpers, manifold, and flowlines during system operation.

Although the operations are time, temperature, and pressure dependent, a typical operating procedure is as follows:

- Operate the flowlines in an underpacked condition during steady state; for example, maintain a sufficient gas void fraction to allow successful depressurization to below the hydrate dissociation pressure at ambient temperatures.
- For a platform shutdown, close the boarding valves and tree as close to simultaneously as possible in order to trap the underpacked condition.
- Design the insulation system to provide enough cooldown time to address facility problems before remedial action is needed, and perform the interventions.
- Inject hydrate inhibitor into the well, tree, jumpers, and manifolds.

- Blow down the flowline to the pressure of fluid below the hydrate-forming pressure.
- Flush flowlines with hot oil prior to restart from a blowdown condition.
- Start up the wells in stages, while injecting hydrate inhibitor. Continue hydrate inhibitor injection until the warm-up period for the well, tree, jumpers, and manifold has passed and enough gas has entered the flowline to permit blowdown.

The systems are normally designed to have a 3-hr no-touch time during which no hydrate prevention actions are required. Blowdown is carried out only during longer, less frequent shutdowns, and about three times per year. The logic charts for start-up and shutdown serve as an outline for the operating guidelines. Figure 12-7 shows a typical logic start-up chart for the cold well start-up of a deepwater field. The start-up logic begins with the pigging of the flowlines, assuming the flowlines have been blown down. The flowlines must be pigged to remove any residual, uninhibited fluids from them. The flowlines are then pressured up to system pressure to avoid any problems caused by a severe pressure drop across the subsea choke and to make manifold valve equalization easier when it is time to switch flowlines. The next step is to start up the well that heats up the fastest, remembering to inject hydrate inhibitor at upstream of the choke while this well is ramping up. When the system is heated, hydrate inhibitor injection can stop. For this case, once the tree has reached 150°F, MeOH injection can be stopped. The temperature of 150°F has been determined to provide 8 hr of cooldown for the tree.

12.3.1. Well Start-Up and Shut-Down

Figure 12-8 shows a simplified typical tree schematic. The subsea tree's production wing valve (PWV) is designated as the underwater safety valve (USV). The production master valve (PMV) is manufactured to USV specifications, but will only be designated for use as the USV if necessary. The well has a remotely adjustable subsea choke to control flow. The subsea choke is used to minimize throttling across the subsea valves during start-up and shutdown.

12.3.1.1. Well Start-Up

The following start-up philosophies are used for cases in which well start-up poses a risk for flowline blockage, particularly if a hydrate blockage is suspected based on the flow assurance study from the design phase:

- Wells will be started up at a rate that allows minimum warm-up time, while considering drawdown limitations. There will be a minimum flow

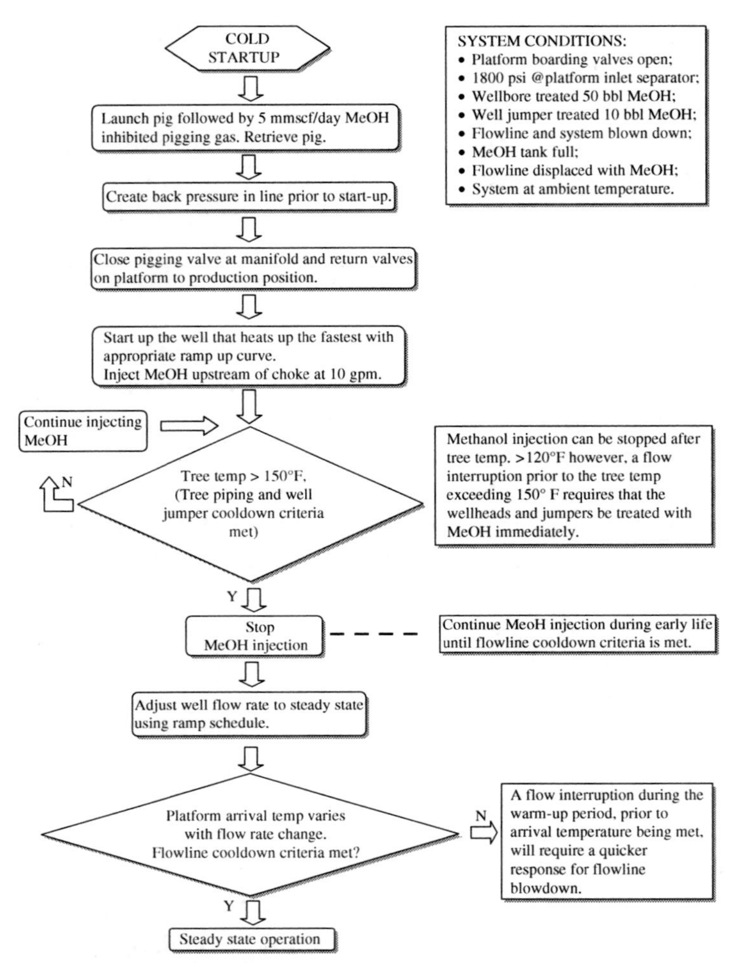

Figure 12-7 Operating Logic Chart of Start-Up for a Cold Well and a Cold Flowline

rate below which thermal losses across the system will keep the fluids in the hydrate-formation region.

- It may not be possible to fully inhibit hydrates at all water cuts, particularly in the wellbore. High water-cut wells will be brought on line without being fully inhibited. Procedures will be developed to minimize the risk of blockage if an unexpected shutdown occurs.
- The system is designed to inject hydrate inhibitor (typically methanol) at the tree during start-up or shutdown. The methanol injected during initial start-up will inhibit the well jumpers, manifold, and flowline if start-up is interrupted and blowdown is not yet possible.

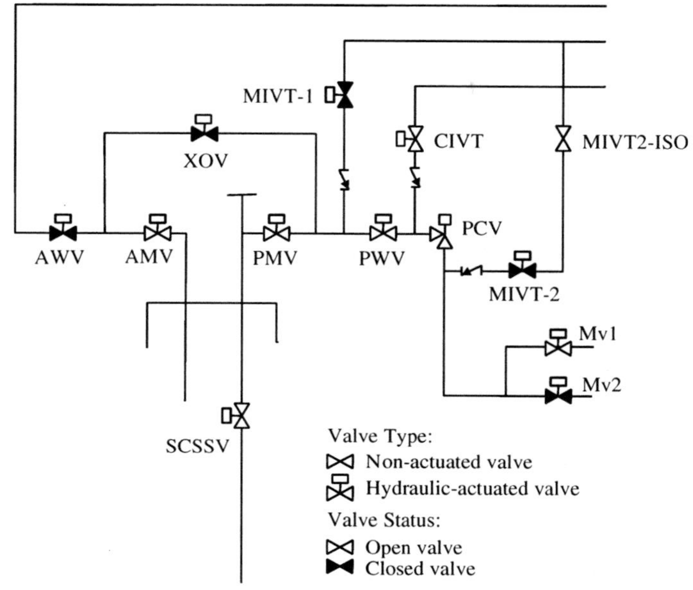

Figure 12-8 Simplified Tree Schematic

12.3.1.2. Well Shut-down

Well shutdown also poses a significant hydrate risk. The following philosophies may be adopted during shutdown operations:

- The subsea methanol injection system is capable of treating or displacing produced fluids with hydrate inhibitor between the manifold and SCSSV following well shutdown to prevent hydrate formation.
- Hydrate prevention in the flowlines is accomplished by blowing the flowline pressure down to less than the hydrate-formation pressure at ambient seabed temperatures.
- Most well shut-downs will be due to short-duration host facility shut-downs. The subsea trees, jumpers, and flowlines are insulated to slow the cooling process and allow the wells to be restarted without having to initiate a full shutdown operation.

12.3.2. Flowline Blowdown

12.3.2.1. Why Do We Blow Down?

The temperature of a flowline system is kept from forming hydrates by the heat from the reservoir fluids moving through the subsea flowlines during steady-state operation. When well shutdown occurs, and the pressure in the system is still high but the temperature of the system will decrease as heat is transferred to the ambient environment, hydrates may form in the flowline. Once the blowdown of the fluid pressures to below the hydrate-formation pressure corresponding to the environmental temperature has been performed, it is safe—from the hydrate-formation point of the view—to leave the system in this condition indefinitely.

12.3.2.2. When Do We Blow Down?

The blowdown is carried out based on several factors:

- Flowline pressure;
- Available thermal energy in the system;
- Ability of the insulation to retain heat.

When the flowline is shut down, the countdown to the hydrate-formation temperature begins. Several hours of no-touch cooldown time is expended before hydrate inhibitor injection or blowdown is required. The base of the risers is the most at risk for hydrate formation in a subsea tie-back system. Riser bases have the least amount of insulation, lowest temperature fluid at the seafloor, and, once the system is shutdown, fluids in the vertical section condense and flow downhill to pool at the riser base. Therefore, this section of the flowline must be treated early to prevent hydrate formation. The hydrate inhibitor is injected at the platform and this will mix with the fluids at the riser base and will effectively lower the temperature at which gas/water interfaces form hydrates. This will allow us to avoid blowdown for several more hours.

Once a blowdown operation has begun, the topside PLC timer is used to determine the length of time to blow the system down. Following a shutdown, all of the flowlines will need to be completely blown down by the end of 12th hour, which is different depending on the project's requirement. Following an extended shutdown that resulted in blowdown of the subsea system, the remaining fluids must be removed from the flowlines before the flowlines can be repressurized. The fluids remaining in the flowlines will have water present and the temperature will be the same as the water temperature. When cold, high-pressure gas is introduced to water, hydrates can form. One option is pigging the flowline to remove residual fluids; sometimes displacement with hot oil is performed without pigging. Prior to beginning a pigging operation, ensure that adequate quantities of methanol are on board the platform for start-up operations and subsequent shutdown or aborted start-up.

REFERENCES

[1] E.G. Hammerschmidt, Gas Hydrate Formation in Natural Gas Pipelines, Oil & Gas Journal vol. 37 (1939). No. 50.

[2] A.A. Kaczmarski, S.E. Lorimer, Emergence of Flow Assurance as a Technical Discipline Specific to Deepwater: Technical Challenges and Integration into Subsea Systems Engineering, OTC 13123, Offshore Technology Conference, Houston, 2001.

[3] E.D. Sloan, Hydrate Engineering, Monograph 21, Society of Petroleum Engineers, Richardson, , Texas, 2000.

[4] F.M. Pattee, F. Kopp, Impact of Electrically-Heated Systems on the Operation of Deep Water Subsea Oil Flowlines, OTC 11894, Offshore Technology Conference, Houston, 2000.

[5] S.E. Lorimer, B.T. Ellison, Design Guidelines for Subsea Oil Systems, presented at Facilities 2000: Facilities Engineering into the Next Millennium, (2000).

CHAPTER 13

Hydraulics

Contents

Subsea Engineering Handbook
ISBN 978-0-12-397804-2, doi:10.1016/B978-0-12-397804-2.00013-8

13.1. INTRODUCTION

To ensure system deliverability of hydrocarbon products from one point in the flowline to another, the accurate prediction of the hydraulic behavior in the flowline is essential. From the reservoir to the end user, the hydrocarbon flow is impacted by the thermal behavior of the heat transfer and phase changes of the fluid in the system. The hydraulic analysis method used and its results are different for different fluid phases and flow patterns. To solve a hydrocarbon hydraulic problem with heat transfer and phase changes, adequate knowledge of fluid mechanics, thermodynamics, heat transfer, vapor/liquid equilibrium, and fluid physical properties for multicomponent hydrocarbon systems is needed. In this chapter, the composition and phase behavior of hydrocarbons are explained first. Then, the hydraulic analyses for single-phase flow and multiphase flow, which include pressure drop versus production flow rate and pipeline sizing, are discussed.

13.2. COMPOSITION AND PROPERTIES OF HYDROCARBONS

13.2.1. Hydrocarbon Composition

The petroleum fluids from reservoirs normally are multiphase and multi-component mixtures, primarily consisting of hydrocarbons, which can be classified into the following three groups:

- Paraffins;
- Naphthenes;
- Aromatics.

In addition to hydrocarbons, water (H_2O), nitrogen (N_2), carbon dioxide (CO_2), hydrogen sulfide (H_2S), salts, and solids are often found in petroleum mixtures.

Table 13-1 lists some typical physical properties of the main components of petroleum. The boiling point of hydrocarbon components increases with

Table 13-1 Physical Properties of Main Petroleum Components [1]

Component	Formula	Boiling Temperature at 1 atm (°C)	Density at 1 atm and 15°C (g/cm³)
		Paraffins	
Methane	CH_4	−161.5	—
Ethane	C_2H_6	−88.3	—
Propane	C_3H_6	−42.2	—
i-Butone	C_4H_{10}	−10.2	—
n-Butane	C_4H_{10}	−0.6	—
n-Pentane	C_5H_{12}	36.2	0.626
n-Hexane	C_6H_{14}	69.0	0.659
i-Octane	C_8H_{18}	99.3	0.692
n-Decane	$C_{10}H_{22}$	174.0	0.730
Naphthenes			
Cyclopentane	C_2H_6	49.5	0.745
Methyl cyclo-pentane	C_2H_6	71.8	0.754
Cyclohexane	C_2H_6	81.4	0.779
Aromatics			
Benzene	C_6H_6	80.1	0.885
Toluene	C_7H_8	110.6	0.867
o-Xylene	C_8H_{10}	144.4	0.880
Naphthalene	$C_{10}H_8$	217.9	0.971
Others			
Nitrogen	N_2	−195.8	—
Carbon dioxide	CO_2	−78.4	—
Hydrogen sulfide	H_2S	−60.3	—

Table 13-2 Typical Composition of a Gas Condensate [2]

Component		Composition (mol %)
Hydrogen sulfide	H_2S	0.05
Carbon dioxide	CO_2	6.50
Nitrogen	N_2	11.71
Methane	C_1	79.06
Ethane	C_2	1.62
Propane	C_3	0.35
i-Butane	i-C_4	0.08
n-Butane	C_4	0.10
i-pentane	i-C_5	0.04
n-Pentane	C_5	0.04
Hexanes	C_6	0.06
Heptanes Plus	C_7+	0.39

an increase in the carbon number of the component formula. If the carbon number of a component is less than 5, the component is in the gas phase at atmospheric pressure. When the mixture contains larger molecules, it is a liquid at normal temperatures and pressures. A typical petroleum fluid contains thousands of different chemical compounds, and trying to separate it into different chemical homogeneous compounds is impractical. A composition of gas condensate is shown in Table 13-2. In the usual composition list of hydrocarbons, the last hydrocarbon is marked with a "+," which indicates a pseudo-component, and lumps together all of the heavier components.

The characterization of a pseudo-component including its molecular weight, density, pseudo-critical properties, and other parameters should be calculated ahead of the flow assurance analysis (see Chapter 12). The characterization should be based on experimental data and analyses. As shown in Table 13-2, hydrocarbon isotropic compounds with the same formula and carbon atom numbers may have very different normal boiling points and other physical properties.

13.2.2. Equation of State

For oil and gas mixtures, the phase behavior and physical properties such as densities, viscosities, and enthalpies are uniquely determined by the state of the system. The equations of state (EOS) for petroleum mixtures are mathematical relations between volume, pressure, temperature, and composition, which are used for describing the system state and transitions between states. Most thermodynamic and transport properties in engineering analyses are derived from the EOS. Since 1873, when the first EOS for representation of real systems was developed by the van der Waals, hundreds of different EOSs have been proposed and they are distinguished by Leland [3] into four families:

1. van der Waals family;
2. Benedict-Webb-Rubin family;
3. Reference-fluid equations;
4. Augmented-rigid-body equations.

The cubic equations of the van der Waals family are widely used in the oil and gas industry for engineering calculations because of their simplicity and relative accuracy for describing multiphase mixtures. The simple cubic equations of state, Soave-Redlich-Kwong (SRK) and Peng-Robinson (PR), are used in flow assurance software (e.g., PIPESIM). They require limited pure component data and are robust and efficient, and usually give broadly similar results.

The SRK Equation [4]

$$p = \frac{RT}{v-b} - \frac{a}{v(v+b)} \tag{13-1}$$

where p is the pressure, T is the temperature, v is the molar volume, R is the gas constant, and a and b are the EOS constants, which are determined by the critical conditions for a pure component:

$$a = 0.42748 /\frac{R^2 T_c^2}{p_c}[1 + m(1 - \sqrt{T_r})]^2$$

$$b = 0.08664 /\frac{RT_c}{p_c}$$

And $m = 0.480 + 1.574\,\omega - 0.176\,\omega^2$, where ω is an acentric factor. For a mixture, a and b are found as follows:

$$a = \sum_i \sum_j z_i z_j a_{ij}$$

$$b = \sum_i z_i b_i$$

where z_i and z_j are the mole fraction of components i and j, respectively, and $a_{ij} = \sqrt{a_i a_j}(1 - k_{ij})$, where k_{ij} is a binary interaction coefficient, which is usually considered equal to zero for hydrocarbon/hydrocarbon interactions, and different from zero for interactions between a hydrocarbon and a nonhydrocarbon, and between unlike pairs of nonhydrocarbons.

PR Equation [5]

$$p = \frac{RT}{v-b} - \frac{a}{v(v+b)+b(v-b)} \tag{13-2}$$

The EOS constants are given by

$$a = 0.45724 \,/\frac{R^2T_c^2}{p_c}\left[1+m(1-\sqrt{T_r}\right]^2$$

$$b = 0.07780 \,/\frac{RT_c}{p_c}$$

and $m = 0.3764 + 1.54226\,\omega - 0.26992\,\omega^2$, where ω is an acentric factor.

13.2.3. Hydrocarbon Properties

Oil and gas are very complex fluids composed of hydrocarbon compounds that exist in petroleum in a wide variety of combinations. Physical properties change with the component composition, pressure, and temperature. No two oils have the same physical properties. In considering hydrocarbon flow in pipes, the most important physical properties are density and viscosity.

13.2.3.1. Density

Dead oil is defined as oil without gas in solution. Its specific gravity γ_o is defined as the ratio of the oil density and the water density at the same temperature and pressure:

$$\gamma_o = \frac{\rho_o}{\rho_w} \tag{13-3}$$

API gravity, with a unit of degree (°) is defined as:

$$^oAPI = \frac{141.5}{\gamma_o} - 131.5 \tag{13-4}$$

where γ_o is the specific gravity of oil at 60°F, and the water density at 60°F is 62.37 lb/ft^3.

The effect of gas dissolved in the oil should be accounted for in the calculation of the *in situ* oil (live oil) density. The oil density can be calculated as follows:

$$\rho_o = \frac{350.4\gamma_o + 0.0764\gamma_g R_s}{5.615B_o} \tag{13-5}$$

where

ρ_o: oil density, lb/ft^3;
γ_o: oil specific gravity;
γ_g: gas gravity;
R_s: dissolved gas, scf/STB;
B_o: formation volume factor, volume/standard volume.

Gas density ρ_g is defined as:

$$\rho_g = \frac{pM}{ZRT} \tag{13-6}$$

where M is the molecular weight of the gas, R is the universal gas constant, p is the absolute pressure, T is the absolute temperature, and Z is the compressibility factor of gas. The gas specific gravity γ_g is defined as the ratio of the gas density and the air density at the same temperature and pressure:

$$\gamma_g = \frac{\rho_g}{\rho_a} = \frac{M}{29} \tag{13-7}$$

For liquids, the reference material is water, and for gases it is air.

13.2.3.2. Viscosity

Dynamic Viscosity

The dynamic viscosity of a fluid is a measure of the resistance to flow exerted by a fluid, and for a Newtonian fluid it is defined as:

$$\mu = \frac{\tau}{dv/dn} \tag{13-8}$$

where τ is the shear stress, v is the velocity of the fluid in the shear stress direction, and dv/dn is the gradient of v in the direction perpendicular to flow direction.

Kinematic Viscosity

$$\nu = \mu/\rho \tag{13-9}$$

where ρ is the density.

The viscosity of Newtonian fluids is a function of temperature and pressure. If the viscosity of a fluid varies with the thermal or time history, the fluid is called non-Newtonian. Most of the hydrocarbon fluids are Newtonian fluid, but in some cases, the fluid in the flowline should be considered to be non-Newtonian. The viscosity of oil decreases with increasing temperature, while the viscosity of gas increases with increasing temperature. The general relations can be expressed as follows for liquids:

$$\mu = A\exp(B/T) \tag{13-10}$$

and as follows for gases:

$$\mu = CT^2 \tag{13-11}$$

where A, B, and C are constants determined from the experimental measurement. The viscosity typically increases with increased pressure for gas. The equation of viscosity can be expressed in terms of pressure as:

$$\mu = \exp(A + B \cdot P) \tag{13-12}$$

The most commonly used unit to represent viscosity is the poise, which can be expressed as dyne · s/cm^2 or g/cm · s. The centipoise or cP is equal to 0.01 poise. The SI unit of viscosity is Pa · s, which is equal to 0.1 poise. The viscosity of pure water at 20°C is 1.0 cP.

The viscosity of crude oil with dissolved gases is an important parameter for the calculation of pressure loss for hydrocarbon flow in pipeline [6]. Pressure drops of high viscous fluid along the pipeline segment may significantly impact deliverability of products. The oil viscosity should be determined in the laboratory for the required pressure and temperature ranges. Many empirical correlations are available in the literature that can be used to calculate the oil viscosity based on the system parameters such as temperature, pressure, and oil and gas gravities when no measured viscosity data are available.

To minimize pressure drop along the pipeline for viscous crude oils, it is beneficial to insulate the pipeline so it can retain a high temperature. The flow resistance is less because the oil viscosity is lower at higher temperatures. The selection of insulation is dependent on cost and operability issues. For most developed fields insulation may not be important. However, for heavy oil, high-pressure drops due to viscosity in the connecting pipeline between the subsea location and the receiving platform, insulation may have a key role to play.

13.3. EMULSION

13.3.1. General

A separate water phase in a pipeline system can result in hydrate formation or an oil/water emulsion under certain circumstances. Emulsions can have high viscosities, which may be an order of magnitude higher than the single-phase oil or water. The effect of emulsions on frictional loss may lead to a significantly larger error in pressure loss than calculated. In addition to affecting pipeline hydraulics, emulsions may present severe problems to downstream processing plants [7].

An emulsion is a heterogeneous liquid system consisting of two immiscible liquids with one of the liquids intimately dispersed in the form of droplets in the second liquid. In most emulsions of crude oil and water, the water is finely dispersed in the oil. The spherical form of the water globules is a result of interfacial tension, which compels them to present the smallest possible surface area to the oil. If the oil and water are violently agitated, small drops of water will be dispersed in the continuous oil phase and small drops of oil will be dispersed in the continuous water phase. If left undisturbed, the oil and water will quickly separate into layers of oil and water. If any emulsion is formed, it will exist between the oil above and the water below. In offshore engineering, most emulsions are the water-in-oil type.

The formation of an emulsion requires the following three conditions:

- Two immiscible liquids;
- Sufficient agitation to disperse one liquid as droplets in the other;
- An emulsifying agent.

The agitation necessary to form an emulsion may result from any one or a combination of several sources:

- Bottom-hole pump;
- Flow through the tubing, wellhead, manifold, or flowlines;
- Surface transfer pump;
- Pressure drop through chokes, valves, or other surface equipment.

The greater the amount of agitation, the smaller the droplets of water dispersed in the oil. Water droplets in water-in-oil emulsions are of widely varying sizes, ranging from less than 1 to about 1,000 μm. Emulsions that have smaller droplets of water are usually more stable and difficult to treat than those that have larger droplets. The most emulsified water in light crude oil, that is, oil above 20° API, is from 5 to 20 vol % water, while emulsified water in crude oil heavier than 20° API is from 10 to 35%. Generally, crude oils with low API gravity (high density) will form a more stable and higher percentage volume of emulsion than will oils of high API gravity (low density). Asphaltic-based oils have a tendency to emulsify more readily than paraffin-based oils.

The emulsifying agent determines whether an emulsion will be formed and the stability of that emulsion. If the crude oil and water contain no emulsifying agent, the oil and water may form a dispersion that will separate quickly because of rapid coalescence of the dispersed droplets. Emulsifying agents are surface-active compounds that attach to the water-drop surface. Some emulsifiers are thought to be asphaltic, sand, silt, shale particles, crystallized paraffin, iron, zinc, aluminum sulfate, calcium carbonate, iron sulfide, and similar materials. These substances usually originate in the oil formation but can be formed as the result of an ineffective corrosion-inhibition program.

13.3.2. Effect of Emulsion on Viscosity

Emulsions are always more viscous than the clean oil contained in the emulsion.

Figure 13-1 shows the tendency of viscosity with the increase of water cut in different emulsion conditions. This relationship was developed by Woelflin [8] and is widely used in petroleum engineering. However, at

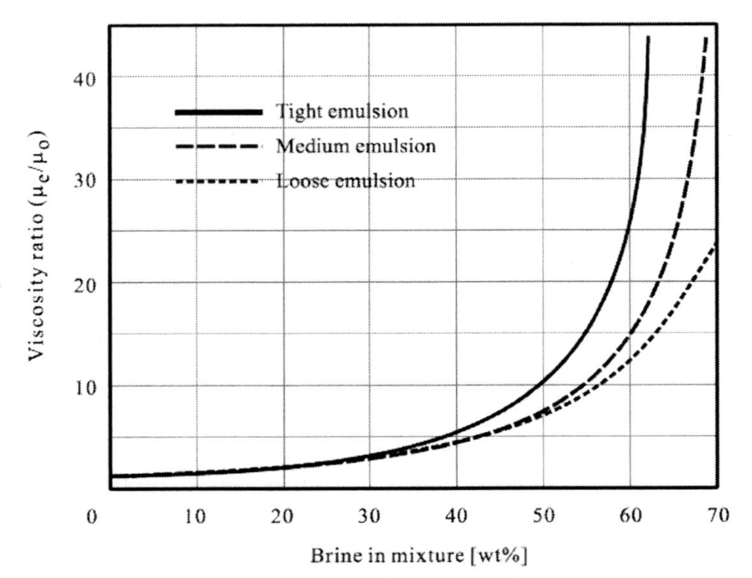

Figure 13-1 Woelflin Viscosity Data

higher water cuts (greater than about 40%), it tends to be excessively pessimistic and may lead to higher pressure loss expectations than are actually likely to occur. Guth and Simha [9] presented a correlation of emulsion viscosity:

$$\mu_e/\mu_o = 1 + 2.5\ C_w \tag{13-13}$$

where μ_e is the viscosity of emulation (in cP or mPa·s), μ_o is the viscosity of clean oil (cP or mPa·s), C_w is the volume fraction of the water phase, it is a ration of water volumetric flow rate compared to the volumetric flow rate of total liquids, including oil, water and other liquids.

This correlation is similar to that of Woelflin in that it predicts the emulsion viscosity to be some multiple of the oil viscosity. The factor is determined solely as a function of the water cut. Up to a water cut of about 40%, the two methods will give almost identical results. Above that, the effective viscosity rises much more slowly with the Guth and Simha [9] correlation, and computed pressure losses will thus be lower at higher water cuts.

Laboratory tests can provide useful information about emulsions, but the form of the emulsion obtained in practice is dependent on the shear history, temperature, and composition of the fluids. Limited comparison of calculated emulsion viscosity with experimental data suggests that the Woelflin correlation overestimates the viscosity at higher water fractions. The following equation, from by Smith and Arnold [10] can be used if no other data are available:

$$\mu_e/\mu_o = 1 + 2.5\ C_w + 14.1 C_w^2 \tag{13-14}$$

13.3.3. Prevention of Emulsion

If all water can be excluded from the oil and/or if all agitation of hydraulic fluid can be prevented, no emulsion will form. Exclusion of water in some wells is difficult or impossible, and the prevention of agitation is almost impossible. Therefore, production of emulsion from many wells must be expected. In some instances, however, emulsification is increased by poor operating practices.

Operating practices that include the production of excess water as a result of poor cementing or reservoir management can increase emulsion-treating problems. In addition, a process design that subjects the oil/water mixture to excess turbulence can result in greater treatment problems. Unnecessary turbulence can be caused by overpumping and poor maintenance of plunger and valves in rod-pumped wells, use of more gas-lift gas than is needed, and pumping the fluid when gravity flow could be used instead. Some operators use progressive cavity pumps as opposed to reciprocating, gear, or centrifugal pumps to minimize turbulence. Others have found that some centrifugal pumps can actually cause coalescence if they are installed in the process without a downstream throttling valve. Wherever possible, pressure drop through chokes and control valves should be minimized before oil/water separation.

13.4. PHASE BEHAVIOR

A multicomponent mixture exhibits an envelope for liquid/vapor phase change in the pressure/temperature diagram, which contains a bubble-point line and a dew-point line, compared with only a phase change line for a pure component. Figure 13-2 shows the various reservoir types of oil and gas systems based on the phase behavior of hydrocarbons in the reservoir, in which the following five types of reservoirs are distinguished:

- Black oils;
- Volatile oils;
- Condensate (retrograde gases);
- Wet gases;
- Dry gases.

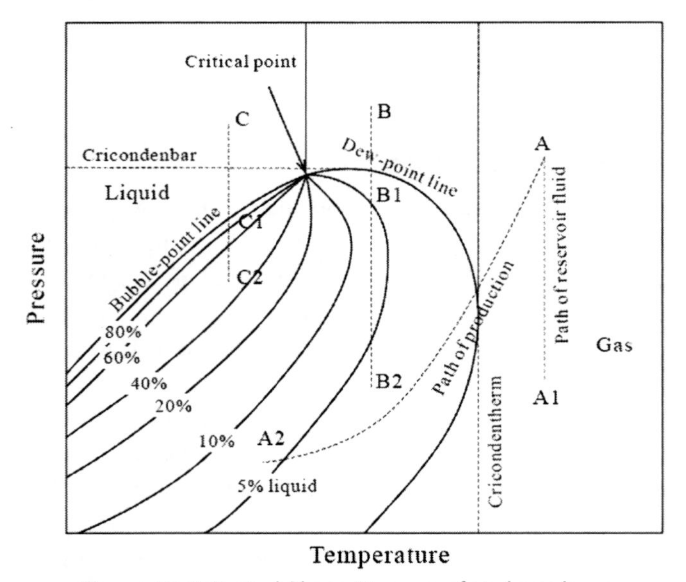

Figure 13-2 Typical Phase Diagram of Hydrocarbons

The amount of heavier molecules in the hydrocarbon mixtures varies from large to small in the black oils to the dry gases, respectively. The various pressures and temperatures of hydrocarbons along a production flowline from the reservoir to separator are presented in curve A–A2 of Figure 13-2. Mass transfer occurs continuously between the gas and the liquid phases within the two-phase envelope.

13.4.1. Black Oils

Black oil is liquid oil and consists of a wide variety of chemical species including large, heavy, and nonvolatile molecules. Typical black oil reservoirs have temperatures below the critical temperature of a hydrocarbon mixture. Point C in Figure 13-2 represents unsaturated black oil; some gases are dissolved in the liquid hydrocarbon mixture. Point C1 in the same figure shows saturated back oil, which means that the oil contains as much dissolved gas as it can take and that a reduction of system pressure will release gas to form the gas phase. In transport pipelines, black oils are transported in the liquid phase throughout the transport process, whereas in production flowlines, produced hydrocarbon mixtures are usually in thermodynamical equilibrium with gas.

13.4.2. Volatile Oils

Volatile oils contain fewer heavy molecules than black oil, but more ethane through hexane. Reservoir conditions are very close to the critical temperature and pressure. A small reduction in pressure can cause the release of a large amount of gas.

13.4.3. Condensate

Retrograde gas is the name of a fluid that is gas at reservoir pressure and temperature. However, as pressure and temperature decrease, large quantities of liquids are formed due to retrograde condensation. Retrograde gases are also called retrograde gas condensates, gas condensates, or condensates. The temperature of condensates is normally between the critical temperature and the cricondentherm as shown in Figure 13-2. From reservoir to riser top, as shown in points B to B2 in the figure, the fluid changes from single gas and partial transfer to condensate liquid and evaporate again as gas, causing a significant pressure drop.

For a pure substance a decrease in pressure causes a change of phase from liquid to gas at the vapor–pressure line; likewise, in the case of

Table 13-3 Hydrocarbon Composition of Typical Reservoirs [11]

	Composition (mol %)				
Component	**Black Oil**	**Volatile Oil**	**Condensate**	**Wet Gas**	**Dry Gas**
CO_2	0.02	0.93	2.37	1.41	0.10
N_2	0.34	0.21	0.31	0.25	2.07
C_1	34.62	58.77	73.19	92.46	86.12
C_2	4.11	7.57	7.80	3.18	5.91
C_3	1.01	4.09	3.55	1.01	3.58
i-C_4	0.76	0.91	0.71	0.28	1.72
n-C_4	0.49	2.09	1.45	0.24	0.0
i-C_5	0.43	0.77	0.64	0.13	0.50
n-C_5	0.21	1.15	0.68	0.08	0.0
C_6	1.61	1.75	1.09	0.14	0.0
C_7^+	56.40	21.76	8.21	0.82	0.0
Total	100.0	100.0	100.0	100.0	100.0

a multicomponent system, a decrease in pressure causes a change of phase from liquid to gas at temperatures below the critical temperature. However, consider the isothermal decrease in pressure illustrated by a line B–B1–B2 in Figure 13-2. As pressure is decreased from point B, the dew-point line is crossed and liquid begins to form. At the position indicated by point B1, the system is 5% liquid. A decrease in pressure has caused a change from gas to liquid. Table 13-3 lists the composition of a typical condensate gas. Curve A–A2 illustrates the variations of pressure and temperature along the flowline from reservoir to the separator on a platform.

13.4.4. Wet Gases

A wet gas exists in a pure gas phase in the reservoir, but becomes a liquid/gas two-phase mixture in a flowline from the well tube to the separator at the topside platform. During the pressure drop in the flowline, liquid condensate appears in the wet gas.

13.4.5. Dry Gases

Dry gas is primarily methane. The hydrocarbon mixture is solely gas under all conditions of pressure and temperature encountered during the production phases from reservoir conditions involving transport and process conditions. In particular, no hydrocarbon-based liquids are formed from the gas although liquid water can condense. Dry gas reservoirs have temperatures above the cricondentherm. Tables 13-3 and 13-4 list the hydrocarbon composition and properties, respectively, of typical reservoirs.

Table 13-4 Hydrocarbon Properties of Typical Reservoirs [11]

Properties	Black Oil	Volatile Oil	Condensate	Wet Gas	Dry Gas
M_7+	274	228	184	130	
γ_7+	0.92	0.858	0.816	0.763	
GOR (scf/STB)	300	1490	5450	105,000	∞
OGR (STB/mmscf)			180	10	0
γ_API	24	38	49	57	
γ_g	0.63	0.70	0.70	0.61	
P_{sat} (psi)	2,810	5,420	5,650	3,430	
ρ_{sat} (lb/ft^3)	51.4	38.2	26.7	9.61	

13.4.6. Computer Models

Accurate prediction of physical and thermodynamic properties is a prerequisite to successful pipeline design. Pressure loss, liquid hold up, heat loss, hydrate formation, and wax deposition all require knowledge of the fluid states.

In flow assurance analyses, the following two approaches have been used to simulate hydrocarbon fluids:

- *"Black-oil" model:* Defines the oil as a liquid phase that contains dissolved gas, such as hydrocarbons produced from the oil reservoir. The "black oil" accounts for the gas that dissolves (condenses) from oil solution with a parameter of R_s that can be measured from the laboratory. This model predicts fluid properties from the specific gravity of the gas, the oil gravity, and the volume of gas produced per volume of liquid. Empirical correlations evaluate the phase split and physical property correlations determine the properties of the separate phases.
- *Composition model:* For a given mole fraction of a fluid mixture of volatile oils and condensate fluids, a vapor/liquid equilibrium calculation determines the amount of the feed that exists in the vapor and liquid phases and the composition of each phase. It is possible to determine the quality or mass fraction of gas in the mixtures. Once the composition of each phase is known, it is also possible to calculate the interfacial tension, densities, enthalpies, and viscosities of each phase.

The accuracy of the compositional model is dependent on the accuracy of the compositional data. If good compositional data are available, selection of an appropriate EOS is likely to yield more accurate phase behavior data than the corresponding black-oil model. This is particularly so if the hydrocarbon liquid is a light condensate. In this situation complex phase effects such as retrograde condensation are unlikely to be adequately handled by the black-oil methods. Of prime importance to hydraulic studies is the viscosity of the fluid phases. Both black-oil and compositional techniques can be inaccurate. Depending on the correlation used, very different calculated pressure losses could result. With the uncertainty associated with viscosity prediction, it is prudent to utilize laboratory-measured values.

The gas/oil ratio (GOR) can be defined as the ratio of the measured volumetric flow rates of the gas and oil phases at meter conditions (multiphase meter) or the volume ratio of gas and oil at the standard condition (1 atm, 60°F) in units of scf/STB. When water is also present, the water cut is generally defined as the volume ratio of the water and total liquid at standard conditions. If the water contains salts, the salt concentrations may be contained in the water phase at the standard condition.

13.5. HYDROCARBON FLOW

13.5.1. General

The complex mixture of hydrocarbon compounds or components can exist as a single-phase liquid, a single-phase gas, or as a multiphase mixture, depending on its pressure, temperature, and the composition of the mixture. The fluid flow in pipelines is divided into three categories based on the fluid phase condition:

- *Single-phase condition:* black oil or dry gas transport pipeline, export pipeline, gas or water injection pipeline, and chemical inhibitors service pipelines such as methanol and glycol lines;
- *Two-phase condition:* oil + released gas flowline, gas + produced oil (condensate) flowline
- *Three-phase condition:* water + oil + gas (typical production flowline).

The pipelines after oil/gas separation equipment, such as transport pipelines and export pipelines, generally flow single-phase hydrocarbon fluid while in most cases, the production flowlines from reservoirs have two- or three-phase fluids, simultaneously, and the fluid flow is then called multiphase flow.

In a hydrocarbon flow, the water should be considered as a sole liquid phase or in combination with oils or condensates, since these liquids basically are insoluble in each other. If the water amount is small enough that it has little effect on flow performance, it may be acceptable to assume a single liquid phase. At a low-velocity range, there is considerable slip between the oil and water phases. As a result, the water tends to accumulate in low spots in the system. This leads to high local accumulations of water and, therefore, a potential for water slugs in the flowline. It may also cause serious corrosion problems.

Two-phase (gas/liquid) models are used for black-oil systems even when water is present. The water and hydrocarbon liquid are treated as a combined liquid with average properties. For gas condensate systems with water, three-phase (gas/liquid/aqueous) models are used.

The hydraulic theory underlying single-phase flow is well understood and analytical models may be used with confidence. Multiphase flow is significantly more complex than single-phase flow. However, the technology to predict multiphase-flow behavior has improved dramatically in the past decades. It is now possible to select pipeline size, predict pressure drop, and calculate flow rate in the flowline with an acceptable engineering accuracy.

13.5.2. Single-Phase Flow

The basis for calculation of changes in pressure and temperature with pipeline distance is the conservation of mass, momentum, and energy of the fluid flow. In this section, the steady-state, pressure-gradient equation for single-phase flow in pipelines is developed. The procedures to determine values of wall shear stress are reviewed and example problems are solved to demonstrate the applicability of the pressure-gradient equation for both compressible and incompressible fluid. A review is presented of Newtonian and non-Newtonian fluid flow behavior in circular pipes. The enthalpy-gradient equation is also developed and solved to obtain approximate equations that predict temperature changes during steady-state fluid flow.

13.5.2.1. Conservation Equations

Mass Conservation

Mass conservation of flow means that the mass in, m_{in}, minus the mass out, m_{out}, of a control volume must equal the mass accumulation in the control volume. For the control volume of a one-dimensional pipe segment, the mass conservation equation can be written as:

$$\frac{\partial \rho}{\partial t} + \frac{\partial (\rho v)}{\partial L} = 0 \qquad (13\text{-}15)$$

where ρ is the fluid density, v is the fluid velocity, t is the time, and L is the length of the pipe segment. For a steady flow, no mass accumulation occurs, and Equation (13-15) becomes:

$$\rho v = \text{constant} \qquad (13\text{-}16)$$

Momentum Conservation

Based on Newton's second law applied to fluid flow in a pipe segment, the rate of momentum change in the control volume is equal to the sum of all forces on the fluid. The linear momentum conservation equation for the pipe segment can be expressed as:

$$\frac{\partial(\rho v)}{\partial t}+\frac{\partial(\rho v^2)}{\partial L}=-\frac{\partial p}{\partial L}-\tau\frac{\pi d}{A}-\rho g\sin\theta \tag{13-17}$$

The terms on the right-hand side of Equation (13-17) are the forces on the control volume. The term $\partial p/\partial L$ represents the pressure gradient, $\tau\pi d/A$ represents the surface forces, and ρg sin θ represents the body forces. Combining Equations (13-16) and (13-17), the pressure gradient equation is obtained for a steady-state flow:

$$\frac{\partial p}{\partial L}=-\tau\frac{\pi d}{A}-\rho g\sin\theta-\rho v\frac{\partial v}{\partial L} \tag{13-18}$$

By integrating both sides of the above equation from section 1 to section 2 of the pipe segment shown in Figure 13-3, and adding the mechanical energy by hydraulic machinery such as a pump, an energy conservation equation, the Bernoulli equation, for a steady one-dimensional flow is obtained as:

$$\frac{p_1}{\gamma_1}+\frac{V_1^2}{2g}+z_1=\frac{p_2}{\gamma_2}+\frac{V_2^2}{2g}+z_2+\sum h_f+\sum h_l-h_m \tag{13-19}$$

where, p/γ is pressure head; $V^2/2g$ is velocity head, z is elevation head, $\sum h_f$ is the sum of the friction head loss between section 1 and 2 caused by

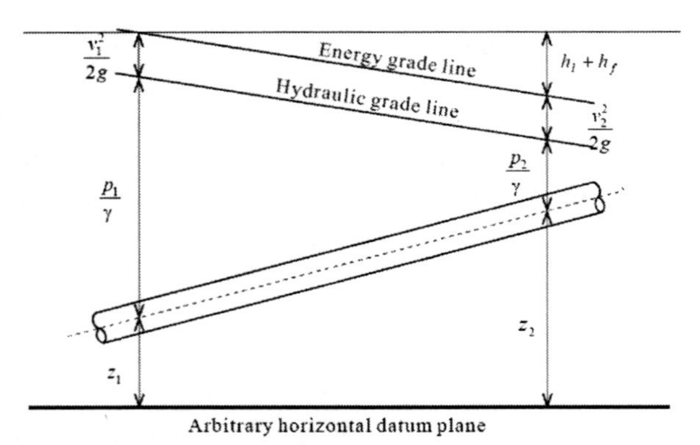

Figure 13-3 Relationship between Energy/Hydraulic Grade Lines and Individual Heads [12]

frictional force, $\sum h_l$ is the sum of the local head loss, and h_m is mechanical energy per unit weight added by hydraulic machinery. The pipe friction head loss is the head loss due to fluid shear at the pipe wall. The local head losses are caused by local disruptions of the fluid stream, such as valves, pipe bends, and other fittings.

The energy grade line (EGL) is a plot of the sum of the three terms in the left-hand side of Equation (13-19), and is expressed as:

$$\text{EGL} = \frac{p}{\gamma} + \frac{V^2}{2g} + z \tag{13-20}$$

The hydraulic grade line (HGL) is the sum of only the pressure and elevation head, and can be expressed as:

$$\text{HGL} = \frac{p}{\gamma} + z \tag{13-21}$$

Figure 13-3 shows the relation of the individual head terms to the EGL and HGL lines. The head losses represent the variation of EGL or HGL between section 1 and 2. In general, in a preliminary design, the local losses may be ignored, but they must be considered at the detailed design stage.

As shown in Equation (13-19), the total pressure drop in a particular pipe segment is the sum of the friction head losses of the pipe segment, valves, and fittings, the local head losses due to valves and fittings, the static pressure drop due to elevation change, and the pressure variation due to the acceleration change.

13.5.2.2. Friction Factor Equation

The Darcy-Weisbach equation is one of the most general friction head loss equations for a pipe segment. It is expressed as:

$$h_f = f\frac{L}{D}\frac{V^2}{2g} \tag{13-22}$$

$$f = f(\text{Re}, \varepsilon/D) \tag{13-23}$$

where f is the friction factor, $\text{Re} = VD\rho/\mu$ is the Reynolds number, and ε/D is the equivalent sand-grain roughness (relative roughness) of the pipe, which is decided by the pipeline material. Figure 13-4 shows the graphical representation of the friction factor at different flow regions, which is also called as Moody diagram.

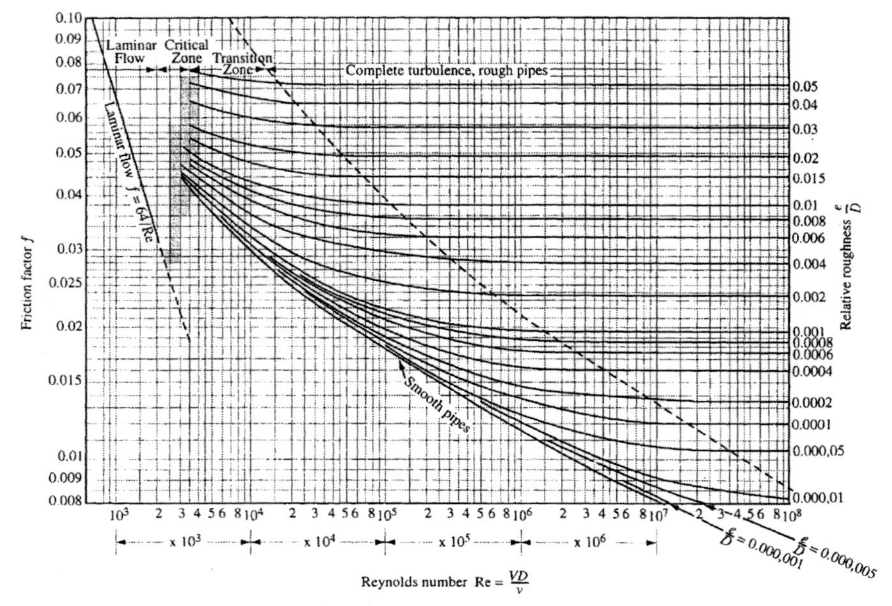

Figure 13-4 Moody Diagram [13]

Table 13-5 lists common values of absolute roughness for several materials. For steel pipe, the absolute roughness is 0.0018 in. unless alternative specifications are specified. The absolute roughness of 0.0018 in. is for aged pipe, but that is accepted in design practice because it is conservative. Flexible pipe is rougher than steel pipe and, therefore, requires a larger diameter for the same maximum rate. For a flexible pipe, the roughness is given by $\varepsilon = ID/250.0$ unless an alternative specification is given. The roughness may increase with use at a rate determined by the material and nature of the fluid. As a general guide, factors of safety of 20% to 30% on the friction factor will accommodate the change in roughness conditions for steel pipe with average service of 5 to 10 years.

Table 13-5 Pipeline Material Roughness [14]

Material	ε, Equivalent Sand-Grain Roughness (in.)
Concrete	0.012−0.12
Case iron	0.010
Commercial or welded steel	0.0018
PVC, glass	0.00006

As shown in Figure 13-4, the function f (Re, ε/D) is very complex. A laminar region and a turbulent region are defined in the Moody diagram based on the flow characteristics.

Laminar Flow

In laminar flow (Re < 2100), the friction factor function is a straight line and is not influenced by the relative roughness:

$$f = 64/\mathrm{Re} \tag{13-24}$$

The friction head loss is shown to be proportional to the average velocity in the laminar flow. Increasing pipe relative roughness will cause an earlier transition to turbulent flow. When the Reynolds number is in the range from 2000 to 4000, the flow is in a critical region where it can be either laminar or turbulent depending on several factors. These factors include changes in section or flow direction. The friction factor in the critical region is indeterminate, which is in a range between friction factors for laminar flow and for turbulent flow.

Turbulent Flow

When the Reynolds number is larger than 4000, the flow inside the pipe is turbulent flow; the fiction factor depends not only on the Reynolds number but also on the relative roughness, ε/D, and other factors. In the complete turbulence region, the region above a dashed line in the upper right part of a Moody diagram, friction factor f is a function only of roughness ε/D. For a pipe in the transition zone, the friction factor decreases rapidly with increasing Reynolds number and decreasing pipe relative roughness. The lowest line in Figure 13-4 is the line representing the friction factor for the smoothest pipe, where the roughness ε/D is so small that it has no effect.

Although single-phase flow in pipes has been studied extensively, it still involves an empirically determined friction factor for turbulent flow calculations. The dependence of this friction factor on pipe roughness, which must usually be estimated, makes the calculated pressure gradients subject to considerable error and summarizes the correlations of Darcy-Weisbach friction factor for different flow ranges. The correlations for smooth pipe and complete turbulence regions are simplified from the one for a transitional region. For smooth pipe, the relative roughness term is ignored, whereas for the complete turbulence region, the Reynolds term is ignored.

Table 13-6 summarizes some Darcy-Weisbach friction factor correlations for both laminar flow and turbulent flow.

Table 13-6 Darcy-Weisbach Friction Factor Correlations [12]

Flow Region	Friction Factor, *f*	Reynolds Range
Laminar	$f = 64/\text{Re}$	$\text{Re} < 2100$
Smooth pipe	$1/\sqrt{f} = 2\log_{10}(\text{Re}\sqrt{f}) - 0.8$	$\text{Re} > 4000$ and $\varepsilon/\text{D} \rightarrow 0$
Transitional, Colebrook-White correlation	$1/\sqrt{f} = 1.14 - 2\log_{10}(\varepsilon/D + 9.35/(\text{Re}\sqrt{f}))$	$\text{Re} > 4000$ and transitional region
Complete turbulence	$1/\sqrt{f} = 1.14 - 2\log_{10}(\varepsilon/D)$	$\text{Re} > 4000$ and complete turbulence

13.5.2.3. Local Losses

In addition to pressure head losses due to pipe surface friction, the local losses are the pressure head loss occurring at flow appurtenances, such as valves, bends, and other fittings, when the fluid flows through the appurtenances. The local head losses in fittings may include:

- Surface fiction;
- Direction change of flow path;
- Obstructions in flow path;
- Sudden or gradual changes in the cross section and shape of the flow path.

The local losses are considered minor losses. These descriptions are misleading for the process piping system where fitting losses are often much greater than the losses in straight piping sections. It is difficult to quantify theoretically the magnitudes of the local losses, so the representation of these losses is determined mainly by using experimental data. Local losses are usually expressed in a form similar to that for the friction loss. Instead of f L/D in friction head loss, the loss coefficient K is defined for various fittings. The head loss due to fitting is given by the following equation:

$$h = K\frac{V^2}{2g} \tag{13-25}$$

where V is the downstream mean velocity normally. Two methods are used to determine the value of K for different fittings, such as a valve, elbow, etc. One method is to choose a K value directly from a table that is invariant with respect to size and Reynolds number.

Table 13-7 gives K values for several types of fittings. In this method the data scatter can be large, and some inaccuracy is to be expected. The other

Table 13-7 Loss Coefficients for Fittings [15]

Fitting	*K*
Globe valve, fully open	10.0
Angle valve, fully open	5.0
Butterfly valve, fully open	0.4
Gate valve, full open	0.2
3/4 open	1.0
1/2 open	5.6
1/4 open	17.0
Check valve, swing type, fully open	2.3
Check valve, lift type, fully open	12.0
Check valve, ball type, fully open	70.0
Elbow, 45°	0.4
Long radius elbow, 90°	0.6

approach is to specify K for a given fitting in terms of the value of the complete turbulence friction factor f_T for the nominal pipe size. This method implicitly accounts for the pipe size. The Crane Company's Technical Paper 410 [15] detailed the calculation methods for the K value for different fittings, and they are commonly accepted by the piping industry.

Manufacturers of valves, especially control valves, express valve capacity in terms of a flow coefficient C_v, which gives the flow rate through the valve in gal/min of water at 60°F under a pressure drop of 1.0 psi. It is related to K by:

$$C_v = \frac{29.9d^2}{\sqrt{K}} \tag{13-26}$$

where d is the diameter of the valve connections in inches.

13.5.3. Multiphase Flow

13.5.3.1. General

Multiphase transport is currently receiving much attention throughout the oil and gas industry, because the combined transport of hydrocarbon liquids and gases, immiscible water, and sand can offer significant economic savings over the conventional, local, platform-based separation facilities. However, the possibility of hydrate formation, the increasing water content of the produced fluids, erosion, heat loss, and other considerations create many challenges to this hydraulic design procedure.

Much research has been carried out on two-phase flow beginning in the 1950s. The behavior of two-phase flow is much more complex than that of single-phase flow. Two-phase flow is a process involving the interaction of many variables. The gas and liquid phases normally do not travel at the same velocity in the pipeline because of the differences in density and viscosities. For an upward flow, the gas phase, which is less dense and less viscous, tends to flow at a higher velocity than the liquid phase. For a downward flow, the liquid often flows faster than the gas because of density differences. Although analytical solutions of single-phase flow are available and the accuracy of prediction is acceptable in industry, multiphase flows, even when restricted to a simple pipeline geometry, are in general quite complex.

Calculations of pressure gradients in two-phase flows require values of flow conditions, such as velocity, and fluid properties, such as density, viscosity, and surface tension. One of the most important factors to consider when determining the flow characteristics of two-phase flows is *flow pattern*. The flow pattern description is not merely an identification of laminar or turbulent flow for a single flow; the relative quantities of the phases and the topology of the interfaces must also be described. The different flow patterns are formed because of relative magnitudes of the forces that act on the fluids, such as buoyancy, turbulence, inertia, and surface tension forces, which vary with flow rates, pipe diameter, inclination angle, and the fluid properties of the phases. Another important factor is *liquid holdup,* which is defined as the ratio of the volume of a pipe segment occupied by liquid to the volume of the pipe segment. Liquid holdup is a fraction, which varies from zero for pure gas flow to one for pure liquid flow.

For a two-phase flow, most analyses and simulations solve mass, momentum, and energy balance equations based on one-dimensional behavior for each phase. Such equations, for the most part, are used as a framework in which to interpret experimental data. Reliable prediction of multiphase flow behavior generally requires use of data or experimental correlations. Two-fluid modeling is a developing technique made possible by improved computational methods. In fluid modeling, the full three-dimensional partial differential equations of motion are written for each phase, treating each as a continuum and occupying a volume fraction that is a continuous function of position,

13.5.3.2. Horizontal Flow

In horizontal pipe, flow patterns for fully developed flow have been reported in numerous studies. Transitions between flow patterns are recorded with visual technologies. In some cases, statistical analysis of pressure fluctuations has been used to distinguish flow patterns. Figure 13-5

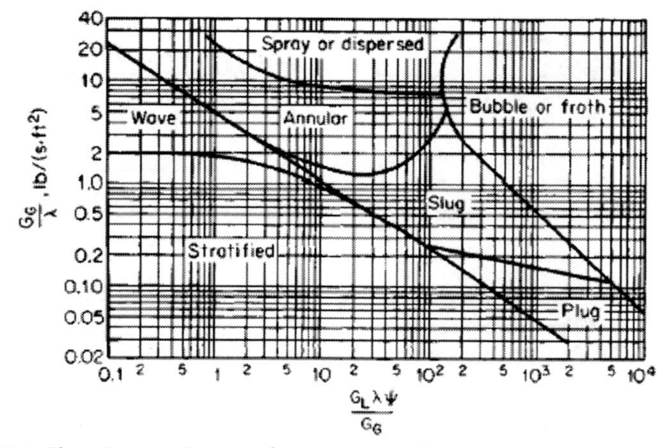

Figure 13-5 Flow Pattern Regions for Two-phase Flow through Horizontal Pipe [16]

shows a typical flow pattern map, which gives an approximate prediction of the flow patterns. Commonly used flow pattern maps such as those of Mandhane and Baker are based on observed regimes in horizontal, small-diameter, air/water systems. Scaling to the conditions encountered in the oil and gas industries is uncertain.

Figure 13-6 shows seven flow patterns for horizontal gas/liquid two-phase flow:

- *Bubble flow:* Gas is dispersed as bubbles that move at a velocity similar to that of liquid and tend to concentrate near the top of the pipe at lower liquid velocities.
- *Plug flow:* Alternate plugs of gas and liquid move along the upper part of the pipe.
- *Stratified flow:* Liquid flows along the bottom of the pipe and the gas flows over a smooth liquid/gas interface.
- *Wavy flow:* Occurs at greater gas velocities and has waves moving in the flow direction. When wave crests are sufficiently high to bridge the pipe, they form frothy slugs that move at much greater than the average liquid velocity.
- *Slug flow:* May cause severe and/or dangerous vibrations in equipment because of the impact of the high-velocity slugs against fittings.
- *Annular flow:* Liquid flows as a thin film along the pipe wall and gas flows in the core. Some liquid is entrained as droplets in the gas core.
- *Spray:* At very high gas velocities, nearly all the liquid is entrained as small droplets. This pattern is also called *dispersed* or *mist flow.*

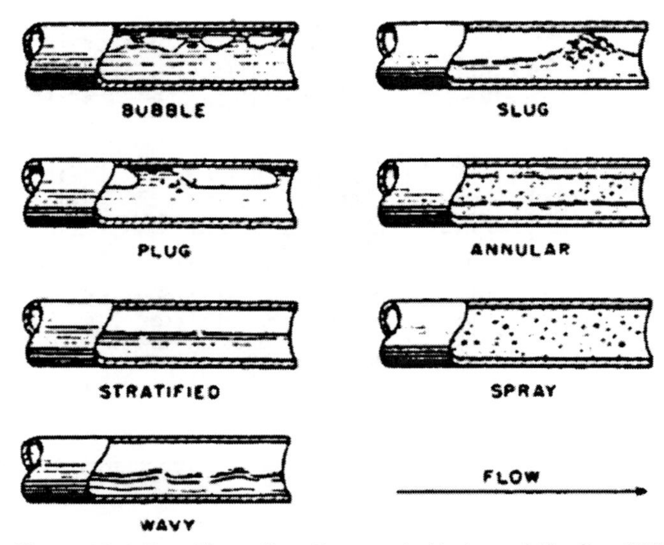

Figure 13-6 Two-Phase Flow Patterns in Horizontal Pipeline [17]

13.5.3.3. Vertical Flow

Two-phase flow in vertical pipe may be categorized into four different flow patterns, as shown in Figure 13-7 and listed here:

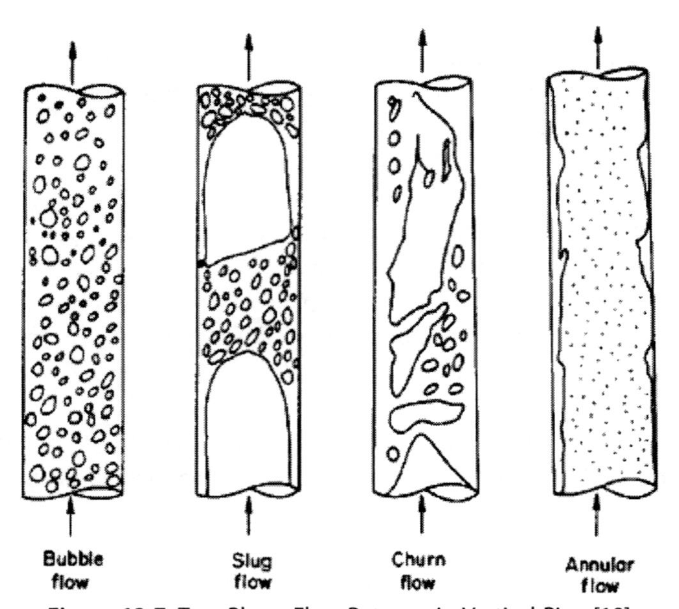

Figure 13-7 Two-Phase Flow Patterns in Vertical Pipe [18]

- *Bubble flow:* The liquid is continuous, with the gas phase existing as randomly distributed bubbles. The gas phase in bubble flow is small and contributes little to the pressure gradient except by its effect on the density.
- *Slug flow:* Both the gas and liquid phases significantly contribute to the pressure gradient. The gas phase in slug flow exists as large bubbles and is separated by slugs of liquid. The velocity of the gas bubbles is greater than that of the liquid slugs, thereby resulting in a liquid holdup that not only affects well and riser friction losses but also flowing density.
- *Churn flow:* The liquid slugs between the gas bubbles essentially disappear and at some point the liquid phase becomes discontinuous and the gas phase becomes continuous. The pressure losses are more the result of the gas phase than the liquid phase.
- *Annular flow:* This type of flow is characterized by a continuous gas phase with liquid occurring as entrained droplets in the gas stream and as a liquid film wetting the pipe wall.

13.5.4. Comparison of Two-Phase Flow Correlations

Many two-phase flow correlations have been developed in past several decades, most of which are based on water/air systems, and every correlation has its limit because of the data used. Most software used in flow assurance has several options for models or correlations for two-phase flow. Baker Jardine [19] conducted a major data matching study to identify the most suitable two-phase flow correlations for oil wells, condensate wells, oil pipelines, and gas/condensate pipelines, and the results of the analysis are summarized in following sections.

13.5.4.1. Description of Correlations

Duns and Ros (D-R)

The Duns and Ros correlation [20] was developed for vertical two-phase flow in wells based on extensive experimental research on oil and air mixtures. Separate correlations were developed for bubble, plug, and froth flows, slug flows, and mist flow regimes. These regions have low, intermediate, and high gas throughputs, respectively.

Orkiszewski (OR)

The Orkiszewski correlation [21] was developed for the prediction of two-phase pressure drops in vertical pipe. The bubble, slug, annular-slug transition, and annular mist flow regimes were considered. The method can accurately predict within 10% the two-phase pressure drops in flowing and gas-lift production wells over a wide range of well conditions based on 148 measured pressure drops.

Hagedorn and Brown (H-B)

The Hagedorn and Brown correlation [22] was developed based on an experimental study of pressure gradients in small-diameter vertical conduits. A 1500-ft experimental well was used to study flow through 1-in., 1.25-in., and 1.5-in. nominal size tubing for widely varying liquid flow rates, gas/liquid ratios, and liquid viscosities.

Beggs and Brill Original (B-BO)

The Beggs and Brill correlation [17] was developed following a study of two-phase flow in horizontal and inclined pipe. The correlation is based on a flow regime map. The model inclines flow both upward and downward at angles of up to $\pm 90°$.

Beggs and Brill Revised (B-BR)

The enhancements to the original method [17] are as follows: (1) An extra flow regime is considered that assumes a no-slip holdup, and (2) the friction factor was changed from the standard smooth pipe model to one that utilizes a single-phase friction factor based on the average fluid velocity.

Mukherjee and Brill (M-B)

The Mukherjee and Brill correlation [23] was developed following a study of pressure drop in two-phase inclined flow. Results agreed well with the experimental data, and correlations were further verified with Prudhoe Bay and North Sea data.

Govier, Aziz, and Fogarasi (G-A)

The Govier, Aziz, and Fogarasi correlation [24] was developed following a study of pressure drop in wells producing gas and condensate. Actual field pressure drop versus flow rate data from 102 wells with gas/liquid ratios ranging from 3,900 to 1,170,000 scf/bbl were analyzed in detail.

No-Slip (NS)

The no-slip correlation assumes homogeneous flow with no slippage between the phases. Fluid properties are taken as the average of the gas and liquid phases, and friction factors are calculated using the single-phase Moody correlation.

OLGAS-89 and OLGAS 92 (O-89, O-92)

OLGAS [25] is based on data from the SINTEF two-phase flow laboratory. The test facilities were designed to operate at conditions similar to field conditions. The test loop was 800 m long and had an 8-in. diameter. Operating pressures were between 20 barg and 90 barg. Gas superficial velocities of up to 13 m/s and liquid superficial velocities of up to 4 m/s were obtained. Pipeline inclination angles between $\pm 1°$ were studied in addition to flow up or down a hill section ahead of a 50-m-high vertical riser. More than 10,000 experiments were run on the test loop. OLGAS considers four flow regimes: stratified, annular, slug, and dispersed bubble flow.

Ansari (AN)

The Ansari model [26] was developed as part of the Tulsa University Fluid Flow Project (TUFFP) research program. A comprehensive model was formulated to predict flow patterns and the flow characteristics of the predicted flow patterns for upward two-phase flow. The model was evaluated by using the TUFFP well data bank that is composed of 1775 well cases, with 371 of them from Prudhoe Bay data.

BJA for Condensates (BJ)

Baker Jardine & Associates have developed a correlation [27] for two-phase flow in gas/condensate pipelines with a no-slip liquid volume fraction of lower than 0.1. The pressure loss calculation procedure is similar in approach to that proposed by Oliemans (see below), but accounts for the increased interfacial shear resulting from the liquid surface roughness.

AGA and Flanigan (AGA)

The AGA and Flanigan correlation [28] was developed for horizontal and inclined two-phase flow of gas/condensate systems. The Taitel Dukler flow regime map is used, which considers five flow regimes: stratified smooth, stratified wavy, intermittent, annular dispersed liquid, and dispersed bubble [29].

Oliemans (OL)

The Oliemans correlation [30] was developed following the study of large-diameter condensate pipelines. The model was based on a limited amount of data from a 30-in., 100-km pipeline operating at pressures of 100 barg or higher.

Gray (GR)

This correlation was developed by H. E. Gray [31] of Shell Oil Company for vertical flow in gas and condensate systems that are predominantly gas phase. Flow is treated as single phase, and water or condensate is assumed to adhere to the pipe wall. It is considered applicable to vertical flow cases where the velocity is below 50 ft/s, the tube size is below 3½ in., the condensate ratio is below 50 bbl/mmscf, and the water ratio is below 5 bbl/mmscf.

Xiao (XI)

The Xiao comprehensive mechanistic model [32] was developed as part of the TUFPP research program. It was developed for gas/liquid two-phase flow in horizontal and near horizontal pipelines. The data bank included large-diameter field data culled from the AGA multiphase pipeline data bank and laboratory data published in the literature. Data included both black oil and compositional fluid systems. A new correlation was proposed that predicts the internal friction factor under stratified flow.

13.5.4.2. Analysis Method

The predictions of pressure drop and flow rate for each of the correlations built in PIPESIM were compared with available experimental data from well/pipeline tests, reservoir fluid analysis, and drilling surveys/pipeline measurements collected by Baker Jardine. The percentage error for data points is calculated by:

$$\% \text{ error} = (\text{calculated value} - \text{measured values})/\text{measured value} \times 100\%$$

Table 13-8 lists the evaluation method with scores for each correlation. The overall scores for pressure drop and flow rate predictions are averaged for each correlation. The score is calculated with the sum of score for each database divided by 3 × the total database number. The evaluation results for each correlation with a database of vertical oil wells, highly deviated oil wells, gas-condensate wells, oil pipelines, and gas-condensate pipelines are demonstrated in Figure 13-8.

Table 13-8 Evaluation of Correlations with Score [33]

Error Range (%)	Evaluation	Score
− 5.0 to + 5.0	Very good	3.0
−10.0 to + 10.0	Good	2.0
−20.0 to + 20.0	Moderate	0
Outside above range	Poor	−3.0

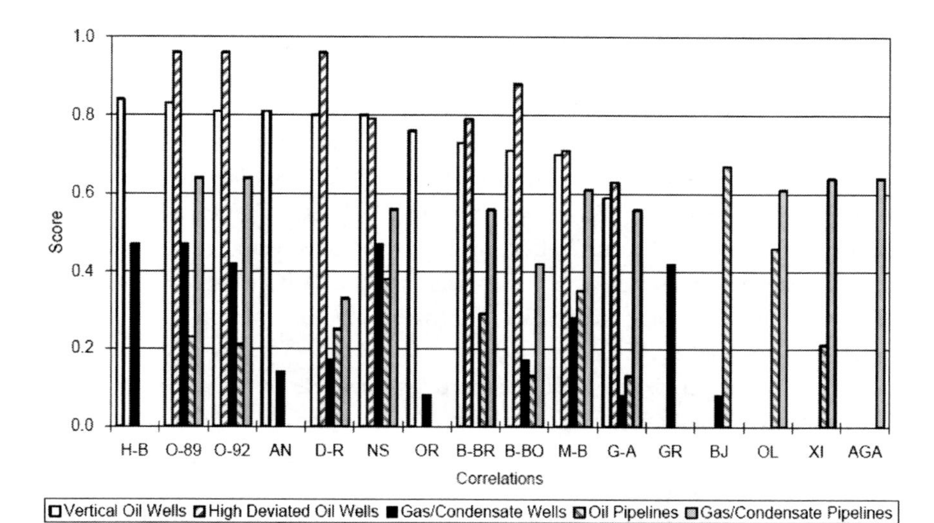

Figure 13-8 Evaluation of the Two-Phase Flow Correlations Used in PIPESIM

The Moody correlation is recommended for single-phase flow in vertical oil flow, horizontal oil flow, vertical gas flow, and horizontal gas flow.

13.6. SLUGGING AND LIQUID HANDLING

13.6.1. General

The occurrence of slug flow in a transportation pipeline can cause many problems in design and operation processes, which include kinetic force on fittings and vessels, pressure cycling, control instability, and inadequate phase separation. Slugging greatly affects the design of receiving facilities. In gas-condensate systems, larger lines result in more liquid being retained in the pipeline at low rates. When the flow rate is increased, much of the liquid can be swept out, potentially overwhelming the liquid handling capability of the receiving facilities. The facilities can be flooded and damaged if the slugs are larger than the slug catcher capacity. Therefore, quantifying the slug size, frequency, and velocity is necessary prior to equipment design.

The pressure at the bottom of the riser can vary if the holdup in the riser is not about the same as that in the line feeding it. If the riser holdup is too large and the gas velocity is too small to provide continuous liquid lift, too much of the liquid reverses and flows downward. Liquid accumulates at the base, causing an unstable pressure situation. This is relieved by large liquid slugs periodically leaving the riser at a high velocity. The changes in liquid amount and the corresponding pressure changes can be dramatic. A large *slug catcher* installation can be provided onshore, but it is not economical to place it on the platform. This is one of the practical reasons why a pipeline section immediately ahead of the riser should be horizontal or have a slightly upward slope of 2° to 5°. The section length probably should be several times the riser height. The upward incline eliminates a possible "sump" effect and serves to decrease pressure/holdup instabilities. Severe slugging in the riser can be enhanced by a negative pipeline inclination just prior to it. Actually, severe slugging is unlikely if there is a positive inclination.

Figure 13-9 shows the effect of the mass flow rate of two-phase flow on flow stability. A higher flow rate helps to decrease the slug and increase flow stability in the flowline. Higher system pressures also increase the tendency for a stable flow. Choking, at the top of the riser can be used to minimize severe slugging. The flow in a riser may differ from that in a wellbore, which has a relatively long horizontal flowline at the end of it. Holdup and surging from that horizontal flowline are transmitted to the relatively short riser. The riser may have to handle far more liquid than a well because the flowline can feed it liquid surges that far exceed those possible by gas-lift or reservoir mechanisms.

In many oil and gas developments that incorporate multiphase flowlines, the possibility of slugs or surges is one of the most important flow assurance

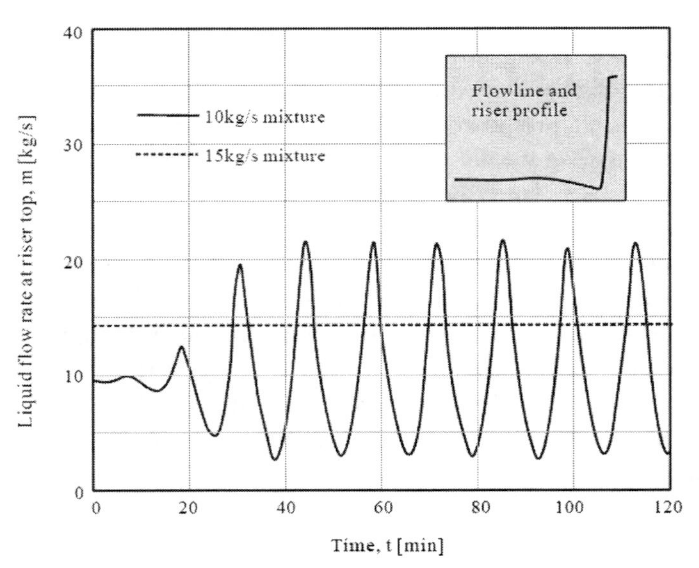

Figure 13-9 Riser Flow Stability versus Flow Rate

concerns due to the excessive demands large changes in oil/gas flow rates place on the processing facilities. Multiphase surges come in three forms:

1. *Hydrodynamic slugs:* formed from the stratified flow regime due to instability of waves at certain flow rates.
2. *Terrain-induced slugs:* caused by accumulation and periodic purging of liquid in elevation changes along the flowline, particularly at low flow rates.
3. *Operationally induced surges:* formed in the system during operation transfer between a steady state and a transient state; for example, during start-up or pigging operations.

13.6.2. Hydrodynamic Slugging

Hydrodynamic slugs are initiated by the instability of waves on the gas/liquid interface in stratified flow under certain flowing conditions. Figure 13-10 shows the formation process of hydrodynamic slugging from a stratified flow. In Figure 13-10a, the gas/liquid interface is lifted to the top of the pipe when the velocity difference between gas phase and liquid phase is high enough. This wave growth is triggered by the Kelvin-Helmholtz instability. Once the wave reaches the top of pipe, it forms a slug (Figure 13-10b). The slug is pushed by the gas and so travels at a greater velocity than the liquid film, and more liquid is then swept into the slug. In Figure 13-10c, entrainment reduces the average liquid holdup in the slug, increasing the turbulence within the slug.

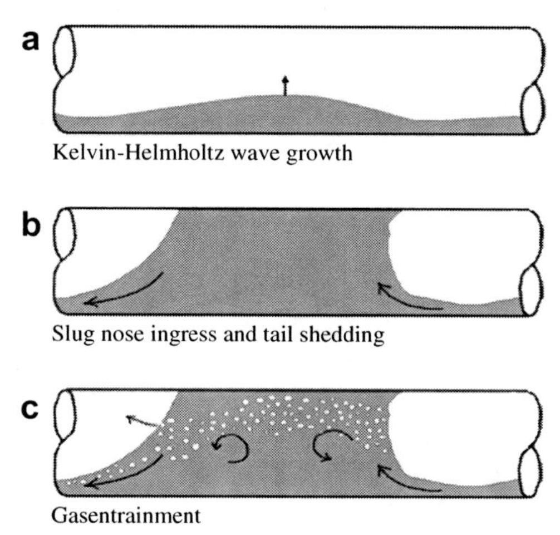

Figure 13-10 Formation of Hydrodynamic Slugging [34]

A main part of the frictional pressure drop in multiphase flow is thought to be due to the turbulent region within the slug. Thus, the size of the turbulent region can have a significant effect on the frictional pressure losses in a pipeline. Two-phase flow pattern maps indicate hydrodynamic slugging, but slug length correlations are quite uncertain. Tracking of the development of the individual slugs along the pipeline is necessary to estimate the volume of the liquid surges out of the pipeline.

Slugging simulations need to be performed over the flow rates, water cuts, and GORs for the field life. The effects of any artificial lift should be included in the simulations. In general, simulation results are presented as liquid and gas flow rates at the separator, slug lengths at the base and top of the riser, and pressure at key locations as a function of time. The key locations in a well are upstream of the well flow control devices and bottomhole. If processing equipment is included in the model, the separator pressure, separator levels, and outlet gas and liquid rates from the separator as a function of time are presented.

In cases where the predicted slugging causes liquid or gas handling problems, the effects of additional choking upstream of the separator should be determined. The evolution of slugs is very sensitive to the pipe inclination and changing the inclination by less than a degree can be sufficient to change the balance, causing a flow regime transition. Thus, peaks and troughs along the pipeline profile of relatively small elevation change may have a very significant effect.

Figure 13-11 compares the flow patterns of a simple horizontal topography and one with some undulations. The multiphase flow in undulate

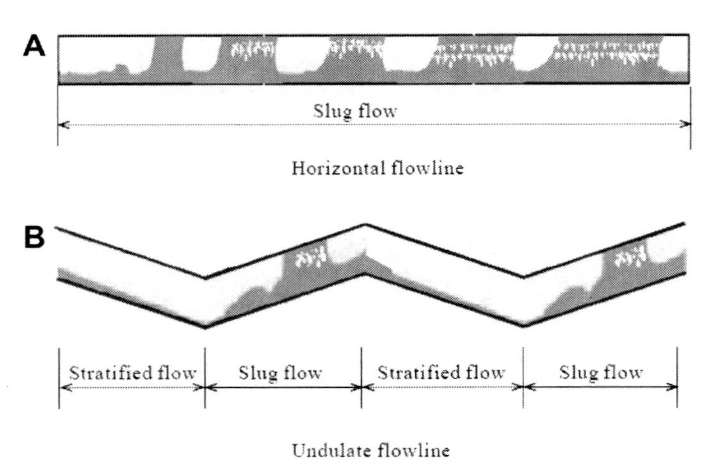

Figure 13-11 Effect of Pipeline Topography to Flow Pattern [34]

flowlines switches between the stratified and slug flow regimes, implying that not only could the slug sizes differ markedly, but the pressure drops could be very different too. Few pipelines have constant inclinations; most undulate following the natural terrain. When modeling multiphase flow in lower flow rates, it is important to represent these undulations as faithfully as possible. At higher flow rates, undulations may not have much impact on predictions.

13.6.3. Terrain Slugging

Figure 13-12 shows the formation process for a terrain slug in a pipeline-riser system:

- *Figure 13-12a:* Low spot fills with liquid and flow is blocked.
- *Figure 13-12b:* Pressure builds up behind the blockage.
- *Figures 13-12c and d:* When the pressure becomes high enough, gas blows liquid out of the low spot as a slug.

Liquid accumulates at the base of the riser, blocking the flow of gas phase. This liquid packet builds up until the gas pressure is sufficiently high to overcome the hydrostatic head and blow the liquid slug from the riser. Slug lengths can be two to three times the riser height.

Terrain slugs can be very severe, causing large pressure variations and liquid surges out of a pipeline. Terrain slugging is a transient situation that requires a dynamic model to predict and describe. When the minimum flow rate is defined as the terrain slugging boundary, a region without severe slugging should be determined as a function of the water cut and including gas lift. In cases where the predicted slugging causes liquid or gas handling problems, the effect of additional choking upstream of the separator should be determined.

It may be difficult to design a slug catcher to cope with the magnitude of terrain slugs. If the transportation system terminates in a vertical riser onto a receiving platform, the passage of the slug from the horizontal pipeline to

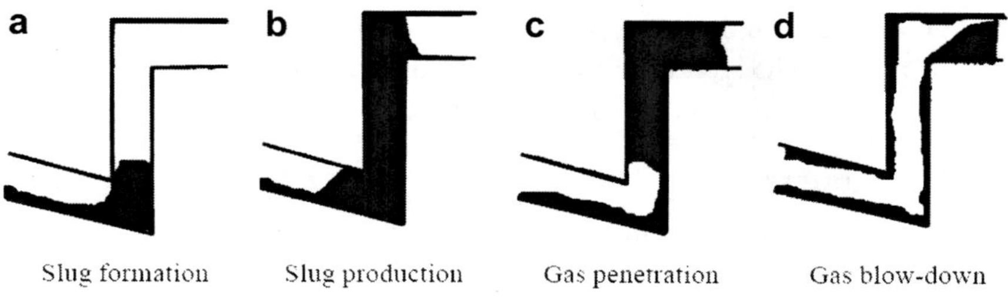

Figure 13-12 Formation of Riser Slugging

the vertical riser results in cyclic flow effects. As the slug decelerates into the vertical riser, the following gas bubble is compressed. Compression continues until sufficient energy is generated to accelerate the slug from the riser. When the wellhead pressure is limited, the vertical riser from the seabed into the platform may form a backpressure due to slugs, and it will limit well production. In such circumstances the vertical pressure loss can be reduced if a slug catcher is located on the seabed, and the gas and liquid phases are separated. The liquid is pumped to the surface with the gas free flowing to the platform through a separate riser. An alternative is to inject gas at the base of the riser, which will lighten the fluid column and minimize vertical pressure losses.

13.6.4. Start-Up and Blowdown Slugging

Slugs form in the start-up operation process because of transformation from a steady state to a transient process. The start-up simulations should be performed starting from shutdown conditions to different representative operating conditions throughout the field life. A range of start-up rates, consistent with reservoir management constraints, should be evaluated. If necessary, artificial lifting to mitigate start-up slugs should be evaluated. For gas lifting, the required gas-lift rate should be determined. If lift gas is unavailable until it is obtained from the production, this operating constraint should be included in the simulations.

13.6.5. Rate Change Slugging

When the flow rate is increased, the liquid holdup in the line decreases. This change in holdup can either exit the line as a steady flow with increased liquid production, or it can come out in the form of a slug, depending on the flow rate change. The rate change slugs can occur in gas/condensate flowlines when the rates are increased. The flowline may be in a steady flow pattern, such as stratified flow, at both the initial and final flow rates but will slug during the transition period until the line reequilibrates at the higher rate.

As with start-up slugs, it is impossible to predict whether slugs will occur when rates are changed using steady-state or hand methods. The flowline must be dynamically simulated using a transient flow program.

13.6.6. Pigging

Pigs are run through pipelines for a variety of reasons, including:

- Liquid inventory control;
- Maintenance and data logging;

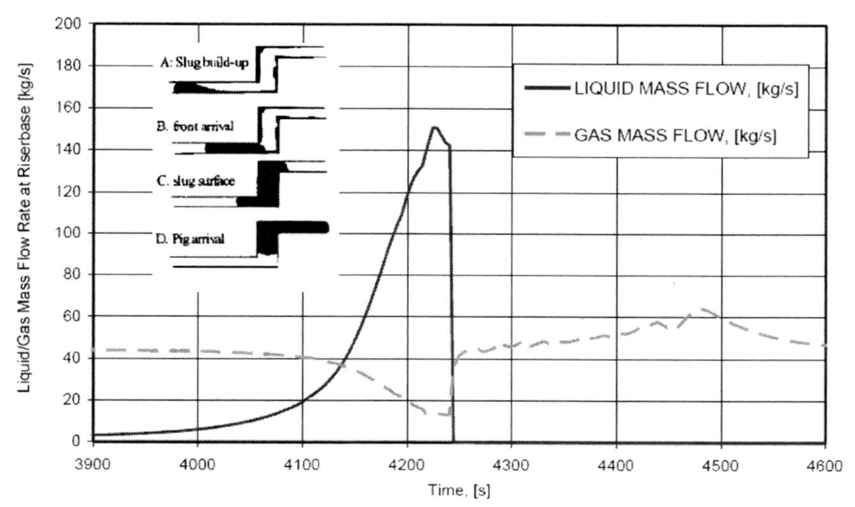

Figure 13-13 Flow Rate Variations Due to Pigging

- Pipeline cleaning and dewaxing;
- Inhibitor application.

Figure 13-13 shows the flow rate variations of liquid and gas at a riser base due to pigging. When the slug is big enough, the full riser base will be filled with liquid phase. The simulations of the pigging operation should be performed where pigging is required. The time for the pigging operation and the pig velocity as a function of time should be reported. For a round trip pigging, the inlet and outlet temperature, pressure, and flow rate as a function of time should be analyzed. For a round trip pigging with liquid, the backpressure necessary to maintain single-phase flow behind the pig should be analyzed.

13.6.7. Slugging Prediction

The slugging prediction can be carried out using PIPESIM (Schlumberger), OLGA (Scandpower Petroleum Technology), ProFES (Aspentech), or TACITE (Simulation Science). Steady-state multiphase simulation software such as PIPESIM can predict hydrodynamic slug distributions along the flowline and riser slugs. However, only OLGA 2000 (standard version), the OLGA 2000 slug tracking module, and ProFES are good at simulating transient multiphase flow and predicting the liquid holdup variations along the flowlines, terrain slugs and start-up/shutdown transient slugs.

Ramp-up slugs are of primary importance for gas/condensate systems where the increased flow rate can sweep out large volumes of liquid. Simulations of flow rate ramps should be performed from turndown rates to intermediate and full production rates over the life of the field. OLGA 2000 slug tracking [35] is generally not required for ramp-ups in gas/condensate systems except in cases with hilly terrain or high liquid loadings (>50 bbl/mmscf). Results may be presented as outlet liquid and gas rates as a function of time.

The production system should be modeled starting at the reservoir using inflow performance relationships provided by reservoir modeling. At the outlet, pressure control is recommended. If this is not possible, a sensitivity study should be performed to ensure that random fluctuations in the outlet pressure do not significantly alter the slugging. The artificial boundary conditions of a constant pressure tend to dampen slugging and should be avoided whenever possible for slugging simulations. True boundary conditions on the flowline/riser can be obtained using a transient pipeline model coupled to a dynamic process model. Integrated dynamic pipeline and process simulation is rarely necessary for the design of wells, flowlines, and risers. However, in certain instances, integrated modeling is advisable for the design and control of the process facilities.

13.6.8. Parameters for Slug Characteristics

The key parameters required for the assessment of the performance of separation and associated downstream facilities are the average slug volume and frequency, and the greatest likely slug volume and its frequency of occurrence. The combination of slug velocity, frequency and liquid holdup is important for the design of pipeline supports. These are the control parameters: V_{sg} (gas superficial velocity), V_{sl} (liquid superficial velocity), D (pipeline diameter), L (pipeline length), ϕ (pipeline inclination), ρ_g (gas density), ρ_l (liquid density), μ_g (gas viscosity), μ_l (liquid viscosity), and σ_l (surface tension).

13.6.9. Slug Detection and Control Systems

Although slug control is very important to avoid facility damage and upsets, control options are limited. The potential impact of slugging on the topside system operation must be addressed and then analyses of the subsea system carried out to assess the effects. Usually a trade-off results between the design of slug catchers and the optimization of the flowline to reduce slugging.

Slugs have been successfully detected using gamma densitometers located on the riser, acoustical measurements, and measurements of pressure at the base of the riser. Slug detection systems should be considered when predicted slugging is expected to give operational difficulties and/or when an advanced control system is to be used for slug mitigation. In this case, the slugging simulations should include advanced controls to test the control algorithm. Flow assurance and process disciplines should demonstrate advanced controls to critically dampen predicted slug volumes and frequency.

13.6.10. Equipment Design for Slug Flow

The design of slug catchers, separators, and control systems in downstream pipelines is based on the presence and severity of slug flow in the system. The parameter estimations of slug volumes, liquid and gas rates exiting the pipeline as a function of time, etc., must be factored into the equipment design for slug flow. These parameters should be calculated for steady-state operation and for a series of transient operation cases: turndown rates, pigging, shutdown and start-up, rate changes, etc.

Transient modeling gives the best estimations of transient slug flow behavior. Some of the transient simulators include separators and control systems as part of the run. In general, the size of slug catching equipment for gas/condensate pipelines will be governed by pigging considerations. For oil-dominated systems, the size of the slug catcher is usually governed by the maximum slug length due to either hydrodynamic or terrain slugs.

13.6.11. Slug Catcher Sizing

Slug catchers should be sized to dampen surges to a level that can be handled by downstream processing equipment. Before dynamic models of the topside facilities are available, the level of acceptable surging is unknown and designers are often forced to make assumptions vis-à-vis surge volumes, such as designing for the "one-in-a-thousand" slug.

The surge volume for gas/condensate requirements is determined from the outlet liquid rates predicted in the ramp-up, start-up, and pigging cases. The required slug catcher size is dependent on liquid handling rate, pigging frequency, and ramp-up rates. An iterative process may be required to identify optimum slug catcher size, pigging frequency, liquid handling rate, and acceptable ramp-up rates. For this optimization, the results of the simulations should be presented as surge volume requirements as a function of liquid handling rate for representative ramp-up rates and pigging frequencies.

Separator volumes for black-oil systems are typically set by separation requirements rather than liquid slug handling capacity. Consequently, the ability of the separator to accommodate slugs from all operations should be confirmed based on the results of the slugging simulations.

13.7. SLUG CATCHER DESIGN

A slug catcher is a piece of process equipment (typically a pressure vessel or set of pipes) that is located at the outlet of production flowlines or pipelines, prior to the remaining production facilities. It is usually located directly at the upstream of the primary production separator as shown in Figure 13-14; in some cases, the primary separator also serves as a slug catcher as shown in Figure 13-15. Slug catchers are used in both oil/gas multiphase production

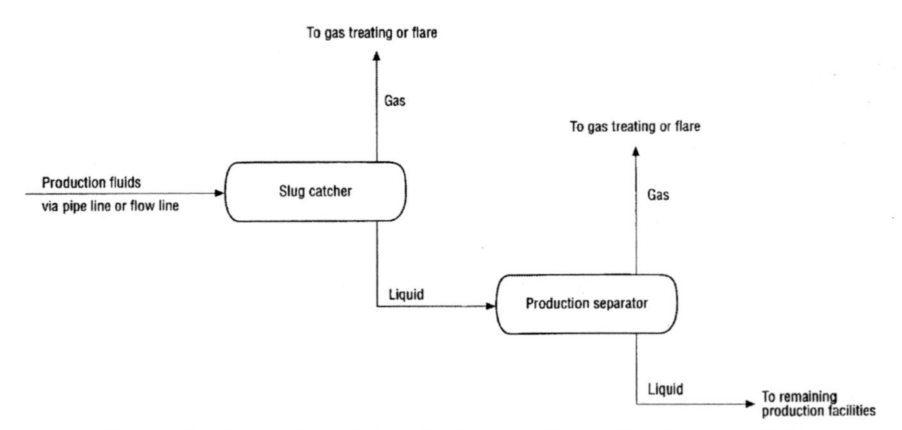

Figure 13-14 Location of Slug Catcher and Production Separator [36]

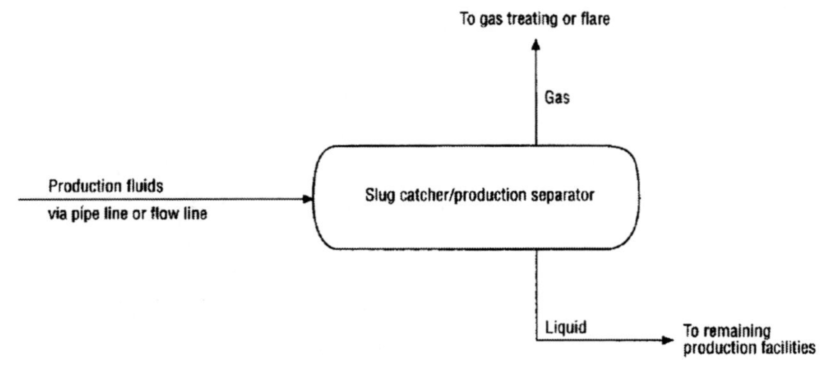

Figure 13-15 Combined Slug Catcher and Production Separator [36]

systems and in gas/condensate systems to mitigate the effects of slugs, which are formed due to terrain, pipeline operation in the slug-flow regime, or pigging. A slug catcher is generally not needed for single-phase liquid lines such as treated oil or produced water because slug flow is not encountered in single-phase operation; however, the need for slug catchers should be evaluated if pigging is expected.

13.7.1. Slug Catcher Design Process

The goal of slug catcher design is to properly size the slug catcher for the appropriate conditions. The process consists of the following steps:

- Determine slug catcher functions.
- Determine slug catcher location.
- Select preliminary slug catcher configuration.
- Compile design data.
- Establish design criteria.
- Estimate slug catcher size and dimensions.
- Review for feasibility; repeat as necessary.

13.7.2. Slug Catcher Functions

The slug catcher functions may be summarized as follows:

- Process stabilization;
- Phase separation;
- Storage.

Each slug catcher can serve one or more functions, and each function is detailed in the following sections.

13.7.2.1. Process Stabilization

Process stabilization is the primary purpose of the slug catcher. In a typical steady-state operation, multiphase production fluids from the flowlines enter the production facilities at constant temperature, pressure, velocity, and flow rate. Process control devices such as pressure control valves and level control valves are used to maintain steady operating conditions throughout the process facilities. During non-steady-state conditions, such as start-up, shutdown, turndown, and pigging, or when slugging during normal operation is expected, the process controllers alone may not be able to sufficiently compensate for the wide variations in fluid flow rates, vessel liquid levels, fluid velocities, and system pressures caused by the slugs.

A slug catcher provides sufficient space to dampen the effects of flow rate surges in order to minimize mechanical damage and deliver an even supply of gas and liquid to the rest of the production facilities, minimizing process and operation upsets.

13.7.2.2. Phase Separation

The second main function of the slug catcher is to provide a means to separate multiphase production fluids into separate gas and liquid streams in order to reduce liquid carryover in the gas stream and gas re-entrainment in the liquid stream. When a slug catcher is provided at the upstream end of a production separator as shown in Figure 13-14, the slug catcher is designed primarily for process stabilization. Gas/liquid separation also occurs, but the efficiency of separation is usually not sufficient to meet oil and gas product specifications. The gas stream may need additional treating to remove entrained liquids prior to treating, compression, or flaring. The liquid may need additional treating for gas/oil/water separation and crude stabilization.

When a separate slug catcher is not provided, the production separator is designed for both process stabilization and efficient gas/liquid separation as shown in Figure 13-15. Production separators, which also function as slug catchers, are generally not designed for gas/oil/water separation. This is because the level surges due to slugging, making it difficult to control the oil/water interface.

13.7.2.3. Storage

Slugs that result from pigging can often be significantly larger than terrain-induced slugs or slugs formed while operating in the slug-flow regime, particularly for gas/condensate systems with long flowlines or pipelines. In these situations, the condensate processing and handling systems may not be sized to quickly process the large slug volume that results from pigging. The slug catcher then acts as a storage vessel to hold the condensate until it can gradually be metered into the process or transported to another location.

13.8. PRESSURE SURGE

13.8.1. Fundamentals of Pressure Surge

An important consideration in the design of single-liquid-phase pipelines is pressure surge, also known as water hammer. Typical surge events in a pipeline or piping system are generally caused by a pump shutdown or a valve closure [37]. The kinetic energy of flow is converted to pressure energy. The velocity of the pressure wave propagation is determined by fluid and pipeline characteristics. Typical propagation velocities range from 1100 ft/s for propane/butane pipelines to 3300 ft/s for crude oil pipelines and up to 4200 ft/s for heavy-wall steel water pipelines.

A rough estimate of the total transient pressure in a pipeline/piping system following a surge event can be obtained from the following equations:

$$v_s = \sqrt{\frac{144Kg/\rho}{1 + [KdC/Et]}} \tag{13-27}$$

where

v_s: speed of sound in the fluid, propagation speed of the pressure wave, ft/s;

K: fluid bulk modulus, 20 ksi;

g: gravitational constant, 32.2 ft/s^2;

ρ: density of liquid, lbf/ft^3;

d: inside diameter of pipe, in.;

E: pipe material modulus of elasticity, psi;

t: pipe wall thickness, in.;

C: constant of pipe fixity; 0.91 for an axially restrained line; 0.95 for unrestrained.

The surge pressure wave travels upstream and is reflected downstream, oscillating back and forth until its energy is dissipated in pipe wall friction. The amplitude of the surge wave, or the magnitude of the pressure surge P_{surge}, is a function of the change in velocity and the steepness of the wave front and is the inverse of the time it took to generate the wave:

$$P_{\text{surge}} = \rho v_s \Delta v/(144g) \qquad \text{for} \quad T < 2L/v_s \tag{13-28}$$

where

Δv: total change in velocity, ft/s (m/s);

$2L/v_s$: propagation time, s;

T: valve closing time, s.

This is only a rough estimation of the surge pressure magnitude for cases limited by the stated time of closure criteria. An accurate analysis of the maximum surge pressure and location of critical points in the system may be obtained from a dynamic simulation by using software such as Pipe Simulator. The surges are attenuated by friction, and the surge arriving at any point on the line is less than at the origin of the surge wave. Nevertheless, when the flow velocity is high and stoppage is complete, or when a pump station is bypassed suddenly, the surge energy generated can produce pressures high enough to burst pipe. Based on the ASME codes, the level of pressure rise due to surges and other variations from normal operations shall not exceed the internal design pressure at any point in the piping system and equipment by more than 10%.

13.8.2. Pressure Surge Analysis

Surge analysis should be performed during a project's early design and planning phases. This analysis will help to ensure the achievement of an integrated and economical design. Surge analysis provides assurance that the selected pumps/compressors, drivers, control valves, sensors, and piping can function as an integrated system in accordance with a suitable control system philosophy. A surge analysis becomes mandatory when one repairs a ruptured pipeline/piping system to determine the source of the problem.

All long pipelines/piping systems designed for high flow velocities should be checked for possible surge pressures that could exceed the maximum allowable surge pressure (MASP) of the system piping or components. A long pipeline/piping system is defined as one that can experience significant changes in flow velocity within the critical period. By this definition, a long pipeline/piping system may be 1500 ft long or 800 miles long.

A steady-state design cannot properly reflect system operation during a surge event. On one project in which a loading hose at a tanker loading system in the North Sea normally operated at 25 to 50 psig, a motor-operated valve at the tanker manifold malfunctioned and closed during tanker loading. The loading hose, which had a pressure rating of 225 psig, ruptured. Surge pressure simulation showed that the hydraulic transient pressure exceeded 550 psig. Severe surge problems can be mitigated through the use of quick-acting relief valves, tanks, and gas-filled surge bottles, but these facilities are expensive single-purpose devices.

13.9. LINE SIZING

13.9.1. Hydraulic Calculations

Hydraulic calculations should be carried out from wells, pipelines, and risers to the surface facilities in line sizing. The control factors in the calculation are fluids (oil, gas, or condensate fluids), flowline size, flow pattern, and application region. Unlike single-phase pipelines, multiphase pipelines are sized taking into account the limitations imposed by production rates, erosion, slugging, and ramp-up speed. Artificial lift is also considered during line sizing to improve the operational range of the system.

Design conditions are as follows:

- Flow rate and pressure drop allowable established: Determine pipe size for a fixed length.
- Flow rate and length known: Determine pressure drop and line size.

Usually either of these conditions requires a trial-and-error approach based on assumed pipe sizes to meet the stated conditions. Some design problems may require determination of maximum flow for a given line size and length, but this is a reverse of the conditions above.

During the FEED stage of a project, the capability of a given line size to deliver the production rate throughout the life of the field is determined by steady-state simulation from the reservoir to the surface facilities throughout the field life. Sensitivities to the important variables such as GOR, water cut, viscosity, and separator pressure should be examined.

The technical criteria for line sizing of pipelines are stated in next section; however, the optimum economic line size is seldom realized in the engineering design. Unknown factors such as future flow rate allowances, actual pressure drops through certain process equipment, etc., can easily overcompensate for what the design predicted when selecting the optimum. In sizing a pipeline, one is always faced with a compromise between two factors. For a given flow rate of a given fluid, piping cost increases with diameter. But pressure loss decreases, which reduces potential pumping or compression costs. There is an economic balance between material costs and pumping costs in downstream flowlines. The optimum pipe size is found by calculating the smallest capitalization/operating cost or using the entire pressure drop available; or increasing velocity to highest allowable.

13.9.2. Criteria

The line sizing of the pipeline is governed by the following technical criteria:

- Allowable pressure drop;
- Maximum velocity (allowable erosional velocity) and minimum velocity;
- System deliverability;
- Slug consideration if applicable.

Other criteria considered in the selection of the optimum line size include:

- *Standard versus custom line sizes:* Generally, standard pipe is less expensive and more readily available. For long pipelines or multiple pipelines for the same size, custom line sizes may be cost effective.
- *Ability of installation:* Particularly in deep water, the technical installation feasibility of larger line sizes may constrain the maximum pipe size.
- *Future production:* Consideration should be given to future production that may utilize the lines.
- *Number of flowlines and risers:* If construction or flow assurance constraints require more than two production flowlines per manifold, optional alternatives including subsea metering processing and bundles should be explored.
- *Low-temperature limits:* Subsea equipment including trees, jumpers, manifolds, and flowlines have minimum temperature specifications. For systems with Joule-Thompson cooling in low ambient temperature environments, operating philosophies and possibly metallurgical sections should be adjusted.
- *High-temperature limits:* Flexible pipe has a maximum temperature limit that depends on the materials of construction, water cut, water composition, and water pH. For systems with flexible pipe, the flow rate may be limited, requiring a smaller pipe.
- *Roughness:* Flexible pipe is rougher than steel pipe and, therefore, requires a larger diameter for the same maximum rate. To smooth flexible pipe, an internal coating may be applied.

13.9.3. Maximum Operating Velocities

Liquid velocity is usually limited because of erosion effects at fittings. Erosion damage can occur in flowlines with multiphase flow because of the continuous impingement of high-velocity liquid droplets. The damage is almost always confined to the place where the flow direction is changed, such as elbows, tees, manifolds, valves, and risers. The erosional velocity is defined as the bulk fluid velocity that will result in the removal of corrosion product scales, corrosion inhibitors, or other protective scales present on the intersurface of a pipeline [38].

The velocity of pipeline fluids should be constrained as follows:

- The fluid velocity in single-phase liquid lines varies from 0.9 to 4.5 m/s (3 to 15 ft/s).

- Gas/liquid two-phase lines do not exceed the erosional velocity as determined from following equation, which is recommended in API RP 14E [39]:

$$V_{max} = C/\rho_m^{0.5} \tag{13-29}$$

where

V_{max}: maximum allowable mixture erosional velocity, m/s (ft/s);
ρ: density of the gas/liquid mixture, kg/m^3 (lb/ft^3);
C: empirical constant, which determined in Table 13-9.
The mixture density, ρ_m is defined as:

$$\rho_m = C_L\rho_L + (1 - C_L)\rho_g \tag{13-30}$$

where

ρ_L: liquid density;
ρ_g: gas density;
C_L : flowing liquid volume fraction, $C_L = Q_L/(Q_L + Q_g)$.

Assuming on erosion rate of 10 mils per year, the following maximum allowable velocity is recommended by Salama and Venkatesh [40], when sand appears in an oil/gas mixture flow:

$$V_M = \frac{4d}{\sqrt{W_s}} \tag{13-31}$$

where

V_M: maximum allowable mixture velocity, ft/s;
d: pipeline inside diameter, in.;
W_s: rate of sand production (bbl/month).

Erosion limits or maximum velocities for flexible pipes should be specified by the manufacturer. The API erosion velocity limit is normally very conservative and the practice of this limit could vary depend on different operators.

Table 13-9 Empirical Constant in the Equation

	Operational Frequency	
Service Type	**Continuous**	**Intermittent**
Two-phase flow without sand	100	125

If possible, the minimum velocity in two-phase lines should be greater than 3 m/s (10 ft/s) to minimize slugging.

13.9.4. Minimum Operating Velocities

The following items may effectively impose minimum velocity constraints:

- *Slugging:* Slugging severity typically increases with decreasing flow rate. The minimum allowable velocity constraint should be imposed to control the slugging in multiphase flow for assuring the production deliverability of the system.
- *Liquid handling:* In gas/condensate systems, the ramp-up rates may be limited by the liquid handling facilities and constrained by the maximum line size.
- *Pressure drop:* For viscous oils, a minimum flow rate is necessary to maintain fluid temperature such that the viscosities are acceptable. Below this minimum, production may eventually shut itself in.
- *Liquid loading:* A minimum velocity is required to lift the liquids and prevent wells and risers from loading up with liquid and shutting in. The minimum stable rate is determined by transient simulation at successively lower flow rates. The minimum rate for the system is also a function of GLR.
- *Sand bedding:* The minimum velocity is required to avoid sand bedding.

13.9.5. Wells

During FEED analysis, the production system should be modeled starting at the reservoir using inflow performance relationships. OLGA 2000 may be used to evaluate the hydraulic stability of the entire system including the wells, flowlines, and risers subject to artificial lift to get minimum stable rates as a function of the operating requirements and artificial lift.

In the hydraulic and thermal analyses of subsea flowline systems, the wellhead pressure and temperature are generally used as the inlet pressure and temperature of the system. The wellhead pressure and temperature are functions of reservoir pressure, temperature, productivity index, and production rate and can be obtained from steady and transient hydraulic and thermal analyses of the well bore. The hydraulic and thermal models may be simulated using the commercial software PIPESIM or OLGA 2000.

The hydraulic and thermal processes of a wellbore are simulated for the following purposes:

- Determining steady-state wellhead pressure and temperature for flowline hydraulics and thermal analyses.
- Determining the minimum production rates that will prevent hydrate formation in the wellbore.

- Determining the minimum production rate for preventing wax deposition in the wellbore and tree. The wellhead temperature is required to be higher than the critical wax deposition temperature in the steady-state flowing condition.
- Determining cooldown time and warm-up time from transient wellbore analysis to prevent hydrate formation in the wellbore.

The transient analyses determine how hydrates can be controlled during startup and shut-in processes. Vacuum insulated tubing (VIT) is one method to help a wellbore out of the hydrate region. Use of 3000 to 4000 ft of VIT results in warm-up of the wellbore to temperatures above the hydrate temperature in typically 2 hr or less. These warm-up rates are rapid enough to ensure that little or no hydrates form in the wellbore. This makes it possible to eliminate the downhole injection of methanol above the subsurface safety valve during warm-up. Instead, methanol is injected upstream of the choke to prevent any hydrates that may be formed during warm-up from plugging the choke. One drawback of VIT is that it cools more quickly than bare tubing since the insulation prevents the tubing from gaining heat from the earth, which also warms up during production. As a result of this quick cooling, it is necessary to quickly treat the wellbore with methanol in the event that a production shut-in occurs during the warm-up period. For a higher reservoir temperature, VIT may not be used in the wellbore, and the cooldown speed of a wellbore is very slow after a process interruption from steady-state production conditions.

13.9.6. Gas Lift

Artificial lift can be used to increase production rates or reduce line sizes. Line sizing may include the affects of gas lift.

For systems employing gas lift, models should extend from the reservoir to the surface facilities incorporating the gas lift gas in addition to the produced fluids. The composition of the gas-lift gas is provided by the topside process but should meet the dew-point control requirement. The preferred location (downhole, flowline inlet, or riser base) is determined based on effectiveness and cost.

The effect of gas lift on the severe slugging boundary may be simulated with OLGA 2000 for the situations, when it is required. Normally, it is set at the riser base. The simulation may decide the minimum flow rate without severe slugging as a function of gas-lift rates and water cuts. When the gas-lift injection point does not include a restriction or valve to choke the gas flows, OLGA 2000 transient simulations may model the transient flow in the gas-lift line by including the gas-lift line as a separate branch to determine maximum variation in mass rate and pressure and still maintain system stability. Maximum production rates throughout the field life are determined as a function of gas-lift rate. In FEED designs, the gas-lift rate and pressure requirements should be recommended, including a measurement and control strategy for the gas-lift system.

REFERENCES

[1] R.C. Reid, J.M. Prausnitz, T.K. Sherwood, The Properties of Gases and Liquids, third ed., McGraw-Hill, New York, 1977.
[2] K.S. Pedersen, A. Fredenslund, P. Thomassen, Properties of Oils and Natural Gases, Gulf Publishing Company (1989).
[3] T.W. Leland, Phase Equilibria, Fluid, Properties in the Chemical Industry, DECHEMA, Frankfurt/Main, 1980. 283–333.
[4] G. Soave, Equilibrium Constants from a Modified Redilich-Kwong Equation of State, Chem. Eng. Sci vol. 27 (1972) 1197–1203.
[5] D.Y. Peng, D.B. Robinson, A New Two-Constant Equation of State, Ind. Eng. Chem. Fundam. vol. 15 (1976) 59–64.
[6] G.A. Gregory, Viscosity of Heavy Oil/Condensate Blends, Technical Note, No. 6, Neotechnology Consultants Ltd, Calgary, Canada, 1985.
[7] G.A. Gregory, Pipeline Calculations for Foaming Crude Oils and Crude Oil-Water Emulsions, Technical Note No. 11, Neotechnology Consultants Ltd, Calgary, Canada, 1990.
[8] W. Woelflin, The Viscosity of Crude Oil Emulsions, in Drill and Production Practice,, American Petroleum Institute vol. 148 (1942) p247.
[9] E. Guth, R. Simha, Untersuchungen über die Viskosität von Suspensionen und Lösungen. 3. Über die Viskosität von Kugelsuspensionen, Kolloid-Zeitschrift vol. 74 (1936) 266–275.
[10] H.V. Smith, K.E. Arnold, Crude Oil Emulsions, in Petroleum Engineering Handbook, in: H.B. Bradley (Ed.), third ed., Society of Petroleum Engineers, Richardson, Texas, 1987.
[11] C.H. Whitson, M.R. Brule, Phase Behavior, Monograph 20, Henry, L. Doherty Series, Society of Petroleum Engineers, Richardson, Texas, (2000).
[12] L.N. Mohinder (Ed.), Piping Handbook, seventh, ed., McGraw-Hill, New York, 1999.
[13] L.F. Moody, Friction Factors for Pipe Flow, Trans, ASME vol. 66 (1944) 671–678.
[14] B.E. Larock, R.W. Jeppson, G.Z. Watters, Hydraulics of Pipeline Systems, CRC Press, Boca Raton, Florida, 1999.
[15] Crane Company, Flow of Fluids through Valves, Fittings and Pipe, Technical Paper No. 410, 25th printing, (1991).
[16] J.P. Brill, H. Mukherjee, Multiphase Flow in Wells, Monograph vol. 17(1999), L. Henry, Doherty Series, Society of Petroleum Engineers, Richardson, Texas.
[17] H.D. Beggs, J.P. Brill, A Study of Two Phase Flow in Inclined Pipes,, Journal of Petroleum Technology vol. 25 (No. 5) (1973) 607–617.
[18] Y.M. Taitel, D. Barnea, A.E. Dukler, Modeling Flow Pattern Transitions for Steady Upward Gas-Liquid Flow in Vertical Tubes, AIChE Journal vol. 26 (1980) 245.
[19] PIPESIM Course, Information on Flow Correlations used within PIPESIM, (1997).

[20] H. Duns, N.C.J. Ros, Vertical Flow of Gas and Liquid Mixtures in Wells, Proc. 6th World Petroleum Congress, Section II, Paper 22-106, Frankfurt, 1963.
[21] J. Qrkifizewski, Predicting Two-Phase Pressure Drops in Vertical Pipes, Journal of Petroleum Technology (1967) 829–838.
[22] A.R. Hagedom, K.E. Brown, Experimental Study of Pressure Gradients Occurring During Continuous Two-Phase Flow in Small-Diameter Vertical Conduits, Journal of Petroleum Technology (1965) 475–484.
[23] H. Mukherjce, J.P. Brill, Liquid Holdup Correlations for Inclined Two-Phase Flow, Journal of Petroleum Technology (1983) 1003–1008.
[24] K.L. Aziz, G.W. Govier, M. Fogarasi, Pressure Drop in Wells Producing Oil and Gas, Journal of Canadian Petroleum Technology vol. 11 (1972) 38–48.
[25] K.H. Beniksen, D. Malnes, R. Moe, S. Nuland, The Dynamic Two-Fluid Model OLGA: Theory and Application, SPE Production Engineering 6 (1991) 171–180. SPE 19451.
[26] A. Ansari, N.D. Sylvester, O. Shoham, J.P. Brill, A Comprehensive Mechanistic Model for Upward Two-Phase Flow in Wellbores, SPE 20630, SPE Annual Technical Conference, 1990.
[27] A.C. Baker, K. Nielsen, A. Gabb, Pressure Loss, Liquid-holdup Calculations Developed, Oil & Gas Journal vol. 86 (No 11) (1988) 55–59.
[28] O. Flanigan, Effect of Uphill Flow on Pressure Drop in Design of Two-Phase Gathering Systems, Oil & Gas Journal vol. 56 (1958) 132–141.
[29] E.A. Dukler, et al., Gas-Liquid Flow in Pipelines, I. Research Results, AGA-API Project NX-28 (1969).
[30] R.V.A. Oliemana, Two-Phase Flow in Gas-Transmission Pipeline, ASME paper 76-Pet-25, presented at Petroleum Division ASME Meeting, Mexico City, (1976).
[31] W.G. Gray, Vertical Flow Correlation Gas Wells API Manual 14BM (1978).
[32] J.J. Xiao, O. Shoham, J.P. Brill, A Comprehensive Mechanistic Model for Two-Phase Flow in Pipelines, SPE, (1990). SPE 20631.
[33] S.F. Fayed, L. Otten, Comparing Measured with Calculated Multiphase Flow Pressure Drop, Oil & Gas Journal vol. 6 (1983) 136–144.
[34] Feesa Ltd, Hydrodynamic Slug Size in Multiphase Flowlines, retrieved from http:// www. feesa.net/flowassurance.(2003).
[35] Scandpower, OLGA 2000, OLGA School, Level I, II.
[36] Deepstar, Flow Assurance Design Guideline, Deepstar IV Project, DSIV CTR 4203b–1, (2001).
[37] J.C. Wu, Benefits of Dynamic Simulation of Piping and Pipelines, Paragon Technotes (2001).
[38] G.A. Gregory, Erosional Velocity Limitations for Oil and Gas Wells, Technical Note No. 5, Neotechnology Consultants Ltd, Calgary, Canada, 1991.
[39] American Petroleum Institute, Recommended Practice for Design and Installation of Offshore Platform Piping System, fifth edition,, API RP 14E, 1991.
[40] M.M. Salama, E.S. Venkatesh, Evaluation of API RP 14E Erosional Velocity Limitations for Offshore Gas Wells, OTC 4485 (1983).

CHAPTER 14

Heat Transfer and Thermal Insulation

Contents

Subsea Engineering Handbook
ISBN 978-0-12-397804-2, doi:10.1016/B978-0-12-397804-2.00014-X

14.1. INTRODUCTION

The thermal performance of subsea production systems is controlled by the hydraulic behavior of the fluid in the pipeline; conversely, it also impacts the hydraulic design indirectly through the influence of temperature on fluid properties such as GOR, density, and viscosity. Thermal design, which predicts the temperature profile along the pipeline, is one of the most important parts of pipeline design, because this information is required for pipeline analyses including expansion analysis, upheaval or lateral buckling, corrosion protection, hydrate prediction, and wax deposition analysis. In most cases, the management of the solids (hydrate, wax, asphaltenes, and scales) produced determines the hydraulic and thermal design requirements. To maintain a minimum temperature of fluid to prevent hydrate and wax deposition in the pipeline, insulation layers may be added to the pipeline.

Thermal design includes both steady-state and transient heat transfer analyses. In steady-state operation, the production fluid temperature decreases as it flows along the pipeline due to the heat transfer through the pipe wall to the surrounding environment. The temperature profile in the whole pipeline system should be higher than the requirements for prevention of hydrate and wax formation during normal operation and is determined from steady-state flow and heat transfer calculations. If the steady flow conditions are interrupted due to a shutdown or restarted again during operation, the transient heat transfer analysis for the system is required to make sure the temperature of fluid is out of the solid formation range within the required time. It is necessary to consider both steady-state and transient analyses in order to ensure that the performance of the insulation coatings will be adequate in all operational scenarios.

The thermal management strategy for pipelines can be divided into passive control and active heating. Passive control includes pipelines insulated by external insulation layers, pipe-in-pipe (PIP), bundle and burial; and active heating includes electrical heating and hot fluid heating.

In addition, if the production fluid contains gas, the fluid can experience a temperature drop due to the Joule-Thompson (JT) effect. The JT effect is primarily caused by pressure-head changes, which predominantly occur in the flowline riser and may cause the flowline temperature to fall below ambient temperatures. JT cooling cannot be prevented by insulation. Thus, the JT effect will not be explicitly discussed other than that its effects are implicitly included in some of the numerical results.

The purpose of this chapter is to discuss thermal behavior and thermal management of a pipeline system, which includes,

- Heat transfer fundamentals;
- Overall heat transfer coefficient calculation;
- Steady-state analysis of thermal design;
- Transient heat transfer analyses of thermal design;
- Insulation or heating management.

14.2. HEAT TRANSFER FUNDAMENTALS

Figure 14-1 shows the three heat transfer modes occurring in nature: conduction, convection, and radiation. Heat is transferred by any one or a combination of these three modes. When a temperature gradient exists in a stationary medium, which may be gas, liquid, or solid, the conduction will occur across the medium. If a surface and a moving fluid have a temperature difference, the convection will occur between the fluid and surface. All solid surfaces with a temperature will emit energy in the form of electromagnetic waves, which is called radiation. Although these three heat transfer modes occur at all subsea systems, for typical pipelines, heat transfer from radiation is relatively insignificant compared with heat transfer from conduction and convection because the system temperature is below 200°C in generally. Therefore, conduction and convection will be solely considered here.

14.2.1. Heat Conduction

For a one-dimensional plane with a temperature distribution $T(x)$, the heat conduction is quantified by the following Fourier equation:

$$q'' = -k \cdot \frac{dT(x)}{dx} \tag{14-1}$$

where

q'': heat flux, Btu/(hr·ft^2) or W/m^2, heat transfer rate in the x direction per unit area;

k: thermal conductivity of material, Btu/(ft-hr-°F) or W/(m-K);

dT/dx: temperature gradient in the x direction, °F/ft or °C/m.

When the thermal conductivity of a material is constant along the wall thickness, the temperature distribution is linear and the heat flux becomes:

$$q'' = -k \cdot \frac{T_2 - T_1}{x_2 - x_1} \tag{14-2}$$

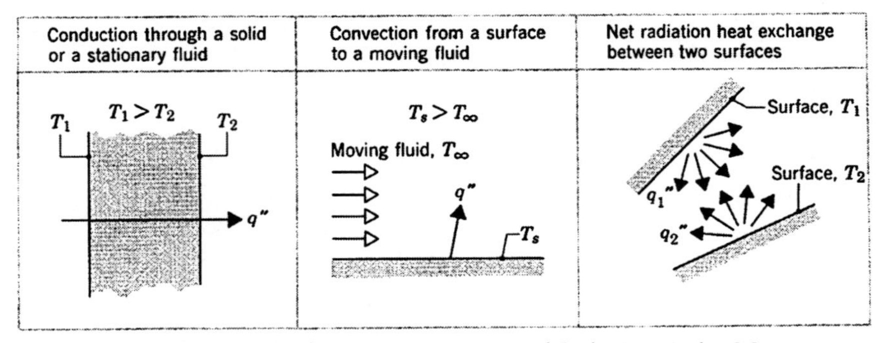

Figure 14-1 Conduction, Convection, and Radiation Modes [1]

Once the temperature distribution is known, the heat flux at any point in the medium may be calculated from the Fourier equation. By applying an energy balance to a 3D differential control volume and temperature boundary condition, the temperature distribution may be acquired from the heat diffusion equation:

$$\frac{\partial}{\partial x}\left(k\frac{\partial T}{\partial x}\right)+\frac{\partial}{\partial y}\left(k\frac{\partial T}{\partial y}\right)+\frac{\partial}{\partial z}\left(k\frac{\partial T}{\partial z}\right)+\dot{q}=\rho c_p\frac{\partial T}{\partial t} \tag{14-3}$$

where

$\dot{q}$: heat generation rate per unit volume of the medium, $\mathrm{Btu/(hr\cdot ft^3)}$ or $\mathrm{W/m^3}$;

ρ: density of the medium, $\mathrm{lb/ft^3}$ or $\mathrm{kg/m^3}$;

c_p: specific heat capacity, $\mathrm{Btu/(lb\cdot {}^\circ F)}$ or $\mathrm{kJ/(kg\cdot K)}$;

x,y,z: coordinates, ft or m;

t: time, sec.

For cylindrical coordinates, the heat diffusion equation may be rewritten as:

$$\frac{1}{r}\frac{\partial}{\partial r}\left(kr\frac{\partial T}{\partial r}\right)+\frac{1}{r^2}\frac{\partial}{\partial \phi}\left(k\frac{\partial T}{\partial \phi}\right)+\frac{\partial}{\partial z}\left(k\frac{\partial T}{\partial z}\right)+\dot{q}=\rho c_p\frac{\partial T}{\partial t} \tag{14-4}$$

where

r,z: radius and axial directions of cylindrical coordinates;

ϕ: angle in radius direction.

For most thermal analyses of flowline systems, where the heat transfer along the axial and circumferential directions may be ignored and, therefore, transient heat conduction without a heat source will occur in the radial direction of cylindrical coordinates, the above equation is simplified as follows:

$$\frac{1}{r}\frac{\partial}{\partial r}\left(kr\frac{\partial T}{\partial r}\right) = \rho c_p \frac{\partial T}{\partial t} \tag{14-5}$$

The temperature change rate ($\partial T/\partial t$) depends not only on the thermal conductivity of material k, but also on the density, ρ, and specific heat capacity, c_p. This equation can be solved numerically. For a steady heat transfer, the right side of equation is equal to zero. The total heat flow per unit length of cylinder is calculated by following equation:

$$q_r = -2\,\pi\,k\frac{T_2 - T_1}{\ln(r_2/r_1)} \tag{14-6}$$

where

r_1, r_2: inner and outer radii of the cylinder medium, ft or m;
T_1, T_2: temperatures at corresponding points of r_1, r_2, °F or °C;
q_r: heat flow rate per unit length of cylinder, Btu/(hr·ft) or W/m.

14.2.2. Convection

Both internal and external surfaces of a subsea pipeline come in contact with fluids, so convection heat transfer will occur when there is a temperature difference between the pipe surface and the fluid. The convection coefficient is also called a film heat transfer coefficient in the flow assurance field because convection occurs at a film layer of fluid adjacent to the pipe surface.

14.2.2.1. Internal Convection

Internal convection heat transfer occurs between the fluid flowing in a pipe and the pipe internal surface; it depends on the fluid properties, the flow velocity, and the pipe diameter. For the internal convection of pipelines, Dittus and Boelter [2] proposed the following dimensionless correlation for fully turbulent flow of single-phase fluids:

$$\mathrm{Nu}_i = 0.0255 \cdot \mathrm{Re}_i^{0.8} \cdot \mathrm{Pr}_i^n \tag{14-7}$$

where

Nu_i : Nusselt number, $\mathrm{Nu}_i = \frac{h_i D_i}{k_f}$;

Re_i : Reynolds number, $\mathrm{Re}_i = \frac{D_i V_f \rho_f}{\mu_f}$;

Pr_i : Prandtl number, $Pr_i = \frac{C_{pf}\mu_f}{k_f}$;

n: 0.4 if the fluid is being heated, and 0.3 if the fluid is being cooled;
h_i : internal convection coefficient, Btu/(ft^2·hr·°F) or W/(m^2·K);
D_i : pipeline inside diameter, ft or m;
k_f : thermal conductivity of the flowing liquid, Btu/(ft·hr·°F) or W/(m·K);
V_f : velocity of the fluid, ft/s or m/s;
ρ_f: density of the fluid, lb/ft^3 or kg/m^3;
μ_f: viscosity of the fluid, lb/(ft·s) or Pa.s;
C_{pf}: specific heat capacity of the fluid, Btu/(lb·°F) or J/(kg·K).

All fluid properties are assumed to be evaluated at the average fluid temperature. This correlation gives satisfactory results for flows with a Reynolds number greater than 10,000, a Prandtl number of 0.7 to 160, and a pipeline length greater than 10D.

If the flow is laminar (i.e., $Re_i < 2100$), h_i may be calculated using Hausen's equation [3] as follows:

$$Nu_i = 3.66 + \frac{0.0668\left(\frac{D_i}{L_o}\right)Re_iPr_i}{1 + 0.4\left[\left(\frac{D_i}{L_o}\right)Re_iPr_i\right]^{2/3}} \tag{14-8}$$

where L_o is the distance from the pipe inlet to the point of interest. In most pipeline cases, $D_i/L_o \approx 0$, therefore Equation (14-8) becomes:

$$Nu_i = 3.66 \tag{14-9}$$

For the transition region ($2100 < Re_i < 10^4$), the heat transfer behavior in this region is always uncertain because of the unstable nature of the flow, especially for multiphase flow in pipeline systems. A correlation proposed by Gnielinski [4] may be used to calculate h_i in this region:

$$Nu_i = \frac{(f/8)(Re_i - 1000)Pr_i}{1 + 12.7(f/8)^{1/2}(Pr_i^{2/3} - 1)} \tag{14-10}$$

where the fiction factor f may be obtained from the Moody diagram, which was discussed in Chapter 13, or for smooth tubes:

$$f = [0.79\ln(Re_i) - 1.64]^{-2} \tag{14-11}$$

This correlation is valid for $0.5 < Pr_i < 2000$ and $3000 < Re_i < 5 \times 10^6$.

Table 14-1 shows typical ranges for internal convection coefficients for turbulent flow. For most pipelines with multiphase flow, an approximate value of internal convection coefficient based on the table generally is adequate.

Tables 14-2 through 14-5 show thermal conductivities and specific heat capacities for a variety of typical oils and gases. These data can be used

Table 14-1 Typical Internal Convection Coefficients for Turbulent Flow [5]

	Internal Convection Coefficient, h_i	
Fluid	**Btu/(ft²·hr·°F)**	**W/(m² K)**
Water	300—2000	1700—11350
Gases	3—50	17—285
Oils	10—120	55—680

Table 14-2 Typical Thermal Conductivities for Crude Oil/Hydrocarbon Liquids [5]

Temperature	0°F (−18°C)		200°F (93°C)	
API Gravity	**Btu/(ft·hr·°F)**	**W/(m·K)**	**Btu/(ft·hr·°F)**	**W/(m·K)**
10	0.068	0.118	0.064	0.111
20	0.074	0.128	0.069	0.119
30	0.078	0.135	0.074	0.128
40	0.083	0.144	0.078	0.135
50	0.088	0.152	0.083	0.144
60	0.093	0.161	0.088	0.152
80	0.103	0.178	0.097	0.168
100	0.111	0.192	0.105	0.182

Table 14-3 Typical Thermal Conductivities for Hydrocarbon Gases [5]

Temperature	50°F (10°C)		100°F (38°C)		200°F (93°C)	
Gas Gravity	**Btu/(ft·hr·°F)**	**W/(m·K)**	**Btu/(ft·hr·°F)**	**W/(m·K)**	**Btu/(ft·hr·°F)**	**W/(m·K)**
0.7	0.016	0.028	0.018	0.031	0.023	0.039
0.8	0.014	0.024	0.016	0.028	0.021	0.036
0.9	0.013	0.022	0.015	0.026	0.019	0.033
1.0	0.012	0.021	0.014	0.024	0.018	0.031
1.2	0.011	0.019	0.013	0.022	0.017	0.029

Table 14-4 Typical Specific Heat Capacities for Hydrocarbon Liquids [5]

Temperature	0°F (−18°C)		100°F (38°C)		200°F (93°C)	
API Gravity	**Btu/(lb·°F)**	**kJ/(kg·K)**	**Btu/(lb·°F)**	**kJ/(kg·K)**	**Btu/(lb·°F)**	**kJ/(kg·K)**
10	0.320	1.340	0.355	1.486	0.400	1.675
30	0.325	1.361	0.365	1.528	0.415	1.738
50	0.330	1.382	0.370	1.549	0.420	1.758
70	0.335	1.403	0.375	1.570	0.430	1.800

Table 14-5 Typical Specific Heat Capacities for Hydrocarbon Gases [5]

Temperature	0°F (−18°C)		100°F (38°C)		200°F (93°C)	
Gas Gravity	Btu/(lb·°F)	kJ/(kg·K)	Btu/(lb·°F)	kJ/(kg·K)	Btu/(lb·°F)	kJ/(kg·K)
0.7	0.47	1.97	0.51	2.14	0.55	2.30
0.8	0.44	1.84	0.48	2.01	0.53	2.22
0.9	0.41	1.72	0.46	1.93	0.51	2.14
1.0	0.39	1.63	0.44	1.84	0.48	2.01

for most purposes with sufficient accuracy if these parameters are not known.

14.2.2.2. External Convection

The correlation of average external convection coefficient suggested by Hilpert [6] is widely used in industry:

$$\mathrm{Nu}_o = C\,\mathrm{Re}_o^{\,m}\,\mathrm{Pr}_o^{\,1/3} \tag{14-12}$$

where

Nu_o : Nusselt number, $\mathrm{Nu}_o = \frac{h_o D_o}{k_o}$;

Re_o : Reynolds number, $\mathrm{Re}_o = \frac{D_o V_o \rho_o}{\mu_o}$;

Pr_{sf} : Prandtl number, $\mathrm{Pr}_o = \frac{C_{p,o}\mu_o}{k_o}$;

h_o: external convection coefficient, Btu/(ft^2·hr·°F) or W/(m^2·K);

D_o : pipeline outer diameter, ft or m;

k_o : thermal conductivity of the surrounding fluid, Btu/(ft·hr·°F) or W/(m·K);

V_o : velocity of the surrounding fluid, ft/s or m/s;

ρ_o: density of the surrounding fluid, lb/ft^3 or kg/m$^{3;}$

μ_o: viscosity of the surrounding fluid, lb/(ft·s) or Pa·s;

$C_{p,o}$: specific heat capacity of surrounding fluid, Btu/(lb·°F) or J/(kg·K);

C, m : constants, dependent on the Re number range, and listed in Table 14-6.

All of the properties used in the above correlation are evaluated at the temperature of film between the external surface and the surrounding fluid.

When the velocity of surrounding fluid is less than approximately 0.05 m/s in water and 0.5 m/s in air, natural convection will have the dominating influence and the following values may be used:

Table 14-6 Constants of Correlation

Re_o	C	m
$4 \times 10^{-1} - 4 \times 10^{0}$	0.989	0.330
$4 \times 10^{0} - 4 \times 10^{1}$	0.911	0.385
$4 \times 10^{1} - 4 \times 10^{3}$	0.683	0.466
$4 \times 10^{3} - 4 \times 10^{4}$	0.193	0.618
$4 \times 10^{4} - 4 \times 10^{5}$	0.027	0.805

$$h_o = \begin{cases} 4\ \mathrm{W/(m^2K)}, & \text{natural convection in air} \\ 200\ \mathrm{W/(m^2K)}, & \text{natural convection in water} \end{cases} \tag{14-13}$$

Note:

- Many of the parameters used in the correlation are themselves dependent on temperature. Because the temperature drop along most pipelines is relatively small, average values for physical properties may be used.
- The analysis of heat transfer through a wall to surrounding fluid does not address the cooling effect due to the Joule-Thompson expansion of a gas. For a long gas pipeline or pipeline with two-phase flow, an estimation of this cooling effect should be made.

14.2.3. Buried Pipeline Heat Transfer

Pipeline burial in subsea engineering occurs for these reasons:

- Placement of rock, grit, or seabed material on the pipe for stability and protection requirements;
- Gradual infill due to sediment of a trenched pipeline;
- General embedment of the pipeline into the seabed because of the seabed mobility or the pipeline movement.

Although seabed soil can be a good insulator, porous burial media, such as rock dump, may provide little insulation because water can flow through the spaces between the rocks and transfer heat to the surroundings by convection.

14.2.3.1. Fully Buried Pipeline

For a fully buried pipeline, the heat transfer is not symmetrical. To simulate a buried pipeline, a pseudo-thickness of the soil is used to account for the asymmetries of the system in pipeline heat transfer simulation software such as OLGA and PIPESIM.

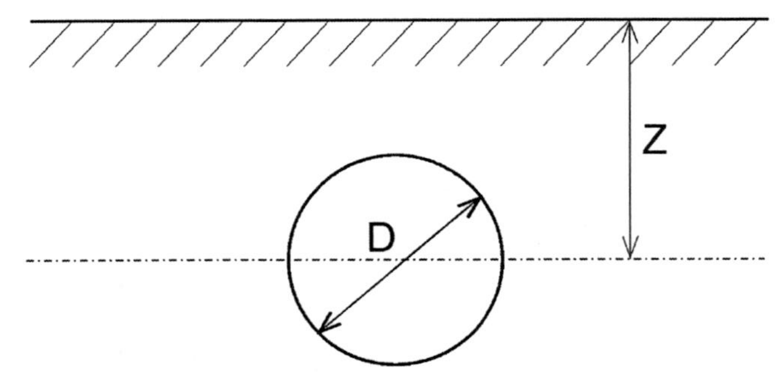

Figure 14-2 Cross Section of a Buried Pipeline

By using a conduction shape factor for a horizontal cylinder buried in a semi-infinite medium, as shown in Figure 14-2, the heat transfer coefficient for a buried pipeline can be expressed as:

$$h_{soil} = \frac{k_{soil}}{\left(\frac{D}{2}\right)\cosh^{-1}\left(\frac{2Z}{D}\right)} \tag{14-14}$$

where

h_{soil}: heat transfer coefficient of soil, Btu/ (ft^2·hr·°F) or W/ (m^2·K);

k_{soil}: thermal conductivity of soil, Btu/(ft·hr·°F) or W/(m·K);

D: outside diameter of buried pipe, ft or m;

Z: distance between top of soil and center of pipe, ft or m.

For the case of $Z > D/2$, $\cosh^{-1}(2Z/D)$ can be simplified as $\ln[(2Z/D) + ((2Z/D)^2 - 1)^{0.5}]$; therefore:

$$h_{soil} = \frac{2k_{soil}}{D \ln\left[\frac{2Z+\sqrt{4Z^2-D^2}}{D}\right]} \tag{14-15}$$

14.2.3.2. Partially Buried Pipeline

The increase in the insulation effect for a partially buried pipeline is not large compared with a fully buried pipeline. Heat flows circumferentially through the steel to the section of exposure. Even exposure of just the crown of the pipeline results in efficient heat transfer to the surroundings due to the high thermal conductivity of the steel pipe. A trenched pipeline (partially buried pipeline) experiences less heat loss than an exposed pipeline but more than a buried pipeline. Engineering judgment must be used for the analysis of trenched pipelines. Seabed currents may be modified to account for the reduced heat transfer, or the heat transfer may be calculated using a weighted average of the fully buried pipe and exposed pipe as follows:

$$h_o = \left(1 - f\right) h_{o,\text{buried}} + f\ h_{o,\text{exposed}} \tag{14-16}$$

where f is the fraction of outside surface of pipe exposed to the surrounding fluid, and h_o is the external heat transfer coefficient, Btu/(ft^2·hr·°F) or W/(m^2·K).

For an exposed pipeline the theoretical approach assumes that water flows over the top and bottom of the pipeline. For concrete-coated pipelines, where the external film coefficient has little effect on the overall heat transfer coefficient, the effect of resting on the seabed is negligible. However, for pipelines without concrete coatings these effects must be considered.

A pipeline resting on the seabed is normally assumed to be fully exposed. A lower heat loss can give rise to upheaval buckling, increased corrosion, and overheating of the coating. It is necessary to take into account the changes in burial levels over field life and, hence, the changes in insulation values.

14.2.4. Soil Thermal Conductivity

Soil thermal conductivity has been found to be a function of dry density, saturation, moisture content, mineralogy, temperature, particle size/shape/arrangement, and the volumetric proportions of solid, liquid, and air phases. A number of empirical relationships (e.g., Kersten [7]) have been developed to estimate thermal conductivity based on these parameters. For a typical unfrozen silt-clay soil, the Kersten correlation is expressed as follows:

$$\kappa_{\text{soil}} = [0.9 \cdot \log(\omega) - 0.2] \times 10^{0.01 \times \rho} \tag{14-17}$$

where

κ_{soil}: soil thermal conductivity, Btu·in./(ft^2·hr·°F);

ω: moisture content in percent of dry soil weight;

ρ: dry density, pcf.

The above correlation was based on the data for five soils and is valid for moisture contents of 7 percent or higher.

Table 14-7 lists the thermal conductivities of typical soils surrounding pipelines. Although the thermal conductivity of onshore soils has been extensively investigated, until recently there has been little published thermal conductivity data for deepwater soil (e.g., Power et al. [8], Von Herzen and Maxwell [9]). Many deepwater offshore sediments are formed

Table 14-7 Thermal Conductivities of Typical Soil Surrounding Pipeline [5]

	Thermal Conductivity, k_{soil}	
Material	**Btu/(ft·hr·°F)**	**W/(m·K)**
Peat (dry)	0.10	0.17
Peat (wet)	0.31	0.54
Peat (icy)	1.09	1.89
Sand soil (dry)	0.25−0.40	0.43−0.69
Sandy soil (moist)	0.50−0.60	0.87−1.04
Sandy soil (soaked)	1.10−1.40	1.90−2.42
Clay soil (dry)	0.20−0.30	0.35−0.52
Clay soil (moist)	0.40−0.50	0.69−0.87
Clay soil (wet)	0.60−0.90	1.04−1.56
Clay soil (frozen)	1.45	2.51
Gravel	0.55−0.72	0.9−1.25
Gravel (sandy)	1.45	2.51
Limestone	0.75	1.30
Sandstone	0.94 −1.20	1.63−2.08

with predominantly silt- and clay-sized particles, because sand-sized particles are rarely transported this far from shore. Hence, convective heat loss is limited in these soils, and the majority of heat transfer is due to conduction [10]. Recent measurements of thermal conductivity for deepwater soils from the Gulf of Mexico [11] have shown values in the range of 0.7 to 1.3 W/(m·K), which is lower than that previously published for general soils and is approaching that of still seawater, 0.65 W/(m·K). This is a reflection of the very high moisture content of many offshore soils, where liquidity indices well in excess of unity can exist and which are rarely found onshore. Although site-specific data are needed for the detailed design most deepwater clay is fairly consistent. Figure 14-3 shows sample thermal conductivity test results for subsea soils worldwide. The sample results prove that most deepwater soils have similar characteristics.

14.3. U-VALUE

14.3.1. Overall Heat Transfer Coefficient

Figure 14-4 shows the temperature distribution of a cross section for a composite subsea pipeline with two insulation layers. Radiation between the internal fluid and the pipe wall and the pipeline outer surface and the environment is ignored because of the relatively low temperature of subsea systems.

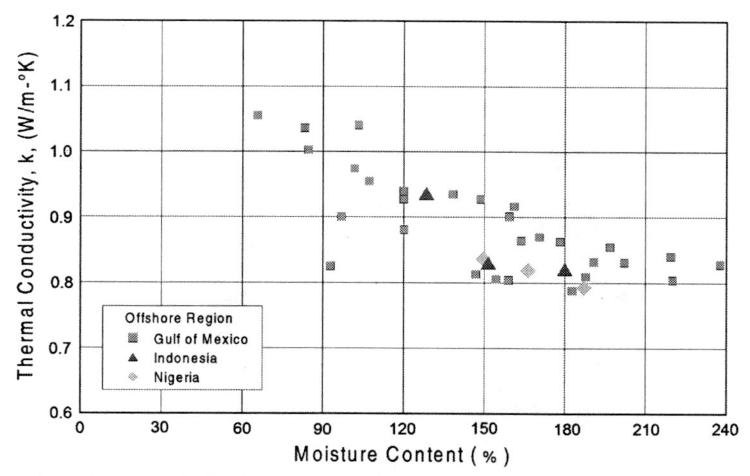

Figure 14-3 Sample Soil Thermal Conductivity Test Results for Offshore Soils Worldwide [12]

Convection and conduction occur in an insulated pipeline as follows:

- Convection from the internal fluid to the pipeline wall;
- Conduction through the pipe wall and exterior coatings, and/or to the surrounding soil for buried pipelines;
- Convection from flowline outer surface to the external fluid.

For internal convection at the pipeline inner surface, the heat transfer rate across the surface boundary is given by the Newton equation:

$$Q_i = A_i \cdot h_i \cdot \Delta T = 2\pi\, r_i\, L\, h_i\, (T_i - T_1) \tag{14-18}$$

where

Q_i : convection heat transfer rate at internal surface, Btu/hr or W;

h_i : internal convection coefficient, Btu/(ft^2·hr·°F) or W/(m^2·K);

r_i : internal radius of flowline, ft or m;

L: flowline length, ft or m;

A_i : internal area normal to the heat transfer direction, ft^2 or m^2;

T_i : internal fluid temperature, °F or °C;

T_1: temperature of flowline internal surface, °F or °C.

For external convection at the pipeline outer surface, the heat transfer rate across the surface boundary to the environment is:

$$Q_o = A_o \cdot h_o \cdot \Delta T_o = 2\pi\, r_o\, L\, h_o\, (T_4 - T_o) \tag{14-19}$$

where

Q_o : convection heat transfer rate at outer surface, Btu/hr or W;

h_o : outer convection coefficient, Btu/(ft^2·hr·°F) or W/(m^2·K);

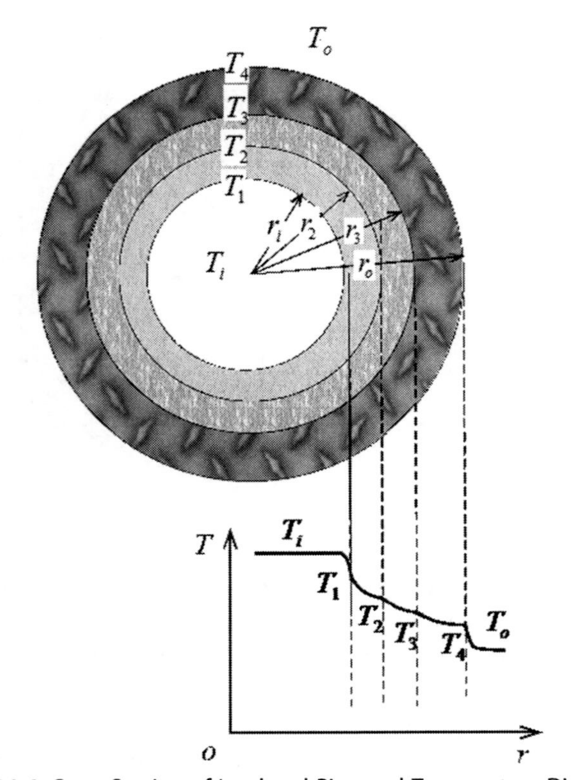

Figure 14-4 Cross Section of Insulated Pipe and Temperature Distribution

r_o : outer radius of flowline, ft or m;

A_o : outer area normal to the heat transfer direction, ft^2 or m^2;

T_o: environment temperature, °F or °C;

T_4: outer surface temperature of flowline, °F or °C.

Conduction in the radial direction of a cylinder can be described by Fourier's equation in radial coordinates:

$$Q_r = -2\pi\, r\, L\, k \frac{\partial T}{\partial r} \tag{14-20}$$

where

Q_r : conduction heat transfer rate in radial direction, Btu/hr or W;

r: radius of cylinder, ft or m;

k: thermal conductivity of cylinder, Btu/(ft·hr·°F) or W/(m·K);

$\partial_T / \partial r$: temperature gradient, °F/ft or °C/m.

Integration of Equation (14-20) gives:

$$Q_r = \frac{2\pi\, L\, k(T_1 - T_2)}{\ln(r_2/r_1)} \tag{14-21}$$

The temperature distribution in the radial direction can be calculated for steady-state heat transfer between the internal fluid and pipe surroundings where heat transfer rates of internal convection, external convection, and conduction are the same. The following heat transfer rate equation is obtained:

$$Q_r = \frac{T_i - T_o}{\frac{1}{2\pi r_i L h_i} + \frac{\ln(r_1/r_i)}{2\pi k_1 L} + \frac{\ln(r_2/r_1)}{2\pi k_2 L} + \frac{\ln(r_o/r_2)}{2\pi k_3 L} + \frac{1}{2\pi r_o L h_o}} \tag{14-22}$$

The heat transfer rate through a pipe section with length of L, due to a steady-state heat transfer between the internal fluid and the pipe surroundings, is also expressed as follows:

$$Q_r = UA(T_i - T_o) \tag{14-23}$$

where

U: overall heat transfer coefficient (OHTC), based on the surface area A, Btu/ (ft^2·hr·°F) or W/ (m^2·K);

A: area of heat transfer surface, A_i or A_o, ft^2 or m^2;

T_o : ambient temperature of the pipe surroundings, °F or °C;

T_i : average temperature of the flowing fluid in the pipe section, °F or °C.

Therefore, the OHTC based on the flowline internal surface area A_i is:

$$U_i = \frac{1}{\frac{1}{h_i} + \frac{r_i \ln(r_1/r_i)}{k_1} + \frac{r_i \ln(r_2/r_1)}{k_2} + \frac{r_i \ln(r_o/r_2)}{k_3} + \frac{r_i}{r_o h_o}} \tag{14-24}$$

and the OHTC based on the flowline outer surface area A_o is:

$$U_o = \frac{1}{\frac{r_o}{r_i h_i} + \frac{r_o \ln(r_1/r_i)}{k_1} + \frac{r_o \ln(r_2/r_1)}{k_2} + \frac{r_o \ln(r_o/r_2)}{k_3} + \frac{1}{h_o}} \tag{14-25}$$

The OHTC of the pipeline is also called the *U-value* of the pipeline in subsea engineering. It is a function of many factors, including the fluid properties and fluid flow rates, the convection nature of the surroundings, and the thickness and properties of the pipe coatings and insulation.

Insulation manufacturers typically use a U-value based on the outside diameter of a pipeline, whereas pipeline designers use a U-value based on the inside diameter. The relationship between these two U-values is:

$$U_o \times \mathrm{OD} = U_i \times \mathrm{ID} \tag{14-26}$$

The U-value for a multilayer insulation coating system is easily obtained from an electrical-resistance analogy between heat transfer and direct current. The steady-state heat transfer rate is determined by:

$$Q_r = UA(T_i - T_o) = (T_i - T_o)/\sum R_i \tag{14-27}$$

where UA is correspondent with the reverse of the cross section's thermal resistivity that comprises three primary resistances: internal film, external film, and radial material conductance. The relationship is written as follows:

$$\frac{1}{UA} = \sum R_i = R_{film,in} + R_{pipe} + \sum R_{coating} + R_{film,ext} \tag{14-28}$$

The terms on the right hand side of the above equation represent the heat transfer resistance due to internal convection, conduction through steel well of pipe, conduction through insulation layers and convection at the external surface. They can be expressed as follows.

$$R_{film,in} = \frac{1}{h_i A_i} \tag{14-29}$$

$$R_{pipe} = \frac{\ln(r_1/r_i)}{2\pi L k_{pipe}} \tag{14-30}$$

$$\sum R_{coating} = \frac{\ln(r_{no}/r_{ni})}{2\pi L k_n} \tag{14-31}$$

$$R_{film,ext} = \frac{1}{h_o A_o} \tag{14-32}$$

wherer_{no} and r_{ni} represent the outer radius and inner radius of the coating layer n, respectively. The term k_n represents the thermal conductivity of coating layer n, Btu/(ft-hr-°F) or W/(m-K). The use of U-values is appropriate for steady-state simulations. However, U-values cannot be used to evaluate transient thermal simulations; the thermal diffusion coefficient and other material properties of the wall and insulation must be included.

14.3.2. Achievable U-Values

The following U-values are lowest possible values to be used for initial design purposes:

- *Conventional insulation:* 0.5 $\text{Btu/ft}^2\cdot\text{hr}\cdot{}^\circ\text{F}$ (2.8 $\text{W/m}^2\cdot\text{K}$);
- *Polyurethane pipe-in-pipe:* 0.2 $\text{Btu/ft}^2\cdot\text{hr}\cdot{}^\circ\text{F}$ (1.1 $\text{W/m}^2\cdot\text{K}$);
- *Ceramic insulated pipe-in-pipe:* 0.09 $\text{Btu/ft}^2\cdot\text{hr}\cdot{}^\circ\text{F}$ (0.5 $\text{W/m}^2\cdot\text{K}$);
- *Insulated bundle:* 0.2 $\text{Btu/ft}^2\cdot\text{hr}\cdot{}^\circ\text{F}$ (1.1 $\text{W/m}^2\cdot\text{K}$).

Insulation materials exposed to ambient pressure must be qualified for the water depth where they are to be used.

14.3.3. U-Value for Buried Pipe

For insulated pipelines, a thermal insulation coating and/or burial provide an order of magnitude more thermal resistance than both internal and external film coefficients. Therefore, the effects of internal and external film coefficients on the U-value of pipelines can be ignored. The U-value of pipelines can be described as:

$$U_i = \frac{1}{\sum \frac{r_i \ln(r_{no}/r_{ni})}{k_n} + \frac{r_i \cosh^{-1}(2Z/D_o)}{k_{soil}}} \quad (14\text{-}33)$$

where the first term in the denominator represents the thermal resistance of the radial layers of steel and insulation coatings, and the second term in the denominator represents the thermal resistivity due to the soil and is valid for $H > D_o/2$. If the internal and external film coefficients are to be included, the external film coefficient is usually only included for nonburied pipelines. For buried bare pipelines, where the soil provides essentially the entire pipeline's thermal insulation, there is a near linear relationship between the U-value and k_{soil}. For buried and insulated pipelines, the effect of k_{soil} on the U-value is less than that for buried bare pipelines.

Figure 14-5 presents the relationship between U-value and burial depth for both a bare pipe and a 2-in. polypropylene foam (PPF)-coated pipe. The U-value decreases very slowly with the burial depth, when the ratio of burial depth to outer diameter is greater than 4.0. Therefore, it is not necessary to bury the pipeline to a great depth to get a significant thermal benefit. However, practical minimum and maximum burial depth limitations are applied to modern pipeline burial techniques. In addition, issues such as potential seafloor scouring and pipeline upheaval buckling need to be considered in designing the appropriate pipeline burial depth.

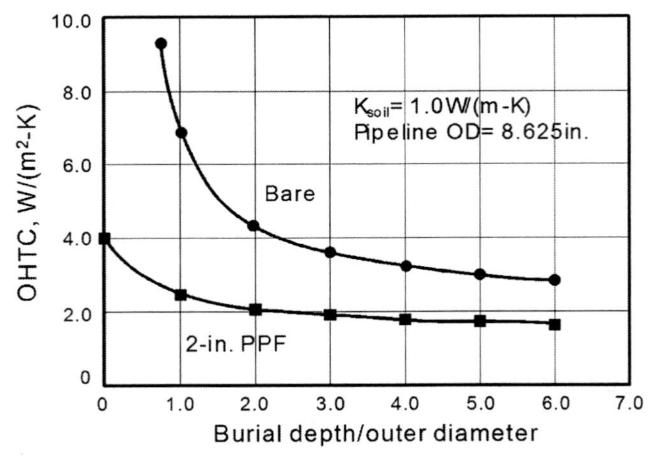

Figure 14-5 U-Value versus Burial Depth for Bare Pipe and 2-In. PPF-Coated Pipe [13]

14.4. STEADY-STATE HEAT TRANSFER

Steady-state thermal design is aimed at ensuring a given maximum temperature drop over a length of insulated pipe. In the steady-state condition, the flow rate of a flowing fluid and heat loss through the pipe wall from the flowing fluid to the surroundings are assumed to be constant at any time.

14.4.1. Temperature Prediction along a Pipeline

The accurate predictions of temperature distribution along a pipeline can be calculated by coupling velocity, pressure, and enthalpy given by mass, momentum, and energy conservation. The complexity of these coupled equations prevents an analytical solution, but they can be solved by a numerical procedure [14].

Figure 14-6 shows a control volume for internal flow for a steady-state heat transfer analysis. In a fully developed thermal boundary layer region of a flowline, the velocity and temperature radial profiles of a single-phase fluid are assumed to be approximately constant along the pipeline. A mean temperature can be obtained by integrating the temperatures over the cross section, and the mean temperature profile of fluid along the pipeline is obtained by the energy conservation between the heat transfer rate through the pipe wall and the fluid thermal energy change as the fluid is cooled by conduction of heat through the pipe to the sea environment, which is expressed as follows:

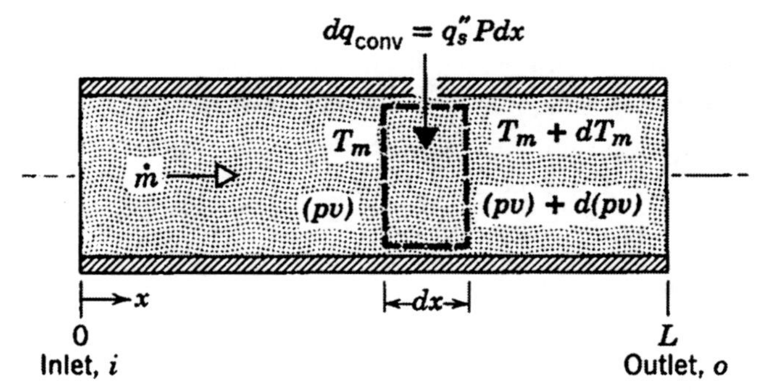

Figure 14-6 Control Volume for Internal Flow in a Pipe [1]

$$UdA(T_m - T_o) = -\dot{m}c_p dT_m \tag{14-34}$$

where

dA : surface area of pipeline, $dA = Pdx = \pi\ Ddx$, ft^2 or m^2;
$\dot{m}$: mass flow rate of internal fluid, lb/s or kg/s;
c_p: specific heat capacity of internal fluid, Btu/(lb·°F) or J/(kg·K);
U: U-value based on pipe diameter D, Btu/ft^2·hr·°F, W/m^2·K;
D: ID or OD, ft or m;
T_m : mean temperature of fluid, °F or °C.

Separating the variables and integrating from the pipeline inlet to the point with a pipe length of x from the inlet, we obtain:

$$\int_{T_{in}}^{T(x)} \frac{dT_m}{T_m - T_o} = \int_0^x -\frac{U\pi Ddx}{\dot{m}c_p} \tag{14-35}$$

where

T_{in} : inlet temperature of fluid, °F or °C;
$T(x)$: temperature of fluid at distance x from inlet, °F or °C.

From integration of Equation (14-35), the following temperature profile equation is obtained:

$$T(x) = T_o + (T_{in} - T_o)\exp\left(\frac{-U\pi Dx}{\dot{m}c_p}\right) \tag{14-36}$$

Here, the thermal decay constant, β, is defined as:

$$\beta = \frac{U\pi DL}{\dot{m}c_p} \tag{14-37}$$

where L is the total pipeline length, ft or m.

However, if two temperatures and their corresponding distance along the route are known, fluid properties in the thermal decay constant are not required. The inlet temperature and another known temperature are used to predict the decay constant and then develop the temperature profile for the whole line based on the exponential temperature profile that is given by the following equation:

$$T(x) = T_o + (T_{in} - T_o)\exp\left(\frac{-x\beta}{L}\right) \tag{14-38}$$

If the temperature drop due to Joule-Thompson cooling is taken into account and a single-phase viscous flow is assumed, the temperature along the flowline may be expressed as:

$$T(x) = T_o + (T_{in} - T_o)\exp\left(\frac{-x\beta}{L}\right) + \frac{[p(x) - p_{in}]/\rho - g[z(x) - z_{in}]}{c_p} \tag{14-39}$$

where

p_{in} : fluid pressure at pipe inlet, Pa;
$p(x)$: fluid pressure at point at distance x from inlet, Pa;
z_{in} : pipeline elevation at pipe inlet, m;
$z(x)$: pipeline elevation at point at distance x from inlet, m;
g: acceleration due to gravity, m/s^2;
ρ: density of fluid, kg/m^3.

14.4.2. Steady-State Insulation Performance

The following insulation methods are used in the field for subsea pipelines:

- External coatings;
- Pipeline burial;
- Pipe-in-pipe (PIP);
- Electrical heating;
- Hot water annulus.

The most popular methods used in subsea pipeines are external coatings, PIP, and pipeline burial.

Table 14-8 presents the U-values for various pipe size and insulation system combinations. Bare pipe has little resistance to heat loss, but buried bare pipe is equivalent to adding approximately 60 mm of external coating of PPF. Combining burial and coating (B&C) achieves U-values near those

Table 14-8 U-Values for Different Subsea Insulation Systems [13]

	U-value [W/(m^2·K)]				
Flowline OD (in.)	**Bare Pipe**	**Coated Pipe t =50 mm PPF k = 0.17 W/(m·K)**	**Buried Pipe H =1.3 m k = 1.0 W/(m·K)**	**Buried and Coated**	**Carrier OD of PIP 8.625 in.* 12.75 in.** 18.0 in.*** **
4.5*	871.0	4.65	4.56	2.52	1.56
8.625**	544.0	4.04	2.87	1.81	1.44
12.75***	362.0	3.81	2.22	1.5	1.02

of PIP. If lower soil thermal conductivity and/or thicker external coating are used, the B&C system can achieve the same level U-value as typical PIPs. Considering realistic fabrication and installation limitations, PIPs are chosen and designed in most cases to have a lower U-value than B&C. However, the fully installed costs of equivalent PIP systems are estimated to be more than double those of B&C systems.

14.5. TRANSIENT HEAT TRANSFER

Transient heat transfer occurs in subsea pipelines during shutdown (cooldown) and start-up scenarios. In shutdown scenarios, the energy, kept in the system at the moment the fluid flow stops, goes to the surrounding environment through the pipe wall. This is no longer a steady-state system and the rate at which the temperature drops with time becomes important to hydrate control of the pipeline.

Pipeline systems are required to be designed for hydrate control in the cooldown time, which is defined as the period before the pipeline temperature reaches the hydrate temperature at the pipeline operating pressure. This period provides the operators with a decision time in which to commence hydrate inhibition or pipeline depressurization. In the case of an emergency shutdown or cooldown, it also allows for sufficient time to carry out whatever remedial action is required before the temperature reaches the hydrate formation temperature. Therefore, it is of interest to be able to predict how long the fluid will take to cool down to any hydrate formation temperature with a reasonable accuracy. When a pipeline is shut down for an extended period of time, generally it is flushed (blown down) or vented to remove the hydrocarbon fluid, because the temperature of the system will eventually come to equilibrium with the surroundings.

The cooldown process is a complex transient heat transfer problem, especially for multiphase fluid systems. The temperature profiles over time may be obtained by considering the heat energy resident in the fluid and the walls of the system, $\sum \dot{m} C_p \Delta T$, the temperature gradient over the walls of the system, and the thermal resistances hindering the heat flow to the surroundings. The software OLGA is widely used for numerical simulation of this process. However, OLGA software generally takes several hours to do these simulations. In many preliminary design cases, an analytic transient heat transfer analysis of the pipeline, for example, the lumped capacitance method, is fast and provides reasonable accuracy.

14.5.1. Cooldown

The strategies for solving the transient heat transfer problems of subsea pipeline systems include analytical methods (e.g., lumped capacitance method), the finite difference method (FDM), and the finite element method (FEM). The method chosen depends on the complexity of the problem. Analytical methods are used in many simple cooldown or transient heat conduction problems. The FDM is relatively fast and gives reasonable accuracy. The FEM is more versatile and better for complex geometries; however, it is also more demanding to implement. A pipe has a very simple geometry; hence, obtaining a solution to the cooldown rate question does not require the versatility of the FEM. For such systems, the FDM is convenient and has adequate accuracy. Valves, bulkheads, Christmas trees, etc., have more complex geometries and require the greater versatility and 3D capability of FEM models.

14.5.1.1. Lumped Capacitance Method

A commonly used mathematical model for the prediction of system cooldown time is the *lumped capacitance model,* which can be expressed as follows:

$$T - T_o = (T_i - T_o)\ e^{-U\pi DLt/\sum mC_p} \tag{14-40}$$

where

T: temperature of the body at time t, °F or °C;

t: time, sec;

T_o : ambient temperature, °F or °C;

T_i : initial temperature, °F or °C;

D: inner diameter or outer diameter of the flowline, ft or m;

L: length of the flowline, ft or m;

U: U-value of the flowline based on diameter D, Btu/ft^2·hr·°F or W/m^2·K;

m: mass of internal fluid or coating layers, lb or kg;

c_p: specific heat capacity of fluid or coating layers, Btu/(lb·°F) or J/(kg·K).

This model assumes, however, a uniform temperature distribution throughout the object at any moment during the cooldown process and that there is no temperature gradient inside of the object. These assumptions mean that the surface convection resistance is much larger than the internal conduction resistance, or the temperature gradient in the system is negligible. This model is only valid for Biot (Bi) numbers less than 0.1. The Bi number is defined as the ratio of internal heat transfer resistance to external heat transfer resistance and may be written as:

$$Bi = \frac{L_c U}{k} \tag{14-41}$$

where

L_c : characteristic length, ft or m;

k: thermal conductivity of the system, Btu/ft·hr·°F or W/m·K.

Because most risers and pipelines are subject to large temperature gradients across their walls and are also subject to wave and current loading, this leads to a Bi number outside the applicable range of the "lumped capacitance" model. A mathematical model is therefore required that takes into account the effects of external convection on the transient thermal response of the system during a shutdown event.

14.5.1.2. Finite Difference Method

Considering the pipe segment shown earlier in Figure 14-4, which has length L, it is assumed that the average fluid temperature in the segment is $T_{f,o}$ at a steady-state flowing condition. The temperature of the surroundings is assumed to be T_a, and constant.

Under the steady-state condition, the total mass of the fluid in the pipe segment is given by:

$$W_f = \frac{\pi}{4} D_i^2 L \rho_f \tag{14-42}$$

where

W_f : fluid mass in the pipe segment, lb or kg;

D_i : pipe inside diameter, in or m;

L: length of the pipe segment, in or m;

ρf: average density of the fluid in the pipe segment when the temperature is $T_{f,o}$, lb/in^3 or kg/m^3.

Note that W_f is assumed to be constant at all times after shutdown. At any given time, the heat content of the fluid in the pipe segment is given by:

$$Q_f = W_f C_{p,f}(T_f - T_{ref}) \tag{14-43}$$

where

$C_{p,f}$: specific heat of the fluid, Btu/(lb·°F) or kJ/(kg °C);

T_{ref} : a constant reference temperature, °F or °C;

T_f : average temperature of fluid in the pipe segment, °F or °C.

The heat content of the pipe wall and insulation layers can be expressed in the same way:

$$Q_{pipe} = W_{pipe} C_{p,pipe}(T_{pipe} - T_{ref})$$

$$Q_{layer} = W_{layer} C_{p,layer}\left(T_{layer} - T_{ref}\right) \tag{14-44}$$

An energy balance is applied to the control volume and is used to find the temperature at its center. It is written in the two forms shown below, the first without Fourier's law of heat conduction. Notice from the equation that during cooldown the quantity of heat leaving is larger than the quantity of heat entering.

In the analysis of heat transfer in a pipeline system, heat transfer along the axial and circumferential directions of the pipeline can be ignored, therefore, the transient heat transfer without heat source occurs only in the radial direction. The first law of thermodynamics states that the change of energy inside a control volume is equal to the heat going out minus the heat entering. Therefore, the heat transfer balance for all parts (including the contained fluid, steel pipe, insulation materials, and coating layers) of the pipeline can be explicitly expressed as in following equation:

$$Q_{i,1} - Q_{i,0} = -(q_{i+1/2} - q_{i-1/2})\Delta t \tag{14-45}$$

where

$Q_{i,0}$: heat content of the part i at one time instant, Btu or J;

$Q_{i,1}$: heat content of the part i after one time step, Btu or J;

$q_{i+1/2}$: heat transfer rate from part i to part $i+1$ at the "0" time instant, Btu/hr or W;

$q_{i-1/2}$: heat transfer rate from part $i-1$ to part i at the "0" time instant, Btu/hr or W;

Δ_t: time step, sec or hr.

If the heat transfer rate through all parts of a pipeline system in the radial direction is expressed with the thermal resistant concept, then:

$$q_{i+1/2} = \frac{A_o(T_{i+1,0} - T_{i,0})}{R_i/2 + R_{i+1}/2} \tag{14-46}$$

$$q_{i-1/2} = \frac{A_o(T_{i,0} - T_{i-1,0})}{R_i/2 + R_{i-1}/2} \tag{14-47}$$

Therefore, the average temperatures for a pipeline with two insulation layers after one time step are rewritten as:

$$T_{f,1} = T_{f,0} - \frac{A_o\Delta t}{W_f C_{p,f}}\left[\frac{T_{f,0} - T_{p,0}}{R_{\text{int}} + R_p/2}\right] \tag{14-48}$$

$$T_{p,1} = T_{p,0} - \frac{A_o\Delta t}{W_p C_{p,p}}\left[\frac{T_{p,0} - T_{f,0}}{R_{\text{int}} + R_p/2} + \frac{T_{p,0} - T_{l1,0}}{R_{l1}/2 + R_p/2}\right] \tag{14-49}$$

$$T_{l1,1} = T_{l1,0} - \frac{A_o\Delta t}{W_{l1} C_{p,l1}}\left[\frac{T_{l1,0} - T_{l2,0}}{R_{l2}/2 + R_{l1}/2} + \frac{T_{l1,0} - T_{p,0}}{R_{l1}/2 + R_p/2}\right] \tag{14-50}$$

$$T_{l2,1} = T_{l2,0} - \frac{A_o\Delta t}{W_{l2} C_{p,l2}}\left[\frac{T_{l2,0} - T_{l1,0}}{R_{l2}/2 + R_{l1}/2} + \frac{T_{l2,0} - T_a}{R_{l2}/2 + R_{surr}}\right] \tag{14-51}$$

The solution procedure of a finite difference method is based on a classical implicit procedure that uses matrix inversion. The main principle used to arrive at a solution is to find the temperature at each node using the energy equation above. Once an entire row of temperatures has been found at the same time step by matrix inversion, the procedure jumps to the next time step. Finally, the temperature at all time steps is found and the cooldown profile is established. The detailed calculation method and definitions are provided in the Mathcad worksheet in the Appendix at the end of this chapter.

To get an accurate transient temperature profile, the fluid and each insulation layer may be divided into several layers and the above calculation method applied. By setting the exponential temperature profile along the pipeline, obtained from the steady-state heat transfer analysis, as the initial temperature distribution for a cooldown transient analysis, the temperature profiles along the pipeline at different times can be calculated. This methodology is easily programmed as an Excel macro or with C^{++} or another computer language, but the more layers the insulation is divided into, the shorter the time step required to obtain a convergent result, because an explicit difference is used in the time domain. The maximum time step may be calculated using following equation:

$$\Delta t_{\max} = \min\left(\frac{W_i C_{p,i} R_i}{A_0}\right) \text{ for all parts or layers } i \tag{14-52}$$

If the layer is very thin, R_i will be very small, which controls the maximum time step. In this case, the thermal energy saved in this layer is very small and may be ignored. The model implemented here conservatively assumes that the content of the pipe has thermal mass only, but no thermal resistance. The thermal energy of the contents is immediately available to the steel. The effect of variations of C_p and k for all parts in the system is of great importance to the accuracy of the cooldown simulation. For this reason, these effects are either entered in terms of fitted expressions or fixed measured quantities. Figure 14-7 shows the variation of measured specific heat capacity for a polypropylene solid with material temperature.

Guo et al. [15] developed a simple analysis model for predicting heat loss and temperature profiles in insulated thermal injection lines and wellbores. The concept of this model is similar to that of the finite difference method described earlier. The model can accurately predict steady-state and transient heat transfer of the well tube and pipeline.

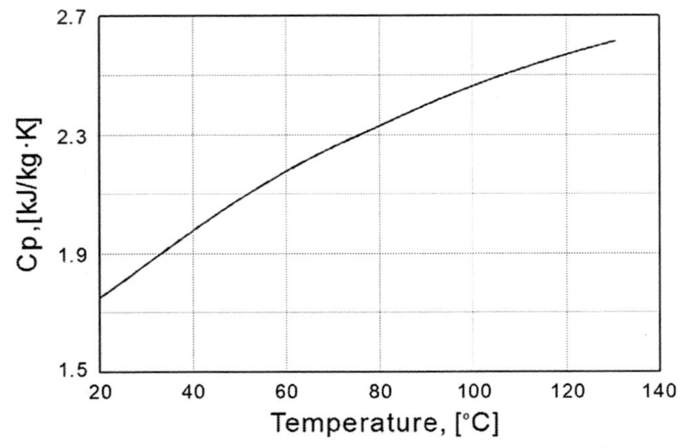

Figure 14-7 Specific Heat Capacity Measurements for Generic Polypropylene Solid

14.5.2. Transient Insulation Performance

Although steady-state performance is generally the primary metric for thermal insulation design, transient cooldown is also important, especially when hydrate formation is possible. Figure 14-8 presents the transient cooldown behavior at the end of a 27-km flowline.

For the case presented in Figure 14-8, the PIP flowline had a slightly lower U-value than that of the B&C flowline. Therefore, the steady-state temperature at the base of the riser is slightly higher for the PIP than the B&C flowline. These steady-state temperatures are the initial temperatures at the start of a transient shutdown. However, the temperature of a B&C system quickly surpasses that of the PIP system and provides much longer cooldown times in the region of typical hydrate formation. For the case considered, the B&C flowline had approximately three times the cooldown time of the PIP before hydrate formation. The extra time available before hydrate formation has significant financial implications for offshore operations, such as reduced chemical injection, reduced line flushing, and reduced topside equipment.

The temperature of the PIP at initial cooldown condition is higher than that of the B&C flowline, but the cooldown curves cross as shown in Figure 14-8. In addition to the U-value, the fluid temperature variation with time is dependent on the fluid's specific heat, C_p, its mass flow rate, and the ambient temperature.

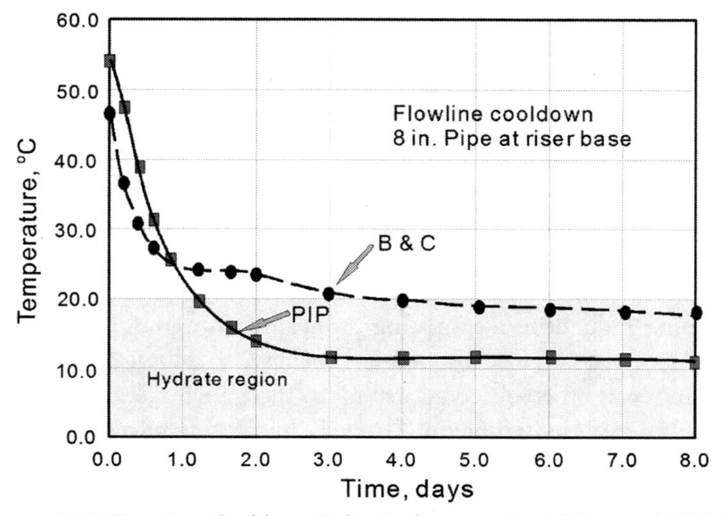

Figure 14-8 Transient Cooldown Behavior between Burial Pipe and PIP [13]

14.6. THERMAL MANAGEMENT STRATEGY AND INSULATION

As oil/gas production fields move into deeper water, there is a growing need for thermal management to prevent the buildup of hydrate and wax formations in subsea systems. The thermal management strategy is chosen depending on the required U-value, cooldown time, temperature range, and water depth. Table 14-9 summarizes the advantages, disadvantages, and U-value range of the thermal management strategies used in subsea systems.

Figure 14-9 summarizes subsea pipeline systems with respect to length and U-value, which concludes:

- U-values of single pipe insulation systems are higher than 2.5 $W/m^2 \cdot K$, and mostly installed using reeled pipe installation methods;
- About 6 different operating PIP systems with lengths of less than 30 km;
- Reeled PIP systems have dominated the 5- to 20-km market with U-values of 1 to 2 $W/m^2 \cdot K$;
- PIP system in development with length of 66 km offers a U-value of 0.5-1.0 $W/m^2 \cdot K$;
- About 3 bundles used on tie-backs of over 10 km, with Britannia also incorporating a direct heating system.

14.6.1. External Insulation Coating System

Figure 14-10 shows a typical multilayer coating system that combines the foams with good thermal insulating properties and polypropylene (PP) shield with creep resistance. These coatings vary in thickness from 25 up to 100 mm or more. Typically thicknesses over about 65 mm are applied in multiple layers.

Coating systems are usually limited by a combination of operating conditions including temperature, water depth, and water absorption. Combinations of temperature and hydrostatic pressure can cause creep and water absorption, with resultant compression of the coating and a continuing reduction in insulating properties throughout the design life. These issues need to be accounted for during the design stage.

Pipelines can be classified according to the pipeline temperature. Table 14-10 shows various temperature ranges with their corresponding insulation materials. Finding suitable thermal insulation for high-temperature subsea pipelines is challenging compared to finding it for lower temperature applications.

Table 14-9 Summary of Thermal Management Strategies for Subsea Systems [16]

Thermal Management Method	Advantages	Disadvantages	U-Value Range (W/m^2 K)
Integral external insulation (including multilayer insulation)	• Suitable for complex geometries • High-temperature resistance • Multilayer systems can tailor configuration to achieve desired properties. • Solid systems have almost unlimited water depth	• Difficult to remove • Limits on level of insulation that can be achieved • Solid systems have almost no buoyancy • Need to allow time to set • Limits on thickness	1.7–5.0
Insulation modules/bundle systems	• Accommodation of auxiliary lines • Can remove during service • Lower cost than PIP systems • Simple implementation to flowlines • Water depth limits lower than PIP or solid integral external insulation (except for bundle systems protected by outer pipe)	• Gaps between modules can lead to convection currents • Large amounts of material required, which can be prohibitive • May be a complex geometry to insulate	2.0–8.0
PIP insulation	• Best insulation properties (except for vacuum system) • Water depth limits defined by capacity of installation vessels	• Expensive to install and fabricate • Some PIP systems cannot be reeled	0.3–1.5
Vacuum systems	• Ultimate insulation properties • Only heat transfer by radiation	• If outer pipe is breached, then all insulation properties lost for pipe section. • Effort required to preserve vacuum • Vacuum level in annulus must be continuously monitored	<0.01

(Continued)

Table 14-9 Summary of Thermal Management Strategies for Subsea Systems [16]—cont'd

Thermal Management Method	Advantages	Disadvantages	U-Value Range (W/m^2 K)
Burial/trenching	• Surrounding soil provides some of the insulation • Provides on-bottom stability for flowlines	• May not be practical to dig trenches for burial • Only applicable where large amounts of pipe are lying on the seabed • Thermal conductivity of soil varies significantly depending on location • Thermal performance uncertain due to variance in soil properties and difficulty in measuring or predicting them	Insulation layers + soil insulating
Electrical heating	• Active system; can control cool-down time during shutdown	• Power requirements may be unfeasible for project • May need to electrically insulate flowline	Depends on power supplied
Hot water/oil systems	• Active system; can control cool-down time during shutdown	• With greater depth longer tie-backs will lose more heat, thus reducing thermal efficiency • Separate return line required	Depends on power supplied
Thin-film/multilayer [17]	• Conjunct with VIT, get a high thermal resistance • Total thickness of layers is less than 3 mm	• With a very low thermal storage, causing a fast cooldown; suitable for trees, jumpers, manifolds, flowlines, risers, and wellbore tubing	Used in BP Marlin, Shell Oregano and Serrano

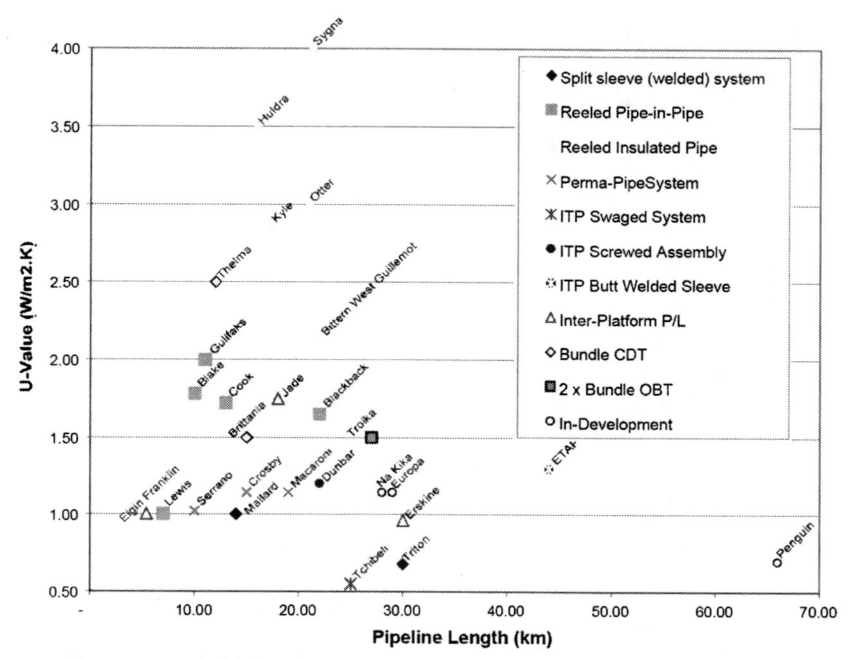

Figure 14-9 Field Development versus Pipeline Length and U-Value [18]

14.6.1.1. Insulation Material

Table 14-11 summarizes the insulation types, limitations, and characteristics of the most commonly applied and new systems being offered by suppliers. The systems have progressed significantly in the past 10 years on issues such as thickness limits, water depth limitations, impact resistance, operating temperature limits, creep with resultant loss of properties, suitability for reel-lay installation with high strain loading, life expectancy, and resultant U-values and submerged weight.

Brief descriptions of external insulation systems for subsea systems are as follows:

- *Polypropylene (PP):* polyolefin system with relatively low thermal efficiency. Applied in three to four layers, can be used in conjunction with direct heating systems. Specific heat capacity is 2000 J/(kg·K);.
- *Polypropylene-foam (PPF), Solid polypropylene (SPP):* Controlled formation of gas bubbles is used to reduce the thermal conductivity; however, as ocean depth and temperature increase, rates of compression and material creep increase. Generally not used on its own.

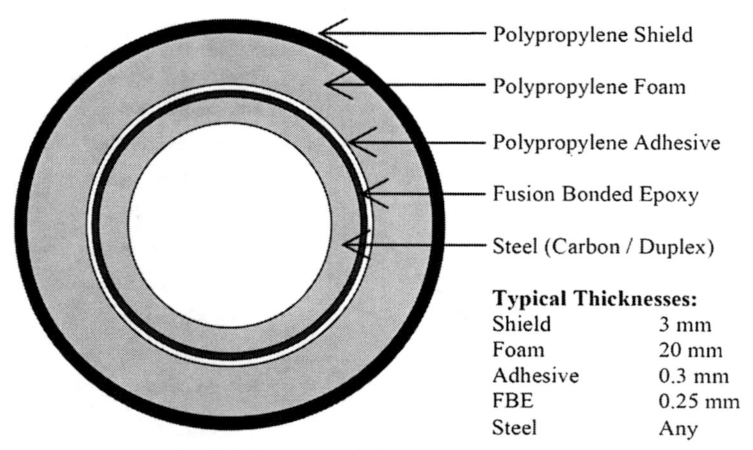

Figure 14-10 Typical Multilayer Insulation System [18]

Table 14-10 Possible Coating Systems for Thermal Insulation of Flowlines [19]

Flowline	Temperature Range (°C)	Insulation System
Low temperature	−10−70	Polyurethane (PU), polypropylene (PP), filled rubbers, syntactic foams based on epoxy and polyurethanes
Medium temperature	70−120	PP, rubber, syntactic epoxy foams
High temperature	120−200	PP systems, phenolic foams, PIP with polyurethane foam or inorganic insulating materials.

- *Polypropylene-reinforced foam combination (RPPF):* A combination of the above two systems incorporating an FBE layer, PPF, and an outer layer of PP to minimize creep and water absorption.
- *Polyurethane (PU):* A polyolefin system with relatively low thermal efficiency; has a relatively high water absorption as the temperature rises over 50°C. Commonly applied with modified stiffness properties to field joints.
- *Polyurethane-syntactic (SPU):* One of the most commonly applied systems over recent years; offers good insulation properties at water depths less than 100 m. Specific heat capacity is 1500 J/(kg·K).
- *Polyurethane-glass syntactic (GSPU):* Similar to SPU but incorporates glass providing greater creep resistance. Specific heat capacity is 1700 J/(kg·K).

Table 14-11 External Insulation Systems [18]

Name of Coating System	Abbreviation	Reelable (yes/no)	Max. Temp. (°C)	Water Absorption	Conductivity Increase at End of Life (%)	Compression at End of Life (%)	Max. Thickness if <100 mm	Density (kg^3/m)	Max. Depth (m)	Conductivity ($W/m^2 \cdot K$)
Polypropylene-solid	PP	Y	145	<0.5%	<1%	<2%	60	900	3000	0.22
Polypropylene-syntactic PPF (e.g., carizite)	SPP	Y	115	<5.0%	<1%	<5%	—	600–800	—	0.13–0.22
Polypropylene-reinforced foam combination	RPPF	Y	115–140	<0.5%	<1%	<0.5%	Multilayer [Note 1]	600–800	600–3000	0.16–0.18
Polyurethane-solid	PU	Y	115	<5.0%	<10%	<2%		1150		0.19– 0.20
Polyurethane-syntactic (plastic beads)	SPU	Y	70–115	<5.0%	<10%	<2%	Multilayer	750–780	100 @ 115°C 300 @ 90°C	0.12–0.15

(Continued)

Table 14-11 External Insulation Systems [18]—cont'd

Name of Coating System	Abbreviation	Reelable (yes/no)	Max. Temp. (°C)	Water Absorption	Conductivity Increase at End of Life (%)	Compression at End of Life (%)	Max. Thickness if <100 mm	Density (kg^3/m)	Max. Depth (m)	Conductivity ($W/m^2 \cdot K$)
Polyurethane-glass syntactic	GSPU	Y	55–90	<2.2%	Variable thickness	<5%	Multilayer	610–830	2000–3000	0.12–0.17
Polyurethane-reinforced foam combination	RPUF	n/a	n/a	n/a	n/a	n/a	Multilayer	448	—	0.080
Phenolic syntactic	PhS	Y	200		<5%	<1%		500	—	0.080
Epoxy syntactic	SEP	N	75–100	<5.0%	<10%	<2%	75	590–720	2000–3000	0.10–0.135
Epoxy syntactic with Mini-Spheres	MSEP	N	75		<10%	<4%		540		0.12

Note 1: For multilayer coatings the properties quoted are for the insulation layer.
Note 2: Value measured at mean product temperature of 50°C.
Note 3: Values are approximate only. All values used in design must be provided by the manufacturer.

- *Phenolic syntactic (PhS), epoxy syntactic (SEP), and epoxy syntactic with mini-spheres (MSEP):* Based on epoxy and phenolic materials, which offer improved performance at higher temperatures and pressures. These materials are generally applied by trowel or are precast or poured into molds. For example, SEP was used on the 6-mile Shell King development in GoM by pouring the material under a vacuum into polyethylene sleeves. Specific heat capacity is 1240 (J/kg K) [20].

PP and PU are the main components of insulation coating systems. The syntactic insulations that incorporate spheres are used to improve insulation strength for hydrostatic pressure.

14.6.1.2. Structural Issues

Installation and Operation Loads

Coating systems that will be installed by the reeling method need to be carefully selected and tested, because some systems have experienced cracking particularly at field joints where there is a natural discontinuity in the coating. This can lead to strain localization and pipe buckling. Laying and operation of subsea pipelines require a load transfer through the coating to the steel pipe and from the steel pipe to the coating. The coating should have a sufficient shear load capacity to hold the steel pipe during the laying process. Thermal fluctuations in the operation process lead to expansion and contraction of the pipe. The thermal insulation coating is required to allow for such changes without being detached or cracked. These requirements have resulted in most thermal insulation systems being based on bonded geometries, comprised of several layers or cast-in shells.

Hydrostatic Loads

Deepwater pipelines are subject to significant hydrostatic loading due to the extreme water depths. The thermal insulation system must be designed to withstand these large hydrostatic loads. The layer of foamed thermal insulation is the weakest member of the insulation system and therefore the structural response of the insulation system will depend largely on this layer. Elastic deformation of the system leads to a reduction in volume of the insulation system and an increase in density and thermal conductivity of the foam layer, which leads to an increase in the U-value of the system. To reduce the elastic deformation of the system, stiffer materials are required. The stiffness of the foams increases with increasing density. Therefore, to reduce elastic deformation of the system, higher density foams are required.

Creep

Thermal insulation operating in deep water depths for long periods (e.g., 20 to 30 years) is subject to significant creep loading. Creep of thermal insulation can lead to some damaging effects, such as structural instability leading to collapse and also loss of thermal performance. Creep of polymer foams leads to a reduction in volume or densification of the foam. The increased density of the material leads to an increased thermal conductivity and, therefore, the effectiveness of the insulation system is reduced. To design effective thermal insulation systems for deepwater applications, the creep response of the insulating foam must be known.

The factors affecting the creep response of polymer foams include temperature, density, and stress level. The combination of high temperatures and high stresses can accelerate the creep process. High-density foams are desirable for systems subject to large hydrostatic loads for long time periods.

The design of thermal insulation systems for deepwater applications is complex and involves a number of thermal and structural issues. These thermal and structural issues can sometimes conflict. To design effective thermal insulation systems, the thermal and structural issues involved must be carefully considered to achieve a balance. For long-distance tie-backs, the need for a substantial thickness of insulation will obviously have an impact on the installation method due to the increase in pipe outside diameter and the pipe field-jointing process. Because most insulation systems are buoyant, the submerged weight of the pipe will decrease. It may be necessary to increase pipe wall thickness to achieve a submerged weight suitable for installation and on-bottom stability.

14.6.2. Pipe-in-Pipe System

A method of achieving U-values of 1 $W/m^2 \cdot K$ or less requires PIP insulation systems, in which the inner pipe carrying the fluid is encased within a larger outer pipe. The outer pipe seals the annulus between the two pipes and the annulus can be filled with a wide range of insulating materials that do not have to withstand hydrostatic pressure.

The key features for application of a PIP system are:

- Jointing method;
- Protection of the insulation from water ingress, water stops/bulkheads;
- Field joint insulation method;
- Response to applied load;
- Offshore fabrication process and lay rate of installation;

- Submerged weight;
- Limitations, cost and properties of insulation materials.

Typical PIP field joints are summarized below:

- Sliding sleeve with fillet welds or fillet weld–butt weld combination;
- Tulip assemblies involving screwed or welded components;
- Fully butt welded systems involving butt-welded half-shells;
- Sliding outer pipe over the insulation with a butt weld on the outer pipe, which has recently been proposed for J-lay installation systems;
- Mechanical connector systems employing a mechanical connector on the inner pipe, outer pipe, or both pipes to minimize installation welding.

For all of these systems, except reel-lay, the lay vessel field-joint assembly time is critical in terms of installation costs; therefore, quadruple joints have been used to speed up the process.

Table 14-12 summarizes some insulation materials used in PIP, their properties, and U-values for a given thickness. For most insulation systems, thermal performance is based mainly on the conductive resistance. However, heat convection and radiation will transfer heat across a gas layer in an annulus if a gas layer exists. The convection and radiation in the gas at large void spaces result in a less effective thermal system than the one that is completely filled with an insulation material. For this reason PIP systems often include a combination of insulation materials. An inert gas such as nitrogen or argon can be filled in the gap or annulus between the insulation and the outer pipe to reduce convection. A near vacuum in the annulus of a PIP greatly improves insulation performance by minimizing convection in the annulus and also significantly reduces conduction of heat through the insulation system and the annulus. The difficulty is to maintain a vacuum over a long period of time.

14.6.3. Bundling

Bundles are used to install a combination of flowlines inside an outer jacket or carrier pipe. Bundling offers attractive solutions to a wide range of flow assurance issues by providing cost-effective thermal insulation and the ability to circulated a heating medium. Other advantages of bundles include:

- Can install multiple flowlines in a single installation
- The outer jacket helps to resist the external hydrostatic pressure in deep water.

Table 14-12 PIP Insulation Materials [18]

Insulation Material	Density (kg/m^3)	Conductivity ($W/m \cdot K$)	Thickness (mm)	Annulus Gap (If Any)	Max. Temperature(°C)	U-Value ($W/m^2 \cdot K$) 16–in. Inner Pipe	Comment
Mineral wool	140	0.037	100	Clearance	700	1.6(40 mm)	Rockwool or Glava, usually in combination with Mylar reflective film
Alumina silicate microspheres	390–420	0.1	no limit	None	1000+	3.9 (100 mm)	Commonly referred to as fly-ash, injected to fill the annulus
Thermal cement	900–1200	0.26	100	None	200	—	Currently being investigated under a JIP to provide collapse resistance with reduced carrier pipe wall thickness
LD PU foam	60	0.027	125	None	147	0.76 (100 mm)	Preassembled as single- or double-jointed system, used on the Erskine replacement
HD PU foam	150	0.035	125	None	147	1.2 (100 mm)	
Microporous silica blanket	200–400	0.022	24	Clearance	900	0.4 (100 mm)	Cotton blanket, calcium-based powder, glass, and titanium fibers
Vacuum insulation panels	60–145	0.006–0.008	10	Clearance	160	0.26 (100 mm)	Foam shells formed under vacuum with aluminium foil and uses gas absorbing "getter" pills to absorb any free gas thereafter

- It may be possible to include heating pipes and monitoring systems in the bundle.

The disadvantages are the bundle length limit and the need for a suitable fabrication and launch site. The convection in the gap between the insulation and pipes needs to be specifically treated.

Bundle modeling is accomplished with these tools:

- FEM Software;
- OLGA 2000 FEMTherm module.

The FEM method is used to easily simulate a wide range of bundle geometries and burial configurations. FEM software may be used to simulate thermal interactions among the production flowlines, heating lines, and other lines that are enclosed in a bundle to determine U-values and cooldown times. The OLGA 2000 FEMTherm module can model bundles with a fluid medium separating the individual production and heating lines, but it is not applicable to cases in which the interstitial space is filled with solid insulation. Once the OLGA 2000 FEMTherm model has been used, the model can be used for steady-state and transient simulations for these components. A comparison of the U-values and cooldown times for the finite element and bundle models is helpful to get a confident result.

14.6.4. Burial

Trenching and backfilling can be an effective method of increasing the amount of insulation, because the heat capacity of soil is significant and acts as a natural heat store. The effect of burial can typically decreases the U-value and significantly increase the pipeline cooldown period as discussed in the Section 14.5.2. Attention to a material's aging characteristics in burial conditions needs to be considered.

Table 14-13 lists a comparison of the impact of pipeline insulation between PIP and B&C systems. The cooldown time for a buried insulated flowline can be greater than that for a PIP system due to the heat capacity of the system. The cost of installing an insulated and buried pipeline is approximately 35% to 50% that of a PIP system. To confidently use the insulating properties of the soil, reliable soil data are required. The lack of this can lead to overconservative pipeline systems.

14.6.5. Direct Heating

Actively heated systems generally use hot fluid or electricity as a heating medium. The main attraction of active heating is its flexibility. It can be used

Table 14-13 Comparison of the Impact of Flowline Insulation between PIP and B&C Systems

PIP Systems	B&C Systems
• Maintains steady-state flow out of hydrate and wax regions for greatest portion of production life • "Threshold" for wax deposition < 5000 bpd • Hydrate formation < 3000 bpd • Insulation hinders hydrate remediation (estimates are of the order of 20 to 22 weeks) • Wax deposition occurs mainly in the riser at low flow rates	• Flowline cools down more slowly than PIP system but starting temperature is lower (lower steady-state temperatures) • Lower insulation characteristic and heat capacity of earth facilitates hydrate remediation (estimates are of the order of 8 to 10 weeks) • Wax deposition is more evenly distributed along the entire flowline at low flow rates • Flow rate "threshold" for wax and hydrate deposition occurs earlier in field life • Wax deposition < 12,000 bpd • Hydrate formation < 5000 bpd

to extend the cooldown time by continuously maintaining a uniform flowline temperature above the critical levels of wax or hydrate formation. It is also capable of warming up a pipeline from seawater temperature to a target operating level and avoids the requirements for complex and risky start-up procedures.

Direct heating by applying an electrical current in the pipe has been used since the 1970s in shallow waters and could mitigate hydrate and wax deposition issues along the entire length of the pipeline; therefore, it has found wide application in recent years. Electrically heated systems have also been recognized as an effective method in removing hydrate plugs within estimated times of 3 days, while depressurization methods employed in deepwater developments can take up to several months. The most efficient electrical heating (EH) systems can provide the heat input as close to the flowline bore as possible and have minimum heat losses to the environment. Trace heating is believed to provide the highest level of heating efficiency and can be applied to bundles or a PIP system. The length capability of an electrical system depends on the linear heat input required and the admissible voltage. Trace heating can be applied for very long tie-backs (tens of kilometers) by either using higher voltage or by introducing intermediate power feeding locations, fed by power umbilicals via step-down transformers.

Table 14-14 summarizes the projects that are currently using EH or have EH as part of their design. It has been reported that there are potential CAPEX savings of 30% for single electrically heated pipelines over a traditional PIP system for lengths of up to 24 km [21]. Beyond this distance, the CAPEX can surpass PIP systems, however operational shutdown considerations could negate these cost differences. OPEX costs may be reduced when offset against reduced chemical use, reduced shutdown and start-up times, no requirement for pipeline depressurization, no pigging requirements, and enhanced cooldown periods.

The disadvantages of EH systems include the requirement for transformers and power cables over long distances, the cost and availability of the power, and the difficulties associated with maintenance. Typical requirements are 20 to 40 W/m for an insulated system of $U = 1.0 \mathrm{W/m^2 \cdot K}$. The method cannot be used in isolation and requires insulation applied to the pipe. The three basic approaches for electrically heated systems consist of the closed and open return systems for wet insulated pipes using either cable or surrounding seawater to complete the electrical circuit, and dry PIP systems.

14.6.5.1. Closed Return Systems

This system supplies DC electrical energy directly to the pipe wall between two isolation joints. The circuit is made complete by a return cable running along the pipeline. Research has shown that the power required for this system is approximately 30% less than that required for an open system. The required power is proportional to the insulation U-value.

14.6.5.2. Open Return Systems (Earthed Current)

Currently implemented on the Åsgard and Huldra fields in the North Sea, this system uses a piggybacked electrical cable to supply current to the pipeline. The return circuit comprises a combination of the pipe wall and the surrounding seawater to allow the flow of electrons. Pipeline anodes are used to limit the proliferation of stray current effects to nearby structures.

14.6.5.3. PIP System

This type of system uses a combination of passive insulation methods and heat trace systems in which direct or alternating current circulates around the inner pipe. Development of optical fiber sensors running the along the inner pipe can provide the operator with a real-time temperature profile. AC systems being offered have been recommended for distances of 20 km.

Table 14-14 Projects Currently Using Electrical Heating Systems [22]

Operator and Project	Line Diameter and Length	Water Depth (m)	Year of Installation	Electrical Heating Method	Mode of Operation	Status
Shell Nakika	10-in. ×16-in. PIP	1900	2003	PIP direct	Remediation	Engineering and qualification testing
Shell Serrano/ Oregano	6-in. ×10-in. PIP; two lines 10 and 12 km	1000	2001	PIP direct	Temperature maintenance for shutdown	In operation
Statoil Huldra	8-in. single pipe; 15 km	300 to 400	2001	Earthed direct	Temperature maintenance of 25°C during shutdown	In operation
Statoil Asgard	8-in. single pipe; 6 lines, 43 km total	300 to 400	2000	Earthed direct	Temperature maintenance of 27°C during shutdown	In operation
Statoil Sleipner	20-in. single pipe; 12.6 km	—	1996	Induction	Temperature maintenance	In operation

The installation would comprise five 20-km sections joined by T-boxes and transformers.

14.6.5.4. Hot Fluid Heating (Indirect Heating)

The use of waste energy from the reception facility to heat and circulate hot water (similar to a shell and tube heat exchanger) has been applied on integrated bundles, for example, Britannia. In this instance the length of the bundle was limited to less than 10 km with the heat medium only applied during pipeline start-up or shutdown. The system would normally have the flexibility to deliver the hot medium to either end of the pipeline first.

To be sufficiently efficient, it is necessary to inject a large flow of fluid at a relatively high temperature. This involves storage facilities and significant energy to heat the fluid and to maintain the flow.

Active heating by circulation of hot fluid is generally more suited to a bundle configuration because large pipeline cross sections are required. There is a length limitation to hot fluid heating because it generally involves a fluid circulation loop along which the temperature of the heating medium decreases.

REFERENCES

[1] F.P. Incropera, D.P. DeWitt, Introduction to Heat Transfer, third ed., John Wiley & Sons, New York, 1996.

[2] F.W. Dittus, L.M.K. Boelter, University of California, Berkeley, Publications on Engineering, vol. 2, p. 443 (1930).

[3] H. Hausen, Darstellung des Warmeuberganges in Rohren durch verallgemeinerte Potezbeziehungen, Z. VDI Beih. Verfahrenstechnik (No. 4) (1943) 91.

[4] V. Gnielinski, New Equations for Heat and Mass Transfer in Turbulent Pipe and Channel Flow, Int. Chemical Engineering vol. 16 (1976) 359–368.

[5] G.A. Gregory, Estimation of Overall Heat Transfer Coefficient for the Calculation of Pipeline Heat Loss/Gain, Technical Note No. 3, Neotechnology Consultants Ltd, 1991.

[6] R. Hilpert, Warmeabgabe von geheizen Drahten und Rohren, Forsch, Gebiete Ingenieurw vol. 4 (1933) 220.

[7] M.S. Kersten, Thermal Properties of Soils, University of Minnesota Eng, 1949. Exp. Station Bull. No. 28.

[8] P.T. Power, R.A. Hawkins, H.P. Christophersen, I. McKenzie, ROV Assisted Geotechnical Investigation of Trench Backfill Material Aids Design of the Tordis to Gullfaks Flowlines, ASPECT 94, 2nd Int. Conf. on Advances in Subsea Pipeline Engineering and Technology, Aberdeen (1994).

[9] R. Von Herzen, A.E. Maxwell, The Measurement of Thermal Conductivity of Deep Sea Sediments by a Needle Probe Method,, Journal of Geophysical Research vol. 64 (No. 10) (1959) 1557–1563.

[10] T.A. Newson, P. Brunning, G. Stewart, Thermal Conductivity of Offshore Clayey Backfill, OMAE2002-28020, 21th International Conference on Offshore Mechanics and Arctic Engineering, Oslo, Norway, 2002. June 23–28.
[11] MARSCO, Thermal Soil Studies for Pipeline Burial InsulationdLaboratory and Sampling Program Gulf of Mexico, internal report completed for Stolt Offshore (1999).
[12] A.G. Young, R.S. Osborne, I. Frazer, Utilizing Thermal Properties of Seabed Soils as Cost-Effective Insulation for Subsea Flowlines, OTC 13137, 2001 Offshore Technology Conference, Houston, Texas, 2001.
[13] K. Loch, Flowline Burial: An Economic Alternative to Pipe-in Pipe, OTC 12034, Offshore Technology Conference, Houston, Texas, 2000.
[14] J.D. Parker, J.H. Boggs, E.F. Blick, Introduction to Fluid Mechanics and Heat Transfer, Addison-Wesley, Reading, Massachusetts, 1969.
[15] B. Guo, S. Duan, A. Ghalambor, A Simple Model for Predicting Heat Loss and Temperature Profiles in Thermal Injection Lines and Wellbores with Insulations, SPE 86983, SPE International Thermal Operations and Heavy Oil Symposium and Western Regional Meeting, Bakersfield, California, 2004, March.
[16] F. Grealish, I. Roddy, State of the Art on Deep Water Thermal Insulation Systems, OMAE2002-28464, 21th International Conference on Offshore Mechanics and Arctic Engineering, Oslo, Norway, 2002. June 23–28.
[17] C. Horn, G. Lively, A New Insulation Technology: Prediction versus Results from the First Field Installation, OTC13136, 2001 Offshore Technology Conference, Houston, Texas, 2001.
[18] J.G. McKechnie, D.T. Hayes, Pipeline Insulation Performance for Long Distance Subsea Tie-Backs, Long Distance Subsea Tiebacks Conference, Amsterdam, 2001, November 26–28.
[19] B. Melve, Design Requirements for High Temperature Flowline Coatings, OMAE2002-28569, 21th International Conference on Offshore Mechanics and Arctic Engineering, Oslo, Norway, 2002. June 23–28.
[20] L. Watkins, New Pipeline Insulation Technology Introduced, Pipeline & Gas Journal (2000). April.
[21] F.M. Pattee, F. Kopp, Impact of Electrically Heated System on the Operation of Deep Water Subsea Oil Flowline, OTC 11894, 2000 Offshore Technology Conference, Houston, Texas, 2000.
[22] S. Cochran, Hydrate Control and Remediation Best Practices in Deepwater Oil Developments, Offshore Technology Conference, Houston, Texas, 2003.

APPENDIX: U-VALUE AND COOLDOWN TIME CALCULATION SHEET

U-VALUE AND COOLDOWN TIME CALCULATION SHEET

JOB NUMBER: CLIENT:
PROJECT NAME:
FLOWLINE:
DATE: 11/19/2004
VERSION: 1.0 DEVELOPER: QIANG BAI
USER: YB CHECKER: QB

Units:

$$F(T) := 32.0 + \frac{9.0}{5.0}\frac{T}{C}\ °F$$

Input Information:

Pipeline Geometry

Pipe outer diameter	$D_o := 0.219$ m	$D_o = 8.62$ in
Wall thickness	$t := 14.3$mm	$t = 0.56$in

Insulation layer Thickness

Insulation layer 1	$t_1 := 0.3$ mm	$t_1 = 0.01$ in
Insulation layer 2	$t_2 := 0.3$ mm	$t_2 = 0.01$ in
Insulation layer 3	$t_3 := 76.2$ mm	$t_3 = 3$ in
Insulation layer 4	$t_4 := 46$ mm	$t_4 = 1.81$ in
Insulation layer 5	$t_5 := 5$ mm	$t_5 = 0.2$in
Length of pipe	$L_p := 1$ m	$L_p = 3.28$ ft

Thermal Conductivity

Fluid	$k_f := 0.35\ W m^{-1} K^{-1}$	$k_f = 0.2 BTU\ ft^{-1}\ hr^{-1}\ °F^{-1}$
Pipe wall	$k_{pipe} := 45\ W m^{-1} K^{-1}$	$k_{pipe} = 26 BTU\ ft^{-1}\ hr^{-1}\ °F^{-1}$
Insulation layer 1	$k_{I1} := 0.3\ W m^{-1} K^{-1}$	$k_{I1} = 0.17 BTU\ ft^{-1}\ hr^{-1}\ °F^{-1}$
Insulation layer 2	$k_{I2} := 0.22\ W m^{-1} K^{-1}$	$k_{I2} = 0.13 BTU\ ft^{-1}\ hr^{-1}\ °F^{-1}$
Insulation layer 3	$k_{I3} := 0.215\ W m^{-1} K^{-1}$	$k_{I3} = 0.12 BTU\ ft^{-1}\ hr^{-1}\ °F^{-1}$
Insulation layer 4	$k_{I4} := 0.175\ W m^{-1} K^{-1}$	$k_{I4} = 0.1\ BTU\ ft^{-1}\ hr^{-1}\ °F^{-1}$
Insulation layer 5	$k_{I5} := 0.215\ W m^{-1} K^{-1}$	$k_{I5} = 0.12 BTU\ ft^{-1}\ hr^{-1}\ °F^{-1}$

Specific Heat Capacity

Fluid	$Cp_{fluid} := 3.054\ 10^3\ J\,kg^{-1}\,K^{-1}$	$Cp_{fluid} = 0.73\,BTU\,lb^{-1}\,°F^{-1}$
Carbon steel	$Cp_{steel} := 2.29\ 10^3\ J\,kg^{-1}\,K^{-1}$	$Cp_{steel} = 0.55\,BTU\,lb^{-1}\,°F^{-1}$
Insulation layer 1	$C_{pl1} := 1.6\ 10^3\ J\,kg^{-1}\,K^{-1}$	$C_{pl1} = 0.38\,BTU\,lb^{-1}\,°F^{-1}$
Insulation layer 2	$C_{pl2} := 1.95\ 10^3\ J\,kg^{-1}\,K^{-1}$	$C_{pl2} = 0.47\,BTU\,lb^{-1}\,°F^{-1}$
Insulation layer 3	$C_{pl3} := 1.95\ 10^3\ J\,kg^{-1}\,K^{-1}$	$C_{pl3} = 0.47\,BTU\,lb^{-1}\,°F^{-1}$
Insulation layer 4	$C_{pl4} := 1.95\ 10^3\ J\,kg^{-1}\,K^{-1}$	$C_{pl4} = 0.47\,BTU\,lb^{-1}\,°F^{-1}$
Insulation layer 5	$C_{pl5} := 2.0\ 10^3\ J\,kg^{-1}\,K^{-1}$	$C_{pl5} = 0.48\,BTU\,lb^{-1}\,°F^{-1}$

Density

Fluid	$\rho_f := 795.8\ kg\ m^{-3}$	$\rho_f = 49.68\,lb\ ft^{-3}$
Carbon steel	$\rho_p := 7850\ kg\ m^{-3}$	$\rho_p = 490.06\,lb\ ft^{-3}$
Insulation layer 1	$\rho_{l1} := 1450\ kg\ m^{-3}$	$\rho_{l1} = 90.52\,lb\ ft^{-3}$
Insulation layer 2	$\rho_{l2} := 900\ kg\ m^{-3}$	$\rho_{l2} = 56.19\,lb\ ft^{-3}$
Insulation layer 3	$\rho_{l3} := 900\ kg\ m^{-3}$	$\rho_{l3} = 56.19\,lb\ ft^{-3}$
Insulation layer 4	$\rho_{l4} := 700\ kg\ m^{-3}$	$\rho_{l4} = 43.7\,lb\ ft^{-3}$
Insulation layer 5	$\rho_{l5} := 890\ kg\ m^{-3}$	$\rho_{l5} = 55.56\,lb\ ft^{-3}$

PROPERTIES OF AMBIENT SURROUNDING

BURIED PIPELINE

Depth of cover to center line	$Z := 0.0\ m$
Thermal conductivity of soil/clay	$k_s := 1.22\ W\,m^{-1}K^{-1}$

SURFACE LAID PIPELINE

If the depth of cover, Z, is set to 0. the ambient condition is sea water.

Density of sea-water	$\rho_{h2o} := 1026\ kg\ m^{-3}$
Viscosity of sea-water	$\mu_{h2o} := 0.00182\ Pa\ s$
Thermal conductivity of seawater	$k_{h2o} := 0.56\ W\,m^{-1}\,K^{-1}$
Velocity of sea-water	$v_{h2o} := 1.6\ ft\,s^{-1}$
Specific heat capacity	$Cp_{h2o} := 4300\ 10^3\ J\ kg^{-1}\ K^{-1}$
Ambient temperature	$T_a := 4.28\ °C$

Internal Fluid Flow Parameters

Reynold number of internal fluid: $Re_f := 1.35\ 10^6$

Prandle number of internal fluid: $Pr_f := 3.5$

Initial fluid temperature $Tf_0 := 87.0\ °C$

Time step $\Delta t := 0.01hr$

Safty operation temperature $T_c := 64\ °C$

Calculations:

Pipe internal diameter $D_i := D_o - 2\ t$ $\quad D_i = 0.19m$

Inside film heat transfer coefficient $h_i := 0.023\ Re_f^{0.8}\ Pr_f^{0.3}\ \frac{k_f}{D_i}$ $\quad h_i = 4.94 \quad 10^3\ \frac{W}{m^2\ K}$

Inside film resistance $R_i := \frac{D_o}{D_i\ h_i}$ $\quad R_i = 2.33 \quad 10^{-4}\ \frac{m^2\ K}{W}$

Pipewall resistance $R_{pipe} := \frac{D_o\ \ln\left(\frac{D_o}{D_i}\right)}{2\ k_{pipe}}$ $\quad R_{pipe} = 3.41 \quad 10^{-4}\ \frac{m^2\ K}{W}$

Insulation 1 resistance $R_{I1} := \frac{D_o\ \ln\left(\frac{D_o + 2\ t_1}{D_o}\right)}{2\ k_{I1}}$ $\quad R_{I1} = 9.99 \quad 10^{-4}\ \frac{m^2\ K}{W}$

Insulation 2 resistance $R_{I2} := \frac{D_o\ \ln\left(\frac{D_o + 2\ (t_1 + t_2)}{D_o + 2\ t_1}\right)}{2\ k_{I2}}$ $\quad R_{I2} = 1.36 \quad 10^{-3}\ \frac{m^2\ K}{W}$

Insulation 3 resistance $R_{I3} := \frac{D_o\ \ln\left(\frac{D_o + 2\ (t_1 + t_2 + t_3)}{D_o + 2\ (t_1 + t_2)}\right)}{2\ k_{I3}}$ $\quad R_{I3} = 0.27\ \frac{m^2\ K}{W}$

Insulation 4 resistance $R_{I4} := \frac{D_o\ \ln\left(\frac{D_o + 2\ (t_1 + t_2 + t_3 + t_4)}{D_o + 2\ (t_1 + t_2 + t_3)}\right)}{2\ k_{I4}}$ $\quad R_{I4} = 0.14\ \frac{m^2\ K}{W}$

Insulation 5 resistance $R_{I5} := \frac{D_o\ \ln\left(\frac{D_o + 2\ (t_1 + t_2 + t_3 + t_4 + t_5)}{D_o + 2\ (t_1 + t_2 + t_3 + t_4)}\right)}{2\ k_{I5}}$ $\quad R_{I5} = 0.01\ \frac{m^2\ K}{W}$

Surrounding resistance

$$f(Z) := \begin{cases} \left(\frac{D_o}{2 \cdot k_s} \cdot \ln\left(\frac{2 \cdot Z}{0.5 \cdot D_o}\right)\right) & \text{if } Z > 0 \cdot m \\ \frac{D_o}{\left[0.0266 \cdot k_{h2o} \cdot \left(\frac{D_i \cdot h_i}{k_f}\right)^{0.8} \cdot \left(\frac{\rho_{h2o} \cdot V_{h2o} \cdot D_o}{\mu_{h2o}}\right)^{0.33}\right]} & \text{otherwise} \end{cases}$$

$$R_{surr} := f(Z) \qquad R_{surr} = 7.03 \times 10^{-4} \frac{m^2 \cdot K}{W}$$

Area for heat transfer $\quad A_o := \pi D_o \cdot L_p$

Weight per unit length

Fluid
$$W_f := \frac{\pi \cdot D_i^2 \cdot \rho_f \cdot L_p}{4}$$

Pipe
$$W_p := \frac{\pi \cdot \left(D_o^2 - D_i^2\right) \cdot \rho_p \cdot L_p}{4}$$

Insulation layer1
$$W_{I1} := \frac{\pi \cdot \left[(D_o + 2 \cdot t_1)^2 - (D_o)^2\right] \cdot \rho_{I1} \cdot L_p}{4}$$

Insulation layer 2
$$W_{I2} := \frac{\pi \cdot \left[(D_o + 2 \cdot t_1 + 2 \cdot t_2)^2 - (D_o + 2 \cdot t_1)^2\right] \cdot \rho_{I2} \cdot L_p}{4}$$

Insulation layer 3
$$W_{I3} := \frac{\pi \cdot \left[(D_o + 2 \cdot t_1 + 2 \cdot t_2 + 2 \cdot t_3)^2 - (D_o + 2 \cdot t_1 + 2 \cdot t_2)^2\right] \cdot \rho_{I3} \cdot L_p}{4}$$

Insulation layer 4
$$W_{I4} := \frac{\pi \cdot \left[(D_o + 2 \cdot t_1 + 2 \cdot t_2 + 2 \cdot t_3 + 2 \cdot t_4)^2 - (D_o + 2 \cdot t_1 + 2 \cdot t_2 + 2 \cdot t_3)^2\right] \cdot \rho_{I4} \cdot L_p}{4}$$

Insulation layer 5
$$W_{I5} := \frac{\pi \cdot \left[(D_o + 2 \cdot t_1 + 2 \cdot t_2 + 2 \cdot t_3 + 2 \cdot t_4 + 2 \cdot t_5)^2 - (D_o + 2 \cdot t_1 + 2 \cdot t_2 + 2 \cdot t_3 + 2 \cdot t_4)^2\right] \cdot \rho_{I5} \cdot L_p}{4}$$

Results:

- **Overall heat transfer coefficient**

U value based on pipeline OD
$$U_o := \frac{1}{R_i + R_{pipe} + R_{I1} + R_{I2} + R_{I3} + R_{I4} + R_{I5} + R_{surr}}$$

$$U_o = 2.38 \frac{W}{m^2 \cdot K}$$

U value based on pipeline ID
$$U_i := U_o \cdot \frac{D_o}{D_i} \qquad U_i = 2.74 \frac{W}{m^2 K}$$

U value based on pipeline ID $U_i := U_o \cdot \frac{D_o}{D_i}$ $U_i = 2.74 \frac{W}{m^2 K}$

- **Cooldown calculation**

Initial temperatures of system for cooldown

Wall

$$Tp_0 := Tf_0 - U_o \cdot \left(R_i + \frac{R_{pipe}}{2} \right) \cdot (Tf_0 - T_a)$$

Insulatioin layer 3

$$Tl3_0 := Tf_0 - U_o \cdot \left(R_i + R_{pipe} + R_{l1} + R_{l2} + \frac{R_{l3}}{2} \right) \cdot (Tf_0 - T_a)$$

Insulatioin layer 4

$$Tl4_0 := Tf_0 - U_o \cdot \left(R_i + R_{pipe} + R_{l1} + R_{l2} + R_{l3} + \frac{R_{l4}}{2} \right) \cdot (Tf_0 - T_a)$$

Insulatioin layer 5

$$Tl5_0 := Tf_0 - U_o \cdot \left(R_i + R_{pipe} + R_{l1} + R_{l2} + R_{l3} + R_{l4} + \frac{R_{l5}}{2} \right) \cdot (Tf_0 - T_a)$$

Inside fluid film heat transfer coefficient for laminar flow
(Velocity of internal fluid is extremely low during cooldown)

$$h_l := 3.66 \cdot \frac{k_f}{D_i} \qquad h_l = 6.73 \frac{W}{m^2 \cdot K}$$

Inside laminar film resistance

$$R_l := \frac{D_o}{D_i \cdot h_l} \qquad R_l = 0.17\, m^2 \frac{K}{W}$$

Cooldown temperature profiles of fluid, pipe and insulation layer 3,4 and 5

$$i := 0 .. 10000$$

$$\begin{pmatrix} Tf_{i+1} \\ Tp_{i+1} \\ Tl3_{i+1} \\ Tl4_{i+1} \\ Tl5_{i+1} \end{pmatrix} := \begin{bmatrix} Tf_i - \frac{A_o \cdot \Delta t}{W_f \cdot Cp_{fluid}} \cdot \left(\frac{Tf_i - Tp_i}{R_l + \frac{R_{pipe}}{2}} \right) \\ Tp_i - \frac{A_o \cdot \Delta t}{W_p \cdot Cp_{steel}} \cdot \left(\frac{Tp_i - Tf_i}{R_l + \frac{R_{pipe}}{2}} + \frac{Tp_i - Tl3_i}{R_{l1} + R_{l2} + \frac{R_{l3}}{2} + \frac{R_{pipe}}{2}} \right) \\ Tl3_i - \frac{A_o \cdot \Delta t}{W_{l3} \cdot C_{pl3}} \cdot \left(\frac{Tl3_i - Tl4_i}{\frac{R_{l3}}{2} + \frac{R_{l4}}{2}} + \frac{Tl3_i - Tp_i}{R_{l1} + R_{l2} + \frac{R_{l3}}{2} + \frac{R_{pipe}}{2}} \right) \\ Tl4_i - \frac{A_o \cdot \Delta t}{W_{l4} \cdot C_{pl4}} \cdot \left(\frac{Tl4_i - Tl3_i}{\frac{R_{l3}}{2} + \frac{R_{l4}}{2}} + \frac{Tl4_i - Tl5_i}{\frac{R_{l4}}{2} + \frac{R_{l5}}{2}} \right) \\ Tl5_i - \frac{A_o \cdot \Delta t}{W_{l5} \cdot C_{pl5}} \cdot \left(\frac{Tl5_i - Tl4_i}{\frac{R_{l5}}{2} + \frac{R_{l4}}{2}} + \frac{Tl5_i - T_a}{R_{surr} + \frac{R_{l5}}{2}} \right) \end{bmatrix}$$

$$f(Tf, T_c) := \begin{array}{|l} i \leftarrow 0 \\ \quad \text{break} \quad \text{if} \quad T_a = Tf_i \\ \text{while} \quad Tf_i \geq T_c \\ \quad i \leftarrow i + 1 \\ i \end{array}$$

This conditional statement is employed to identify whether the critical condition (i.e. WAT or HFT) is breached.

$$\text{Critical\_Time} := f(Tf, T_c) \cdot \Delta t$$

Critical time is the required time for fluid coolling down to the critical temperature.

Cooldown Time : Critical_Time = 19.55hr

$$\text{Time}_i := i \cdot \frac{\Delta t}{hr}$$

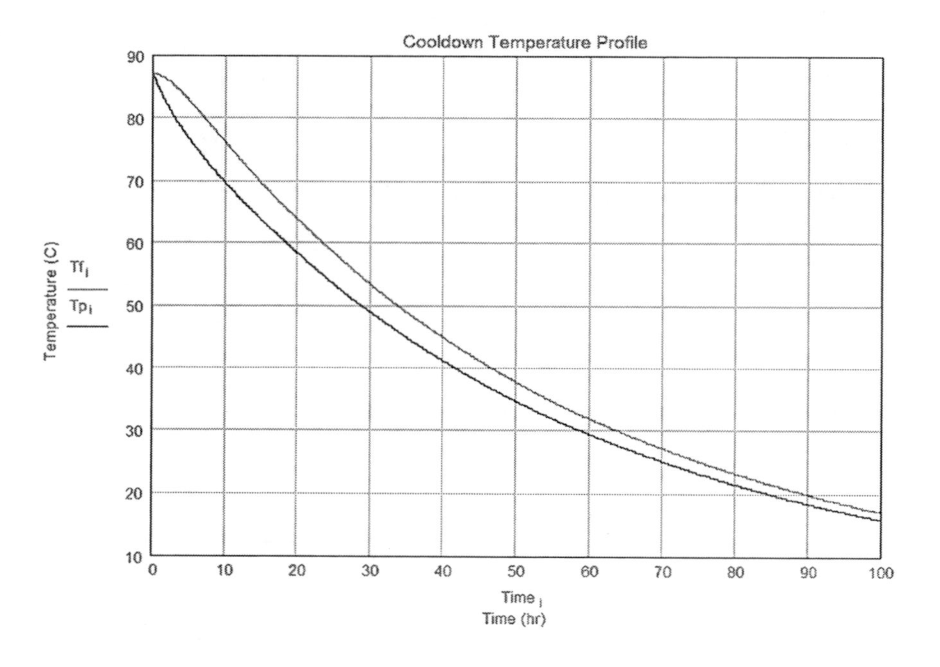

CHAPTER 15

Hydrates

Contents

15.1. INTRODUCTION

Natural gas hydrates are crystalline compounds formed by the physical combination of water molecules and certain small molecules in hydrocarbons fluid such as methane, ethane, propane, nitrogen, carbon dioxide,

Subsea Engineering Handbook
ISBN 978-0-12-397804-2, doi:10.1016/B978-0-12-397804-2.00015-1

and hydrogen sulfide. Hydrates are easily formed when the hydrocarbon gas contains water at high pressure and relatively low temperature.

As development activities in deeper waters increase, gas hydrates have become one of the top issues in the offshore industry. Deepwater operations are being conducted all over the world in locations such as the Gulf of Mexico (GoM), West Africa (WA), and the North Sea (NS). Gas hydrate problems manifest themselves most commonly during drilling and production processes. Hydrates may appear anywhere and at any time in an offshore system when there is natural gas, water, and suitable temperature and pressure. Figure 15-1 shows a sketch of areas where hydrate blockages may occur in a simplified offshore deepwater system from the well to the platform export flowline. Hydrate blockages in a subsea flowline system are most likely to be found in areas where the direction of flow changes in the well, pipeline, and riser parts of a system.

Hydrates are rarely found in the tubing below the downhole safety valve and the pipelines after the platform. In general, hydrocarbon fluids are at higher pressure and higher temperature before the downhole safety valve, such that the fluid temperature is higher than the hydrate formation temperature corresponding to the local pressure. On the platform, the processes of separation, drying, and compression are usually carried out. First the multiphase hydrocarbons fluids are separated into gas, oil, and water phases; then, the gas is dried and dehydrated; last, the gas is compressed for export. After gas and water are separated on the platform, hydrates cannot form in a gas export flowline without the presence of water.

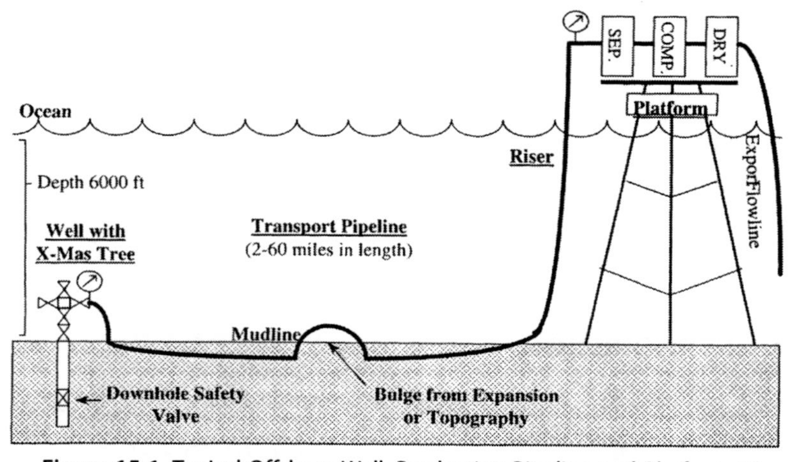

Figure 15-1 Typical Offshore Well, Production Pipeline and Platform [1]

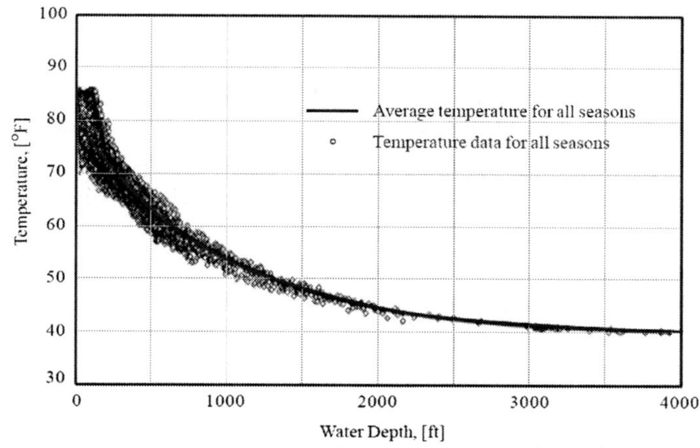

Figure 15-2 Seawater Temperatures of GoM for All Seasons at Different Water Depths [2]

Figure 15-2 shows typical temperature variations with water depth in the GoM. At the water surface, the temperature deviation in all seasons is about 15°F. When water depth is deeper than 3000 ft, the temperature deviations are very small and water temperature at the seabed becomes approximately constant at 40°F. In the wellhead and Christmas tree, many valves are used to control the production operation. Hydrates may possibly occur during the shutdown/start-up operation in the well and Christmas tree because the ambient temperatures are typically around 40°F in deepwater, even though they may not form during normal operation at steady-state conditions in which the flow rate and temperature of hydrocarbon fluid are higher. In most deepwater production, the ambient water temperature is the main factor causing hydrate formation in the subsea system, but numerous examples also exist of hydrates forming due to Joule-Thomson (JT) cooling of gas, where the gas expands across a valve, both subsea and on the platform. Without insulation or another heat control system for the flowline, the fluids inside a subsea pipeline will cool to the ambient temperature within a few miles from the well. The cooling rate per length depends on the fluid composition, flow rate, ambient temperature, pipe diameter, and other heat transfer factors.

The development of hydrates should be avoided in offshore engineering because they can plug flowlines, valves, and other subsea devices. Hydrates are of importance in deepwater gas developments because ambient temperatures are low enough to be in the hydrate formation region at operating pressures. The presence of a certain amount of water in the hydrocarbon systems can be troublesome due to the formation of hydrates. When temperature and pressure are in the hydrate formation region, hydrates grow as long as water and small molecule hydrocarbons are present. Hydrate crystals can develop into flow blockages, which can be time consuming to clear in subsea equipment or flowlines and cause safety problems. Lost or delayed revenue and costs associated with hydrate blockages can be significant due to vessel intervention costs and delayed production. Thus, hydrate prevention and remediation are important design factors for deepwater developments.

The following topics about hydrates are discussed in this chapter:

- Fundamental knowledge of hydrates;
- Hydrate formation process and consequences;
- Hydrate prevention techniques;
- Hydrate remediation;
- Hydrate control design philosophies;
- Thermal inhibitor recovery.

15.2. PHYSICS AND PHASE BEHAVIOR

15.2.1. General

Natural gas hydrates are crystalline water structures with low-molecular-weight guest molecules. They are often referred to as clathrate hydrates. The presence of the gas molecules leads to stability of the crystalline structure, allowing hydrates to exist at much higher temperatures than ice. Natural gas hydrates typically form one of three crystal structures, depending primarily on the sizes of the guest molecules. They are metastable minerals whose formation, stability, and decomposition depend on pressure, temperature, composition, and other properties of the gas and water. Hydrate formers include nitrogen, carbon dioxide, hydrogen sulfide, methane, ethane, propane, iso-butane, *n*-butane, and some branched or cyclic C5–C8 hydrocarbons. Figure 15-3 shows one of the typical hydrate crystal structures found in oil and gas production systems.

Natural gas hydrates are composed of approximately 85 mol% water; therefore, they have many physical properties similar to those of ice. For instance, the appearance and mechanical properties of hydrates are comparable to those of ice. The densities of hydrates vary somewhat due to the nature of the guest molecule(s) and the formation conditions, but are generally comparable to that of ice. Thus, hydrates typically will float at the

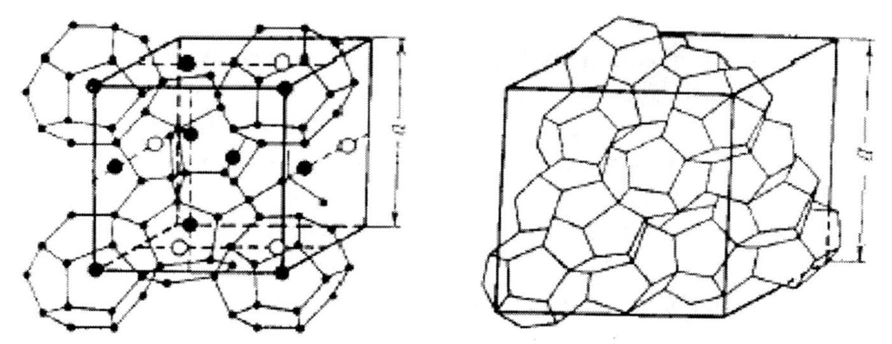

Figure 15-3 Hydrate Crystal Structures in Oil and Gas Production Systems [2]

water/hydrocarbon interface. However, in some instances, hydrates have been observed to settle on the bottom of the water phase. If a hydrate plug breaks from the pipe walls, it can be pushed down along the flowline by the flowing of hydrocarbon fluid like an ice bullet, potentially rupturing the flowline at a restriction or bend.

Four components are required to form gas hydrates: water, light hydrocarbon gases, low temperature, and high pressure. If any one of these components is absent, then gas hydrates will not form. Hydrate problems can appear during normal production, but transient operations are often more vulnerable. For instance, during a shut-down, the temperature of the subsea line drops to that of the surrounding environment. Given sufficient time under these high pressures and low temperatures, hydrates will form.

The extent to which the gas, oil, and water partition during shutdown somewhat limits the growth of hydrates; although direct contact between the gas phase and the water phase is not needed for hydrate formation, an intervening oil layer slows transport of the hydrate-forming molecules. Additionally, hydrates typically form in a thin layer at the water/oil interface, which impedes further contact between the water and gas molecules. Even if the flowlines do not plug during shutdown, when the well is restarted, the agitation breaks the hydrate layer and allows good mixing of the subcooled water and gas. Rapid hydrate formation often leads to a blockage of flow at low spots where water tends to accumulate. Plugging tendency increases as the water cut increases, because there is a higher likelihood that sufficient hydrate particles will contact each other and stick together. Other typical locations include flow restrictions and flow transitions, as occurs at, for instance, elbows and riser bases.

Hydrate plugs may also occur in black oil subsea systems. Most deepwater black oil systems are not producing significant volumes of water. As water cuts rise, the incidence of hydrate plugs in black oil lines will certainly increase. Some black oils have a tendency not to plug, even when hydrates are formed. The hydrates remain small particles dispersed in the liquid phase and are readily transported through the flowline. These back oils will eventually plug with hydrates if the water cut gets high enough.

15.2.2. Hydrate Formation and Dissociation

Hydrate formation and dissociation curves are used to define pressure/temperature relationships in which hydrates form and dissociate. These curves may be generated by a series of laboratory experiments, or more commonly, are predicted using thermodynamic software such as Multi-Flash or PVTSIM based on the composition of the hydrocarbon and aqueous phases in the system. The hydrate formation curve defines the temperature and pressure envelope in which the entire subsea hydrocarbons system must operate in at steady state and transient conditions in order to avoid the possibility of hydrate formation [3].

Figure 15-4 shows an example of these curves, which shows the stability of natural gas hydrates as a function of pressure and temperature. To the right of the dissociation curve is the region in which hydrates do not form; operating in this region is safe from hydrate blockages. To the left of hydrate formation curve is the region where hydrates are thermodynamically stable

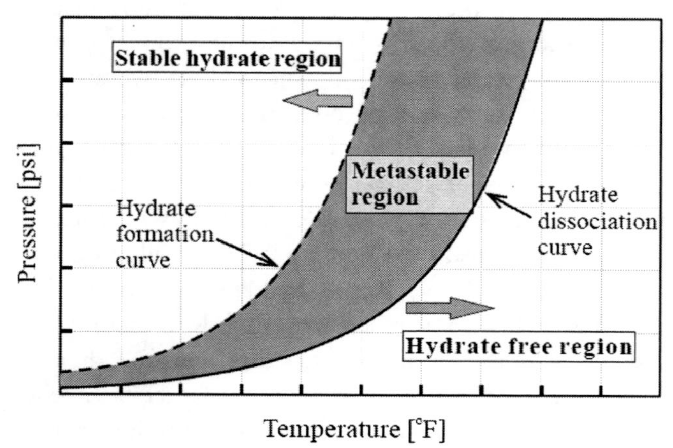

Figure 15-4 Hydrate Formation and Dissociation Regions

and have the potential to form [4]. This does not mean that hydrates will necessarily form or that formed hydrates will cause operational difficulties. The stability of hydrates increases with increasing pressure and decreasing temperature. There is often a delay time or the temperature must be lowered somewhat below the hydrate stability temperature in order for hydrates to form. The subcooling of a system is often used when discussing gas hydrates, which is defined as the difference between hydrate stability temperature and the actual operating temperature at the same pressure. If the system is operating at 40°F and 3000 psi, and the hydrate dissociation temperature is 70°F, then the system is experiencing 30°F of subcooling.

The subcooling of a system without hydrate formation leads to an area between the hydrate formation temperature and the hydrate dissociation temperature, called the metastable region, where hydrate is not stable. Some software packages attempt to predict this metastable region. Regularly operation within this metastable region is risky. While such differences in hydrate formation and dissociation temperatures are readily observed in the laboratory, the quantitative magnitude of this hysteresis is apparatus and technique dependent.

Hydrates may not form for hours, days, or even at all even if a hydrocarbon system containing water is at a temperature and pressure conditions close to the hydrate dissociation curve. A certain amount of "subcooling" is required for hydrate formation to occur at rates sufficient to have a practical impact on the system. Figure 15-5 shows the variation of hydrate formation time with the subcooling of hydrate formation temperature. When subcooling increases, the hydrate formation time decreases exponentially. In

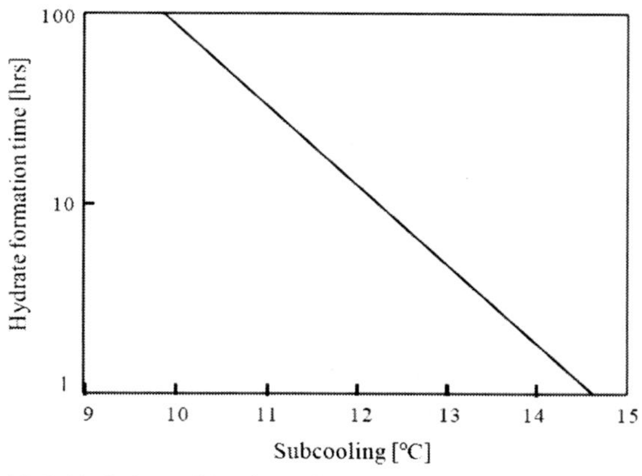

Figure 15-5 Variation of Hydrate Formation Time with Subcooling [5]

general, subcooling higher than 5°F will cause hydrate formation to occur at the hydrocarbon/water interface in flowlines.

A thermodynamic understanding of hydrates indicates the conditions of temperature, pressure, and composition at which a hydrate may form. However, it does not indicate where or when a hydrate plug will form in the system. Hydrate plugs can form in just a few minutes, or take several days to block production. There are two mechanisms of plug formation, one in which hydrates slowly build up on the bottom of the pipe, gradually restricting flow, and the other in which hydrates agglomerate in the bulk fluid, forming masses of slush that bridge and eventually block the flow. Both mechanisms have been observed in the field, although the latter is believed to be more prevalent. The mechanics of plug formation are not yet well understood, although it is known that certain geometries, such as flow restrictions at chokes, are prone to hydrate plug formation.

Control of hydrates relies on keeping the system conditions out of the region in which hydrates are stable. During oil production operations, temperatures are usually above the hydrate formation temperature, even with the high system pressures at the wellhead (on the order of 5000 to 10,000 psi). However, during a system shutdown, even well-insulated systems will fall to the ambient temperatures eventually, which in the deep GoM is approximately 38 to 40°F. Many methods are available for hydrate formation prediction. Most of them are based on light gas hydrocarbon systems and vary in the complexity of the factors utilized within the computational procedures. The Peng-Robinson method is one typical equation of state (EOS) method that is currently extensively utilized to predict hydrate boundaries.

Knowledge about hydrates has significantly improved in the past 10 years. Hydrate disassociation can be predicted within 1 to 3° with the exception of brines that have a high salt concentration. The hydrate disassociation curves typically provide conservative limits for hydrate management design. The effects of thermodynamic hydrate inhibitors, methanol and ethylene glycols, can be predicted with acceptable accuracy.

When the temperature and pressure are in the hydrate region, hydrates grow as long as water and light hydrocarbons are available and can eventually develop blockages. Clearing hydrate blockages in subsea equipment or flowlines poses safety concerns and can be time consuming and costly. Hydrate formation is typically prevented by several methods including controlling temperature, controlling pressure, removing water, and by shifting thermodynamic equilibrium with chemical inhibitors such as methanol or monoethylene glycol, low-dosage hydrate inhibitors.

15.2.3. Effects of Salt, MeOH, and Gas Composition

The hydrate dissociation curve may be shifted toward lower temperatures by adding a hydrate inhibitor. Methanol, ethanol, glycols, sodium chloride, and calcium chloride are common thermodynamic inhibitors. Hammerschmidt [6] suggested a simple formula to roughly estimate the temperature shift of the hydrate formation curve:

$$\Delta T = \frac{KW}{M(100 - W)} \quad \text{15-1}$$

where

ΔT: temperature shift, hydrate depression, °C;

K: constant, which is defined in Table 15-1;

W: concentration of the inhibitor in weight percent in the aqueous phase;

M: molecular weight of the inhibitor divided by the molecular weight of water.

The Hammerschmidt equation was generated based on more than 100 natural gas hydrate measurements with inhibitor concentrations of 5 to 25 wt% in water. The accuracy of the equation is 5% average error compared with 75 data points. The hydrate inhibition abilities are less for substances with a larger molecular weight of alcohol, for example, the ability of methanol is higher than that of ethanol and glycols. With the same weight percent, methanol has a higher temperature shift than that of glycols, but MEG has a lower volatility than methanol and MEG may be recovered and recycled more easily than methanol on platforms.

Figure 15-6 shows the effect of typical thermodynamic inhibitors on hydrate formation at 20 wt% fraction. Salt, methanol, and glycols act as thermodynamic hydrate inhibitors that shift the hydrate stability curve to the left. Salt has the most dramatic impact on the hydrate stability temperature. On a weight basis, salt is the most effective hydrate inhibitor

Table 15-1 Constant of Equation (15-1) for Various Inhibitors

Inhibitor	*K*-Value
Methanol	2335
Ethanol	2335
Ethylene glycol (MEG)	2700
Diethylene glycol (DEG)	4000
Triethylene glycol (TEG)	5400

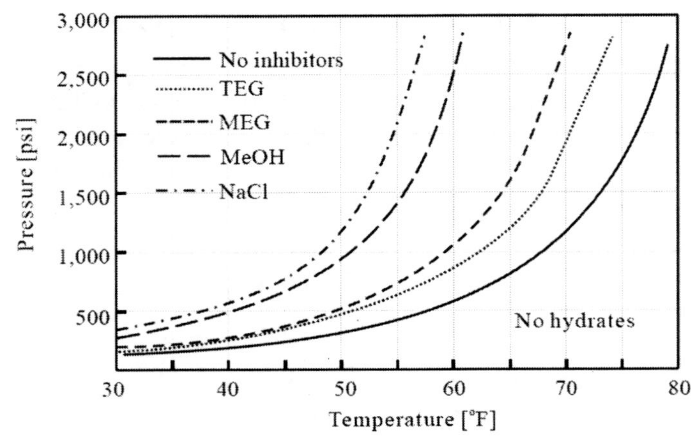

Figure 15-6 Effect of Thermodynamic Inhibitors on Hydrate Formation

and so accounting correctly for the produced brine salinity is important in designing a hydrate treatment plan. In offshore fields, MEG found more application than DEG and TEG because MEG has a lower viscosity and has more effect per weight.

Figure 15-7 shows the effect of salts in fluid on hydrate formation. Increasing salt content in the produced brine shifts the hydrate curve to lower temperatures at the same pressure. The solubility of salt to water has a limit based on the temperature.

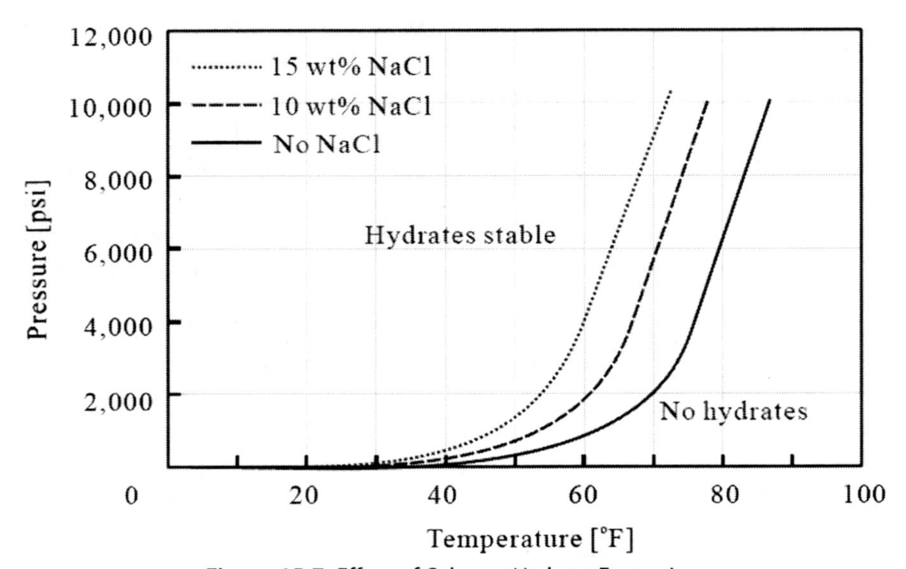

Figure 15-7 Effect of Salts on Hydrate Formation

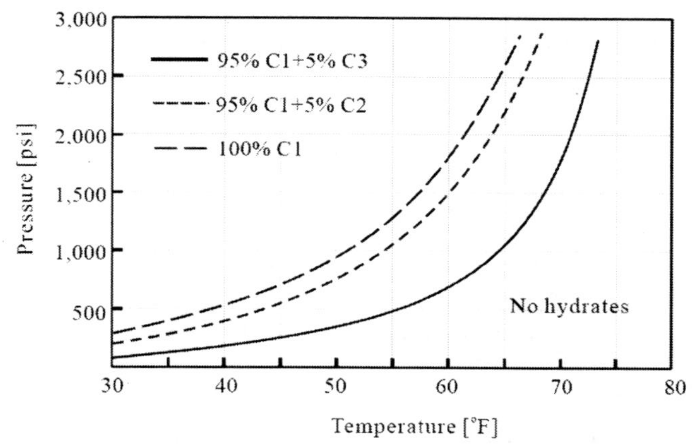

Figure 15-8 Effect of Gas Composition on Hydrate Formation

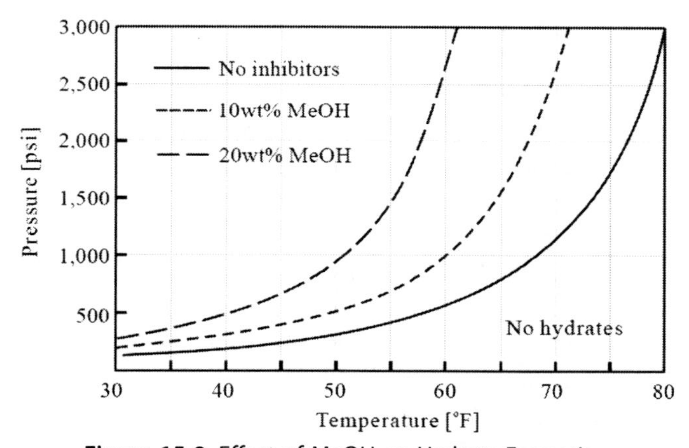

Figure 15-9 Effect of MeOH on Hydrate Formation

Figure 15-8 shows the effect of gas composition on the hydrate formation curve. More small molecular components results in a lower hydrate formation at the same pressure.

Figure 15-9 shows the effect of weight percentage of methanol on the hydrate formation curve. More weight percentage of methanol leads to a greater temperature shift of the hydrate formation curve.

15.2.4. Mechanism of Hydrate Inhibition

Two types of hydrate inhibitors are used in subsea engineering: thermodynamic inhibitors (THIs) and low-dosage hydrate inhibitors (LDHIs).

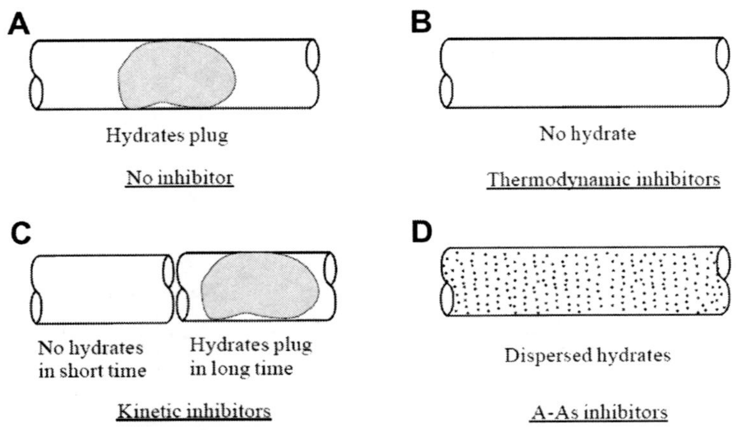

Figure 15-10 Mechanism of Hydrate Inhibition

The most common THIs are methanol and MEG, even though ethanol, other glycols (DEG, TEG), and salts can be effectively used. They inhibit hydrate formation by reducing the temperature at which hydrates form. This effect is the same as adding antifreeze to water to lower the freezing point. Methanol and MEG are the most commonly used inhibitors.

LDHIs include anti-agglomerates and kinetic inhibitors. LDHIs have found many applications in subsea systems in recent years. LDHIs prevent hydrate blockages at significantly lower concentrations, for example, less than 1 wt%, than thermodynamic inhibitors such as methanol and glycols. Unlike thermodynamic inhibitors, LDHIs do not change the hydrate formation temperature. They either interfere with formation of hydrate crystals or agglomeration of crystals into blockages. Anti-agglomerates can provide protection at higher subcooling temperatures than can kinetic hydrate inhibitors. However, low-dosage hydrate inhibitors are not recoverable and they are expensive. The differences in hydrate inhibition mechanism between LDHIs and THIs are shown in Figure 15-10.

LDHIs are preferred for regular operations because they reduce volumes and can work out to be cheaper. For transient events, the volumes required are not usually that large, so there is not much benefit in LDHIs, and methanol becomes the preferred inhibitor.

15.2.4.1. Thermodynamic Inhibitors

THIs inhibits hydrate formation by reducing the temperature at which hydrates form by changing the chemical potential of water. This effect is the same as adding antifreeze to water to lower the freezing point. THIs includes methanol, glycols, and others. In general, methanol is vaporized into the gas phase of a pipeline, and then dissolves in any free water accumulation to prevent hydrate formation. Hydrate inhibition occurs in the aqueous liquid, rather than in the vapor or oil/condensate. Although most of the methanol dissolves in the water phase, a large amount of methanol remains in the vapor or oil/condensate phase; therefore, the proportions of methanol dissolved in the vapor or oil/condensate liquid phases are usually counted as an economic loss.

15.2.4.2. Low-Dosage Hydrate Inhibitors

Kinetic inhibitors (KIs) are low-molecular-weight water-soluble polymers or copolymers that prevent hydrate blockages by bonding to the hydrate surface and delaying hydrate crystal nucleation and/or growth. They are dissolved in a carrier solvent and injected into the water phase in pipelines [7]. These inhibitors work independently of water cuts, but are limited to relatively low subcooling temperatures (less than 20°F), which may not be sufficient for deepwater applications. For greater subcooling, KIs must be blended with a thermodynamic inhibitor. Additionally, the inhibition effect of KIs is time limited and, thus, their benefit for shut-down is limited. KIs have been applied in the North Sea and the Gulf of Mexico. Long-term shutdowns will require depressurization, which complicates the restart process, and methanol without KIs will be required for restarts. KIs are generally environmentally friendly.

Anti-agglomerates (AAs) are surfactants, which cause the water phase to be suspended as small droplets. When the suspended water droplets convert to hydrates, the flow characteristics are maintained without blockage. They allow hydrate crystals to form but keep the particles small and well dispersed in the hydrocarbon liquid. They inhibit hydrate plugging rather than hydrate formation. AAs can provide relatively high subcooling up to 40°F, which is sufficient for deepwater applications and have completed successful field trials in deepwater GoM production systems. AA effectiveness can be affected by type of oil or condensate, water salinity, and water cut. For deepwater gas developments, AAs can only be applied where there is sufficient condensate, such that the *in situ* water cut is less than 50%. Methanol may still be required for shutdown and restart. AAs have toxicity issues and may transport microcrystals of hydrate into and remain in the condensed/oil phase.

15.3. HYDRATE PREVENTION

Gas subsea systems typically contain small quantities of water, which allows them to be continuously treated with methanol or glycol to prevent hydrate formation. These inhibitors prevent the formation of hydrates by shifting the hydrate stability curve to lower temperatures for a given pressure. If the systems produce too much water, it is difficult to economically treat with methanol. As a result, the system designs have to incorporate insulation of almost all components of a system and develop complex operating strategies to control hydrate formation during transient activities such as system start-up and shutdown.

Hydrate prevention techniques for subsea systems include [8]:

- Thermodynamic inhibitors;
- Low-dosage hydrate inhibitors (LDHIs);
- Low-pressure operation;
- Water removal;
- Insulation;
- Active heating.

15.3.1. Thermodynamic Inhibitors

The most common method of hydrate prevention in deepwater developments is injection of thermodynamic inhibitors, which include methanol, glycols, and others. Methanol and MEG are the most commonly used inhibitors, though ethanol, other glycols, and salts can be effectively used.

Inhibitor injection rates, whether methanol or MEG, are a function of water production and inhibitor dosage. Inhibitor dosage is a function of design temperature and pressure and produced fluid composition. Water production multiplied by the methanol dosage is the inhibitor injection rate, which will change throughout field operating life due to typically decreasing operating pressures and increasing water production.

Selection of the hydrate inhibitor is an important decision to be made in the early FEED design and can involve a number of criteria:

- Capital costs of topside process equipment, especially for regeneration;
- Capital costs of subsea equipment;
- Topside weight/area limitations;
- Environmental limits on overboard discharge;
- Contamination of the hydrocarbon fluid and impacts on downstream transport/processing;
- Safety considerations;

- System operability;
- Local availability of inhibitor.

Technical advantages and disadvantages of methanol and MEG are compared in Table 15-2.

Methanol and MEG are both effective inhibitors if sufficient quantities are injected; for deep water, inhibitor dosages of 0.7 to 1 bbl of inhibitor per barrel of water are generally used. Methanol can provide higher hydrate temperature depression but this effect is typically countered by high losses to the hydrocarbon liquid and gas phases. The selection of inhibitor is often based on economics, downstream process specifications, environmental issues, and/or operator preferences.

Costs for inhibition systems are driven by up-front capital costs, which are dominated by the regeneration system and also by makeup costs for inhibitor loss. Methanol is cheaper per unit volume, but has greater makeup requirements. Additionally, a methanol regeneration system may be as much as 50% less expensive than a MEG regeneration system. The methanol system starts out cheaper, but, with increasing field life, becomes more expensive due to methanol makeup costs.

The risks of using thermodynamic inhibitors include:

- Underdose, particularly due to not knowing water production rates;
- Inhibitor not going where intended (operator error or equipment failure);

Table 15-2 Comparison of Methanol and MEG [9]

Inhibitor	Advantages	Disadvantages
Methanol	• Move hydrate formation temperature more than MEG in a mass basis • Less viscous • Less likely to cause salt precipitation • Relative cost of regeneration system is less than for MEG • Approximate GoM cost of 1.0 $/gal	• Losses of methanol to gas and condensate phases can be significant, leading to a lower recovery (<80%) • Impact of methanol contamination in downstream processing • Low flash point • Environmental limitations on overboard discharge
MEG	• Easy to recover with recovery of 99% • Low gas and condensate solubility • Approximate GOM cost of 2.5$/gal	• High viscosity, impacts umbilical and pump requirements • Less applicable for restarts, stays with aqueous phase at bottom of pipe • More likely to cause salt precipitation

- Environmental concerns, particularly with methanol discharge limits;
- Ensuring remote location supply;
- Ensuring chemical/material compatibility;
- Safety considerations in handling methanol topside.

15.3.2. Low-Dosage Hydrate Inhibitors

While development of AAs and KIs continues, cost per unit volume of LDHIs is still relatively high, but is expected to decrease as their use increases. A potentially important advantage is that they may extend field life when water production increases.

15.3.3. Low-Pressure Operation

Low-pressure operation refers to the process of maintaining a system pressure that is lower than the pressure corresponding to the ambient temperature based on the hydrate dissociation curve. For deep water with an ambient temperature of 39°F (4°C), the pressure may need to be 300 psia (20 bar) or less. Operation at such a low pressure in the wellbore is not practical because pressure losses in a deepwater riser or long-distance tie-back would be significant.

By using subsea choking and keeping the production flowline at a lower pressure, the difference between hydrate dissociation and operating temperatures (i.e., subcooling) is reduced. This lower subcooling will decrease the driving force for hydrate formation and can minimize the inhibitor dosage.

15.3.4. Water Removal

If enough water can be removed from the produced fluids, hydrate formation will not occur. Dehydration is a common hydrate prevention technique applied to export pipelines. For subsea production systems, subsea separation systems can reduce water flow in subsea flowlines. The advantage of applying subsea separation is not only hydrate control, but also increasing recovery of reserves and/or accelerating recovery by making the produced fluid stream lighter and easier to lift. Another benefit is reduced topside water handling, treatment, and disposal.

As a new technology, subsea water separation/disposal systems are designed to separate bulk water from the production stream close to subsea trees on the seafloor. Basic components of such a system include a separator, pump to reinject water, and water injection well. Additional components include instrumentation, equipment associated with controlling the pump and separator, power transmission/distribution equipment, and chemical injection. Water cut leaving the separator may be as high as 10%. Operating experience on the Troll Pilot has shown water cuts of 0.5 to 3%. Because these systems do not remove all free water, and water may condense farther downstream, subsea bulk water removal does not provide complete hydrate protection. These systems need to be combined with another hydrate prevention technique, for example, continuous injection of a THI or LDHI. The main risk associated with subsea water separation systems is reliability.

15.3.5. Thermal Insulation

Insulation provides hydrate control by maintaining temperatures above hydrate formation conditions. Insulation also extends the cooldown time before reaching hydrate formation temperatures. The cooldown time gives operators time either to recover from the shutdown and restart a warm system or prepare the system for a long-term shutdown.

Insulation is generally not applied to gas production systems, because the production fluid has low thermal mass and also will experience JT cooling. For gas systems, insulation is only applicable for high reservoir temperatures and/or short tie-back lengths. One advantage of an insulated production system is that it can allow higher water production, which would not be economical with continuous inhibitor injection. However, shutdown and restart operations would be more complicated. For example, long-term shutdowns will probably require depressurization. An overview of insulation requirements was described in the Chapter 14.

15.3.6. Active Heating

Active heating includes electrical heating and hot fluid circulation heating in a bundle. In flowlines and risers, active heating must be applied with thermal insulation to minimize power requirements.

15.3.6.1. Electrical Heating

Electrical heating (EH) is a very fast developing technology and has found applications in the offshore fields including Nakika, Serrano, Oregano, and Habanero in the GoM, and Asgard, Huldra, and Sliepner in the North Sea [10]. Advantages of electrical heating include eliminating flowline depressurization, simplifying restart operations, and providing the ability to quickly remediate hydrate blockages.

Electrical heating techniques include:

- Direct heating, using the flowline as an electrical conductor for resistance heating;
- Indirect heating, using an EH element installed on the outer surface of flowline.

15.3.6.2. Hot Fluid Circulation Heating in a Pipe Bundle

Hot fluid heating has many of the same advantages as electrical heating. Instead of using electricity for supplying heat, however, hot fluid, typically inhibited water, circulates in the bundles to provide heat to the production fluids. Examples of such bundles include Statoil Asgard and Gullfaks South, Conoco Britannia, and BP King. These bundles can be complex in design, with thermal and mechanical design, fabrication, installation, life cycle, and risk issues that need to be addressed.

Active heating techniques provide a good level of protection. With active heating, hydrate control is simply a matter of power, insulation, and time. Active heating can increase the operating flexibility of a subsea production system, such that concerns including water cut, start-up, and operating flow rate and depressurization times are of lesser importance.

Electrically heated flowlines and low-dosage hydrate inhibitors are two developing technologies in hydrate prevention for reducing the complexity of the design and operation of subsea systems. Electrically heated flowline technology reduces hydrate concerns in subsea systems. Instead of relying on the lengthy process of blowdown for hydrate remediation, electrical heating provides a much faster way to heat the flowline and remove the plug. The other potential advantages of electrical heating are covered in Chapter 14. Low-dosage hydrate inhibitors reduce the volume of chemicals that must be transported and injected into the subsea system. Methanol treatment rates, for hydrate control, are on the order of one barrel of methanol for each barrel of produced brine. The low-dosage hydrate inhibitors may be able to accomplish the same task at dosage rates of less than 1%. This leads to a reduction in umbilical size and complexity. Note, however, that the hydrate inhibitors must be injected continuously to prevent hydrate formation.

15.4. HYDRATE REMEDIATION

Like the kinetics of hydrate formation, hydrate dissociation is a poorly understood subject and applying laboratory observations to field predictions has proven difficult. Part of the reason is the complicated interplay of flow, heat transfer, and phase equilibria. The dissociation behavior of hydrate depends on the hydrate size, porosity, permeability, volume of occluded water, "age" of the deposit, and local conditions such as temperature, pressure, fluids in contact with the plug, and insulation layers over the pipeline.

Two factors combine to make hydrate plugs exceedingly difficult to remove: It takes a large amount of energy to dissociate the hydrate, and heat transfer through the hydrate phase is slow. Hydrates also concentrate natural gas: 1 ft^3 of hydrates can contain up to 182 ft^3 of gas. This has significant implications for safety in depressurizing hydrate plugs. Hydrate dissociation is highly endothermic. If heat transfer through the pipeline insulation layer from the surroundings is low, the temperature near a dissociating hydrate can drop rapidly. In addition, as gas evolves during hydrate dissociation, JT cooling of the expanding gas is also possible. By either of these mechanisms, additional hydrates and/or ice can form during the dissociation process. For a more complete discussion of gas hydrate structures and properties, the reader is referred to the book by Sloan [1].

Although the design of a unit is intended to prevent hydrate blockages, industry operators must include design and operational provisions for remediation of hydrate blockages. A *hydrate blockage remediation plan* should be developed for a subsea system where hydrate formation is an issue. This tells operators how to spot when a blockage might be occurring and what to do about it. The state of the art would be to have an "online" or "real-time" system using a calculation engine such as OLGA to continuously predict temperatures and pressures in the pipeline and raise an alarm if hydrate formation conditions are detected. Such a system may also be able to pinpoint the most likely blockage location.

Hydrate remediation techniques are similar to hydrate prevention techniques, which include:

- Depressurization from two sides or one side, by reducing pressure below the hydrate formation pressure at ambient temperature, will cause the hydrate to become thermodynamically unstable.
- Thermodynamic inhibitors can essentially melt blockages with direct hydrate contact.
- Active heating is used to increase the temperature to above the hydrate dissociation temperature and provide significant heat flow to relatively quickly dissociate a blockage.
- Mechanical methods such as drilling, pigging, and scraping have been attempted, but are generally not recommended. Methods include

inserting a thruster or pig from a surface vessel with coiled tubing through a workover riser at launchers, and melting by jetting with MEG.

- Replace the pipeline segment.

15.4.1. Depressurization

Depressurization is the most common technique used to remediate hydrate blockages in production systems. Rapid depressurization should be avoided because it can result in JT cooling, which can worsen the hydrate problem and form ice. From both safety and technical standpoints, the preferred method to for dissociating hydrates is to depressurize from both sides of the blockage. If only one side of a blockage is depressurized, then a large pressure differential will result across the plug, which can potentially create a high-speed projectile.

When pressure surrounding a hydrate is reduced below the dissociation pressure, the hydrate surface temperature will cool below the seabed temperature, and heat influx from the surrounding ocean will slowly melt the hydrate at the pipe boundary. Lowering the pressure also decreases the hydrate formation temperature and helps prevent more hydrates from forming in the rest of the line. Because most gas flowlines are not insulated, hydrate dissociation can be relatively fast due to higher heat flux from pipeline surfaces, as compared to an insulated or buried flowline.

The depressurization of the flowlines, a process known as blowdown, creates many operational headaches. Not only does the host facility have to handle large quantities of gas and liquid exiting the flowlines, it must also be prepared to patiently wait until the plug dissociates. Because multiple plugs are common, the process can be extremely long and much revenue is lost. Some subsea system configurations, such as flowlines with a number of low spots, can be extremely difficult to blow down. The best policy is to operate, if at all possible, in a manner that prevents hydrates from forming in the first place. Depressurization may not be effective due to production system geometry; a sufficiently high liquid head in the riser or flowline may prevent depressurization below hydrate conditions. In this case, some methods may be needed to reduce the liquid head. If additional equipment is needed to perform depressurization or remediation, equipment mobilization needs to be factored into the total downtime. System designers need to evaluate the cost/benefit of including equipment in the design for more efficient remediation versus using higher remediation times.

15.4.2. Thermodynamic Inhibitors

Thermodynamic inhibitors can be used to melt hydrate blockages. The difficulty of applying inhibitors lies in getting the inhibitor in contact with the blockage. If the injection point is located relatively close to the blockage, as may be the case in a tree or manifold, then simply injecting the inhibitor can be effective. Injecting inhibitor may not always help with dissociating a hydrate blockage, but it may prevent other hydrate blockages from occurring during remediation and restart.

If the blockage can be accessed with coiled tubing, then methanol can be pumped down the coiled tubing to the blockage. In field applications, coiled tubing has reached as far as 14,800 ft in remediation operations, and industry is currently targeting lengths of 10 miles.

15.4.3. Active Heating

Active heating can be used to remediate hydrate plugs by increasing the temperature and heat flow to the blockage; however, safety concerns arise when applying heat to a hydrate blockage. During the dissociation process, gas will be released from the plug. If the gas is trapped within the plug, then the pressure can build and potentially rupture the flowline. Heat evenly applied to a flowline can provide safe, effective remediation.

Active heating can remediate a blockage within hours, whereas depressurization can take days or weeks. The ability to quickly remediate hydrate blockages can enable less conservative designs for hydrate prevention.

15.4.4. Mechanical Methods

Pigging is not recommended for removing a hydrate plug because the plug can become compressed, which will compound the problem. If the blockage is complete, it will not be possible to drive a pig through. For a partial blockage, pigging may create a more severe blockage.

Coiled tubing is another option for mechanical hydrate removal. Drilling a plug is not recommended because it can cause large releases of gas from the blockage. Coiled tubing can be inserted through a lubricator. Coiled tubing access, either at the host or somewhere in the subsea system, should be decided early in the design phase.

15.4.5. Safety Considerations

Knowledge of the location and length of a hydrate blockage is very important in determining the best approach to remediation, although the methodology is not well defined, This information facilitates both safety considerations in terms of distance from the platform and time necessary to dissociate the blockage.

When dissociating a hydrate blockage, operators should assume that multiple plugs may exist both from safety and technical standpoints. The following two important safety issues should be kept in mind:

- Single-sided depressurization can potentially launch a plug like a high-speed projectile and result in ruptured flowlines, damaged equipment, release of hydrocarbons to the environment, and/or risk to personnel.
- Actively heating a hydrate blockage needs to be done such that any gas released from the hydrate is not trapped.

15.5. HYDRATE CONTROL DESIGN PHILOSOPHIES

Steady-state temperature calculations from the flow assurance process are used to indicate the flow rates and insulation systems that are needed to keep the system above the hydrate formation temperature during normal operation. Transient temperature calculations are used to examine the conditions of transient action such as start-up and shutdown process. It is essential for each part of the system to have adequate cooldown time. Dosing of hydrate inhibitor is the way hydrate formation is controlled when system temperatures drop into the range in which hydrates are stable during the transient actions. Dosing calculations of thermodynamic hydrate inhibitor indicate how much inhibitor must be added to the produced water at a given system pressure. The inhibitor dosing requirements are used to determine the requirements for inhibitor storage, pumping capacities, and number of umbilical lines in order to ensure that the inhibitor can be delivered at the required rates for treating each subsea device during start-up and shut-down operations.

15.5.1. Selection of Hydrate Control

Injection rates of thermodynamic inhibitors are a function of water production and inhibitor dosage. Inhibitor dosage is a function of design temperature and pressure, and produced fluid composition. Water production multiplied by the methanol dosage is the inhibitor injection rate, which will change throughout field operating life due to typically decreasing operating pressures and increasing water production. Hydrate control strategies for gas and oil systems are different. Gas systems are designed for continuous injection of a hydrate inhibitor. Water production is small, typically only the water of condensation. Inhibitor requirements are thus relatively small, of the order of 1 to 2 bbl MeOH/mmscf. Oil systems produce free water. Continuous inhibition is generally too expensive so some alternative control system is adopted. Insulation is often used to control hydrates during normal production, and some combination of blowdown and methanol injection is used during start-up and shutdown. But insulation is costly, and the more serious drawback may be lengthening of the hydrate plug remediation time, because insulation limits the heat transfer required to melt the hydrate solid, for cases in which hydrates have formed.

Table 15-3 provides a summary of the applications, benefits, and limitations of the three classes of chemical inhibitors.

Table 15-3 Summary of Chemical Inhibitors Applications, Benefits, and Limitations [11]

Thermodynamic Hydrate Inhibitors	Kinetic Hydrate Inhibitors	Anti-Agglomerate Inhibitors
Applications		
• Multiphase • Gas and condensate • Crude oil	• Multiphase • Gas and condensate • Crude oil	• Multiphase • Condensate • Crude oil
Benefits		
• Robust and effective • Well understood • Predictable • Proven track record	• Lower OPEX/CAPEX • Low volumes (<1 wt%) • Environmentally friendly • Nontoxic • Tested in gas systems	• Lower OPEX/CAPEX • Lower volumes (<1 wt%) • Environmentally friendly • Nontoxic • Wide range of subcooling
Limitations		
• Higher OPEX/ CAPEX • High volumes (10 to 60 wt %) • Toxic/hazardous • Environmentally harmful • Volatile losses to vapor	• Limited subcooling (<10°C) • Time dependency • Shutdowns • System-specific testing • Compatibility	• Time dependency? • Shutdowns? • Restricted to lower water cuts • System-specific testing • Compatibility

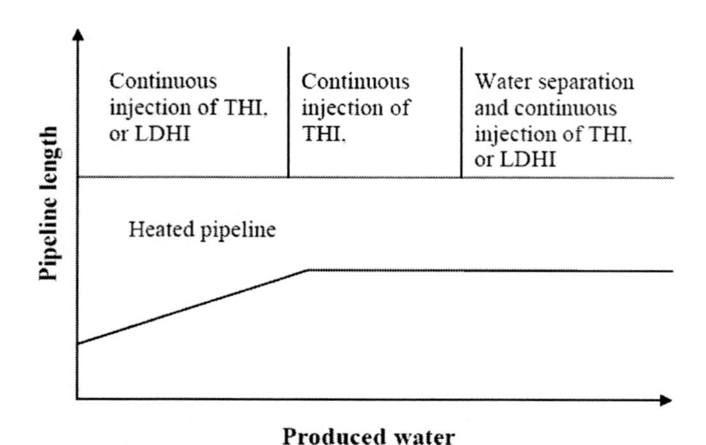

Figure 15-11 Hydrate Control Method for Different Water Cut and Pipeline Length

Figure 15-11 shows the relationship between hydrate control options for different water cuts and pipeline lengths. LDHIs offer the ability to treat production with higher water rates because they are injected at lower quantities. Subsea separation of bulk water in combination with either thermodynamic or low-dosage inhibitors also can enable developments with potentially high water production. Basically, THIs, LDHIs, low-pressure operation, subsea processing (water separation), and thermal management are used in fields for hydrate prevention in subsea systems.

The selection of hydrate mitigation and remediation strategies is based on technical and economic considerations and the decision is not always clear-cut. While continuous injection of THIs is expected to remain the most economic and technically feasible approach to hydrate control, LDHIs or subsea processing will offer advantages for some developments. Whatever the hydrate control strategy, these decisions are critical in the early design stage because of the many impacts on both the subsea and topside equipment selection and design.

The main benefits of the thermodynamic hydrate inhibitors are their effectiveness, reliability (provided sufficient quantities are injected), and proven track records. However, these benefits are outweighed by significant limitations, including the high volumes, high associated costs (both CAPEX and OPEX), and their toxicity and flammability. In addition, they are harmful to the environment and significant disposal into the environment is prohibited.

Kinetic hydrate inhibitors are injected in much smaller quantities compared to thermodynamic inhibitors and therefore offer significant potential costs savings, depending on the pricing policies of major chemical suppliers. They are also typically nontoxic and environmentally friendly. Moreover, considerable field experience is now available following a number of successful trials. However, they have some important limitations, including restrictions on the degree of subcooling (typically only guaranteed for less than 10°C) and problems associated with residence times, which as implications for shutdowns. In addition, the effectiveness of KHIs appears to be system specific, meaning that testing programs are required prior to implementation. Unfortunately adequate testing can require appreciable quantities of production fluids, which may not be available, particularly for new field developments. Furthermore, KHIs can interact with other chemical inhibitors (e.g., corrosion inhibitors), and testing programs need to account for this too. Finally, there are no established models for predicting the effectiveness of KHIs, which presents difficulties for field developers considering the application of these chemicals.

The benefits and limitations of anti-agglomerates are largely similar to those for KHIs, although AAs do not have the same subcooling limitations. However, there is uncertainty about the effectiveness of AAs under shut-down or low flow rate conditions and it is postulated that agglomeration may still proceed. In addition, the one major limitation of AAs compared to KHIs or THIs is that they are limited to lower water cuts due the requirement for a continuous hydrocarbon liquid phase. Finally, compared to both THIs and KHIs, field experience with AAs appears to be lacking, which is reflected by the relatively small number of publications available in the open literature.

Thermal management can assist with maintaining some room for response by assisting with adequate temperatures for both hydrate and paraffin control. Passive thermal management is by insulation or burial. Pipe-in-pipe and bundled systems can extend the production cooldown time before continuous hydrate inhibition is required. However, there may not be sufficient thermal capacity to provide the necessary cooldown time for a shutdown greater than 4 to 8 hours. Phase change material insulation systems that provide heat "storage" are beginning to appear as commercial systems. Burial can provide extended cooldown times due to the thermal mass of the soil. Warm-up and cooldown times can be optimized with insulation and thermal mass.

Both electrical and heating media systems belong to the active thermal management process. Heating media systems require pipe-in-pipe or bundle flowline designs to provide the flow area required for the circulation of the heating media. When design, installation, and corrosion management issues are successful, the heating media systems can be reliable. However, the heat transported is still limited by the carrier insulation and heat input at the platform.

15.5.2. Cold Flow Technology

Cold flow technology is a completely new concept that is designed to solve hydrate blockage problems during steady-state operating conditions. It has been recently developed and is owned by SINTEF. It is different from the chemical-based technologies (THIs, LDHIs) and insulation/heating technologies for hydrate protection, and concerns the cost-effective flow of oil, gas, and water mixtures in deepwater production pipelines, from wellhead to processing facility without using chemicals to prevent hydrate or wax deposition. In cold flow technology the hydrate formation occurs under controlled conditions in specialized equipment. The formed hydrate particles flow easily with the bulk fluid mixture as slurry and do not form wall deposits or pipeline blockage. Cold flow technology is environmentally friendly because of the envisaged reduction in the use of bulk and specialty chemicals. This technology provides an exciting challenge and could result in significant economic savings.

Figure 15-12 shows a flow diagram for the cold flow hydrate technology process. The flow enters from a wellbore (WELL), and leaves through a cold flow pipeline (CFP). This system has several main process units: a wellhead unit (WHU), a separator unit (SU), a heat exchanger unit (HXU), and a reactor unit (RU). The term G/L refers to gas/liquid. Depending on the water cut and gas/oil ratio (GOR) and other fluid properties, more than one set of SU+HXU+RU may be required.

Figure 15-13 is a schematic of the SINTEF-BP cold flow project, which develops a very robust process for long-distance transportation of an

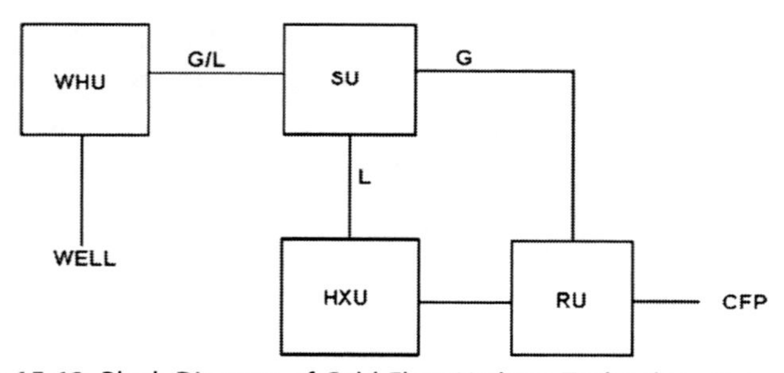

Figure 15-12 Block Diagram of Cold Flow Hydrate Technology Process [12]

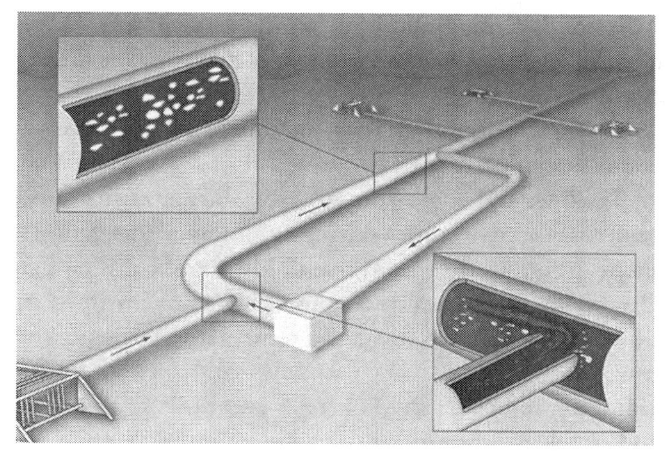

Figure 15-13 Schematic of a Cold Flow Project [13]

unprocessed well stream containing water by converting the water to a very stable and transportable gas hydrate. This process has been successfully proven in a 1-in. flow loop facility, operated with a variety of field fluids. This technology can reduce CAPEX by 15 to 30% for subsea tie-backs.

15.5.3. Hydrate Control Design Process

The hydrate control design process in subsea hydrocarbons system is summarized as follows:

- Determine operating conditions: pressure, temperature, GOR, water cut.
- Obtain good representative samples of oil, gas, and water.
- Measure chemical composition and phase behavior.
- Analyze reservoir fluids to determine hydrate formation conditions.
- Perform hydrate prediction calculations.
- Estimate the effects of insulation and thermodynamic inhibitors.
- Determine thermodynamic hydrate inhibitor dosing and sizes of umbilical and inhibitor storage; consider the use of LDHIs.

15.5.4. Hydrate Control Design and Operating Guidelines

The guidelines utilized for hydrate control in subsea hydrocarbon system designs and operations are summarized as follows:

- Keep the entire production system out of the hydrate formation envelope during all operations. Current knowledge is not sufficient to design a system that can operate in the hydrate region without hydrate or blockage formation.

- Inject thermodynamic inhibitors at the subsea tree to prevent the formation of hydrates in the choke and downstream during transient operations.
- Use LDHIs only for transient start-up/shutdown operations and not for continuous operation.
- Insulate flowlines and risers from heat loss during normal operation and to provide cooldown time during shutdown. Insulation of subsea equipment (trees, jumpers, and manifolds) should also be done.
- Consider wellbore insulation to provide fast warm-up during restart operations and to increase operating temperatures during low flow rate operation.
- Determine minimum production rates and flowing wellhead temperatures and check consistency with technical and economic criteria.
- Establish well and flowline start-up rates to minimize inhibitor injection while assuring that the system warms in an acceptable amount of time.
- Ramp-up well production rates sufficiently fast to outrun hydrate blockage formation in wellbores.
- Provide system design and operating strategies to ensure the system can be safely shut down.
- Monitor water production from individual wells.
- Locate SCSSVs at a depth where the geothermal temperature is higher than the hydrate temperature at shut-down pressure.
- Remediate hydrate blockages via depressurization or heating.

15.6. RECOVERY OF THERMODYNAMIC HYDRATE INHIBITORS

Thermodynamic hydrate inhibitors (methanol, glycols) are widely used as a primary hydrate inhibitor in subsea systems. For projects producing oil, the amount of methanol is limited in terms of the oil's quality. The issue of an oxygenated solvent limit for glycols in oil is still under discussion. Even though the recovery units of methanol and glycol are improving, the units require appreciable heat to recover the THIs, and scaling in the methanol and glycol stills can generate operations challenges, especially when the produced water chemistry is not available from the reservoir appraisal during the design phase. Glycol recovery units can be designed to remove the salts that have traditionally limited the glycol quality. To reduce the methanol and glycol to the requirements of oil quality, crude washing requires large volumes of water that must be treated to seawater injection quality. The

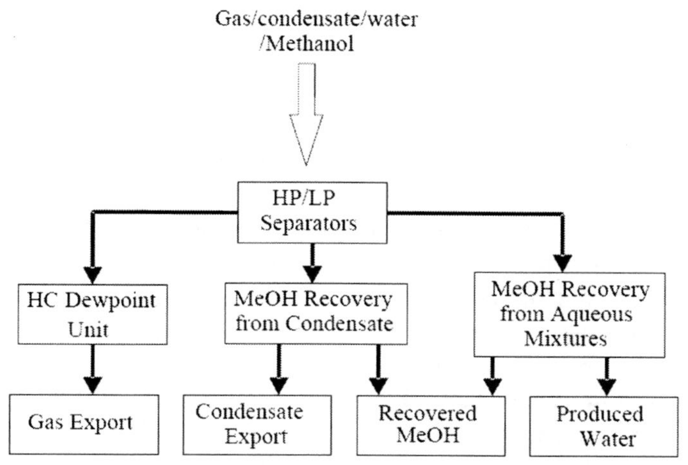

Figure 15-14 MeOH Recovery Flowchart of Offshore Gas Facility

recovery units and wash units have a significant weight and can affect the project's design. These units are large and heavy.

Figure 15-14 shows a typical flowchart for an offshore gas production facility with methanol as the hydrate inhibitor. As shown in the figure, the fluids from the subsea flowlines enter the separation vessels and are then distributed in three separate phases:

- Gas export stream to onshore gas plant in vapor phase;
- Condensate or oil export stream to shore in liquid hydrocarbon phase;
- Produced water stream in aqueous phase.

The methanol in the hydrocarbon vapor phase is recovered by adsorption, and the methanol in the hydrocarbon liquid phase is recovered by water wash, mechanical separation, or a combination of the two. Most of the injected methanol is in the aqueous phase. A methanol tower is used to recover the methanol from the aqueous phase. More than 96% of the methanol injected can be recovered if good engineering judgment and experience are applied.

The aqueous phase mixture contains most of the methanol injected to the system. Methanol is a highly polar liquid and is fully miscible with water; therefore, the recovery of methanol from water is achieved by distillation rather than phase separation.

Methanol is fully miscible with water, while the solubility of methanol in hydrocarbons is very small. Therefore, water (the solvent) can be used to extract methanol (the solute) from the hydrocarbon condensate (the feed) efficiently. Because water is the solvent, this extraction process is called water wash.

Methanol losses in the hydrocarbon liquid phase are difficult to predict. Solubility of methanol is a strong function of both the water phase and hydrocarbon phase compositions. The recovery of the dispersed and dissolved methanol from the feed to the condensate stabilizer can be achieved by mechanical separation via coalescence, liquid–liquid extraction, or a combination of the two depending on the feed characteristics, weight, space, and cost of the entire subsea topside operation. Mechanical separation can separate the dispersed methanol from the hydrocarbon liquid phase to a certain extent. Liquid–liquid extraction can recover the smaller droplets of dispersed methanol and the dissolved methanol from the hydrocarbon liquid phase. However, liquid–liquid extraction is more costly than mechanical separation. The solvent needs to be regenerated and reused by the extraction tower. It increases the flow rate and the heating and cooling duties of the methanol distillation tower. For subsea operation, the treating facility's space requirements, weight, and cost have to be considered together to determine which system to use to recover the methanol from the liquid hydrocarbon phase.

Methanol in the hydrocarbon vapor phase can be recovered by adsorption. If the hydrocarbon vapor passes through a cryogenic system, the methanol in the hydrocarbon vapor phase condenses into the liquid phase and can be recovered.

REFERENCES

[1] E. Sloan, Offshore Hydrate Engineering Handbook, SPE Monograph vol. 21 (2000).
[2] S.E. Lorimer, B.T. Ellison, Design Guidelines for Subsea Oil Systems, Facilities 2000: Facilities Engineering into the Next Millennium (2000).
[3] B.T. Ellison, C.T. Gallagher, S.E. Lorimer, The Physical Chemistry of Wax, Hydrates, and Asphaltene, OTC 11963 (2000).
[4] B. Edmonds, R.A.S. Moorwood, R. Szczepanski, X. Zhang, Latest Developments in Integrated Prediction Modeling Hydrates, Waxes and Asphaltenes, Focus on Controlling Hydrates, Waxes and Asphaltenes, IBC, Aberdeen, 1999, October.
[5] B. Ellision, C.T. Gallagher, Baker Petrolite Flow Assurance Course, Texas, Houston, 2001.
[6] E.G. Hammerschmidt, Possible Technical Control of Hydrate Formation in Natural Gas Pipelines, Brennstoff-Chemie vol. 50 (1969) 117–123.
[7] A.P. Mehta, P.B. Hebert, E.R. Cadena, J.P. Weatherman, Fulfilling the Promise of Low Dosage Hydrate Inhibitors: Journey from Academic Curiosity to Successful Field Implementation, OTC 14057 (2002).
[8] S. Cochran, Hydrate Control, Remediation Best, Practices in Deepwater Oil Developments, OTC 15255 (2003).
[9] S. Cochran, R. Gudimetla, Hydrate Management: Its Importance to Deepwater Gas Development Success, World Oil vol. 225 (2004) 55–61.

[10] F.M. Pattee, F. Kopp, Impact of Electrically-Heated Systems on the Operation of Deep Water Subsea Oil Flowlines, OTC11894, Offshore Technology Conference, Houston, Texas, 2000. May.
[11] P.F. Pickering, B. Edmonds, R.A.S. Moorwood, R. Szczepanski, M.J. Watson, Evaluating New Chemicals and Alternatives for Mitigating Hydrates in Oil & Gas Production, IIR Conference, Aberdeen, Scotland, 2001.
[12] J.S. Gudmundsson, Cold Flow Hydrate Technology, 4th International Conference on Gas Hydrates, Yokohama, Japan (2002). May.
[13] D. Lysne, Ultra Long Tie-Backs in Arctic Environments with the SINTEF-BP Cold Flow Concept, Oil and Gas Developments in Arctic and Cold Regions, U.S.–Norway Oil & Gas Industry Summit, Houston, Texas, 2005. March.

CHAPTER 16

Wax and Asphaltenes

Contents

16.1. INTRODUCTION

Waxes or paraffins are typically long-chain, normal alkane compounds that are naturally present in crude oil. When the temperature drops, these compounds

Subsea Engineering Handbook
ISBN 978-0-12-397804-2, doi:10.1016/B978-0-12-397804-2.00016-3

can come out of the oil and form waxy and elongated crystals. If the control of wax deposition is not effective, the waxy deposits can build up significantly with time and cause disruption of production, reduction of throughput, and even complete blockage of the flowlines. Subsea production facilities and pipelines are very susceptible to wax deposits and asphaltene precipitates induced by the lower temperature and decreasing pressure environment.

Asphaltenes are a component of the bitumen in petroleum and are usually black, brittle coal-like materials. They also exist as a thick black sludge. Asphaltenes can flocculate and form deposits under high shear and high velocity flow conditions. Asphaltenes are insoluble in nonpolar solvents, but are soluble in toluene or other aromatics-based solvents. Asphaltene deposits are very difficult to remove once they occur. Unlike wax deposits and gas hydrates, asphaltene formation is not reversible. Frequently asphaltene deposition occurs with wax deposition, and makes the combined deposits very hard and sticky and difficult to remove.

16.2. WAX

16.2.1. General

Wax varies in consistency from that of petroleum jelly to hard wax with melting points from near room temperature to more than 100°C. Wax has a density of around 0.8 g/cm^3 and a heat capacity of around 0.140 W/(m·K). Paraffinic hydrocarbon fluids can cause a variety of problems in a production system ranging from solids' stabilized emulsions to a gelled flowline. Although the pipeline is thermal insulated, it will ultimately cool to the ambient temperature given sufficient time in a shutdown condition. Problems caused by wax occur when the fluid cools from reservoir conditions and wax crystals begin to form. Wax is a naturally occurring substance in most crude oils and, depending on its characteristics, it may cause problems. Wax deposition on the pipeline walls increases as the fluid temperature in the pipeline decreases. The deposits accumulate on the pipe walls and over time may result in a drastic reduction of pipeline efficiency. The temperature at which crystals first begin to form is called the cloud point or wax appearance temperature (WAT). At temperatures below the cloud point, crystals begin to form and grow. Crystals may form either in the bulk fluid, forming particles that are transported along with the fluid, or deposit on a cold surface where the crystals will build up and foul the surface. WAT is calculated from a proprietary thermodynamic model calibrated to field data and using input PVT tests and proprietary high-temperature gas chromatography measurements. Systems with pipe wall temperatures below the WAT need a wax management strategy. It is important to note that the WAT indicates the temperature at which deposition will begin, but it does not indicate the amount of wax that will be deposited or the rate at which it will be deposited. Deposition rate measurements should be made in previous studies and will be used in future modeling. The pour point temperature is the temperature at which the oil stops flowing and gels.

Gel formation and deposition are the main problems that wax may cause in a production system. A crude oil gel forms when wax precipitates from the oil and forms a 3D structure spanning the pipe. This does not occur while the oil is flowing because the intermolecular structure is destroyed by shear forces. However, when the oil stops flowing, wax particles will interact, join together, and form a network resulting in a gel structure if enough wax is out of solution.

Wax deposition results in flow restrictions or possibly a blockage of a pipeline. Complete blockage of flow due to deposition is rare. Most pipeline blockages occur when a pig is run through a pipeline after deposition has occurred and a significant deposit has built up. In this situation the pig will continue to scrape wax from the pipe wall and build up a viscous slug or candle in front of the pig. However, if the candle becomes too large there will be insufficient pressure for the pig to move. When this occurs the pig becomes stuck and mechanical intervention to remove the candle will be necessary before the pig can be moved.

In subsea systems, the following issues caused by waxes should be addressed by flow assurance analyses:

- Deposition in flowlines is gradual with time but can block pipelines.
- Gelation of crude oil can occur during shutdown.
- High start-up pressures and high pumping pressures occur as a result of the higher viscosity.
- Insulation for pipeline increases capital expenses.
- Wax inhibitors increase operational expenses.
- Pigging operation in offshore environment is more difficult than that in onshore.
- Wax handling in surface facilities requires a higher separator temperature.

16.2.2. Pour Point Temperature

The pour point is an indicator of the temperature at which oil will solidify into a gel. At the pour point, the fluid still "pours" under gravity; at the next lower measurement temperature (3°C lower), the fluid has gelled and does not pour. In environments with ambient and fluid temperatures above the pour point, no pour point strategy is necessary. In environments where the ambient or fluid temperature is at or below the pour point temperature, a specific pour point strategy is required to maintain flow ability of the fluid. So the environmental temperature could be analyzed, in general, such that the analysis was conservative, 100-year extreme minimum seabed temperatures were used along the pipeline route.

The pour point temperature can be measured using ASTM methods and estimated using high-temperature gas chromatography (HTGC) date correlations.

There are currently at least five ASTM pour point protocols, most of which are designed for petroleum products rather than crude oils. The two most widely used are ASTM D97-09 [1] and D5853-09 [2]. In general, these tests require 100 to 250 mL of fluid. ASTM D97 is an older oil-products pour point protocol that requires heating to a prescribed temperature (60°C) and cooling in a series of baths until the oil gels or solidifies. Shell has devised a "mini" D97 pour point that uses a 30-mL sample. The "mini" pour point has been calibrated using the ASTM D97 method. The repeatability (95% confidence limits, same lab) of the D97 pour points (measured on fuel oils) is reported by ASTM to be 5°F (2.8°C); the reproducibility (95% confidence limits, different labs) is 12°F (6.7°C).

ASTM D5853 [2] is designed specifically for crude oils and recognizes the potentially strong effect of thermal history on the gelling temperature. Two separate heating and cooling protocols are employed in order to see the effects of two substantially different thermal histories. The minimum pour point protocol requires heating to 105°C and cooling in air for 20 min at room temperature before entering a series of cooling baths. This protocol lowers the measured pour point in two ways: (1) It ensures that all wax is in solution before cooling is started, and (2) it cools relatively quickly, potentially "outrunning" the wax kinetics and reaching a lower temperature before the gel forms. The maximum pour point protocol requires heating to 60°C or less, cooling in air for 24 hr, and a short reheat to 45°C followed by the cooling baths. This protocol raises the measured point by allowing a long time for "seed crystals" to form at room temperature, which in turn decreases the time for gel formation. ASTM D5853 is difficult to adapt to small volumes; therefore, we have no "mini" technique for ASTM D5853. The repeatability of D5853 using crude oils is reported by ASTM to be 6 to 12°F (3.3 to 6.6°C), but the reproducibility is as high as 32 to 40°F (17.8 to 22.2°C).

16.2.3. Wax Formation

The wax deposit is complex in nature, comprised of a range of normal paraffins of different lengths, some branched paraffins, and incorporated oil. The buildup of wax over time can eventually reach a point where flow rates are restricted. Wax deposits in oil production flowlines are primarily comprised of paraffins of C^{7+} in length. The deposition of paraffins is controlled by temperature. As the temperature in a system drops, paraffins that are in the liquid phase begin to come out of solution as solids. Wax deposits form at the wall of the pipe where the temperature gradient is at its highest.

In addition to wax deposition, the formation of sufficient wax solids can cause oil to "gel" at sufficiently low temperature during a shutdown of the system. Once this occurs, it is difficult or even impossible to restart flow in the system due to the very high viscosity of the gelled oil. The wax properties of oils have been characterized by cloud point and pour point measurements. The cloud point essentially measures the point at which wax first visibly comes out of solution as oil is cooled at atmospheric pressure. The pour point is the temperature at which the oil ceases to flow when the vessel is inverted for 5 sec. These measurements give a general understanding of the temperatures at which wax deposition will become a problem and when crude oil gelling will become a problem.

The key to wax deposition prediction is a precise analysis of the concentration of the normal paraffins in an oil sample, which is carried out using the HTGC technique. The paraffin composition data are used to construct a thermodynamic model for prediction of wax deposition rates in the flowline as well as for predicting the cloud point and pour point dependence on pressure. The thermodynamic model may be combined with the model of flowline using software such as HYSYS, PIPESIM, or OLGA to predict where wax deposits will occur, how fast wax will accumulate, and the frequency at which the line must be pigged.

In contrast to hydrates, wax deposits slowly and can be controlled by controlling the system temperature and temperature differential at the pipe wall. Cloud points for deepwater GoM crude oils generally fall in the range of 80 to 100°F. If the system is operated at a temperature approximately 10 to 20°F above the cloud point, wax will not deposit. A "rule of thumb" for the deposition temperature that has been frequently used is *cloud point + 15°F.* Although this can usually be achieved for the wellbore and subsea tree, it is often not possible to operate at this high a temperature in the flowline. The

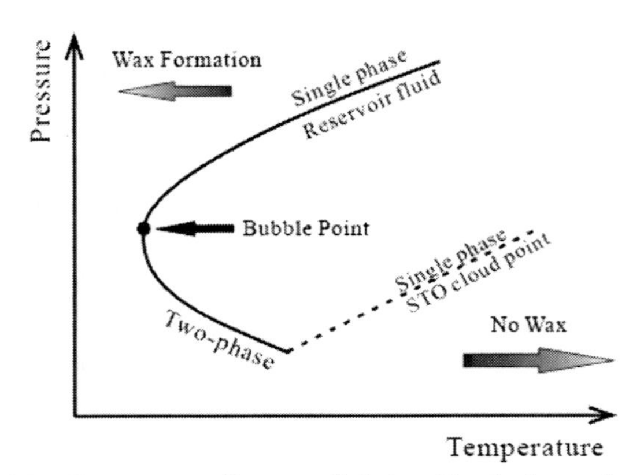

Figure 16-1 Temperature/Pressure Relationships in Formation of Wax

host arrival temperature may be limited due to processing concerns or by the temperature rating of equipment. During late life, the reservoir temperature may have dropped to the point at which host arrival temperatures are substantially below the cloud point leading to significant deposition. Figure 16-1 shows the relationship of temperature and pressure to wax formation. The left side of the curve shows the area where waxes will deposit, whereas the right side of the curve indicates where waxes will not form.

Wax formed in wellbores can only be removed by a process such as coiled tubing, which is prohibitively expensive for subsea wells. As a result, it is important to control the temperature of the well tubing, tree, and other components that cannot be pigged (such as jumpers) above the point at which wax deposition occurs. This critical wax temperature may be chosen as cloud point + 15°F. The flowing wellhead temperatures and pressures from flow assurance analysis are used to check that the tubing and tree remain above this critical wax deposition temperature. The need to prevent wax deposition in the wellbore, tree, and jumpers may set a minimum late life production rate, based on the temperature predictions from flow assurance.

High pour point oils present a potentially serious problem somewhat similar to that for formation of hydrates. Wax formation during shutdown can be sufficient to make line restart difficult or even impossible. Even though temperatures may be well above the pour point during steady-state operation, it is impossible to know when a shutdown may occur.

Wax deposition will occur once the oil has fallen below the cloud point if there is a negative temperature gradient between the bulk oil and a surface. However, the Deepstar research [3] has shown that wax deposition also can occur above the dead oil cloud point in some systems. Therefore, to prevent

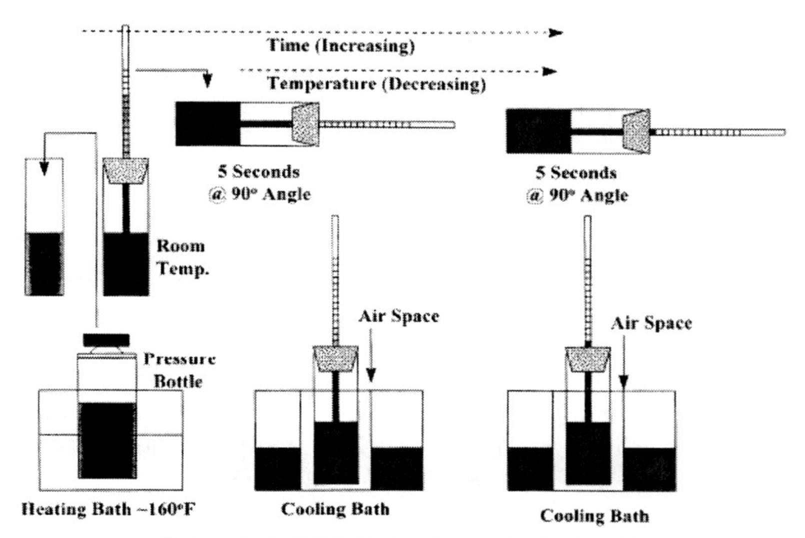

Figure 16-2 ASTM D5853 Pour Point Testing [4]

wax deposition, the system temperature should remain greater than 9°C above the dead oil cloud point. Experimentally the tendency of oil to deposit and the rate of deposition can be measured by placing a cold surface in contact with warm oil. Experimental systems include cold fingers, coaxial shearing cells, and pipe loops. The cold finger consists of a test tube–shaped metal finger cooled by flowing chilled fluid through the finger, and a heated stirred container for an oil sample. The coaxial shearing cell is similar to a cold finger but the finger rotates to create uniform shear on the surface. A pipe loop is a pipe-in-pipe heat exchanger in which the cold fluid is pumped through the shell side and the oil is heated and pumped through the tube side. It might be expected that the pipe loop would be the preferred method due to its geometric similarity to a pipeline. However, if an actual field sample is to be used, none of the methods properly simulate the system since dynamic similitude is impossible to achieve without building a system of the same size. However, each method can be used to measure the wax flux to the surface. With careful analysis the data from any laboratory method can be used to make predictions about field deposition.

As oil cools far below the cloud point it may begin to gel. While the amount of wax out of solution needed to form a gel structure varies considerably, 2% paraffin is used as a useful rule of thumb. The typical method of measuring a crude oil pour point is ASTM D5853-09 shown in Figure 16-2, which specifies the conditioning required for obtaining reproducible values. Because the cooling rate is not controlled in this method, results may vary from lab to lab, but in general the results are accurate to within ±5°F.

16.2.4. Gel Strength

The pour point is used to determine whether or not a crude oil will form a gel; the gel strength is defined as the shear stress at which gel breaks. The gel strengths measured in a rheometer are typically of the order of 50 dyne/cm^2. However, the sample in a rheometer will yield uniformly due to the small gap. In a pipe the sample will not yield uniformly. A pressure front will propagate through the oil and the oil near the inlet will yield and begin to flow when a flowline is restarted. Because this problem is not tractable due to the compressibility and non-Newtonian behavior of a crude oil gel, predictions continue to be made from rheometer measurements or short pipe sections.

Start-up pressure predictions can be made using following equation:

$$\Delta P = 4\tau_y|_{\text{wall}}\frac{L}{D} \tag{16-1}$$

where $\Delta T$is the pressure drop, $\tau_y$is the yield stress, L is the length of the pipe, and D is the diameter of the pipe. Equation (16-1) assumes that the whole flowline is gelled, will yield simultaneously, and that the yield at the wall is the appropriate parameter to consider. This method typically overpredicts the actual restart pressures.

For export pipelines laboratory pour point methods are very good indicators of potential problems since the export pipelines and blowdown flowlines are full of dead crude. If a flowline were not blown down, then the gel situation would improve due to an increase in light components in the oil.

16.2.5. Wax Deposition

The deposition tendency and rate can also be predicted adequately by calculating the rate of molecular diffusion of wax to the wall by the following equation:

$$\frac{dm}{dt} = -\rho D_m A \frac{dC}{dr} \tag{16-2}$$

where

m: mass of deposit, kg;

ρ: density of wax, kg/m^3;

D_m: molecular diffusion constant, m^2/s;

A: deposition area, m^2;

C: concentration of wax, %;

r: radial position, m.

The radial concentration gradient can easily be calculated if broken into two components by applying the chain rule as shown in following equation:

$$\frac{dm}{dt} = -\rho D_m A \frac{\partial C}{\partial T} \frac{dT}{dr} \quad (16\text{-}3)$$

where T is temperature. The concentration gradient may be calculated from the wax concentrations predicted by a thermodynamic model for a range of temperatures.

16.2.6. Wax Deposition Prediction

Waxes are more difficult to understand than pure solids because they are complex mixtures of solid hydrocarbons that freeze out of crude oils if the temperature is low enough. Waxes are mainly formed from normal paraffins but isoparaffins and naphthenes are also present and some waxes have an appreciable aromatic content. The prediction of wax deposition potentials comprises two parts: the cloud point temperature or wax appearance temperature, and the rate of deposition on the pipe wall. The cloud point is the temperature below which wax crystals will form in oil, and the rate of deposition determines the wax buildup rate and pigging frequency requirement. The cloud point temperature can be satisfactorily predicted using thermodynamic models. These models require detailed hydrocarbon compositions. Figure 16-3 shows the comparison between experimental data and model simulation. The tendency of the pressure effect is accurately predicted.

The prediction of deposition rates depends on a thermodynamic model for the amount of wax in oil and a diffusion rate model [5]. The accuracy of the predictions is not satisfactory at present, but some industrial JIPs (e.g., University of Tulsa) are investigating to improve the accuracy. The wax models used include these:

- Tulsa University model;
- Olga wax module;
- BPC model.

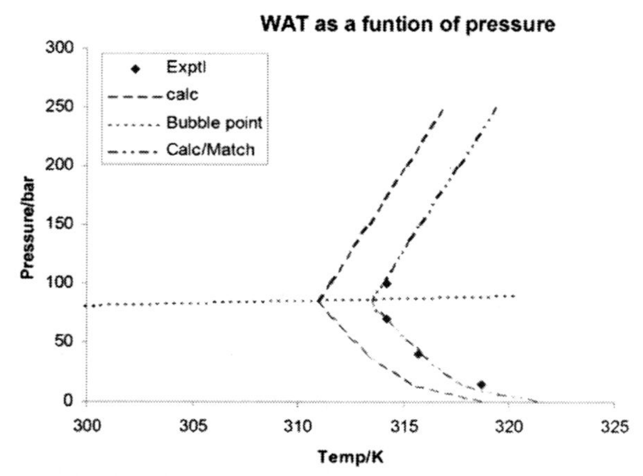

Figure 16-3 Comparison of WAT between Experimental Data and Calculation Results at Different Pressures [6]

In the wax prediction process, the pipeline temperature profile for various insulation systems with different flow velocities should be analyzed and compared with the WAT and pour point temperature to identify the location of wax appearance. Also whether the pipeline is buried or not should be considered due to the thermal insulation properties of a buried pipeline.

16.3. WAX MANAGEMENT

16.3.1. General

Several methods of wax control and management are practiced by production operations, but the transportation of crude oil over a long distance in subsea system demands significant planning and forethought. The wax management strategy generally is based on one or more of the following methods:

- Flowline pigging;
- Thermal insulation and pipeline heating;
- Inhibitor injection;
- Coiled-tubing technology.

The most common method of wax control is flowline pigging if the wax has formed. The solid deposit is removed by regularly removing the wax layer by the scouring action of the pig. Chemical inhibitors can also help control wax deposition, although these chemicals are not always effective and tend to be expensive. In cases where flowline pigging of the production lines is not practical, particularly for subsea completions, wax deposition is controlled by maintaining fluid temperatures above the cloud point for the whole flowline.

Coiled-tubing technology has become an important means of conducting well cleanup procedures. This technology involves the redirection of well production to fluid collection facilities or flaring operations while the coiled tubing is in the well. Heavy coiled tubing reels are placed at the wellhead by large trucks, the well fluids are diverted, and high-pressure nozzles on the end of the coiled tubing are placed in the well. Tanker trucks filled with solvent provide the high-pressure pumps with fluids that are used to clean the well tubing as the coiled tubing is lowered into the well. The value of this method is apparent in many areas of the world, since certain integrated production companies maintain a fleet of coiled tubing trucks that remain busy a large percentage of the time.

16.3.2. Thermal Insulation

Good thermal insulation can keep the fluid above the cloud point for the whole flowline and thus eliminate wax deposition totally. Although line heaters can be successfully employed from the wellhead to other facilities, the physical nature of the crystallizing waxes has not been changed. This can be a problem once the fluids are sent to storage, where the temperature and fluid movement conditions favor the formation of wax crystals and lead to gels and sludge.

16.3.3. Pigging

The rate of deposition can be reduced by flowline insulation and by the injection of wax dispersant chemicals, which can reduce deposition rates by up to five times. However, it must be emphasized that these chemicals do not completely stop the deposition of wax. Therefore, it is still necessary to physically remove the wax by scouring the flowline. To facilitate pigging, a dual-flowline system with a design that permits pigging must be built. Pigging must be carried out frequently to avoid the buildup of large quantities of wax. If the wax deposit becomes too thick, there will be insufficient pressure to push the pig through the line as the wax accumulates in front of it. Pigging also requires that the subsea oil system be shut down, stabilized by a methanol injection and blowdown, and finally restarted after the pigging has been completed. This entire process may result in the loss of 1 to 3 days of production. The deposition models created based on the fluids analysis work and the flow assurance calculations are the key to establishing pigging intervals that are neither too frequent to be uneconomical or too infrequent to run the risk of sticking the pig in the flowline. OLGA models the "slug" of wax pushed ahead of the pig.

16.3.4. Inhibitor Injection

Chemical inhibition is generally more expensive than mechanical pigging, although the cost comparison depends on pigging frequency requirements, chemical inhibition effectiveness, and many other factors [7]. Chemical inhibitors can reduce deposition rates but rarely can eliminate deposition altogether. Therefore, pigging capabilities still have to be provided as a backup when chemical inhibition is used. The chemicals must match the chemistry of the oil, at the operating conditions, to be effective. Testing of inhibitor effectiveness is absolutely necessary for each application. The tests should be carried out at likely operating conditions. The chemical inhibitors for wax prevention include:

- *Thermodynamic wax inhibitor (TWI):* Suppresses cloud point, reduces viscosity and pour point, requires high volume.
- *Pour point depressants:* Modify wax crystal structure, reduce viscosity and yield stress, but do not reduce rate of wax deposition.
- *Dispersants/surfactants:* Coat wax crystals to prevent wax growth; alter wetting characteristics to minimize wax adhesion to pipe wall or other crystals.
- *Crystal modifiers:* Co-crystallize with wax, reduce deposition rate, but do not prevent formation, modify wax crystals to weaken adhesion and prevent wax from forming on pipe wall, inhibit agglomeration; suitable for steady state and shutdown, reduce viscosity/pour point, no universal chemical—performance is case specific, high cost, pigging still required, inject above cloud point.

16.4. WAX REMEDIATION

Wax remediation treatments often involve the use of solvents, hot water, a combination of hot water and surfactants, or hot oil treatments to revitalize production. The following methods are used for removal of wax, paraffin, and asphaltenes:

- Heating by hot fluid circulation or electric heating;
- Mechanical means (scraping);
- NGS, nitrogen generating system, thermo-chemical cleaning;

- Solvent treatments;
- Dispersants;
- Crystal modifiers.

16.4.1. Wax Remediation Methods

16.4.1.1. Heating

Removal of wax by means of a hot fluid or electric heating works well for downhole and for short flowlines [8]. The hydrocarbon deposit is heated above the pour point by the hot oil, hot water, or steam circulated in the system. It is important for the hydrocarbons to be removed from the wellbore to prevent redeposition. This practice, however, has a drawback. The use of hot oil treatments in wax-restricted wells can aggravate the problem in the long run, even though the immediate results appear fine.

16.4.1.2. Mechanical Means

This method is only suitable for cleaning a flowline that is not completely plugged. The wax is cleaned by mechanically scraping the inside of the flowline by pigging. The effectiveness of the pigging operation can vary widely depending on the design of the pigs and other pigging parameters. The pigging strategies in subsea systems, and the pigging requirements for subsea equipment, flowlines, platforms, and FPSO design have been discussed by Gomes et al. [9]. Coiled tubing is another effective mechanical means used in wax remediation.

16.4.1.3. Nitrogen-Generating System

A nitrogen-generating system (NGS), introduced by Petrobras in 1992, is a thermochemical cleaning method. The NGS process combines thermal, chemical, and mechanical effects by controlling nitrogen gas generation to comprise the reversible fluidity of wax/paraffin deposits. Such an exothermal chemical reaction causes the deposits to melt.

16.4.1.4. Solvent Treatments

Solvent treatments of wax and asphaltene deposition are often the most successful remediation methods, but are also the most costly. Therefore, solvent remediation methods are usually reserved for applications where hot oil or hot water methods have shown little success. When solvents contact the wax, the deposits are dissolved until the solvents are saturated. If they are not removed after saturation is reached, there is a strong possibility that the waxes will precipitate, resulting in a situation more severe than that prior to treatment.

16.4.1.5. Dispersants

Dispersants do not dissolve wax but disperse it in the oil or water through surfactant action. They may also be used with modifiers for removal of wax deposits. The dispersants divide the modifier polymer into smaller fractions that can mix more readily with the crude oil under low shear conditions.

16.4.1.6. Crystal Modifiers

Wax crystal modifiers are those chemically functionalized substances that range from polyacrylate esters of fatty alcohol to copolymers of ethylene and vinyl acetate. Crystal modifiers attack the nucleating agents of the hydrocarbon deposit and break it down and prevent the agglomeration of paraffin crystals by keeping the nucleating agents in solution.

Chemicals are available that can be tailored to work with a particular crude oil composition, but tests should be carried out on samples of the crude oil to be sure that the chemical additives will prevent wax deposition. The combined hot water and surfactant method allows the suspension of solids by the surfactant's bipolar interaction at the interface between the water and wax. An advantage of this method is that water has a higher specific heat than oil and, therefore, usually arrives at the site of deposition with a higher temperature.

16.4.2. Assessment of Wax Problem

The process of assessment for a wax problem can be summarized as follows [10]:

- Obtain a good sample;
- Cloud point or WAT based on solid-liquid-equilibria;
- Rheology: viscosity, pour point, gel strength;
- Crude oil composition: standard oil composition, HTGC;
- Wax deposition rates: cold finger or flow loop;
- Wax melting point;
- Consider the use of wax inhibitors.

16.4.3. Wax Control Design Philosophies

Wax control guidelines for the subsea devices and flowlines can be summarized as follows [5]:

- Design the subsea system to operate above the WAT by thermal insulation.
- Operate the well at sufficiently high production rates to avoid deposition in the wellbore and tree.

- Remove wax from flowlines by pigging, and pig frequently enough to ensure that the pig does not stick.
- Utilize insulation and chemicals to reduce pigging frequency.
- Identify and treat high pour point oils continuously.

For the pour point the strategies can be summarized as follows:

- In steady-state operation: heat retention (pipeline insulation) to maintain temperatures above the pour point;
- For planned shutdown and start-ups, injection of PPD;
- For unplanned shutdown, focus on restarting the system within the cooldown time of pipeline insulation; if this is not possible, use export pumps to move the gelled plug as early as possible. The required cool-down time has yet to define by operations.

In steady and transient states, the strategies can be summarized as follows:

- In steady-state operation, heat retention (pipeline insulation) is used to maintain temperatures above WAT as far along the pipeline as reasonably possible, especially in the deepwater section. Regular operational pigging will be needed throughout life to remove wax deposition.
- In transient operations, wax deposition is considered a long-term issue, so short durations (i.e., during start-ups) of low temperatures will not be addressed.

16.5. ASPHALTENES

16.5.1. General

Asphaltenes are a class of compounds in crude oil that are black in color and not soluble in *n*-heptane. Aromatic solvents such as toluene, on the other hand, are good solvents for asphaltenes. From an organic chemistry standpoint, they are large molecules comprised of polyaromatic and heterocyclic aromatic rings, with side branching. Asphaltenes originate with the complex molecules found in living plants and animals, which have only been partially broken down by the action of temperature and pressure over geologic time. Asphaltenes carry the bulk of the inorganic component of crude oil, including sulfur and nitrogen, and metals such as nickel and vanadium. All oils contain a certain amount of asphaltene. Asphaltenes only become a problem during production when they are unstable. Asphaltene stability is a function of the ratio of asphaltenes to stabilizing factors in the crude such as aromatics and resins. The factor having the biggest impact on asphaltene stability is pressure. Asphaltenes may also be destabilized by the

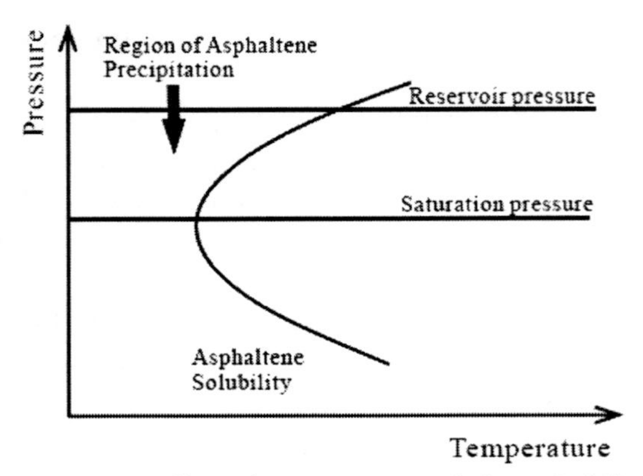

Figure 16-4 Effect of Pressure on Asphaltene Stability

addition of acid or certain types of completion fluids and by the high temperatures seen in the crude oil refining process.

In general, asphaltenes cause few operational problems since the majority of asphaltic crude oils have stable asphaltenes. Typically problems only occur downstream due to blending or high heat. Crude oils with unstable asphaltenes suffer from some severe operational problems, most of which are fouling related and affect valves, chokes, filters, and tubing. Asphaltenes become unstable as the pressure of the well decreases and the volume fraction of aliphatic components increases. If the aliphatic fraction of the oil reaches a threshold limit, then asphaltenes begin to flocculate and precipitate. This pressure is called the *flocculation point*. Figure 16-4 shows the effect of pressure on asphaltene stability. On the left side of the curve, asphaltenes are unstable, whereas to the right side of the curve, asphaltenes are stable.

There are currently no standard design and operating guidelines for the control of asphaltenes in subsea systems. Some experience has been gained from asphaltene control programs used for onshore wells. Approaches have varied from allowing the wellbore to completely plug with asphaltenes, then drilling the material out, to utilizing periodic solvent washes with coiled tubing to remove material. Relatively few operators have chosen to control asphaltene deposition with dispersants, possibly due to the expense of doing so and variable results.

16.5.2. Assessment of Asphaltene Problem

One method of characterizing oil is with a SARA (saturates, aromatics, resins, and asphaltenes) analysis. This method breaks the oil down into four pseudo-components or solubility classes and reports each as a percentage of the total. The four pseudo-components are saturates, aromatics, resins, and asphaltenes. The asphaltene fraction is the most polar fraction and is defined as aromatic soluble and *n*-alkane insoluble. Asphaltenes are condensed polyaromatic hydrocarbons that are very polar. Whereas wax has a hydrogen-to-carbon ratio of about 2, asphaltenes have a hydrogen-to-carbon mole ratio of around 1.15. The low hydrogen content is illustrated in Figure 16-5, which shows hypothetical asphaltene molecules.

Asphaltenes have a density of approximately 1.3 g/cm^3. In oil production systems the asphaltenes are often found mixed with wax. There are two mechanisms for fouling that occur during formation. The first involves acid; the second is adsorption to formation material. Acidizing is one of the most common well treatments and can cause severe damage to a well with asphaltic crude oil. The acid causes the asphaltenes to precipitate, sludge, and form rigid film emulsions that severely affect permeability, often cutting production by more than 50%. Formation materials, particularly clays, contain metals that may interact with the asphaltenes and cause the chemisorption of the asphaltenes to the clay in the reservoir. A SARA screen, aliphatic hydrocarbon titration, or depressurization of a bottomhole sample is used to determine if asphaltenes are unstable in a given crude. One

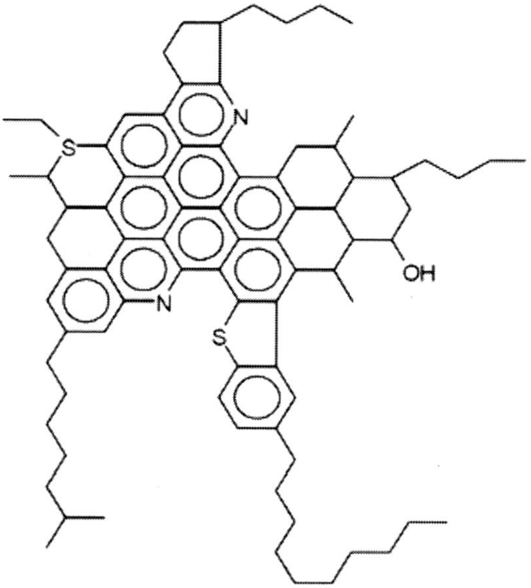

Figure 16-5 Hypothetical Asphaltene Molecules [11]

SARA screen is the colloidal instability index (CII) 4. The CII is the ratio of the unfavorable components to the favorable components of the oil as shown in following equation:

$$\mathrm{CII} = \frac{S + As}{R + Ar} \qquad 16\text{-}4$$

where S is the percentage of saturates, As is the percentage of asphaltenes, R is the percentage of resins, and Ar is the percentage of aromatics in the oil. If the CII is greater than 1, then the amount of unfavorable components exceeds the amount of favorable components in the system and the asphaltenes are likely to be unstable.

Several aliphatic hydrocarbon titrations are available that can be used to assess the stability of asphaltenes in a dead crude oil. One method in particular involves the continuous addition of an aliphatic titrant to oil and the measurement of the optical density of the solution. In this method the precipitation point of asphaltenes is detected by monitoring changes in transmission of an infrared laser. The instrument is referred to as the asphaltene precipitation detection unit (APDU). As the hydrocarbon is added to the oil, the optical density decreases and the laser transmittance through the sample increases. At the point when enough titrant has been added to the sample that asphaltenes become unstable and precipitate, the optical density drops dramatically. This point is called the APDU number, which is defined as the ratio of the volume of titrant to the initial mass of crude oil.

Depressurization of a live bottomhole sample provides the most direct measurement of asphaltene stability for production systems. During depressurization, the live oil flocculation point or the pressure at which asphaltenes begin to precipitate in the system is determined by monitoring the transmittance of an infrared laser that passes through the sample. Onset of flocculation will produce a noticeable reduction of light transmittance. If oil has a flocculation point, then the asphaltenes are unstable at pressures between the flocculation points to just below the bubble point. Many oils are unstable only near the bubble point, which has led many engineers to believe that problems only occur at the bubble point. However, some oils have an instability window of several thousand pounds per square inch. Figure 16-6 depicts an asphaltene solubility curve. The onset or flocculation point, saturation pressure, asphaltene saturation point, and reservoir fluid asphaltene concentration are indicted with dashed lines, whereas the unstable region is indicated by a shaded area between the asphaltene

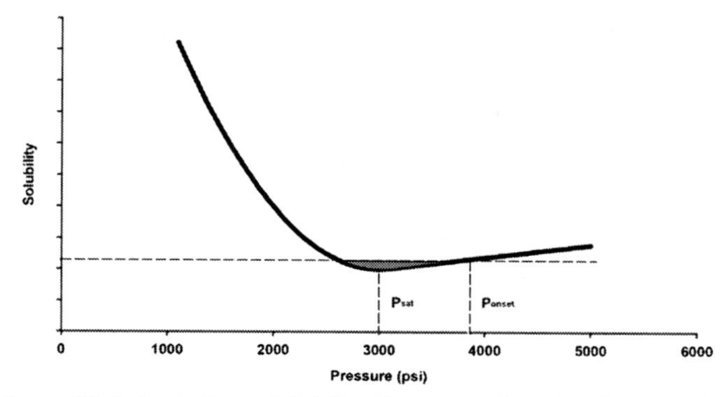

Figure 16-6 Asphaltene Solubility Curve as a Function Pressure [11]

saturation point and the solubility curve. As shown on the curve the region of instability is from 2600 to 3800 psi.

16.5.3. Asphaltene Formation

Asphaltene solids are typically black coal-like substances. They tend to be sticky, making them difficult to remove from surfaces. In addition, asphaltene solids tend to stabilize water/oil emulsions, complicating oil separation and water treatment at the host. It is felt that asphaltenes will be a problem for a relatively small fraction of deepwater projects. Current research efforts have focused on improving screening tests for asphaltene and figuring out how screening test results relate to field problems. Three screening tools are used. The first is a test known as the P-value test, which involves the titration of crude oil with cetane, the normal paraffin with a carbon chain length of 16. Additions of normal paraffins tend to destabilize asphaltenes in the crude oil. The stability of the oil increases with the amount of cetane that can be added before visible amounts of asphaltenes come out of solution. The SARA screen test examines the stability of asphaltenes by determining the concentration of the primary components of crude oil, saturated hydrocarbons, aromatics, resins, and asphaltenes. The ratio of saturates to aromatics and asphaltenes to resins is computed and used to determine the stability of the oil. The PVT screen utilizes two values available from a PVT analysis, the *in situ* density and the degree of undersaturation (difference between reservoir pressure and bubble point) to make a general assessment of asphaltene stability. In general, increasing under saturation and decreasing *in situ* density are associated with decreasing asphaltene stability.

16.5.4. Asphaltene Deposition

After the bubble point has been reached, the mass of precipitated asphaltenes and the mass of asphaltenes deposited in the cell are measured. These two measurements provide a means to assess the likelihood of problems due to deposition. We have observed that only a small percentage of total asphaltenes adheres to a surface during an experiment. Based on these observations we would expect only about 5% of the total asphaltenes to play a role in asphaltene deposition. The mass of deposited asphaltenes can be used to predict the mass of deposition in a production system. Although this number is likely to be an overestimate, it is a good number for design purposes and for contingency planning.

Asphaltene problems occur infrequently offshore, but can have serious consequences on project economics. Because asphaltene deposition is most likely as the produced fluid passes through the bubble point, the deposition often occurs in the tubing. Subsea systems designed to mitigate asphaltene problems generally rely on bottomhole injection of an inhibitor with provision to solvent treat the wellbore when required. In the absence of inhibitors, monthly tubing cleanouts are not uncommon onshore. This frequency would be intolerable in a subsea system, making effective inhibitors essential to the design of the project.

16.6. ASPHALTENE CONTROL DESIGN PHILOSOPHIES

In subsea wells, direct intervention with coiled tubing is very expensive and is not a viable means of control. Therefore, the strategy that has been proposed for control of subsea wells utilizes a combination of techniques to minimize deposition. This strategy is as follows [12]:

- Inject an asphaltene dispersant continuously into the wellbore (injection must be at the packer to be effective).
- Install equipment to facilitate periodic injection of an aromatic solvent into the wellbore for a solvent soak.
- Be financially and logistically prepared to intervene with coiled tubing in the wellbore to remove deposits.
- Control deposition in the flowline with periodic pigging with solvents.

This strategy requires the installation of additional umbilical lines for delivery of asphaltene dispersant and large volumes of solvent, as well as a downhole line for injection of dispersant immediately above the packer. These hardware requirements add considerable project cost. Currently, there are no models for asphaltene deposition as a function of system pressure or other parameters. Flow assurance modeling is helpful in understanding the pressure profile in the subsea system, especially where the bubble point is reached. Because the bubble point is typically the pressure at which asphaltenes are least stable, deposition problems would be expected to be the worst at this location.

If steady-state flow cannot be reached in a reasonable time due to the high viscosity of the fluids in the pipeline or difficulty in passing the gel segments through the outlet choke, consider displacing the pipeline contents with diluent hydrocarbon such as diesel or perhaps water. Another similar approach is to chemically treat the produced fluids being introduced at the pipeline inlet. Once the pipeline is filled with diesel, water, or treated fluids, flow can be accelerated rapidly due to the low viscosity and no concern about gel segments. If the gel does not break some options to consider are:

- Use a coiled-tubing system to push a tool into the pipeline and flush out the gel. Commercial coiled tubing equipment is available with an extended reach up to several miles.
- Generate pressure pulses in the gel. Because the gels will compress somewhat, a pressure pulse will move the exposed end of the gel and break a portion of it. With successive pressure pulses, a small segment of the gel can be broken with each pulse up to a limit.
- An easy first approach is to apply pressure to the gel and wait. There have been field reports of the gel breaking after several hours of exposure to pressure.

To plan, design, and operate a subsea pipeline to transport high paraffinic crudes, the following activities are recommended:

- Measure key properties of the crude including cloud point, pour point, gel strength, effectiveness of pour point depressant chemicals, and viscosity as a function of temperature, shear rate, and dissolved gas.
- Assess the thermal conditions to be experienced by the crude for steady-state flow and transient flow during start-up, shutdown, and low production rates.
- Based on crude properties and thermal data, develop operational plans for shutdown, start-up, and low-flow situations. A key decision for start-up is whether to break the gel with pressure or to employ more costly means to prevent gel formation.
- If gel breaking or warm-up is required for restart, the necessary flow control equipment and possibly a static mixer at the pipeline outlet will be required.

Asphaltene design and control guidelines are still in the early stages of development. There is no predictive model for asphaltene deposition, as there is for wax and hydrates. Asphaltene design and control are usually considered with wax design and the following steps and analyses are carried out [10]:

1. Define samples to be taken and analyses to be performed.
2. Perform wax modeling to provide:
   - WAT of live fluids;
   - Location of deposition;
   - Rates of deposition;
   - Amount of wax deposited;
   - Pipeline design parameters.
3. Assess asphaltene stability under producing conditions.
4. Provide chemical and thermal options.
5. Determine pigging frequency.
6. Design cost-effective solutions for prevention and remediation of wax and asphaltenes.

REFERENCES

[1] ASTM D97-09, Standard Test Method for Pour Point of Petroleum Products, American Society for Testing and Materials, West Conshohocken, PA, 2009.

[2] ASTM D5853-09, Standard Test Method for Pour Point of Crude Oils, American Society for Testing and Materials, West Conshohocken, PA, 2009.

[3] Pipeline Deepstar, Wax Blockage Remediation, Deepstar III Project, DSIII CTR 3202, Radoil Tool Company (1998).

[4] J.R. Becker, Crude Oil, Waxes, Emulsions and Asphaltenes, Pennwell Publishing, Tulsa, Oklahoma, 1997.

[5] S.E. Lorimer, B.T. Ellison, Design Guidelines for Subsea Oil Systems, Facilities 2000: Facilities Engineering into the Next Millennium (2000).

[6] B. Edmonds, R.A.S. Moorwood, R. Szczepanski, X. Zhang, Latest Developments in Integrated Prediction Modeling Hydrates, Waxes and Asphaltenes, Focus on Controlling Hydrates, Waxes and Asphaltenes, IBC, Aberdeen, 1999, October.

[7] Y. Chin, Flow Assurance: Maintaining Plug-Free Flow and Remediating Plugged Pipelines, Offshore vol. 61 (Issue 2) (2001).

[8] Y. Chin, J. Bomba, Review of the State of Art of Pipeline Blockage Prevention and Remediation Methods, Proc. 3rd Annual Deepwater Pipeline & Riser Technology Conference & Exhibition, 2000.

[9] M.G.F.M. Gomes, F.B. Pereira, A.C.F. Lino, Solutions and Procedures to Assure the Flow in Deepwater Conditions, OTC 8229, Offshore Technology Conference, Houston, Texas, 1996.

[10] B. Ellision, C.T. Gallagher, Baker Petrolite Flow Assurance Course., 2001.

[11] B.T. Ellison, C.T. Gallagher, S.E. Lorimer, The Physical Chemistry of Wax, Hydrates, and Asphaltene, OTC 11960, Offshore Technology Conference, Houston, 2000.

[12] H. James, I. Karl, Paraffin, Asphaltenes Control Practices Surveyed, Oil & Gas Journal (1999, July 12) 61–63.

CHAPTER 17

Subsea Corrosion and Scale

Contents

Subsea Engineering Handbook
ISBN 978-0-12-397804-2, doi:10.1016/B978-0-12-397804-2.00017-5

17.1. INTRODUCTION

In most subsea developments, oil and gas products are transported from subsea wells to platforms in multiphase flow without using a separation process. Corrosion, scale formation, and salt accumulation represent increasing challenges for the operation of subsea multiphase pipelines. Corrosion can be defined as a deterioration of a metal, due to chemical or electrochemical interactions between the metal and its environment. The tendency of a metal to corrode depends on a given environment and the metal type.

The unprotected buried or unburied pipelines that are exposed to the atmosphere or submerged in water are susceptible to corrosion in external pipe surfaces, and without proper maintenance, the pipeline will eventually corrode and fail, because corrosion can weaken the structure of the pipeline, making it unsafe for transporting oil, gas, and other fluids. A strong adhesive external coating over the whole length of the pipeline will tend to prevent corrosion. However, there is always the possibility of coating damage during handling of the coated pipe either during shipping or installation. Cathodic protection is provided by sacrificial anodes to prevent the damaged areas from corroding.

The presence of carbon dioxide (CO_2), hydrogen sulfide (H_2S), and free water in the internal production fluid can cause severe corrosion problems in oil and gas pipelines. Internal corrosion in wells and pipelines is influenced by temperature, CO_2 and H_2S content, water chemistry, flow velocity, oil or water wetting, and the composition and surface condition of the steel. Corrosion-resistant alloys such as 13% Cr steel and duplex stainless steel are often used in the downhole piping of subsea structures. However, for long-distance pipelines, carbon steel is the only economically feasible alternative and corrosion has to be controlled so as to protect the flowline both internally and externally.

Scale is a deposit of the inorganic mineral components of water. Solids may precipitate and deposit from the brine once the solubility limit or capacity is exceeded. The solid precipitates may either stay in suspension in the water or form an adherent scale on a surface such as a pipe wall. Suspended scale solids may cause problems such as formation plugging. Adherent scale deposits can restrict flow in pipes and damage equipment such as pumps and valves. Corrosion and microbiological activity are often accelerated under scale deposits.

The purpose of this chapter is to evaluate the effects of corrosion and scale deposits on subsea oil and gas pipelines and describe protection methods. The evaluation focuses on the following three aspects:

- Pipeline internal corrosion and protection;
- Pipeline external corrosion and protection;
- Scale.

17.2. PIPELINE INTERNAL CORROSION

Two types of corrosion can occur in oil and gas pipeline systems when CO_2 and H_2S are present in the hydrocarbon fluid: sweet corrosion and sour corrosion. Sweet corrosion occurs in systems containing only carbon dioxide or a trace of hydrogen sulfide (H_2S partial pressure < 0.05 psi). Sour corrosion occurs in systems containing hydrogen sulfide above a partial pressure of 0.05 psia and carbon dioxide.

When corrosion products are not deposited on the steel surface, very high corrosion rates of several millimeters per year can occur. This "worst case" corrosion is the easiest type to study and reproduce in the laboratory. When CO_2 dominates the corrosivity, the corrosion rate can be reduced substantially under conditions where iron carbonate can precipitate on the steel surface and form a dense and protective corrosion product film. This occurs more easily at high temperatures or high pH values in the water phase. When H_2S is present in addition to CO_2, iron sulfide films are formed rather than iron carbonate, and protective films can be formed at lower temperature, since iron sulfide precipitates much easier than iron carbonate. Localized corrosion with very high corrosion rates can occur when the corrosion product film does not give sufficient protection, and this is the most feared type of corrosion attack in oil and gas pipelines.

17.2.1. Sweet Corrosion: Carbon Dioxide

CO_2 is composed of one atom of carbon with two atoms of oxygen. It is a corrosive compound found in natural gas, crude oil, and condensate and produced water. It is one of the most common environments in the oil field industry where corrosion occurs. CO_2 corrosion is enhanced in the presence of both oxygen and organic acids, which can act to dissolve iron carbonate scale and prevent further scaling.

Carbon dioxide is a weak acidic gas and becomes corrosive when dissolved in water. However, CO_2 must hydrate to carbonic acid H_2CO_3, which is a relatively slow process, before it becomes acidic. Carbonic acid causes a reduction in the pH of water and results in corrosion when it comes in contact with steel.

Areas where CO_2 corrosion is most common include flowing wells, gas condensate wells, areas where water condenses, tanks filled with CO_2, saturated produced water, and pipelines, which are generally corroded at a slower rate because of lower temperatures and pressures. CO_2 corrosion is enhanced in the presence of both oxygen and organic acids, which can act to dissolve iron carbonate scale and prevent further scaling.

The maximum concentration of dissolved CO_2 in water is 800 ppm. When CO_2 is present, the most common forms of corrosion include uniform corrosion, pitting corrosion, wormhole attack, galvanic ringworm corrosion, heat-affected corrosion, mesa attack, raindrop corrosion, erosion corrosion, and corrosion fatigue. The presence of carbon dioxide usually means no H_2 embrittlement.

CO_2 corrosion rates are greater than the effect of carbonic acid alone. Corrosion rates in a CO_2 system can reach very high levels (thousands of mils per year), but it can be effectively inhibited. Velocity effects are very important in the CO_2 system; turbulence is often a critical factor in pushing a sweet system into a corrosive regime. This is because it either prevents formation or removes a protective iron carbonate (siderite) scale.

CO_2 corrosion products include iron carbonate (siderite, $FeCO_3$), iron oxide, and magnetite. Corrosion product colors may be green, tan, or brown to black. This can be protective under certain conditions. Scale itself can be soluble. Conditions favoring the formation of a protective scale are elevated temperatures, increased pH as occurs in bicarbonate-bearing waters, and lack of turbulence, so that the scale film is left in place. Turbulence is often the critical factor in the production or retention of a protective iron carbonate film. Iron carbonate is not conductive. Therefore, galvanic corrosion cannot occur. Thus, corrosion occurs where the protective iron carbonate film is not present and is fairly uniform over the exposed metal. Crevice and pitting corrosion occur when carbonate acid is formed. Carbon dioxide can also cause embrittlement, resulting in stress corrosion cracking.

17.2.1.1. Corrosion Predictions

CO_2 corrosion of carbon steel used in oil production and transportation, when liquid water is present, is influenced by a large number of parameters, some of which are listed below:

- Temperature;
- CO_2 partial pressure;
- Flow (flow regime and velocity);
- pH;
- Concentration of dissolved corrosion product ($FeCO_3$);
- Concentration of acetic acid;
- Water wetting;
- Metal microstructure (welds);
- Metal prehistory.

The detailed influence of these parameters is still poorly understood and some of them are closely linked to each other. A small change in one of them may influence the corrosion rate considerably.

Various prediction models have been developed and are used by different companies. Among them are the de Waard et al. model (Shell), CORMED (Elf Aquitaine), LIPUCOR (Total), and a new electrochemically based model developed at IFE. Due to the complexity of the various corrosion controlling mechanisms involved and a built-in conservatism, the corrosion models often overpredict the corrosion rate of carbon steel.

The Shell model for CO_2 corrosion is most commonly used in the oil/gas industry. The model is mainly based on the de Waard equation [1] published in 1991. Starting from a "worst case" corrosion rate prediction, the model applies correction factors to quantify the influence of environmental parameters and corrosion product scale formed under various conditions. However, the first version of the model was published in 1975, and it has been revised several times, in order to make it less conservative by including new knowledge and information. The original formula of de Waard and Milliams [1] implied certain assumptions that necessitated the application of correction factors for the influence of environmental parameters and for the corrosion product scale formed under various conditions.

CO_2 corrosion rates in pipelines made of carbon steel may be evaluated using industry accepted equations that preferably combine contributions from flow-independent kinetics of the corrosion reaction at the metal surface, with the contribution from the flow-dependent mass transfer of dissolved CO_2.

The corrosion rate calculated from the original formula with its correction factors is independent of the liquid velocity. To account for the effect of flow, a new model was proposed that takes the effect of mass transport and fluid velocity into account by means of a so-called resistance model:

$$V_{cr} = \frac{1}{\frac{1}{V_r} + \frac{1}{V_m}} \tag{17-1}$$

where V_{cr} is the corrosion rate in mm/year, V_r is the flow-independent contribution, denoted as the reaction rate, and V_m is the flow-dependent contribution, denoted as the mass transfer rate.

In multiphase turbulent pipeline flow, V_m depends on the velocity and the thickness of the liquid film, whereas V_r depends on the temperature, CO_2 pressure, and pH. For example, for pipeline steel containing 0.18% C and 0.08% Cr, the equations for V_r and V_m for liquid flow in a pipeline are:

$$\log(V_r) = 4.93 - \frac{1119}{T_{mp} + 273} + 0.58 \cdot \log(p_{CO_2}) \tag{17-2}$$

where T_{mp} is the pipeline fluid temperature in °C, and the partial pressure p_{CO_2} of CO_2 in bar. The partial pressure p_{CO2} can be found by:

$$p_{CO_2} = n_{CO_2} \cdot p_{opr} \tag{17-3}$$

where n_{CO_2} is the fraction of CO_2 in the gas phase, and p_{opr} is the operating pressure in bar.

The mass transfer rate V_m is approximated by:

$$V_m = 2.45 \cdot \frac{U^{0.8}}{d^{0.2}} \cdot p_{CO_2} \tag{17-4}$$

where U is the liquid flow velocity in m/s and d is the inner diameter in meters.

17.2.1.2. Comparison of CO_2 Corrosion Models

The corrosion caused by the presence of CO_2 represents the greatest risk to the integrity of carbon steel equipment in a production environment and is more common than damage related to fatigue, erosion, or stress corrosion cracking. NORSOK, Shell, and other companies and organizations have developed models to predict the corrosion degradation.

NORSOK's standard M-506 [2] can be used to calculate the CO_2 corrosion rate, which is an empirical model for carbon steel in water containing CO_2 at different temperatures, pH, CO_2 fugacity and wall shear

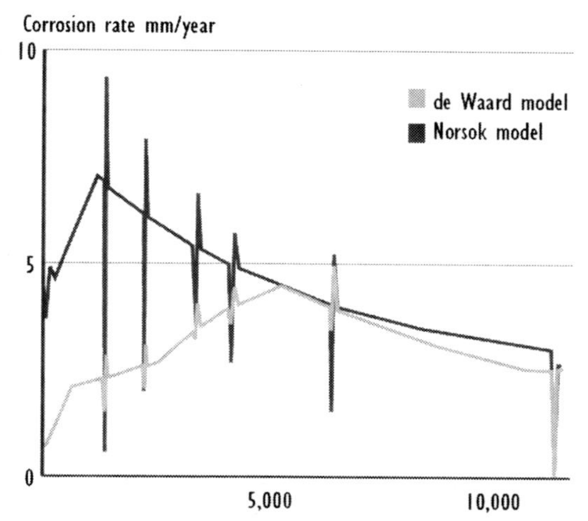

Figure 17-1 Predicted Corrosion Rate in a Subsea Pipeline [3]

stress. The NORSOK model covers only the corrosion rate calculation where CO_2 is the corrosive agent. It does not include additional effects of other constituents, which may influence the corrosivity (e.g. H_2S), which commonly appears in the production flowlines. If such a constituent is present, the effect must be evaluated separately. None of the de Waard models includes the H_2S effect.

Figure 17-1 shows an example of corrosion rate prediction in a subsea gas condensate pipeline. Here, two of the most commonly used corrosion prediction models were combined with a three-phase fluid flow model in order to calculate corrosion rate profiles along a pipeline. This can help to identify locations where variations in the flow regime, flow velocity, and water accumulation may increase the risk of corrosion damage. For this pipeline, the temperature was 90°C at the inlet and 20°C at the outlet, and the decrease in predicted corrosion rates toward the end of the pipeline is mainly a result of the decreasing temperature. The lower corrosion rates close to the pipeline inlet are due to the effect of protective corrosion films at high temperature, which is predicted differently by the two corrosion models used. The peaks in predicted corrosion rates result from variation in flow velocity due to variations in the pipeline elevation profile.

17.2.1.3. Sensitivity Analysis for CO_2 Corrosion Calculation

Table 17-1 presents the base case for the following sensitivity analysis. These data are based on the design operating data for a 10–in. production flowline.

Table 17-1 Base Case for Sensitivity Analysis

Parameter	Units	Base Case
Total pressure	bar	52
Temperature	°C	22.5
CO_2 in gas	mol%	0.5
Flow velocity	m/s	2.17
H_2S	ppm	220
pH		4.2
Water cut		50%
Inhibitor availability		50%

Total System Pressure and CO_2 Partial Pressure

An increase in total pressure will lead to an increase in corrosion rate because p_{CO_2} will increase in proportion. With increasing pressure, the CO_2 fugacity f_{CO_2} should be used instead of the CO_2 partial pressure p_{CO_2} since the gases are not ideal at high pressures. The real CO_2 pressure can be expressed as:

$$f_{CO_2} = a * p_{CO_2} \tag{17-5}$$

where a is fugacity constant that depends on pressure and temperature:

$$a = 10^{P(0.0031-1.4/T)} \quad \text{for P} \leq 250 \text{ bar}$$

$$a = 10^{250(0.0031-1.4/T)} \quad \text{for P} < 250 \text{ bar}$$

Figures 17-2 and 17-3 present the effect of total pressure and CO_2 partial pressure, respectively, on the corrosion rate. With increasing total pressure and CO_2 partial pressure, the corrosion rate is greatly increased.

System Temperature

Temperature has an effect on the formation of protective film. At lower temperatures the corrosion product can be easily removed by flowing liquid. At higher temperatures the film becomes more protective and less easily washed away. Further increases in temperature result in a lower corrosion rate and the corrosion rate goes through a maximum [1]. This temperature is referred to as the *scaling temperature*. At temperatures exceeding the scaling temperature, corrosion rates tend to decrease to close to zero, according to de Waard. Tests by IFE Norway revealed that the corrosion rate is still increasing when the design temperature is beyond the scaling temperature [4]. Figure 17-4 shows the effect of temperature on the corrosion rate,

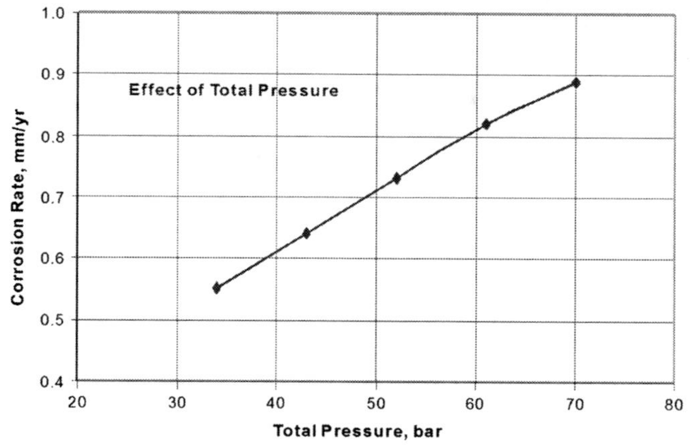

Figure 17-2 Effect of Total Pressure on Corrosion Rate

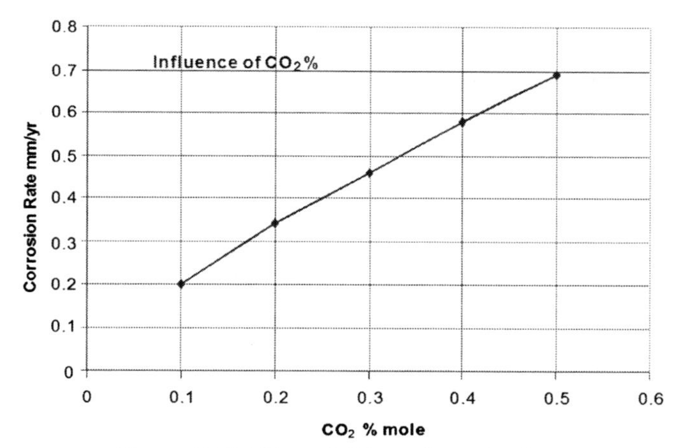

Figure 17-3 Effect of CO_2 on Corrosion Rate

where the total pressure is 48 bara and the pH is equal to 4.2. The corrosion rate increases with increasing temperature, when the temperature is lower than the scaling temperature.

H_2S

H_2S can depress pH when it is dissolved in a CO_2 aqueous solution. The presence of H_2S in CO_2/brine systems can reduce the corrosion rate of steel when compared to the corrosion rate under conditions without H_2S at temperatures of less than 80°C, due to the formation of a metastable iron sulfide film. At higher temperatures the combination of H_2S and chlorides

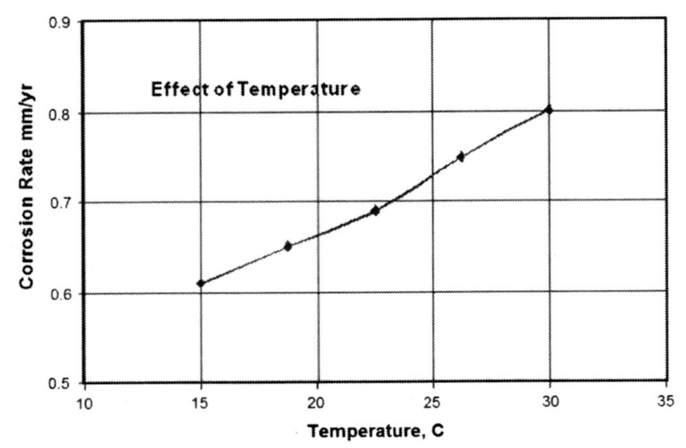

Figure 17-4 Effect of Temperature on Corrosion Rate

will produce higher corrosion rates than just CO_2/brine system because a protective film is not formed.

H_2S at levels below the NACE criteria for sulfide stress corrosion cracking (per the MR0175 NACE publication [5]) reduces general metal loss rates but can promote pitting. The pitting proceeds at a rate determined by the CO_2 partial pressure; therefore, CO_2-based models are still applicable at low levels of H_2S. Where the H_2S concentration is greater than or equal to the CO_2 value, or greater than 1 mol%, the corrosion mechanism may not be controlled by CO_2 and, therefore, CO_2-based models may not be applicable.

pH

pH affects the corrosion rate by affecting the reaction rate of cathodes and anodes and, therefore, the formation of corrosion products. The contamination of a CO_2 solution with corrosion products reduces the corrosion rate. pH has a dominant effect on the formation of corrosion films due to its effect on the solubility of ferrous carbonate. An increase in pH slows down the cathodic reduction of H^+. Figure 17-5 presents the relationship between pH and corrosion rate. In a solution with a pH of less than 7, the corrosion rate decreases with increasing pH.

Inhibitors and Chemical Additives

Inhibitors can reduce the corrosion rate by presenting a protective film. The presence of the proper inhibitors with optimum dosage can maintain the

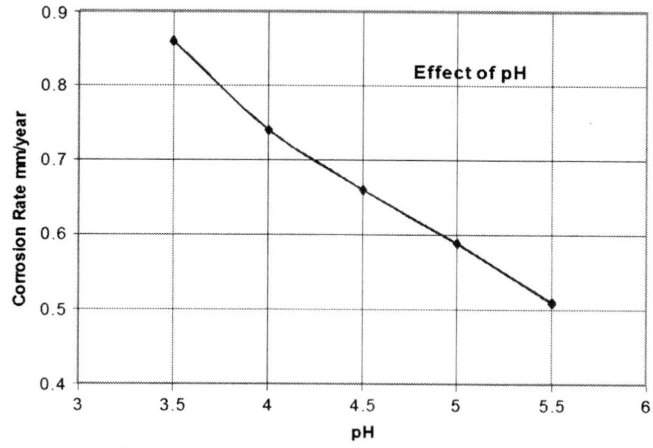

Figure 17-5 Effect of pH on Corrosion Rate

corrosion rate at 0.1 mm/year. Use of inhibitor can greatly decrease corrosion rates and, hence, increase pipeline life.

The impingement of sand particles can destroy the inhibitor film and, therefore, reduce inhibitor efficiency. Inhibitors also perform poorly in low-velocity lines particularly if the fluids contain solids such as wax, scale, or sand. Under such circumstances, deposits inevitably form at the 6 o'clock position, preventing the inhibitor from reaching the metal surface. Flow velocities below approximately 1.0 m/s should be avoided if inhibitors are expected to provide satisfactory protection; this will be critical in lines containing solids.

Inhibitor Efficiency versus Inhibitor Availability

When inhibitors are applied, there are two ways to describe the extent to which an inhibitor reduces the corrosion rate: inhibitor efficiency (IE) and inhibitor availability (IA). A value of 95% for IE is commonly used. However, inhibitors are unlikely to be constantly effective throughout the design life. For instance, increased inhibitor dosage or better chemicals will increase the inhibitor concentration. It may be assumed that the inhibited corrosion rate is unrelated to the uninhibited corrosivity of the system and all systems can be inhibited to 0.1 mm/year. The corrosion inhibitor is not available 100% of the time and therefore corrosion will proceed at the uninhibited rate for some periods. Figure 17-6 shows the inhibited corrosion rate under different inhibitor availabilities. This figure is based on the assumed existence of corrosion

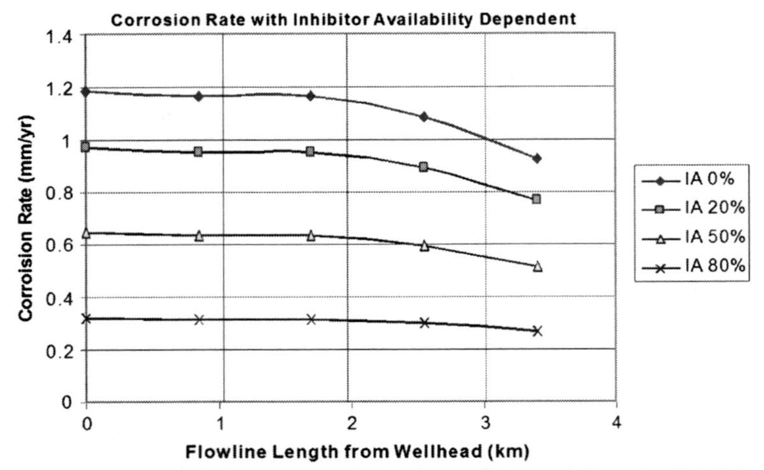

Figure 17-6 Inhibited Corrosion Rate under Different Inhibitor Availabilities

inhibitors that are able to protect the steel to a corrosion rate CR_{mit} (typically 0.1 mm/year) regardless of the uninhibited corrosion rate CR_{unmit}, taking into consideration the percentage of time IA the inhibitor is available.

Chemical Additives

Glycol (or methanol) is often used as a hydrate preventer on a recycled basis. If glycol is used without the addition of a corrosion inhibitor, there will be some benefit from the glycol. De Waard has produced a glycol correction factor. However, if glycol and inhibitor are both used, little additional benefit will be realized from the glycol and it should be ignored for design purpose.

Methanol is batch injected during start-up until flowline temperatures rise above the hydrate formation region and during extended shutdown.

Single-Phase Flow Velocity

Single-phase flow refers to a flow with only one component, normally oil, gas, or water, through a porous media. Fluid flow influences corrosion by affecting mass transfer and by mechanical removal of solid corrosion products. The flow velocity used in the corrosion model is identified as the true water velocity. Figure 17-7 shows that the corrosion rate increases consistently with increased flow rate at low pH.

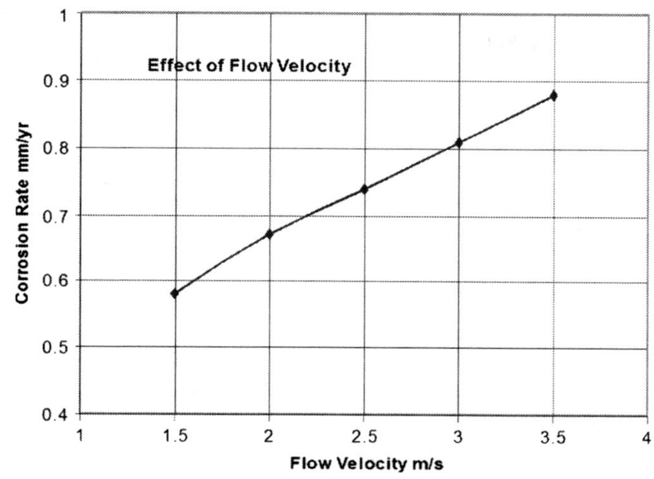

Figure 17-7 Effect of Flow Velocity on Corrosion Rate

Multiphase Flow

Multiphase flow refers to the simultaneous flow of more than one fluid phase through a porous media. Most oil wells ultimately produce both oil and gas from the formation and often also produce water. Consequently, multiphase flow is common in oil wells. Multiphase flow in a pipeline is usually studied by the flow regime and corresponding flow rate. Because of the various hydrodynamics and the corresponding turbulence, multiphase flow will further influence the internal corrosion rate, which is significantly different from that of single-phase flow in a pipeline in terms of corrosion.

Water Cut

The term *water cut* refers to the ratio of water produced compared to the volume of total liquid produced. CO_2 corrosion is mainly caused by water coming in contact with the steel surface. The severity of the CO_2 corrosion is proportional to the time during which the steel surface is wetted by the water phase. Thus, the water cut is an important factor influencing the corrosion rate. However, the effect of the water cut cannot be separated from the flow velocity and the flow regime.

Free-Span Effect

Pipeline spanning can occur on a rough seabed or a seabed subjected to scour. The evaluation of the allowable free-span length should be considered in order to avoid excessive yielding and fatigue. The localized

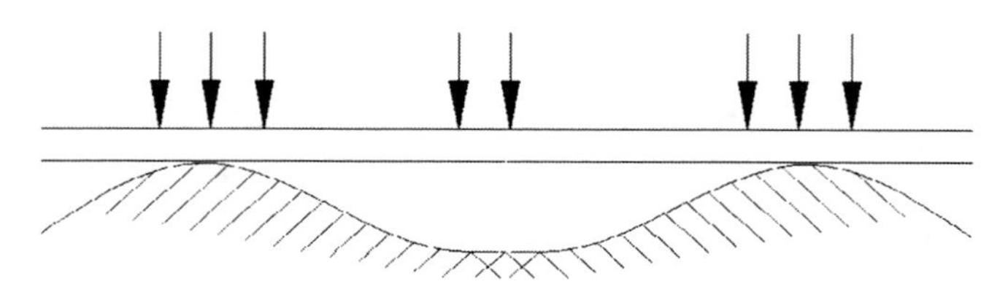

Figure 17-8 Effect of Free Spans on Corrosion Defect Development

reduction of wall thickness influences the strength capacity of the pipeline and, therefore, the allowable free-span length.

Figure 17-8 shows that the middle point of a free span contains additional accumulated waters and marine organisms that may accelerate corrosion development. The flow regime and flow rates will change. The corrosion defect depth in the region close to the middle point will most likely be deeper.

17.2.2. Sour Corrosion: Hydrogen Sulfide

Hydrogen sulfide is a flammable and poisonous gas. It occurs naturally in some groundwater. It is formed from decomposing underground deposits of organic matter such as decaying plant material. It is found in deep or shallow wells and also can enter surface water through springs, although it quickly escapes to the atmosphere. Hydrogen sulfide often is present in wells drilled in shale or sandstone, or near coal or peat deposits or oil fields.

Hydrogen sulfide gas produces an offensive "rotten egg" or "sulfur water" odor and taste in water. In some cases, the odor may be noticeable only when the water is initially turned on or when hot water is run. Heat forces the gas into the air, which may cause the odor to be especially offensive in a shower. Occasionally, a hot water heater is a source of hydrogen sulfide odor. The magnesium corrosion control rod present in many hot water heaters can chemically reduce naturally occurring sulfates to hydrogen sulfide.

Hydrogen sulfide (H_2S) occurs in approximately 40% of all wells. Wells with large amounts of H_2S are usually labeled sour; however, wells with only 10 ppm or above can also be labeled sour. Partial pressures above 0.05 psi H_2S are considered corrosive. The amount of H_2S appears to increase as the well ages. H_2S combines with water to form sulfuric acid (H_2SO_4), a strongly corrosive acid. Corrosion due to H_2SO_4 is often referred to as *sour corrosion*. Because hydrogen sulfide combines easily with water, damage to stock tanks below water levels can be severe.

Water with hydrogen sulfide alone does not cause disease. However, hydrogen sulfide forms a weak acid when dissolved in water. Therefore, it is a source of hydrogen ions and is corrosive. It can act as a catalyst in the absorption of atomic hydrogen in steel, promoting sulfide stress cracking (SSC) in high-strength steels. Polysulfides and sulfanes (free acid forms of polysulfides) can form when hydrogen sulfide reacts with elemental sulfur.

The corrosion products are iron sulfides and hydrogen. Iron sulfide forms a scale at low temperatures and can act as a barrier to slow corrosion. The absence of chloride salts strongly promotes this condition and the absence of oxygen is absolutely essential. At higher temperatures the scale is cathodic in relation to the casing and galvanic corrosion starts. The chloride forms a layer of iron chloride, which is acidic and prevents the formation of an FeS layer directly on the corroding steel, enabling the anodic reaction to continue. Hydrogen produced in the reaction may lead to hydrogen embrittlement. A nuisance associated with hydrogen sulfide includes its corrosiveness to metals such as iron, steel, copper, and brass. It can tarnish silverware and discolor copper and brass utensils.

17.2.3. Internal Coatings

The primary reason for applying internal coatings is to reduce friction, which enhances flow efficiency. Besides, the application of internal coatings can improve corrosion protection, precommissioning operations, and pigging operations. Increased efficiency is achieved through lowering the internal surface roughness since the pipe friction factor decreases with a decrease in surface roughness. In actual pipeline operation the improved flow efficiency will be observed as a reduction in pressure drop across the pipeline.

The presence of free water in the system is one cause of corrosion in an inner pipeline. An effective coating system will provide an effective barrier against corrosion attack. The required frequency of pigging is significantly reduced with a coated pipeline. The wear on pig disks is substantially reduced due to the pipe's smoother surface.

The choice of a coating is dictated by both environmental conditions and the service requirements of the line. The major generic types of coatings used for internal linings include epoxies, urethanes, and phenolics. Epoxy-based materials are commonly used internal coatings because of their broad range of desirable properties, which include sufficient hardness, water resistance, flexibility, chemical resistance, and excellent adhesion.

17.2.4. Internal Corrosion Inhibitors

Corrosion inhibitors are chemicals that can effectively reduce the corrosion rate of the metal exposed to the corrosive environment when added in small concentration. They normally work by adsorbing themselves to form a film on the metallic surface [6].

Inhibitors are normally distributed from a solution or by dispersion. They reduce the corrosion process by either:

- Increasing the anodic or cathodic polarization behavior;
- Reducing the movement or diffusion of ions to the metallic surface;
- Increasing the electrical resistance of the metallic surface.

Inhibitors can be generally classified as follows [6]:

- Passivating inhibitors;
- Cathodic inhibitors;
- Precipitation inhibitors;
- Organic inhibitors;
- Volatile corrosion inhibitors.

The key to selection of an inhibitor is to know the system and anticipate the potential problems in the system. The system conditions include water composition (such as salinity, ions, and pH), fluid composition (percentage water versus hydrocarbon), flow rates, temperature, and pressure. Application of the inhibitors can be accomplished by batch treatments, formation squeezes, continuous injections, or a slug between two pigs.

Inhibitor efficiency can be defined as:

$$\text{Inhibitor efficiency } (\%) = 100^{*}\ (CR_{\text{uninhibited}} - CR_{\text{inhibited}})/CR_{\text{uninhibited}}$$

where $CR_{\text{uninhibited}}$ is the corrosion rate of the uninhibited system and $CR_{\text{inhibited}}$ is the corrosion rate of the inhibited system. Typically the inhibitor efficiency increases with an increase in inhibitor concentration.

17.3. PIPELINE EXTERNAL CORROSION

Infrastructures such as steel pipelines are susceptible to corrosion. This section deals with coatings and external corrosion protection such as cathodic protection (CP). The preferred technique for mitigating marine corrosion is use of coatings combined with CP. Coatings can provide a barrier against moisture reaching the steel surface and therefore provide a defense against external corrosion. However, in the event of the failure of coatings, a secondary CP system is required.

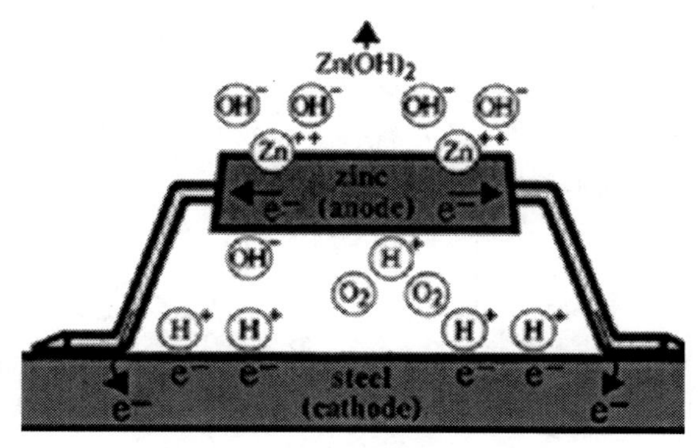

Figure 17-9 Galvanic Corrosion

Corrosion is the degradation of a metal through its electrochemical reaction with the environment. A primary cause of corrosion is due to an effect known as galvanic corrosion. All metals have different natural electrical potentials. When two metals with different potentials are electrically connected to each other in an electrolyte (e.g., seawater), current will flow from the more active metal to the other, causing corrosion to occur. The less active metal is called the cathode, and the more active, the anode. In Figure 17-9, the more active metal, zinc, is the anode and the less active metal, steel, is the cathode. When the anode supplies current, it will gradually dissolve into ions in the electrolyte, and at the same time produce electrons, which the cathode will receive through the metallic connection with the anode. The result is that the cathode will be negatively polarized and, hence, protected against corrosion.

17.3.1. Fundamentals of Cathodic Protection

Carbon steel structures exposed to natural waters generally corrode at an unacceptably high rate unless preventive measures are taken. Corrosion can be reduced or prevented by providing a direct current through the electrolyte to the structure. This method is called cathodic protection (CP), as showed in Figure 17-10.

The basic concept of cathodic protection is that the electrical potential of the subject metal is reduced below its corrosion potential, such that it will then be incapable of corroding. Cathodic protection results from cathodic polarization of a corroding metal surface to reduce the corrosion rate. The anodic and cathodic reactions for iron corroding in an aerated near neutral electrolyte are:

$$Fe \rightarrow Fe^{2+} + 2e^{-} \qquad (17\text{-}6)$$

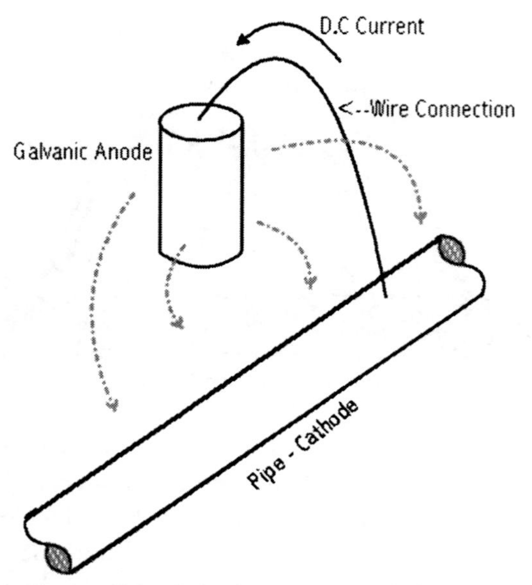

Figure 17-10 Cathodic Protection of a Pipeline

and

$$O_2 + 2H_2O + 4e^- \rightarrow 4OH^-, \quad (17\text{-}7)$$

respectively. As a consequence of Equation (17-7), the pH of the seawater immediate close to a metal surface increases. This is beneficial because of the precipitation of solid compounds (calcareous deposits) by the reactions:

$$Ca^{2+} + HCO_3^- + OH^- \rightarrow H_2O + CaCO_3 \quad (17\text{-}8)$$

and

$$Mg^{2+} + 2OH^- \rightarrow Mg(OH)_2 \quad (17\text{-}9)$$

These deposits decrease the oxygen flux to the steel and, hence, the current necessary for cathodic protection. As a result, the service life of the entire cathodic protection system is extended.

Offshore pipelines can be protected as a cathode by achieving a potential of –0.80 $V_{Ag/AgCl}$ or more negative, which is accepted as the protective potential (E_c^o) for carbon steel and low-alloy steel in aerated water. Normally, it is the best if the potentials negative to −1.05 $V_{Ag/AgCl}$ are avoided because these can cause a second cathodic reaction [7]:

$$H_2O + e^- \rightarrow H + OH^- \quad (17\text{-}10)$$

which results in
- Wasted resources;
- Possible damage to any coatings;
- The possibility of hydrogen embrittlement.

Cathodic protection systems are of two types: impressed current and galvanic anode. The latter has been widely used in the oil and gas industry for offshore platforms and marine pipeline in the past 40 years because of its reliability and relatively low cost of installation and operation. The effectiveness of cathodic protection systems allows carbon steel, which has little natural corrosion resistance, to be used in such corrosive environments as seawater, acid soils, and salt-laden concrete.

17.3.2. External Coatings

Oil and gas pipelines are protected by the combined use of coatings and cathodic protection. The coating systems are the primary barrier against corrosion and, therefore, are highly efficient at reducing the current demand for cathodic protection. However, they are not feasible for supplying sufficient electrical current to protect a bare pipeline. Cathodic protection prevents corrosion at areas of coating breakdown by supplying electrons.

Thick coatings are often applied to offshore pipelines to minimize the holidays and defects and to resist damage by handling during transport and installation. High electrical resistivity retained over long periods is a special requirement, because cathodic protection is universally used in conjunction with coatings for corrosion control. Coatings must have good adhesion to the pipe surface to resist disbondment and degradation by biological organisms, which abound in seawater. Pipe coating should be inspected both visually and by a holiday detector set at the proper voltage before the pipe is lowered into the water. Periodic inspection of the pipeline cathodic protection potential is used to identify the coating breakdown areas.

Coatings are selected based on the design temperature and cost. The principal coatings, in rough order of cost, are:
- Tape wrap;
- Asphalt;
- Coal tar enamel;
- Fusion bonded epoxy (FBE);
- Cigarette wrap polyethylene (PE);
- Extruded thermoplastic PE and polypropylene (PP).

The most commonly used external coating for offshore pipeline is the fusion bonded epoxy (FBE) coating. FBE coatings are thin-film coatings, 0.5 to 0.6 mm thick. They consist of thermosetting powders that are applied to a white metal blast-cleaned surface by electrostatic spray. The powder will melt on the preheated pipe (around 230°C), flow, and subsequently cure to form thicknesses of between 250 and 650 μm. The FBE coating can be used in conjunction with a concrete weight coating. The other coating that can be used with a concrete coating is coal tar enamel, which is used with lower product temperatures.

The external coating can be dual layer or triple layer. Dual-layer FBE coatings are used when additional protection is required for the outer layer, such as protection from high temperatures or abrasion resistance. For deepwater flowlines the high temperature of the internal fluid dissipates rapidly, reaching ambient within a few miles. Therefore, the need for such coatings is limited for steel catenary riser (SCRS) at the touchdown area where abrasion is high and an additional coating with high abrasion resistance is used. Triple-layer PP coating consists of an epoxy or FBE, a thermoplastic adhesive coating, and a PP top coat. The PE and PP coatings are extruded coatings. These coatings are used for additional protection against corrosion and are commonly used for dynamic systems such as steel catenary risers and where the temperature of the internal fluid is high. These pipe coatings are also frequently used in reel pipelines. The field joint coating for the three-layer systems is more difficult to apply and takes a longer time. However, in Europe, PE and PP coatings are preferred because of their high dielectric strength, water tightness, thickness, and very low CP current requirement.

17.3.3. Cathodic Protection

Cathodic protection is a method by which corrosion of the parent metal is prevented. For offshore pipelines, the galvanic anode system is generally used. This section specifies parameters to be applied in the design of cathodic protection systems based on sacrificial anodes.

17.3.3.1. Design Life

The design life t_r of the pipeline cathodic protection system is to be specified by the operator and shall cover the period from installation to the end of pipeline operation. It is normal practice to apply the same anode design life as for the offshore structures and submarine pipelines to be protected because maintenance and repair of CP systems are very costly.

17.3.3.2. Current Density

Current density refers to the cathodic protection current per unit of bare metal surface area of the pipeline. The initial and final current densities, i_c (initial) and i_c (final), give a measure of the anticipated cathodic current density demands to achieve cathodic protection of bare metal surfaces. They are used to calculate the initial and final current demands that determine the number and sizing of an anode. The initial design current density is necessarily higher than the average final current density because the calcareous deposits developed during the initial phase reduce the current demand. In the final phase, the developed marine growth and calcareous layers on the metal surface will reduce the current demand. However, the final design current density should take into account the additional current demand to repolarize the structure if such layers are damaged. The final design current density is lower than the initial density.

The average (or maintenance) design current density is a measure of the anticipated cathodic current density, once the cathodic protection system has attained its steady-state protection potential. This will simply imply a lower driving voltage, and the average design current density is therefore lower than both the initial and final design value.

Table 17-2 Summary of Recommended Design Current Densities for Bare Steel

Organization	Location	Water Temp. (°C)	Design Current Density (mA/m^2)		
			Initial	Mean	Final
NACE	Gulf of Mexico	22	110	55	75
	U.S. West Coast	15	150	90	100
	N. North Sea	0–12	180	90	120
	S. South Sea	0–12	150	90	100
	Arabian Gulf	30	130	65	90
	Cook Inlet	2	430	380	380
	Buried/mud zone	All	10–30	10–30	10–30
DNV	Tropical	>20	150/130	70/60	90/80
	Subtropical	12–20	170/150	80/70	110/90
	Temperate	7–12	200/180	100/80	130/110
	Arctic	<7	250/220	120/100	170/130
	Buried/mud zone	all	20	20	20
ISO	Nonburied	>20	—	70/60	90/80
		12–20	—	80/70	110/90
		7–12	—	100/80	130/110
		<7	—	120/80	170/130
		All	20	20	20

Note: DNV and ISO format: "(depths less than 30 m)/(depth greater than 30 m)."

Table 17-2 gives the recommended design current densities for the cathodic protection systems of nonburied offshore pipelines under various seawater conditions in different standards. For bare steel surfaces fully buried in sediments, a design current density of 20 mA/m^2 is recommended irrespective of geographical location or depth.

17.3.3.3. Coating Breakdown Factor

The coating breakdown factor describes the extent of current density reduction due to the application of a coating. The value $f_c = 0$ means the coating is 100% electrically insulating, whereas a value of $f_c = 1$ implies that the coating cannot provide any protection.

The coating breakdown factor is a function of coating properties, operational parameters and time. The coating breakdown factor f_c can be described as follows:

$$f_c = k_1 + k_2 \cdot t \tag{17-11}$$

where t is the coating lifetime, and k_1 and k_2 are constants that are dependent on the coating properties.

Four paint coating categories have been defined for practical use based on the coating properties in DNV [9]:

Category I: one layer of primer coat, about 50-μm nominal dry film thickness (DFT);

Category II: one layer of primer coat, plus minimum one layer of intermediate top coat, 150- to 250-μm nominal DFT;

Category III: one layer of primer coat, plus minimum two layers of intermediate/top coats, minimum 300-μm nominal DFT;

Category IV: one layer of primer coat, plus minimum three layers of intermediate top coats, minimum 450-μm nominal DFT.

The constants k_1 and k_2 used for calculating the coating breakdown factors are given in Table 17-3.

For cathodic protection design purposes, the average and final coating breakdown factors are calculated by introducing the design life t_r:

$$f_c(\text{average}) = k_1 + k_2 \cdot t_r/2 \tag{17-12}$$

$$f_c(\text{final}) = k_1 + k_2 \cdot t_r \tag{17-13}$$

17.3.3.4. Anode Material Performance

The performance of a sacrificial anode material is dependent on its actual chemical composition. The most commonly used anode materials are Al

Table 17-3 Constants (k_1 and k_2) for Calculation of Paint Coating Breakdown Factors [10]

	Coating Category			
	I	II	III	IV
	$k_1 = 0.1$	$k_1 = 0.05$	$k_1 = 0.02$	$k_1 = 0.02$
Depth (m)	k_2	k_2	k_2	k_2
0–30	0.1	0.03	0.015	0.012
>30	0.05	0.02	0.012	0.012

Table 17-4 Design Electrochemical Efficiency Values for Al- and Zn-Based Sacrificial Anode Materials [9]

Anode Material Type	Electrochemical Efficiency (Ah/kg)
Al-based	2000 (max 25°C)
Zn-based	700 (max 50°C)

Table 17-5 Design Closed-Circuit Anode Potentials for Al- and Zn-Based Sacrificial Anode Materials [9]

Anode Material Type	Environment	Closed-Circuit Anode Potential (V rel. Ag/AgCl Seawater)
Al-based	Seawater	−1.05
	Sediments	−0.95
Zn-based	Seawater	−1.00
	Sediments	−0.95

and Zn. Table 17-4 gives the electrochemical efficiency ϵ of anode materials applied in the determination of the required anode mass. The closed circuit anode potential used to calculate the anode current output should not exceed the values listed in the Table 17-5.

17.3.3.5. Resistivity

The salinity and temperature of seawater have an influence on its resistivity. In the open sea, salinity does not vary significantly, so temperature becomes the main factor. The resistivities of 0.3 and 1.5 $\Omega \cdot m$ are recommended to calculate anode resistance in seawater and marine sediments, respectively, when the temperature of surface water is between 7 and 12°C [9].

17.3.3.6. Anode Utilization Factor

The anode utilization factor indicates the fraction of anode material that is assumed to provide a cathodic protection current. Performance becomes unpredictable when the anode is consumed beyond a mass indicated by the utilization factor. The utilization factor of an anode is dependent on the detailed anode design, in particular, the dimensions and location of anode cores. Table 17-6 lists anode utilization factors for different types of anodes [9].

17.3.4. Galvanic Anode System Design

17.3.4.1. Selection of Anode Type

Pipeline anodes are normally of the half-shell bracelet type (see Figure 17-11). The bracelets are clamped or welded to the pipe joints after application of the corrosion coating. Stranded connector cables are be used for clamped half-shell anodes. For the anodes mounted on the pipeline with concrete, measures should be taken to avoid the electrical contact between the anode and the concrete reinforcement.

Normally, bracelet anodes are distributed at equal spacing along the pipeline. Adequate design calculations should demonstrate that anodes can provide the necessary current to the pipeline to meet the current density requirement for the entire design life. The potential of a pipeline should be polarized to –0.8 $V_{Ag/AgCl}$ or more negative. Figure 17-11 shows the potential profile of a pipeline protected by galvanic bracelet anodes.

Because the installation expense is the main part of CP design, larger anode spacing can reduce the overall cost. However, the potential is not evenly distributed along the pipeline. The pipeline close to the anode has a more negative potential. The potential at the middle point on the pipeline between two anodes is more positive and must be polarized to –0.80 $V_{Ag/AgCl}$ or more negative in order to achieve cathodic protection

Table 17-6 Design Utilization Factors for Different Types of Anodes

Anode Type	Anode Utilization Factor
Long 1, slender stand-off	0.90
Long 1, flush-mounted	0.85
Short 2, flush-mounted	0.80
Bracelet, half-shell type	0.80
Bracelet, segmented type	0.75

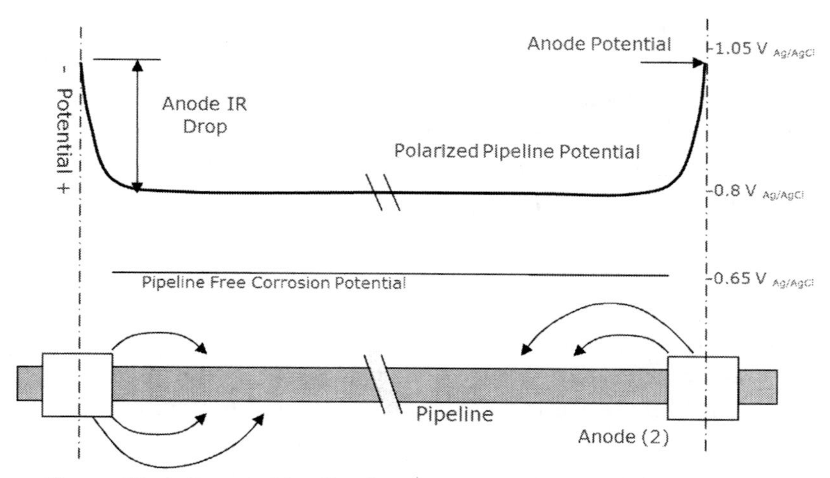

Figure 17-11 Potential Profile of a Pipeline Protected by Bracelet Anodes

for the whole pipeline. Increased anode spacing results in a bigger mass per anode, causing an uneven potential distribution. The potential close to the anode could be polarized to be more negative than $-1.05\ V_{Ag/AgCl}$, which should be avoided because of reaction 1.5. [10] Figure 17-12 illustrates the anticipated potential attenuation for situations using large anode spacing [10].

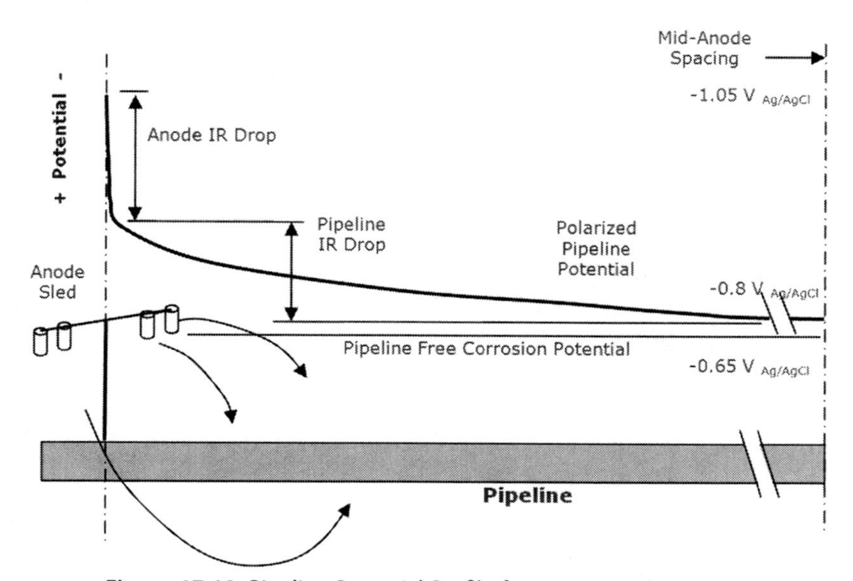

Figure 17-12 Pipeline Potential Profile for Large Anode Spacing

17.3.4.2. CP Design Practice

Offshore pipeline CP design includes the determination of the current demand I_c, required anode mass M, and number and current output per anode I_a. The current demand is a function of cathode surface area A_c, a coating breakdown factor f_c, and current density i_c, and can be expressed as follows [9]:

$$I_c = A_c \cdot f_c \cdot i_c \tag{17-14}$$

where i_c depends on water depth, temperature, seawater versus mud exposure, and whether or not the mean or final life of the CP system is being evaluated. Current density i_c is normally in the range of 60 to 170 mA/m^2 [9]. Because the initial polarization period preceding steady-state conditions is normally quite short compared to the design life, the mean (time-averaged) design current density i comes very close to the steady-state current density. Therefore, it is used to calculate the minimum mass of anode material necessary to maintain cathodic protection throughout the design life. Correspondingly, M can be calculated as:

$$M = \frac{8760 \cdot i \cdot {}_m T}{u \cdot C} \tag{17-15}$$

where u is a utilization factor, C is anode current capacity, and T is design life. The cathode potential is assumed to be spatially constant. Therefore, the current output per anode can be calculated by:

$$I_a = \frac{\phi_c - \phi_a}{R_a} \tag{17-16}$$

where ϕ_c and ϕ_a are the closed-circuit potential of the pipe and anode, respectively, and R_a is the anode resistance.

17.3.4.3. Anode Spacing Determination

Bethune and Hartt [11] have proposed a new attenuation equation to modify the existing design protocol interrelating the determination of the anode spacing L_{as}, which can be expressed as:

$$L_{as} = \frac{(\phi_c - \phi_a)}{\phi_{\text{corr}} - \phi_c} \cdot \frac{\alpha \cdot \gamma}{2\pi \cdot r_p \cdot R_a} \tag{17-17}$$

where,

ϕ_{corr}: free corrosion potential;

α: polarization resistance;

γ: reciprocal of coating breakdown factor f;

r_p: pipe radius.

This approach makes certain assumptions:

- Total circuit resistance is equal to anode resistance.
- All current enters the pipe at holidays in the coating (bare areas).
- The values of ϕ_c and ϕ_a are constant with both time and position. The ISO standards recommend that the distance between bracelet anodes not exceed 300 m [12].

17.3.4.4. Commonly Used Galvanic Anodes

The major types of galvanic anodes for offshore applications are slender stand-off, elongated flush mounted, and bracelet (Figure 17-13). The type of anode design to be applied is normally specified by the operator and should take into account various factors, such as anode utilization factor and current output, costs for manufacturing and installation, weight, and drag forces exerted by ocean current. The slender stand-off anode has the highest current output and utilization factor among these commonly used anodes.

17.3.4.5. Pipeline CP System Retrofit

Cathodic protection system retrofits become necessary as pipeline systems age. An important aspect of such retrofitting is determination of when such action should take place. Assessment of cathodic protection systems is normally performed based on potential measurements. As galvanic anodes waste, their size decreases; this causes a resistance increase and

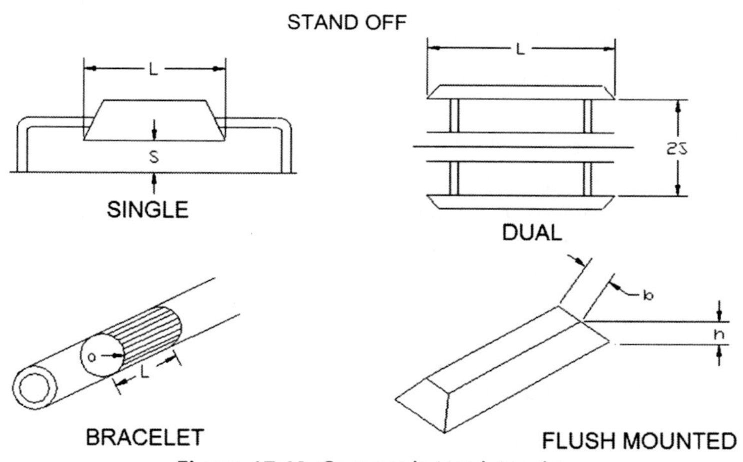

Figure 17-13 Commonly Used Anodes

a corresponding decrease in polarization. Models have been constructed for potential change that occurs for a pipeline protected by galvanic bracelet anodes as these deplete. Anode depletion is time dependent in the model.

Bracelet anodes have been used for cathodic protection of marine pipelines, especially during the "early period" (roughly 1964 to 1976), when many oil companies had construction activities in the Gulf of Mexico. According to recent survey data, many of these early anode systems have been depleted or are now being depleted. Retrofitting of old anode systems on pipelines installed in the 1960s and 1970s and even newer ones is required since these are still being used for oil transportation. Anodes can be designed as multiples or grouped together to form an anode array (anode sled) (see Figure 17-13). Anode arrays typically afford a good spread of protection on a marine structure. They are a good solution for retrofitting old cathodic protection systems.

17.4. SCALES

Scales are deposits of many chemical compositions as a result of crystallization and precipitation of minerals from the produced water. The most common scale is formed from calcium carbonate. Scale is one of the most common and costly problems in the petroleum industry. This is because it interferes with the production of oil and gas, resulting in an additional cost for treatment, protection, and removal. Scale also results in a loss of profit that makes marginal wells uneconomical. Scale deposition can be minimized using scale inhibition chemicals. Anti-scale magnetic treatment methods have been studied for the past several decades as an alternative. Acid washing treatments are also used for removal of scale deposits in wells.

The solubility of individual scale species is dependent on the equilibrium constants of temperature and pressure; the activity coefficients, which are dependent on concentrations plus the temperature and pressure of each individual species; the bulk ionic strength of the solution; and the other ionic species present. Once the solution exceeds the saturation limit, scale will begin to precipitate.

17.4.1. Oil Field Scales

Oil field scales are generally inorganic salts such as carbonates and sulfates of the metals calcium, strontium, and barium. Oil field scales may also be the complex salts of iron such as sulfides, hydrous oxides, and carbonates. Oil field scales may form in one of following two ways:

- Due to the change of temperature or pressure for brine during production, the solubility of some of the inorganic constituents will decrease and result in the salts precipitating.
- When two incompatible waters (such as formation water rich in calcium, strontium, and barium and seawater rich in sulfate) are mixed. Scales formed under these conditions are generally sulfate scales.

Table 17-7 lists the most common water-formed scales in oil field waters. Table 17-8 lists the common causes for the formation of these deposits.

17.4.1.1. Calcium Carbonate

Calcium carbonate, the most common scale in oil and gas field operations, occurs in every geographical area. Calcium carbonate precipitation occurs when calcium ion is combined with either carbonate or bicarbonate ions as follows,

$$Ca^{2+} + CO_3^{2-} \rightarrow CaCO_3 \tag{17-18}$$

$$Ca^{2+} + 2(HCO^-) \rightarrow CaCO_3 + CO_2 + H_2O \tag{17-19}$$

The preceding equations show that the presence of CO_2 will increase the solubility of $CaCO_3$ in brine. Increasing CO_2 also makes the water more acidic and decreases the pH. The calcium carbonate scaling usually occurs with a pressure drop, for example, at the wellbore. This reduces the partial pressure of CO_2, thereby increasing the pH and decreasing the $CaCO_3$ solubility. The solubility of calcium carbonate decreases with increasing temperature.

17.4.1.2. Calcium Sulfate

The precipitation of calcium sulfate is given by the reaction

$$Ca^{2+} + SO4^{2-} \rightarrow CaSO_4 \tag{17-20}$$

This scale may occur in different forms. Gypsum ($CaSO_4$•$2H_2O$) is the most common scale in oil field brines. It is associated with lower temperatures. Anhydrite ($CaSO_4$) may occur at high temperatures. Theoretically, anhydrite would be expected to precipitate above 100°F in preference to gypsum because of its lower solubility. However, gypsum may be found at temperatures as high as 212°F. With the passage of time, gypsum will dehydrate to anhydrite.

Table 17-7 Common Oil Well Scale Deposits—Solubility Factors [13]

Deposit	Chemical Formula	Mineral Name	Most Important Factors Affecting Solubility							
			Partial Pressure CO	pH	Total Pressure	Temp.	Total Salinity	Corrosion	H_2S	O_2
Calcium carbonate	$CaCO_3$	Calcite	X	X		X	X			
Calcium sulfate	$CaSO_4 \bullet 2H_2O$	Gypsum			X	X	X			
	$CaSO_4$	anhydrive								
Barium sulfate	$SrSO_4$	Barite			X	X	X			
Iron carbonate	$FeCO_3$	Celestite	X	X		X	X	X		
Iron sulfide	FeS	Trolite		X				X	X	
Iron oxide	Fe_2O_3	Hematite		X				X		X
	Fe_3O_4	Magnetite		X				X		X
Sodium chloride	NaCl	Halite				X	X			
Magnesium Mhydroxide	$Mg(OH)_2$	Brucite		X		X				
Silicates	Variable			X		X	X			

Table 17-8 Common Oil Well Scale Deposits—Causes and Removal Chemicals [13]

Deposit	Occurrence	Chemical Formula	Mineral Name	Most Frequent Causes of Scale Deposit	Removal Chemical
Calcium carbonate	Common	$CaCO_3$	Calcite	Mixing brines, changes in temperature and pressure	15% HCL
Calcium sulfate	Common	$CaSO_4 \bullet 2H_2O$ CaSO	Gypsum Anhydrive	Same as above	Converting solutions EDTA type dissolvers
Barium sulfate	Common	$BaSO_4$	Barite	Mixing of brines	?
Strontium sulfate	Not common	$SrSO_4$	Celestite	Mixing brines, changes in temperature and pressure	?
Iron carbonate	Common	$FeCO_3$	Siderite	Same as above	Sequestered acid
Iron sulfide	Common	FeS	Trolite	Corrosion by sour crude or H_2S Gas	Same as above
Iron oxide	Common	Fe_2O_3 Fe_3O_4	Hematite Magnetite	Reaction of oxygen with dissolved ferrous ion	Same as above
Sodium chloride	Not common	NaCl	Halite	Evaporation of water and addition of MeOH for hydrate control	Water or 1–3% HCl
Magnesium hydroxide	Not common	$Mg(OH)_2$	Brucite	Excessive amounts of oxygen enter the well or alkaline fluids in well, high temperature	15% HCl
Silicates		Variable		Cooling of hot brine high in dissolved silica	HCl:HF acid mixtures

A common mechanism for gypsum precipitation in the oil field is a reduction in pressure (e.g., at the production wellbore). The solubility increases with higher pressure because, when the scale is dissolved in water, the total volume of the system decreases.

17.4.1.3. Barium Sulfate

This scale is especially troublesome. It is extremely insoluble and almost impossible to remove chemically. Barium sulfate scaling is likely when both barium and sulfate are present, even in low concentrations.

$$Ba^{2+} + SO4^{2-} \rightarrow BaSO_4 \quad (17\text{-}21)$$

Barium sulfate scale is common in North Sea and GoM reservoirs. These fields often have barium in the original formation brine. Seawater injection (high sulfate concentration) for secondary oil recovery causes the scale problem. As the water flood matures and the seawater breaks through, these incompatible waters mix and a barium scale forms.

Generally, barium sulfate solubility increases with temperature and salinity. Similar to gypsum, $BaSO_4$ solubility increases with an increase in total pressure and is largely unaffected by pH.

17.4.1.4. Strontium Sulfate

Strontium sulfate is similar to barium sulfate, except fortunately its solubility is much greater:

$$Sr^{2+} + CO_4^{2-} \rightarrow SrSO_4 \quad (17\text{-}22)$$

Strontium sulfate solubility increases with salinity (up to 175,000 mg/L), temperature, and pressure. Again, pH has little effect. Pure strontium sulfate scale is rare except for some fields in the Middle East. $SrSO_4$ deposits in producing wells where the strontium-rich formation water mixes with the sulfate-rich injected seawater.

17.4.2. Operational Problems Due to Scales

Scale deposits are not restricted to any particular location in the production system, although some locations are more important than others in terms of ease and cost of remedial treatment. The following are areas or events where scale formation is possible in production systems.

17.4.2.1. Drilling/Completing Wells

Scale can cause problems at this early stage if the drilling mud and/or completion brine is intrinsically incompatible with the formation water. For example, allowing a seawater-based mud to contact a formation water rich in barium and strontium ions would be undesirable, similar to allowing a high calcium brine to contact a formation water rich in bicarbonate.

17.4.2.2. Water Injection

Scale problems may be encountered when new water injection wells are commissioned if the injection water is intrinsically incompatible with the formation water. For example, seawater injection into an aquifer rich in strontium and/or barium ions could cause problems.

17.4.2.3. Water Production

As soon as a well begins to produce water, the risk of carbonate scale formation arises, assuming that the produced water has a tendency to precipitate carbonate scale. The severity of the problem will depend on the water chemistry, the rate of drawdown, and other factors such as pressure and temperature.

17.4.2.4. HP/HT Reservoirs

HP/HT reservoirs have some potentially unique scaling problems due to the following characteristics:

- Total dissolved solids (TDS) up to 300,000+ ppm;
- Reservoir temperatures in excess of 350°F (175°C);
- Reservoir pressures in excess of 15,000 psi.

Examples are the Eastern Trough Area Project and Elgin/Franklin reservoirs in the North Sea.

17.4.3. Scale Management Options

Scale can be managed in several ways:

- Prevent deposition by using scale inhibitors, etc.
- Allow scale to form, but periodically remove it.
- Use pretreatments that remove dissolved and suspended solids.

The typical way of preventing scale deposition in oil field production is through the use of scale inhibitors.

17.4.4. Scale Inhibitors

Scale inhibitors are chemicals that delay or prevent scale formation when added in small concentrations in water that would normally create scale deposits. Use of these chemicals is attractive because a very low dosage (several ppm) can be sufficient to prevent scale for extended periods of time for either surface or downhole treatments. The precise mechanism for scale inhibitors is not completely understood but is thought to be following:

- Scale inhibitors may adsorb onto the surface of the scale crystals just as they start to form. The inhibitors are large molecules that can envelop these microcrystals and hinder further growth. This is considered to be the primary mechanism.
- Many oil field chemicals are designed to operate at oil/water, liquid/gas, or solid/liquid interfaces. Since scale inhibitors have to act at the interface between solid scale and water, it is not surprising that their performance can be upset by the presence of other surface active chemicals that compete for the same interface. Before deployment, it is important to examine in laboratory tests the performance of a scale inhibitor in the presence of other oil field chemicals.
- Because these chemicals function by delaying the growth of scale crystals, the inhibitor must be present before the onset of precipitation. Suspended solids (nonadherent scales) are not acceptable. This suggests two basic rules in applying scale inhibitors: (1) The inhibitor must be added upstream of the problem area. (2) The inhibitor must be present in the scaling water on a continuous basis to stop the growth of each scale crystal as it precipitates.

17.4.4.1. Types of Scale Inhibitors

The common classes of scale inhibitors include:

- Inorganic polyphosphates;
- Organic phosphates esters;
- Organic phosphonates;
- Organic polymers.

17.4.4.2. Scale Inhibitor Selection

Following are criteria for the selection of scale inhibitor:

- Efficiency;
- Stability;
- Compatibility. The inhibitor must not interfere with other oil field chemicals nor be affected by other chemicals.

The detailed factors in the selection of scale inhibitor candidates for consideration in the performance tests include:

- *Type of scale:* The best scale-inhibitor chemistry based on the scale composition should be selected.
- *Severity of scaling:* Fewer products are effective at high scaling rates.
- *Cost:* Sometimes the cheaper products prove to be the most cost effective; sometimes the more expensive products do.

- *Temperature:* Higher temperatures and required longer life limit the types of chemistry that are suitable.
- *pH:* Most conventional scale inhibitors perform less effectively in a low-pH environment.
- *Weather:* The pour point should be considered if the inhibitor will be used in a cold-climate operation.
- *Chemical compatibility:* The scale inhibitor must be compatible with other treatment chemicals, such as oxygen scavengers, corrosion inhibitors, and biocides.
- *Application technique:* This is most important if the inhibitor is to be squeezed into the formation.
- *Viscosity:* This is important when considering long umbilical applications such as in remote subsea fields.

17.4.5. Scale Control in Subsea Field

17.4.5.1. Well

The production wells of a HP/HT reservoir in the G0M have a potential for barium sulfate scale deposition in the formation at the near-wellbore location or within the tubing. These deposits occur due to mixing of injection seawater and formation water. Sulfates in the injected seawater react with naturally occurring barium in the formation water to induce barium sulfate scale. Barium sulfate is not soluble in acid, so prevention rather than remediation by acid treatment is the key.

- If left untreated, barium sulfate scale is likely to form downhole and possibly in the formation after produced water breakthrough. These areas are not treatable with continuous downhole chemical injection. For this reason the treatment method will consist of periodic batch scale squeeze treatments into each production well. It is estimated that each production well will require treatment once per year.
- The barium sulfate scale control will depend on accurate well testing and analysis of produced fluids to detect whether there is adequate scale inhibitor in the near-wellbore area in order to prevent the formation of the scale. Well tests and fluid analysis will be required for each well at a frequency of once every 2 weeks.

17.4.5.2. Manifold and Pipeline

Both barium sulfate and calcium carbonate scale formation may occur at the manifold and in the pipelines due to comingling of incompatible produced waters from different reservoirs. Deposition in these areas will be controlled by scale inhibitor injection at the subsea tree. Scale inhibitor injection is required upstream of the subsea choke.

REFERENCES

[1] C. de Waard, U. Lotz, D.E. Milliams, Predictive Model for CO_2 Corrosion Engineering in Wet Natural Gas Pipelines, Corrosion vol. 47 (no12) (1991), pp. 976–985.
[2] Norwegian Technology Standards Institution, CO_2 Corrosion Rate Calculation Model, NORSOK Standard No. M-506, 2005.
[3] R. Nyborg, Controlling Internal Corrosion in Oil and Gas Pipelines, Oil and Gas Review(Issue 2) (2005).
[4] A. Dugstad, L. Lunde, K. Videm, Parametric Study of CO_2 Corrosion of Carbon Steel, Corrosion/94, NACE International, paper no.14, Houston, TX, 1994.
[5] National Association of Corrosion Engineers, Metals for Sulfide Stress Cracking and Stress Corrosion Cracking Resistance in Soul Oilfield Environments, NACE MR 0175 (December 2003).
[6] www.corrosion-doctors.org/Inhibitors/lesson11.htm.
[7] D.A. Jones, Principles and Prevention of Corrosion, first ed., McMillan, New York, 1992, pp. 437–445.
[8] E.D. Sunde, Earth Conduction Effects in Transaction Systems, Dover Publishing, New York, 1968, pp. 70–73.
[9] Det Norsk Veritas, Cathodic Protection Design, DNV RP B401, 1993.
[10] W.H. Hartt, X. Zhang, W. Chu, Issues Associated with Expiration of Galvanic Anodes on Marine Structures, Paper 04093, Corrosion (2004).
[11] K. Bethune, W.H. Hartt, A Novel Approach to Cathodic Protection Design for Marine Pipelines: Part II -Applicability of the Slope Parameter Method, presented at Corrosion, paper no.00674, 2000.
[12] International Organization for Standardization, Pipeline Cathodic Protection-Part 2:Cathodic Protection of Offshore Pipelines, ISO/TC 67/SC 2 NP 14489, Washington, D.C, 1993.
[13] DeepStar, Flow Assurance Design Guideline, 2001.

CHAPTER 18

Erosion and Sand Management

Contents

Subsea Engineering Handbook
ISBN 978-0-12-397804-2, doi:10.1016/B978-0-12-397804-2.00018-7

18.1. INTRODUCTION

The fluid associated with hydrocarbon production from wells is a complex multiphase mixture, which may include the following substances,

- Hydrocarbon liquids: oil, condensate, bitumen;
- Hydrocarbon solids: waxes, hydrates;
- Hydrocarbon gases: natural gas;
- Other gases: hydrogen sulfide, carbon dioxide, nitrogen and others;
- Water with dilute salts;
- Sand and other particles.

Erosion has been long recognized as a potential source of problems in hydrocarbon production systems. Many dangerous elbow failures due to erosion have occurred on production platforms, drilling units, and other subsea equipment in the past decades. The inherently variable nature of the erosion process makes it very difficult to develop best practice recommendations for the design and operation of hydrocarbon production systems with elbows. The potential mechanisms that could cause erosion damage are:

- Particulate erosion (sand erosion);
- Liquid droplet erosion;
- Erosion-corrosion;
- Cavitation.

Particulate erosion due to sand (sand erosion) is the most common source of erosion problems in hydrocarbon systems, because small amounts of sand entrained in the produced fluid can result in significant erosion and erosion-corrosion damage. Even in "sand-free" or clean service situations where sand production rates are as low as a few pounds per day, erosion damage could also be very severe at high production velocities. Sand erosion can also cause localized erosion damage to protective corrosion scales on pipe walls and result in accelerated erosion-corrosion damage.

However, all other mechanisms are equally aggressive under the right conditions. Erosion is affected by numerous factors, and small or subtle changes in operational conditions can significantly affect the damage it causes. This can lead to a scenario in which high erosion rates occur in one production system, but very little erosion occurs in other seemingly very similar systems. Detection of erosion as it progresses is also difficult, which makes erosion management difficult, especially for those unfamiliar with the manner in which erosion occurs.

The hydrodynamics of the multiphase mixture of hydrocarbons within a pipeline has a very important influence on a number of physical phenomena, which determine the corrosion and erosion performance of the system. In many pipeline applications, particularly those involving gas fields and gas production systems, liquid droplets are present as the dispersed phase. The impingement of these droplets onto surfaces can cause erosion.

Velocity is an important parameter for hydrocarbon production flow. Excessive fluid velocities caused by incorrect pipe sizing and/or process design can have a detrimental effect on the pipeline and fittings and the effectiveness of any chemical inhibitors. The phenomena involved are complex, involving aspects of the hydrodynamics of the pipeline, the chemistry of the multiphase mixture within the pipeline, and the materials used to manufacture the pipeline.

18.2. EROSION MECHANISMS

There are two primary mechanisms of erosion. The first is erosion caused by direct impingement. Normally, the most severe erosion occurs at fittings that redirect the flow such as at elbows and tees. The particles in the fluid can possess sufficient momentum to traverse the fluid streamlines and impinge the pipe wall. The other mechanism is erosion caused by random impingement. This type of erosion occurs in the straight sections of pipe even though there is no mean velocity component directing flow toward the wall. However, turbulent fluctuations in the flow can provide the particles with momentum in the radial direction to force them into the pipe wall, but the turbulent fluctuations are a random process, therefore, it is termed *random* impingement. These two mechanisms can cause different types of erosion based on the fluid compositions, velocity, and configurations of piping systems.

Venkatesh [1] provides a good overview of erosion damage in oil wells. Regardless of the erosion mechanism, the most vulnerable parts of production systems tend to be components in which:

- The flow direction changes suddenly.

- High flow velocities occur that are caused by high volumetric flow rates.
- High flow velocities occur that are caused by flow restrictions.

Components and pipe systems upstream of the primary separators carry multiphase mixtures of gas, liquid, and particulates and are consequently more likely to suffer from particulate erosion, erosion-corrosion, and droplet erosion. Also the vulnerability of particular components to erosion heavily depends on their design and operational conditions. However, the following list is suggested as a rough guide to identify which components are most vulnerable to erosion (the first on the list being most likely to erode):

- Chokes;
- Sudden constrictions;
- Partially closed valves, check valves, and valves that are not full bore;
- Standard radius elbows;
- Weld intrusions and pipe bore mismatches at flanges;
- Reducers;
- Long radius elbows, miter elbows;
- Blind tees;
- Straight pipes.

18.2.1. Sand Erosion

The rate of sand erosion is determined by following factors [2]:

- Flow rate of sand and the transport manner;
- Velocity, viscosity, and density of the flowing fluid;
- Size, shape, and hardness of the particles (sand);
- Configuration of the flow path such as straight tubing, elbow, or tee.

The effect of configuration of the flow path has already be described in the last section (section 18.2), only the effects of these factors are detailed in following programs.

18.2.1.1. Flow Rate of Sand and Manner of Transport

The nature of sand and the way in which it is produced and transported determines the rate of erosion within a production system. The sand production rate of a well is determined by a complex combination of geological factors, and it can be estimated by various techniques. Normally, new wells produce a large amount of sand as they "clean up." Sand production then stabilizes at a relatively low level before increasing again as the well ages and the reservoir formation deteriorates. The sand production rate is not stable, and if a well produces less than 5 to 10 lb/day, it is often regarded as being sand free. However, this does not eliminate the possibility that erosion may be taking place.

The sand transport mechanism is an important factor for controlling sand erosion. Gas systems generally run at high velocities (>10 m/s), making them more apt to erosion than liquid systems. However, in wet gas systems sand particles can be trapped and carried in the liquid phase. Slugging in particular can periodically generate high velocities that may significantly enhance the erosion rate. If the flow is unsteady or the operational conditions change, sand may accumulate at times of low flow and be washed out through the system when high flows occur. The flow mechanisms may act to concentrate sand, increasing erosion rates in particular parts of the production system.

18.2.1.2. Velocity, Viscosity, and Density of the Fluid

The sand erosion rate is highly dependent on, and proportional to, the sand impact velocity. When the fluid velocity is high enough, the sand impact velocity will be close to the fluid velocity, and erosion will be an issue. Therefore, erosion is likely to be the worst where the fluid flow velocity is the highest. Small increases in fluid velocity can cause substantial increases in the erosion rate when these conditions are satisfied. In dense viscous fluids particles tend to be carried around obstructions by the flow rather than impacting on them. While in low-viscosity, low-density fluids, particles tend to travel in straight lines, impacting with the walls when the flow direction changes. Sand erosion is therefore more likely to occur in gas flows, because gas has a low viscosity and density, and gas systems operate at higher velocities.

18.2.1.3. Sand Shape, Size, and Hardness

Sand sizes in hydrocarbon production flow depend on the reservoir geology, the size of the sand screens in the well, and the breakup of particles as they travel from the reservoir to the surface. Without sand exclusion measures, such as downhole sand screens, particle sizes typically range between 50 and 500 microns. With sand exclusion in place particles larger than 100 microns are usually excluded. A sand particle density of about 2600 kg/m^3 is generally accepted.

Particle size influences erosion primarily by determining how many particles impact on a surface. Figure 18-1 shows the paths of particles as they are carried through an elbow. The flow paths depend on the particle weight and the amount of drag imparted on the particles by the fluid as

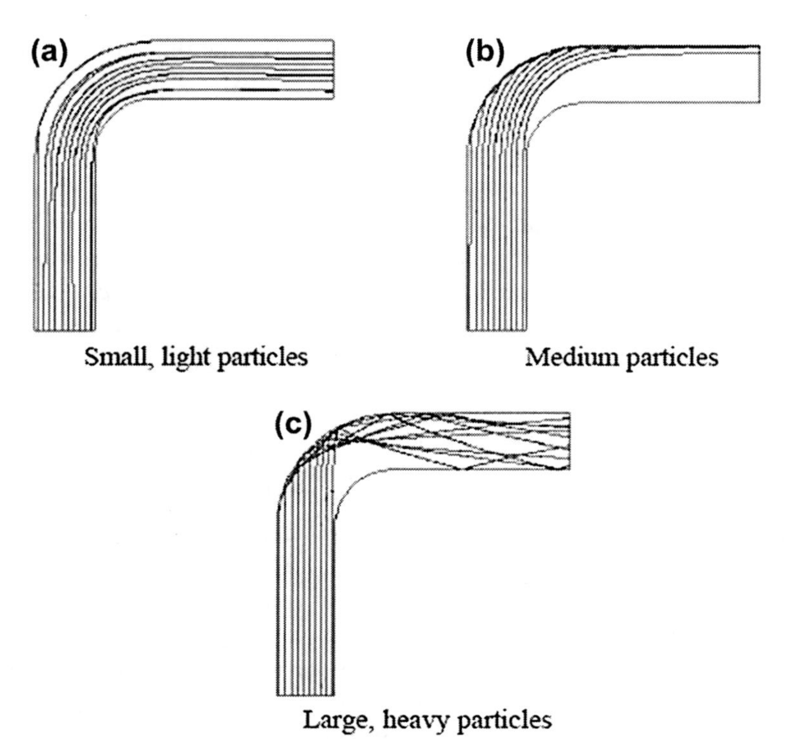

Figure 18-1 Paths of Different Sized Particles through an Elbow [2]

they pass through the elbow. Figure 18-1a shows particle paths typically seen for small sand grains (of the order of 10 microns) in a liquid flow. Figure 18-1b is representative of typically sized sand grains (of the order of 200 microns) in liquid flows, and Figure 18-1c is representative of typically sized sand in gas flows. Small light particles require very little drag to change direction. Therefore, they tend to follow the flow as shown in Figure 18-1a. This figure also represents a case of particles flowing in a highly viscous, dense fluid. Figure 18-1c illustrates the path of large heavy particles in an elbow, the large heavy particles have a relatively high momentum and they will hardly be deflected by the fluid flow; therefore, they tend to travel in straight lines, bouncing off the elbow walls as they go. Figure 18-1c illustrates the case where particles are flowing in a low-density, low-viscosity fluid.

Normally, hard particles cause more erosion than soft particles, and sharp particles do more damage than rounded particles. However, it is not clear whether the variability of sand hardness and sharpness causes a significant difference between the erosion rate in production systems associated with different wells or fields.

18.2.2. Erosion-Corrosion

Erosion often causes localized grooves, pits, or other distinctive patterns in the locations of elevated velocity. Corrosion is usually more dispersed and identifiable by the scale or rust it generates. Erosion-corrosion is a combined effect of particulate erosion and corrosion. The progression of the erosion-corrosion process depends on the balance between the erosion and corrosion processes. Erosion-corrosion can be avoided by ensuring that operating conditions do not allow either erosion or corrosion.

In a purely corrosive flow, without particulates in it, new pipe system components typically corrode very rapidly until a brittle scale develops on the surfaces exposed to the fluid. After this scale has developed, it forms a barrier between the metal and the fluid that substantially reduces the penetration rate. In this case, very low-level erosion is also taking place simultaneously with corrosion. In highly erosive flows, in which corrosion is also occurring, the erosion process dominates, and scale is scoured from exposed surfaces before it can influence the penetration rate. Corrosion therefore contributes little to material penetration. At intermediate conditions erosion and corrosion mechanisms can interact. In this case, scale can form and then be periodically removed by the erosive particles.

18.2.3. Droplet Erosion

Droplet erosion occurs in wet gas or multiphase flow systems in which droplets can form. The erosion rate is dependent on a number of factors including the droplet size, impact velocity, impact frequency, and liquid and gas density and viscosity. It is very difficult to predict the rate of droplet erosion because most of these values are unknown in the field conditions.

Experimental results indicate that under a wide range of conditions, the material lost by droplet erosion varies with time. Initially, the impacting droplets do not cause erosion due to the existence of protective layers on the surface. However, after a period of time, rapid erosion sets in and the weight loss becomes significant and will increase linearly with time.

The hydrodynamics of the multiphase mixture within the pipeline also affects the degree of wetting of the pipe walls and the distribution of corrosion inhibitors injected into the pipeline system. Above a certain velocity, the inhibitor film will be removed, leading to increased rates of corrosion. Currently, to determine the critical velocity, the following empirical correlations are used:

$$\nu_{\text{erosion}} = \frac{C}{\sqrt{\rho}} \tag{18-1}$$

where

ν_{erosion}: velocity of erosion, ft/s;

ρ: mixture density, lb/ft^3;

C: constant defined by Table 18-1.

Table 18-1 shows different erosion velocities proposed by various researchers. From the API specification [3], the maximum gas velocity should not exceed 15 m/s (50 ft/s) in order to prevent stripping of the film-forming corrosion inhibitor from the inside surface of the pipe. However, there is a suggestion that the API specification is overly conservative. The suggestion in DNV-RP-O501 [4] that droplet erosion and liquid impingement erosion are unlikely to occur in steel components at velocities below 70 to 80 m/s is probably more realistic although it is not clear where these values come from.

18.2.4. Cavitation Erosion

When liquid passes through a restrictive low-pressure area, cavitation can be generated. If the pressure is reduced below the vaporization pressure of the liquid, bubbles are formed. These bubbles then collapse and generate shock waves. These shock waves can damage a pipe system. Cavitation is rare in oil and gas production systems because the normal operating pressure is

Table 18-1 Erosion Velocity for Gas/Condensate Systems with Various Ratios of Condensate and Gas (CGRs)

			API 14E [3]	S&V [6]
	Mixture Density (kg m^{-3})/(lb ft^{-3})		$c = 100$	$c = 300$
CGR [bbl/mmscf]	kg m^{-3}	lb ft^{-3}	m/s	m/s
1.0	64.0	4.0	15.3	45.7
10.0	68.5	4.3	14.7	44.2
50.0	78.5	4.9	13.8	41.3
100.0	90.5	5.7	12.8	38.5

generally much higher than the liquid vaporization pressures. Evidence of cavitation can be sometimes found in chokes, control valves, and pump impellers, but is unlikely to occur in other components.

The onset of cavitation in equipment or components with flow constrictions can be predicted by calculating a cavitation number K, defined as below:

$$K = \frac{2(P_{\min} - P_{vap})}{\rho \cdot v^2} \tag{18-2}$$

where

$P_{\min}$: minimum pressure occurring in the vicinity of the restriction, Pa;

P_{vap}: vapor pressure of the liquid, Pa;

ρ: density of the liquid, kg/m^3;

v: flow velocity through the restriction, m/s.

A cavitation number of less than 1.5 indicates that cavitations may occur.

18.3. PREDICTION OF SAND EROSION RATE

Knowing a facility's erosion risk is of the utmost importance for well management optimization against sand production. Normally, the sand erosion models consider the erosion process into three stages. First, the fluid flow in the facilities is modeled or in some way approximated. This flow prediction is then used to derive the drag forces imparted by the fluid on the particles (sands); hence, the trajectories of a large number of particles are predicted. Computational fluid dynamics (CFD) software has been used to model the fluid flow and particle trajectories. It has been shown to be good at predicting erosion locations and erosion scar shapes. Second, the damage due to the individual particles' impact on a wall is calculated using a material-specific empirical or a theoretically derived impact damage model. Last, the average impact damage of a large number of particles can then be used to predict the distribution and depth of erosion damage on a surface. However, the physical process is only partially understood and the prediction of the critical conditions has to rely on empirical or semi-empirical models.

On the basis of experimental results, the sand erosion rate can be summarized in relation to the kinetic energy of sand fragments through the following parameters:

- Fluid velocity;
- Fluid density;

- Size of the sand fragments;
- Sand production rate;
- Pipe or conveyance diameter;
- System geometry;
- Metal hardness (resistance to erosion).

The sand erosion rate is expressed in the following form based on the impact damage model:

$$E = A \cdot V_p^n \cdot F(\alpha) \tag{18-3}$$

where,

E: sand erosion rate, kg of material removed/kg of erodant;
V_p: particle impact velocity;
A: a constant depending on the material being eroded and other factors;
α: particle impact angle;
$F(\alpha)$: material dependent function of the impact angle, which is between 0 and 1.0;
n: material dependent index.

18.3.1. Huser and Kvernvold Model

Huser and Kvernvold [5] developed the following impact damage equation:

$$E = m_p \cdot K \cdot V_p^n \cdot F(\alpha) \tag{18-4}$$

where,

m_p: mass flow of particles impacting on an area;
K, n: constants given for steel and titanium grade materials and Glass-reinforced plastics (GRP) piping.

The values for K, n, and $F(\alpha)$ are derived from sand-blasting tests on small material samples.

Figure 18-2 shows the angle relationship $F(\alpha)$ used by Huser and Kvernvold. This model has been used with both CFD particle models and with various empirical particle models. DNV RP -0501 [5] compiles these empirical particle models for the calculation of sand erosion in straight pipes, around welds, and in elbows, tees, and reducers.

18.3.2. Salama and Venkatesh Model

Salama and Venkatesh [6] developed a method similar to that of Huser and Kvernvold, although they simplified their model by making a conservative assumption that all sand impacts occur at about 30°, therefore, $F(\alpha)$ is set to

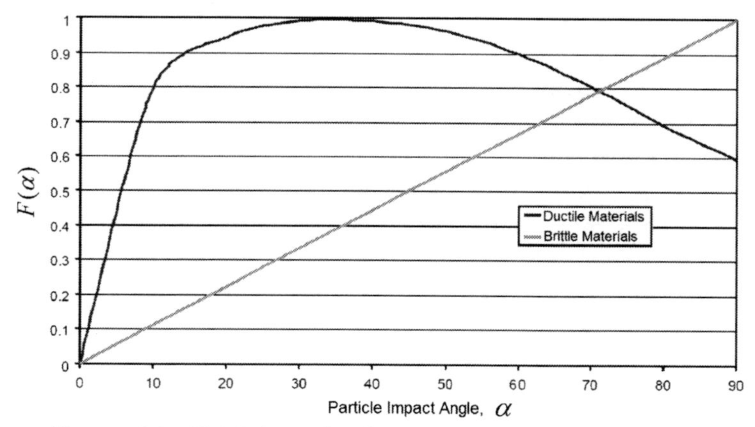

Figure 18-2 $F(\alpha)$ Relationship for Ductile and Brittle Materials [5]

1. This approximation is reasonable for gas flows, but does not account for the particle drag effects in liquid flows.

$$Er = \frac{S_k \cdot W \cdot V^2}{D^2} \tag{18-5}$$

where:

Er: erosion rate, thousandths of inch per year;

W: sand flow rate, lb/day;

V: fluid velocity, ft/sec;

D: pipe diameter, in.;

S_k: geometry dependant constant; $S_k = 0.038$ for short radius elbows and $S_k = 0.019$ for ells and tees.

The predicted erosion rate based on the above equation is an average of 44% greater than the measured value in comparison with experimental air/sand tests of elbows. This method is consistently conservative. Svedeman and Arnold [7] also suggested using this equation, but they gave values of S_k for gas systems, with an S_k of 0. 017 for long radius elbows and an S_k of = 0.0006 for plugged tees.

18.3.3. Salama Model

Salama [8] updated the Salama and Venkatesh model [6] for gas/liquid systems and developed a new equation as follows:

$$E_p = \frac{1}{S_p} \cdot \frac{V_m^2 \cdot d}{D^2 \cdot \rho_m} \tag{18-6}$$

where

E_p: erosion rate, mm/kg;
D: pipe diameter, mm;
d: particle diameter, microns;
S_p: geometrical constant
V_m: mixture velocity (m/s) defined by $V_m = V_{\text{liquid}} + V_{\text{gas}}$
ρ_m: mixture density defined by $\rho_m = (\rho_{\text{liquid}} \cdot V_{\text{liquid}} + \rho_{\text{gas}} \cdot V_{\text{gas}})/V_m$

Particle size and liquid effects are included in the equation. By calibrating against experimental data from Weiner and Tolle [9] and Bourgoyne [10], the geometrical constants are given as follows:

$S_p = 2000$ for elbows (1.5D and 5D);
$S_p = 12{,}000$ for ells;
$S_p = 25{,}000$ for plugged tees (liquid/gas) and 500,000 for plugged tees (gas).

18.3.4. Tulsa ECRC Model

A range of particle models have been developed by the University of Tulsa Erosion Corrosion Research Center (ECRC) based on a significant amount of research work on pipe component erosion. Those models are utilized with an impact damage model of the form:

$$E = F_M \cdot F_S \cdot m_p \cdot V_p^n \cdot F(\alpha) \tag{18-7}$$

In the above equation the coefficient of F_M accounts for the variation in material hardness. McLaury and Shirazi [11] give typical values of F_M for a number of different steels ranging from 0.833 to 1.267, suggesting a ±25%

Table 18-2 Material Properties and Erosion Ratio Coefficients for Noncarbon Steels

Material Type	Hardness Brinell B	Material Factor* $F_M \times 10^6$	Material Factor** $F_M \times 10^6$
1018	210	0.833***	1.066
13 Cr annealed	190	1.267	1.622
13 Cr heat treat	180	1.089	1.394
22 Cr 5 Ni duplex	217	0.788	1.009
316 Stainless steel	183	0.918	1.175
Incoloy 825	160	0.877	1.123

Notes:
*For VL in m/s.
**For VL in ft/s.
***For carbon steel materials, $F_M = 1.95\times10^{-5}/B^{-0.59}$ (for VL in m/s) where B is the Brinell hardness factor.

in erosion resistance between different steels. These values have been derived from impingement tests. Table 18-2 summarizes material properties of hardness and material factor for noncarbon steels.

The coefficient of F_S accounts for sand sharpness: 1.0, 0.53, and 0.2 for sharp, semirounded, and rounded grains, respectively. A value of 0.53 is used to represent production systems.

18.4. THRESHOLD VELOCITY

The commonly used practice for controlling sand erosion in gas and oil producing wells is to limit production velocity; the critical velocity is called fluid threshold velocity, below which an allowable amount of erosion occurs.

The API RP14E guideline [3] limits the production rates for avoiding erosional damage and the recommended velocity limitation described by the Equation (18-8). The recommended value for the constant C is 100 for continuous service and 125 for intermittent service. When sand is present, API RP 14E suggests that the value of C should be smaller than 100 but does not indicate what the value should be. Furthermore, the sand properties, fluid viscosity, etc., cannot be considered in API RP 14E.

$$V_{\text{erosion}} = \frac{c}{\sqrt{\rho}} \tag{18-8}$$

The recently developed methods for predicting threshold velocities are based on the penetration rates in elbow geometry because the sections with this geometry are more susceptible to erosion damage than a straight pipe section.

18.4.1. Salama and Venkatesh Model

Salama and Venkatesh [6] developed a model for predicting the penetration rate for an elbow. Their suggested equation is:

$$h = 496920 \cdot \left(q_{sd} v^2 p / T \cdot d^2 \right) \tag{18-9}$$

where

h: penetration rate, mil/yr;

q_{sd}: sand production rate, ft^3/D;

v_p: particle impact velocity, ft/s;
T: hardness, psi;
d: pipe diameter, μm.

By assuming the hardness of T is equal to 1.55×10^5 psi and allowing the penetration rate of h to equal mil/yr, Salama and Venkatesh obtained an expression for the erosion velocity of threshold velocity for sand erosion:

$$v_e = \frac{1.73 \cdot d}{\sqrt{q_{sd}}} \tag{18-10}$$

The authors also suggested that this equation be used for gas flows only and indicated that particle-impact velocity in gas flows with low density and viscosity nearly equals the flow stream velocity. They noted that this equation is not valid for liquid flows because the threshold velocity given by this equation actually represents the particle-impact velocity, which is generally lower than the flow stream velocity.

18.4.2. Svedeman and Arnold Model

On the basis of Salama and Venkatesh's work, Svedeman and Arnold [8] suggested the following formula for predicting a threshold velocity based on the penetration rate of 5 mil/yr:

$$v_e = K_s \left(\frac{d}{\sqrt{q_{sd}}} \right) \tag{18-11}$$

where
v_e: erosional-threshold velocity, ft/s;
K_s: empirical constant based on data from [2].

18.4.3. Shirazi et al. Model

Shirazi et al. [12] proposed a method for calculating the penetration rates for various pipe geometries, in which the following expression is used for calculating the maximum penetration rate in elbows for carbon steel material:

$$h = A \cdot F_s \cdot F_p \cdot \left(q_{sd} \cdot \rho_p \cdot v_p^{1.73} / \mathrm{B}^{0.59} \cdot d^2 \right) \tag{18-12}$$

where
h: the penetration rate, mil/yr;
A:coefficient dependent on pipe material;
F_s: sand sharpness factor as listed in Table 18-3;

Table 18-3 Sand Sharpness Factor, F_s

Description	F_s
Sharp corners, angular	1.0
Semi-rounded corners	0.53
Rounded, spherical glass beads	0.20

F_p: penetration factor (where F_p of 3.68 in./lb·m is obtained from experimental data for steel elbows and tees);
ρ_p: particle density;
v_p: particle impact velocity;
B: Brinell hardness.

This method accounts for many of the physical variables in the flow and erosion processes and includes a way to predict the maximum penetration rate for sand erosion. The capabilities of the method are evaluated by comparing predicted penetration rates with experimental data found in the literature. A major difference between this method and the earlier work lies in that this method is developed to finding the "characteristic impact velocity of the particles" on the pipe wall, v_p. This characteristic impact velocity of the particles depends on many factors, including pipe geometry and size, sand size and density, flow regime and velocity, and fluid properties.

Given an allowable penetration rate (of 5 or 10 mil/yr), a threshold flow stream velocity can be readily calculated and combined with the procedure of impact velocity particles using the iterative solution procedure as shown in the following section.

18.4.4. Particle Impact Velocity

Generally, the erosion rate is proportional to the particle impact velocity, so there is a necessity to analyze the characteristic impact velocity in detail.

A simple model is used to describe the characteristic of impact velocity of particles. Figure 18-3 shows the procedure presented conceptually. In this procedure, a concept "stagnation zone," which represents the fluid layer that the sand particles must penetrate, is defined. Also a characteristic length, called the equivalent stagnation length L, is used to represent this distance in the simpler direct impingement model. Figure 18-4 illustrates erosion testing results of the equivalent stagnation length, which can be used to calculate L for elbow geometry.

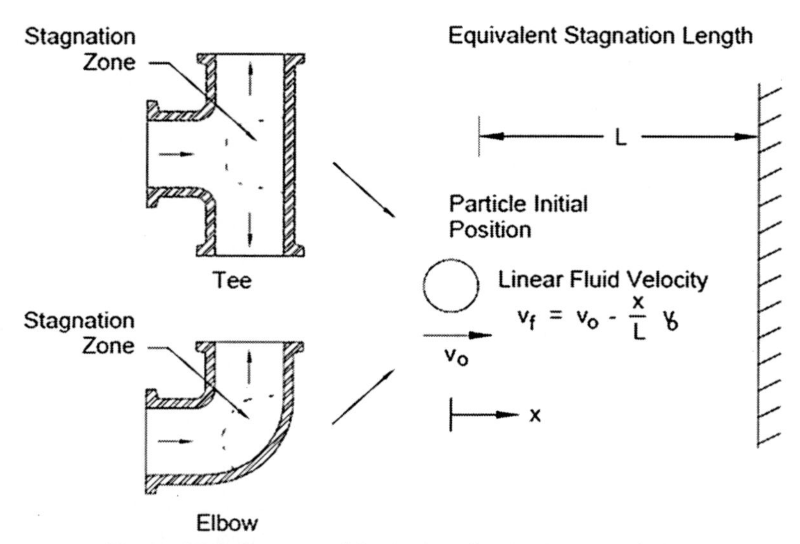

Figure 18-3 Concept of Equivalent Stagnation Length [11]

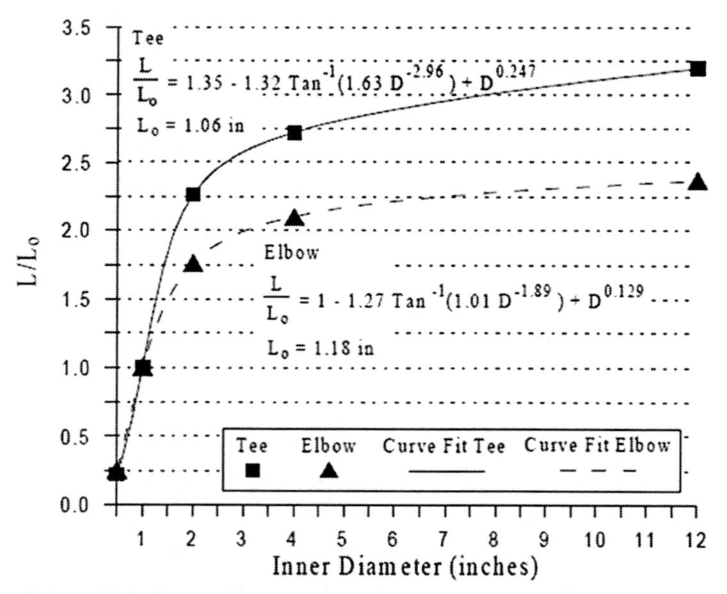

Figure 18-4 Stagnation Length versus Pipe Diameter for Elbow [11]

The behavior of the particles in the stagnation region mainly depends on:

- Pipe-fitting geometry;
- Fluid properties;
- Sand properties.

A simplified particle-tracking model is used to compute the characteristic impact velocity of the particles. This model assumes that the article is traveling through an inside diameter flow field that is assumed to have a linear velocity in the direction of the particle motion and uses a simplified drag-coefficient model.

The impact velocity depends on:

- Flow stream velocity;
- Characteristic length scale;
- Fluid density and viscosity;
- Particle density and diameter.

These parameters are combined into three dimensionless groups related to one another as shown in Figure 18-5. The dimensionless groups are defined as follows:

Particle Reynolds number, N_{Re}

$$N_{\mathrm{Re}} = \rho_f \cdot v \cdot d_p / \mu_f \tag{18-13}$$

where

ρ_f: density of fluid;
v: velocity of fluid;

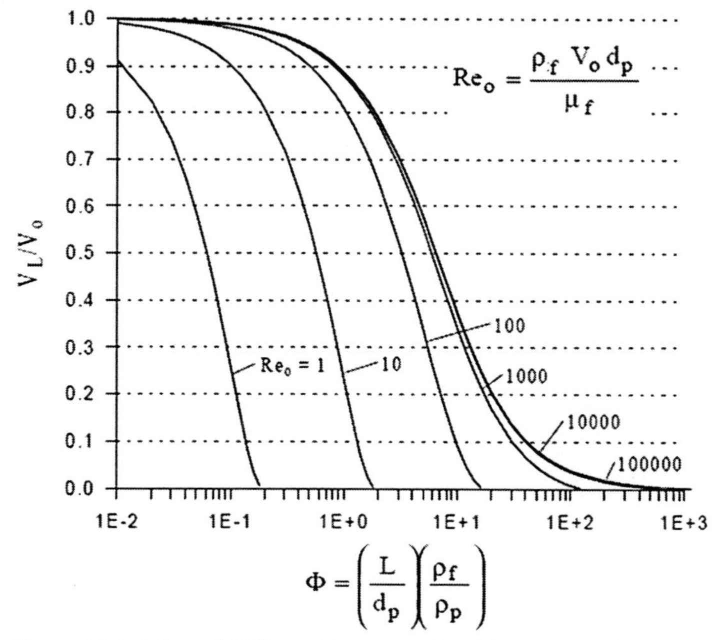

Figure 18-5 Effect of Different Factors on Particle Impact Velocity [11]

d_p: particle diameter;
μ_f: viscosity of fluid.

Dimensionless parameter, Φ

$$\Phi = L \cdot \rho_f / d_p \cdot \rho_p \tag{18-14}$$

Figure 18-5 contains much useful information about how various parameters affect v_p and sand erosion. For example, it shows how fluid and sand properties affect v_p. Once v_p is determined, it is used in Equation (18-11) to calculate the erosion and penetration rates for a specific geometry, such as an elbow.

18.4.5. Erosion in Long Radius Elbows

In this model, the erosion condition in a long radius elbow has been studied on the basis of a standard elbow mechanistic model. To extend the mechanistic model to be able to predict the penetration rate in long radius elbows, a new term called the *elbow radius factor* is introduced [13]. The elbow radius factor ($\mathrm{ERF}_{r/d}$) is defined as follows:

$$\mathrm{ERF}_{r/d} = \frac{Pn_L}{Pn_{std}} \tag{18-15}$$

where Pn_L is the maximum penetration rate in the long radius elbow, and Pn_{std} is the maximum penetration rate in a standard elbow. The introduction of elbow radius factor preserves the accuracy of the mechanistic model for standard elbows and extends it to predict penetration rates in long radius elbows.

$$\mathrm{ERF}_{r/d} = e^{-\left(\frac{\rho_f^{0.4}\mu_f^{0.65}}{d_p^{0.3}} \cdot 0.215 + 0.03\rho_f^{0.25} + 0.12\right)\left(\frac{r}{d} - C_{std}\right)} \tag{18-16}$$

where,

$\mathrm{ERF}_{r/d}$: elbow radius factors for long radius elbows;
C_{std}: r/d of a standard elbow; C_{std} is set equal to 1.5;
ρ_f: fluid density;
μ_f: fluid viscosity;
d_p: particle diameter;
r: radius of curvature of the elbow.

This equation accounts for the elbow radius curvature effect in different carrier fluids and sand particle size. Note that this equation is based on a sand particle density of $165.41\mathrm{b/ft}^3$.

The model did not investigate the effects of turbulent fluctuation on the erosion predictions because direct impingement is the dominant erosion mechanism for elbows. However, as the radius of curvature increases significantly, the long radius elbow becomes closer to a straight section of pipe and the random impingement mechanism can become important.

18.5. EROSION MANAGEMENT

Erosion management includes the erosion monitoring and erosion mitigation methods.

18.5.1. Erosion Monitoring

The following methods are used for erosion monitoring on steel pipes or special tab erosion [14]:

- Ultrasonic gauges are used to clamp to the external surface of the pipe. They send out an ultrasonic pulse to measure the thickness and the material loss from which to determine the erosion severity. The method is sensitive to the noise from other sources; also the primary limitation of this method is that it only checks a limited local region of the pipe.
- Weight-loss coupons made of the same or similar material as the pipe being monitored are installed and periodically retrieved and weighed. They provide only discrete monitoring and are unsuitable for subsea engineering equipment.
- Electrical resistance probes measure the accumulated erosion as an increase in electrical resistance on a known cross section. Calibration and temperature changes are of concern.
- Electrochemical probes determine the erosion rate through measurement of the linear polarization resistance between electrodes through a conductive electrolyte flowing inside the pipe. This method is suitable only for conductive liquids such as water, or oil systems with high water cuts.

18.5.2. Erosion Mitigating Methods

A number of measures can be taken to mitigate erosion [3], as discussed next.

18.5.2.1. Reduction of Production Rate

Reducing the production rate includes reducing the flow velocity and sand production rate. However, this has adverse financial implications.

18.5.2.2. Design of Pipe System

Minimizing the flow velocity and avoiding sudden changes (e.g., at elbow, constrictions, and valves) in the flow direction should be given much attention in order to reduce the severity of any erosion. Blind tees are generally perceived as being less prone to erosion than elbows, so the use of full-bore valves and blind tees in place of elbows can reduce erosion problems. Also, the flow regime has an impact on erosion problems and slugging flows can be particularly damaging; therefore, slug catchers may be appropriate for reducing the severity of any erosion.

18.5.2.3. Increasing Wall Thickness

Thick-walled pipes are often used to increase the wear life of a pipe system. However, the thick wall thickness reduces the pipe bore, which in turn elevates flow velocities and increases the erosion rate, particularly with small-bore pipe systems.

18.5.2.4. Specialized Erosion-Resistant Materials

Generally, in oil and gas production systems nearly all of the components will be made of ductile metals, although other materials such as plastic and rubber may also be used. Material properties have a significant effect on erosion problems. If erosion problems are suspected, specialized erosion-resistant materials such as tungsten carbide can be used.

The primary factor of ductile materials in controlling erosion is their hardness. Consequently, steels are more resistant than other softer metals. In vulnerable components, specialized materials such as tungsten carbides, coatings, and ceramics are often used. These materials are generally hard and brittle and have a super erosion resistance to steel (often orders of magnitude better). However, some coated materials' resistance may rapidly reduce once the coating or its substrate fails.

Brittle materials erode in a different manner. Impacts on brittle materials abrade the surface, and erosion increases linearly with the impact angle, until reaching a maximum for perpendicular impacts.

18.6. SAND MANAGEMENT

Sand management is an operating concept in which traditional sand control means are not normally applied and production is managed through monitoring and control of well pressures, fluid rates and sand influx [14]. Sand management has proven to be an effective tool in North Sea oil and gas production wells.

Sand management has proven to be workable, and has led to the generation of highly favorable well skins because of self-cleanup associated with the episodic sand bursts that take place. These low skins have, in turn, led to higher productivity indexes, and each of the wells where sand management has been successful has displayed increased oil or gas production rates. Furthermore, expensive sand control devices are avoided and the feasibility of possible future well interventions is guaranteed.

Different analysis and design tools are necessary for evaluating the sand production probability, for quantifying risk reduction, and for establishing practical operational criteria for safe and optimum production. Such design tools include the capacity to predict:

- Sand production onset;
- Sand quantities and production rate;
- Equipment erosion risks;
- Conditions of sand transported inside the production flow line.

Another essential tool is sand monitoring technology that allows for real-time quantitative sand influx tracking.

18.6.1. Sand Management Philosophy

Classical sand control techniques, such as gravel packing, use of wire-wrapped or expandable screens, frac-and-pack, and chemical consolidation, are based on a sand exclusion philosophy: Absolutely no sand in the production facilities can be tolerated. Alternatively, in the absence of means of totally excluding sand influx, the traditional approach is to reduce the production rate to minimize the amount of entering sand.

The most extensive field validation of the reliability and cost effectiveness of sand management is possessed by the Canadian heavy-oil wells. This approach is a combination of techniques that define and extend the safety limits. The too conservative approaches in which no sand is permitted into the production system are avoided or delayed.

18.6.2. The Sand Life Cycle

Risk management requires reliable analysis of the "sand life cycle," starting with predicting formation conditions conducive to sanding, and ending with the ultimate disposal of the produced material at the surface. These techniques are based on:

- An extensive data acquisition of the field;

- Theoretical modeling of the involved physical processes;
- Active monitoring and follow-up on production data;
- Well testing to optimize production rates.

Also, the techniques will help the production engineer optimize the design and provide risk assessment throughout the well's production life.

18.6.2.1. Sand Detachment

Sand detachment is a mixed hydromechanical process, which releases sandstone fragments from the formation near the well, can be viewed as a mixed hydromechanical process. Many models have been established to predict the sanding initiation conditions.

First, due to excessive drawdown or reservoir pressure depletion, the production stratum fails in compression or extension from excessive local stresses at the free surface near the wellbore. Alternatively, sanding may result from formation weakening, perhaps from fatigue effects related to repeated well shut-downs, or from water breakthrough and related capillary or chemical cohesion loss.

Second, the yielded material is destabilized and fluidized by hydrodynamic forces from the fluid flow into the well. In addition, the force varies with time along with local geometry, so sand cannot flow constantly and is likely to be produced as bursts, which has been verified by small-scale laboratory experiments. Transient pressure gradient effects that result from well shut down and start-up and relative permeability changes are the major reasons for the episodic increases in the sand influx, also these are the best known causes for increased forces acting on the sand in the vicinity of wellbore.

18.6.2.2. Sand Transport

Once sand is detached, it follows the fluid through the perforations and into the well. Then gravity and hydrodynamic forces will act on the grains and sand fragments. The effects of sand, including the probability of transport to the surface, blocking of perforation tunnels, or settling into the well sump or horizontal well section, depends on the balance of the following factors:

- Fluid rheology and density;
- Local flow velocity;
- Local geometrical obstructions;
- Sand fragment size;
- Well inclination.

In particular, sand may sediment and be remobilized later as conditions change (e.g., velocity changes, water cut increases) in long horizontal wells.

These events may often be interpreted as sanding because of formation failure, rather than as a well cleanup process.

18.6.2.3. Sand Erosion

Sand erosion including the area of tubing, flow lines, and chokes is significantly related to the sand transport process. The kinetic energy of the moving particles is transferred to the steel when they impinge on a surface, causing abrasive steel removal. Generally, sand flow rate and sand fragment velocity are the two main factors determining the erosion risk. For heavy crude fields, the velocity is low and the risk is also low; however, HP/HT gas condensate fields, gas expansion and acceleration near the wellhead dramatically increase the erosion risk.

Erosion risk is a major technical and economical constraint because it may lead to severe safety problems. Erosion forces the production rate to be kept below a limit that is considered safe.

18.6.2.4. Surface Sand Deposition

Once sand passes the wellhead, it passes through the surface lines, or, for subsea wells, through the sea line, to deposit in the separator, which must be cleaned and flushed from time to time according to the expected average sand rate. The oil-contaminated sand that is produced is collected and sent for ultimate disposal. For subsea engineering equipment, it has been dumped into the sea in the past, but this practice is not viewed as an option in future operations.

18.6.3. Sand Monitoring

Sand monitoring is a critical aspect of sand management. Sand monitors are used when erosion problems are suspected. Current sand monitoring methods are discussed next.

18.6.3.1. Volumetric Methods

Volumetric methods include the following issues:

- Sand traps can be installed to capture sand at tees or bends usually. However, sand traps are not a real-time method because they need to be disassembled to measure the sand production. These techniques have not proven effective, because the majority of the produced sand is normally not captured. (North Sea experience indicates a recovery of 1% to 10%.)
- Fluid sampling, including centrifugation for water and sand cuts after the primary separator, includes measurements of bottom sediment and

water, which are carried out during appraisal well testing or during normal production. However, this method cannot be guaranteed due to the much remaining in the primary separator.

- Another method that has been used quite extensively in the Adriatic Sea on gas wells consists of dismounting the sand separator, jetting it clean of all sand, and quantifying all produced solids. However, the accuracy and practicality (i.e., the time and manpower required to dismount, jet, and remount the separator) limit its application.
- Use of an in-line sand cyclone is a new method used on some North Sea platforms. Sand is effectively separated from the produced fluid and stored in a tank. The load cells or other devices on the tank allow for the measurement of sand accumulation in real time.

18.6.3.2. Acoustic Transducers

Acoustic transducers have proven to be an effective sand monitoring tool. Installed in the flow system, a transducer includes the following items:

- An impact probe installed in the flow line to detect sand grain impacts;
- An acoustic collar that can be used to capture the information about impact of the sand grains against the wall of pipe or the choke throat.

18.6.4. Sand Exclusion and Separation

Downhole sand screens and gravel packs are often used to stop sand from entering the production system. Typically, sand screens prevent particles larger than 100 microns from entering the production stream. However, a balance should be struck between reducing the productivity by including a sand screen and having to choke back an unprotected well to avoid excessive sand production. In addition, even very small particles can generate a significant degree of erosion; therefore, sand screens and gravel packs cannot guarantee erosion-free operation.

Sand separation can be conducted at three principal levels:

- The primary separation facility topside;
- The subsea separation module (for offshore installations);
- In the downhole separator in conjunction with oil/water separation.

Surface modules include high-pressure horizontal baffle-plate separators, vertical gravitational separators, and centrifugal segregators. These devices can be very effective at protecting chokes in particular. In subsea and downhole separation, water-wet sand will normally be separated by gravity and stored or reinjected. A specially designed sand cyclone or sand centrifuge may be required for oil-wet sand.

Table 18-4 Evaluation of Different Sand Prevention Methods

Control Method	Major Short-comings
Chemical consolidation	• Some permeability reduction • Placement and reliability issues • Short intervals only
Screens, slotted liners, special filters	• Lack of zonal isolation • High placement and workover costs • Longevity of devices • Plugging and screen collapse • Screen erosion • Potential damage during installation
Inside-casing gravel packing	• PI reduction • Placement and workovers difficult • High cost of installation • Positive skin development
Open-hole gravel packing	• PI reduction • Complexity of operation • Necessity for extensive underreaming in most cases • Costs of installation
Propped fracturing, including frac-and-pack, stress frac, and use of resin-coated sand	• Permeability recovery • Risks of tip screen-out during installation • Directional control and tortuosity issues (in inclined wells) • Fracture containment control • Proppant flowback on production
Selective perforating	• Problematic in relatively homogeneous formations • Need for formation strength data • Reduces inflow area
Oriented perforating	• Necessity for full stress mapping • Theoretical analysis required • Perforation tool orientation needed • Limited field validation available
Production rate control	• Erosion of facilities • Sand monitoring required • Separation and disposal required • Potential for lost production

18.6.5. Sand Prevention Methods

Table 18-4 evaluates and lists disadvantages of the different methods used for dealing with sand production. In general, sand control represents high-cost/low-risk solution. Sand management leads to a low-cost solution, but it also involves active risk management.

18.7. CALCULATING THE PENETRATION RATE: EXAMPLE

This example illustrates the calculation process for determining the maximum penetration rate for a long radius elbow. The calculation includes the following steps:

- Determine the elbow radius factor ($ERF_{r/d}$).
- Calculate the particle impact velocity;
  - Equivalent stagnation length L
  - Dimensionless parameters, N_{Re} and Φ;
  - v_p/v based on Figure 18-5.
- Compute the penetration rate in a standard elbow.
- Determine the penetration rate in long radius elbow.

In this example, a 2-in. pipe with a Schedule 40, ASTM 234 Grade WPB seamless elbow with an ID of 2.067 in. was used. The r/d ratio of the elbow was 3.0, and the liquid used was clay/mud at a velocity of 25 ft/s; the sand rate is 1800 ft^3/D. Detailed input data for the example is summarized in Table 18-5.

18.7.1. Elbow Radius Factor

The elbow radius factor (ERFr/d) is calculated according to Equation (18-16). In this calculation, the unit of density is lb/ft^3 and the viscosity of μ_f is cp. The calculated ERFr/d is 0.304.

Table 18-5 Input Data for Calculating Penetration Rate

Parameter	Description and Units	Value
A	Coefficient dependent on pipe material	0.9125
r/d	Long radius elbow	3
F_s	Sand sharpness factor (Table 18-3)	0.53
F_p	Penetration factor	3.68
qsd	Sand rate (ft^3/D)	1800
ρ_p	Particle density (lbm/ft^3)	165.4
ρ_f	Fluid density (lbm/ft^3)	65.4
v	Fluid velocity (ft/s)	25
d	Pipe diameter (in.)	2
ID	Elbow diameter (in.)	2.067
μ_f	Fluid viscosity($lbm/ft \cdot s$)	3.36×10^{-3}
B	Brinell hardness	120
d_p	Particle diameter (μm)	350

18.7.2. Particle Impact Velocity

The particle impact velocity is calculated based on the methods illustrated in Section 18.4.4.

Step 1: Equivalent Stagnation Length *L*

Calculation of the equivalent stagnation length L is based on Figure 18-4. For $L_0 = 1.18$ in., Equation (18-17), shown in Figure 18-4, is used:

$$\frac{L}{L_0} = 1 - 1.27\arctan(1.01d^{-1.89}) + d^{0.129} \tag{18-17}$$

and $L/L_0 = 1.78$ is calculated. Therefore, L =, so $L = 1.78 \times 1.18 = 2.1$ in.

Step 2: Compute N_{Re} and Φ

The dimensionless numbers N_{Re} and Φ are calculated using parameters with consistent units. These parameters are used for Φ: $L = 2.1$ in., $d_p = 0.0138$ in. (350 μm), $\rho_f = 65.4$ lbm/ft^3, and $\rho_p = 165.4$ lbm/ft^3. We obtain $\Phi = 60.285$. For N_{Re}, $\rho_f = 65.4$ lbm/ft^3, $v = 25$ ft/s, $d_p = 1.148 \times 10^{-3}$ ft (350 μm), and $\mu f = 3.36 \times 10^{-3}$ lbm/ft·s. We obtain $N_{Re} = 558.77$.

Step 3: Determine the v_p/v of a Valve

The v_p/v is obtained from Figure 18-5 based on the given N_{Re} and Φ. For this example, $v_p/v = 0.0175$. Then, v_p is calculated with $v_p = (v_p/v) \times v = 0.0175 \times 25 = 0.4375$ ft/s.

18.7.3. Penetration Rate in Standard Elbow

In this example, semirounded corner sand is assumed. From Equation (18-12), with $A = 0.9125$, $F_s = 0.53$, $F_p = 3.69$ in./lbm, $v_p = 0.4375$ ft/s, $q_{sd} = 1800$ ft^3/D, and $d_{elbow} = 2.067$ in./1.0 in. $= 2.067$:

$$h := 0.9125 \cdot 0.53 \cdot 3.68 \frac{1800 \cdot 165.4 \cdot 0.4375^{1.73}}{120^{0.59} \cdot 2.067^2}$$

$h = 1761$ mil/yr.

18.7.4. Penetration Rate in Long Radius Elbow

Equation (18-15) is rewritten as the following equation:

$$P_{nL} = \text{ERF}_{r/d} \cdot Pn_{std} \tag{18-18}$$

Then, $P_{nL} = 0.304 \times 1761 = 535.344$ mil/yr.

REFERENCES

[1] E.S. Venkatesh, Erosion Damage in Oil and Gas Wells, Proc. Rocky Mountain Meeting of SPE, Billings, MT (1986) 489–497. May 19-21.

[2] N.A. Barton, Erosion in Elbows in Hydrocarbon Production System: Review Document, Research Report 115, HSE, ISBN 0 7176 2743 8, 2003.

[3] American Petroleum Institute, Recommended Practice for Design and Installation of Offshore Production Platform Piping Systems, fifth ed., API- RP-14E, 1991.

[4] Det Norsk Veritas, Erosive Wear in Piping Systems, DNV- RP- O501 (1996).

[5] A. Huser, O. Kvernvold, Prediction of Sand Erosion in Process and Pipe Components, Proc. 1st North American Conference on Multiphase Technology, Banff, Canada, pp. 217–227 (1998).

[6] M.M. Salama, E.S. Venkatesh, Evaluation of API RP 14E Erosional Velocity Limitation for Offshore Gas Wells, OTC 4485, Offshore Technology Conference, Houston, Texas, 1983.

[7] S.J. Svedeman, K.E. Arnold, Criteria for Sizing Multiphase Flow Lines for Erosive/ Corrosive Service, SPE 26569, 68th Annual Technical Conference of the Society of Petroleum Engineers, Houston, Texas, 1993.

[8] M.M. Salama, An Alternative to API 14E Erosional Velocity Limits for Sand Laden Fluids, OTC 8898, pp. 721 –733, Offshore Technology Conference, Houston, Texas (1998).

[9] P.D. Weiner, G.C. Tolle, Detection and Prevention of Sand Erosion of Production Equipment. API OSAPR Project No 2, Research Report, Texas A&M University, College Station, Texas, 1976.

[10] T. Bourgoyne, Experimental Study of Erosion in Diverter Systems. SPE/IADC 18716, Proc SPE/IADC Drilling Conference, New Orleans, 28 February - 3 March, pp. 807–816, 1989.

[11] B.S. McLaury, S.A. Shirazi, Generalization of API RP 14E for Erosive Service in Multiphase Production, SPE 56812, SPE Annual Technical Conference and Exhibition, Houston, Texas, 1999.

[12] S.A. Shirazi, B.S. McLaury, J.R. Shadley, E.F. Rybicki, Generalization of the API RP 14E Guideline for Erosive Services, SPE28518, Journal of Petroleum Technology, August 1995 (1995) 693–698.

[13] B.S. McLaury, J. Wang, S.A. Shirazi, J.R. Shadley, E.F. Rybicki, Solid Particle Erosion in Long Radius Elbows and Straight Pipes, SPE 38842, SPE Annual Technical Conference and Exhibition, San Antonio, Texas, 1997.

[14] J. Tronvoll, M.B. Dusseault, F. Sanfilippo, F.J. Santarelli, The Tools of Sand Management, SPE 71673, 2001, SPE Annual Technical Conference and Exhibition held in New Orleans, Louisiana, 2001.

PART *Three*

Subsea Structures and Equipment

CHAPTER 19

Subsea Manifolds

Contents

Subsea Engineering Handbook
ISBN 978-0-12-397804-2, doi:10.1016/B978-0-12-397804-2.00019-9

19.1. INTRODUCTION

Subsea manifolds have been used in the development of oil and gas fields to simplify the subsea system, minimize the use of subsea pipelines and risers,

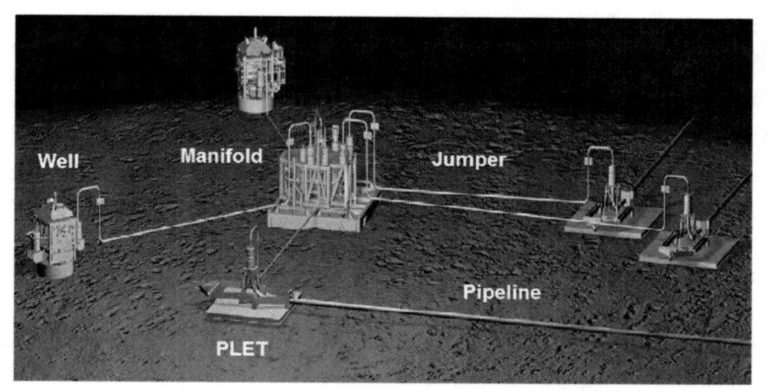

Figure 19-1 Typical Subsea Structure System [1]

and optimize the fluid flow of production in the system. The manifold is an arrangement of piping and/or valves designed to combine, distribute, control, and often monitor fluid flow. Subsea manifolds are installed on the seabed within an array of wells to gather product or to inject water or gas into wells as shown in Figure 19-1. There are numerous types of manifolds, ranging from a simple pipeline end manifold (PLEM/PLET) to large structures such as an entire subsea process system. A PLEM is one of the most common manifolds in this range and it is detailed in the next section separately because it is directly connected to pipelines and its installation considerations are key factors in the design.

A subsea manifold system is structurally independent from the wells. The wells and pipelines are connected to the manifold by jumpers. The subsea manifold system is mainly comprised of a manifold and a foundation. The manifold support structure is an interface between the manifold and foundation. It provides necessary facilities to guide, level, and orient the manifold relative to the foundation. The connection between the manifold and the manifold supporting structure allows the manifold to retrieve and reinstall with sufficient accuracy so as to allow for the subsequent reuse of the production and well jumpers.

A manifold is a structural frame with piping, valves, control module, pigging loop, flow meters, etc. The foundation provides structural support for the manifold. It may be either a mudmat with skirt or a pile foundation, depending on seabed soil conditions and manifold size. The functions and purposes of subsea manifolds can be summarized as follows:

- Provide an interface between the production pipeline and well.
- Collect produced fluids from individual subsea wells.

- Distribute production fluids, inject gas, and inject chemicals and control fluid.
- Distribute electrical and hydraulic systems.
- Support manifold wing hubs, pipeline hubs, and umbilical hubs.
- Support and protect all pipe work and valves.
- Provide lifting points for the manifold system during installation and retrieval.
- Provide a support platform for ROVs during ROV operations.
- Provide sea-fastening interfaces.

Subsea manifold systems are often configured for the following specific functions:

- Production and/or test manifolds, for controlling flow of individual wells into production and test headers, which is connected with pipelines;
- Gas injection manifolds, for injecting gas into the riser base to decrease slugs in the production flow;
- Gas lift manifolds, for injecting gas into the tubing to lighten the fluid column along the tubing in order to increase oil production;
- Water injection manifolds, for supplying water to the last valve in the well before the shutdown valve to increase oil production;
- Choke or kill manifolds, for controlling well operations.

This chapter discusses the issues and standards related to the design, manufacture, and testing of subsea manifolds. Requirements for individual components such as valves, actuators, and piping will also be detailed. The discussion includes the following contents:

- Manifold components;
- Manifold design and analysis;
- Foundation and suction pile;
- Installation and maintenance.

19.1.1. Applications of Manifolds in Subsea Production Systems

Subsea manifolds are an integral part of many subsea developments that provides effective gathering and distribution points for subsea systems. The manifolds are designed in various configurations, such as template or cluster manifolds with internal or external pigging loops, horizontal or vertical connectors. Two applications of manifolds in subsea production systems are described in the following sections along with their functions.

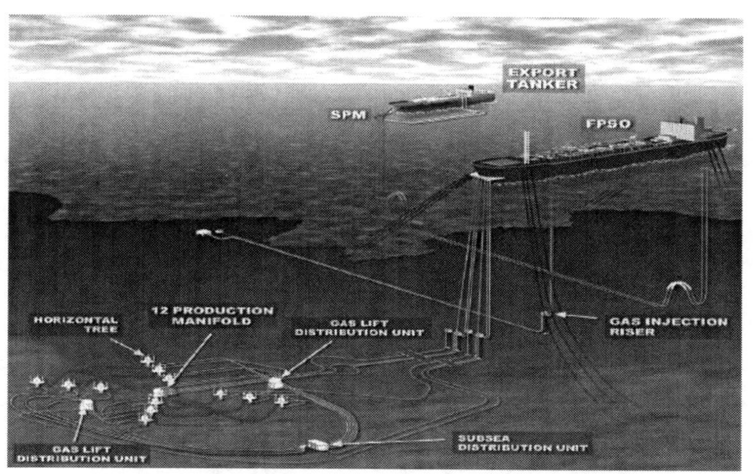

Figure 19-2 Field Arrangement of Kuito Field Phase 1A [2]

19.1.1.1. Kuito Field Phase 1A

The manifold used in Kuito Field Phase 1A is a clustered manifold system with 12 production wells [2]. The production manifold is tied back with three flowlines to an FPSO. Each well is connected to the manifold with a 4-in. (10.2cm) flexible well jumper. The field arrangement of Kuito Field Phase 1A is illustrated in Figure 19-2.

The production manifold has an 11-in. (0.28m) production header designed for round-trip pigging and a 6-in. (0.15m) test header for individual testing of each well. There is a hydraulically actuated ball valve in line with each header. The 11-in. (0.28m) ball valve isolates the two production flowlines and allows pigging of the production lines. The 6-in. (0.15m) ball valve isolates the test header from the production headers. Twenty-four 4 1/16-in. (0.10m) hydraulically actuated gate valves allow each well to be connected to either the production or test header.

The manifold structure is permanently installed on a 30-in. (0.76m) pile. A guide mast, installed in the pile prior to installation of the manifold, provides guidance during installation of the manifold structure and reduces the vertical loads during landing.

19.1.1.2. The Gemini Field

The Gemini Field is located in Mississippi Canyon Block 292 at a water depth of 3400 ft (1000m), and it is a subsea development tied back to the Viosca Knoll 900 Platform.

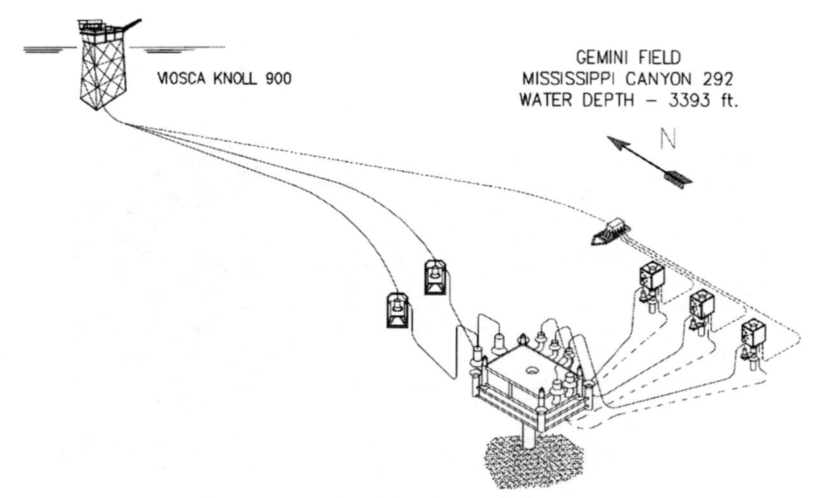

Figure 19-3 Gemini Field Arrangement [3]

The subsea system consists of three subsea wells connected to a four-slot manifold using 4-in. (0.10m)-diameter flexible jumpers. The manifold is supported on the seafloor using a single conductor. Two 12-in. (0.30m) rigid jumpers connect the manifold to two 12-in. (0.30m) flowlines that extend 27.5 miles(44.3km) to the platform. Details of the Gemini Field arrangement are shown in Figure 19-3.

The subsea manifold gathers the production from three wells and directs it to one of the two flowlines to the platform. The manifold consists of dual 12-in. (0.30m) headers and a removable pigging loop. One of the 12-in. (0.30m) headers contains an ROV-operated isolation valve, whereas the other includes a remotely operated fail-safe closed valve that serves as the pigging isolation valve while the system is in normal production. The manifold has the ability to accept production from up to four subsea wells with four 4-in. nominal connection hubs. Each hub is piped to a dual valve block, which provides the ability to access either header. The valves are remotely operated through the control pod on the tree associated with the connection.

The manifold was designed to be a moon-pool deployed and retrieved using the Ocean Star semisubmersible drilling rig and it is supported on the seabed using a single 36-in. (0.91m) suction pile.

19.1.2. Trends in Subsea Manifold Design

Continuous innovations in subsea manifold design have been achieved in past two decades as the water depth of oil/gas field developments has increased. The diverless subsea manifolds present quite a few more challenges than the manifolds used in shallow water. Operating needs and costs are the main factors pushing these innovations. The development of subsea manifolds has gone through the following stages based on the applications of Petrobras' subsea manifolds in deepwater [4]:

19.1.2.1. First Stage

The manifolds of the first stage were designed to be operated in water depths up to 1000 m (3280.8ft). The manifolds have an independent subsea support base with four lift points at the top of frame. A leveling system is designed to compensate up to 2° degrees of inclination, and normally four retrievable chokes and valve modules are set, with each one for two different wells. The manifolds at this stage were designed to guarantee reliability and installation in any soil conditions, but the design resulted in a large, heavy manifold, which made installation difficult and expensive. Figure 19-4a shows a first-stage manifold being installed.

19.1.2.2. Second Stage

The next stage of manifold design was diverless manifolds being built that incorporated the experience acquired from the first-stage subsea manifolds. Figure 19-4b shows a typical manifold from the second stage. A comparison of the manifolds in the first and second stages is summarized in Table 19-1.

The weight of the manifolds in the second stage was reduced, but also the size. No subsea base installations were required because geomechanical analysis of the soil confirmed that no special foundation was necessary. The manifold structure itself has a mudmat with skirts around its base and is designed to compensate the seabed slope. No additional leveling system was

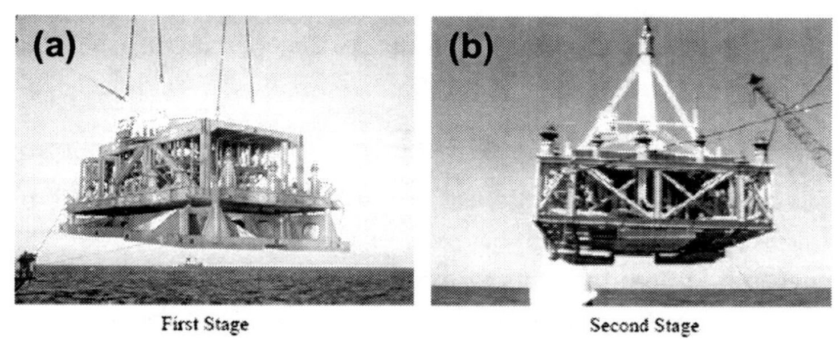

Figure 19-4 Typical Manifolds of the First Stage and the Second Stage [4]

Table 19-1 Comparison of Manifolds in the First Stage and the Second Stage [4]

Characteristics	First Stage	Second Stage
Dimensions (m)	19(L) × 14(W) × 7(H)	12(L) × 12(W) × 6(H)
Weight (tonne)	400	160
Well slots	8	8
Production capacity (m^3/d)	8000	8000
Price (relative value)	100%	70%
ROV operation	Docked on the structure	Landed on the top of the manifold
Control system	Electrohydraulic multiplexed	Electrohydraulic multiplexed
Retrievable modules	Control module and valves/chokes	Control module
Connection system	Vertical	Vertical
Subsea base	Independent	Fixed
Lift point	Four	One central
Leveling system	Yes	No

required. Besides the reduction in size and weight, another reason for the decreased installation cost was the introduction of a central point, which enabled the manifold to be lifted and hung by drilling risers or wire during the offshore transport and deployment phases.

19.1.2.3. Third Stage

The manifolds in the third stage incorporated a retrievable choke module and the provision of a cradle for one subsea multiphase flow meter (MFM) after installation. In the manifolds designed during this stage, the valves were set at the highest or lowest positions of the pipes to avoid contact with gas and water caused by oil segregation during the period during which any valve is closed. They also saw the development of a new procedure for removal of the water before any well enters into production.

19.2. MANIFOLD COMPONENTS

Manifold components such as headers, valves, chokes, hubs for pipeline connections, hubs for a multiphase meter module, and hydraulic and electrical lines are designed to operate without maintenance during a project's lifetime. Furthermore, components such as choke modules, subsea control modules, and multiphase meter modules are designed to be retrievable.

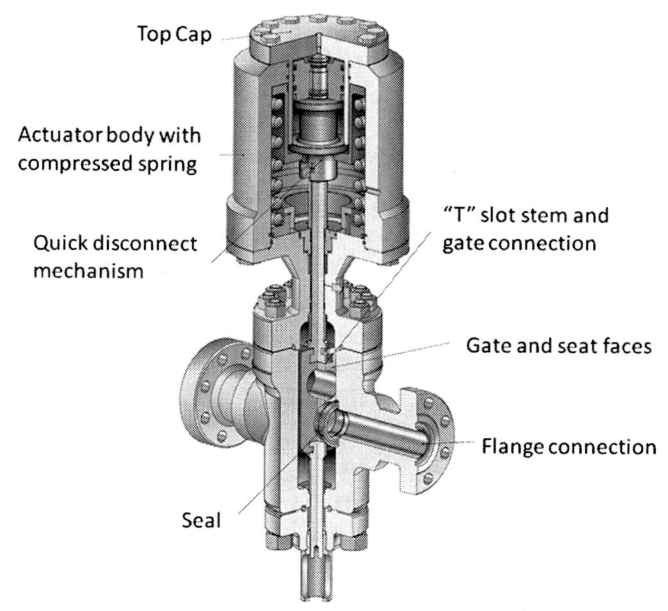

Figure 19-5 Magnum Subsea Gate Valve with Fail-Safe Actuator [5]

19.2.1. Subsea Valves

Subsea manifold valves are mounted within the piping system to control the production and injection fluids. The reliability of subsea manifolds is strongly dependent on subsea valves because the flow is directly through them. In a diverless subsea manifold, these valves are hydraulically actuated, on/off types and are used to direct the flow to the manifold production header or to the manifold test header and they are also used in chemical injection lines. In the manifolds designed to allow the passing of pigs, valves are used to provide the connection between the main headers. Figure 19-5 illustrates a Magnum subsea gate valve with fail-safe actuator, which can be used in water depths up to 2500 m (8200 ft).

One of the main design requirements of the subsea system is that all remotely controlled valves should be located within retrievable manifold frames or modules. The selection and location of the valves on the manifolds is of prime importance. Design specifications of valve type, size, pressure rating, and actuator function are required prior to detailing manifold layouts. Every type of valve has its strengths and weaknesses. For some applications, a ball valve provides a number of benefits over an API 6A gate valve that would potentially make it the best choice. Some of the features to consider are size, weight, height, speed of operation, seal wear, weldability, depth sensitivity, ROV intervention, fluid displacement during operation, and low-pressure sealing.

19.2.1.1. Valve Type

Gate valves or ball valves are two typical valves used in the manifolds. Gate valves have a long history of use in subsea blowout preventer (BOP) stacks, trees, and manifolds and are considered relatively reliable devices because both the valve and the valve actuators have been through extensive development with proven field use and design improvements. Figure 19-6 illustrates two types of subsea gate valves. Figure 19-6a shows a WOM (Worldwide Oilfield Machine, Inc.) subsea gate valve with actuator, compensator, and ROV bucket. The hydraulic actuator is designed with a fail-safe model and spring returns with the ROV. The mechanical ROV is for backup. Figure 19-6b shows a WOM subsea gate valve with only an ROV bucket. Both valves are designed, built, and tested based on API 6A [6] and 17D [7], which can be used up to a water depth of 13,000 ft (4000m).

Ball valves also are proven items and their use in deeper water depths is increasing. In some deepwater applications, ball valves can provide

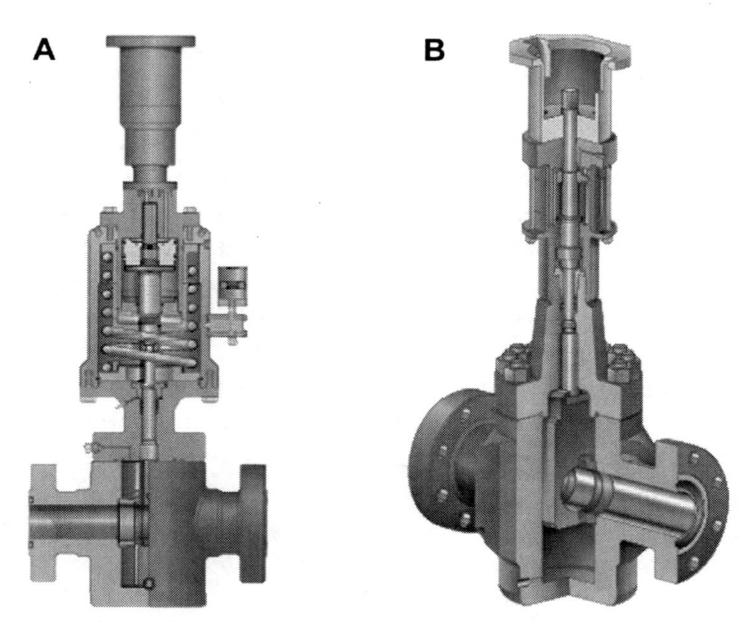

Figure 19-6 WOM Subsea Gate Valve with ROV Bucket [5]

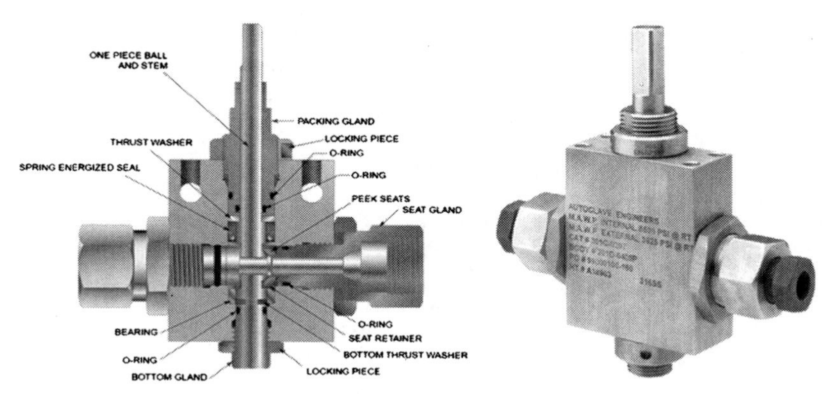

Figure 19-7 Typical Two-Way Subsea Ball Valve [8]

operational and cost advantages over gate valves, and improvements in nonmetallic seals and coatings are raising the reliability of ball valves. Ball valves were initially used downstream by the gas industry in gas pipeline valves. At that time, pipeline gate valves were the standard valves used in liquid pipelines [9]. Even today, gate valves are frequently specified for liquid pipelines, and ball valves are specified for gas pipelines. When gas wells were completed in the Gulf of Mexico in the 1960s, ball valves were installed in pipelines both as isolation valves and as terminal valves to tie in lateral lines from future wells and platforms. In the late 1970s, ball valves were installed in the North Sea and encountered problems due to the more challenging conditions of the sea. Later, ball valves were installed in subsea projects as emergency shutdown (ESD) valves to prevent gas in a pipeline from flowing back to a platform in the event of a major leak.

Figure 19-7 shows a typical two-way subsea ball valve from Autoclave Engineers that is designed to facilitate operation by an ROV. The valve design incorporates additional O-ring seals, which prevent the ingress of seawater into the valve. Seawater would adversely affect the operation of the valve and also contaminate the process fluid. The valve can be used in water depths to 12,500 ft (3800m) with maximum internal pressures of 20 ksi.

19.2.1.2. Size and Pressure

Manifold valves are usually required to carry the same pressure rating as Christmas tree valves in terms of the possible failure of the choke or valves on the tree. Normally, ball valves can more easily accommodate the size,

Table 19-2 Comparison of Typical API 6A Size with Pipe Size [10]

Nominal Pipe Size	Standard Valve Bore*	Pipe ID**
2 in.	1-13/16 in. or 2-1/16 in.	1.689 in.
3 in.	2-9/16 in.	2.624 in.
4 in.	4-1/16 in.	3.438 in.
6 in.	5-1/8 in.	5.189 in.
8 in.	7-1/16 in.	6.813 in.
10 in.	9 in.	8.500 in.
12 in.	11 in.	10.126 in.
16 in.	13-5/8 in.	12.814 in.

Notes:
*Standard valve bore as per API 6A, 10K flanges.
**Pipe ID for the nominal pipe size and a standard pipe schedule of 160.

whereas gate valves are more suited for the pressure. When people think of deepwater valve applications, API 6A gate valves typically come to mind because the first ventures into subsea at any depth have been exploration drilling. Typical BOP chokes and kill stack valves are 3 1/16-in. (0.078m), 10,000- or 15,000-psi API 6A gate valves. For the API 6A gate valves, size is a real challenge, while the increasing need for 10-ksi ball valves has been the biggest development effort for ball valves. Gate valves are preferred for all smaller sizes in which they are available, and ball valves are preferred for sizes 10 in. (0.25m) and larger.

Size has revealed another anomaly in subsea manifolds for all valves. Specifications are calling for 5000- or 10,000-psi valves that meet API 6A or 17D requirements in pipeline sizes that do not exist in either specification. A 5 1/8-in. (0.13m) or 6 3/8-in. (0.16m) 6A valve size defines the bore, but pipe size is defined by the outside diameter. A manifold specification calling for 8-in. (0.20m), 10,000-psi valves will likely be met with a valve having a 6 3/8–in. (0.16m) bore. The pipe grade and wall thickness must be known to properly select the valves. Table 19-2 compares some typical API 6A sizes with a heavy-wall pipe.

19.2.2. Chokes

The choke is a kind of valve used to control the flow of the well by adjusting the downstream pressure in a production manifold or upstream pressure in an injection manifold to allow commingled production/injection. For diverless subsea manifolds, hydraulic-actuated variable chokes are used. These chokes can be residents at the manifold or installed at retrievable modules.

19.2.3. Control System

Subsea valves are normally actuated by means of a direct control system or electrohydraulic multiplexed system. The use of one type or another depends on the distance of the subsea manifold to the host platform and the number of functions to be controlled. A multiplexed control system provides a quick response to valve actuation and greatly reduces the number of umbilicals needed.

The control system has a significant impact on the acquisition cost of a subsea manifold, especially the multiplexed control system. The impact on maintenance costs is also considerable because an unexpected failure can lead to an interruption of the well's production.

19.2.4. Subsea Modules

The main objective of the subsea module is to house components with a high failure probability in a place that permits the recovery of all components either by divers, ROVs, or a specific running tool, thus saving time during maintenance operations.

One important aspect to be considered in subsea manifold modularization is any increase in cost, weight, and equipment complexity. Modularization increases equipment availability, but it also increases the number of connections, which in turn increases the chances of leakage. Therefore, it is essential for a highly reliable seal component to be used in these modules and retrievable cartridges.

19.2.5. Piping System

The overall weight of the manifold is dependent on the piping configuration adopted. Normally only functional requirements are established by operators, so manufacturers are not given the arrangement to be adopted, , which leaves the design team relatively free to design the piping system. The use of flanged components or welded ones is basically dependent on manufacturers' criteria and must be considered in the analysis of a manifold's designed lifetime.

19.2.6. Templates

A template is a subsea structure on the seabed that provides guidance for drilling or other equipment. It is also the structural framework that supports other equipment, such as manifolds, risers, wellheads, drilling and completion equipment, and pipeline pull-in and connection equipment. The structure should be designed to withstand any loads, such as from thermal expansion of the wellheads and the pipelines.

Production from the templates may flow to floating production systems, platforms, shore, or other remote facilities. The template is typically used to group several subsea wells at a single seabed location. Templates may be of a unitized or modular design. Typical templates are described next. Actual templates may combine features of more than one of these types.

19.2.6.1. Modular Template

A modular template is one that is installed as one unit or as modules assembled around a base structure, such as a well.

19.2.6.2. Multiwell/Manifold Template

A template with multiple wells that incorporates a manifold system is illustrated in Figure 19-8.

19.2.6.3. Manifold Template/Center

A manifold template is one that is used to support a manifold for produced or injected fluids. Wells would not be drilled through such a template, but may be located near it or in the vicinity of the template. This template would be similar to the one shown in Figure 19-8 with the well-drilling guides omitted.

19.2.6.4. Well Spacer/Tie-Back Template

A multiwell template that is used as a drilling guide to predrill wells prior to installing a surface facility is referred to as a well space/tie-back template. The wells are typically tied back to the surface facility during completion. The wells could also be completed subsea with individual risers back to the surface as shown in Figure 19-9. If wells are to be drilled through the template, it should provide a guide for drilling, landing capability for the first casing string, and sufficient space for landing a BOP stack. If subsea trees are to be installed on the template, it should provide proper mechanical guidance for positioning of the trees and sufficient room for all installation and intervention operations.

19.2.6.5. Riser-Support Template

A riser-support template supports a marine production riser or loading terminal and serves to react to loads on the riser throughout its service life. This type of template may also include a pipeline connection capability. Figure 19-10 shows a combination template with wells, manifold, and production riser support.

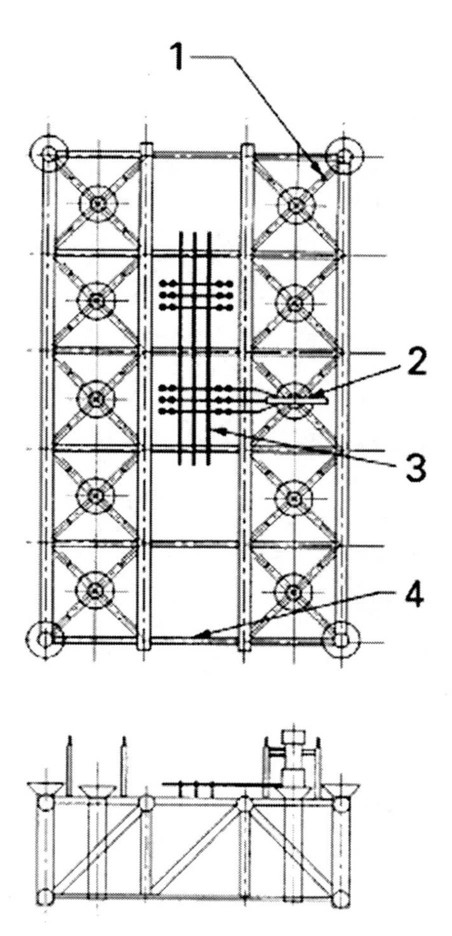

Key

1 Tree guide post receptacle (typical, if required)
2 Tree
3 Manifold header and valves
4 Pipeline connection bay

Figure 19-8 Multiwell/Manifold Template [11]

Templates designed to provide underwater wellhead support should incorporate a well bay and associated structure capable of withstanding drilling loads. The design should provide sufficient rigidity to maintain required spatial tolerances between components when under deflection.

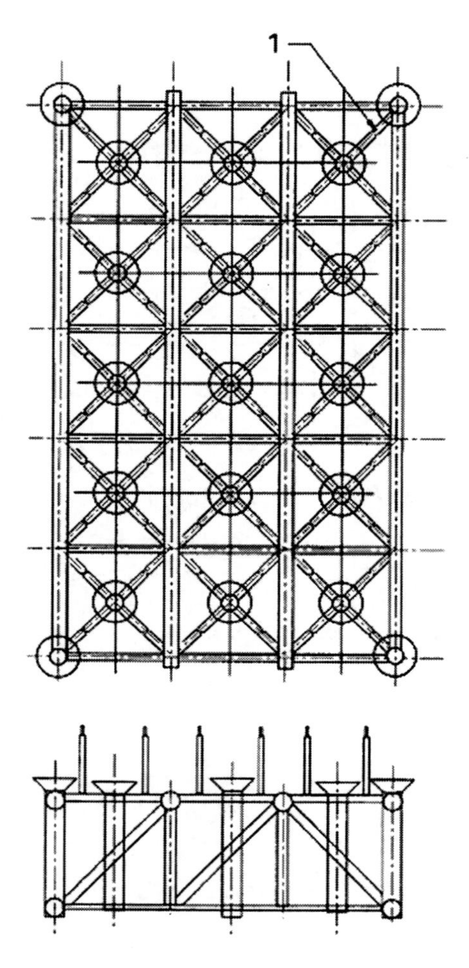

Figure 19-9 Well Spacer/Support Template [11]

Design loads may also include weight from surface casings due to inadequate cementing and drilling riser loads caused by vessel offsets.

The template should withstand pipeline installation forces, snag loads, and any loads induced by thermal expansion. If the template cannot be practically designed for pipeline snag loads, a protective breakaway device should be considered. The template should provide a sufficiently strong foundation to be able to transfer design loads into the seabed. The template

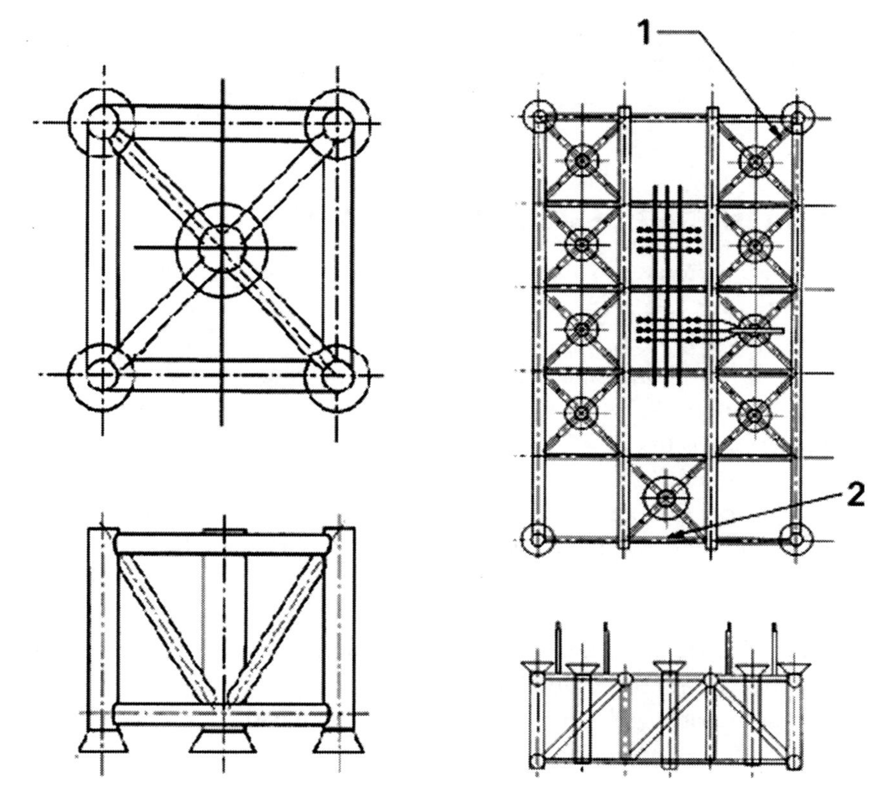

Figure 19-10 Manifold Template with Production Riser [11]

should be capable of withstanding the applicable loads due to the use of maintenance equipment.

In seismically active regions, the effects of earthquakes on the template should be considered. The design should allow for thermal expansion loads of elements heated by the produced fluid stream. The possibility of accidental snagging of the structure by trawling equipment, anchors, or other foreign objects, and the resultant loading, should be considered. The structural shape and strength should be designed to accommodate the maximum snag load without damaging critical components.

If the template is to be recovered at the end of the project, a recovery method should be considered. During installation and retrieval of the template, the vertical center of buoyancy should be maintained well above the center of gravity. Placement of any sacrificial anodes should be considered, avoiding areas subjected to accumulation of drill cuttings.

19.3. MANIFOLD DESIGN AND ANALYSIS

The manifold should provide a sufficient amount of piping, valves, and flow controls to, in a safe manner, collect produced fluids or distribute injected fluids such as gas, water, or chemicals. The manifold should have sufficiently sized bores in piping and valves to allow pigging. If wells are to be completed on the template, the manifold should provide for the connection of the Christmas tree.

The manifold may provide for mounting and protection of equipment needed to control and monitor production/injection operations. The manifold may also include a distribution system for hydraulic or electrical supplies for the control system.

Where wells are incorporated into the template and manifold, the addition of spare well slots should be considered. If satellite wells are to be tied in, the template and manifold design should provide this capability considering pipeline and umbilical connections, pull-in loads, valving, and pigging capabilities. The initial design considerations should consider the types of accidental damage loads to which the system may be subjected.

A maintenance approach should be considered early in the design of a template/manifold system. Specifically, the following items should be considered:

- Maintenance method;
- Retrievable components;
- Access space for divers, ROVs, or other maintenance equipment;
- Height above seabed for visibility;
- System safety with components removed;
- Identification of failing components.

Manifold design and analyses should address the following issues:

- Steel frame structures and painting design;
- Pipework and valve design;
- Connection equipment and control equipment;
- CP design;
- Flow assurance and hydraulics.

19.3.1. Steel Frame Structures Design

The manifold should be designed with the ability to be installed together with the template and as a separate module. Cluster manifolds should be based on an integrated template and protection structure with a separately retrievable manifold.

19.3.1.1. Structural Design

Manifold modules belong to subsea structures. Therefore, their structural design follows relevant standards for subsea structures, such as ISO 13819-1 [12] and ISO 13819-2 [13].

The design pressure for a steel manifold piping system or the nominal wall thickness for a given design pressure may be determined according to ASME B31.8, Chapter VIII [14], or DNV OS F-101 [15]. Formulas used for wall thickness design, flexibility analysis, and code stress checks may be determined from the same documents. For example, the formula from ASME B31.8 is:

$$P = \frac{2St}{D} FET \tag{19-1}$$

where

P: design pressure;

D: nominal outside diameter of pipe;

S: specified minimum yield strength;

t: nominal wall thickness;

F: design factor;

E: longitudinal join factor;

T: temperature derating factor.

19.3.1.2. Structural Analysis

The manifold system may include valves, hubs, connectors for pipeline and tree interfaces, chokes for flow control, and through-flowline (TFL) diverters. It may also include control system equipment, such as a distribution system for hydraulic and electrical functions. Figure 19-11 shows the front and plan views for a typical tie-in manifold.

Structural Frame

Figure 19-12 shows an ISO view of a production manifold structural frame. The manifold structural frames satisfy the following requirements:

- Be designed to the requirements of API RP2A [16].
- Be fabricated to requirements of AWS D1.1 [17].
- Provide support and protection for header piping and control tubing.

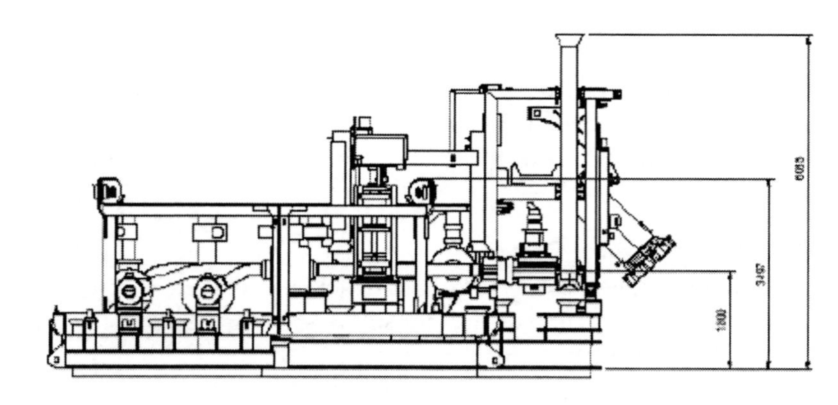

FRONT VIEW

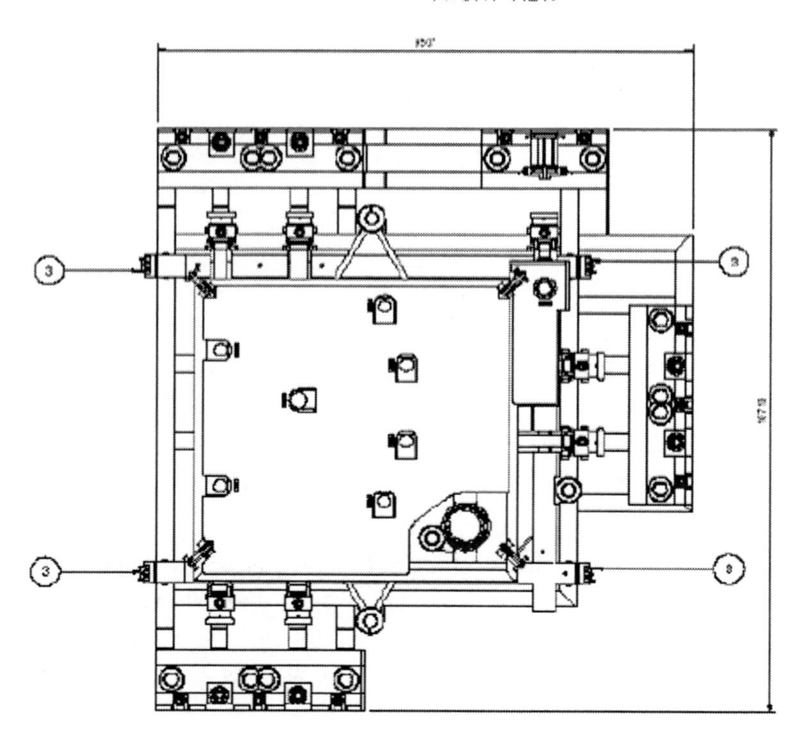

PLAN VIEW

Figure 19-11 Front and Plan Views for a Typical Tie-In Manifold

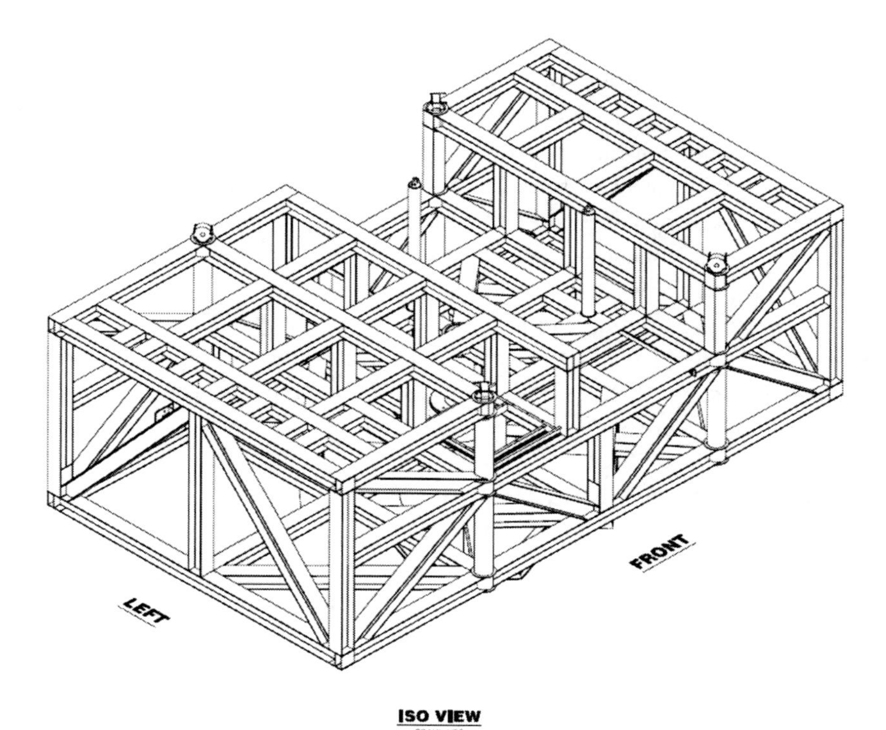

Figure 19-12 ISO View of a Manifold Structure Frame

- Provide the means for guidance and alignment of the manifold onto the pile foundation.
- Consist of a primary structural frame and ancillary protective structures, support structures, alignment and guidance systems, and installation and handling facilities.
- Provide sufficient open space for installing the manifold piping and valves, and be strong enough to resist all anticipated loads and deflections.
- Be designed with the bottom of the frame suitable for supporting the static weight of the manifold on the ground/deck.
- Provide a four-point lift arrangement suitable for a four-legged flexible wire rope sling. The lift points shall be strong enough to support the weight of the manifold in its as-installed configuration. All lift points will be designed in accordance with API 17D [7].
- Contain all anodes necessary to cathodically protect the manifold system from external corrosion, including any anodes needed to protect the system from external drainage, such as half of the total area of attached jumpers and flying leads.

Interfaces

Figure 19-13 shows a typical production manifold with interfaces for UTA (Umbilical Termination Assembly), ROV (remotely operated vehicle), and others. The following interfaces in the design of the manifold are considered:

- Well/flowline jumper interface to manifold hub including spacing for and access to pipeline end connector running tools;
- Pressure cap installation and retrieval;
- ROV access for valve functioning/isolation;
- ROV access to flying lead junction plates;
- Operational interfaces associated with installation, lifting and handling (padeyes, slings, bridles, spreader bars, installation vessel, etc.);
- Pile foundation top.

19.3.2. Manifold Piping Design

The piping system, header, and branch piping should be designed based on DNV OS-F101/ASME31.8, Chapter VIII, and comprises the entire piping: straight pipes, bends, tees, and reducers on the header and branches.

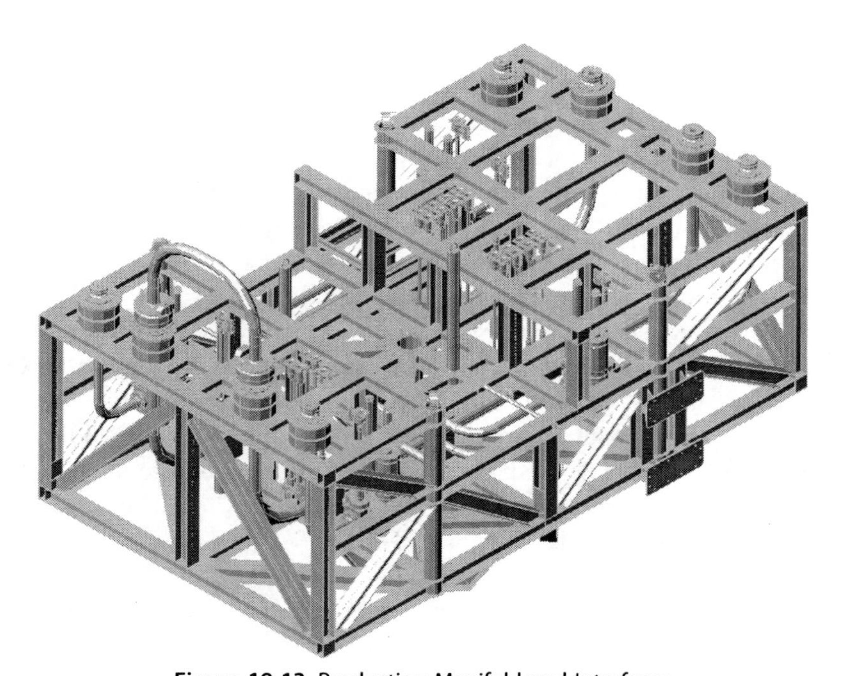

Figure 19-13 Production Manifold and Interfaces

The following issues should be considered for the piping stress analysis:

- Internal pressure;
- Hydrotesting;
- Thermal loads;
- Operating with jumper loads;
- Flowline jumper connection loads;
- Well jumper connection loads;
- Environmental loads;
- External corrosion;
- Internal corrosion/erosion;
- Piping supports to accommodate all anticipated loading, deflections, and vibrations.

Piping system should be designed to satisfy the requirements for internal pressure, thermal loads, hydrostatic collapse, and external operational loads, and fabricated to ASME Section IX requirements. Piping stress analysis should be performed using a finite element software package, such as CAESAR II, Ansys, or Abaqus, to confirm that the manifold piping system is fit for the intended purpose for its entire life span.

To conduct a detailed finite element analysis for a piping system such as that shown in Figure 19-14, beam elements are used to simulate the piping system. Figure 19-15 shows the analysis results of the maximum stress in the whole system under operational loads. The displacements of piping at valve positions also should be checked to ensure they are within the movement limits of the valve base.

Pipe fittings in the manifold piping system should be in accordance with ASME B16.9 if practicable. If nonstandard fittings are used, the fitting should be checked in accordance with ASME B31.8, Appendix F, and ASME Boiler and Pressure Vessel Code, Section VIII, Division 2 [18]. Figure 19-16 shows an 8.625-in. (0.22m) × 8.625-in. (0.22m) × 6.625-in. (0.17m) header reducing tee used in a piping system. The linearized stresses at three sections of the tee are evaluated for the hydrotest conditions and operating cases under external and internal pressure loads as well as thermal expansion loads based on the methods of the ASME Boiler and Pressure Vessel Code, Section VIII, Division 2.

Figure 19-17 shows the stress distribution in a tee section for evaluation. Based on the stress criteria in *Mandatory Design Based on Stress Analysis* of Section VII, Division 2. ASME "Design by Analysis" approach is based on the Tresca (shear stress) failure theory with a linear elastic material property. The design by analysis requires the most severe combination of loadings satisfy

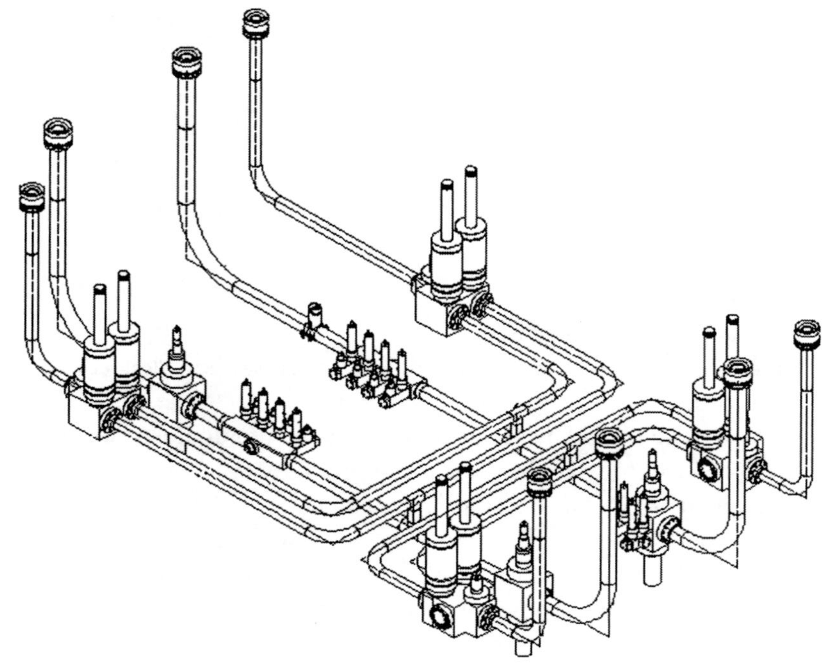

Figure 19-14 Manifold Piping System

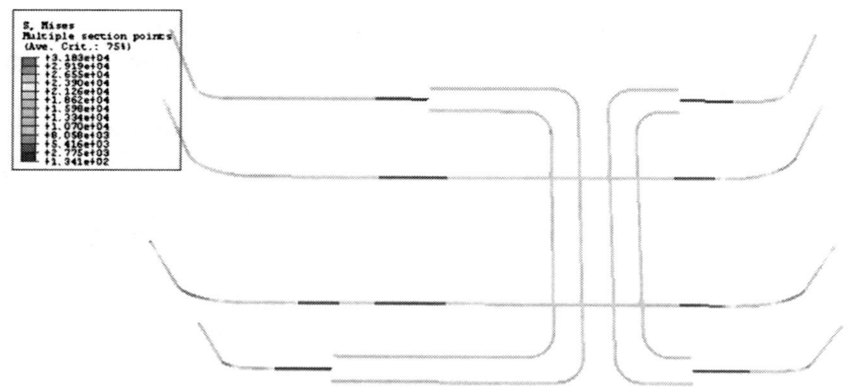

Figure 19-15 Von Mises Stress Distribution of Piping System in the Operating Condition

limits of stress intensities. In the code, the general membrane stress is an average primary stress across a solid section excluding geometry discontinuities. The local membrane stress is the average primary stress across a solid section including discontinuities. The bending stress is a component of primary stress proportional to the distance from the centroid of a solid section.

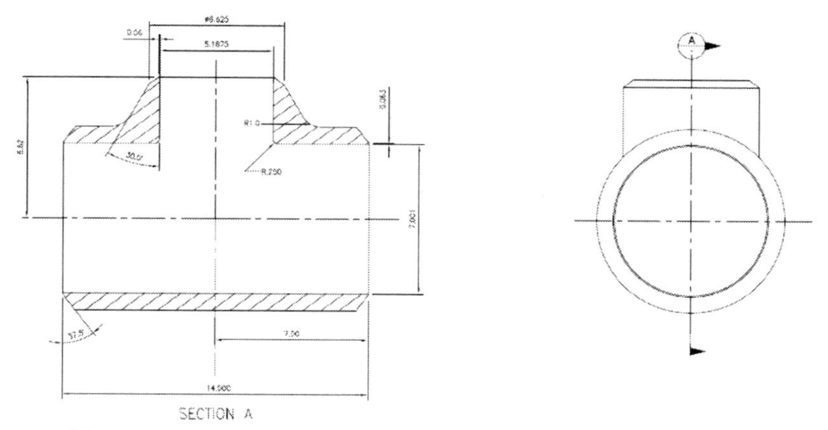

Figure 19-16 8.625-in. × 8.625-in. × 6.625-in. Header Reducing Tee

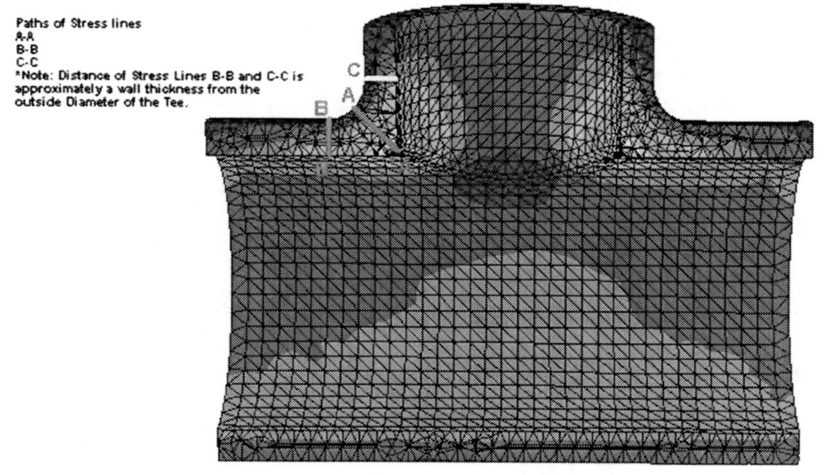

Figure 19-17 Stress Distribution in Middle Section of Tee

- The primary general membrane stress allowable:
  $P_L < k \times S_m$, where $k = 1$, Sm is the allowable stress intensity – typically the lesser of 2/3 the yield stress (SY) or 1/3 the ultimate tensile stress (SUT).
- The primary local membrane stress allowable:
  $P_L < 1.5 \times k \times S_m$.
- The primary membrane stress plus bending stress allowable:
  $P_L + P_B < 1.5 \times k \times S_m$
- The primary stress plus secondary stress allowable:
  $P_L + P_B + Q < 3.0 \times k \times S_m$

For the hydrotest case, the stress criteria are as follows:

- The maximum primary membrane stress intensity is $0.9\ S_Y$.
- The allowable primary membrane plus bending stress:

  $P_m + P_B < 1.35S_Y$ for $P_m < 0.67S_Y$

  $P_m + P_B < 2.15S_Y - 1.2P_m$ for $0.67 < P_m < 0.9S_Y$.

Three sections are chosen for the stress evaluation and the positions of the sections are shown in Figure 19-17, where section A–A crosses the area of wall thickness discontinuity.

19.3.3. Pigging Loop

The manifold piping loop is designed to allow passage of pigs through the main headers to provide round-trip pigging of the flowlines from the production platform.

The pigging loop should satisfy the following requirements:

- Is subsea installable and retrievable without guidelines using a suitable spreader beam with rigging?
- Conforms to the same standards as the manifold header piping.
- Has a full-bore gate valve that will allow a pig to pass without hang-ups. This valve should be hydraulically operated.
- Provides facilities to inject chemicals/methanol through two parallel hydraulically operated gate valves with integral check valves mounted to either side of the full-bore gate valve.
- Is of the same material and size as the production manifold pipeline header piping.
- Is insulated to the same requirements as the production manifold pipeline header piping, with the exception of the pipeline end connectors.
- Is constructed with bends having a minimum radius of five times the nominal pipe diameter (5D).
- Mounts on the inlet hubs of the production manifold.
- Is not required to provide anodes for its own cathodic protection. Electrical continuity to the production manifold will be demonstrated.
- Is coated with three-part epoxy paint with the exception of stainless steel tubing, unless otherwise specified.

Pigging loop system design should consider the following:

- Piping size;
- Bend radius;
- Internal protrusions;
- Valve types;

Figure 19-18 Manifold Pigging Loop

- Pig launcher/receiver;
- Pig location determination.

Figure 19-18 shows a pigging loop designed for a production manifold. The loop has two 5D bends and an ROV panel for the connector.

19.3.4. Padeyes

Padeyes should be designed in accordance with the industry practice using a design safety factor of typically 4 or greater based on minimum specified ultimate material strength at the maximum rated pickup angle during installation [7]. This factor of safety is to be applied to the design lift. The design lift includes the total weight to be lifted multiplied by the dynamic load factor (DAF = 1.33), which can be found in the *Lloyd's Rules for Fixed Offshore Installations* [19], Part 4, Chapter 3, or 1.3 in DNV Rules for Marine Operations, Part II, Chapter 5, Section 2.2.2; the weight inaccuracy factor (WF = 1.1), which can be found in DNV Rules for Marine Operations, Part I, Chapter 3, Section 3.5; and the sling factor (SKL = 1.3), which can be found in DNV Rules for Marine Operations, Part II, Chapter 5, Section 3.1. The out-of-plane force is considered by using 8 degrees as the transverse angle in the calculations.

Padeye Limits at Design Load Based on Yield Stress, σ_y

- Bearing $< 0.9\ \sigma_y$
- Shear $< 0.4\sigma_y$
- Tensile $< 0.6\sigma_y$

Padeye Limits at Design Load Based on Ultimate Strength σ_u (API 17D, s.f. = 4)

- Bearing $< 0.25\ \sigma_u$
- Shear $< 0.25 * 0.57\sigma_u$
- Tensile $< 0.25\sigma_u$

Allowable values for weld material are the same as base material for full penetration and groove welds.

19.3.5. Control Systems

The production control system and its components should be designed according to ISO 13628-6 [20]. For polymer-based hoses, material selection should be based on a detailed evaluation of all fluids to be handled, but it should not be used for pure methanol service (with less than 5% water); see API Spec 17E [21].

The annulus bleed system is exposed to a mixture of fluids, such as production fluid, methanol, completion fluid, and pressure-compensating fluid.

A hose qualification program should be carried out, including testing of candidate materials in stressed conditions representative of actual working pressures, unless relevant documentation exists.

For umbilicals, the electric cable insulation material should also be qualified for all relevant fluids. The materials selected for the electrical termination should be of similar type, in order to ensure good bonding between different layers. The material selection for metals and polymers in electrical cables in the outer protection (distribution harness) and in connectors in distribution systems should have qualified compatibility with respect to dielectric fluid/pressure-compensation fluid and seawater.

19.3.6. CP Design

Cathodic protection should be used for all submerged, metallic materials, except for materials that are immune to seawater corrosion. A surface coating should also be used for components with complex geometries and where its use will result in a cost-effective design.

Cathodic design should be carried out in accordance with DNV RP B401 [22], *Cathodic Protection Design*, or ANSI/NACE RP 0176 [23], *Cathodic Protection of Steel Fixed Offshore Structures*. The design should ensure reliable electrical continuity to each individual element for the defined design life, including continuity through the sealine termination.

Welded anode connections are recommended for subsea applications. Flanged and screwed connections should be avoided where possible. The electrical continuity to the cathodic protection system should be verified by actual measurements for all components and parts not having a welded connection to an anode.

Any components permanently exposed to ambient seawater and for which efficient cathodic protection cannot be ensured should be fabricated from seawater-resistant materials. Exceptions are components where corrosion can be tolerated, that is, where pressure containment or structural integrity will not be compromised.

Material selection should take into account the probability for, and consequences of, component failure.

The following materials are regarded as corrosion resistant when submerged in seawater at ambient temperature:

- Alloy 625 and other nickel alloys with equal or higher PRE value (PRE = % Cr + [3.3 × % Mo] + [16 × % Ni]);
- Titanium alloys with suitable performance under cathodic protection (should be documented for the relevant operating conditions);
- GRP;

Stainless steels Type 6Mo and Type 25 Cr duplex are borderline cases and not considered as fully seawater resistant for temperatures above 15°C (60°F). These materials should not be used for threaded connectors without cathodic protection.

The location and number of CP inspection points for intervention and risk of hydrogen-induced cracking should also be evaluated and determined. Hydrogen-induced stress cracking (HISC) issues should be addresses for duplex stainless steel pipework exposed to cathodic protection. The duplex stainless steels are susceptible to HISC when exposed to elevated stresses in conjunction with cathodic protection potentials lower than typically −850 mV relative to the Ag/AgCl/ seawater reference electrode. DNV- RP- F112 should be incorporated into the piping design code when designing duplex stainless steel pipework.

19.3.7. Materials for HP/HT and Corrosion Coating

19.3.7.1. Materials for HP/HT Manifolds

Materials for HP/HT manifolds should meet these requirements:

- All piping pipe is made of duplex stainless steel alloy.
- All tees, crosses, elbows, and flanges are made of duplex stainless steel alloy.
- The production pipeline end connectors and hubs are made of materials based on their UNS designation (unified numbering system designation).
- The manifold frame is made from carbon steel.

19.3.7.2. Material Evaluation

For the materials used in subsea structures, manifolds, piping, and other components having importance for the safety and operability of the subsea production system, the following factors apply to the materials selection:

- Materials with good market availability and documented fabrication and service performance;
- Design life;
- Operating conditions;
- Mass reduction;
- Experience with materials and corrosion protection methods from conditions with similar corrosivity;
- System availability requirements;
- Minimization of the number of different material types considering costs, interchangeability, and availability of relevant spare parts;
- Inspection and corrosion-monitoring possibilities;
- Effect of external and internal environment, including compatibility of different materials;
- Evaluation of failure probabilities, failure modes, criticalities, and consequences;
- Environmental issues related to corrosion inhibition and other chemical treatments.

The materials to be used should normally fulfill the following requirements:

- The material should be listed by the relevant design code for use.
- The material should be standardized by recognized national and international standardization bodies.
- The material should be readily available on the market.
- The material should be readily weldable, if welding is relevant.
- The material preferably has a past experience record for the particular application.

19.3.7.3. Metallic Materials

Corrosivity Evaluation in Hydrocarbon Systems

Evaluation of corrosivity should include at the least:

- CO_2 content;
- H_2S content;
- Oxygen content and content of other oxidizing agents;
- Operating temperature and pressure;
- Acidity, pH;
- Halogenide concentration/water chemistry;
- Velocity.

The evaluation of CO_2 corrosion should be based on an agreed-on corrosion prediction model or previous experience from the same field.

Risk for "sour" conditions during the lifetime should be evaluated. Requirements for corrosion-resistant alloys in sour service should comply with ANSI/NACE MR 0175, *Sulfide Stress Cracking Resistant Metallic Materials for Oil Field Equipment*, with amendments given in this part of ISO 13628.

Corrosivity Evaluation in Water Injection Systems

Water injection systems are used for injection of deaerated seawater, raw untreated seawater, and produced water including aquifer water.

Corrosivity evaluations for deaerated injection seawater should be based on the maximum operating temperature appropriate for the given geographical area. For carbon steel submarine injection flowlines used for low corrosive services, the minimum corrosion allowance should be 3 mm (0.12 in.).

All components that may contact injection water should be resistant against well-treatment chemicals or well-stimulation chemicals if backflow situations can occur. For carbon steel piping, the maximum flow velocity should be evaluated considering the corrosivity and erosivity of the system.

19.3.8. Hydrate Prevention and Remediation

As subsea field development expands into deepwater environments, operators need to consider the potential risk of gas hydrate formation in well-bores, subsea pipelines, and subsea equipment during both drilling and production operations. Hydrates can also form in certain sections of subsea equipment exposed to stagnant fluids under normal or transient flow conditions. The hydrate management philosophy for subsea systems is as follows:

- No continuous inhibition of the subsea system is required in flowing conditions.
- No part of the fluid system is allowed to enter the hydrate formation domain during flowing and shutdown conditions.
- During start-up operations, the cold production fluid released by the well is inhibited at the wellhead until the production temperature reaches a temperature high enough to ensure sufficient cooldown time if another shutdown occurs.

Remediation of hydrate plugs in subsea equipment can be difficult or impossible. In shallow-water field developments, the pipeline and subsea equipment could potentially be depressurized to allow a hydrate plug to dissociate. In deepwater, the hydrostatic head of liquids in the pipeline may be generally high enough to keep the hydrates stable. While the pipeline could be displaced, the tree piping, jumper piping, and manifold branch valves are typically not designed for pigging or circulation.

One of the challenges is to apply thermal insulation to subsea equipment in optimal locations to achieve the most benefits. To design thermal insulation for subsea equipment with complex geometries, thermal finite element analysis (FEA) is typically performed. Although a certain minimum thermal insulation thickness is required to achieve the required cooldown time, it is sometimes not practical to insulate all of the surfaces due to access, manufacturing assembly process, equipment testing requirement, contingency procedures for disassembly and repair, etc.

Figure 19-19 illustrates a subsea tree that was modeled using a 3D FEA method that was capable of modeling the complex geometry. The tree body, the valve blocks, the choke, and tree connectors have irregular exterior surfaces. Based on the 3D thermal analysis, the thermal insulation thickness can be optimized to avoid cold spots to meet the cooldown requirements.

Subsea gate valves have a complex internal geometry and flat exterior surfaces unlike the typically round cross section of upstream and downstream piping. Areas that trap fluids inside the valve cavity are prone to hydrate blockage if the valve is not properly insulated. Furthermore, the valve actuators are not typically insulated to avoid overheating the actuator sealing elements and the hydraulic fluid. The actuator behaves like a large fin and dissipates a lot of heat unless the valve assembly is properly analyzed. Three-dimensional thermal FEA is useful for understanding the heat flow and, hence, designing the insulation around the valve body. Figure 19-20

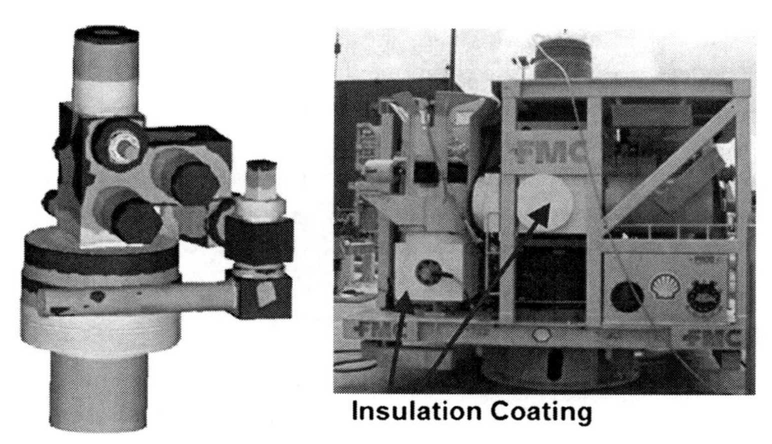

Figure 19-19 Insulation Design Based on 3D Thermal FEA [24]

shows the results of temperature distribution in a group of valves in a manifold.

The insulation on any part of a manifold piping system is usually of sufficient thickness to meet the cooldown time required for hydrate management. This insulation coupled with high fluid temperatures may exceed the qualification temperature of the electronic components for instruments. Care should be taken to select the proper instruments and

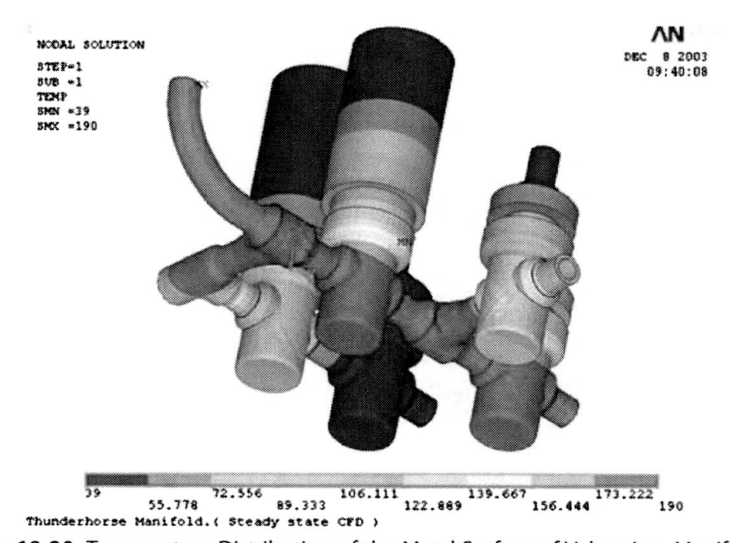

Figure 19-20 Temperature Distribution of the Metal Surface of Valves in a Manifold [24]

determine the insulation thickness around those electronic components. Either the electronics for the components should be requalified at a higher temperature or sections of the insulation have to be cut away to provide the cooling for the electronics. The connecting jumpers between the tree and manifold, and between the manifold and flowline, would appear to be reasonably straightforward to insulate, and do not suffer from the potential cold spots at the manifold due to valves.

19.4. PILE AND FOUNDATION DESIGN

The increasingly more robust and specialized anchoring systems with piles for drilling and production systems have found wide applications in the exploration and exploitation of hydrocarbons offshore from shallow water to deepwater. The anchoring systems were designed based on soil conditions and load considerations and can be categorized into three groups: (1) mudmat, (2) suction pile, and (3) pile. Figure 19-21 shows a comparison of these three groups based on the foundation aspect ratios of L/D (length/diameter). The foundation used in subsea equipment may be either a mudmat, a skirt, or a suction pile depending on the seabed soil conditions, while piles are usually used for anchoring the drilling and production unit. The mudmat configuration relies on the footprint of the structure to support the combined weight of the manifold system and attached

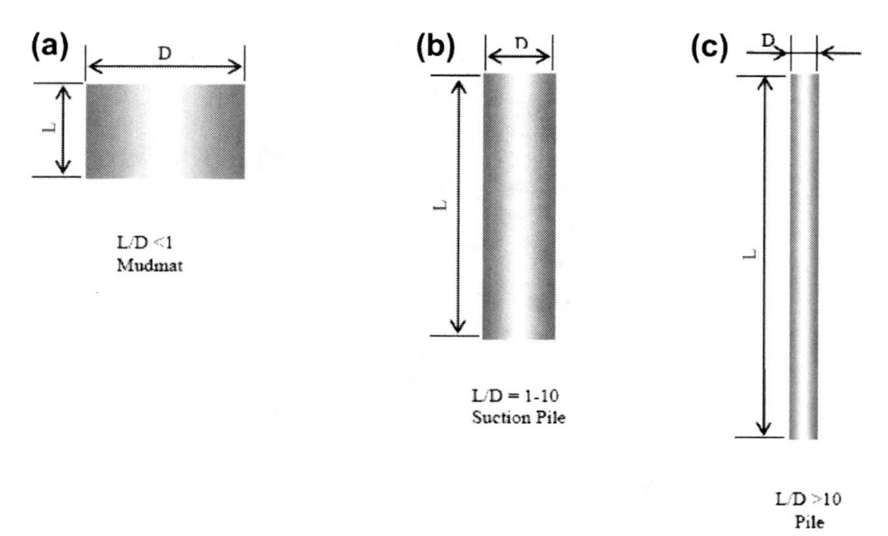

Figure 19-21 Aspect Ratios of Piles

components such as jumpers. A skirt that penetrates deeper into the seabed than the mudmat resists horizontal forces and some portion of the vertical load. The mudmat foundation is described in the chapter on PLETs.

The change in the suction pile aspect ratio (characteristic shape of suction piles) was inevitable based on the soil conditions due to the requirement of a big driving force. Based on the required driving force, the suction pile aspect ratio may range between 2:1 for stiff clay and 7:1 for very soft clay [25]. The full penetration in soil of a suction pile is critical.

Since the early 1980s, when the idea of suction installation was first introduced and applied, the suction pile has found wide applications in deepwater development due to its lower fabrication costs and installation time compared to conventional foundation options. Suction anchors are generally used as the main foundation and anchoring system in deepwater fields because of the following advantages [26]: (1) fixed location on the seabed, which is important in subsea developments; (2) simple installation procedures with no need for proof load testing at the site; (3) no special limitation on the water depth for application or installation: and (4) the capacity of suction anchors can be defined more precisely than that of drag anchors.

Figure 19-22 shows a typical suction pile and its installation mechanism. A suction anchor is a cylindrical unit with an open bottom. It is equipped with a valve at the top. Suction piles are installed by applying differential pressure by the way of pumping out water from the interior of the pile. The differential pressure constitutes the driving force necessary to overcome the resistance due to the penetrated soil. In the first step of installation, the suction pile (anchor) penetrates to soil under its self-weight, with free evacuation of the water located inside the skirts. In the second step, an additional driving force is created by pumping out the water entrapped between the anchor top and the soil plug. After reaching the final penetration depth, the valve is generally closed in order to increase the pullout resistance. Installation by suction is less efficient in cohesionless soils than in clay soil. A suction pile works well in homogeneous soils, such as clean sand and particularly very soft to soft clay, which is commonly used in deepwater applications.

Suction piles are widely used in subsea structures, such as subsea manifolds and well protection structures, and it has been also employed as an alternative solution for pipeline installation and for stopping pipeline walking under cyclic thermal loads, as shown in Figure 19-23. A piled foundation may use either a single pile or multiple piles.

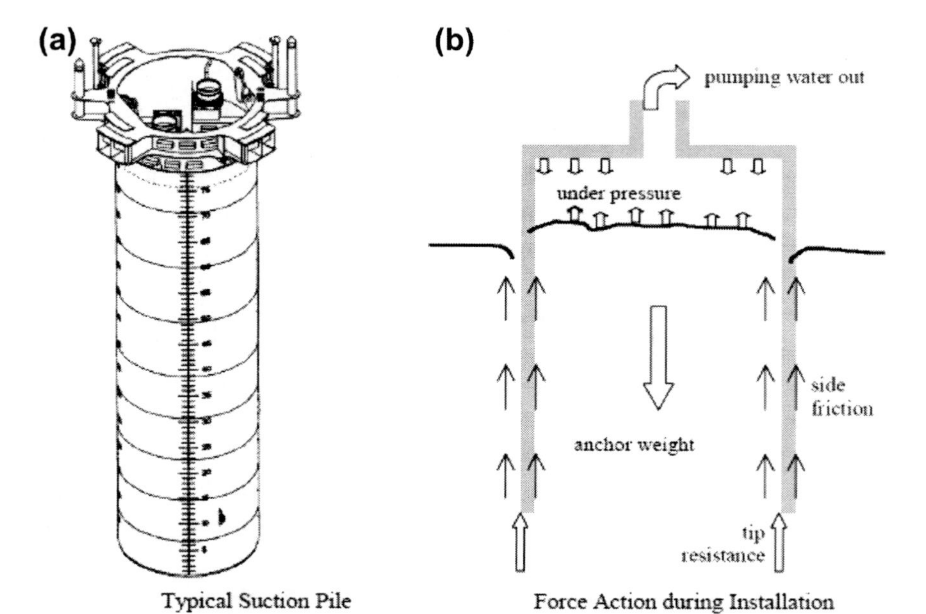

Figure 19-22 Suction Pile and Its Installation Mechanism [27]

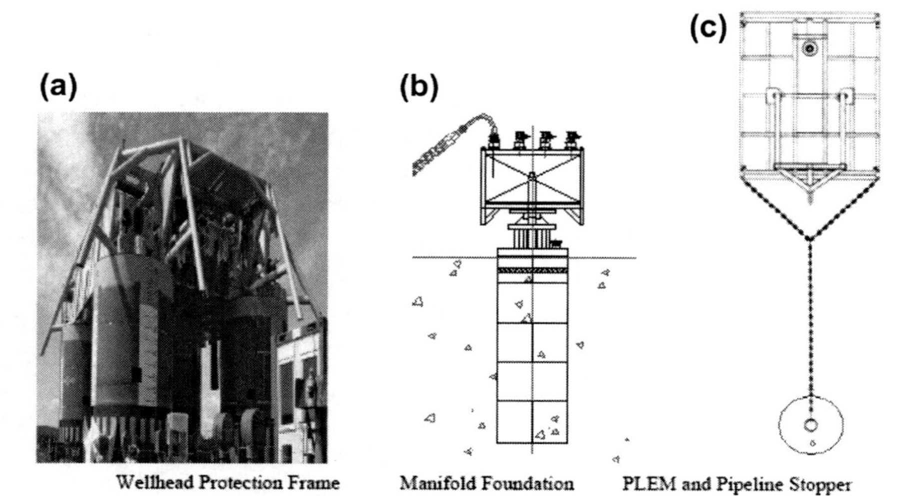

Figure 19-23 Suction Pile Used in Subsea Structure

This section provides a general description of the design of suction piles and foundations, including:

- *Design methodology:* selection of appropriate design code and relative safety factors;
- *Geotechnical profiles:* interpretation and selection of design soil profiles and geotechnical parameters;
- *Pile capacity and sizing:* calculation of in-place resistance and pile capacities for pile sizing;
- *Detailed structure design:* verification of suction pile responses under the design loadings;
- *Installation and retrieval:* analyses of installation/retrieval.

19.4.1. Design Methodology

The API RP design codes ensure adequate foundation safety using the working stress design method. The partial coefficient method is used in DNV codes, by which the target safety level for the structure is reached by applying partial coefficients to the characteristic loads and soil parameters and taking into account the reliability of the design data.

Suction pile design includes preliminary pile sizing and final structure design. In the preliminary pile sizing, the pile aspect ratio (L/D) versus steel weight should be optimized prior to the final suction pile design. Suction pile sizing assumes that the pump-in suction pressure is acceptable for the available soil strength (i.e., soil plug stability check) and that the pump-out pressure is acceptable for the available soil strength.

The suction pile foundation design should account for the loads in the lift, transportation, installation, penetration, and in-place conditions:

- Seabed slope, installation tolerances, and effects from possible scouring;
- Soil bear capacity and bear stress should satisfy API RP 2A-WSD [16];
- Stability analysis of the manifold should include overturning, sliding, and shearing resistance;
- Suction loads due to repositioning or levelling;
- Invasion of soil into pile sleeves should be prevented;
- For foundation and skirt systems, arrangements should be made for air escape during splash zone transfer and water escape during seabed penetration. Lift stability and wash-out of soil should be taken into account;
- Structures with skirt foundation should be designed for self-penetration;
- Skirt-system facilities for suction and pumping should, where required, be included to allow for final penetration, leveling, and breaking out

prior to removal. The suction and pump systems should be operated in accordance with the selected intervention strategy;
- Settlement of the structures for initial settlement and long-term deformation (during installation and lifetime) should be accounted for.

Designers must ensure that the pump capacity will be sufficient to provide adequate flow at the required pressure in order to allow efficient operations. This must be verified by testing the pump in submerged condition.

19.4.1.1. Codes and Standards

The following is a list of codes and standards that can be used for the design of suction piles:
- API RP 2SK: *Recommended Practice for Design and Analysis of Station keeping Systems for Floating Structures;*
- API RP 2A-LRFD: *Recommended Practice for Planning, Designing and Constructing Fixed Offshore Platforms;*
- API RP 2A-WSD: *Recommended Practice for Planning, Designing and Constructing Fixed Offshore Platforms;*
- AISC *Manual of Steel Construction—LRFD;*
- API Bulletin 2U: *Stability Design for Cylindrical Plates;*
- API Bulletin 2V: *Design of Flat Plate Structures;*
- DNV, *Rules for Classification of Fixed Offshore Installations*, 1995;
- DNV-RP-B401, *Recommended Practice—Cathodic Protection Design;*
- DNV-RP-E303, *Geotechnical Design and Installation of Suction Anchors in Clay.*

19.4.2. Design Loads

19.4.2.1. Permanent Loads

The permanent load will not vary in magnitude, position, and direction in the consideration period. It includes the dead load of the structure and any long-term static load applied to the subsea structure.

19.4.2.2. Live and Dynamic Loads

A live load is applied to the foundation during installation or the considered working period. It may vary in magnitude, position, and direction.

19.4.2.3. Environmental Loads

The influence of environmental factors such as current and wave action can result in significant hydrodynamic loads on installations, which are typically transferred as dynamic loads to the foundations.

19.4.2.4. Accidental Loads

An accidental load considers the load due to fishing gear snagging or impact of dropped objects.

19.4.3. Geotechnical Design Parameters

Reliable information concerning the soil layering and properties should be collected. The soil investigation program should include high-quality geophysical and geotechnical surveys. The geophysical survey allows geohazards to be detected and also provides for a regional geological overview. The geotechnical survey should allow for the collection of high-quality soil samples in as nearly an undisturbed state as feasible. *In situ* geotechnical tests (such as cone penetrometer tests, *in situ* shear vane tests) should be performed. The depth of the geotechnical borings should exceed the anchor's penetration. The number of these borings should be defined as a function of the soil variability. One boring should be performed at each anchor location when lateral variability of the soil properties is expected.

The following main soil properties are needed for the design of suction anchors:

- Index properties;
- *In situ* stresses and stress history;
- Undrained shear strength;
- Drained characteristics;
- Consolidation characteristics;
- Interface strength and thixotropy.

19.4.3.1. Index Properties

The index properties (including water content, unit weight, plasticity index, grain size distribution, and organic content) are needed to characterize the soil. The unit weight, as deduced from laboratory tests, should be in agreement with the bulk density of the soil evaluated by core logging (gamma-ray absorption). Core logging before liner opening using nondestructive techniques is the best way to assess the quality of the samples as shown in Figure 19-24.

19.4.3.2. In Situ Stresses

The *in situ* effective stresses and stress history are mainly needed to establish the consolidation stresses for laboratory tests. The *in situ* stresses are evaluated from the following geotechnical parameters: unit weight, pore pressure,

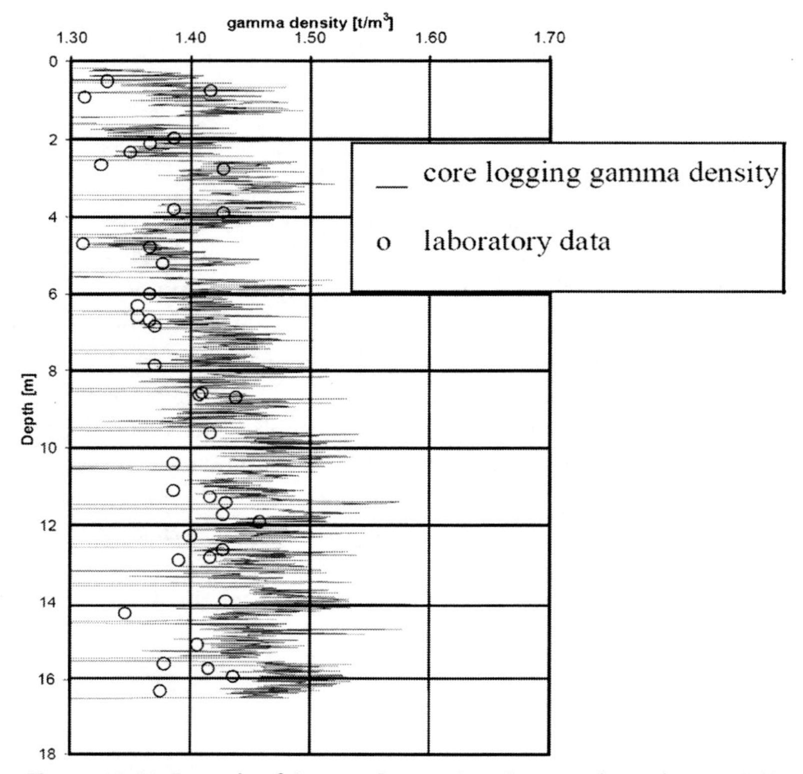

Figure 19-24 Example of Gamma Density Log Compared to Lab Tests [27]

and coefficient of earth pressure at rest. The stress history is commonly expressed by an overconsolidation ratio (OCR).

19.4.3.3. Undrained Shear Strength

The undrained shear strength is a key parameter for the design. The variation of this parameter with the stress path should be assessed. Anisotropic triaxial compression (CAUc), triaxial extension (CAUe), and direct simple shear tests (DSS) are necessary to derive the undrained shear strength profiles. The *in situ* cone penetration tests (CPTs) generally provide the trend in the variation of the undrained shear strength with depth. The *in situ* shear vane allows *in situ* measurement of the undrained shear strength.

The undrained cyclic shear strength has to be evaluated to calculate the capacity under cyclic loads. The undrained cyclic shear strength should be obtained for various stress paths. The remolded shear strength is an important parameter because it is used to calculate the penetration resistance of the skirt walls. The remolded shear strength profile may be deduced from laboratory tests (fall-cone, laboratory vane tests) and *in situ* tests (mainly from *in situ* vane tests).

A typical example of undrained shear strength profiles obtained on a field is shown in Figure 19-25. The figure shows a satisfactory agreement between these different sets of data.

19.4.3.4. Drained Characteristics

The drained friction angle and its associated cohesion have to be evaluated when the suction anchor is required to support long-term loading.

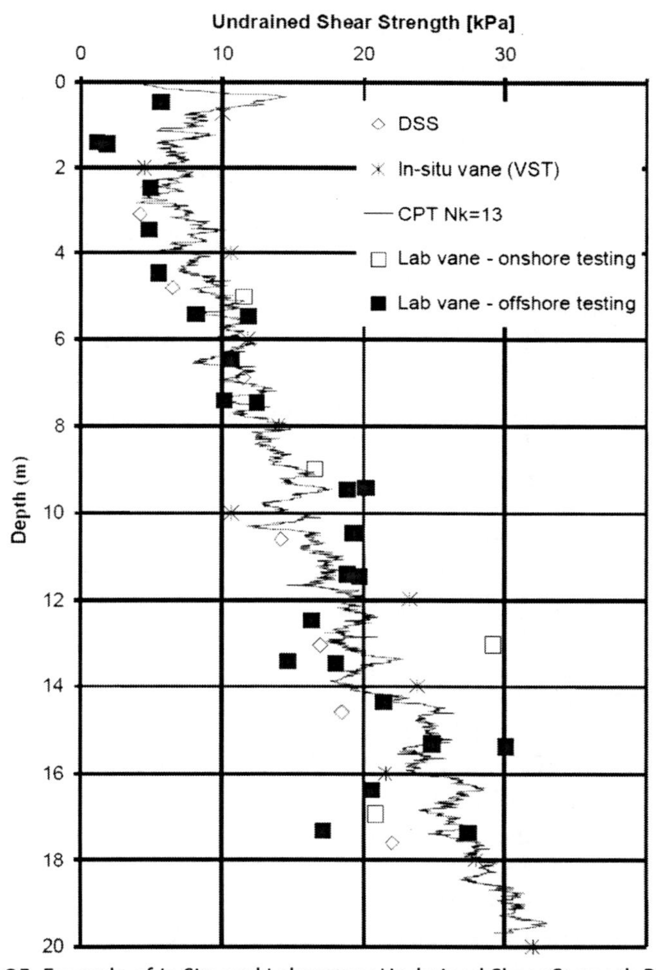

Figure 19-25 Example of *In Situ* and Laboratory Undrained Shear Strength Profiles [27]

19.4.3.5. Consolidation Characteristics

The consolidation characteristics include compressibility modulus, Poisson's ratio, coefficient of permeability and the OCR. These characteristics are needed for anchor capacity calculations under long-term loads.

19.4.3.6. Thixotropy

Thixotropy is the increase in the undrained shear strength of remolded clay with time and without any consolidation effect.

19.4.4. Suction Pile Sizing—Geotechnical Design

19.4.4.1. Pile Axial, Lateral, and Torsional Capacities

Suction anchors can be designed for different loading conditions: (1) mooring applications for floating production units; in this case the suction anchor should resist to an inclined mooring load where the vertical component may be high when fiber ropes are used; (2) anchoring for TLPs and other structures such as riser towers where the suction anchor should support uplift loading; and (3) foundations for subsea structures where the anchor should resist to a compression load combined with horizontal load and moment.

For the suction pile foundations used to support the manifolds, the foundations should have sufficient capacities to withstand all loads from the manifold system, including compression, lateral, moment, and torsion under both subsea equipment installation and operating conditions. The safety factors are determined in the design basis. As an example, the following safety factors (FOS) were used in a GoM application:

Operating Conditions

FOS = 1.9 for the axial failure;
FOS = 1.9 for the lateral failure;
FOS = 1.5 for the torsional failure.

Installation Conditions

FOS = 1.5 for all type of failures.

The safety factor of 1.9 for the axial failure and lateral failure includes a load factor of 1.3 for dead loads and a resistance factor of 0.7 for soil axial capacity. The safety factor for the torsional failure is reduced to 1.5 due to the fact of the consequence of torsional failure is less severe than axial and lateral failures. However, the negative impact of torsional loading on the axial capacity and lateral capacity of the system's foundation should be considered during the design stage. A safety factor against all type of installation loading is reduced to 1.5 due to the nature of their temporary, short period of presentation, which also agrees with API RP 2A, Section 6.3.4.

Suction piles should be designed to provide adequate lateral and axial capacities. The tasks include:

- Foundation stability;
- Penetration resistance and the accompanying required suction for final penetration;
- Retrieval resistance and the accompanying required overpressure;
- Vertical settlements of the manifold;
- Maximum horizontal displacements;
- Maximum responses due to the impact when landing the suction pile anchor and the interface guide base at the seabed.

19.4.4.2. Stability Analyses

Stability calculations are performed for the following two separate conditions:

- Vertical load in combination with torsion moment, utilizing the skin friction and plugged end bearing of the suction pile anchor;
- Horizontal load in combination with an overturning moment, utilizing the shear strength of the soil surrounding the anchor.

The combined vertical and torsion resistances are solved by computing the allowable skin friction according to API and the maximum value applicable for suction-assisted penetration beyond the self-penetration depth.

19.4.4.3. Penetration Analysis

A penetration analysis involves the calculation of skirt penetration resistance, underpressure needed to achieve the target penetration depth, allowable underpressure (underpressure giving either large soil heave inside the skirts or cavitations in the water), and soil heave inside the caisson as shown in Figure 19-22b. The penetration analysis is commonly performed using the general principles given in Andersen and Jostad [28]. Such an analysis includes the following parameters:

- Penetration resistance;
- Self-weight penetration of the anchor;
- Required underpressure as a function of depth;
- Allowable underpressure as a function of depth;
- Soil heave as function of depth;
- Maximum penetration depth.

The penetration resistance R_{TOT} is conventionally evaluated using the following expression:

$$R_{TOT} = R_{side} + R_{tip} \quad (19\text{-}2)$$

where R_{side} is the resistance along the sides of the walls due to the side shear, and R_{tip} is the bearing capacity at the skirt tip. The effect of inner stiffeners must be included in the penetration resistance, and the penetration resistance may be higher if there are sand layers or boulders in the clay.

The required underpressure needed to penetrate the skirts is calculated by the following expression:

$$\Delta u = (R_{TOT} - W')/A_{in} \quad (19\text{-}3)$$

where W' is the submerged weight of the suction anchor during installation and A_{in} is the inside suction anchor area. The allowable underpressure with respect to large soil heave within the cylinder due to bottom heave at the skirt tip level can be calculated by bearing capacity. In shallow water the allowable pressure should not exceed the cavitation pressure.

The required suction should be less than the allowable suction, which is the critical suction that may induce failure of the soil plug inside the suction anchor.

In the conventional approach, the resistance along the sides of the walls (R_{side}) is calculated considering that the external (f_{ext}) and the internal (f_{int}) frictions are governed by the same expression:

$$f_{ext} = f_{int} = \alpha . Su_{DSS} \quad (19\text{-}4)$$

where α is an adhesion factor commonly taken as the inverse of the sensitivity value (the ratio between the intact and the remolded shear strengths), and Su_{DSS} is an average DSS shear strength over the penetration depth.

Recent installation of suction anchors equipped with inside stiffeners in highly plastic clays showed that the conventional approach gives high penetration resistance compared to the measurements. It is believed that a mixture of remolded soil and water is trapped within the stiffener zone, leading to low internal strength during installation not taken into account by the conventional approach. Several analyses are required to establish a mechanism to address the bearing resistance of the internal stiffeners and the friction generated on the inner skirt wall: (1) special laboratory tests to investigate the behavior of the soil between internal ring stiffeners, (2) finite element analyses with large displacement formulations, and (3) site-specific calibration.

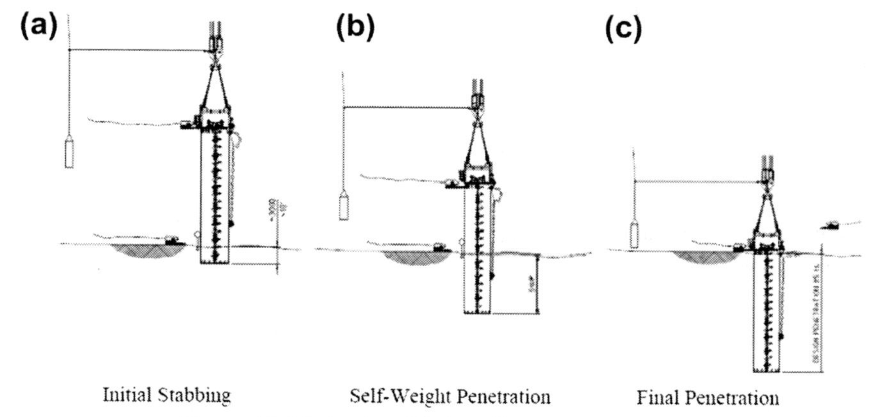

Figure 19-26 Suction Pile Installation Sequence [30]

19.4.4.4. Installation Impact Response

The impact response analysis of lowering the suction pile to the seabed is to check the limiting heave motions of the pile close to the seabed. A range of crane lowering speeds should be used considering the available water evacuation areas, suction pile anchor geometry, and prevailing soil conditions. The check criteria are to avoid soil-bearing capacity failures and to avoid high hydrodynamic pressures within the suction pile anchor that could influence its structural design [29]. Figure 19-26 illustrates the landing sequence of suction pile installation.

19.4.5. Suction Structural Design

The suction pile should be designed to withstand the following loads:

- Maximum loads applied and equilibrated by the soil reactions;
- Maximum negative pressure (underpressure) required for pile embedment;
- Maximum internal pressure (overpressure) required for pile extraction;
- Maximum loads imposed on the pile during lifting, handling, launching, lowering, recovery, etc.

The maximum horizontal and vertical loads should be used for the global structural design of piles. A structural finite element model may be used for the global structural pile analysis to ensure that the pile wall structure and appurtenances have adequate strength in highly loaded areas. The structural components of the suction pile should be designed in accordance with the applicable provisions of API RP 2A, AISC, and API

Bulletins 2U and 2V. In general, cylindrical shell elements should be designed in accordance with API RP 2A or API Bulletin 2U, flat plate elements in accordance with API Bulletin 2V, and all other structural elements in accordance with API RP 2A or AISC, as applicable.

In API RP 2A and AISC, allowable stress values are expressed, in most cases, as a fraction of the yield stress or buckling stress. In API Bulletin 2U, allowable stress values are expressed in terms of critical buckling stresses. In API Bulletin 2V, allowable stresses are classified in terms of limit states. Two basic limit states are considered in API Bulletin 2V: ultimate limit states and serviceability limit states. Ultimate limit states are associated with the failure of the structure, whereas serviceability limit states are associated with adequacy of the design to meet its functional requirements. For the purposes of suction pile design, only the ultimate limit state is considered in design.

19.4.5.1. Allowable Stresses and Usage Factors

For structural elements designed in accordance with API RP 2A or AISC, the safety factors recommended in API RP 2A and AISC should be used for normal design conditions. For extreme design conditions, the allowable stresses may be increased by one-third.

For structural elements analyzed using finite element techniques, the von Mises (equivalent) stress should not exceed the maximum permissible stress as shown here:

$$\sigma = \eta_0 \sigma_y \tag{19-5}$$

where η_0 is the basic usage factor, and σ_y is the material yield strength. The basic usage factor η_0 is 0.8 for the maximum in-place loading condition and 0.6 for normal operating, transportation, lifting, lowering, and recovery conditions. The permissible stresses are based on the fiber stresses for simple beam analyses, and the membrane or midthickness stresses for FEAs using plate elements. For laterally loaded plates also exposed to in-plane (e.g., membrane) stresses, the surface von Mises stress computed at the middle of the plate field (e.g., midway between stiffeners and/or girders) should not exceed the following:

$$\sigma_p = (\eta_0 + 0.1)\,\sigma_y$$

The nominal elastic stress calculated in the middle of the plate field due to lateral pressure acting alone should not exceed $\eta_0 \sigma_y$.

19.4.5.2. Buckling Checks

Hydrostatic buckling calculations should be performed in order to check the capacity of the shell wall of a pile during pile embedment for local shell buckling. Although the buckling of the shell wall is not a concern for operating conditions due to the soil supporting the structure, it is a potential concern during installation. When the pile is penetrating the soil and is subjected to differential embedment pressures, there is a potential for local buckling due to axial and hoop stress interactions.

The design minimum buckling collapse pressure of the pile should have an adequate safety margin to ensure the shell of pile is strong enough to resist the maximum possible suction pressure. In addition, an adequate safety margin between the design minimum buckling pressure and the required suction pressure for embedment of the pile should be applied.

A closed, unstiffened cylinder under hydrostatic pressure will buckle between end supports by forming a pattern of circumferential lobes. The lobes formed around the circumference, or the buckling mode, are a function of the unsupported shell length-to-diameter ratio L/D and cylinder D/t ratio. Buckling may be purely elastic or a combination of elastic deformation and elastoplastic deformation. In the case of suction piles, the D/t and L/D ratios are such that elastic buckling is the predominant mechanism.

Two analytical approaches are available for determining the buckling capacity:

- Design code-based buckling capacity analysis methods using an estimated unsupported length of shell;
- Finite element analysis of buckling behavior incorporating full soil/structure interactions to explicitly model the degree of soil support to the shell of a suction pile.

A number of design codes provide analysis methods to determine the design buckling strength of a suction pile. The methodology detailed in API RP 2A for combined axial compression, hydrostatic pressure, and bending may be used to check the pile wall for local buckling during the embedment process. The semiempirical technique used in API RP 2A is appropriate for pile diameter–to–wall thickness ratios of less than 300 ($D/t < 300$), the methodology must be adapted to account for multiple wall thickness cans between ring frames. This is accomplished by using a weighted average wall thickness in the calculations that is a function of effective buckling length. The effective buckling length will decrease as the pile penetrates the soil and the part of the pile sticking up above the mudline decreases. As the pile penetrates deeper into the soil, the soil will progressively support the pile wall and increase the buckling capacity by forcing the wall structure into a higher mode of buckling failure.

19.4.5.3. Transportation Analysis

A transportation analysis should be carried out for the following procedures:

- The pile structure is handled and transferred to the transportation barge.
- During transportation to the installation site, the piles in the transportation barge are normally supported at two locations with cradles. The cradles will take the roll forces, while pitch plates welded to the pile and the barge deck will take pitch forces.
- Before installation into the seabed, the pile structure is upended from a horizontal to a vertical position.

The loads considered in the transportation analysis should include the self-weight of the suction pile and inertial loads developed from a motion analysis. A dynamic load factor of 2.0 should be used in the design calculations according to API RP 2A-WSD. The calculations for padeye design can be carried out by hand, and the padeye support structures should be checked by an FEA method for the simulation of all pile structures. Lift analysis should be performed to check the structural integrity of the suction pile and lift attachments under installation and removal onto the transport barge and pile lowering and recovery.

19.5. INSTALLATION OF SUBSEA MANIFOLD

In deepwater fields the contribution of installation activities to project costs and schedule is higher than for shallower developments. The risks associated with installation are also higher. The metocean condition of deepwater is a key factor for the installation of subsea structures. Relatively gentle conditions in offshore West Africa may not mean installation operations are easier than in the GoM because the persistent swell is likely to result in ideal conditions for vessel motion resonance. High currents in offshore Brazil are likely to be a dominating factor.

A subsea development may have more than 30 wells, which requires an extensive installation program of manifolds, PLETs, jumpers, and suction pile foundations. Excluding the flowlines, which are small lift weight components having compact dimensions, we must still consider the large number of individual items that require a long installation program and the large number of heavier manifolds, which also present installation challenges.

19.5.1. Installation Capability

In the deepwater installation, a number of challenges come up due to the increasing of water depths in the field development. These challenges potentially constrain the installation capacity for subsea structures in deepwater. The capabilities of the subsea installation system are mainly depending on the limitations of the following components of the system:

- Lifting and lowering system, which includes vessel, lift line, and overboarding/lift line deployment system;
- Load control and positioning system, including motion compensation system, buoyancy hook/payload control/positioning, and communications.

The lifting and lowering system directly related to the weight of the loads to be lowered to the deep seabed, the dynamic responses that can augment these loads, and the decrease in the capability of the lifting systems. The load control and positioning system related to placing the load in the desired location, at the correct compass heading, and at a stable attitude on the seabed. The effects of these issues are discussed in following sections.

19.5.1.1. Lifting and Lowering System

Steel wire ropes with multifall lowering systems are very well understood and durable, but they are limited in their application to very deep water Figure 19-27 shows the relationship of lift capacity with deployment depth. As the water depth increases, the ratio of the weight of the cable to the weight of the payload increases quickly. At 3000 m (9840ft) the weight of a 5-in. (0.13m) wire rope is about the same as its 170-tonne payload. At a depth of about 6000 m (19700ft), the safe working load (SWL) of the steel wire rope is entirely used up by its self-weight, leaving zero payload capacity.

The potential for resonance, associated with the natural period of the payload on the line and with surface vessel excitation, can give rise to very large dynamic loads. It is also difficult to estimate the hydrodynamic added mass for a complex payload shape. Fiber rope is also used in deepwater hardware installations. Fiber options include aramids, polyester, and high-modulus polyethylene for offshore applications. The fiber ropes have several attractive properties such as lower self-weight, small allowable bend radii, lower axial stiffness, a large damping ability to absorb heave created by surface waves, and the ability to be repaired. Petrobas used a polyester rope to successfully install a manifold in deep water of 2000 m (6560ft). However, fiber rope has potential problems related to stretch, creep, and its relatively low melting point.

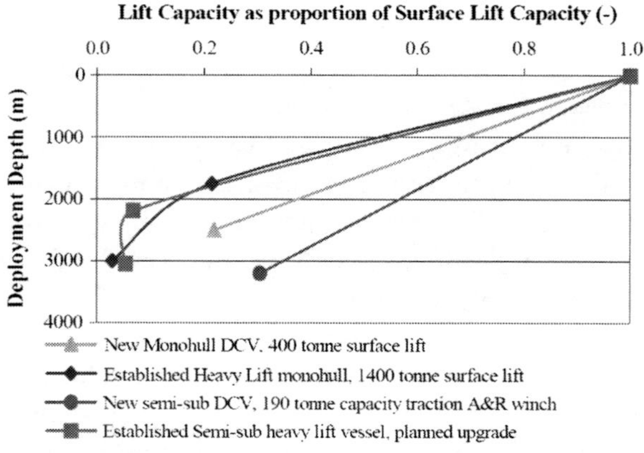

Figure 19-27 Degradation of Lift Capability with Depth [31]

19.5.1.2. Load Control and Positioning System

Buoyancy units may have an important role to play in reducing the static lifting line tensions. However, buoyancy units for large subsea components to be installed in deep water are not easy to design in such a way that they are manageable and economic. The load control and stability problems of buoyancy are difficult to solve; in particular, the inertia and hydrodynamic loading of the system are increased, which contributes to undesirable dynamic effects.

Very significant dynamic effects can result when lowering heavy weights on long lines. The excitation caused by the motions of the surface vessel can be amplified with large oscillations and high dynamic tensile loads in the lifting line. Motions in the heave direction may be only lightly damped, and the added mass of the load can be very significant. For example, a suction anchor consisting of a flooded cylinder with closed top will have an added mass that is many times its weight in air due to the water trapped inside and entrained around it. When combined with dynamic magnification caused by oscillations due to surface waves, the line tension of 460 tonnes is obtained for a suction anchor with a weight in air of 44 tonnes.

The shape of the item to be installed, which in turn determines the added mass, can therefore be crucial to this dynamic response and to the ability to install it. It can be shown that for lowering into deep water there will nearly always be a depth at which a resonant response will occur. It is important for this resonant region to be passed through relatively quickly and for it not to occur at full depth where careful control is required for placement of the payload on the seabed. Modeling methods, as shown in the next section, have been developed to predict the behavior of these dynamic responses so that design and planning of the lowering operations can attempt to minimize and avoid them.

19.5.1.3. Motion Compensation

For the installation of subsea structures in deep water, installation contractors encounter limitations when using equipment without heave compensation systems. Limitations include:

- Excessive dynamic amplification of the load during lowering, with the risk of overloading or even rupturing of the cable;
- Unstable situations during the landing of the subsea structure and its positioning on the seabed;
- ROV assistance operations with respect to the moving load.

Active and passive motion compensation systems are mainly used in subsea installations. Many cranes for subsea installation are equipped with an active heave compensation (AHC) system to compensate for the vertical heave motion at the crane jib tip by means of a powerful computer system and a high-speed winch system. A passive heave compensator (PHC), such as a cranemaster, is another kind of heave compensator that can alleviate the high hook loads. In essence, the cranemaster is a damped spring system set above the subsea structure using compacted gas over an oil accumulator acting as an isolator. Two types of PHC systems for subsea structure installation are introduced here because of their impressive heave compensation effects.

Heave-Compensated Landing System

The heave-compensated landing system (HCLS) used by Delmar consists of the following equipment: a compensating buoy system, compensating belly chain, and pendant wire. Figure 19-28 shows an example of the HCLS installation of subsea structure with an anchor handling vessel (AHV) [32]. The buoyancy system suspends the subsea structure, keeping the system close to a neutrally buoyant condition in the vertical direction. The subsea contracture is lowered to the seafloor in a controlled manner by paying out wire from the AHV, which transfers more chain and thus more weight to the buoyancy system. As more chain is added, the added weight overrides the buoyant forces, causing the subsea structure to slowly sink. The length and configuration of the chain are the compensating portion of the system. HCLS has a very good heave compensation effect, but it is difficult and complicated to get the subsea equipment overboard from the barge.

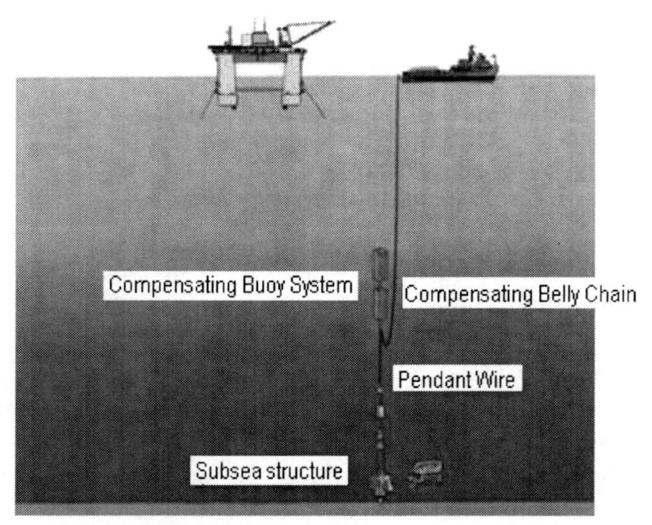

Figure 19-28 Installation of Subsea Structure with HCLS [32]

Polyester Rope

Polyester rope is one of the most popular ropes used as anchor and mooring lines in the mooring systems of the offshore industry. It is very close to nylon in strength but stretches very little. Polyester rope has highly elastic properties and a good damping effect. Typical mechanical properties for polyester rope can be found in the Orcaflex software database, which is based on a catalog published by Marlow Ropes Ltd.

19.5.2. Installation Equipment and Installation Methods

The installation of a subsea structure/manifold requires careful planning and coordination with workboats, a crane barge or floating drilling vessel, and acoustic and/or electrical location equipment. The choice of installation vessel is based on vessel availability, existing mooring equipment, adequate crane capacity, and suitable deck space for transportation.

19.5.2.1. Crane Barge

Figure 19-29 illustrates a manifold installation from a crane barge. The foundation suction pile is 60.3 tonne, while the manifold is 88.5 tonne. The manifold is landed and locked to the suction pile. A ROV is used to monitor the direction and position of the manifold during the entire installation procedure.

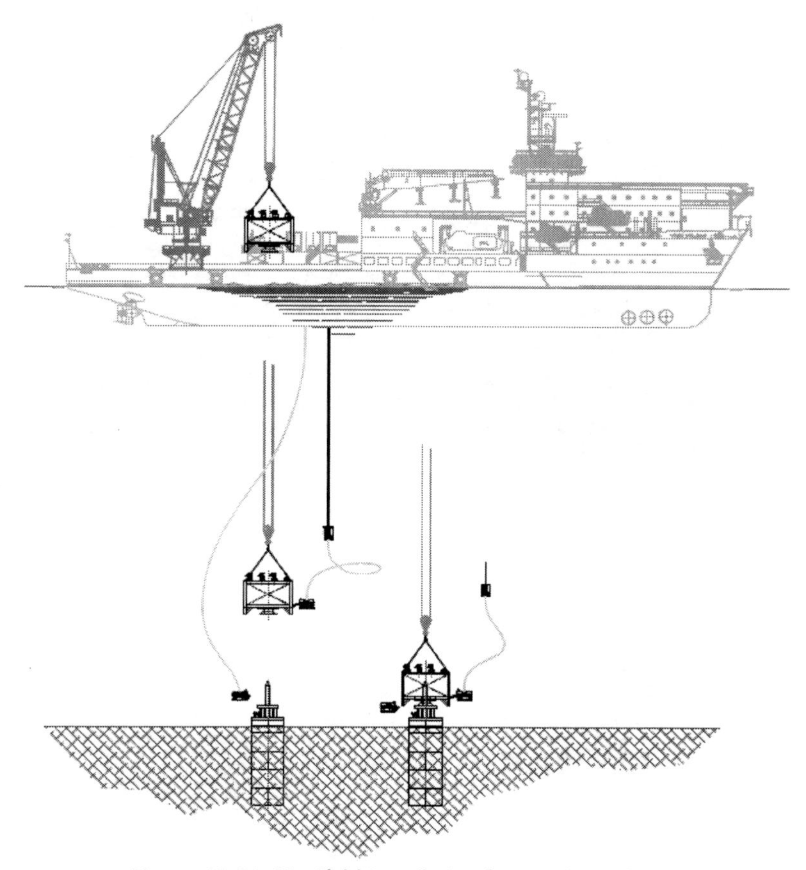

Figure 19-29 Manifold Installation from a Crane Barge

In water depths up to 1000 m (3280.8ft), manifolds are normally installed by cable from a crane barge or drilling riser directly, depending on the wire length and crane capacity. For water depths greater than 1000 m (3280.8ft), for which the length of the crane cable is not long enough to land the subsea structure directly on the seabed, an A&R wire and winch may be used to land the structure, and the load is transferred from the crane to the A&R wire at a water depth of approximately 100 m (328.1m). Figure 19-30 shows the installation of a temple using THIALF in the Ormen Lange Field at a water depth of 820 m (2690ft).

19.5.2.2. DP Rig and Drilling Riser

Figure 19-31 shows a manifold installation from a DP rig using a drilling riser. Because the manifold size was bigger than the moon-pool size on the

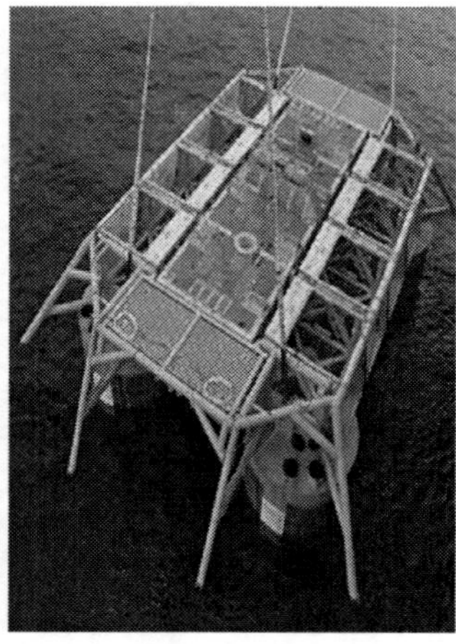

Figure 19-30 Installation of Temple Using THIALF in the Ormen Lange Field [33]

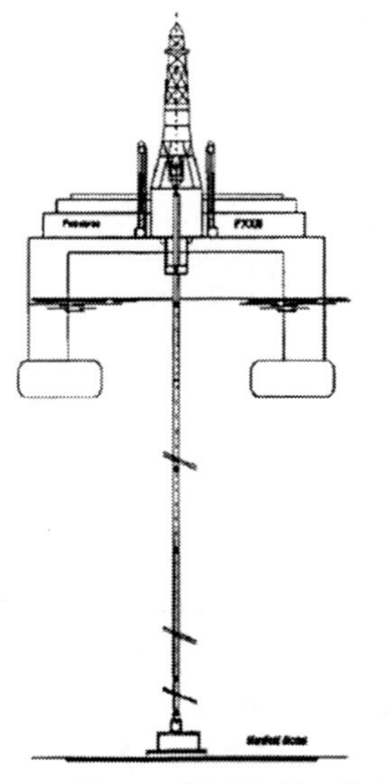

Figure 19-31 Manifold Installation Using Drilling Riser [4]

available rig, the manifold could not be lowered through the moon-pool directly, so a means to transport the manifold from the shipyard to the field was required. The rig was maneuvered so that it was placed over the transportation barge, allowing the riser connection and the lift operation. The manifold was then fixed by six stoppers and prior welded underneath the moon-pool to avoid movement during the transport and installation phases.

The installation methods by crane barge and by DP rig using drilling risers in a water depth of 500 m (1640ft) were compared based on the analysis of Petrobras projects. Analyzing the results and considering the barge's daily cost as around 80% of the rig's daily cost, the following conclusions were made:

- Installation using a crane barge is not economically feasible in crowded areas where a lot of equipment and flowlines have already been installed at the seabed or in water depths where the mooring system demands special resources.
- Time spent waiting on the weather for an operational window and the mooring system deployment task represents around 35% of the total manifold installation cost when the installation is made by a crane barge. The installation of stoppers in the drilling rig, underneath the moon-pool, represents 30% of the total manifold installation cost.
- The total cost of the installation by drilling riser is 25% lower when compared with the total cost of the installation by crane barge and there is a potential optimization that can be reached.
- The option of using a crane barge is still attractive because such a resource is promptly available and an installation by rig may impact the drilling schedule and completion of other activities.

19.5.2.3. Other Installation Methods

Installation of Sheave Manifold

The Roncador manifold project located in the ultradeep Campos Basin offshore from Brazil marks the first time a subsea manifold has been installed at a water depth exceeding 1600 m (5250ft). The Roncador manifold was installed at water depth of 1885 m (6184ft) using steel cables and support vessels.

Figure 19-32 shows the installation procedure of the sheave manifold, in which an SS rig provided heave motion compensation. AHTS 1 lifted the manifold together with the SS rig and AHTS 2 oriented the manifold.

Pendulous Method for Manifold Installation

Figure 19-33 shows the pendulous procedure for P52 manifold installation by a barge and an AHTS. The manifold has these dimensions: 16.5 m (54.4ft) (L) × 8.5 m (27.9ft) (W) × 5.2 m (17.1ft) (H), with a weight in air of 280 tonne at water depth of 1900 m (6233.6ft). Buoyancy sets are used to decrease the load on the rope.

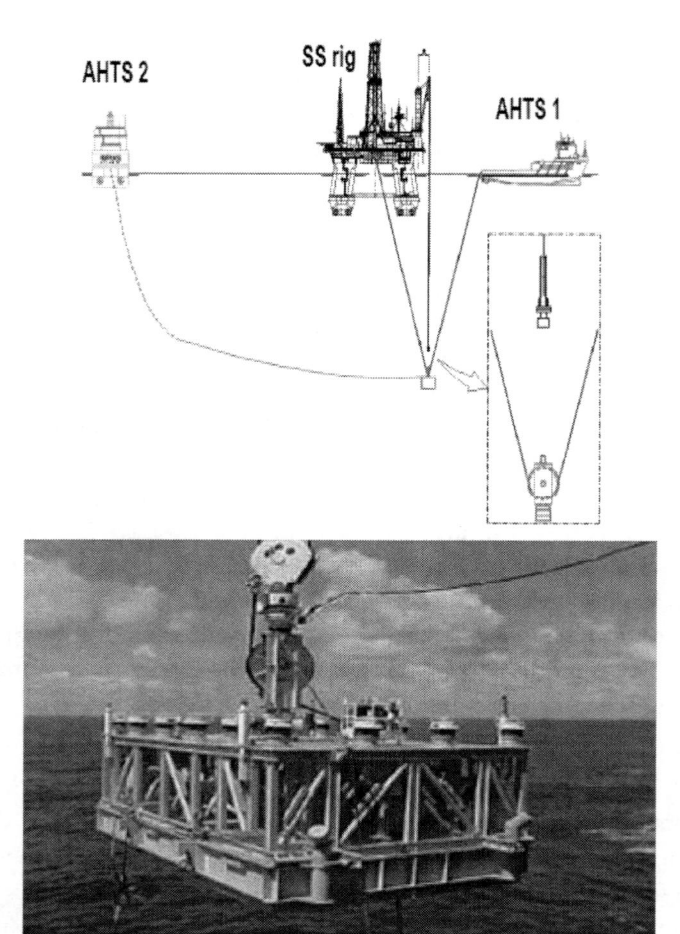

Figure 19-32 Installation of Sheave Manifold [34]

Selection of Installation Method

The installation method and equipment selected for the subsea structure should ensure safe and reliable operation in accordance with the selected intervention strategy. The subsea production system should fulfill the following requirements:

- Installation equipment (temporary and permanent) should not cause obstructions and restrict intervention access.
- Disconnection of lifting slings and lifting beams/frames/arrangements used during installation should be according to the selected intervention strategy. A backup system may be provided.

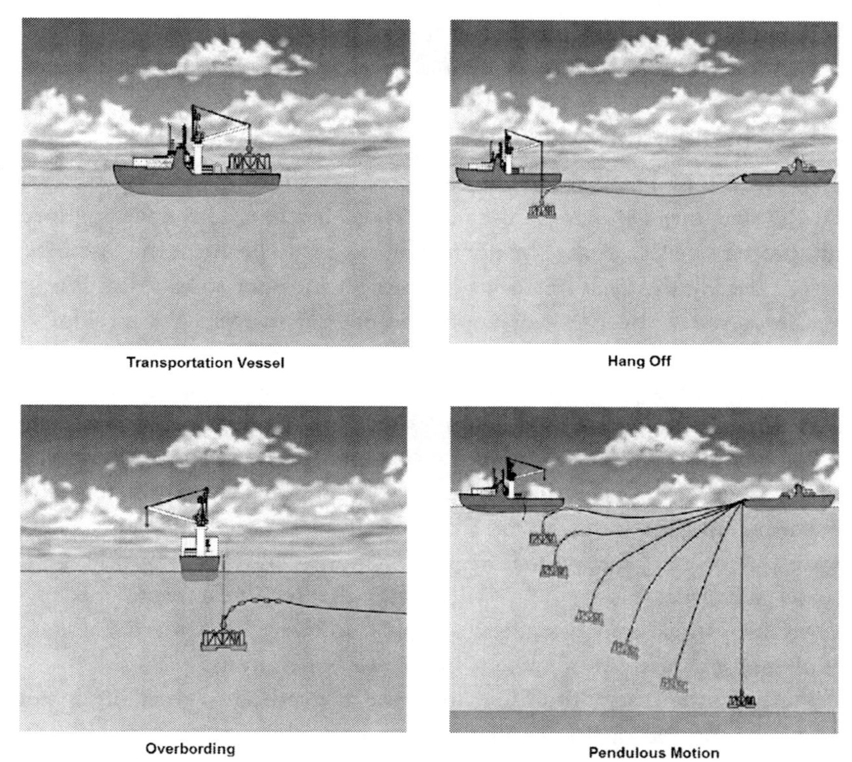

Figure 19-33 Pendulous Method for P52 Manifold Installation [34]

- The installation system should not represent any hazard to the permanent works during installation, release, reconnection, and removal.
- Lifting/installation arrangements should be designed to minimize the lifting height.

The installation of a subsea system should satisfy the following issues:

- Be video -recorded during installation operations.
- Use installation tools with a fail-safe design.
- Allow flushing of hydraulic circuits subsequent to connection of interfaces.
- Where possible, not be dependent on unique installation vessels.
- Have position indicators on all interface connections.
- Be installable utilizing a minimum number of installation vessels.
- Require installation within a defined practical weather window that is consistent with the specific type of installation equipment and vessel to be used.

- Require a minimum number of special installation tools.
- Facilitate fully reversible sequential installation techniques/operations.

19.5.3. Installation Analysis

Subsea manifold design should be subjected to a thorough analysis to ensure that the structure can handle the installation, leveling, and lowering forces with proper safety factors. The normal practice is to perform the installation analysis during the final phases of the project in order to establish limiting weather criteria for the marine operations. However, if a preliminary dynamic analysis is conducted earlier, during the conceptual or design phase, a preliminary assessment of the main factors of dynamic loading may allow integration of structural design criteria and operational requirements. Lifting analysis can help to quantify dynamic forces and thus identify the critical stages of the operation. In cases where the hydrodynamic forces will determine the sea state limitation, an analysis can potentially enable an expansion of the operational weather window and optimization of the project schedule. Figure 19-34 illustrates the technical issues related to deepwater installations. Orcaflex is typical software used for the dynamic simulation and analysis of various installation operations.

During installation, the lifted structure is exposed to dynamic loading due to the motion of the installation vessel and the direct action of waves. The installation procedure involved in the lifting operations normally includes following steps:

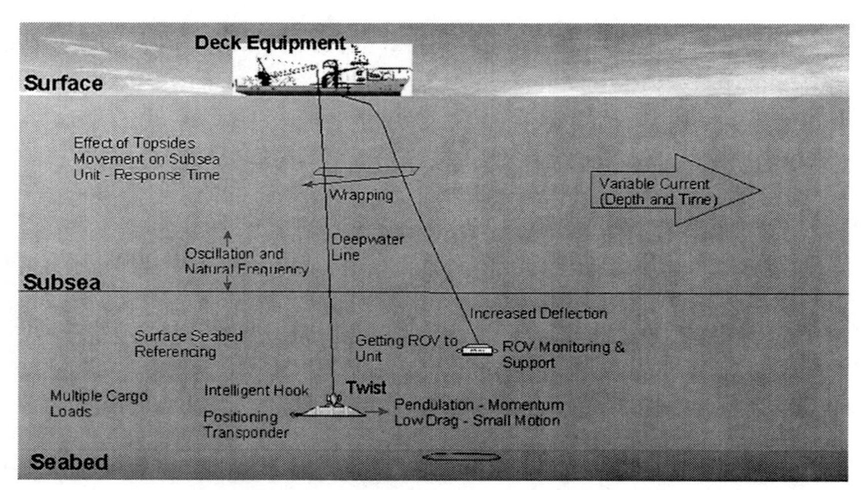

Figure 19-34 Deepwater Installation Issues [35]

- *Barge lift:* lifting from the deck of the crane vessel;
- *Splash zone:* lowering through the splash zone;
- *Transferring to A&R wire:* load from the manifold is transferred from the crane to A&R wire due to water depth;
- *Lowering in deep water:* lowering analysis at critical resonance depth;
- *Landing on the seabed:* the velocity and heave amplitude of the installed manifold should be controlled within the installation criteria to land the manifold safely.

The characteristics of these installation steps are summarized in the following subsections.

19.5.3.1. Barge Lift

The main purpose of the analysis is to determine the minimum crane lifting speed required to avoid recontact of the manifold with the barge deck while lifting. This minimum crane lifting speed could be obtained by using the criteria that the crane lifting velocity should be greater than the relative vertical velocity between the manifold base and the barge to avoid recontact. The pendulum motion of a structure due to crane tip movement is one of the limiting criteria. Bumper frames and tugger lines can be used to control the motion of the structure.

19.5.3.2. Splash Zone

The motions of the installation vessel combined with the motion of surface waves will create a significant load on the subsea structure when it is lowered through the splash zone. It is important to determine the wave loads during the passage through the splash zone and the added mass and damping when the subsea structure is submerged to the seabed. However, the added mass and drag coefficients are immersion dependent.

The installed structure should be exposed to extreme direct wave loading in the analysis. The added mass associated with the installed structure at different submerged volumes may be estimated using the radiation/diffraction software Diodore, while the drag coefficients are estimated using CFD software. The slamming forces are estimated separately using the DNV code [36].

19.5.3.3. Transferring to A&R Wire

After lowering through the splash zone, the installed structure is lowered further to a water depth of about 100 m (328.1ft) to transfer the load from the crane to A&R wire if the crane wire is not long enough for the water depth. Once it reaches the transfer depth, the payout on the crane wire is stopped, the A&R winch wire is deployed and then connected with an ROV to the lowering yoke, and the load transfer is made from the crane main hook wire to the A&R winch.

The hydrodynamic coefficients of the installed structure are kept the same as those of the final phase of the splash zone analysis. The tensions on the yoke and the motions of the installed structure during the upending procedure are checked.

19.5.3.4. Lowering in Deep Water

After transferring the structure load to the A&R winch, A&R wire is used to lower the structure to the seabed. To determine the peak vertical motions of the subsea structure during the lowering procedure, the resonance depth has to be calculated. In a regular wave analysis, the resonance occurs where the wave period corresponding to the maximum heave motions of the vessel in a particular wave period range matches the natural period of the subsea structure lowering system. The natural period of the lowering system depends on the length of the A&R wire, so at a certain length of A&R wire resonance occurs and the corresponding depth is identified as the resonance depth. For the calculation of the natural period of the lowering system, the reader is referred to Section 5.3.5.1 of DNV-RP-H103 [36]. The hydrodynamic parameters of the structure are the same as for the last step. Dynamic loading is mainly due to crane tip motion.

19.5.3.5. Landing on the Seabed/Subsea Structure

The hydrodynamic coefficients used in the landing analysis are same as those used in the lowering analysis. The maximum allowable touchdown velocity should be specified in the installation criteria. Installation may be performed in heave compensation mode to satisfy the criteria. The hydrodynamic parameters of the structure may be influenced by the seabed. Dynamic loading is mainly due to crane tip motion.

REFERENCES

[1] B. Rose, Flowline Tie-in Systems, SUT Subsea Awareness Course, Houston, 2008.
[2] C. Davison, P. Dyberg, P. Menier, Fast-Track Development of Deepwater Kuito Field, Offshore Angola, OTC 11873, Offshore Technology Conference, Houston, Texas, 2000.
[3] E. Coleman, G. Isenmann, Overview of the Gemini Subsea Development, OTC 11863, Offshore Technology Conference, Houston, Texas, 2000.
[4] M.T.R. Paula, E.L. Labanca, C.A.S. Paulo, Subsea Manifolds Design Based on Life Cycle Cost, OTC 12942, Offshore Technology Conference, Houston, Texas, 2001.

[5] Gate Valves & Actuators, http://www.magnum-ss.com/ss-gatevalve.html.
[6] American Petroleum Institute, Specification for Wellhead and Christmas Tree Equipment, seventh ed., API Spec 6A, 2000.
[7] American Petroleum Institute, Specification for Subsea Wellhead and Christmas Tree Equipment, API Spec 17D, 1992.
[8] Autoclave Ball Value, http://www.autoclave.com/products/ball_valves/index.html.
[9] American Petroleum Institute, Pipeline Valves, twenty-second ed., API Spec 6D, (2002).
[10] D.R. Mefford, Deep Water Subsea Ball Valves, Cameron, http://www.c-a-m.com/, (2010).
[11] American Petroleum Institute, Recommended Practice for Design and Operation of Subsea Production Systems, API-RP- 17A (2002).
[12] International Organization for Standardization, Petroleum and Natural Gas Industries - Offshore structures - Part 1: General requirements, ISO 13819-1, first ed., (1995).
[13] International Organization for Standardization, Petroleum and Natural Gas Industries - Offshore structures - Part 2: Fixed steel structure (Interim standard), ISO 13819-2, first ed., (1995).
[14] American Society of Mechanical Engineers, Gas Transmission and Distribution Piping Systems, ASME B31.8, 2010.
[15] Det Norske Veritas, Submarine Pipeline Systems, DNV-OS- F101 (2003).
[16] American Petroleum Institute, Recommended Practice for Planning, Designing and Constructing Fixed Offshore Platforms-Working Stress Design, twenty first ed, API-RP-2A-WSD, 2002.
[17] American Welding Society (AWS), Structural Welding Code – Steel, AWS D1.1, 2008 ed.
[18] American Society of Mechanical Engineers, Boiler and Pressure Vessel Code, Section VIII, Div. 3, ASME (2007).
[19] Lloyd's Register of Shipping, Rules and Regulations for the Classification of Fixed Offshore Installations, Lloyd's Register (1990).
[20] International Organization for Standardization, Petroleum and Natural Gas Industries – Design and Operation of Subsea Production Systems – Part 6: Subsea Production Control Systems, ISO 13628-6, (2006).
[21] American Petroleum Institute, Specification for Subsea Umbilicals, third ed., API Spec 17E, 2003.
[22] Det Norske Veritas, Cathodic Protection Design, DNV RP-401, (1993).
[23] NACE International, Corrosion Control of Steel-Fixed Offshore Platforms Associated with Petroleum Production, NACE Standard RP 0176-03, Houston, 2003.
[24] D. Janoff, N. McKie, J. Davalath, Prediction of Cool Down Times and Designing of Insulation for Subsea Production Equipment, OTC 16507, Offshore Technology Conference, Texas, Houston, 2004.
[25] A. Eltaher, Y. Rajapaksa, K.T. Chang, Industry Trends for Design of Anchoring Systems for Deepwater Offshore Structures, OTC 15265, Offshore Technology Conference, Houston, Texas, 2003.
[26] J.-L. Colliat, Anchors for Deepwater to Ultra-deepwater Moorings, OTC 14306, Offshore Technology Conference, Houston, Texas, 2002.
[27] H. Dendani, Suction Anchors: Some Critical Aspects for Their Design and Installation in Clayey Soils, OTC 15376, Offshore Technology Conference, Houston, Texas, 2003.
[28] K.H. Andersen, H.P. Jostad, Foundation Design of Skirted Foundations and Anchors in Clay, OTC 10824, Offshore Technology Conference, Texas, Houston, 1999.
[29] P. Sparrevik, Suction Pile Technology and Installation in Deep Waters, OTC 14241, Offshore Technology Conference, Houston, Texas, 2002.

[30] A. Couch, et al., Independence Installation, OTC 18585, Offshore Technology Conference, Houston, Texas, 2007.
[31] R.G. Standing, B. Mackenzie, R.O. Snell, Enhancing the Technology for Deepwater Installation of Subsea Hardware, OTC 14180 (2002).
[32] J. Soliah, M. Guinn, Subsea Equipment Installations Utilizing Anchor Handling Vessels, Deepwater Technology (October 2003) 25–27.
[33] T. Bernt, E. Smedsrud, Ormen Lange Subsea Production System, OTC 18965, Offshore Technology Conference, Houston, Texas, 2007.
[34] J. Mauricio, et al., Development of Subsea facilities in the Roncador Field (P-52), OTC 19274, Offshore Technology Conference, Houston, Texas, 2008.
[35] S.J. Rowe, B. Mackenzie, R. Snell, Deepwater Installation of Subsea Hardware, Proc. 10th Offshore Symposium, Houston, Texas, 2001.
[36] Det Norske Veritas, Modeling and Analysis of Marine Operations, DNV-RP-H103 (2009).

CHAPTER 20

Pipeline Ends and In-Line Structures

Contents

20.1. INTRODUCTION

As oil/gas field developments move further away from existing subsea structures, it becomes advantageous to consider subsea tie-ins of their export

Subsea Engineering Handbook
ISBN 978-0-12-397804-2, doi:10.1016/B978-0-12-397804-2.00020-5

systems with existing deepwater pipeline systems that offer spare transport capacity. This necessitates incorporating pipeline end manifolds (PLEMs) at both pipeline ends to tie in the system. A PLEM is a subsea structure (a simple manifold) set at the end of pipeline that is used to connect a rigid pipeline with other subsea structures, such as manifolds or trees, through a jumper. It is also called a pipeline end termination (PLET), especially when serving as a support for one pipeline valve and one vertical connector. Figure 20-1 shows PLEMs used in the Anadarko GC518 field [1], which tied back to the Marco Polo TLP. The in-line structure is a simple manifold set at the middle of the pipeline, which is connected in line with the pipeline and used as a tee connector to divide or combine pipelines.

The deepwater PLEM with its mudmat and up-looking hub was designed as an economical and reliable method for terminating pipelines of all sizes. The mudmat is favored for economy and position compliance. The PLEM must be remotely installable and designed to support ROV execution. PLEMs are installed with the pipeline end from the installation barge and are lowered into their final position. A PLEM can also be a platform for a range of optional components such as valves, taps, and instrumentation. After installation, the PLEM can be accessed for repair or maintenance by removing the pipeline connection jumper and recovering the PLEM to the surface.

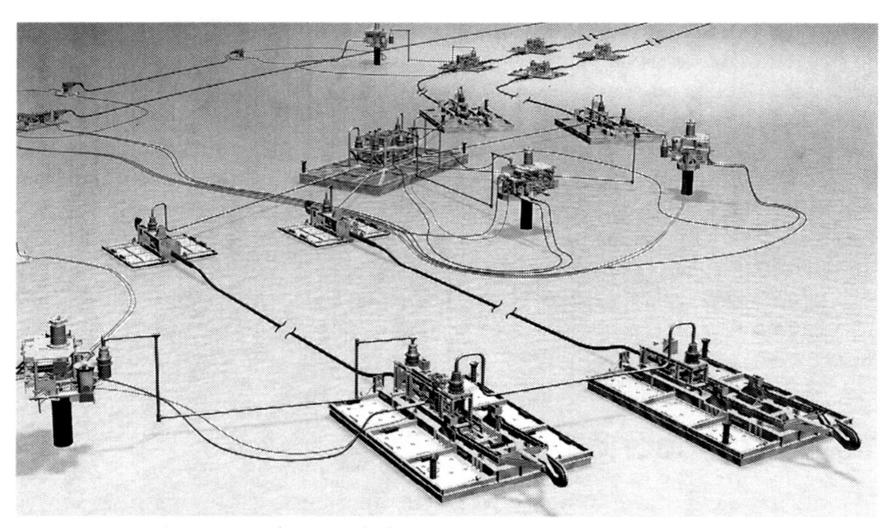

Figure 20-1 PLEMs Used in Anadarko Project, Tied Back to the Marco Polo TLP [1]

20.1.1. PLEM General Layout

The purpose of the PLEM is to provide an installation structure to attach these piping components and then lower them to the seafloor in the desired orientation. The structure must not only withstand the installation loads while being lowered to the seafloor, but also retrieval to the surface should a failure occur. Figure 20-2 shows a typical configuration for a PLEM. The PLEM can be any configuration, but generally is made up of the following assemblies:

- Piping system and components;
- Structural frame, which supports the piping and components (such as wye, valves, hubs);
- Foundation (mudmat), which distributes the loads to the seabed while minimizing settlement;
- Installation yoke, which is hinged to the structural frame, to minimize torsion-related rotation of the PLEM pipeline assembly during installation and laydown of the PLEM onto the seabed.

The PLEM foundation provides support for the structural frame of the piping system. In addition to their normal functions, PLEMs are designed to accommodate reasonable thermal expansion of the pipeline during

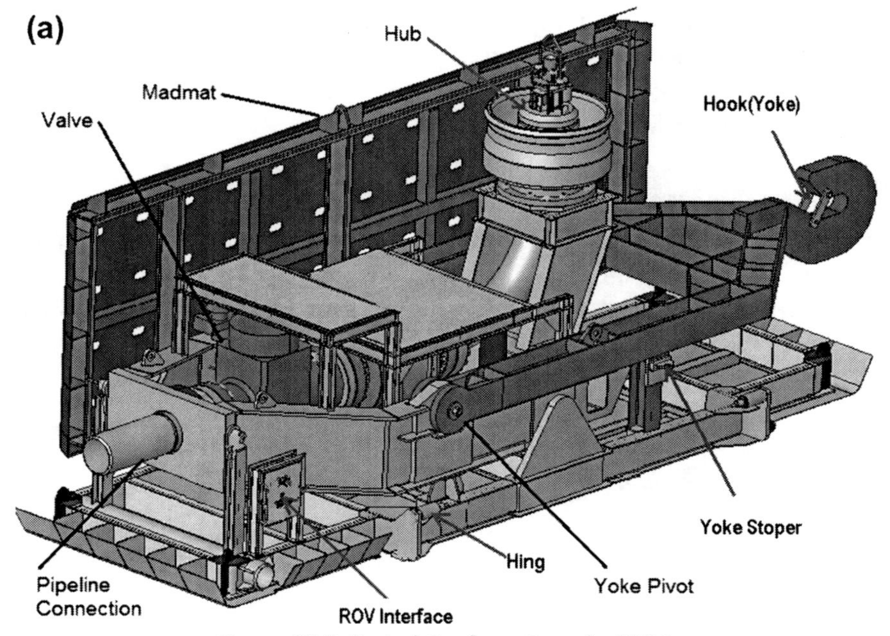

Figure 20-2 Typical Configuration of a PLEM

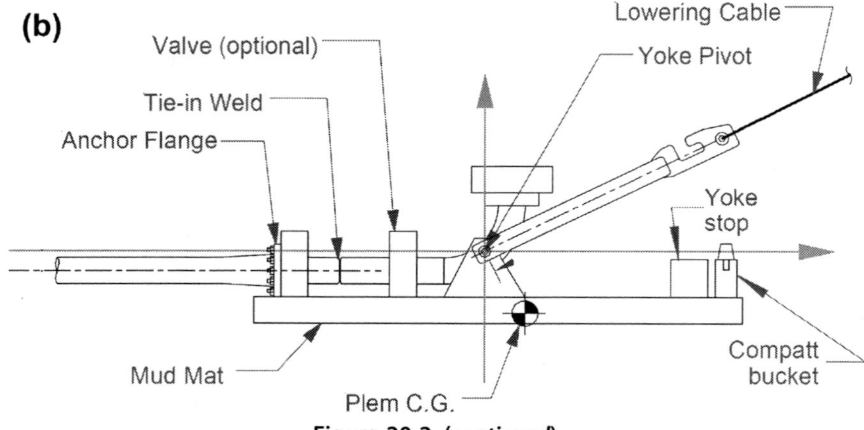

Figure 20-2 (*continued*).

operation. To accommodate for this movement, the design is based on a stationary mudmat foundation that supports a moving pipe support frame, which allows the pipeline end to slide on the frame. The preliminary foundation center section for a foldable mudmat is selected with a width and length that is based on the nominated installation vessel-handling capabilities. The foundation section geometry is further developed by providing folding wings on each side of the center section to obtain the required foundation area to support the anticipated loads on the PLEM.

20.1.2. Components of PLEMs

Figure 20-3 shows a 12-in. pipeline that transports gas to a 16-in. pipeline via two PLEMs. Collet connectors are used for both PLEMs and connected by a rigid jumper [2]. The PLEMs provide the supporting structure for the piping systems and collet connectors. Detailed configurations for both PLEMs are illustrated in Figure 20-4. In the following sections, the 12- and 16-in. PLEMs shown in the figure are used to explain the components of a PLEM.

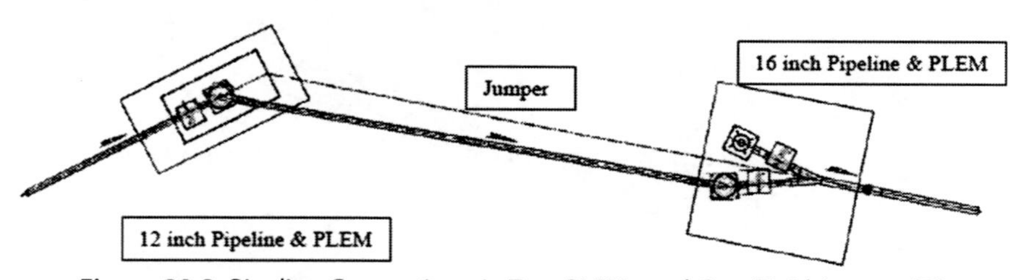

Figure 20-3 Pipeline Connection via Two PLEMs and One Rigid Jumper [3]

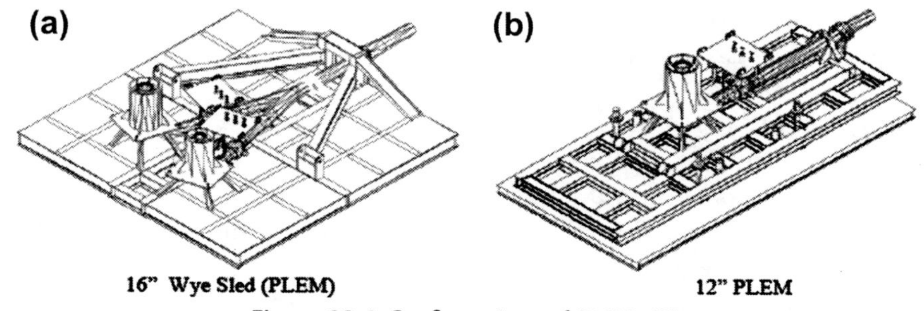

Figure 20-4 Configurations of PLEMs [3]

20.1.2.1. 16-in. PLEM

Piping System and Components

The primary components of the 16-in. PLEM are 16-in. ANSI 1500 class and 12-in. ANSI 1500 class male hubs for the diverless collet connectors and a 16-in. piggable wye. The gas export flowline utilizes a 12-in. connector, leaving the 16-in. connector free for future tie-ins and system expansion. In the piping system, 16- and 12-in. ANSI 1500 subsea ball valves are used for both PLEMs to isolate each pipeline segment. Incorporated in the ball valves are complete ROV interface panels allowing not only operation of the valves but also maintenance. Each ROV interface allows for both lubrication and sealant injection to the valve.

With the piggable wye oriented horizontally and the use of different sized collect connector hubs and valves, the piping was arranged such that the center of gravity was along the centerline of the pipeline, which would minimize the rotation tendency of the PLEM during installation. Rotation of the PLEM was critical in this example because the installation contractor was held to a 3-degree verticality requirement for both collect connector hubs.

Structural Components

The structural components may be grouped into two functional components: installation loaded members and the in-place loaded members. The members subjected to the in-place loads attach and support the piping components to the structural components. The in-place loads consisted of the jumper loads developed in the operating condition, the dead loads of all components, and an applied pipeline torque. Installation experience has shown that the pipeline will tend to rotate as it is retrieved from the seabed, particularly for S-lay pipeline installation. Because the PLEM will achieve a fixed orientation on the seabed, there will be some residual pipeline torque after pipeline is installed.

20.1.2.2. 12-in. PLEM

Piping System and Components

The main components of the 12-in. PLEM are a 12-in. ANSI 1500 class ball valve and a 12-in. ANSI 1500 class male hub for the diverless collet connector. While the design pressure of the piping is 2220 psig, the use of ANSI 1500 class valves allows the pipeline to be hydrostatically tested with the valve in the closed position. The pressure cap should be set on the collect connector hub without leakage. The components are connected by the piping system.

Structural Components

The piping components are set at the top of the frame with the hinge of yoke pivoting, whereas the mudmat attached to the bottom remains horizontal as the pipe is lowered from the J-lay tower. Once the PLEM is laid on the seabed, a temporary cable holds the PLEM in place while the pipeline normal lay is initiated. After the vessel has finished laying pipe, the structure will hinge down and lock in a final, horizontal position. The PLEM is laid first on the seabed and is called the first-end PLEM. The 12-in. PLEM is 16 ft (4.87m) wide by 32 ft (9.75m) long and weighs approximately 34.5 tonns.

20.2. PLEM DESIGN AND ANALYSIS

The objective of this section is to summarize the typical method used for PLEM design and analysis, but it is not a cookbook approach, because PLEMs have different design criteria and requirements, and the configurations of PLEM are based on each project's requirements.

20.2.1. Design Codes and Regulations

The design codes and regulations used for PLEM design should be approved by clients depending on the project's requirements. The following codes and regulations are often used for the design and structural analysis of PLEM in the deepwater of GOM (*Note:* The latest edition of the codes should be used):

- AISC, *Steel Construction Manual, Allowable Stress Design (ASD);*
- API RP 2A-WSD, *Recommended Practice for Planning, Designing and Construction: Fixed Offshore Platforms—Working Stress Design (WSD);*

- API 2RD, *Design of Risers for Floating Production Systems (FPSs) and Tension-Leg Platforms (TLPs);*
- API 5L, *Specification for Line Pipe;*
- ASME B31.8, *Gas Transmission and Distribution Piping Systems.*

The following specifications are used for the reference:

- DNV RP B401, *Cathodic Protection Design;*
- Classification Notes No. 30.4, *Foundations,* for the design of the mudmat;
- NACE RP 0176, *Corrosion Control of Steel Fixed Platforms Associated with Petroleum Production;*
- NACE TM 0190, *Impressed Current Test Method for Laboratory Testing of Aluminum Anodes;*
- NACE No.2 ISSSP-SP10, *Near-White Metal Blast Cleaning;*
- NACE RP0387, *Metallurgical and Inspection Requirements for Cast Sacrificial Anodes for Offshore Applications;*
- NACE 7L198, *Design of Galvanic Anode Cathodic Protection System for Offshore Structures;*
- API 1104, *Pipeline and Pipe Welding,* 18th ed.;
- AWS D1.1-02, *Structural Welding Code—Steel;*
- ISP 8501-1, *Preparation of Steel Substrates before Application of Paints and Related Products.*

20.2.2. Design Steps

The design procedure for a PLEM may be divided into the following four steps:

1. Architecture design, which includes geometry and piping configuration, foundation selection (pile or mudmat), installation requirement (yoke, hinges), and fabrication and installation limitations;
2. Initial sizing, which includes determining the principal load path, configuring the primary structural components (forging, yoke, etc.) and the layout of structure frames and supports to support valves and pressure/temperature components that facilitate installation and maintenance, and sizing the mudmat based on loads, structure dead weight, and soil data;
3. Drafting a model for analysis, which includes developing a 3D PLEM system with AutoCAD 3D software and saving the 3D model in a SAT format file for finite element detailed analysis; meanwhile a SACS/StruCAD model is developed to determine the structure steel frame member's sizes;

4. Stress analysis, which includes importing the SAT file into ABAQUS/ ANSYS finite element software, meshing the 3D model, applying the loads in the operating and installation conditions, then applying a stress analysis and checking stress to satisfy the design criteria.

20.2.3. Input Data

The following data are required for a PLEM design:
- Geotechnical data;
- Specifics of the equipment, including weight and center of gravity (CG) of each item;
- Thermal loads and maximum anticipated excursion due to expansion;
- Jumper installation and operation loads;
- Jumper impact velocity;
- ROV access requirements;
- Vessel limitations for handling;
- Installation methodology;
- Maximum tension during installation;
- Maximum installation sea conditions;
- Specifics of hubs, connectors, and miscellaneous tools;
- Specifics on ROV-activated valves where applicable.

20.3. DESIGN METHODOLOGY

The PLEMs should be designed to facilitate installation and allow them to land properly on the seabed. The following design and analyses should be performed during PLEM design:
- Structural design and analysis;
- Mudmat design and analysis;
- PLEM installation analysis.

20.3.1. Structure

The structural analysis software SACS/StruCAD, which is widely used in the analysis of subsea steel structures, is one of the most useful tools for the analysis of a PLEM's steel frame structure. The SACS/StruCAD analysis provides loads, stresses, and deflections for all structural members of the PLEMs in various load conditions. The structural design of a PLEM should be performed in accordance with the latest editions of API-RP-2A [4] and AISC's *Steel Construction Manual, Allowable Stress Design (ASD)* [5].

The structural analyses for a PLEM should be carried out for lifting, installation, and in-place load conditions. Under installation conditions, the controlling design load on the PLEM is the tension load from the pipeline. The tension load includes the weight of the PLEM plus the tension in the pipeline with a dynamic factor of 1.5.

If the PLEM is designed with a yoke to support the installation, the yoke may be designed based on the forces obtained from pipelay analysis for the pipeline in a flooded condition. The yoke design also considers the irretrievability loads. Designs of the yoke and PLEM are coordinated with the installation contractor for adaptability with the installation vessel capability.

The PLEM analysis for structural integrity should include the following conditions:

- Fabrication;
- Lifting;
- Transportation;
- Installation;
- In-place operating conditions, structural loads, and fatigue damage;
- Thermal and pressure-induced expansion of pipelines in the operating condition;
- Piping requirements, including bends, tees, reducers, and valves.

The PLEM is designed with padeyes and miscellaneous installation aids as required and has a sufficient number of padeyes for handling during various phases of installation. The design should ensure that the CG of the PLEM assembly, including the yoke, is near the geometric center of the mudmat, in line with the center of the pipeline (or slightly below) and below the pivot point of the yoke and slightly to the rear of the yoke pivot to ensure that the PLEM will stabilize with the correct orientation for landing. The center of the yoke pin will be located in line with the flowline as shown in Figure 20-5. The pivot is the local coordinate origin on the PLEM for dimensions and calculations. The offsets e_{pipe} and e_{PLEM} are negative numbers. The PLEM is welded to the pipeline to minimize the cost and potential leak paths.

The equipment size and location for a PLEM should take into consideration the need for ROV operability and visibility. The structural frame and mudmat for the PLEM are coated and cathodically protected for the design field life. The cathodic protection design and analysis may be carried out based on DNV RP B401 [6], "Cathodic Protection Design."

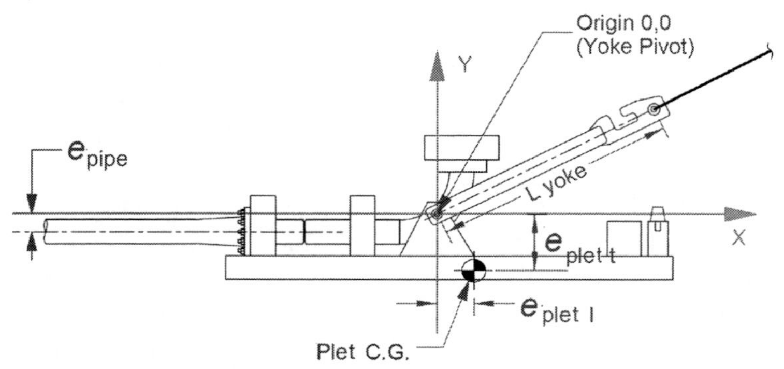

Figure 20-5 Coordinates and CG Position of Second-End PLEM

20.3.2. Mudmat

A preliminary mudmat area for the PLEM is estimated based on the soil bearing capacity and the estimated weights of the hubs, valves, piping, jumper loads, and mat.

The soil shear strength profiles are developed based on site-specific soil data obtained from drop cores. The design shear strength is obtained as the depth-averaged value over the shear strength profile at a depth equal to half the PLEM width.

The foundation bearing stresses are calculated for the following anticipated jumper loading conditions[7]:

- Jumper installation;
- Jumper operating;
- Jumper operating, future jumper installation;
- Future jumper operating.

The mudmat has a shallow foundation and may be designed to cover the following issues per API-RP2A WSD or DNV Classification Notes No. 30.4, "Foundations"[8]:

- Settlement;
- Eccentric loading;
- Overturning and cyclic effects;
- Safety factors for bearing, sliding, torsional, and overturn failures.

Mudmats with skirts are designed to keep the PLEM stability and to reduce settlement and movement on the seabed when in clay soil. However, PLEMs without skirts are often used in shallow water on sandy soil so it can move about on the seabed, because the penetration on sand is very small.

20.3.3. PLEM Installation

The pipe in a PLEM must not be overstressed at any time during different load conditions. The PLEM must be stable during installation and configured to land on the seabed in the correct orientation. The PLEM used to initiate pipelay is termed a first-end PLEM. A PLEM installed on completion of the pipelay is termed a second-end PLEM. The second-end PLEM has a yoke that pivots near the pipe centerline and the PLEM center of gravity.

The following problems could occur during installation and should be considered during the design procedure:

- Center of gravity too high due to late or unplanned equipment additions;
- Extra measures required to land upright due to pipe torsion;
- Bent pipe due to lowering too far with the PLEM held inverted by pipe torsion.

The installation of a second-end PLEM starts with the abandonment configuration of the pipeline. The pipeline can be laid by using an S-lay or J-lay process. On completion of pipe laying, the pipeline is fitted with an abandonment and recovery (A&R) head to prevent flooding and for attachment of the A&R wire. There are two reasons for not attaching the PLEM at this point:

- It is not practical to attach and maneuver the PLEM structure through the pipelay stinger or J-lay tower.
- It is prudent to lay the end of the pipe on the bottom and assess the unconstrained top-of-pipe orientation before attaching the PLEM.

The PLEM weight, balance, and geometry are all optimized during design to ensure that the PLEM has an intrinsic tendency to land in the correct orientation. Nevertheless, experience has shown that it is essential to attach the PLEM according to the observed natural top-of-pipe.

Figure 20-6 shows the installation sequence for a second-end PLEM. In the Figure 20-6a, the pipeline is lowered to the seabed with an A&R wire. The cut length and top-of-pipe are assessed with an ROV inspection. In Figure 20-6b, the pipeline is recovered to the vessel in a J-mode (pipeline is suspended without an overbend in this configuration) and set in a hang-off receptacle or slips so that the A&R head can be removed from the pipeline end to the attached PLEM. In Figure 20-6c, with the PLEM welded, the entire assembly is lifted from the support and lowered to the seabed. If all goes well, the PLEM sled gently lands upright on the seabed.

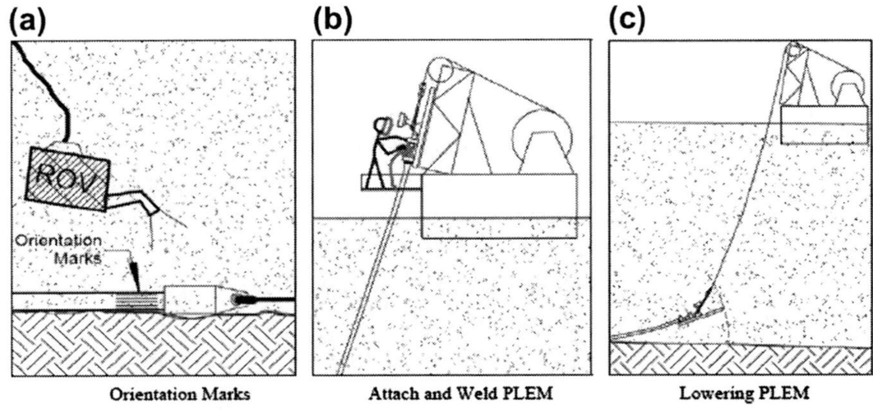

Figure 20-6 Installation Sequence for a Second-End PLEM [9]

The rigging to lower the PLEM is connected to a yoke that applies a lift force to a pivot near the centerline of the pipe and above the CG of the PLEM. The pivot is located so as to:

- Control bending load on the pipeline throughout the installation sequence.
- Direct the force to correct the orientation of the PLEM as the PLEM approaches the seabed.

20.4. FOUNDATION (MUDMAT) SIZING AND DESIGN

The PLEMs foundation provides a support for the PLEM structure. A preliminary mudmat area or dimensions for the PLEM are estimated based on the soil bearing capacity and the estimated weights of the hubs, valves, piping, jumper loads, and mudmat. The mudmat may be designed with skirts for stability and to reduce settlement and movement on the seabed. The dimensions will be adjusted with consideration given to the limitations imposed by the installation vessel, fabrication shop, and overland and sea transportation situations. The design should be analyzed for structural integrity in terms of:

- Fabrication;
- Lifting;
- Transportation;
- Installation;
- In-place operating conditions, structural loads, and fatigue damage;
- Thermal expansion of the pipelines.

Knowledge of the soil properties at the PLEM target area is required to properly size the mudmat, to determine stability, and to estimate short-term and long-term settlement. These soil data are more important for a mudmat than the deep soil sample required for pile design. The foundation design for piles was detailed in the chapter on subsea manifolds, Chapter 19.

20.4.1. Load Conditions

General load conditions for mudmat sizing are as follows:

- For the PLEM landing condition, the submerged weight, including the hubs, hub end caps, piping, and the support frame are applied to the foundation.
- For the jumper installation case, the loads from the jumper, connector tool, and ROV are added.
- For the operation case, the ROV and connector tool are removed and the jumper operating loads introduced.
- For the maintenance case, ROV loads are added to the operation case. For two-hub PLEMs, loads on one hub are kept as operating but ROV loads are added to the second hub.
- For the case of operation and maintenance in the thermal expanded positions, the location of the hubs and loads are transferred to the expanded position on the mudmat foundation.

The foundations should be designed to support the PLEMs with the design loads as indicated, within limits of the required safety factors in accordance with API RP2A (WSD) and a minimum factor of safety for soil bearing of 2.0.

20.4.2. Mudmat Analysis

The design of mudmat foundations should include the following issues per API RP 2A-WSD [4]:

1. Stability, including failure due to overturning, bearing, sliding, or combinations thereof;
2. Static foundation deformations, including possible damage to components of the structure and its foundation or attached facilities;
3. Dynamic foundation characteristics, including the influence of the foundation on structural response and the performance of the foundation itself under dynamic loading;
4. Hydraulic instability such as scour or piping due to wave pressures, including the potential for damage to the structure and for foundation instability;

5. Installation and removal, including penetration and pull-out of shear skirts or the foundation base itself and the effects of pressure build-up or drawdown of trapped water underneath the base.

20.4.2.1. Overturning Capacity

The objective of checking the overturning capacity for a shallow foundation is to ensure that, under the applied loads, the foundation will be stable if placed on a hard surface. Overturning could take place either around the longitudinal edge or the transverse edge of the foundation. The governing overturning direction will yield the lowest safety factor. The following formula is used to calculate the overturning safety factor of the shallow foundation around one of its edges:

$$\text{Safety factor}_{\text{overturning}} = \frac{F_z L}{M} \tag{20-1}$$

in which M is the resultant overturning moment in the overturning direction, F_z is the resultant vertical force acting at the geometric center of the mudmat, and L is the distance from the geometric center to the rotating axis.

Load eccentricity decreases the ultimate vertical load that a footing can withstand. This effect is accounted for in bearing capacity analysis by reducing the effective area of the footing according to empirical guidelines. The calculation method of effective area is detailed in the section title "C6.13 Stability of Shallow Foundations" of API-RP-2A [4].

The recommendations pertaining to the design issues for a shallow foundation are given in Sections 6.13 through 6.17 of API-RP-2A [4] and include:

- Stability of shallow foundations;
- Static deformation of shallow foundations (short-term and long-term deformation);
- Dynamic behavior of shallow foundations;
- Hydraulic instability of shallow foundations;
- Installation and removal of shall foundations.

The following sections provide a detailed design example for a mudmat based on DNV Classification Notes No. 30.4, "Foundations" [8].

20.4.2.2. Penetration Resistance of Skirts

The height of the skirts is designed to guarantee that with the occurrence of the maximum estimated consolidation settlement, the structures will retain their ability to be displaced freely through the mudline.

The skirt penetration resistance in sand, Q_p, is estimated according to DNV code [8]:

$$Q_P = q_t A_t + \tau_w A_w \tag{20-2}$$

where

q_t = skirt tip pressure, $q_t = k_p\, q_{c,\mathrm{CPT}}$;
t_w = skirt wall friction $t_w = \mathrm{k_f}\, \mathrm{q}_{c,\mathrm{CPT}}$;
A_t and A_w = skirt tip and skirt wall area;
$q_{c,\mathrm{CPT}}$ = point resistance from CPT boring according to [8];
$k_p = 0.6$ and $k_f = 0.003$ for highest expected penetration resistance;
$k_p = 0.3$ and $k_f = 0.001$ for most probable penetration resistance.

The following relationship is an example of soil survey results at the PLEM location:

$$q_{c,\mathrm{CPT}} = 40 \times z \quad [\mathrm{MPa}] \quad \text{for} \quad z < 0.5 \text{ m}$$

20.4.2.3. Bearing Capacity during Installation

The mudmat minimum dimensions required to keep the structure supported at the mudline are a function of the ultimate vertical capacity of the soil, considering the effect of the resultant load eccentricities (related to its length and width), as a reduction of the mudmat area (effective area concept).

Assuming that the subsoil contains sand, the vertical bearing capacity is checked according to:

$$Q_v = 0.5^* N_r \gamma' B A + N_q p' A \tag{20-3}$$

where

N_γ, N_q = load bearing factors;
γ': effective unit weight of sand;
B: width of mudmat;
A: mudmat foundation area;
p': effective vertical stress.

For the case of clayey soil, the vertical bearing capacity may in principle be found by:

$$Q_v = N_c S_u A \tag{20-4}$$

where

N_c: load bearing factor (function of several factors);
S_u: average undrained shear strength along slip surface;
A: mudmat foundation area.

20.4.2.4. Bearing Capacity during Operation

Potential soil failure modes due to trawler loading and thermal expansion loading include failure from lateral sliding and deep-seated failure. A short description of calculation principles for the different failure modes is given below.

Lateral Sliding

The lateral soil resistance against pure sliding of the foundation may be calculated by:

$$Q_h = r\tan(\varphi)W \tag{20-5}$$

where, $r\tan(\varphi)$ is the friction between foundation and seabed and W is the vertical load on the mudmat.

If a rock berm is placed on the mudmat, additional lateral capacity (earth pressure) is obtained according to:

$$Q_v = P_p - P_a \tag{20-6}$$

where

$$P_p - P_a = (\chi_p + \chi_a)^{1/2}\gamma' H_r B$$

and

$\chi p + \chi_a$: earth pressure coefficients;
γ': effective unit weight of rock;
H_r: height of rock dump;
B: equivalent width of mudmat.

Deep-Seated Vertical Failure

A foundation's stability is based on the limiting equilibrium methods ensuring equilibrium between the driving and the resisting forces, according to Janbu et al. [10]. The foundation base has been idealized to account for load eccentricity according to the principle of plastic stress distribution over an effective base area.

In addition, a verification of the final rotation (loss of horizontality) of the structures should be done, considering the overturning moment envelope that will act during the entire life of the structure.

20.4.2.5. Settlement Analyses

The settlement of the structures is analyzed according to Janbu's method [11]. The soil strains ε are calculated by:

$$\varepsilon = \Delta\sigma'_v/M \tag{20-7}$$

where $\Delta\sigma'_v$ is the vertical effective stress increase, and M is the deformation modulus. Distribution of additional vertical soil stress with depth is based on the recommendation given by Janbu [11].

The vertical consolidation settlement δ is then found by:

$$\delta = \int_0^H \varepsilon \; dz \tag{20-8}$$

where H is soil layer thickness.

20.5. PLEM INSTALLATION ANALYSIS

The installation analysis for a pipeline system with a PLEM should check pipeline tensile, bending, shear, and hoop stresses. The installation operations can be divided into the following main steps:

- First-end PLEM from going overboard through the splash zone;
- Initiation of first-end PLEM;
- Normal pipe laying (including installation of buoyancy);
- Abandonment of second-end PLEM;
- Abandonment and recovery (contingency operations).

The following analyses should be executed to ensure pipeline integrity during operations:

- Static analysis for each of the above steps. The main objective of this analysis is to determine the requirements for each step during the installation (vessel movement, pipe payout, etc.);
- Regular wave dynamic analysis for each of the above steps for both empty and flooded conditions. The most critical steps of the installation defined by the static analysis should also be subjected to regular wave dynamic analysis. The main objective of this dynamic analysis is to define the critical environmental conditions for the installation window. If regular wave results are found to be too conservative, an irregular wave analysis may be considered as an alternative.
- A local buckling check should be used to verify that the pipeline will resist combined loads (axial force, bending moment, external or internal pressure) and that no local buckling will occur along the catenary for each of the steps.

The critical steps for fatigue occurring during installation are whenever the pipe is clamped for several hours to weld the PLEM.

OFFPIPE and Orcaflex are popular tools for the analysis of offshore pipeline and PLEM installations. OFFPIPE offers a recognized pipeline installation simulation that is simple, fast, and gives results with reasonable accuracy. However, it was developed based on the DOS operating system without a Windows interphase to illustrate the installation procedure. Compared with OFFPIPE, Orcaflex is widely utilized for subsea installation analyses because of its friendly interface. Orcaflex is a fully 3D, nonlinear, time-domain finite element program capable of dealing with arbitrarily large deflections of the pipe from the initial configuration. A simple lumped mass element is used, which greatly simplifies the mathematical formulation and allows for quick and efficient development of the program to include additional force terms and constraints on the system in response to new engineering requirements. The program is designed for the static and dynamic analysis of subsea structures, pipeline, and cable systems in an offshore environment.

During the design phase of a PLEM, there is not usually enough information and data available for the installation analysis using the software just described. Some simplified methods have been developed based on MathCAD or Excel worksheets. One of the typical methods described below is mainly based on a paper by Gilchrist [12], where further details are given.

20.5.1. Second-End PLEM

20.5.1.1. Static Force Balance of a PLEM

To assess the design of a PLEM, the static force balance of the PLEM is evaluated to predict the behavior of the PLEM during different steps of the installation procedure. For static force analysis, the following simplifications are assumed:

- Dynamic forces due to vessel motions and PLEM lowering movements are negligible.
- Pipe and cable shear loads are negligible. The suspended pipe and cable are modeled as catenaries. This assumption tends to slightly overpredict pipe touchdown bending strain and slightly underpredict pipe top tension.
- The PLEM is always aligned with the centerline of the top of the suspended pipeline.
- Wave and current loads to the PLEM and pipeline are ignored.

When the PLEM mudmat is at the water surface through the splash zone, wave and current loads will cause high flowline stresses at the PLEM bulkhead. Even through the maximum stress is in general below the allowable value, the installation contractor should consider reducing the installation sea state and avoiding pipeline welding or other activities when the PLEM is at the water surface to avoid a girth weld fatigue failure. As the

PLEM is positioned deeper, the wave-induced velocity dissipates and steady current becomes the dominant contributor to the hydraulic load. The dynamic effects are negligible.

The relationship of free body forces at the second-end PLEM with a cable in the top and pipe in the bottom during installation is shown in Figure 20-7. where,

W_s: submerged weight of PLEM;

T_{oc}: bottom tension of cable;

T_p: top tension of pipeline;

θ_{oc}: bottom angle of cable;

θ_p: top angle of pipe.

Summations of forces in the x and y directions are:

$$\Sigma F_x = T_{oc}\cdot\cos(\theta_{oc}) - T_p\cdot\cos(\theta_p) = 0 \tag{20-9}$$

$$\Sigma F_y = T_{oc}\cdot\sin(\theta_{oc}) - T_p\cdot\sin(\theta_p) - W_s = 0 \tag{20-10}$$

The force vector diagram in Figure 20-7 is a graphical representation of Equations (20-9) and (20-10).

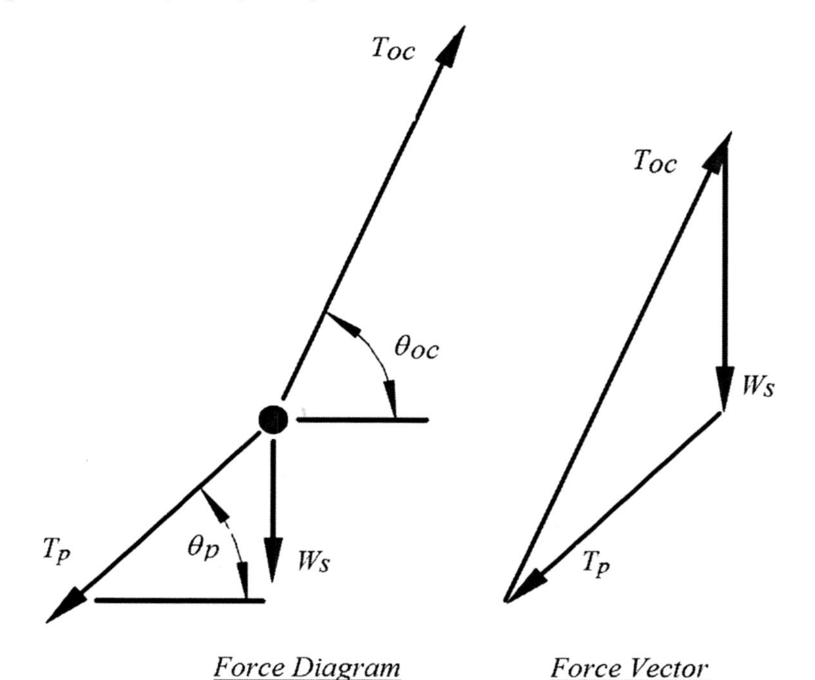

Figure 20-7 Static Force Balance of Second-End PLEM

The top tension and top angle of pipe for a suspended pipe without shear at the top can be solved using catenary equations. The above equations can be rearranged to give formulas for the cable bottom tension and angle:

$$T_{oc} = \sqrt{T_p^2 + 2 \cdot T_p \cdot W_s \cdot \sin(\theta_p) + W_s^2} \tag{20-11}$$

$$\theta_{oc} = \arctan\left[\frac{\sin(\theta_p) + W_s/T_p}{\cos(\theta_p)}\right] \tag{20-12}$$

20.5.2.2. Bending Load on a Pipe with the PLEM Correctly Oriented

Figure 20-8 shows the forces and moment balances at the pivot point of the second-end PLEM, in which the summary of moments due to forces is resisted by a balancing moment in the suspended pipe, M_{pipe}. The pipe must be capable of providing the reaction moment without overstressing during the installation procedures.

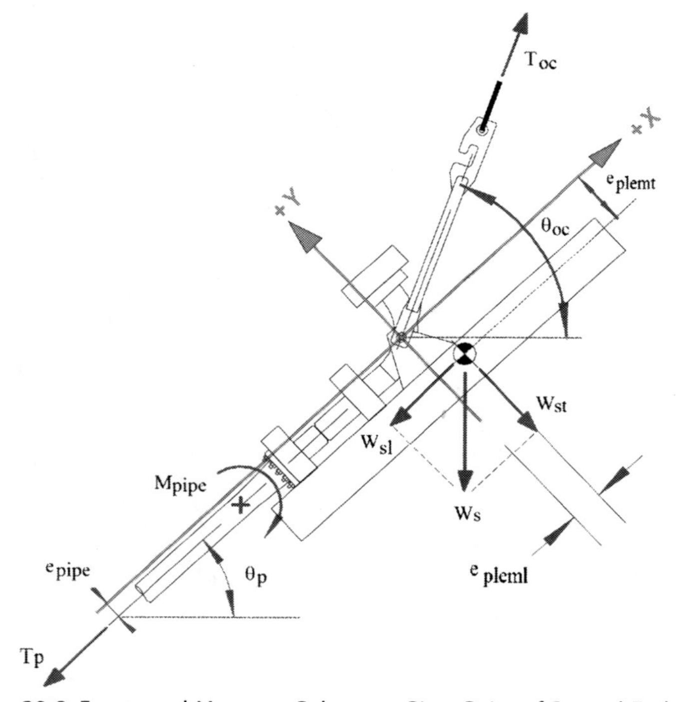

Figure 20-8 Forces and Moments Balance at Pivot Point of Second-End PLEM

The sum of moments in the PLEM is zero because the only loads considered are vector forces acting through the PLEM pivot. The moments occurring due to eccentricity of forces between the pivot, pipe axis, and PLEM center of gravity are calculated in the following equations:

1. Moment about the pivot due to pipe top tension is expressed as:

$$M_{Tp} = -T_p \cdot e_{\text{pipe}} \tag{20-13}$$

where e_{pipe} is the pipe eccentricity from the pivot, and is negative here.

2. Moment about the pivot due to PLEM weight is broken down into two components:

- Moment due to the transverse weight component times the longitudinal eccentricity:

$$W_{st} = W_s \cdot \cos(\theta_p) \tag{20-14}$$

$$M_{\text{pleml}} = W_{st} \cdot e_{\text{pleml}} \tag{20-15}$$

where e_{pleml} is the PLEM CG longitudinal eccentricity from the pivot, and W_{st} is the transverse component of the submerged weight of the PLEM.

- Moment due to the longitudinal weight component times transverse eccentricity:

$$W_{s1} = W_s \cdot \sin(\theta_p) \tag{20-16}$$

$$M_{\text{plemt}} = -W_{st} \cdot e_{\text{plemt}} \tag{20-17}$$

where e_{plemt} is the PLEM CG transverse eccentricity from the pivot, and W_{sl} is the longitudinal component of the submerged weight of the PLEM.

The reaction at the pivot is a force vector only; a pivot has no moment capacity. There is no moment due to cable tension because it is a vector that never has any eccentricity with respect to the pivot. The total moment about the pivot must sum to zero. Therefore,

$$M_{\text{pipe}} = M_{\text{pleml}} + M_{\text{plemt}} + M_{Tp} \tag{20-18}$$

Pipe bending moments arise from the eccentricity of the pivot from the centerline of the pipe (pipe tension is applied in line with the pipe) and eccentricity of the PLEM CG from the pivot. The bending load on the pipe to balance the moment load about the pivot should be evaluated through the running sequence. The pivot and CG of the PLEM must be located to avoid exceeding the moment capacity of the pipe. To keep the stress of pipe in the allowable range, the pivot should be located above and aft of the PLEM CG and above the centerline of the pipe. At the beginning, the pipe bending moment is dominated by the moment arising from the eccentricity of the pipe centerline from the pivot. At the end, the pipe reaction moment is dominated by the moment arising from the longitudinal eccentricity of the PLEM CG from the pivot.

The durability of the system can be increased by fitting the PLEM with a tailpiece of heavy wall pipe (one or two joints), which will have greater moment capacity either to increase the safety factor or to allow greater eccentricity of the pivot. The moment force on the pipe from Equation (20-18) is conservative because deflection of the pipe due to the moment force tends to reduce the pipe eccentricity and that in turn reduces the moment load on the pipe.

20.5.1.3. Righting Torsion with the PLEM Misoriented by 90 Degrees

It is essential for the PLEM to land in the correct orientation, which is ensured by designing a PLEM with a righting torsion to force the unit upright as it approaches the seabed if it does not land in the correct orientation. Righting torsion is zero when the PLEM is correctly oriented and a maximum when the PLEM is misoriented by 90 degrees. In this section, the PLEM is assumed to be out of orientation a full 90 degrees to calculate the maximum righting torsion.

Figure 20-9 illustrates the PLEM rotated 90 degrees out of correct orientation in which the righting torsion is applied. Two forces act to rotate the PLEM upright:

- The transverse component W_{st} of the PLEM's submerged weight, W_s. The yoke weight should be included in the PLEM submerged weight and the CG calculations.
- The transverse component T_{oct} of A&R cable bottom tension T_{oc}, an upward force applied at the yoke lift eye.

The axis for the righting torsion is the centerline of the pipe. Only transverse load components W_{st} and T_{oc} time their eccentricities to contribute to the righting torsion. Longitudinal loads do not contribute to the righting torsion. The value of W_{st}, the transverse component of PLEM submerged weight, is taken from Equation (20-14).

The transverse component of cable tension is:

$$T_{oct} = T_{oc} \cdot \sin(\theta_{oc} - \vartheta_p) \quad (20\text{-}19)$$

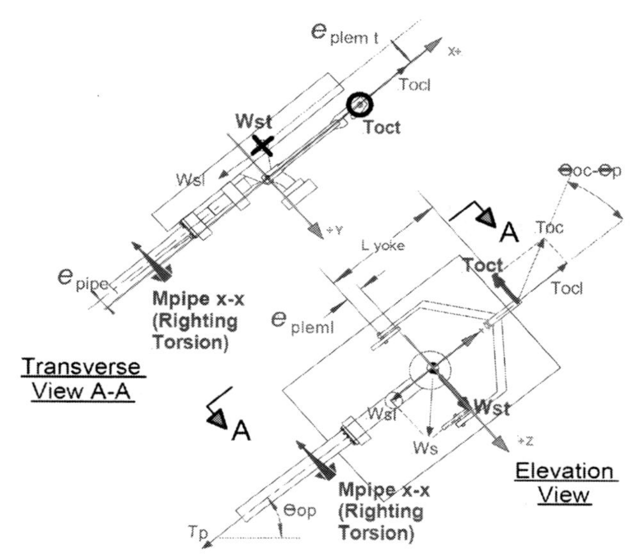

Figure 20-9 Righting Torsion for PLEM That Is Misoriented by 90 Degrees

The righting torsions about the pipe axis are:

$$M_{\text{pipe } x-x \text{ plem}} = -W_{st} \cdot (e_{\text{plem t}} - e_{\text{pipe}}) \tag{20-20}$$

$$M_{\text{pipe } x-x \text{ cable}} = -T_{oct} \cdot e_{\text{pipe}} \tag{20-21}$$

$$M_{\text{pipe } x-x} = M_{\text{pipe } x-x \text{ plem}} + M_{\text{pipe } x-x \text{ cable}} \tag{20-22}$$

where T_{oc} is the cable bottom tension, transverse component, and $M_{x\text{-}x}$ is the pipe torsional moment. Note that the transverse loads and consequently the righting torsion are small when the pipe end is near vertical. It is not unusual for a PLEM to rotate one or more times during the descent. The x-x subscript indicates that the axis of the moment is parallel to the x axis of the PLEM.

Righting torsion can be increased by increasing the PLEM weight and increasing the transverse eccentricity of the CG. The most efficient way of doing this is to add the thickness of the mudmat. To adjust the longitudinal location of the PLEM CG, the fore and aft plates can be of different thicknesses. Increasing the pivot eccentricity from the pipe centerline is usually not an option because pipe bending stress at the start of the installation is dominated by the contribution of T_p. e_{pipe} as shown in Equation (20-13). In deepwater situations, the PLEM's submerged weight is always much lower than initial top-of-pipe tension, so there is more scope to change the eccentricity of the PLEM CG.

The yoke's lateral load is also at a maximum when the PLEM is misoriented by 90 degrees. The lateral design load for the yoke and pivots is expressed as:

$$F_{\text{yoke lateral}} = T_{oct} \tag{20-23}$$

20.5.1.4. Bending Load on a Pipe with PLEM Misoriented by 90 Degrees

Figure 20-10 illustrates the 90-degree misoriented case, in which the moment in the pipe is different from that in the correctly oriented running case.

The bending moment due to cable eccentricity is:

$$M_{\text{pipe } y-y(90 \text{ degree out of orientation})} = W_{st} \cdot (L_{\text{yoke}} - e_{\text{plem l}}) \tag{20-24}$$

Shortening the yoke length can reduce this moment load. However, a long yoke is beneficial at the end of the lowering sequence when the yoke lifts.

Force couples due to pipe eccentricity, and PLEM CG eccentricity from the yoke still exists in the x-y plane. The moment z-z due to pipe eccentricity from the yoke pivot is the same as previously noted in Equation (20-13):

$$M_{\text{pipe } z-z(\text{top tension})} = -T_p \cdot e_{\text{pipe}} \tag{20-25}$$

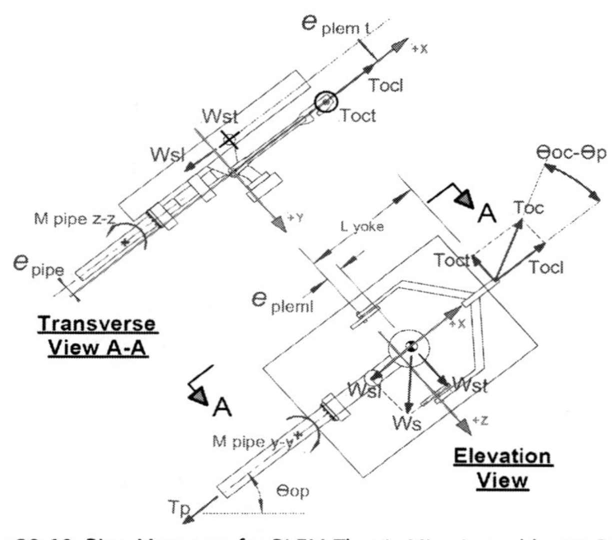

Figure 20-10 Pipe Moments for PLEM That Is Misoriented by 90 Degrees

The moment z-z due to the longitudinal component of sled weight is:

$$M_{\text{pipe } z-z(\text{plem})} = -W_{sl} \cdot e_{\text{plem t}} \quad (20\text{-}26)$$

The total moment z-z is:

$$M_{\text{pipe } z-z} = M_{\text{pipe } z-z(\text{top tension})} + M_{\text{pipe } z-z(\text{plem})} \quad (20\text{-}27)$$

The combined pipe moment with the PLEM running 90 degrees misoriented is the vector sum of the z-z and y-y moments:

$$M_{\text{pipe}(90^\circ\text{misoriented})} = \sqrt{M_{z-z}^2 + M_{y-y}^2} \quad (20\text{-}28)$$

This equation can be used to determine when the lowering should be stopped if the PLEM is out of orientation and the yoke has failed to lift. If the PLEM does not right, the situation must be reviewed with particular attention paid to actual CG location, uprighting torsion calculations, and pipe torque. If pipe torque is the problem, the PLEM is best recovered and the PLEM reoriented about the pipe.

20.5.2. First-End PLEMs

Figure 20-11 illustrates a first end PLEM, which can be recognized as a variant of a second-end PLEM in that the yoke and mudmat functions are combined.

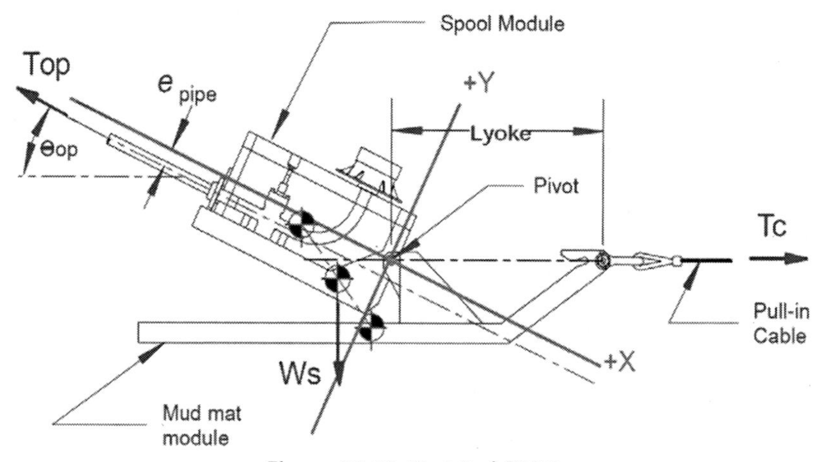

Figure 20-11 First End PLEM

A typical installation procedure for the first-end PLEM is:

- Preinstall pull-in pile and lay cable out to an accessible stand-off location.
- At the standoff location, lower the PLEM on the pipe.
- Connect the pull-in cable to the PLEM.
- Pull the PLEM into position while adding pipe and maintaining the PLEM clearance from the seabed.

With the pull-in cable pulling horizontally, the static forces acting at the PLEM are as shown in Figure 20-12. The forces in Figure 20-12 are analogous to Figure 20-7 for the second-end PLEM except the pipe and cable have swapped positions. The equations are:

$$\Sigma F_x = T_{oc} - T_{op} \cdot \cos\theta_{op} = 0 \quad (20\text{-}29)$$

$$\Sigma F_y = W_s - T_{op} \cdot \sin\theta_{op} = 0 \quad (20\text{-}30)$$

The force vector polygon in Figure 20-12 is a graphical representation of Equation (20-29) and Equation (20-30). There is only one solution for pipe bottom tension and pipe bottom angle in this state of equilibrium. The equations for pipe bottom tension and pipe bottom angle are expressed as follows:

$$T_{op} = \sqrt{T_{oc}^2 + W_s^2} \quad (20\text{-}31)$$

$$\theta_{op} = \arctan(W_s / T_{oc}) \quad (20\text{-}32)$$

Fixing these two variables in turn defines the suspended pipe span. The suspended pipe span can be solved using catenary equations. The analysis is carried to the conclusion of the pull-in by decreasing the PLEM submerged weight W_s in steps once the PLEM is over the target location. When the cable bottom tension is equal to the pipe bottom tension, then normal pipelay has been achieved.

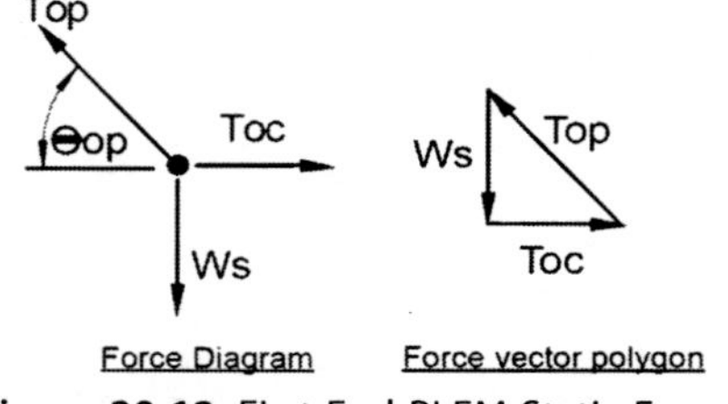

Figure 20-12 First-End PLEM Static Forces

20.5.3. Stress Analysis for Both First- and Second-End PLEMs

Pipe outer surface stresses due to the bending moment loads can be calculated using the following formula:

$$\sigma_{\text{bending}} = \frac{M \cdot D}{2I} \tag{20-33}$$

where M is the moment, I is the moment of inertia for pipe section, and D is the pipe outer diameter.

Pipe stress resulting from pipe-top tension is shown as:

$$\sigma_{\text{tension}} = \frac{T}{\text{A}_\text{s}} \tag{20-34}$$

Pipe stress due to hydrostatic pressure is compressive:

$$\sigma_{\text{hydrostatic}} = -\frac{P_e \cdot A_e}{A_s} \tag{20-35}$$

where P_e is the pipe external pressure (hydrostatic pressure); A_e is the external area of the pipeline, $A_e = \pi D^2/4$; and A_s is the pipeline cross-section area of steel.

The maximum pipe outer surface stress in the pipe is the sum of all three of the above stresses:

$$\sigma_{\text{pipe}} = \sigma_{\text{bending}} + \sigma_{\text{tension}} + \sigma_{\text{hydrostatic}} \tag{20-36}$$

20.5.4. Analysis Example of Second End PLEM

Modeling of the second-end PLEM running can be done in a stepwise fashion based on Equations (20-9) to (20-28) using a spreadsheet such

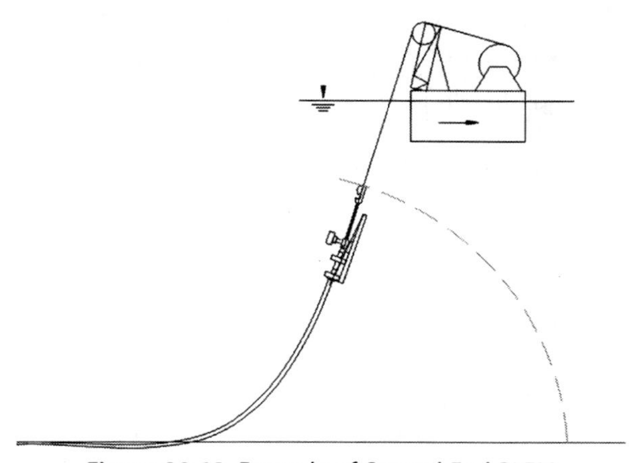

Figure 20-13 Example of Second-End PLEM

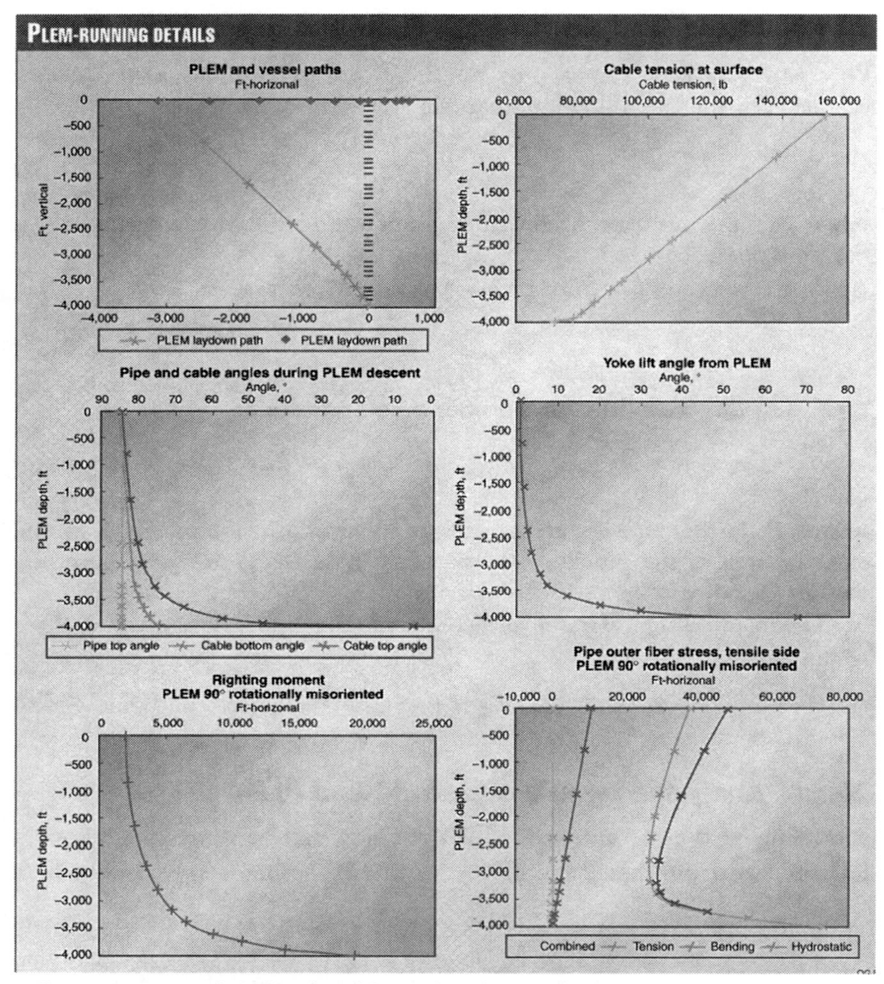

Figure 20-14 Analysis Results of Second-End PLEM for Various Water Depths [13]

as Microsoft Excel or MathCAD. Figure 20-13 shows an example problem.

The input data include:

PLEM: $W_s = 20{,}000$ lb; $L_{yoke} = 6$ ft; $e_{plem\ l} = 0.5$ ft; $e_{plem\ t} = -1.0$ ft

Pipe: OD = 6.625 in.; W.T. = 0.719 in.; $e_{pipe} = 0.25$ ft

Cable: OD = 2.875 in.; $W_c = 12.5$ lb/ft

Water depth = 4,000 ft; seawater density = 64 lb/cu ft

Figure 20-14 shows the analysis results for the example of a second-end PLEM.

REFERENCES

[1] M. Faulk, FMC ManTIS (Manifolds & Tie-in Systems), SUT Subsea Awareness Course, Houston, 2008.
[2] G. Corbetta, BRUTUS: The Rigid Spoolpiece Installation System, OTC 11047, Offshore Technology Conference, Houston, Texas, 1999.
[3] J.K. Antani, W.T. Dick, D. Balch, T. Van Der Leij, Design, Fabrication and Installation of the Neptune Export Lateral PLEMs, OTC 19688, Offshore Technology Conference, Houston, Texas, 2008.
[4] American Petroleum Institute, Recommended Practice for Planning, Designing and Constructing Fixed Offshore Platforms–Working Stress Design, API RP 2A-WSD (2007).
[5] American Institute of Steel Construction, Manual of Steel Construction: Allowable Stress Design, nineth ed., AISC, Chicago, 2002.
[6] DET NORSKE VERITAS, Cathodic Protection Design, DNV RP B401 (1993).
[7] K.H. Andersen, H.P. Jostad, Foundation Design of Skirted Foundations and Anchors in Clay, OTC 10824, Offshore Technology Conference, Houston, Texas, 1999.
[8] DET NORSKE VERITAS, Foundations, DNV, Classification Notes No. 30.4 (1992).
[9] K.C. Dyson, W.J. McDonald, P. Olden, F. Domingues, Design Features for Wye Sled Assemblies and Pipeline End Termination Structures to Facilitate Deepwater Installation by the J-Lay Method, OTC 16632, Offshore Technology Conference, Houston, Texas, 2004.
[10] N. Janbu, L.O. Grande, K. Eggereide, Effective Stress Stability Analysis for Gravity Structures, BOSS'76, Trondheim, Vol. 1 (1976) 449–466.
[11] N. Janbu, Grunnlag i geoteknikk, Tapir forlag, Trondheim, Norway (in Norwegian). (1970).
[12] R.T. Gilchrist, Deepwater Pipeline End Manifold Design, Oil & Gas Journal, special issue (1998, November 2).
[13] D. Wolbers, R. Hovinga, Installation of Deepwater Pipelines with Sled Assemblies Using the New J-Lay System of the DCV Balder, OTC 15336, Offshore Technology Conference, Houston, Texas, 2003.

CHAPTER 21

Subsea Connections and Jumpers

Contents

Subsea Engineering Handbook
ISBN 978-0-12-397804-2, doi:10.1016/B978-0-12-397804-2.00021-7

21.1. INTRODUCTION

In subsea oil/gas production system, a subsea jumper is a short pipe connector used to transport production fluid between two subsea components, for example, tree and manifold, manifold and manifold, or manifold and export sled. It may also connect other subsea structures such as PLEM/PLETs and riser bases. In addition to production fluid, it can also be a pipe by which water/chemicals are injected into a well.

21.1.1. Tie-In Systems

Subsea fields have been developed using a variety of tie-in systems in the past decades. Different types of horizontal and vertical tie-in systems and associated connection tools are used for the tie-in of flowlines, umbilicals, and other applications. For flowlines, the subsea tie-in systems are used to connect tree to manifold, tree to tree, and pipeline end to tree or manifold. For subsea control systems, the subsea tie-in systems are used to connect umbilicals to tree or manifold.

The horizontal tie-in connection systems are mainly used in shallow water with diver-based installation, such as expansion spools used for connecting pipeline with a fixed riser nearby the platform. In recent years, subsea engineering has undergone a fundamental change in that it has gone from being a diver-dominated activity to the use of remote systems for the construction of deepwater field developments. Horizontal connecting rigid spools are used in deep water with the aid of ROVs. Most of the horizontal connection systems are based on mechanically bolted flange joints, whereas the vertical spool connection system uses guideline-deployed inverted U- or M-shaped rigid spools with collet connectors.

21.1.1.1. Vertical Tie-In Systems

Vertical jumpers, mainly adopted in the Gulf of Mexico, are typical vertical tie-in systems, which are generally characterized by an inverted U-shaped rigid spool, and use mechanical collet connectors at each end as shown in Figure 21-1a. The vertically connected jumper system consists of a pipe spool between two vertically oriented downward connectors (hydraulic or mechanical type) with metal-to-metal seals for piping connections between subsea facilities (tree to manifold, manifold to PLET, etc.). The actuated half of the connector is part of the retrievable jumper assembly with the mating hub attached to the subsea equipment. The jumper connectors are landed onto upward hubs, made up, and tested to verify seal integrity.

Vertical connections are installed directly onto the receiving hub during tie-in. Because the vertical connection system does not require a pull-in capability, it simplifies the tool functions, provides a time efficient tie-in operation, and reduces the length of rigid spools. Stroking and connection are carried out by the connector itself, or by the ROV-operated connector actuation tool (CAT) system as shown in Figure 21-1b.

21.1.1.2. Horizontal Tie-In Systems

Figure 21-2 shows a horizontal tie-in system, which can be used for both first-end and second-end tie-in of flowlines. The termination head is hauled into the tie-in point by use of a subsea winch. Horizontal tie-in may be made up by clamp connectors operated from a tie-in tool, by integrated hydraulic connectors operated through a ROV, or by nonhydraulic collet connectors with assistance from a CAT and ROV. Horizontal connection leaves the flowline/umbilical in a straight line, which is easily protected if overtrawling from fishermen occurs.

The installation procedure for a horizontal tie-in system is illustrated in Figure 21-3 and described here:

- The spool of the horizontal tie-in system hooked up to a spreader beam is deployed and lowered to within a few meters above the target areas on the subsea structures as shown in Figure 21-2.

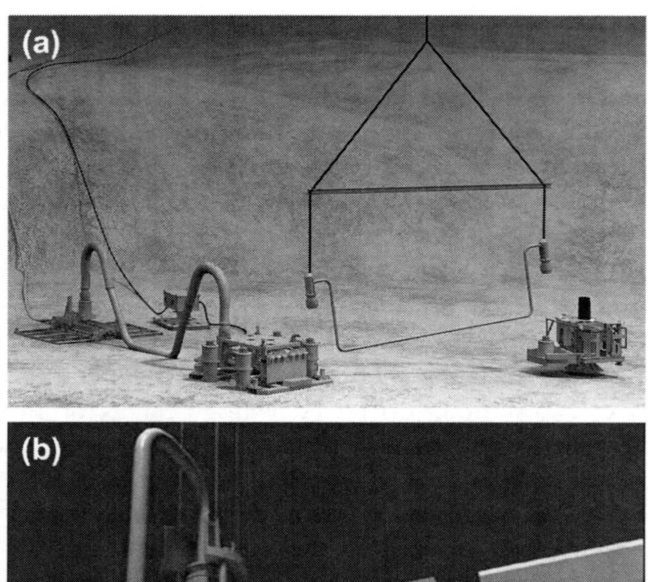

Figure 21-1 Vertical Tie-In Systems [1]

- The spool is lowered until the stab on the first termination head enters the stab receptacle on the tie-in porch as shown in Figure 21-3a. The second termination head will align horizontally as the spool continues to be lowered until the stab enters the stab receptacle and lands on the tie-in porch.
- The CAT is landed and locked on the first termination head by the ROV as shown in Figure 21-3b.
- The termination head is leveled and locked in the horizontal position. The protection caps are removed from the connector and the inboard hub as shown in Figure 21-3c.

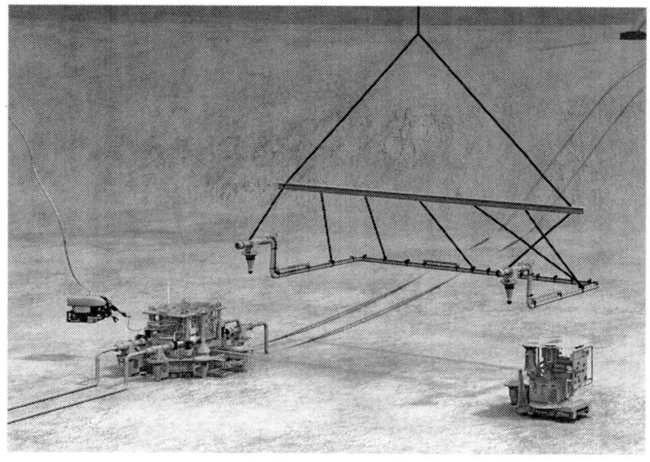

Figure 21-2 Horizontal Tie-in System [1]

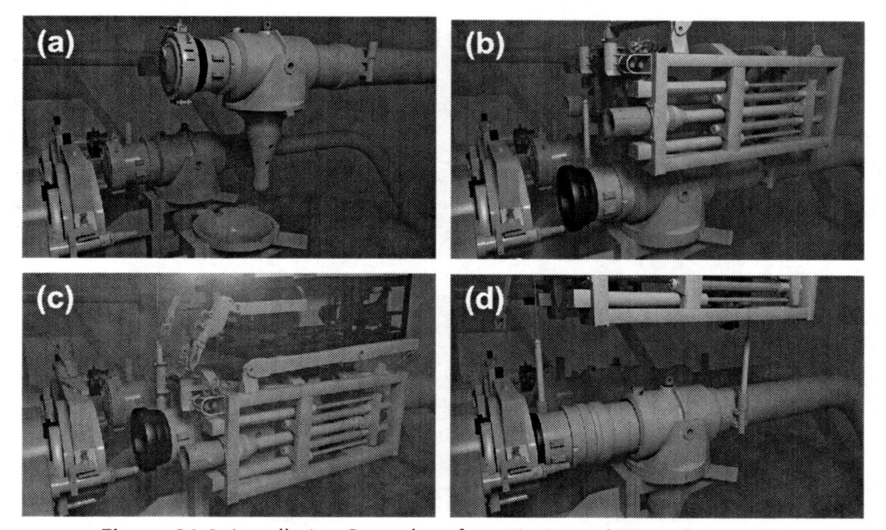

Figure 21-3 Installation Procedure for a Horizontal Tie-In System [1]

- The termination head is stroked against the inboard hub and the connector is closed as shown in Figure 21-3d. A pressure test is carried out to verify the integrity of the connector seal. The CAT is unlocked and lifted from the termination head and inboard hub. The connection procedure is repeated to connect the second termination head to the inboard hub without returning the CAT to the surface vessel.

Figure 21-4 Well Pu-PG to Manifold Jumper [2]

Figure 21-4 shows a well-to-manifold jumper that uses a horizontal tie-in system. This system was used in the water injection system of BP's Greater Plutonio project.

21.1.1.3. Comparison between Horizontal and Vertical Tie-In Systems

Table 21-1 summarizes a comparison between horizontal and vertical tie-in connection systems. The main technical disadvantages of the horizontal tie-in system over the vertical system include increased offshore operation time and increased complexity and reliance on ROVs. However, these disadvantages are largely offset by the advantages of reduced subsea structure size, simplicity of the flowline connector, and lower vessel/weather dependence [3].

21.1.2. Jumper Configurations

A typical jumper consists of two end connectors and a pipe between the two connectors. If the pipe is a rigid pipe, the jumper is called a rigid jumper; otherwise, if the pipe is flexible, the jumper is called a flexible jumper.

Figure 21-5 shows some rigid jumper configurations. For the rigid pipe jumpers, the M-shaped style and inverted U-shaped style are two commonly used styles. There is also the horizontal Z-shaped style and so on. Jumper configurations are dictated by design parameters, interfaces with subsea equipment, and the different modes in which the jumper will operate. The configurations in Figure 21-5a and c use bends to connect straight pipes, while that in Figure 21-5b uses elbows to connect rigid pipes. The extended part of the elbows is designed for the erosion of sand. For the case shown here, the production flow should move from right to left.

The subsea rigid jumpers between various components on the seabed are typically rigid steel pipes that are laid horizontally above the seabed. After

Table 21-1 Comparison between Horizontal and Vertical Tie-In Systems

Evaluation Issue	Horizontal Connection	Vertical Connection
Tie-in/connection required equipment	Relatively complex ROV to deploy and operate all equipment.	Relatively simple deployment equipment, and reliance on ROV to perform tasks is relatively low.
Duration of tie-in	Long.	Short.
Connector complexity and size	Simple; API/ANSI flange or clamp connector. Connector weight is comparatively low.	Complex; collet connector required either with integral stroking mechanism or separate running tool. Connector weight is on the order of several tons.
Metrology and fabrication accuracy	Medium level of accuracy required, since connection system can elastically deform spool for alignment if required.	High level of accuracy required since system has no means of correcting for inaccuracy other than by the connector.
Possibility of snagging by anchors and others	Relatively low, since the system stays close to seabed.	Relatively problematic, since higher profile is more exposed to a snag.
Vessel requirements	Relatively low-specification DP vessel with deck space and crane capacity for spool.	Relatively high-specification DP with stable RAOs (Response Amplitude Operator) to operate with guidelines and maximize weather window.
	Crane height may be low.	Due to vertical nature of spool, crane height may need to be relatively high.
Seal change	Relatively simple; push back and replace seal with ROV seal.	Depends on collet connector brand. Some are relatively involved: recover connector and flowline to surface/deck, change seal, and redeploy flowline. Some may be replaced with ROV tool after lifting spool clear.
Weather dependence	Very low, since operation is independent of vessel motion.	Relatively high due to dependence on guidelines.

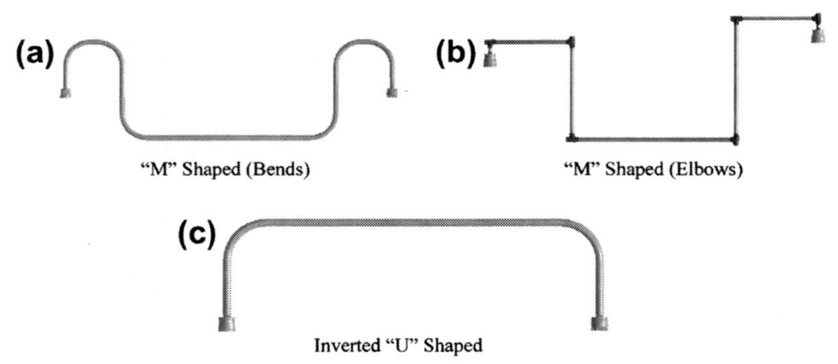

Figure 21-5 Rigid Jumper Configurations

the subsea hardware is installed, the distances between the components to be connected are measured or calculated. Then the connecting jumper is fabricated to the actual subsea metrology for the corresponding hub on each component, in which the pipes are fabricated to the desired length and provided with coupling hubs on the ends for the connection between the two components. Once the jumper has been fabricated, it is transported to the offshore location for the deployment of subsea equipment. The jumper will be lowered to the seabed, locked onto the respective mating hubs, tested, and then commissioned.

If the measurements are not precisely made or the components moved from their originally planned locations, new jumpers may need to be fabricated. The distance and the orientation between the subsea components must be known in advance before the flowline jumpers can be fabricated because the lengths will be critical. Also, a small change in the jumper configuration should be considered when the jumper is lowered below the water surface and the dead load of jumper changes due to the buoyancy: otherwise, the jumper's dimensions may change and the jumper installation may fail. After the installation, if one of the components needs to be retrieved or moved, it is a time-consuming task to disconnect the flowline jumper from the component.

Figure 21-6 shows a configuration for a subsea flexible jumper. It consists of two end connectors and a flexible pipe between the two connectors. The subsea flexible jumpers are usually used for transporting fluid between two subsea components. They are also used to separate the rigid riser from the vessel to effectively isolate the riser from the fatigue due to the motions of the FPSO. For example, both rigid and flexible jumpers are used in a FSHR (Free Standing Hybrid R*iser*) system as

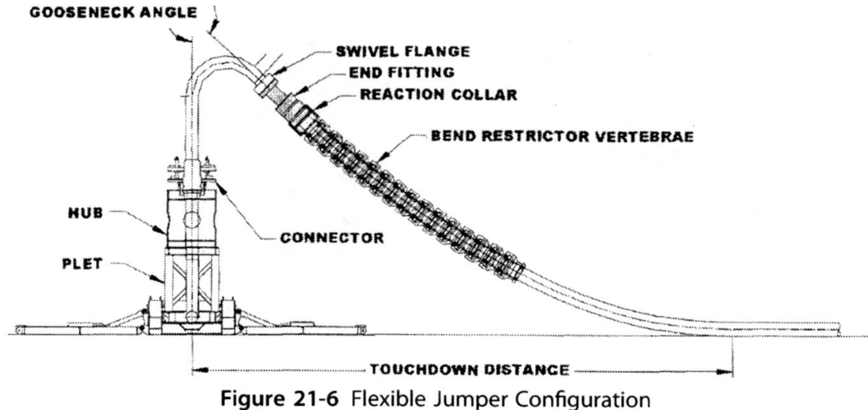

Figure 21-6 Flexible Jumper Configuration

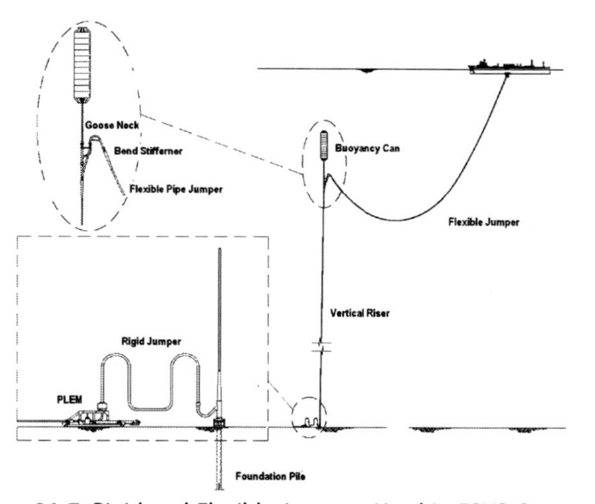

Figure 21-7 Rigid and Flexible Jumpers Used in FSHR System [4]

shown in Figure 21-7. The FSHR consists of a vertical steel pipe (riser) tensioned by a near-surface buoyancy can with a flexible jumper connecting the top of the riser and the FPSO. An M-shaped rigid jumper is used to connect the vertical riser and a PLEM.

21.2. JUMPER COMPONENTS AND FUNCTIONS

21.2.1. Flexible Jumper Components

A typical flexible jumper consists of two end connectors and a flexible pipe between the two connectors. Figure 21-8 shows the components of a Coflexip flexible jumper.

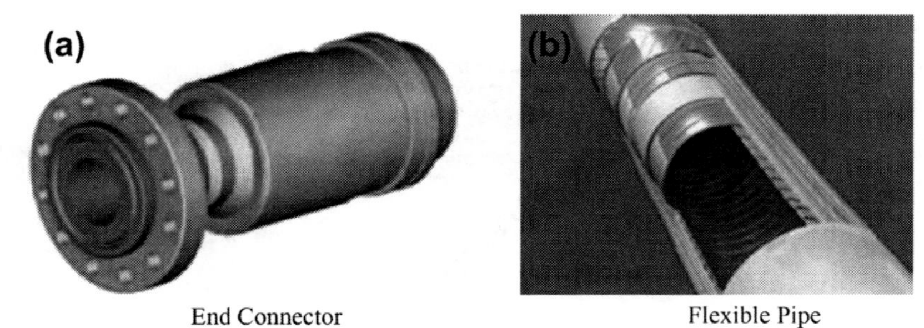

Figure 21-8 Flexible Jumper Components [5]

The Coflexip subsea flexible jumper consists of a number of functional layers, namely: (1) a stainless steel internal casing, which is designed to resist external pressure; (2) a thermoplastic sheath, which creates a fluid seal; (3) a helically wound steel wire, which is designed to resist internal pressure; (4) an axial armored wire, which is for tensile load reinforcement; (5) an external polymer sheath for protection from the environment; and (6) an optional external stainless steel carcass for additional protection.

21.2.1.1. End Fittings of Flexible Pipe

The standard end fittings installed on the subsea flexible jumper are produced with high specification carbon steel coated with ultra-high-corrosion-resistant and damage-tolerant Nikaflex® cladding. Standard terminations are available with API flanges, Grayloc®, Cameron hubs, or hammer unions.

21.2.1.2. Corrosion Resistance

The subsea jumpers for HP/HT flowlines should use the corrosion resistance of high-performance materials technology to provide a high-pressure, high-temperature, corrosion-resistant pipe. The pipes are designed and specified to resist the worst operational conditions. Especially all pipe structures are well adapted to resist corrosive agents such as water, H_2S, CO_2, aromatics, acids, and bases without loss of mechanical design life integrity.

21.2.1.3. Advantages of Flexible Jumpers

The advantages of flexible jumpers can be summarized as follows:

- Flexible;
- H_2S and corrosion resistant;
- High-temperature resistance;

- Suitable for dynamic or static service;
- Excellent fatigue design life;
- Can be pigged;
- High resistance to collapse;
- Highly simplified length specification;
- Low U-value.

The disadvantages of flexible jumpers are expansive and smaller diameter for a higher internal pressure.

21.2.2. Rigid Jumper Components

The jumper connection system is configured with steel pipe and mechanical connections at each end for connection to the subsea facility's piping. The metal seal is retained and sheltered within the connector during installation and retrieval. The connector design is capable of resisting the design loads due to the combined effects of internal pressure, external bending, torsion, tensile, thermal, and installation loads.

In the design of a rigid jumper system, the following items should be considered: (1) The jumper system is installable by a vessel; (2) the jumper components are independently retrievable from the connector receiver (or support structure); and (3) all parts of the system are analyzed with respect to reliability and expected failure rates. The system is designed to minimize failures.

For the design of components for a rigid jumper system, the following items should be considered:

- Materials for the components should be selected so as to minimize potential galling or damage to sealing surfaces during assembly, installation, and maintenance.
- A redundancy philosophy for all parts of the system should be analyzed with respect to safety, cost, and reliability.
- Jumper components of the same design should be interchangeable.
- Standardized parts should be used throughout the system.
- All system components should be compatible with the intended fluids (e.g., all produced, injected, and completed fluids, as applicable) and designed to operate without failure or maintenance for the design life.
- All system components are designed to operate in seawater at the design depth rating.

The rigid jumper connection system may consist of the following items:

- Preformed curved steel piping;
- Two connector assemblies, each with an integral or retrievable actuator;

- A soft landing system either integral to the connectors or provided for in the connector actuator (running tool);
- Two connector receivers;
- Hub end closures;
- Jumper measurement tool;
- Fabrication jigs/test stands and lifting tools such as spreader bar and rigging;
- Shipping/transportation stands;
- Testing equipment;
- Running tool;
- Metal seal replacement tool;
- Insulation;
- Connector actuator (running tool);
- Hub cleaning tool;
- Vortex-induced vibration (VIV) suppression devices (if required).

21.2.3. Connector Assembly

The connector assembly may consist of the following components:

- Connector receiver (hub and support structure);
- Connector;
- Connector actuator;
- Soft landing system;
- Connector override tool.

These subsea components are summarized in the following subsections.

21.2.3.1. Connector Receiver

The connector receiver is mounted on the subsea equipment (such as manifold, PLET, and tree) and includes the connector mating hub, connector alignment system, piping, and support structure. Metal-to-metal seal surfaces for each connector receiver are inlaid with a corrosion-resistant alloy.

The connector receiver should include the following components: (1) an alignment system for guiding the connector onto the mating hub (The alignment system should provide coarse alignment during the initial landing of the jumper connectors onto the connector receivers and align the connector and mating hub during final lowering within the makeup tolerances of the connector.), (2) a measurement tool interface, (3) two high-quality inclinometers (90 degrees to one another) to measure the inclination of each hub directly, and (4) padeyes for attachment of a guideline by an ROV.

The connector receiver should be designed with the following requirements in mind:

- The connector receiver should be designed to be a self-contained assembly that can be independently tested prior to integration into the subsea equipment (tree, manifold, PLET, etc.).
- The connector receiver should be field weldable to the subsea equipment (e.g., piping weld).
- The connector should be able to transfer jumper loads and connector override loads into the subsea structure (tree frame, manifold structure, PLET, etc.).
- The connector should protect critical surfaces on the mating hub during installation and retrieval.
- The connector should be designed to accept hub end caps.

21.2.3.2. Connector

The connector should be used at the end of the jumper piping to lock and seal the mating hub on the connector receiver. The connector should be equipped with the following components: (1) a mechanism (e.g., collet fingers, or dogs) designed to resist lateral and longitudinal forces that may be encountered in the process of aligning and final lowering, prior to makeup of the connection (the connector is designed to accommodate the design loads); (2) metal-to-metal seal surfaces inlaid with corrosion resistant alloy; the seal surfaces shall be relatively insensitive to contaminants or minor surface defects and maintain seal integrity in the presence of the maximum bending moments and/or torsional moments; (3) a metal seal with an elastomeric backup capable of multiple makeups; (4) mechanical position indicators to indicate lock/unlock operations that are clearly readable by an ROV; and (5) a mechanical release override or a hydraulic secondary release system.

The connector should be designed to satisfy the following requirements:

- The connector should not permanently deform the mating hub during connection.
- The connector should protect seals and seal surfaces during deployment and retrieval.
- The connector should retain the metal-to-metal seal during jumper installation and retrieval. The metal seal should be capable of being replaced by an ROV without bringing the jumper to the surface.
- The connector should be welded to the jumper pipe. The mechanical unlatch load should not exceed the allowable stresses in the connector receiver structure and piping.

- The connector should provide for an external pressure test of the metal-to-metal seal after makeup using an ROV hot stab. The external pressure test is at least 1.25 times the ambient hydrostatic head. A volumetric compensator is included in the seal test circuit. This circuit is constructed with welded fittings wherever possible.

21.2.3.3. Connector Actuator

The connector actuator may be integral with the connector or independently retrievable from the connector. The connector actuator should include the following components: (1) ROV handles, which are positioned to observe critical surfaces while guiding the stabbing operation; (2) an ROV panel with pressure gauges to monitor tool functions, hot stab receptacles, etc., for operation of the soft landing system, connector makeup, and seal test (all control and seal test tubing is to be welded whenever possible; separate ROV hot stabs should be provided for each function); (3) indicators that allow the ROV to monitor tool hydraulic functions; (4) a mechanical release (override); and (5) an optional hydraulic secondary release. Note, however, that even if the hydraulic release is used, the mechanical release is still required.

The connector actuator should be designed considering the following requirements:

- The connector actuator should be designed to preclude accidental unlocking from impact loads, vibration, thermal loads, and any other loads affecting the locking mechanism. A secondary lock should be provided where self-locking mechanisms are used.
- The independently retrievable actuator should be positively locked onto the connector and protect the connector during running, retrieval, and topside handling operations.

Figure 21-9 illustrates two typical collet connector assemblies, which includes the connector, connector receiver, and connector actuator: Cameron CVC system [6] and FMC max tie-in system [1]. The connector assemblies are connected with rigid pipe and are widely used in the connection of subsea structures.

21.2.3.4. Soft Landing System

A soft landing system for retrieval and installation should be provided to minimize the impact loads. This system is an integral part of the connector or connector actuator. It is designed to absorb impact loads while landing the jumper on the connector receivers and to maintain the separation

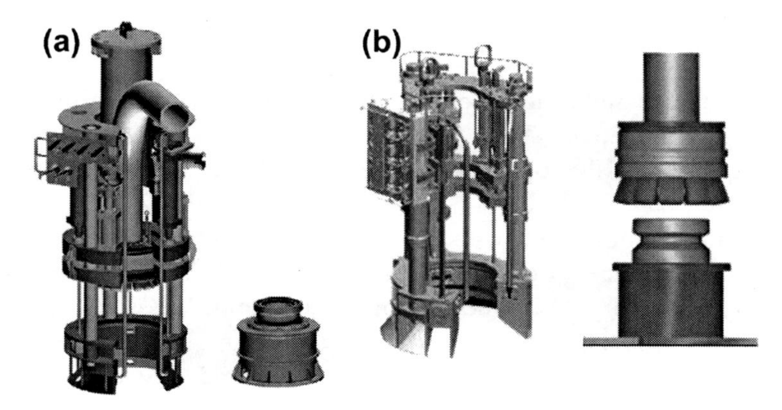

Cameron CVC System [1] FMC Max Tie-in System [6]

Figure 21-9 Typical Connector Assemblies (Connector, Connector Receiver, and Connector Actuator)

between the sealing surfaces when the jumper lands on the connector receiver structure. The soft landing system supports the weight of the jumper to isolate the jumper from vessel motion as the connectors are lowered onto the mating hubs. The system pulls the connector and mating hub faces together prior to locking the connector on to the hub by lowering or raising the connector from the mating hub in a controlled manner. It has mechanical position indicators to indicate a lowering operation that are clearly readable by an ROVand allows each end of the jumper to be lowered or raised independently of the other.

21.2.3.5. Connector Override Tool

The connector override tool should be capable of releasing the mechanical connections if the main actuating device malfunctions.

21.2.4. Jumper Pipe Spool

The jumper pipe spool should include an assembly of straight pipes and bends between the end connections configured to provide compliance during installation and operation. The jumper pipe spool should satisfy the following requirements:

- The pipe should be designed to satisfy the requirements of ASME B31 and API RP 1111, whichever is more stringent.
- Full welds should be required for straight pipes and bends between end connections. Pipe welds should be in accordance with ASME B31.3/ ASME Section IX.

- A minimum of 3D pipe bends should be used; 5D pipe bends for pigging purposes.
- Should have sufficient flexibility to accommodate measurement and fabrication tolerances.
- Should satisfy the bending and torsional limits of the connection system.
- Should accommodate the end movements due to pipeline expansion.
- The jumper pipe spool should be fully assembled. The final welds may be performed at a shore base or offshore after jumper measurements have been determined.
- Vortex suppression devices should be incorporated as required.
- Eccentric gravity loads should be minimized.
- Thermal insulation should be incorporated if required.

21.2.5. Hub End Closure

A hub end closure is provided for each hub unless the hub is designed to be equipped with other components (for example, pig catcher, pigging loop) during installation. The hub end closure is secured to the hub prior to installation of the subsea equipment. In the design of a hub end closure, a mechanism for aligning with the hub for installation by an ROV and the interface for the jumper measurement tools should be included. The hub end closure is designed to be recovered or deployed using a lift line and an ROV. In addition, the main sealing area should be protected.

When a pressure-containing end closure is specified, it should meet the following additional requirements:

- The hub closure should have a minimum pressure rating equal to the connector and hub.
- Seal surfaces should be inlaid with corrosion-resistant alloy. The seal surfaces are insensitive to contaminants or minor surface defects.
- The hub closure should be able to be used during subsea equipment hydrotesting.
- The hub should include a provision for a secondary override. If a mechanical pressure cap is used, an ROV jacking tool will be provided to remove the cap from the hub without damaging the hub, piping, or adjacent equipment.
- It acts as a secondary pressure barrier when the jumper is removed and the closure is reinstalled.

- The hub end closure seal may seal on the seal surface used by the connector metal seal.
- It has an inhibitor injection port and pressure relief port operated via an ROV hot stab interface. The design should allow the determination of the internal pressure prior to cap removal and for injection/circulation of inhibitor.

21.2.6. Fabrication/Testing Stands

The fabrication/testing stands should include the following components: (1) pressure caps and jumper test hubs; (2) ports for quick, easy hydrotesting of the jumper system after fabrication and assembly; (3) a removable hub if more than one size of connector is to be tested on the fixture; and (4) access ladders and work platforms.

The fabrication/testing stands should be designed to satisfy the following items:

- The stand should be capable of being used at a shore base or on the installation vessel.
- The stand should be usable as fabrication/testing or shipping stands.
- It should support the jumper in the installed configuration at the fabrication site or while en route to the installation site.
- It should provide for angular and height adjustments of the mating hubs for jumper fabrication based on field measurements or for modeling misalignment tolerances to verify jumper system functionality.
- It should not limit connector operation and access.
- It should be more rigid than the hub support structures on the subsea equipment.
- It should allow for welding down for transport of a jumper offshore, and quick release of jumper tie-downs offshore during installation. Jumper tie-down release is performed at the deck level.

21.2.6.1. Transportation/Shipping Stands

Figure 21-10 show a transportation/shipping stand with a rigid jumper on a barge. The transportation/shipping stands are similar to the fabrication/testing stands. They are designed with the following items in mind:

- The stand should support the jumper in the installed configuration at the fabrication site or while en route to the installation site.
- The stand should not limit connector operation and access.
- It should have access ladders and work platforms if required.

Figure 21-10 Transportation/Shipping Stand

- It should allow for welding down for transport of a jumper offshore.
- It should allow for quick, easy release of jumper tie-downs offshore during installation. Jumper tie-down release should be performed at deck level.
- It should be designed to accommodate shipping of different jumper sizes.
- It should allow for the jumper to be transported with a new ring gasket.
- It should use a guide funnel to facilitate installation of a jumper.

21.2.6.2. Testing Equipment

Testing equipment for jumper fabrication and testing should be configured for offshore use. The test equipment as a minimum includes the following:

- Hydraulic power unit with flying lead, fittings, hot stabs, hydraulic fluid, etc., to operate a soft landing system, connector actuator, and external seal test, or as required to perform a complete test on the jumper connection system onshore;
- Hydraulic water pump capable of achieving the required hydrotest pressure of the jumper system.

21.2.6.3. Running Tool

The running tool should be capable of installing and retrieving the jumper from an installation vessel with the assistance of an ROV. The running tool is designed to consist of a spreader bar arrangement with lift slings attached to the jumper. Normally, it includes (1) a ROV- friendly rigging (ROV-releasable shackles) for removal during jumper installation and attachment during jumper recovery (e.g., safety hooks); (2) jumper lift slings of sufficient length to allow rigging to be attached and ready for lifting and installation with the running tool in the transportation position; (3) attachment points for steadying lines to facilitate rigging operations; and (4) guide funnels/guide arms for temporary mounting for contingencies.

In the design of the running tool, the following items should be considered:

- The tool should be able to install and recover the jumper with or without the use of guidelines.
- It should be able to run on drill pipe or lowering crane wire.
- It should not damage subsea facilities during jumper deployment and recovery.
- It should allow for hook up to a jumper.
- It should be configured so that the jumper connector rotation will be minimized during landing of the jumper assembly.
- It should allow for disassembly for truck transportation.

21.2.6.4. Jumper Measurement Tool

The normal jumper measurement tool includes taut line and acoustics systems. Both systems are utilized when performing subsea measurement of jumpers and the taut line system is utilized for surface fabrication.

21.2.6.5. Metal Seal Replacement Tool

The metal seal replacement tool should be designed to replace a metal seal without having to bring the jumper to the surface. The operation may utilize the soft landing system and actuator to execute the replacement.

The metal seal removal tool positively locks onto the metal seal and extracts it from the connector. The tool protects the extracted seal so that it may be inspected on retrieval.

21.2.6.6. Hub Cleaning Tool

Figure 21-11 shows a hub cleaning tool on the top of a hub. The hub cleaning tool is used after removal and inspection of the end closure and prior to deployment of the jumper. It consists of a plug with cleaning pads for cleaning debris and grease from the seal surfaces (seal profile) of the connector hub.

Figure 21-11 Hub Cleaning Tool

21.2.6.7. Vortex-Induced Vibration Suppression Devices

The requirement for VIV suppression devices is determined based on a vibration analysis.

21.2.6.8. ROV-Deployable Thickness Gauge

On each connector end, where erosion is most prevalent, a measurement funnel and plug if required are provided. The arrangement will allow an ultrasonic probe to be installed for reading pipe wall thickness with an ROV at this location. A baseline reading is taken and recorded for each jumper.

21.3. SUBSEA CONNECTIONS

The subsea connections include flowline connections and umbilical control connections. The types of subsea flowline connection may be divided into four types of connections: welded, flanged, clamp hub, and mechanical connections. The welded connection is normally used for subsea connections in very shallow water, and the welding procedure is carried out in a one-atmospheric chamber to achieve a dry environment. For the other three types of subsea connections, the primary purpose of the connection method is to create a pressure-tight seal that resists the loads associated with subsea environments. The connection between the sealine and the connection point is generally made after sealine end alignment is complete. The mechanical connectors use either a mandrel or hub style interface, and the actuating tools are hydraulically or mechanically actuated. For deep water, all seals experiencing hydrostatic pressure should have bidirectional capability. In the following subsections we discussed several types of sealine connectors that typify the numerous options available.

21.3.1. Bolted Flange

The flanged connection utilizes metal gasket designs to allow the flanges to make face-to-face contact. The most typical flanges used are either APl 6A/17D [7] type or ANSI/ASME B16.5 [8] type flanges, and the design of bolted flanges is covered in ANSI/ASME B16.5 and ISO 13628-4 [9]. They are commonly used in shallow water (<1000 ft) with the aid of a diver. The designs make use of metal ring joint gaskets, which compress when the bolts are tightened. Special consideration should be given to these gaskets for underwater applications. Some gaskets tend to trap water behind the gasket when made up underwater, resulting in improper sealing of the gasket and flange connection. The gasket ring grooves should be specified with welded inlays to provide a corrosion-resistant surface finish. Welded inlays are not relevant when corrosion-resistant alloy (CRA) materials are used for flanges.

Figure 21-12 shows a cross section of a bolted flange connection, which may permit a limited degree of initial misalignment. However, rotational alignment is restricted because of the bolt-hole orientation.

Swivel flanges are used to facilitate bolt-hole alignment. Figure 21-13a shows swivel flange assemblies of the Taper-Lok type. The misalignment flange designs, such as those by Hydrotec and Taper-Lok, have been used in

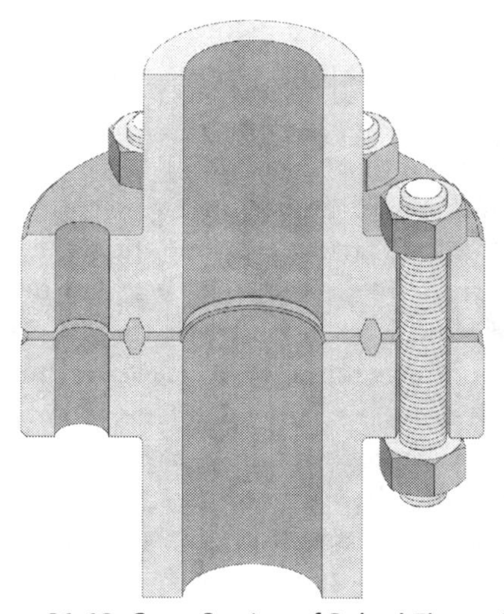

Figure 21-12 Cross Section of Bolted Flange [10]

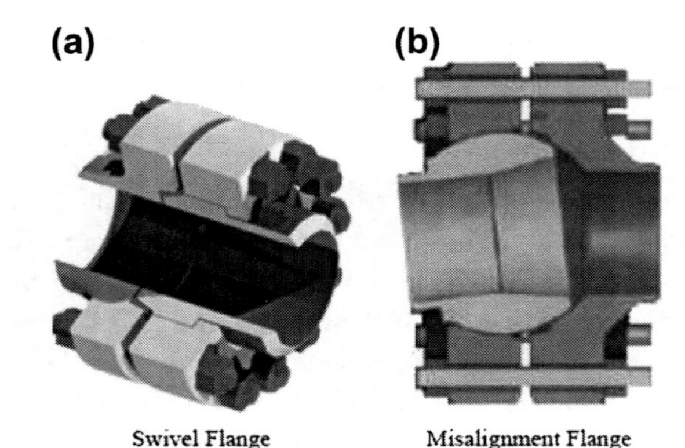

Figure 21-13 Misalignment and Swivel Flange Assemblies of Taper-Lok Type

subsea pipeline connections. The flanges usually include at least one swivel flange at each interface flange to allow for ease of makeup. The flange fasteners are typically tightened through the use of some type of stud tensioning system. The standard Taper-Lok misalignment assembly shown in Figure 21-13 allows up to 10 degrees of both axial and angular misalignment of piping.

A bolted flange has the following characteristics:

- Bolts and nuts are used to preload two flanges.
- A metal gasket is compressed between the two flanges to create a seal.
- It is commonly used at shallow diver depths.
- It can be made up with specialized ROV-operated tools.
- Flanges must be closely aligned before makeup.
- Swivel flanges can account for rotational misalignment.

21.3.2. Clamp Hub

A clamp hub connector is similar in principle to the bolted flange connector. Clamp hub connectors may use the same metal ring gaskets as bolted flange connectors or use proprietary gasket designs. The clamping device forces the mating hubs together as the clamping device is tightened. Figure 21-14 illustrates a typical clamp hub connection. The clamped hub connections are generally faster to make up than bolted flange connections, because fewer bolts are required. Rotational alignment is unnecessary since the mating hubs do not have bolt holes, except for multibore hubs. On the other hand, most clamped hubs do not permit the amount of initial misalignment that bolted flange connections may provide.

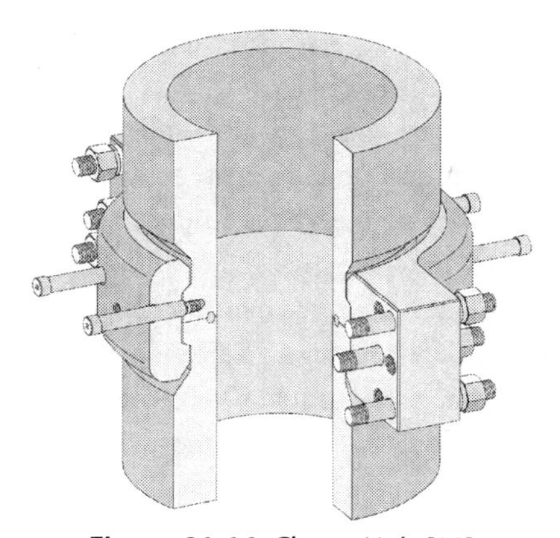

Figure 21-14 Clamp Hub [10]

Clamp hubs are often used for subsea connections, because they are small and have a reduced number of bolts to handle, install, and tension. They can, however, be difficult to disconnect on occasion. Since the clamp hubs are pulled together by the tensioning of the bolts (wedging the clamp halves on the tapered hub profile), residual stress and friction resists removal of the clamp halves, even with the bolts removed. Divers often have to hammer or use a special tool, to strip the clamp halves of the hubs, particularly if they have been subsea for a considerable length of time. The characteristics of clamp hubs are summarized as follows:

- The clamping device preloads two hubs as the device is tightened.
- A metal gasket is compressed between the two hubs to create a seal.
- It can be made up by divers or ROV-operated tools
- It is faster to make up than a bolted flange because it has fewer bolts.
- Hubs must be closely aligned before makeup. Rotational alignment is not critical.

21.3.3. Collet Connector

Collet connectors are the most widely used for jumper spool style connections. The connection principle is similar to that for a hub and clamp except that a series of longitudinally segmented *collets* replaces the clamp. The collets are activated by an annular locking cam ring. The cam ring slides axially along the collet length to either open or close the connector. The driving angle on the cam ring is typically self-locking to prevent incidental unlocking of the connector. A collet connector consists of a body and hub around in which individual "fingers" or collets are arranged in a circular pattern and attached to the body, engaging the hub to form a fully structural connection as shown in Figure 21-15.

The engagement between the hub and collets is either mechanically or hydraulically actuated. Funnels or guide posts provide coarse alignment for the mating hubs. After the collet connector is landed, an ROV installs the hot stab and provides hydraulic power to the stroking cylinders, pulling the hubs together at a controlled rate. The open collets provide initial makeup alignment to protect the metal seal. The standard design accommodates up to ±2° angular misalignment and ±1.5–in. axial offset. Once the hubs are mated, hydraulic collet actuating cylinders close the collet fingers to complete the connection. The collets impart a preload that gives the connection the same strength characteristics as the pipe. The seal is pressure tested after makeup via the ROV control panel. The characteristics of collet connectors are summarized as follows:

- Collet segments are driven around a connector body and hub by an actuator ring.
- A metal gasket is compressed between the body and hub to create a seal.

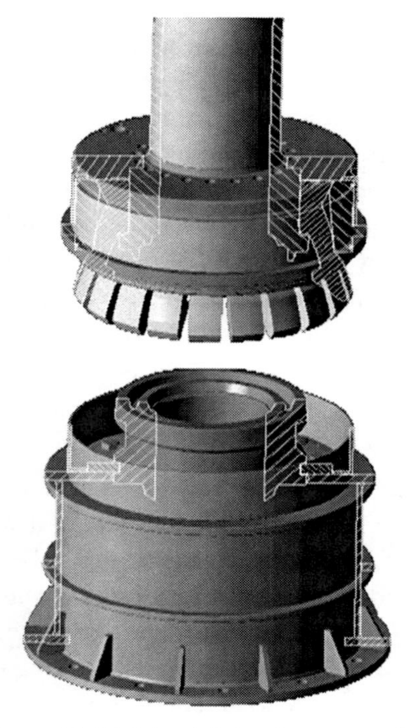

Figure 21-15 Pair of Collet Connectors [10]

- It functions by means of integral hydraulics or by an ROV-operated tool.
- It can align misaligned hubs.
- Rotational alignment is not critical.

21.3.4. Dog and Window Connector

Dog and window flowline connectors are similar in principle to standard wellhead connectors where a number of "dog" segments are collapsed into a corresponding groove in the mating hub to lock the connection. The dogs are captured by "windows" within the connector assembly. The number of "dog" segments varies between manufacturers and can be as simple as a longitudinal cut in a fully circumferential ring. A locking piston is driven forward to collapse the dogs, setting the seal, and preloading the connection.

Figure 21-16 illustrates a typical dog and window connector. Its characteristics are summarized as follows:

- The locking dogs held in a window of the connector body are driven inward around a hub by an actuator ring.
- A metal gasket is compressed between the body and hub to create a seal.
- It functions by means of integral hydraulics or by an ROV-operated tool.
- Rotational alignment is not critical.

21.3.5. Connector Design

The sealine connector choice and subsequent design should consider factors such as water depth, intervention method, type of connection point, sealine installation method, and misalignment tolerance compatibility with the alignment method. In addition, the choice and design of the connector can be influenced by the factors discussed next.

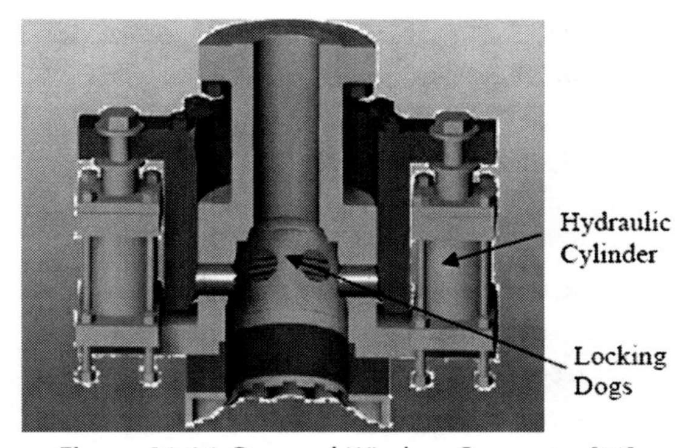

Figure 21-16 Dog and Window Connector [10]

21.3.5.1. Connector Stresses

Residual stresses in the line itself and the sealine connector resulting from a particular alignment method and the additional axial movement required for end connection should be analyzed in conjunction with operating stresses to determine if the combined stress is within allowable limits.

21.3.5.2. Makeup Requirements

Makeup requirements for connectors should be reviewed to ensure that (1) the connector will deform or deflect the gasket to result in a seal; (2) there is enough preload in the connector to offset the installation and operating loads, which could otherwise break the gasket seal; and (3) there is enough axial clearance and access for seal replacement.

The sealine connector design should be such that after makeup it will not lose its sealing capability under cyclic pressures, temperatures, or natural vibration loading and design external loads.

21.3.5.3. Testing

The sealine connectors should be tested in plant to the hydrostatic test pressure stipulated for the sealine. In some cases, the connector may be part of and tested with the subsea facility. In such cases the connector should be tested to the hydrostatic test pressure stipulated for the subsea facility. If TFL (through flow line) is specified, each made-up connector should be drifted in accordance with API RP 17C [11]. Additional in-plant testing may be required to verify makeup preloads, fit, and functional performance of locking devices and hydraulic actuation devices.

The end-connector equipment is designed to provide some testing means to verify that the gasket has formed an adequate seal and the connector has been fully actuated or clamped together after it has been installed in subsea. In the evaluation of the connection methods, the following factors should be considered:

- Sealing reliability;
- Ruggedness or resistance to damage;
- Ease of recovery;
- Installation cost;
- Template piping interface fabrication difficulty;
- Resistance to operating loads;
- Initial equipment costs.

To choose the connection method, the criterion is assigned to each factor according to its importance. Each connection method is ranked against the other methods on a scale from one to three within each criterion.

21.4. DESIGN AND ANALYSIS OF RIGID JUMPERS

21.4.1. Design Loads

The design parameters used to analyze a rigid pipe jumper are:

- Working pressure (external hydrostatic pressure not considered);
- Temperature of the product inside the pipe, which may constitute derating the minimum yield strength of the per code;
- Insulation thickness and density required to prevent hydrate formation;
- Height of the jumper's mating hubs off the mud line.

The design of the jumper system should take into account all loads, which may be imposed as a result of the following issues:

- Fabrication, assembly, and testing loads;
- Loadout and transportation;
- Installation, including keelhauling, cross hauling, lowering to seabed, and landing. The jumper system will be assumed filled with water during deployment;
- Hydrostatic load;
- Hydrotest load;
- Fabrication and measurement tolerances;
- Thermal and pressure loads, including effect of thermal/pressure cycling;
- Wave and current forces;
- Flowline operational loads (flowline end movement due to pressure, temperature, etc.);
- ROV impact loads;
- Subsidence loads (differential settlement);
- Vibration fatigue (from unsteady flow or VIVs from currents);
- Other operation-induced loads, including drill riser/BOP movement to tree/well jumper during workover.

21.4.2. Analysis Requirements

The jumper analysis should consider the analyses discussed next as a minimum.

21.4.2.1. Tolerance Analysis

A tolerance analysis should be performed to determine maximum allowable tolerances between mating components and subassemblies. It demonstrates that repeatable interfaces can be achieved and that connectors are fully interchangeable with mating hubs. It also determines the optimum jumper geometry for meeting functional requirements.

21.4.2.2. ROV Access Analysis

The contractor should perform an ROV accessibility analysis to demonstrate that the ROV has clear access to all ROV panels, mechanical overrides, hydraulic hot stabs, position indicators, jumper connections, etc.

21.4.2.3. Loading Analysis

The design must be confirmed through analysis of the following loading conditions:

- *Transportation:* Transportation criteria should be determined during the detailed design phase. The jumpers should be analyzed for transportation to the field.
- *Offshore lifting:* Analyses should be performed with appropriate load factors taking into account the dynamics of the installation vessel.
- *Installation:* An installation analysis should be performed.
- *Thermal expansion:* All thermal expansion loads should be considered to establish design loads for the manifold structure and the jumper system.
- *Local stress analyses:* Local stress analyses should be carried out for all joints, lifting points, and high stressed welds.

21.4.2.4. Thermal Analysis

A thermal analysis should be performed to confirm the expected time required to reach hydrate formation temperature in the jumper system. The analysis must include the jumper and the connector assembly.

21.4.3. Materials and Corrosion Protection

A coating system should be used for protection against corrosion. Careful attention should be given to the grounding of subassemblies to maintain continuity throughout the jumper connection system. The use of dissimilar metals in the system should be avoided.

21.4.4. Subsea Equipment Installation Tolerances

Tie-in connections are either vertical or horizontal based on system selection. Designed for water depths exceeding 10,000 ft (3000 m) and

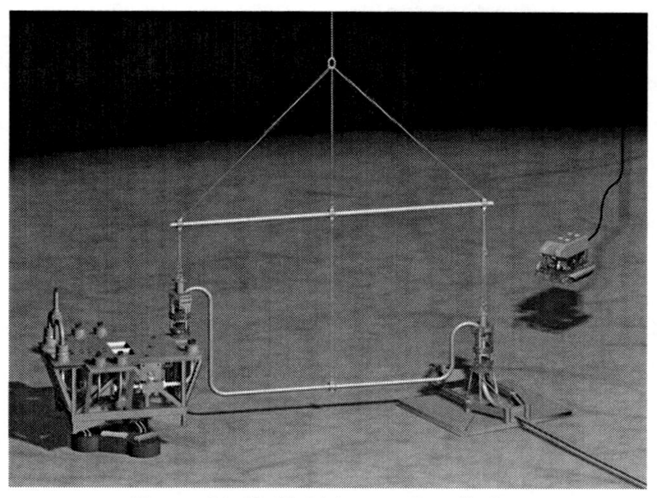

Figure 21-17 Rigid Jumper Installation

working pressures to 15,000 psi, all jumpers or spool pieces should be installed using guideline-less techniques as shown in Figure 21-17. Jumpers or spool pieces are installed after onshore construction and testing to mate to previously installed equipment, based on subsea metrology data. The installation tolerances of the subsea equipment may not exceed the tolerance of 2 degrees off verticality for any vertically oriented hub. Installation tolerances should be determined from the tolerance analysis of the well and flowline jumpers.

21.5. DESIGN AND ANALYSIS OF A FLEXIBLE JUMPER

Figure 21-18 shows a 4-in. flexible jumper used to connect a subsea tree and manifold. The flexible jumper has found wide application as a connector between subsea structures. Flexible jumpers can be used to connect trees and pipelines (PLET) or manifolds. Its flexibility allows the connection to be made such that one side of the connection has sufficient movement.

Figure 21-19 and 21-20 show, respectively, typical Oil States Industries (OSI) and Vetco Gray (VG) jumper gooseneck assemblies and components of a flexible jumper.

In the following sections, jumper in-place analyses and installation analyses are detailed for a GoM project.

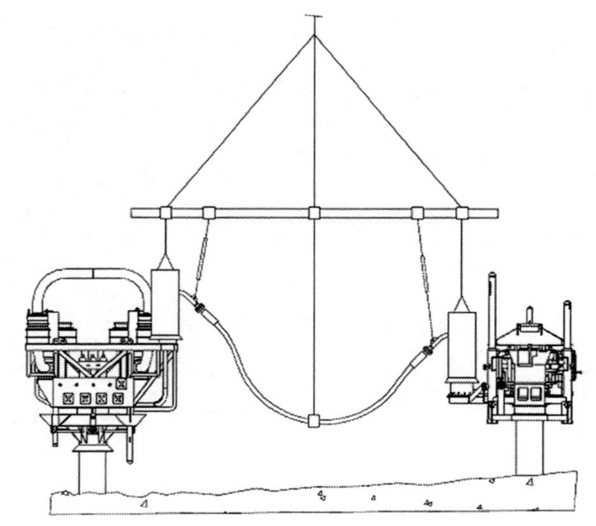

Figure 21-18 Flexible Jumper between Tree and Manifold [12]

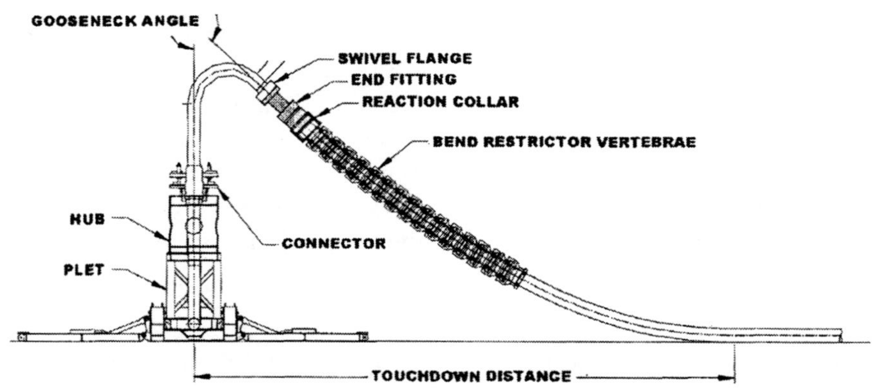

Figure 21-19 Typical Oil States Industries Jumper Gooseneck Assembly of a Flexible Jumper

21.5.1. Flexible Jumper In-Place Analysis

21.5.1.1. Allowable Jumper Loads for Jumper In-Place Analysis

The allowable jumper loads for a jumper in-place analysis are affected by the following issues:

- Maximum tension of flexible jumper. To ensure the integrity of supporting structures and the collet connectors of subsea structures, the horizontal tension should be limited to a maximum value (for example, 5 kips). The jumper loads and moments due to the maximum horizontal tension should be below the allowable limits.

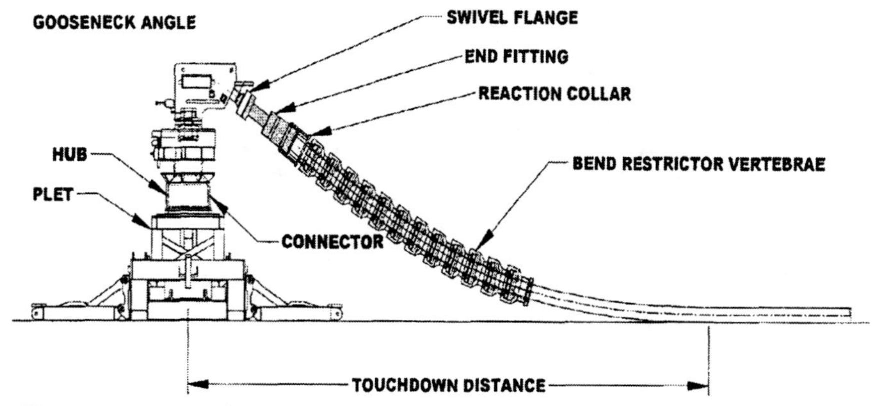

Figure 21-20 Typical Vetco Gray Jumper Gooseneck Assembly of a Flexible Jumper

- The minimum allowable bend radius should not below 1.25 × MBR (minimum bend radius).
- The flexible jumper should experience a positive bottom tension at all times (i.e., no axial compressive load is permitted).
- The induced maximum jumper loads should ensure the integrity and stability of adjacent structures at all times.

21.5.1.2. Analysis Methodology

During installation, and throughout the design life of jumpers, loads will be transferred from the jumpers to the PLET collet connector and manifold structures. The global static analysis is performed to ensure the integrity of jumpers during installation and operating conditions, and to assess the induced load transfer from jumpers to the adjacent structures.

The jumper static design process includes the design steps shown in the flowchart of Figure 21-21.

Figure 21-22 shows the flowchart for system configuration design. Based on the flowchart, the following sections describe the analysis methodology adopted during the system configuration design.

Static Jumper Analysis

All static analyses are 2D or 3D load cases, which account for all functional loads. In the analysis of the flexible jumper, the following external forces are considered:

- Uniformly distributed submerged weight of jumper and bend restrictor;
- Concentrated load due to gooseneck, end connector, running tool, etc.;

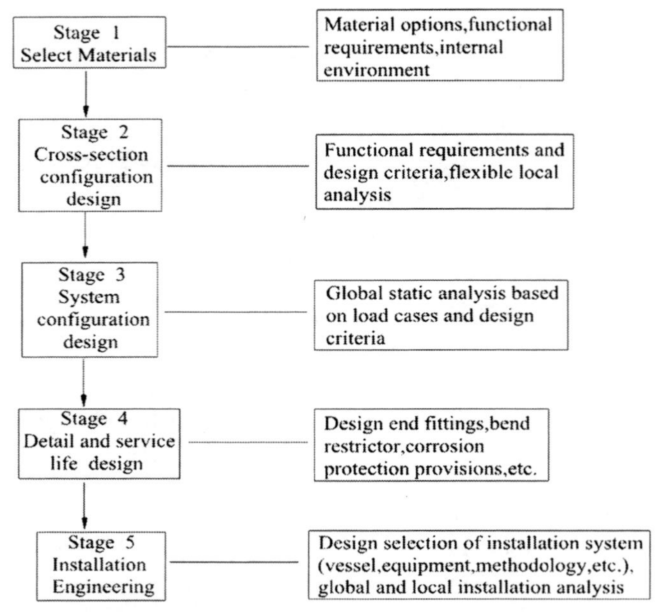

Figure 21-21 Typical Static Application Design Flowchart

- Seabed foundation in the vertical plane, which is assumed to be a continuous elastic foundation;
- External hydrostatic pressure.

Free-End Catenary Analysis

The free-end catenary analysis is performed to define the gooseneck angle. The term *free-end* refers to the top end of the jumper, which is fixed at a predetermined elevation although it is free to rotate according to the natural catenary configuration of the jumper. The free-end catenary analysis is performed for the operating condition, which is the long running load case that the jumper will experience during its design life.

For each jumper, the free-end catenary analysis assumes that the lower end of the jumper is resting on the seabed under a constant horizontal bottom tension. The predetermined elevation of the top end reflects the supporting structure height, with the nodal rotational degree of freedom unrestrained to allow the top end of the jumper to follow the natural catenary configuration of the free hanging string. For a range of bottom tension, the above analytical modeling allows study of jumper behavior to establish a range of touchdown locations and associated gooseneck angles.

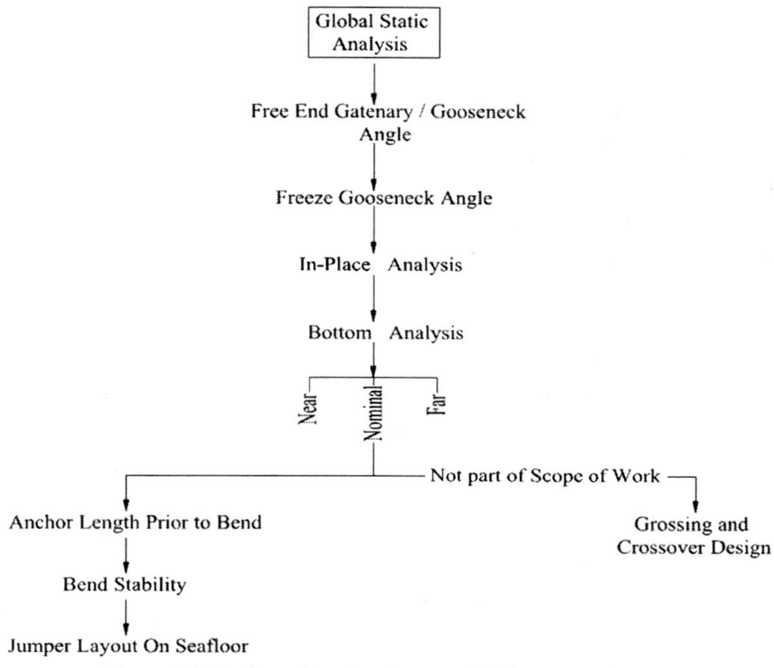

Figure 21-22 Flowchart for System Configuration Design

Based on this study, the gooseneck angle can be optimized to meet loading requirements. Based on the optimized gooseneck angle, the nominal bottom tension is defined for the operating condition.

In-Place Jumper Analysis

The in-place jumper analysis is performed to ensure that the gooseneck angle defined in the free-end catenary analysis provides a favorable solution for all other loading conditions. The in-place jumper analysis is also performed to provide a range of bottom tensions for which the integrity of the jumper and the supporting structure is maintained.

Based on the axial soil resistance, it is concluded that jumper lengths, in most cases, are not long enough to fully or partially restrain the jumper in the longitudinal direction. Any variation in axial tension, therefore, is anticipated to freely travel from the first end to the second end of the jumper. For this reason, it is recommended that the entire jumper, including first- and second-end elevations, be modeled when performing in-place analysis.

As minimum the following three extreme cases are investigated:

- Near position;
- Far position;
- Maximum out-of-plane excursion.

The *near position* is defined as the loading direction that results in the touchdown point (TDP) moving toward the supporting structure. In this situation, bottom tension is at a minimum and the jumper bend radius at the touchdown location is approaching the MBR criteria limitation. The *far position* is defined as the loading direction that results in the TDP moving away from the supporting structure. For this load case, the minimum bend radius usually occurs at the jumper/end-fitting connection interface. The induced moment and top tension are high for this loading condition. The transverse loading direction results in the TDP moving in the out-of-plane direction to the jumper.

The following criteria are used to establish jumper extreme positions:

- The minimum allowable bend radius should not below 1.25 × MBR.
- The flexible jumper should experience a positive bottom tension at all times (i.e., no axial compressive load is permitted).
- The induced maximum jumper loads should ensure the integrity and stability of adjacent structures at all times.
- A pipe rigidity transition zone of 1.5 × OD should be assumed after the jumper/end-fitting interface.

Jumper Curve Stability Analysis

The jumper stability analysis for the most conservative residual on-bottom tension will be performed to evaluate the minimum stable curve radius.

A lateral stability check is performed for each lateral bend using the following relationship:

$$R_{\min} = SF \frac{DAF\ T_R}{\mu\ W_s}$$

where

$R_{\min}$: minimum stable radius, m;
SF: 1.5 (safety factor against sliding);
DAF: 1.0 (dynamic amplification factor);
T_R: maximum residual laying tension in the bend, kN;
μ : lateral friction factor of the jumper/seabed interface
W_s: jumper submerged weight, kN/m.

Flexural Anchor Reaction Length Prior to Bend

The jumper curve stability analysis described above ensures jumper stability along the curve, during the laying operation. However, prior to a lateral bend, adequate anchorage is required to initiate the bend. Depending on the lateral jumper/seabed soil resistance, and the bend radius, a straight section of flexible pipe is defined to resist the induced moment and reaction loads at the bend tangential point. This straight section is referred to as the flexural anchor reaction (FAR) length and is evaluated using the following relationship:

$$L_{\min} = SF\sqrt{\frac{2EI}{\mu RW_s}}$$

where

$L_{\min}$: minimum flexural reaction length, m;
SF: 1.2 (safety factor);
EI: flexural rigidity of the jumper;
R: minimum stable radius, m;
μ : lateral friction factor of the jumper/seabed interface;
W_s: jumper submerged weight, kN/m.

21.5.2. Flexible Jumper Installation

The finite element program Orcaflex is commonly used to simulate the flexible jumper installation. The software is developed by Orcina Ltd. The following installation steps should been analyzed in the installation analysis:

Step 1: Lowering of the jumper;
Step 2: Obtaining an installation configuration;
Step 3: Connecting to PLET;
Step 4: Connecting to another PLET or manifold.

The main parameters of the analysis and the items that should be investigated are:

- Wave height – constant throughout the analysis;
- Wave period – constant throughout the analysis;
- Wave type – Stokes fifth-order waves may be used in the analysis for conservatism;
- Current – assumed to be constant throughout the analysis;
- Vessel heading is maintained at 180 degrees or head seas, unless noted otherwise;
- Layback range (horizontal distance from overboard point to TDP);
- Curvature/bending radius;
- Tension in the jumper and deployment wire;

- Bending moment (BM) applied to bend restrictors (BRs).

Data required for the installation analysis include:

- Water depth;
- Current profile;
- Wave data;
- Seabed friction factors.

21.5.2.1. Design Criteria

The installation design criteria for a flexible jumper are listed below. The installation analysis is performed to ensure that these criteria could be safely and effectively achieved.

- MBR of flexible jumpers must not be exceeded.
- Tension at the PLETs and manifolds should not exceed a given value, for example, 5 kips.
- The bending moment (BM) applied to the BRs must remain within allowable limits.
- Connector hub angle must remain $\pm 4°$ of vertical when connecting.
- Maximum load on the lifting adapter should not be exceeded.
- Load capacity of the cranes should not be exceeded.
- Maximum load on the A&R wire should be limited to allowable value, for example, 45 tonne.
- There should be no compression in the jumpers.

21.5.2.2. Installation Steps

The installation steps for a flexible jumper for the project in the deep water of the GoM are given to show the procedure used for an installation analysis.

Step 1: Lowering of the Jumper

Analysis starts off with the jumper attached to crane hooks and hanging over the side of the vessel. This is the position from which the crane wires begin to pay out equally on both sides to begin lowering the jumper into the water. Figure 21-23 shows the lowering of the first end of the flexible jumper.

Step 2: Obtaining an Installation Configuration

This step involves lowering the jumper by simultaneously paying out equal amounts of crane wire on both sides into; the crane hooks reach a certain depth. After this depth is reached, only the crane wire attached to the first end is allowed to pay out further to lead into step 3. Figure 21-24 shows the installation configuration of a flexible pipeline after step 2.

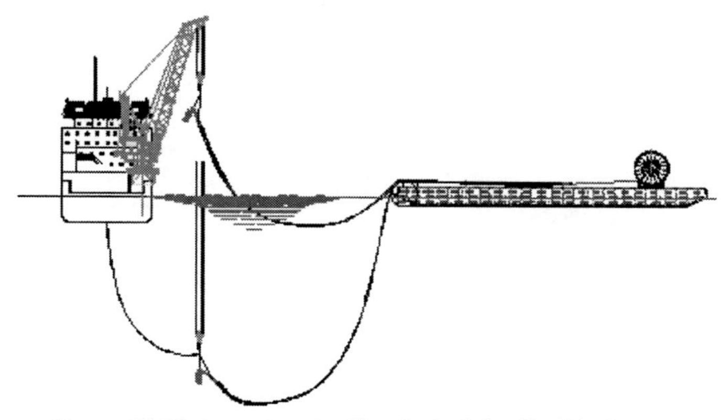

Figure 21-23 Lowering the First End of the Flexible Jumper

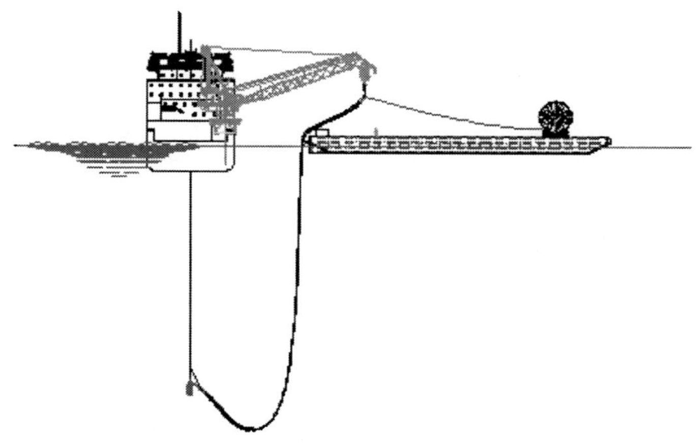

Figure 21-24 Installation Configuration after Step 2

Step 3: Connecting to First-End

This step starts when the wire attached to the first end is paid out to reach the depth at which the first end can be connected to the applicable PLET/manifold. This is done while the other wire is held at the position where it was at the end of step 2. For some of the longer jumpers, the crane separation on board the vessel may need to be increased to around 70 m. The vertical separation of the two connectors has to be increased so as to assist in obtaining the vertical orientation of the connector being attached to the first end. Higher bending moments and curvatures are expected in the jumper due to this difference in elevation. Figure 21-25 shows the connection of the first end of the jumper to the PLET.

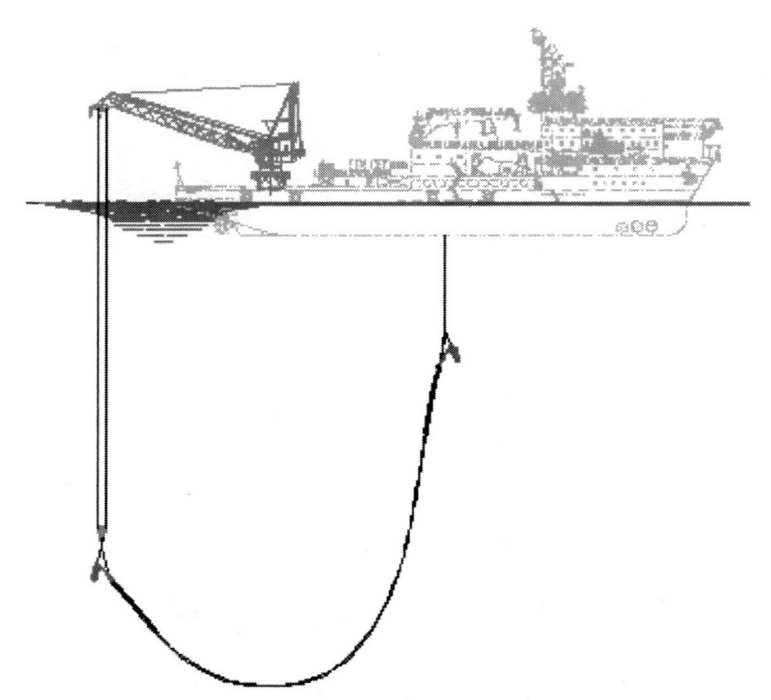

Figure 21-25 Connection of First End of the Jumper

Step 4: Connecting to Second-End

This step begins with the first end of the jumper being rigidly connected to the PLET/manifold. Thereafter, 10 m of the second-end crane wire is paid out (step 4, substep 1) to achieve touchdown and establish a layback. The subsequent steps involve paying out crane wire at the second end until the hub is within 10 m of its connection point and lowered further toward

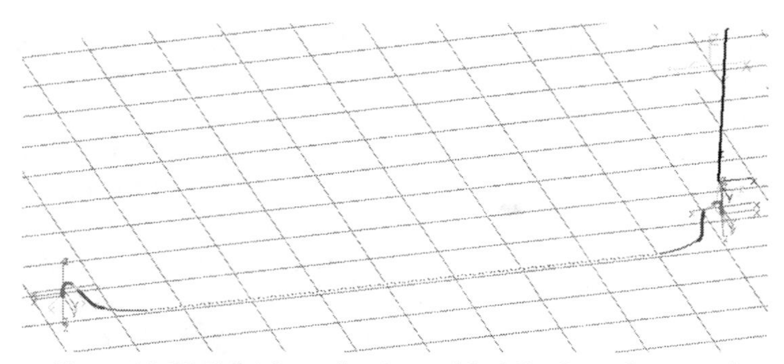

Figure 21-26 Finial Step after Second End Has Been Connected

the second-end PLET/manifold in two steps (substeps 2 and 3), to be followed by the second end reaching the final connection position in the final step (substep 4). Figure 21-26 illustrates the finial step after the second end has been connected.

REFERENCES

[1] FMC Technologies, Subsea Tie-In Systems, http://www.fmctechnologies.com/subsea.
[2] T. Oldfield, Subsea, Umbilicals, Risers and Flowlines (SURF): Performance Management of Large Contracts in an Overheated Market; Risk Management and Learning, OTC 19676, Offshore Technology Conference, Houston, Texas, 2008.
[3] G. Corbetta, D.S. Cox, Deepwater Tie-Ins of Rigid Lines: Horizontal Spools or Vertical Jumpers? 2001, SPE Production & Facilities, 2001.
[4] F.E. Roveri, A.G. Velten, V.C. Mello, L.F. Marques, The Roncador P-52 Oil Export System Hybrid Riser at an 1800m Water Depth, OTC 19336, Offshore Technology Conference, Houston, Texas, 2008.
[5] Technip, COFLEXIP Subsea and Topside Jumper Products, www.technip.com.
[6] Cameron, Cameron Vertical Connection (CVC) System, http://www.c-a-m.com.
[7] American Petroleum Institute, Specification for Subsea Wellhead and Christmas Tree Equipment, API Spec 17D (1992).
[8] American Society of Mechanical Engineers, Pipe Flanges and Flanged Fittings, ASME/ANSI B16.5 (1996).
[9] International Organization for Standardization, Petroleum and Natural Gas Industries - Design and Operation of Subsea Production Systems - Part 4: Subsea Wellhead and Tree Equipment, ISO 13628-4, (1999).
[10] B. Rose, Flowline Tie-in Systems, SUT Subsea Awareness Course, Houston, 2008.
[11] American Petroleum Institute, TFL (Through Flowline) Systems, second ed., API-RP-17C, 2002.
[12] E. Coleman, G. Isenmann, Overview of the Gemini Subsea Development, OTC 11863, Offshore Technology Conference, Houston, Texas, 2000.

CHAPTER 22

Subsea Wellheads and Trees

Contents

Subsea Engineering Handbook
ISBN 978-0-12-397804-2, doi:10.1016/B978-0-12-397804-2.00022-9

22.1. INTRODUCTION

Subsea wellheads and Xmas trees are one of the most vital pieces of equipment in a subsea production system. The subsea wellhead system performs the same general functions as a conventional surface wellhead. It supports and seals casing strings and also supports the BOP stack during drilling and the subsea tree after completion.

A subsea Xmas tree is basically a stack of valves installed on a subsea wellhead to provide a controllable interface between the well and production facilities. It is also called a Christmas tree, cross tree, X-tree, or tree. Subsea Xmas tree contains various valves used for testing, servicing, regulating, or choking the stream of produced oil, gas, and liquids coming up from the well below. The various types of subsea Xmas trees are used for either production or water/gas injection. Configurations of subsea Xmas trees can be different according to the demands of the various projects and field developments.

Subsea wellhead systems and Xmas trees are normally designed according to the standards and codes below:

- API 6A, *Specification for Wellhead and Christmas Trees Equipment;*
- API 17D, *Specification for Subsea Wellhead and Christmas Tree Equipment;*

- API RP 17A, *Recommended Practice for Design and Operation of Subsea Production Systems;*
- API RP 17H, *Remotely Operated Vehicle (ROV) Interfaces on Subsea Production System;*
- API RP 17G, *Design and Operation of Comlpetion/Workover Risers;*
- ASME B31.3, *Process Piping;*
- API 5L, *Specification for Line Pipe;*
- ASME B31.8, *Gas Transmission and Piping System;*
- ASME BPVC VIII, *Rules for Construction of Pressure Vessels, Divisions 1 and 2;*
- AWS D1.3, *Structural Steel Welding Code;*
- DNV RP B 401, *Cathodic Protection;*
- NACE MR-0175, *Petroleum and Natural Gas Industries—Material for Use in H_2S-Containing Environments in Oil and Gas Production.*

22.2. SUBSEA COMPLETIONS OVERVIEW

Prior to the start of production, a subsea well is to be completed after drilling and temporarily suspended. Subsea completion is the process of exposing the selected reservoir zones to the wellbore, thus letting the production flow into the well. Two completion methods are commonly and widely used in the industry, as illustrated in Figure 22-1:

- *Open hole completion:* Open hole completions are the most basic type. This method involves simply setting the casing in place and cementing it above the producing formation. Then continue drilling an additional hole beyond the casing and through the productive formation. Because this hole is not cased, the reservoir zone is exposed to the wellbore.
- *Set-through completion:* The final hole is drilled and cemented through the formation. Then the casings are perforated with tiny holes along the wall facing the formations. Thus, the production can flow into the well hole.

The completion design includes the tubing size, completion components and equipment, and subsea Xmas tree configuration. Components of subsea completion equipment include the subsea wellheads and the subsea tubing hanger/tree systems, which will be discussed in the following sections.

22.3. SUBSEA WELLHEAD SYSTEM

The main function of the subsea wellhead system is to serve as a structural and pressure-containing anchoring point on the seabed for the drilling and

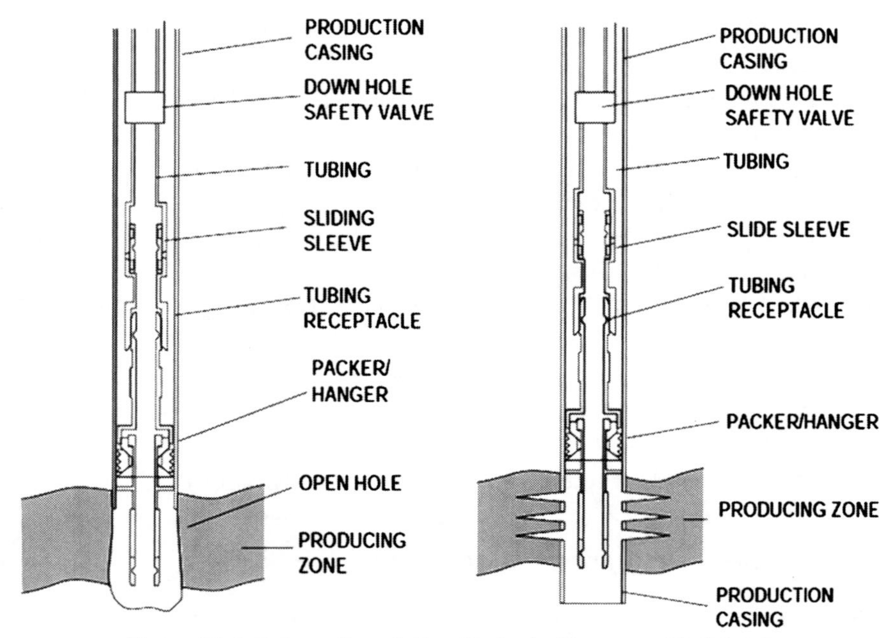

Figure 22-1 Subsea Completion Methods *(Courtesy of Dril-Quip)*

completion systems and for the casing strings in the well. The wellhead system incorporates internal profiles for support of the casing strings and isolation of the annulus. In addition, the system incorporates facilities for guidance, mechanical support, and connection of the systems used to drill and complete the well. Figure 22-2 illustrates the main building blocks of a subsea wellhead system.

22.3.1. Function Requirements

The subsea wellhead system should:

- Provide orientation of the wellhead and tree system with respect to the tree-to-manifold connection.
- Interface with and support the Xmas tree system and blowout preventer (BOP).
- Accept all loads imposed on the subsea wellhead system from drilling, completion, and production operations, inclusive of thermal expansion. Particular attention should be given to the horizontal tree concept where the BOP is latched on top of the Xmas tree.
- Ensure alignment, concentricity, and verticality of the low-pressure conductor housing and high-pressure wellhead housing.

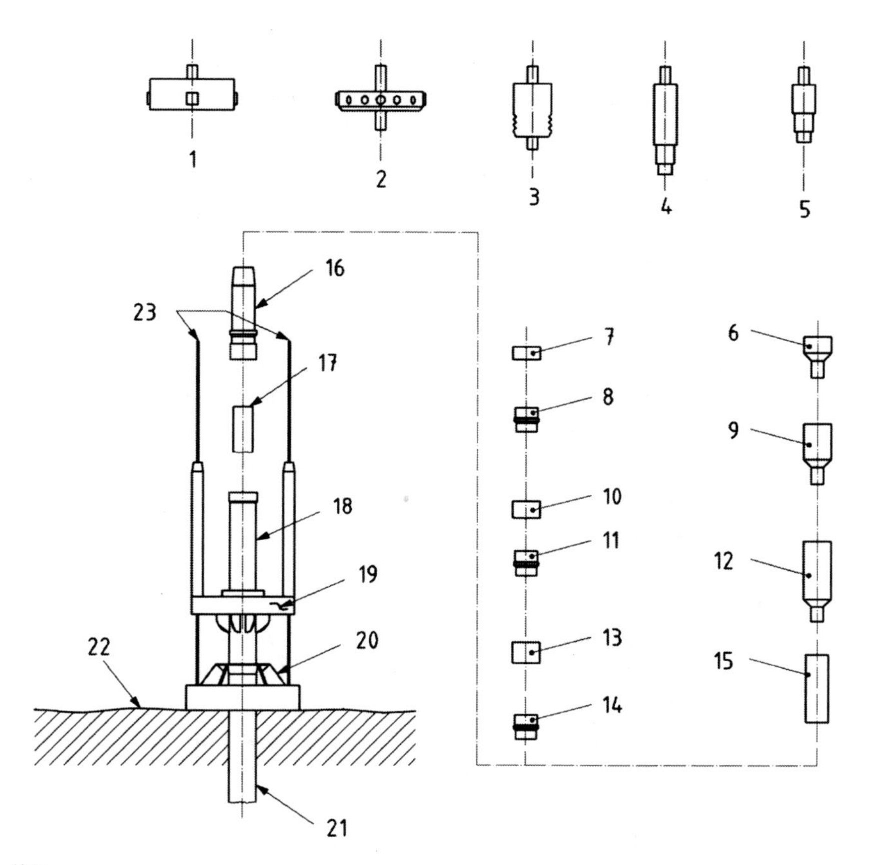

Key

1 TGB running tool
2 762 mm (30 in) housing running tool
3 high-pressure housing running tool
4 casing hanger running tool (drillpipe or fullbore)
5 test tool
6 177,8 mm (7 in) wear bushing
7 244,5 mm × 177,8 mm (9 5/8 in × 7 in) annulus seal assembly
8 177,8 mm (7 in) casing hanger
9 244,5 mm (9 5/8 in) wear bushing
10 339,7 mm × 244,5 mm (13 3/8 in × 9 5/8 in) annulus seal assembly
11 244,5 mm (9 5/8 in) casing hanger
12 339,7 mm (13 3/8 in) wear bushing
13 508,0 mm × 339,7 mm (20 in × 13 3/8 in) annulus seal assembly
14 339,7 mm (13 3/8 in) casing hanger
15 housing bore protector
16 high-pressure wellhead housing
17 casing [normally 508,0 mm (20 in)]
18 low-pressure conductor housing [normally 762,0 mm (30 in)]
19 PGB
20 TGB
21 762,0 mm (30 in) conductor casing
22 sea floor
23 guidelines

Figure 22-2 Subsea Wellhead System Building Blocks *(Courtesy of API RP17A)*

- Be of field proven design, as far as possible, and designed to be installed with a minimum sensitivity to water depth and sea conditions.

22.3.2. Operation Requirements

The subsea wellhead system should:

- Provide the ability to install the following equipment in the same trip: the production guide base (PGB), the 36-in. conductor, and the low-pressure conductor housing. The assembly should be designed to be preinstalled in the moon-pool prior to being run subsea.
- Allow for jetting operations for the casing pressure and for the drill and cement as a contingency case.
- Include provision for efficient discharge of the drill cuttings/cement returns associated with the drilling operations.
- Provide a bore protector and wear bushings to protect the internal bores of the wellhead system components during drilling, completion, and retrieval operations.
- Ensure that all seals and locking arrangements can be tested *in situ*.
- Ensure that the complete packoff/seal assembly can be retrieved and replaced in the even of a failed test.
- Ensure that all permanent seals are protected during the running phase and remotely energized after landing.
- Be designed such that the running string with wellhead tools and components will not snag or be restricted when running in or being pulled out of the hole.
- Provide tooling that allows for seal surfaces to be cleaned after cement operation and prior to setting seal assemblies without pulling the running string; that is, the tool should allow cleaning of seal surfaces by circulation prior to pack offsetting.
- Be designed to allow for landing of the casing hanger and installation of the seal assembly and removal of the same (in case of failure) in a single trip. Multipurpose tools should as far as possible be used to avoid pulling of the running string for tool change-outs.
- Allow for large enough flow-by areas, and particle size, at the casing hanger and casing hanger running tool level (to be compared with the clearance between ID of the previous casing and the OD of the collars of the attached casing).
- Be designed to allow for testing of the BOP without having to pull the wear bushing.

- Provide guidance for equipment entering the well during drilling, completion, and subsequent operations.
- Allow for safe and efficient retrieval of all installed equipment during permanent abandonment of the well.
- Be designed to allow access for both work class and inspection ROVs. ROV grab bars should be included wherever an ROV operation is defined to provide stabilized working conditions for the ROV.

22.3.3. Casing Design Program

For subsea wellhead system design, it is imperative to consider casing growth, which will affect the wellhead load intensively. Generally, the casing will connect with a tubing hunger with a screw thread on the top and be fixed with cement. It is a comparatively simple structure. The most important parameters for casing design are wall thickness and length. Based on guaranteeing the intensity and reliability, there is a growing need to consider conversative materials and resources with the increase in operating costs and withstand the cyclic swing of oil prices. Subsea oil development is not significant to survival when oil prices are very low.

There are many typical casing design examples to refer to. See Figures 22-3 and 22-4, which are typical wellhead casing schemes used in North Sea. As the oil explorations move into deepwater drilling of high-pressure and high-temperature wells, it has become more and more popular and necessary to increase the scope of the optimization by encompassing more design parameters into the analysis. Consequently, numerous variables can be taken into account within the design spectrum. However, usually it is imperative to integrate all of the subsea components in the analysis of the casing design, which we will elaborate on in wellhead reliability analysis section.

Normally, a lot of different design parameters are proposed under the same conditions. To counter this problem, a dimensionless parameter called the wellhead growth index (WHI) has been developed, which greatly aids the ability to determine the severity of the design and a means of describing the severity of wellhead growth, without sacrificing any rigor. WHI encapsulates the annuli fluid expansion and wellhead growth and it provides a simple practical way to view the casing movement and fluid expansion in the annuli during the course of drilling and also during the production phase of the well. It is defined as the ratio of the annulus fluid expansion of the casing to the actual volume of the exposed segment above the top of the cement [1].

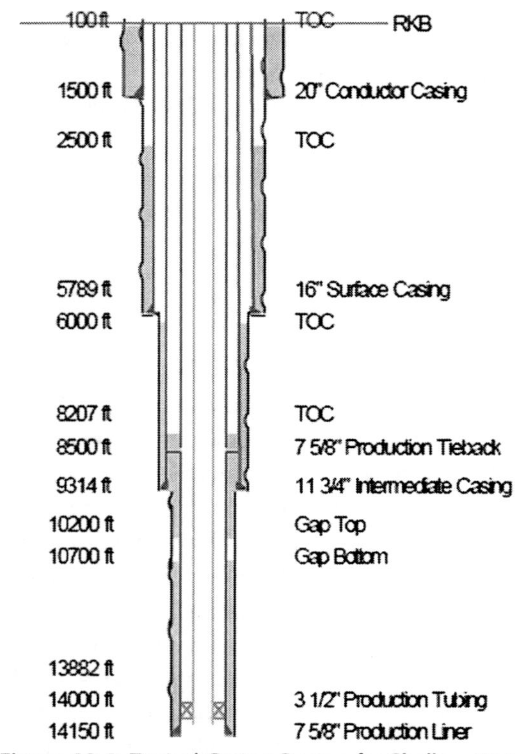

Figure 22-3 Typical Casing Design for Shallow Water

The annulus fluid expansion includes the unconstrained volume change and the annulus volume change due to annulus pressures. Wellhead growth gives an estimate of the circumferential and axial strain on the casings. With the circumferential and lateral strain, the total volume of the expansion of all casing string for all casing segments is given by:

$$\Delta v = \sum_{j-1}^{m} \sum_{i-1}^{n} \left[\frac{\pi}{4} (2d\Delta dl + d^2 \Delta l) + v_a \right]_{i,j} \tag{22-1}$$

The total area of the annulus cross section for each casing string is given by

$$a = \sum \sum \frac{\pi}{4} (D^2) \mid_{i,j} \tag{22-2}$$

where:

d = casing diameter, in

D = annulus gap between the casings, in

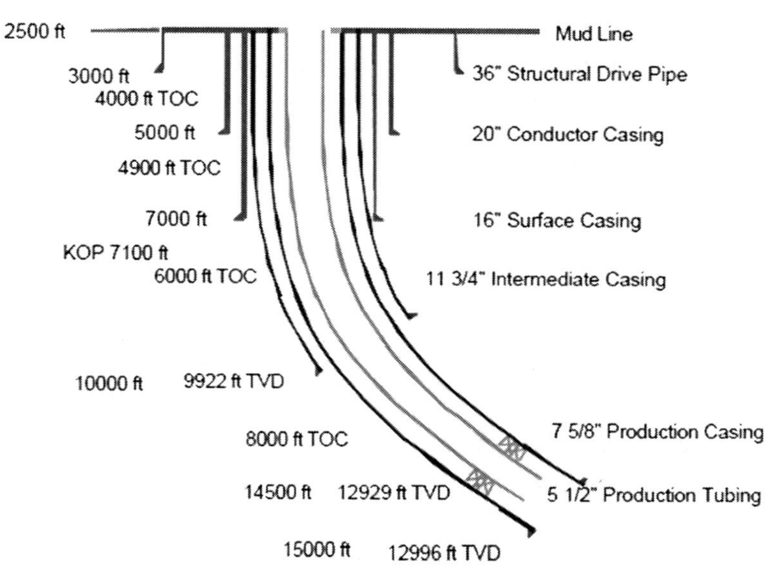

Figure 22-4 Typical Casing Design for Deep Water

l = segment length of the exposed casing, ft
n = number of exposed casing sections,
m = number of casings,
v = annulus volume, ft^3
v_2 = volumetric change due to annulus pressures
Δd = change in the casing diameter, in
Δl = wellhead growth, in
Δv = change in the annulus volume, ft^3
WHI = wellhead growth index.

Using Equations (22-1) and (22-2) with approximations, the wellhead growth index for multiple casing string is given by

$$\text{WHI} = \frac{\sum_{j-1}^{m} \sum_{i-1}^{n} \left[\frac{\pi}{4}(2d\Delta dl + d^2\Delta l) + v_a\right]_{i,j}}{\sum\sum \frac{\pi}{4}(D^2)\big|i, j^l} \tag{22-3}$$

where:

d	= casing diameter, in
D	= annulus gap between the casings, in
l	= segment length of the exposed casing, ft
n	= number of exposed casing sections,
m	= number of casings,
v	= annulus volume, ft^3
v_2	= volumetric change due to annulus pressures
Δd	= change in the casing diameter, in
Δl	= wellhead growth, in
Δv	= change in the annulus volume, ft^3
WHI	= wellhead growth index.

WHI gives a quantitative predictive capability for interpreting the calculation results. The higher the value of WHI, the higher the severity of the casing design involved. Calculation of WHI at different stages of the casing design will aid in comparing the relative rigorousness of the overall casing design.

22.3.4. Wellhead Components

A subsea wellhead system mainly consists of wellhead housing, conductor housing, casing hungers, annulus seals, and guide base (TGB and PGB). The high-pressure wellhead housing is the primary pressure-containing body for a subsea well, which supports and seals the casing hangers, and also transfers external loads to the conductor housing and pipe, which are eventually transferred to the ground.

Figure 22-5 illustrates the typical $18^3/_4$-in. subsea wellhead components.

22.3.4.1. Wellhead Housing

The wellhead housing is the primary housing supporting both the intermediate and production casing strings. In API 17D [2], very detailed profiles are introduced. Figure 22-6 is a schematic of a typical wellhead housing.

Two kinds of subsea analyses are necessary to consider in the wellhead housing design procedure: load stress analysis and thermal analysis. The hanger landing shoulder will sustain loads from the tubing hanger. Normally, a finite element analysis (FEA) and riser fatigue analysis will be performed to verify the design capacities.

In addition, thermal analysis is performed to determine the temperature profiles through the system so that temperature derating can be accounted

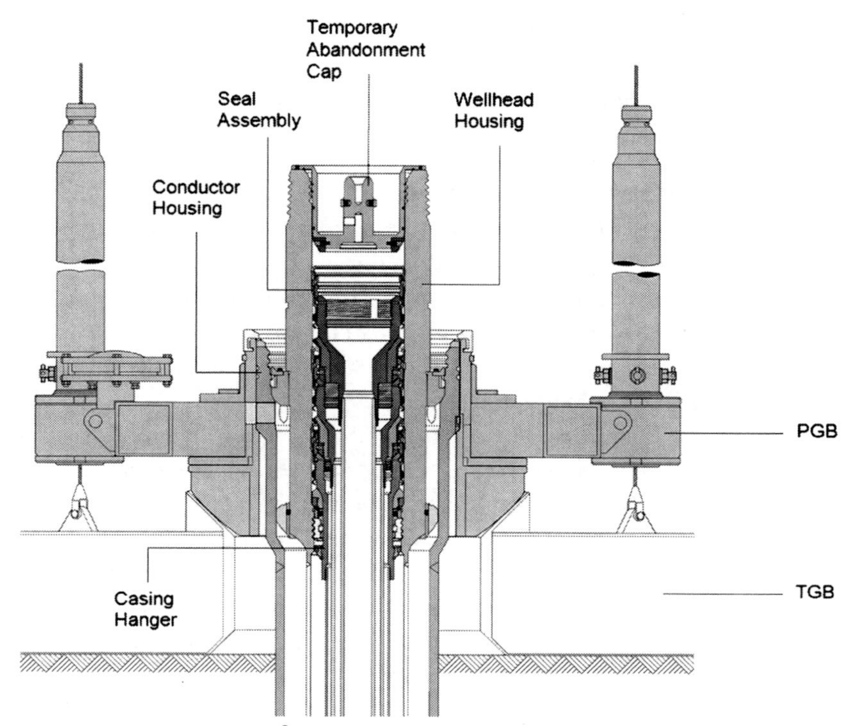

Figure 22-5 Typical $18^{3}/_{4}$-in.Subsea Wellhead System *(Courtesy of Dril-Quip)*

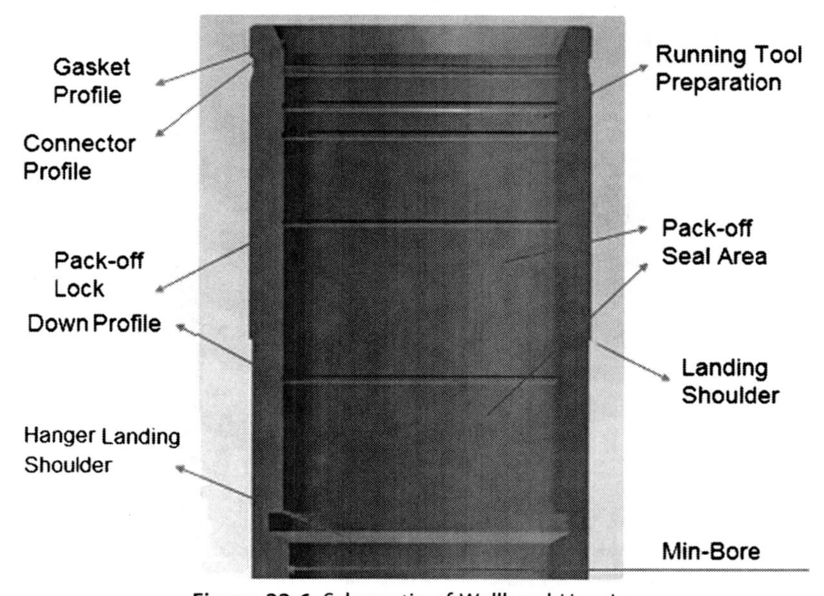

Figure 22-6 Schematic of Wellhead Housing

for as appropriate. A prototype wellhead should be tested to the test pressure as well as loaded with simulated casing loads and BOP test pressure with hangers in place simulating the real production environment in total without experiencing any permanent deformation.

22.3.4.2. Intermediate Casing Hanger

The intermediate casing hanger for the system lands in the first hanger position in the lower portion of the wellhead. Figure 22-7 illustrates the profile for an intermediate casing hanger. The casing hanger can nominally be for either a 16- or 13-$^5/_8$-in. casing. The casing hanger features an expanding load ring that lands into the wellhead seat segments to suspend the casing and BOP pressure end loads.

The analysis is performed on existing field-proven hangers of similar designs to compare stress levels. Reliability data also will be collected from similar equipment and lessons learned should be incorporated into the design. Further reliability work is performed during FMECAs (Failure Mode effect and Criticality Analysis). Finally, through testing, the hanger is positively proof tested to the design loads, pressures, and combined loads in the exact sequence in which they would be applied in the field, without experiencing any permanent deformation.

22.3.4.3. Production Casing Hanger

The production casing hanger for the system lands in the second hanger position. The casing hanger can nominally be either for an 11$^3/_4$- or 10$^3/_4$-in.

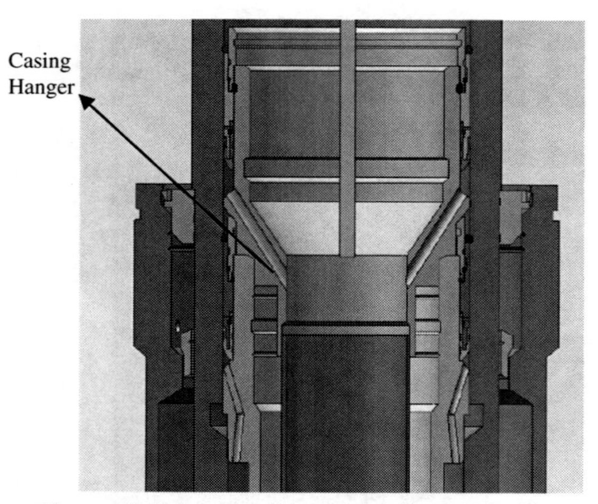

Figure 22-7 Intermediate Casing Hanger Profile

casing. The casing hanger features an expanding load ring that lands into the second set of wellhead seat segments to support the casing and BOP pressure end loads. Using the approach above, detailed stress analysis and classical calculations should be performed in fashion similar to the analysis performed for the intermediate hanger.

22.3.4.4. Lockdown Bushing

The lockdown bushing is used to permanently hold the production casing hanger in place so that the annulus seal assembly locked to the hanger does not move and get damaged during start-up/shutdown operations. This system's lockdown bushing has a rated lockdown capacity of 3.2 million pounds. It is installed using a related tool using full-open water operations.

The design approach for this piece of equipment is handled as same as other components of wellhead. Design calculations and finite element analysis are performed in conjunction with the gathering of reliability lessons learned and FMECA to confirm the integrity of the design up-front. Tests should be performed to confirm that the lockdown bushing and tool can definitely function as designed, and load testing is performed to confirm the load capacity.

22.3.4.5. Metal-to-Metal Annulus Seal Assembly

The metal-to-metal annulus seal assembly is used to seal off the casing string annulus pressure from the bore pressure to isolate geological formations from one another. The typical profile is showed in Figure 22-8.

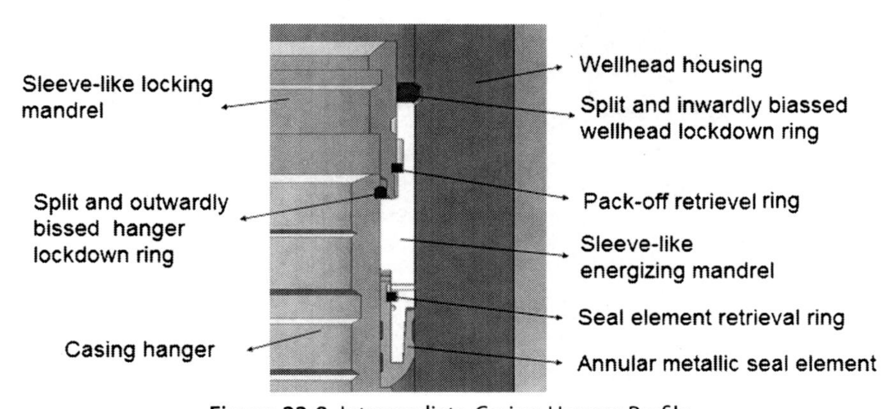

Figure 22-8 Intermediate Casing Hanger Profile

The metal-to-metal assembles usually needs practice testing to confirm that it could withstand the high pressure and temperature.

22.3.4.6. Elastomeric Annulus Seal Assembly

The elastomer seal assembly is used in an emergency when the primary metal seal fails to function should the bore of the wellhead or casing hanger have a deep scratch. The elastomer seal assembly seals in a different vertical location in the wellhead, which should seal away from the damaged area.

22.3.4.7. Casing Hanger Running Tools

The intermediate and production casing hanger running tools run the casing hangers and set the annulus. These tools normally are designed using the same technology, lessons learned, and in many cases the same parts as the standard 15-ksi running tool, as shown in Figure 22-9.

22.3.4.8. BOP Test Tool

The BOP test tool is designed with an approach that is similar that for other components and tools in this system. It is used to test the BOP in terms of future formation pressure that the operator is currently drilling into, and is also used to run and retrieve wear bushings, as shown in Figure 22-10.

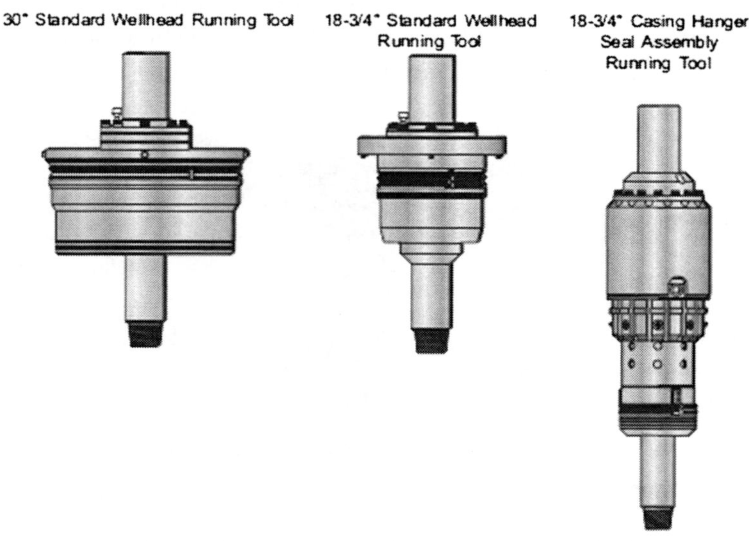

Figure 22-9 Casing Hanger Running Tool *(Courtesy of Dril-Quip)*

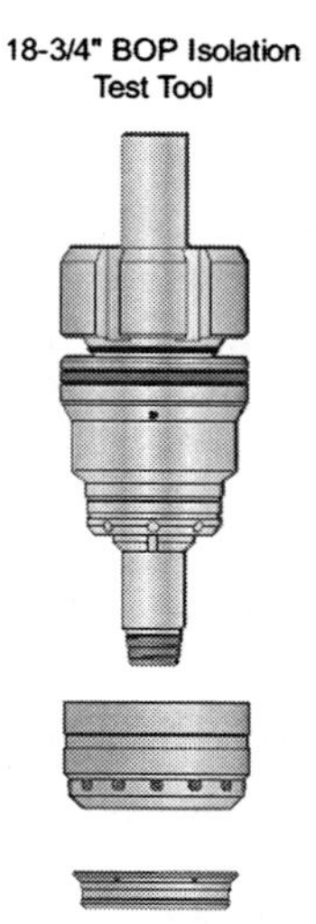

Figure 22-10 BOP Isolation Test Tool *(Courtesy of Dril-Quip)*

22.3.4.9. Isolation Test Tool

The isolation test tool is used to test the pack-off per MMS (Mineral Management Service) requirements while simultaneously isolating the BOP (Blowout Preventer) stack/riser. The tool operates with a simple straight-in/straight-out approach. Once set in position with the weight down, drill string pressure is applied up to a target test pressure of 20 ksi.

22.3.4.10. OD Wear Bushing and OD BOP Test Tool

The $13^5/_8$-in. OD wear bushing and $13^5/_8$-in. OD BOP test tool are key tools to expediting completions while the $18^3/_4$-in., 20-ksi BOP is being developed for the industry. This wear bushing and running tool combination is run through a $13^5/_8$-in., 20-ksi BOP that latches either to the wellhead or 20-ksi tubing head; they allow for BOP tests. The wear bushing protects all seal surfaces including those on the production casing hanger while drilling/logging operations are performed.

22.3.5. Wellhead System Analysis

An analysis approach based on a simple linear elastic model (SLEM) of multistring wellbore systems is described in this section. This approach can be used to facilitate understanding and analysis of various complex wellhead load events in terms of a simple linear model. Although the simple linear model is limited because nonlinear effects due to buckling are not considered, wellhead loads and displacement behave as a linear system to a good first-order approximation in many realistic situations. Loads on conductor and surface casings are considered in particular, since many surface/rig events tend to impact these strings more directly. Also, they must bear the primary load burden due to their greater relative stiffness and tendency to displace linearly.

22.3.5.1. Basic Theory and Methodology

A brief review is presented of the SLEM for multistring wellhead displacements and loads for a free-standing wellhead system. Use of the SLEM in a systematic analysis of wellhead load events is also discussed.

Recall that an OCTG casing or tubing string typically operates within the material's linearly elastic region. The relation of axial stress strain is governed by Hooke's law and the string material elastic modulus, E. Since the tubular string functions as a prismatic bar, Hooke's law can be expressed as follows:

$$\delta = \frac{PL}{EA} \tag{22-4}$$

where δ is the resultant displacement subject to an applied axial load P on a string of free length L and cross-sectional area A.

This may be expressed in terms of a stiffness or spring constant k as is familiar for linear elastic springs:

$$P = k \cdot \delta \tag{22-5}$$

where stiffness k is given by:

$$k = \frac{EA}{L} \tag{22-6}$$

For an offshore platform or jack-up well with a free-standing wellhead structure, the wellhead is free to move vertically, and all casing and tubing strings landed in the wellhead are subject to uniform wellhead displacement. The system is statically indeterminate and must be analyzed as a composite system. For a wellbore with n strings linked at the wellhead (not including downhole liners or outer casings not in contact with the wellhead), the composite system stiffness k_{sys} is the sum of the stiffness from each string:

$$k_{sys} = \frac{E_1 \cdot A_1}{L_1} + \frac{E_2 \cdot A_2}{L_2} + \cdots + \frac{E_n \cdot A_n}{L_n} = \sum_{i=1}^{n} k_n \tag{22-7}$$

Note that for a casing or tubing string $i = q$ composed of w sections with changes in geometry or material, the composite stiffness of that particular string is given by the following equation:

$$\frac{1}{k_q} = \frac{1}{k_{q,1}} + \frac{1}{k_{q,2}} + \cdots + \frac{1}{k_{q,w}} = \sum_{z=1}^{w} \frac{1}{k_{q,z}} \tag{22-8}$$

Each string landed into the wellhead contributes an axial load, P_i. To satisfy mechanical equilibrium, the sum of all axial loads at the wellhead must be zero:

$$\sum_{i=1}^{n} p_i = 0 \tag{22-9}$$

If m static wellhead loads W_j, such as the weight of the wellhead, BOPs, or Christmas trees and also upward forces applied by rig tension systems, are applied to the system, then the equation of equilibrium now requires that the sum of all string axial loads balance the net static load:

$$\sum_{i=1}^{n} P_i + \sum_{j=1}^{m} W_j = 0 \tag{22-10}$$

When any load W is applied to the system, the wellhead and all strings landed in the wellhead will undergo a uniform displacement, δ_{sys}, which is determined by the system stiffness and the applied load (upward displacement is positive):

$$\delta_{sys} = -W/k_{sys} \tag{22-11}$$

Based on this uniform displacement, the applied load W is thus distributed onto each string in proportion to its relative individual stiffness. As a result, the axial load of each string is changed by an incremental load ΔP_i:

$$\Delta P_i = k_i \cdot \delta_{sys} \tag{22-12}$$

To model the change in wellhead displacement and actual wellhead loads for each string throughout the life of the well, the preceding equations and conditions must be applied to each step of the well construction process as well as subsequent states during production operations. The global datum point for wellhead displacement is the flange height of the outer casing string in its initial free-standing state. The methodology is simplified by utilizing the initial and subsequent load states for each string considered in isolation from the overall wellbore system. An axial load result can be calculated using a single-string model based on a fixed nominal wellhead condition. Each single-string load can then be added to the multistring wellhead system and the appropriate redistribution of axial loads can be determined based on the discussion and equations above. The initial axial load $P_{i,o}$ contributed by each string added to the system corresponds to the hook load when it is landed in the wellhead. This is calculated from cumulative buoyed weight based on its nominal length, tubular weight, mud and slurry densities, and wellbore deviation. In addition, the landed weight of each string may include overpull or slackoff.

For any new load state S such as a well life production operation, each string undergoes a change in axial load $\Delta P_{i,S,}$ When changing in operation conditions such as temperature or pressures.

For example, after commencement of production, the wellbore heats up and any given casing will tend to expand axially due to a net increase in temperature relative to the initial state. Similarly, during a cold injection or stimulation operation, each string will tend to go into increased tension due a thermal contraction. The resultant change in axial load and the associated unconstrained axial displacement for each string considered in isolation may be calculated using a standard single-string force/displacement model such as in Mitchell [3] (1996).

As each new string is added to the system, the wellhead undergoes a uniform displacement as discussed above and the new string landed weight will be distributed among the outer strings. If the cement has set before the new string is landed, the new string will also "slump" somewhat to bear a portion of its own weight. Likewise, for any changes in the string state during operation, the subsequent changes in axial load are redistributed across the multistring system based on relative stiffness.

A procedural method based on the foregoing discussion of simple linear elasticity, the SLEM can be summarized as follows:

1. For operational load step S, identify the strings $i = 1$ to n already installed or to be landed in the current step and calculate the current composite system stiffness:

$$k_s = \sum_{i=1}^{n} k_n \tag{22-13}$$

2. For each string $i = 1$ to n, determine the load change ΔP relative to $P_{i,o}$ based on single-string analysis; for a string to be landed in the current step S, define $\Delta P_{i,S} = P_{i,o}$.
3. Identify static wellhead loads $j = 1$ to m to be applied in the current load step S: $W_{1,S}$, $W_{2,S}$, ..., $W_{m,S}$.
4. Calculate the current incremental wellhead system displacement δ_S as follows:

$$\delta s = \left[\sum_{i=1}^{n} \Delta P_{i,s} + \sum_{j=1}^{m} W_{j,s} \right] / k_s \quad (22\text{-}14)$$

5. For each string, calculate the final redistributed multistring axial load based on the current load step:

$$P_{i,S} = P_{i,S-1} + \Delta P_{i,S} - \delta_s \cdot k_i \quad (22\text{-}15)$$

22.3.5.2. Static Wellhead Loading

In the first sensitivity study, incremental wellhead displacements were calculated for a range of arbitrary static wellhead loads using both SLEM Equation (22-11) and also state-of-the-art stress simulation software with advanced numerical modeling of multistring system behavior. There is nearly exact agreement between the SLEM solution and the numerical model, which does account for any nonlinear effects such as buckling. This indicates that the multistring system reacts in an essentially linear fashion to static wellhead loads. This result holds even for loads on the order of the axial rating of the outer casing.

This is consistent with the general observations noted above, in that the majority of the load is distributed onto the outer casing, which is either unbuckled or for which buckling strain is negligible relative to elastic strain.

Table 22-1 summarizes the system stiffness values used in the SLEM calculation. In this example, the outer surface casing accounts for 92.1% of the composite system stiffness.

22.3.5.3. Thermal Induced Loading

A second sensitivity study was carried out to investigate the effects of thermal induced loads. Heating of the wellbore tends to induce buckling of the inner casings and tubing. This is because the outer casing will tend to

Table 22-1 Buckling Sensitivity Example: SLEM Multistring System Values

String No.	Section	*A* (in.^{2})	*E* (psi)	L_ft (ft)	*K* (lbf/in.)		K/K_{sys} (%)	K_{sys} (lbs/in.)
1	1	38.0427	30E+06	500	190214		92.1%	206529
2	1	20.7677	30E+06	5000	10384		5.0%	
3	1	31.5342	30E+06	1000	78836	4094	2.0%	
	2	15.5465	30E+06	9000	4318			
4	1	8.4494	30E+06	11500	1837		0.9%	

restrain upward well growth driven primarily by the compressive force of the inner strings; as a consequence it will tend to go into tension. Production and injection operations were defined appropriately to focus on thermal effects as opposed to production pressures.

The initial operating state is steady-state production followed by a long duration of shutdown at roughly constant SITHP. After the wellbore returns to geostatic temperatures, kill operations are undertaken that cool the wellbore even more and also reduce wellhead pressure to a minimal value. The kill is then continued as a long-term, low-rate mud injection.

Figure 22-11 shows a plot of cumulative wellhead displacement versus time over the duration of the shutdown and injection operations using SLEM and the numerical modeling software. The datum reference point is the as-built wellhead elevation after landing all strings. The SLEM matches the overall trend of wellhead movement, and the difference in magnitude with the numerical model is negligible.

If the conductor is added to the system by assuming the wellhead to be fixed to the outer flange, then the overall system becomes stiffer. The increase in system stiffness results in a greatly reduced range of movement and a tighter clustering of the different model results.

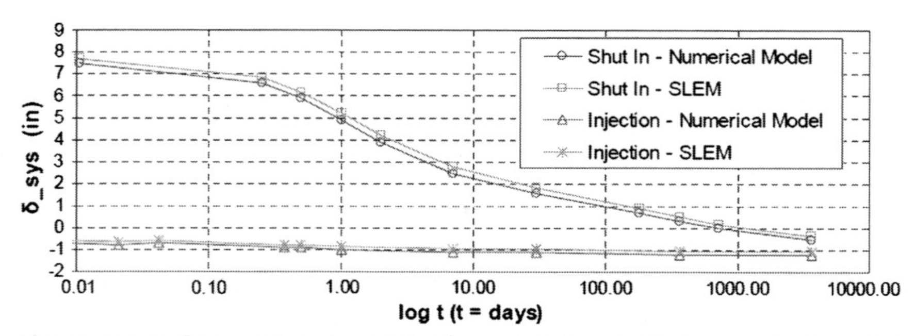

Figure 22-11 Comparative Sensitivity-Thermal Induced Displacement: Numerical Model versus SLEM

22.3.5.4. Wellhead System Reliability Analysis

When drilling a well in deep water, the wellhead will bear the intricate forces from the environment and the drilling operation, which will affect the integral reliability of the wellhead system. Normally an FEA model is built to analyze the wellhead system reliability in terms of taking all factors into consideration, including loading from the marine environment and the drift of a drilling vessel or platform, and nonlinear response between casing string and soil stratum. Figure 22-12 illustrates the scheme of a wellhead system subjected to different loads.

Figure 22-13 shows the force diagram for a wellhead during the drilling operation, where F_x is the sum of the external force on the riser in the x direction, F_y is the sum of the external force on the riser in the y direction,

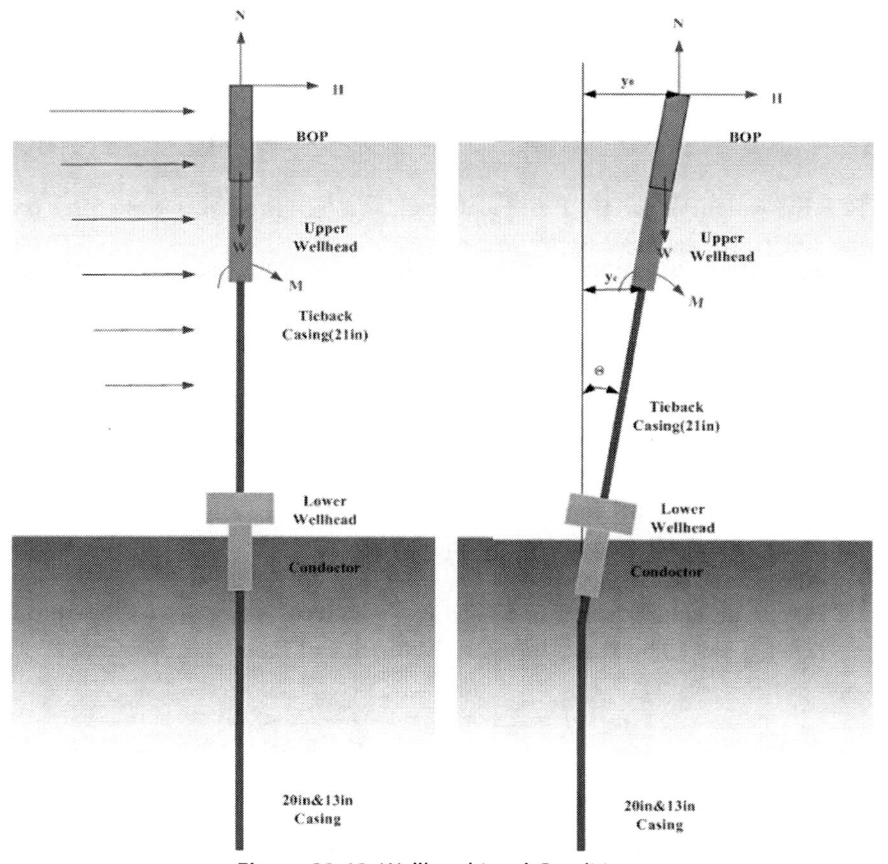

Figure 22-12 Wellhead Load Conditions

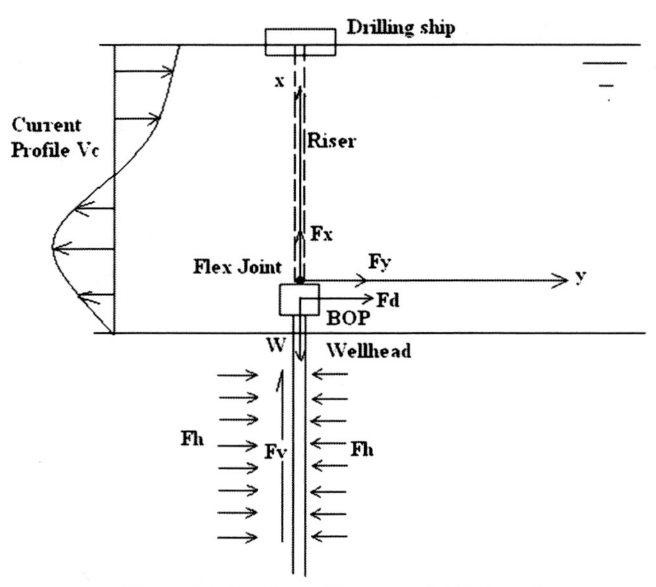

Figure 22-13 Force Diagram of Wellhead

W is the weight of the BOP and wellhead, and F_d is the direct wave force on the BOP and wellhead upper seabed.

The casing string's displacement equation can be explained as follows:

$$\frac{d^2}{d^2x}\left[EI(x)\frac{d^2y}{d^2x}\right] + \frac{d}{dx}\left[N(x)\frac{dy}{dx}\right] + Dc(x)p(x,y) = q(x) \qquad (22\text{-}16)$$

where

$EI(x)$: bending stiffness of combination of casing string, cement ring, etc., kNm^2;

$q(x)$: unit external force, kN/m;

$Dc(x)$: external diagram of casing string, m;

$N(x)$: axial force, kN;

$P(x,y)$: unit (area) horizontal soil force, which is determined by Equations (22-17) and (22-18) according to Matlock and Reese [4], kN/m^2.

$$\begin{cases} p_u = 3C_u + \gamma X = \dfrac{\varepsilon C_u X}{D_c}, X < X_r \\ p_u = 9C_u, X \geq X_r \\ X_r = \dfrac{6C_u D_c}{\gamma D_c + \varepsilon C_u} \end{cases} \qquad (22\text{-}17)$$

$$\begin{cases} \dfrac{P}{P_u} = 0.5\left(\dfrac{y}{y_{50}}\right)^{\frac{1}{3}}, \dfrac{y}{y_{50}} < 8 \\ \dfrac{P}{P_u} = 1.0, \dfrac{y}{y_{50}} \geq 8 \end{cases} \quad (22\text{-}18)$$

where

P_u : critical soil force;

x: distance under the seabed;

y: horizontal displacement of casing string;

C_u : undrained shear strength of soil;

ε: a coefficient 0.25 to 0.5 ;

γ: submerged weight of soil.

After obtaining the displacement of the casing string, the bending moment can also be obtained and used to judge the stability of the wellhead.

22.3.6. Guidance System

22.3.6.1. Guide Base Options

Guidance onto the wellhead system is dependent on the actual type of development concept (template or satellite well) by means of the following guide base options:

- *Concept 1:* multiwell template with production manifold (integral or modular) and template-mounted production guide base;
- *Concept 2:* individual wells connected to subsea manifold center;
- *Concept 3:* two slots structures;
- *Concept 4:* individual wells in daisy chain configuration.

Three different types of guide base are proposed for these four concepts:

- Twin production guide base;
- Template-mounted guide base;
- Single-well or cluster production guide base.

22.3.6.2. General Requirements

The production guide base is an element attached to and installed with the low-pressure housing of the wellhead. All guide bases should comply with the following general requirements:

- All guide bases should orient and lock to the 30-in. conductor housing incorporating an antirotation device.
- In all type of wells, it should provide orientation to the tree to be in the right position for connecting the flowlines and hydraulic and electric lines.

- The ability to reuse some of the existing exploration and appraisal wells is to be included within the design either as part of the overall design or as part of a specific modification.
- The guide base and receptacle should be designed to withstand all vertical and horizontal loads associated with deployment, installation and retrieval of all modules including BOPs, without any permanent deformation. The base should tolerate landing and retrieval of modules with angular misalignment relative to the centerline of the wellhead.
- The guide base should be designed to prevent snagging of an ROV and associated umbilical.
- All guide bases should be designed for a 36-in. conductor pipe to be run through the guide base at the surface.
- The production guide base should be installed with the conductor pipe and latched and locked to the 30-in. housing. It should provide an antirotation device. The locking and antirotation mechanism should not be affected by any cutting or cement, during drilling and cementing on the associated two first phases of the well. The PGB and the template, if to be considered, should facilitate the evacuation of cuttings and cement from the next drilling phase, in order to preserve the previously installed equipment.
- The orientation device will allow the guide base to be installed in multiple orientation positions.
- The configuration should allow retrieving the tree with the jumper connector remaining in place.
- The guide base should be designed to accommodate rigs that have a funnel-down guidance arrangement. However, a funnel-up configuration should be proposed as an option.
- The guide base and low-pressure housing should be designed to provide a support to suspend the conductor pipe in the moon-pool prior to running.
- The conductor pipe should be jetted and the guide base arrangement should allow for installing and retrieving all of the tools used for this purpose.
- Drilling and cementing the conductor pipe is part of the contingency plan. The guide base design should cope with such operations.
- The structure and the conductor locking mechanism with the antirotation device should accommodate all combined loads resulting from the drilling and production operation such as bending and fatigue loads and thermal and pressure effects.

- A device (bull's-eye for example) should be installed on the guide base to check the conductor pipe verticality at the installation.
- The guide base should include all equipment needed for ROV intervention or connection tools.
- The guide base will be equipped with a cathodic system, which should be dimensioned to cater to both the guide base and the well (well drain) for the entire life of the well.
- The guide base should be designed to avoid any clash with the different BOP planned to be used in the project in any orientation.
- A clump weight should be added to the guide base in order to handle it in a horizontal position.
- A dedicated tool, drill pipe should be provided for retrieving/reorienting/reinstalling the guide base. This system will be equipped with a secondary backup release which is operated by ROVs.
- If relevant, the PGB should include retrievable protection caps for vertical mandrels. These should be installable and retrievable by ROV. These caps provide temporary protection of bores, seal areas, and hydraulic couplers against dropped objects and environmental impact, per ISO 13628-4 [5]. The protection caps should be used during transportation and storage on land and offshore.

22.3.6.3. Twin Production Guide Base (Twin-PGB)

The twin-PGB should provide a second well slot adjacent to an existing wellhead, by means of a base structure that can be landed and locked to the existing well conductor housing by use of the guide base locking profile. It should satisfy the same functional and design requirements as the PGB defined above, with the following additions/modifications:

- The twin-PGB includes a second well slot with the same functionality as a TMGB.
- The twin-PGB includes a two-well manifold with piping arrangement, ROV-operable isolation valves, hubs for vertical tree connection, and horizontal inboard tie-in facilities for production lines and service umbilical, as well as an electrical connection system between the inboard hub and the trees.
- The twin-PGB will not be installed with the conductor pipe.

22.3.6.4. Template-Mounted Guide Base (TMGB)

In a template configuration, the function of the guide base will be limited to the orientation and the elevation of the Xmas tree. Providing all of the functions listed before are included in the design, the PGB, in that special case could be installed with the template. The TMGB should include the equipment necessary to facilitate the interface between the Xmas tree and the template/manifold, such as:

- Equipment for guidance;
- Alignment and suspension of the wellhead;
- Flowline/service umbilical inboard connection (if relevant).

22.3.6.5. Single-Well or Cluster Production Guide Base (SWPGB)

In addition to the above general requirements and the PGB requirements, the SWPGB should satisfy the following:

- In a cluster or single-well configuration, the production guide base should orient the tree and support the tree-to-jumper connection. In such a case, the PGB should be designed so that it can be either retrieved to the surface or reoriented.
- The PGB should provide a support for deepwater subsea telemetry to perform the measurements needed for the jumper installation.

22.4. SUBSEA XMAS Trees

22.4.1. Function Requirements

Typical function requirements for subsea Xmas trees include:

- Direct the produced fluid from the well to the flowline (called production tree) or to canalize the injection of water or gas into the formation (called injection tree).
- Regulate the fluid flow through a choke (not always mandatory).
- Monitor well parameters at the level of the tree, such as well pressure, annulus pressure, temperature, sand detection, etc.
- Safely stop the flow of fluid produced or injected by means of valves actuated by a control system.
- Inject into the well or the flowline protection fluids, such as inhibitors for corrosion or hydrate prevention.

22.4.2. Types and Configurations of Trees

22.4.2.1. Vertical Xmas Tree

The master valves are configured above the tubing hanger in the vertical Xmas tree (VXT). The well is completed before installing the tree. VXTs are applied commonly and widely in subsea fields due to their flexibility of

Figure 22-14 Xmas Vertical Tree *(Courtesy of FMC)*

installation and operation. Figure 22-14 shows a vertical Xmas tree being lowered subsea.

Figure 22-15 illustrates the schematic of a typical vertical tree. The production and annulus bore pass vertically through the tree body of the tree. Master valves and swab valves are also stacked vertically. The tubing hanger lands in the wellhead, thus the subsea Xmas tree can be recovered without having to recover the downhole completion.

22.4.2.2. Horizontal Xmas Tree

Another type of subsea Xmas tree developed rapidly in recent years is the horizontal tree (HXT). Figure 22-16 shows a horizontal tree made by FMC.

Figure 22-17 shows the schematic of a horizontal tree. The valves are mounted on the lateral sides, allowing for simple well intervention and tubing recovery. This concept is especially beneficial for wells that need a high number of interventions. Swab valves are not used in the HXT since they have electrical submersible pumps applications.

The key feature of the HXT is that the tubing hanger is installed in the tree body instead of the wellhead. This arrangement requires the tree to be installed onto the wellhead before completion of the well.

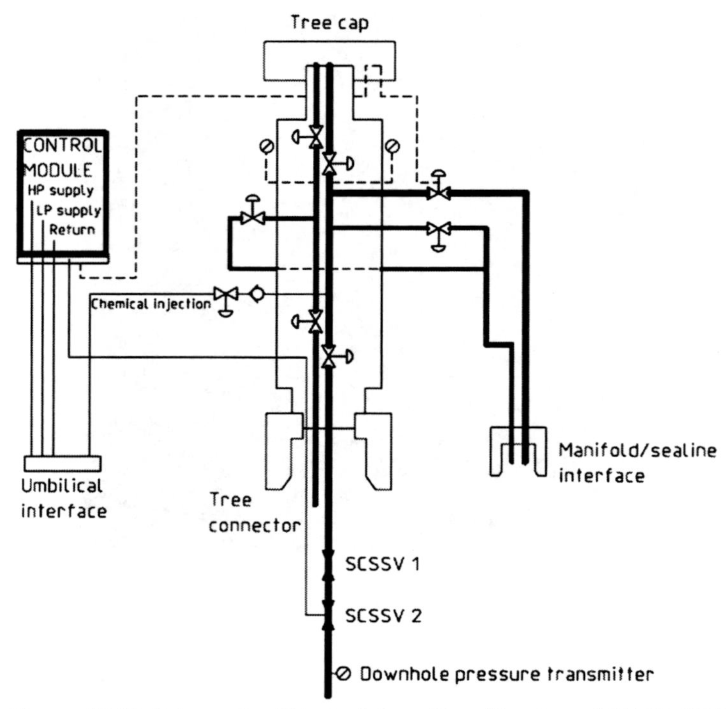

Figure 22-15 Schematic of Vertical Xmas Tree *(Courtesy of API RP 17A)*

Figure 22-16 Horizontal Xmas Tree *(Courtesy of FMC)*

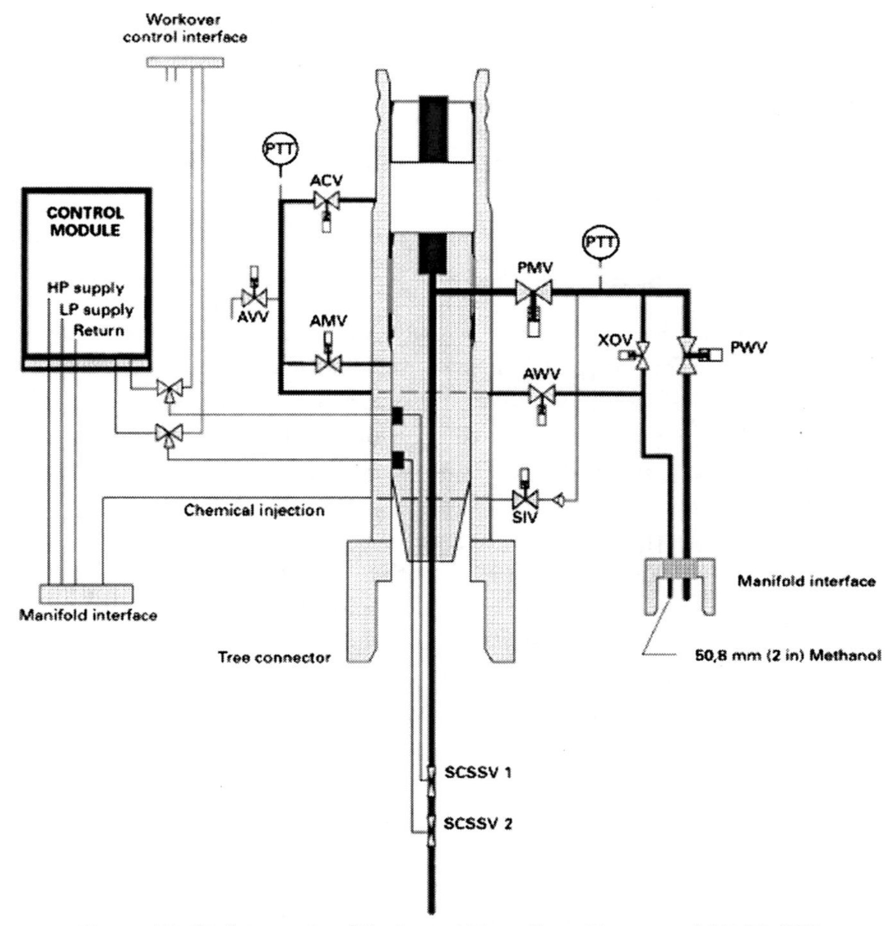

Figure 22-17 Schematic of Horizontal Xmas Tree *(Courtesy of API RP 17A)*

22.4.2.3. Selection Criteria

In the selection of a horizontal tree (HXT) or a vertical tree (VXT), the following issues should be considered:

- The cost of an HXT is much higher than that of a VXT; typically the purchase price of an HXT is five to seven times more.
- A VXT is larger and heavier, which should be considered if the installation area of the rig is limited.
- Completion of the well is another factor in selecting an HXT or VXT. If the well is completed but the tree has not yet been prepared, a VXT is needed. Or if an HXT is desired, then the well must be completed after installation of the tree.

- An HXT is applied in complex reservoirs or those needing frequent workovers that require tubing retrieval, whereas a VXT is often chosen for simple reservoirs or when the frequency of tubing retrieval workovers is low.
- An HXT is not recommended for use in a gas field because interventions are rarely needed.

22.4.3. Design Process

The designs of subsea trees vary in many ways: completion type (simple, diver assist, diverless, or guideline-less), purpose of the tree (production or injection), service conditions (H_2S, CO_2, or H_2S and CO_2), and so on. These parameters will affect the selection of the tree type, materials, and component arrangement. A typical design process for a subsea tree is shown in Figure 22-18.

The design requirements include the requirements of function, performance, working capabilities, and the cost of the product, which is referred to as its *economy*. These are basic requirements when designing a product. Design requirements are shown in Figure 22-19. After the functions and requirements are clearly known by the designer, the tree type can be determined and the schematic diagram drawn. Draft structural drawings of the tree are usually done during this phase.

Component design is almost the most important point in subsea Xmas tree design. Components of a typical subsea Xmas tree include:

- Tubing hanger system;
- A tree connector to attach the tree to the subsea wellhead;
- The tree body, a heavy forging with production flow paths, designed for pressure containment. Annulus flow paths may also be included in the tree body;
- Tree valves for the production bore, the annulus, and ancillary functions. The tree valves may be integral with the tree body or bolted on;
- Valve actuators for remotely opening and closing the valves. Some valves may be manual and will include ROV interfaces for deep water;
- Control junction plates for umbilical control hook-up;
- Control system, including the valve actuator command system and pressure and temperature transducers. The valve actuator command system can be simple tubing or a complex system including a computer and electrical solenoids depending on the application;
- Choke (optional) for regulating the production flow rate;

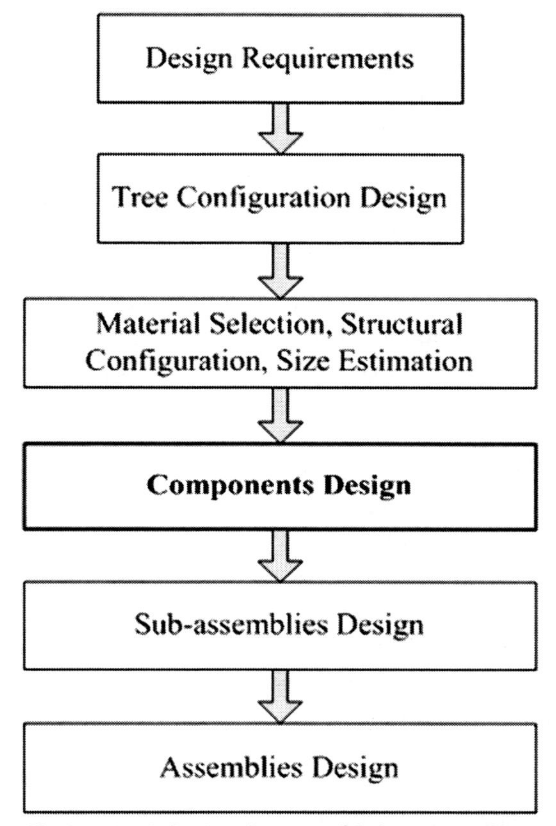

Figure 22-18 Design Process of Subsea Xmas Tree

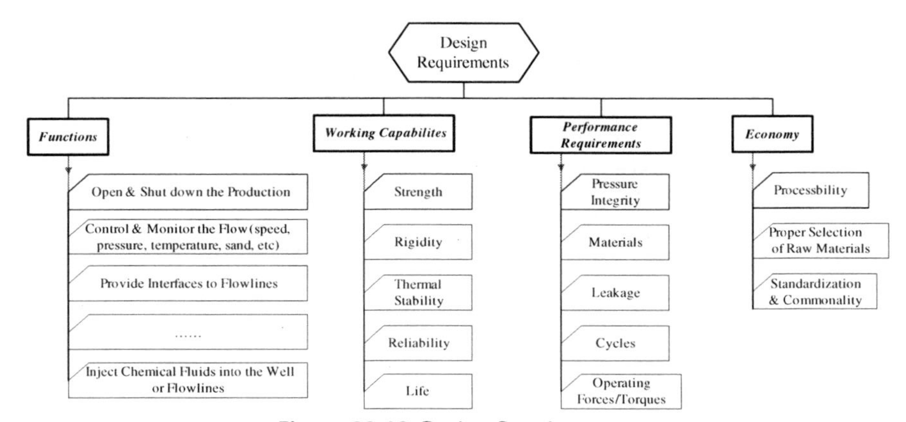

Figure 22-19 Design Requirements

- Tree piping for conducting production fluids, crossover between the production bore and the annulus, chemical injection, hydraulic controls, etc.;
- Tree guide frame for supporting the tree piping and ancillary equipment and for providing guidance for installation and intervention;
- External tree cap for protecting the upper tree connector and the tree itself. The tree cap often incorporates dropped object protection or fishing trawler protection.

The design should consider the components although some of them may be a system or subassembly. Calculations for, drawings, and reports for the components are completed in this phase. Subassembly and assembly design includes a report that shows the assembly procedures and the assembly drawings. Design procedures are shown in Figure 22-20.

22.4.4. Service Conditions

Pressure ratings of subsea Xmas trees are standardized to 5000 psi (34.5 MPa), 10,000 psi (69.0 MPa), and 15,000 psi (103.5 Mpa). Recently 20,000-psi (138-Mpa) subsea Xmas trees have been applied successfully in subsea fields.

Table 22-2 shows the standard temperature ratings per API Specification 6A [6]. Subsea equipment should be designed and rated to operate throughout a temperature range of 35 to 250°F (Rating V) according to the API Specification 17D [2].

Choosing materials classes is the responsibility of the user. All pressure-containing components should be treated as "bodies" for determining material requirements from Table 22-3. However, other wellbore pressure

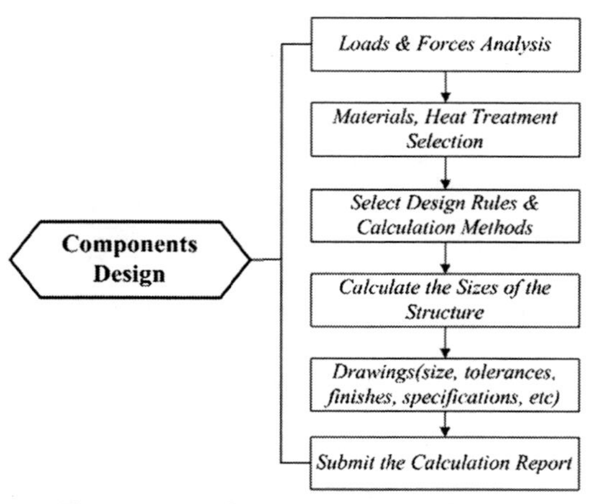

Figure 22-20 Component Design Process

Table 22-2 Standard Temperature Ratings [46]

Temperature Classification	Operating Range			
	Minimum (°C)	Maximum (°F)	Minimum (°C)	Maximum (°F)
K	−60	82	−75	180
L	−46	82	−50	180
P	−29	82	−20	180
R	Room temperature		Room temperature	
S	−18	66	0	150
T	−18	82	0	180
U	−18	121	0	250
V	2	121	35	250

boundary penetration equipment, such as grease/bleeder fittings and lockdown screws, should be treated as "stems." Metal seals should be treated as pressure-controlling parts.

Equipment should be designed to use materials based on the material class required, as shown in Table 22-4. If the mechanical properties can be met, stainless steels may be used in place of carbon and low-alloy steels, and corrosion-resistant alloys may be used in place of stainless steels.

22.4.5. Main Components of Tree

22.4.5.1. General

Components of a subsea Xmas tree vary according to the specific design requirements of specific subsea fields. However, typical VXT components, as illustrated in Figure 22-21, include the following:

- Tree cap;
- Upper production master valve (UPMV);

Table 22-3 Material Class Rating [2]

Retained Fluids	Relative Corrosivity of Retained Fluid	Partial Pressure of CO_2 (psia) (MPa)	Recommended Material Class
General service	Noncorrosive	<7 (0.05)	AA
General service	Slightly corrosive	7 to 30 (0.05 to 0.21)	BB
General service	Moderately to highly corrosive	>30 (0.21)	CC
Sour service	Noncorrosive	<7 (0.05)	DD
Sour service	Slightly corrosive	7 to 30 (0.05 to 0.21)	EE
Sour service	Moderately to highly corrosive	>30 (0.21)	FF
Sour service	Very corrosive	>30 (0.21)	HH

Table 22-4 Material Requirements [6]

Material Class	Minimum Material Requirements	
	Body, Onnet, End, and Outlet Connections	**Pressure-Controlling Parts, Stems, and Mandrel Hangers**
AA – General service	Carbon or low-alloy steel	Carbon or low-alloy steel
BB – General service	Carbon or low-alloy steel	Stainless steel
CC – General service	Stainless steel	Stainless steel
DD – Sour service[a]	Carbon or low-alloy steel[a]	Carbon or low-alloy steel[b]
EE – Sour service[a]	Carbon or low-alloy steel[a]	Stainless steel[b]
FF – Sour service[a]	Stainless steel[a]	Stainless steel[b]
HH – Sour service[a]	CRAs[b]	CRAs[b]

[a]As defined by NACE MR 0175 [7].
[b]In compliance with NACE MR 0175 [7].

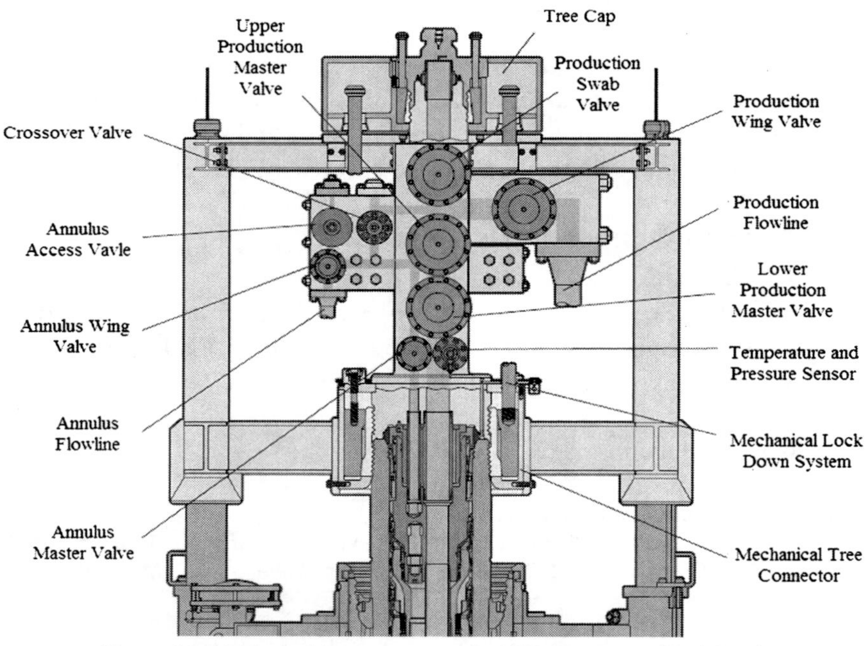

Figure 22-21 Typical Components of a VXT *(Courtesy of Dril-Quip)*

- Lower production master valve (LPMV);
- Production wing valve (PWV);
- Production swab valve (PSV)
- Crossover valve (XOV);
- Annulus master valve (AMV);

- Annulus access valve (AAV) or annulus swab valve (ASV);
- Annulus wing valve (AWV);
- Pressure and temperature sensors (PT, TT, PTT, etc.);
- Tree connector;
- Conventional tubing hanger system.

Typical components of a horizontal Xmas tree, as illustrated in Figure 22-22, are as follows:

- Tree debris cap;
- Tree body;
- Internal tree cap (or upper crown plug);
- Crown plug (or lower crown plug);
- Production master valve (PMV);
- Production wing valve (PWV);
- Annulus access valve (AAV);
- Annulus master valve (AMV);
- Annulus wing valve (AWV);
- Crossover valve (XOV);
- Sensors (PT, TT, PTT, etc.);
- Tree connector;
- Tubing hanger system.

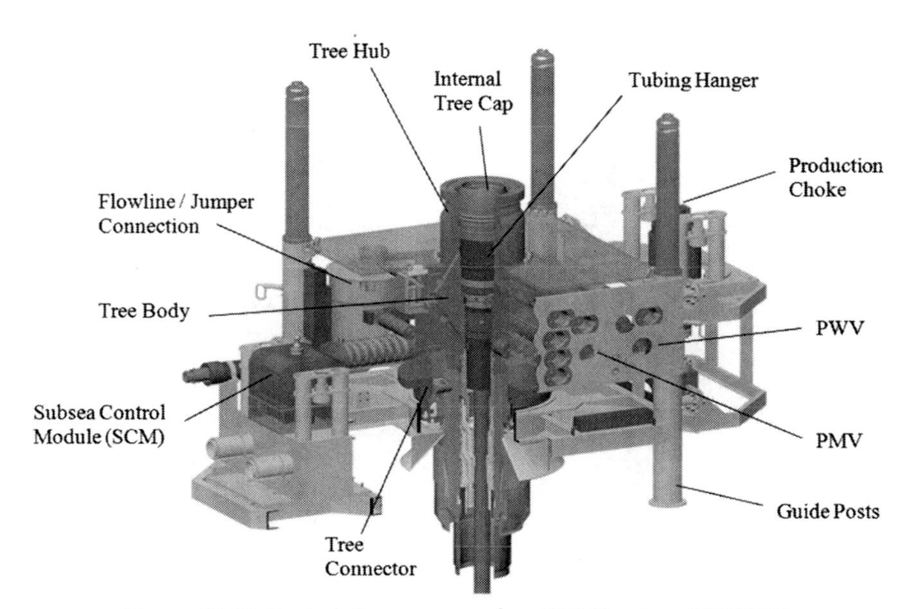

Figure 22-22 Typical Components of an HXT *(Courtesy CAMERON)*

The main components that vary between VXT and HXT are as follows:

- *Tree body:* The tree body in a HXT is normally designed to be an integrated spool. The PMV is located in this tree body, as well as the annulus valves. The PWV is usually designed to be integrated into a production wing block, which can be easily connected to the tree body by flange methods. This design results in components that are interchangeable between the HXTs in the industry. In addition, the tubing hanger system is located in the tree body.
- *Tubing hanger system:* A VXT utilizes a conventional tubing hanger, which has a main production bore and an annulus bore. The tubing hanger is located in the wellhead. However, in an HXT, the tubing hanger is a monobore tubing hanger with a side outlet through which the production flow will pass into the PWV. Because the TH in the HXT is located in the tree body, it needs the crown plugs as the barrier method. An internal tree cap is the second barrier located above the crown plug. If dual crown plugs are designed in a TH system, an internal tree cap is not used.
- *Tree cap:* The tree cap in a VXT system has the functions of providing the control interfaces during workover and sealing the tree from seawater ingress. An HXT, in contrast, has internal tree caps and tree debris caps.

These differences are illustrated in Figure 22-23.

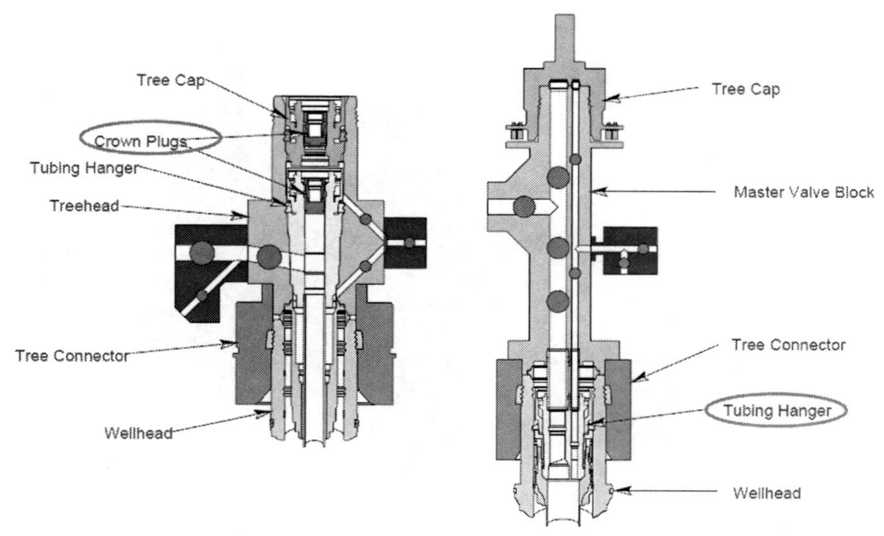

Figure 22-23 Differences between VXTs and HXTs *(Courtesy of Vetco Gray)*

22.4.5.2. Tubing Hanger

Tubing hangers (TH) system is designed to suspend and seal the tubing strings in the production bore of subsea well. TH shall be able to install through the BOP stack and lock onto the internal profile of either the casing hanger or the tree bore. Specific design considerations with regard to the tubing hanger system include as follows:

- Number, size and mass of tubing strings to be supported;
- Type of threaded connection for the tubing;
- Number and size of control ports and pressure rating for downhole safety valves and others as required;
- Requirement for electrical connectors for downhole monitoring and/or control;
- Orientation, if required, relative to a given datum for corresponding interface with the tree;
- Type of riser, integral riser or individual tubing tieback strings used for installation and for wireline work;
- Protection of control ports from debris/fluid contamination.

Tubing Hanger Types

Subsea tubing hangers can be grouped into the following two types as illustrated in Figure 22-24:

- Dual bore tubing hanger
- Monobore tubing hanger

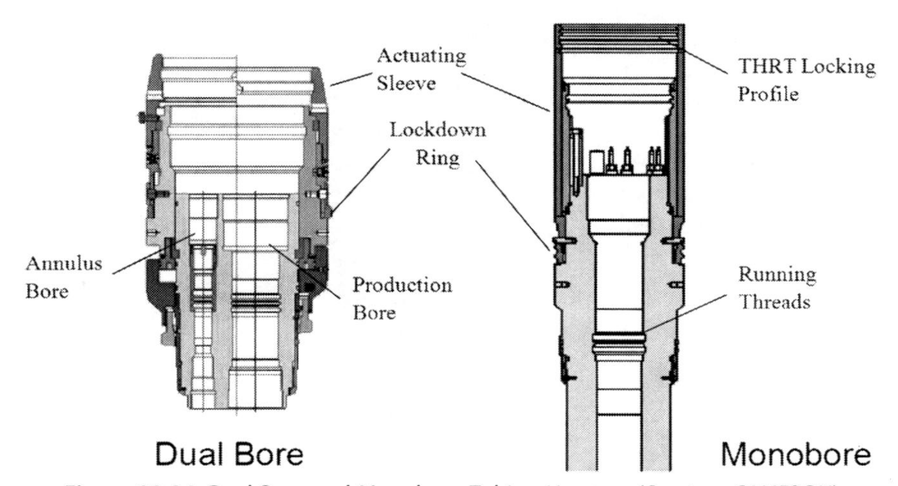

Figure 22-24 Dual Bore and Monobore Tubing Hangers *(Courtesy CAMERON)*

In a dual completion, where tubing strings are run to two separate zones, a dual bore tubing hanger is required for the independent suspension and sealing of the tubing strings. Conventional tree also requires access to both the production bore and annulus bores.

The monobore tubing hanger has production bore only. The lower part was usually designed to be threaded in order to run the tubing strings easily. The monobore tubing hangers are normally used for horizontal Xmas trees. The annulus fluid can be bled off through a concentric port and then through the annulus master valve (AMV) in the tree.

In a horizontal tree system, the tubing hanger configuration is normally of the concentric type, with a production outlet beside (see Figure 22-25). The tubing hanger assembly consists of the hanger body, lockdown sleeve, locking dogs, gallery seals, pump down seal, electrical penetrator receptacle, bottom dry mate connector, and pup joint. There are four basic sizes of tubing hanger: $3^1/_2$, 4, 5, and 7 in. nominal.

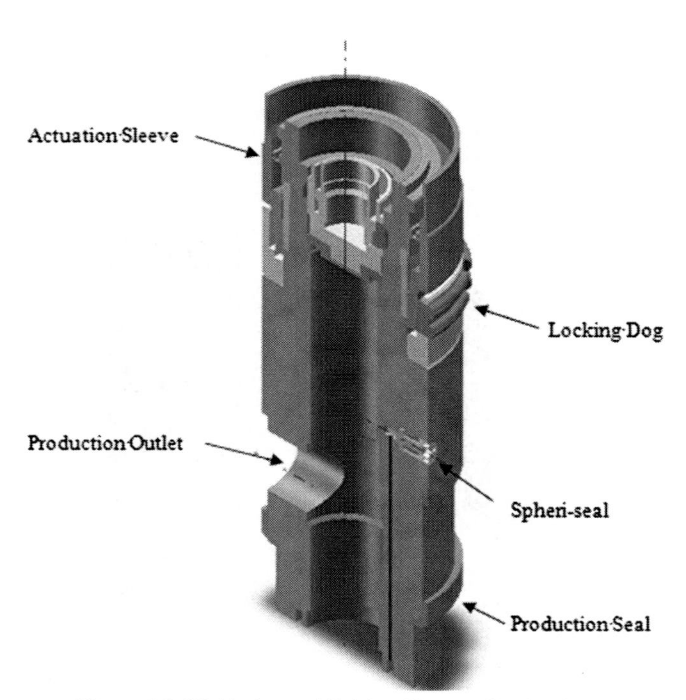

Figure 22-25 Horizontal Tubing Hanger Section View

Penetration Configuration

Figure 22-26 shows a typical tubing hanger penetration configuration. The number of control ports through the tubing hanger depends on:

- How many SCSSV there are and whether they are balanced or unbalanced. A balanced SCSSV requires two control lines, whereas an unbalanced valve needs only one.
- Downhole pressure/temperature monitors (electric connector at the tubing hanger/tree extension sub interface). An electric cable extends below to the bottom of the tubing string where a sensing device is located.

For example, a typical configuration would include:

- Downhole injection: 1 Hyd.
- SCSSV: 2 Hyd.
- Smart well hydraulics: 2 Hyd.
- Soft landing: 1 Hyd.
- Downhole gauges: 1 or 2 Elec.

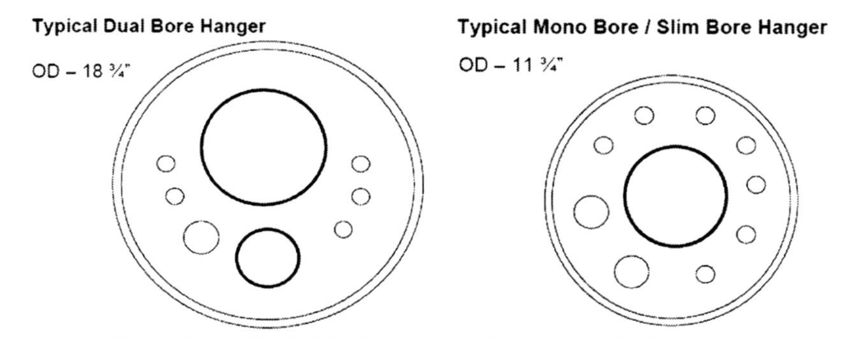

Figure 22-26 Typical Tubing Hanger Penetration Configurations

Tubing Hanger Running Tool

The tubing hanger running tool (THRT) is used to run the tubing hanger (TH) into the tree body. The THRT should have a balanced piston design with equivalent cross-sectional areas to prevent annulus pressure from acting to unlock the THRT from the TH and to ensure that the THRT remains locked even upon loss of hydraulic pressure during installation or workover operations. The system design should be configured to prevent hydraulic locking of the seals during installation/retrieval of the THRT to/from the TH and slick joint. A THRT is illustrated in Figure 22-27.

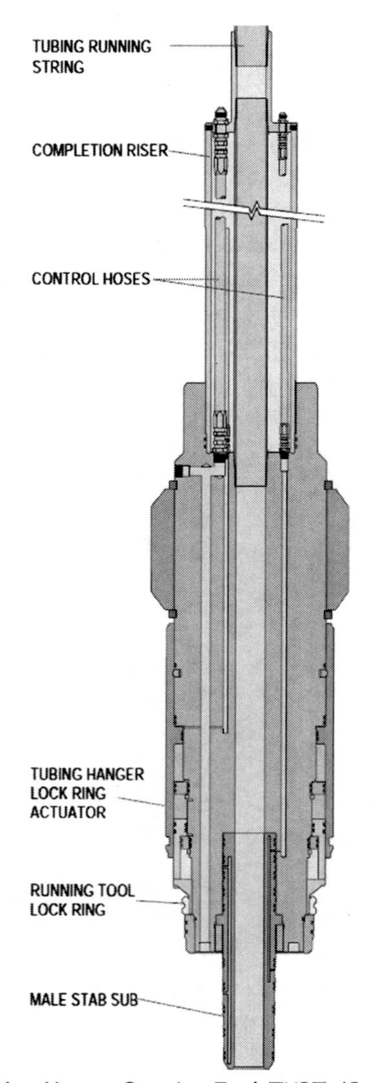

Figure 22-27 Tubing Hanger Running Tool (THRT) *(Courtesy of Dril-Quip)*

22.4.5.3. Tree Piping

Tree piping is defined as pipeworks, fittings and valves in a subsea Xmas tree. The design of tree piping shall consider:

- Allowable stresses at working pressure and testing pressure;
- External loading;
- Tolerances;

- Corrosion/Erosion allowance;
- Temperature;
- Wall thickness due to bending.

The tree flowloops may be fabricated using forged fittings or pre-bent pipe sections, or may be formed in a continuous piece of pipe.

22.4.5.4. Flowline Connector

A flowline connector is used to connect subsea flowlines and umbilicals via a jumper to the subsea Xmas tree. In some cases, the flowline connector also provides the means for disconnecting and removing the tree without retrieving the subsea flowline or umbilical to the surface. Figure 22-28 shows a horizontal flowline connector system.

Flowline connectors generally come in three types: manual connectors operated by divers or ROVs, hydraulic connectors with integral hydraulics, or mechanical connectors with the hydraulic actuators contained in a separate running tool. The flowline connector system may utilize various installation methods, such as first-end or second-end connection methods. It may be either diverless or diver assisted and may utilize guidelines/guideposts to provide guidance and alignment of the equipment during installation.

The flowline connector support frame reacts to all loads imparted by the flowline and umbilical. Tree valves and tree piping are protected from flowline/umbilical loads, which could damage these components. Alignment of critical mating components is provided and maintained during installation. Trees can be removed and replaced without damage to critical

Figure 22-28 Flowline Connector *(Courtesy of FMC)*

mating components. The flowline connector support frame is designed to allow landing a BOP stack on the wellhead housing after the flowline connector support frame is installed.

22.4.5.5. Tree Connectors

Tree connector is the mechanism to join and seal a subsea tree to a subsea wellhead. It may require diver assistance for installation, or be hydraulically actuated to permit remote operation. Typical types of the tree connectors are illustrated in Figure 22-29:

- H-4 type connector
- Collet type connector

The H-4 type connector is a hydraulic actuated connector developed by Vetco Gray and applied for the H-4 wellhead profiles. The hydraulic piston is designed to provide for releasing force 25% greater than locking force for the same operating pressure, using the primary release only. If the secondary release system is used together, the final releasing force is 125% greater than the locking force.

The collet type connector is applied for the hub type tree. This concept is designed to ensure that the bottom collet connection portion would be stronger than any corresponding API connection used on the top. The collet connector has internal porting for actuating the operating cylinders. Guide arms are attached for centering, and the clamp segments form

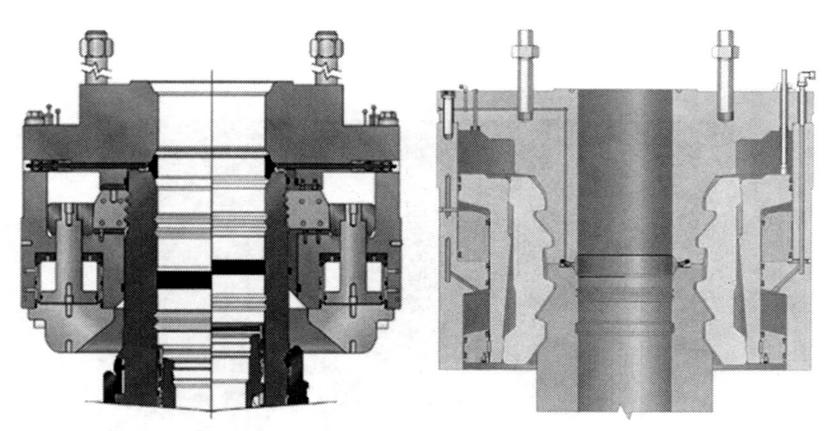

H-4 connector landing on mandrel type wellhead
Collet connector locked to hub type wellhead

Figure 22-29 H-4 Connector and Collet Connector *(Courtesy VetcoGray, CAMERON)*

a funnel for centering the collet connector as it approaches a mating hub. When the collet connector is locked down, the clamp segments are supported the full 360° around the mating hubs on a friction locking taper by the clamp actuator ring.

22.4.5.6. Tree Valves

Subsea Xmas tree contains various valves used for testing, servicing, regulating, or choking the stream of produced oil, gas, and liquids coming up from the well below. Figure 22-30 shows a typical tree valve arrangement and configuration.

The production flow coming from the well below passes through the downhole safety valve (DHSV), which will shut down if it detects an accident, leak, or overpressure occurring.

Production master valves (PMVs) provide full opening during normal production. Usually these valves are high-quality gate valves. They must be capable of holding the full pressure of the well safely for all anticipated

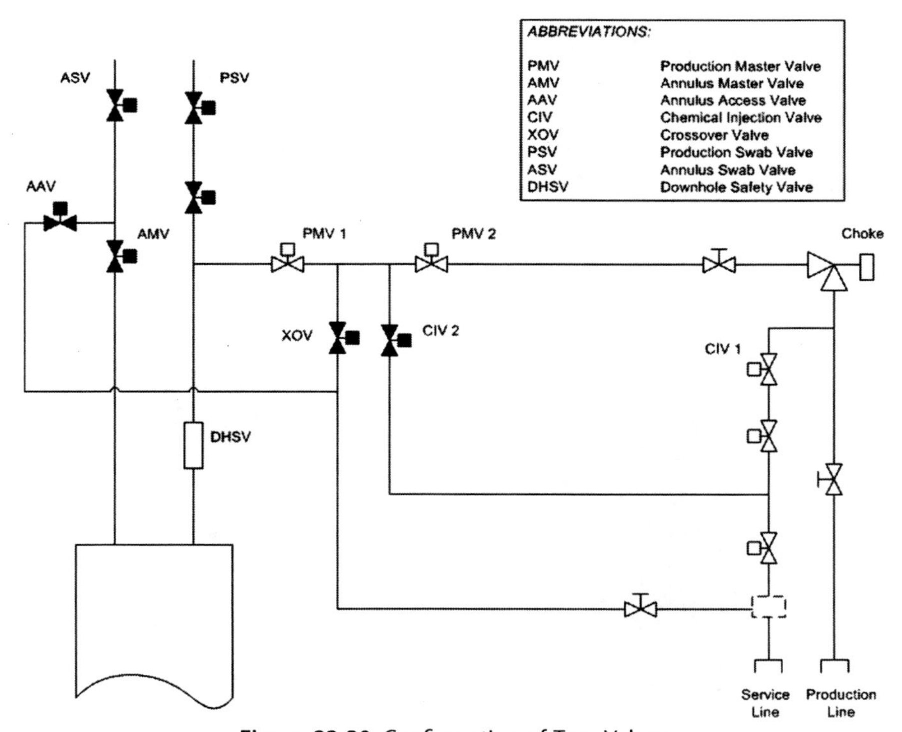

Figure 22-30 Configuration of Tree Valves

purposes, because they represents the second pressure barrier (the first is the DHSV). A production choke is used to control the flow rate and reduce the flow pressure.

The annulus master valve (AMV) and annulus access valve (AAV) are used to equalize the pressure between the upper space and lower space of the tubing hanger during the normal production (i.e., when the DHSV is open).

Located in the crossover loop, a crossover valve (XOV) is an optional valve that, when opened, allows communication between the annulus and production tree paths, which are normally isolated. An XOV can be used to allow fluid passage for well kill operations or to overcome obstructions caused by hydrate formation.

The production swab valve (PSV) and annulus swab valve (ASV) are open when interventions in the well are necessary.

Subsea Xmas tree valves should be designed, fabricated, and tested in accordance with API 17D [2], API 6A [6], and API 6D [9]. The valves can be both bolted on or built in.

22.4.5.7. Production Choke

A production choke is a flow control device that causes pressure drop or reduces the flow rate through an orifice. It is usually mounted downstream of the PWV in a subsea Xmas tree in order to regular the flow from the well to the manifold. It can also be mounted on the manifold. Figure 22-31 shows the subsea choke in a subsea Xmas tree.

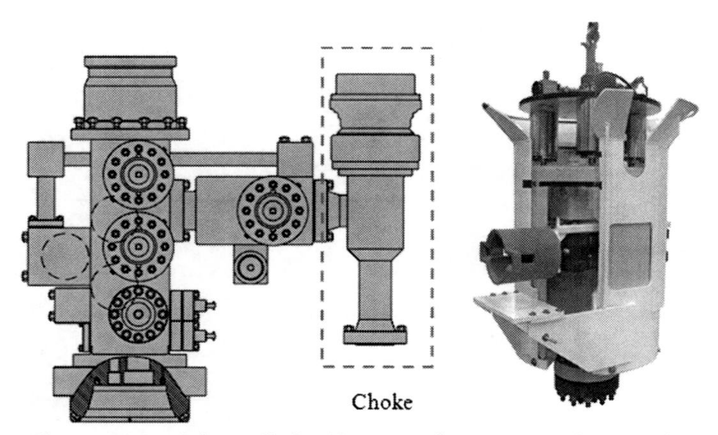

Figure 22-31 Subsea Choke *(Courtesy of Cameron and MasterFlo)*

The two most widely used choke types are positive chokes and adjustable chokes. The adjustable choke can be locally adjusted by a diver or adjusted remotely from a surface control console. They normally have a rotary stepping hydraulic actuator, mounted on the choke body. This adjusts the size of orifice at the preferred value. Chokes have also been developed to be installed and retrieved by ROV tools without using a diver. In addition, the insert-retrievable choke leaves the housing in place, while the internals and the actuator are replaceable units.

Trims/Orifices Types

Typical orifices used are of the disk type or needle/plug type. The disk type acts by rotating one disk and having one fixed. This will ensure the necessary choking effect. The needle/plug type regulates the flow by moving the insert and thereby providing a gap with the body. The movement is axial. Figure 22-32 shows all of the trim/orifice types per ISO 13628-4 [5].

Choke Design Parameters

Several measurements must be known in order to select the proper choke for a subsea production system: how fast the flow is coming into the choke, the inlet pressure P_1 of the flow, the pressure drop that occurs crossing the orifice, and the outlet or downstream pressure P_2 of the flow, as shown in Figure 22-33.

Choke sizing is determined by coefficient value (Cv), which takes into account all dimensions as well as other factors, including size and direction changes, that affect fluid flow in a choke. The Cv equals number of gallons of per minute that will pass through a restriction (orifice) with a pressure drop of 1 psi at 60C. This Cv calculation normally follows Instrument Society of America (ISA) guidelines.

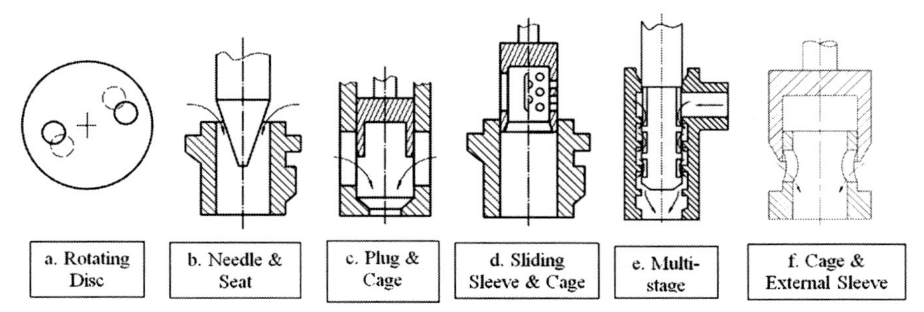

Figure 22-32 Trim Types

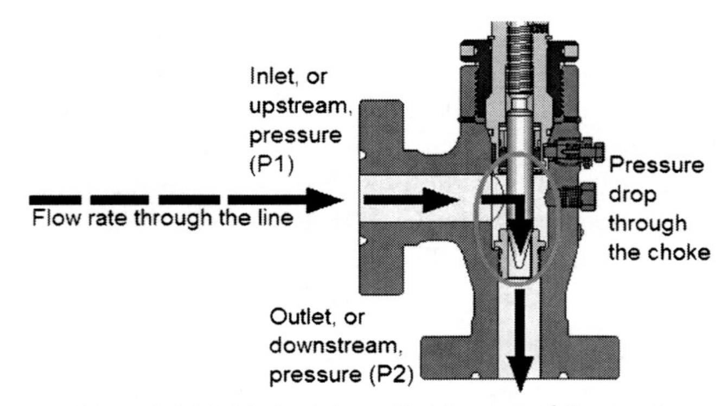

Figure 22-33 Choke Schematic *(Courtesy of Cameron)*

Pressure is maintained through the tree piping as P1. When the flow crosses the orifice of the choke, the pressure drops. But soon the pressure will recover to a level (P2). The process is illustrated in the Figure 22-34.

The pressure drop is determined by the equation $\Delta P = P_1 - P_2$ (inlet pressure minus outlet pressure). The ΔP ratio, ΔPR, is considered the most important parameter for evaluating and ensuring the success of the subsea field development project. This ratio is determined as $\Delta PR = \Delta P/P_1$, which used to measure the capacity and recovery of the choke. The higher the value of ΔPR, the higher the potential damage to the choke trim or body. Normally a special review of the trim is required if ΔPR is beyond 0.6.

22.4.5.8. Tree Cap

Tree caps are designed to both prevent fluid from leaking from the wellbore into the environment and small dropped objects from getting into the mandrel. Designs are very different between HXTs and VXTs. Tree caps are usually designed to be recoverable for easy maintenance. The debris cap covers the top of the tree spool. It is installed, locked, unlocked, released, and recovered via ROV-assisted operations. See Figure 22-35.

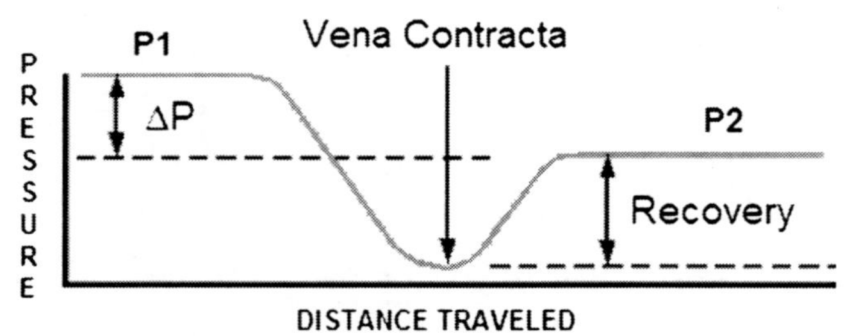

Figure 22-34 Pressure Drop in a Choke *(Courtesy of Cameron)*

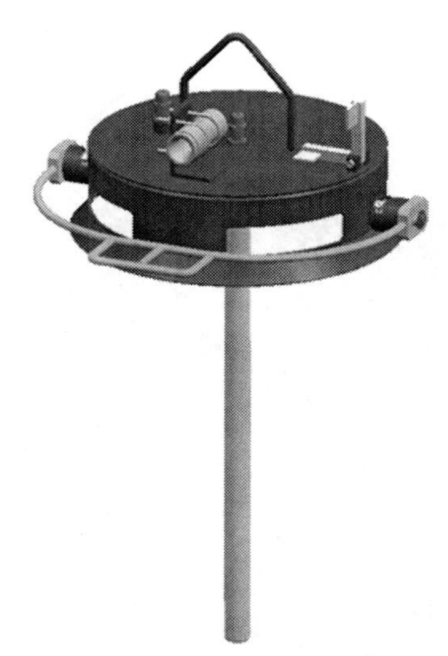

Figure 22-35 Tree Debris Cap

An internal tree cap is designed to latch onto the spool body above the tubing hanger and seal off the area above the tubing hanger to the maximum rated working pressure. It is installed through the marine riser and latches full within the bore of the horizontal tree and should provide primary metal-to-metal and secondary elastomeric seals to isolate the internal tree from the environment. Figure 22-36 illustrates a configuration for an ROV-operated internal tree cap.

22.4.5.9. Tree Frame

The tree frame is designed to protect critical components on the tree from objects falling from the surface. It also provides structural mounting for:

- Tree body;
- Tree valves;
- Subsea control module (SCM);
- Choke;
- Tree piping;
- Flowline connectors;
- Tree connector;
- Flying leads and connections;
- ROV panel;
- Anodes.

Guidance and orientation systems are designed for the tree frame in order to land the tree on the production guide base or a template. The tree

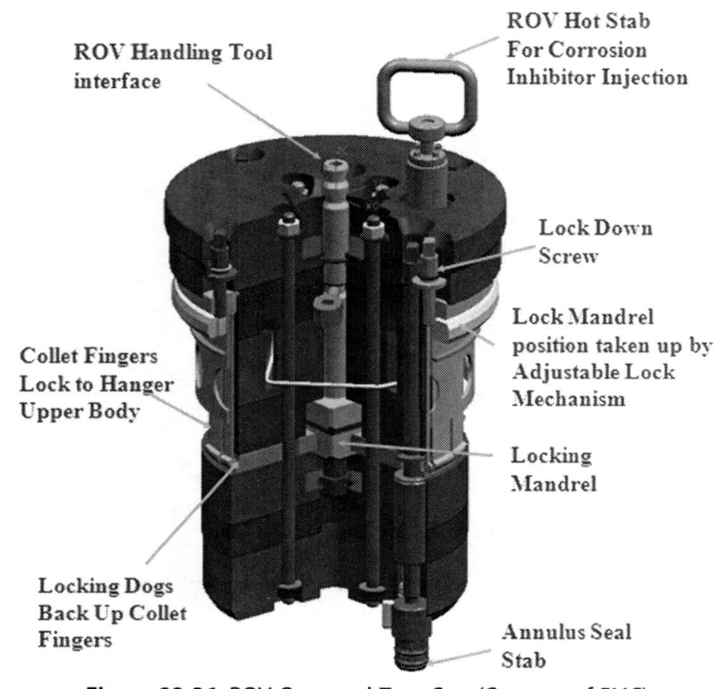

Figure 22-36 ROV-Operated Tree Cap *(Courtesy of FMC)*

frame is designed to protect the tree components during handling on the surface and subsea running and retrieving operations. Its strength and entire weight are calculated and checked to ensure these operations can be completed successfully.

The subsea tree frame must be designed with no snag points or sharp edges that may cut or entangle the ROV tether or control umbilical.

22.4.6. Tree-Mounted Controls

22.4.6.1. Subsea Control Module (SCM)

The subsea control module is located on the subsea Xmas tree, or manifold. The SCM normally controls all hydraulically actuated tree and downhole safety valves as well as monitors all tree pressure and temperature and sand detection sensors. Most of the SCMs are designed to be capable of being installed or retrieved by guidelineless installation using a single lift wire with ROV assistance. Figure 22-37 shows a SCM.

The SCM contains all control valves, hydraulic pressure monitoring transducers and electronics. The subsea control module shall be filled

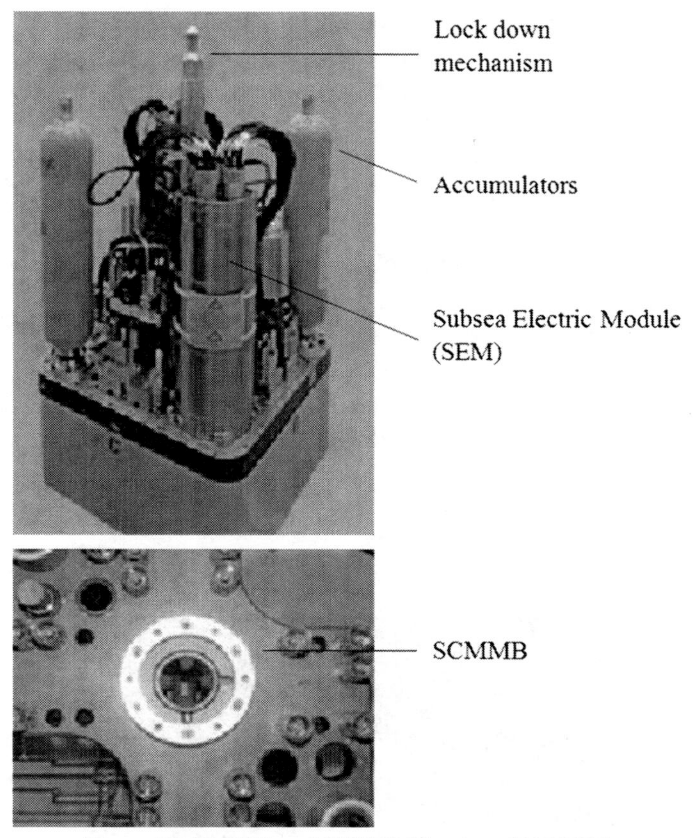

Figure 22-37 SCM and SCMMB (*Courtesy CAMERON*)

with dielectric fluid and pressure compensated to provide a secondary barrier to ingress from sea water.

The SCM is normally landed and locked onto the SCM mounting base (SCMMB). Hydraulic and electrical connectors/couplers are distributed on the base for the SCM to allow the connection to be made or broken.

The following hydraulic connections are normally required:

- Lower pressure hydraulic supply
- High pressure hydraulic supply

22.4.6.2. Pressure, Temperature Transmitters and Sand Detectors

Subsea pressure/temperature transmitters shall be constructed to withstand the pressure and temperature underwater, which require particular considerations for the sensors design, construction and interface.

All the transmitters have to be constructed from corrosion resistant materials such as 316SS. Parts of the unit in contact with process fluids should be covered with resistant materials to match the environmental conditions, such as Inconel 625.

The pressure and temperature range should be same as these of the tree. The transmitters normally designed for API type flange mount as shown in Figure 22-38.

Typical transmitters used for subsea structures measurements are:

- Pressure Transmitter (PT);
- Temperature Transmitter (TT);
- Combined Pressure Temperature Transmitter (PTT);
- Downhole Pressure/Temperature Transmitter (DHPTT).

Transmitters are typically installed at the following locations:

- Annulus line;
- Downstream of production choke;
- Upstream of production choke.

Figure 22-38 PTT Located on a Subsea Xmas Tree

22.4.7. Tree Running Tools

Tree running tools are designed for subsea deployments such as subsea tree, jumper, wellhead, and etc. All of the running tools are normally hydraulic actuated for deepwater applications. As if using the torque setting, it will be very difficult to count the turn number of the drill pipe due to the flexibility in the pipe. Hydraulic tools can provide hydraulic signals applying to the locking positions, which will be utilized to confirm the correct position of the mechanism. However, it is necessary to design a fail-safe system to avoid the hydraulic fails while the equipment is deployed subsea.

Figure 22-39 shows a tree running tool, which typically consists of a connector interfacing the top of the tree. It is often combined with a lower riser package containing a set of safety valves to control the well during installation/workover operations.

22.4.8. Subsea Xmas Tree Design & Analysis

22.4.8.1. Chemical Injection

Chemical injection is designed according to the reservoir type and the fluid characteristics. The final objective of chemical injection is to make sure that it produces fluids economically from the reservoir to a production facility through the whole life cycle in the field development. The chemical injections are performed at start-up and during production. Figure 22-40 shows the chemical injection point of a subsea Xmas tree.

Chemical injection design issues include:

- Sizing of chemical injection core;
- Viscosities at seabed temperatures;

Figure 22-39 Tree Running Tool *(Courtesy of Dril-Quip)*

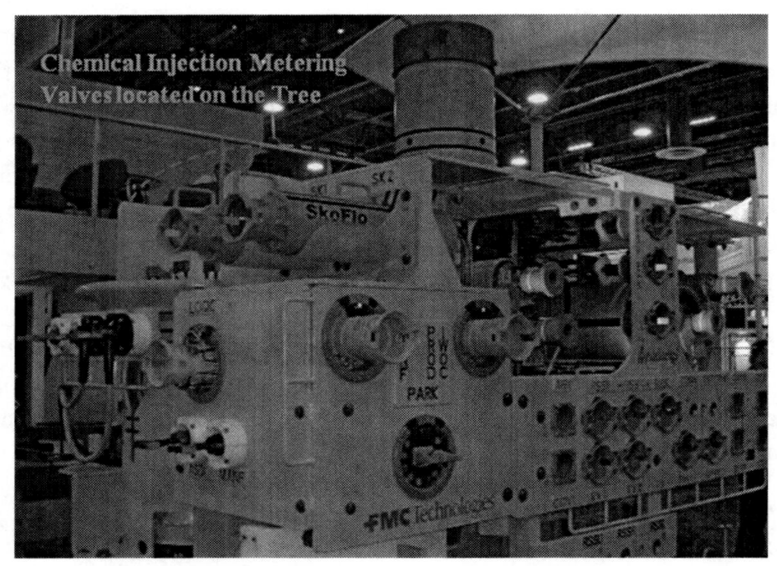

Figure 22-40 Chemical Injection Point at Tree *(Courtesy FMC Technologies)*

- Material compatibilities;
- Miscibility of chemicals;
- Chemical permeabilities;
- Metering valves & reliability.

22.4.8.2. Cathodic Protection

Design of cathodic protection (CP) for subsea Xmas tree system shall be carried out in accordance with DNV RP B401 or NACE MR-0174. The main design parameters and calculations are:

- Protect criteria for specific materials;
- Current density;
- Surface to be protected;
- Coating breakdown factors;
- Anodes mass and number to be applied.

For subsea wellheads, trees, and other structures, different materials are often used in the system. For CRA and high strength steel materials, the risk of hydrogen induced cracking is of main concern when the cathodic protection is applied.

To apply CP to a subsea Xmas tree system, the following design features are recommended:

- All submerged metallic components are connected electrically to the base housing to ensure the cathodic protection of the complete assembly.

Items such as pressure caps which cannot be fully or easily connected electrically shall be analyzed individually and have independent protection. Surface areas of all submerged components are calculated as the inputs of the sacrificial anode calculations.
- All submerged components exposed to seawater, except for the stainless steel control tubing, junction plates, control couplers, etc., are coated with a subsea three coat epoxy system.
- To achieve a cost-effective corrosion control program for each subsea structure, it may be beneficial to allow a certain amount of the structure to remain uncoated. The repair of minor coating damage may be eliminated if the cathodic protection system design accounts for the additional bare surface area. The bare or uncoated area shall be protected by the inclusion of additional galvanic anodes.

Detailed design and calculation of Current Demand, Selection of Anodes, Anode Mass and Number are designed according to DNV RP B401.

22.4.8.3. Insulation and Coating

The trees and wellhead, as well as well jumpers, manifolds, flowline jumpers, and associated equipment, require corrosion coatings and thermal insulation to enable sufficient cooldown time in the event of a production stoppage.

The main objectives of thermal insulation are:
- Have sufficient time to confidently perform the preservation sequence at any operation condition.
- Avoid dramatic consequences of hydrate formation with associated production losses.
- Solve the shutdown problem and avoid the burden of the launching preservation sequence with associated production losses

The insulation system includes a layer of corrosion coating suitable for working temperature on the steel surface. This corrosion coating is applied in accordance with the manufacturer's specifications. Areas that require insulation are specified in the engineering drawings. Areas that are not to be insulated because insulation will be detrimental to the function of the components are marked or adequately protected during installation process.

22.4.8.4. Structural Loads

The tree connector, tree body, tree guide frame, and tree piping must be designed to withstand internal and external structural loads imposed during installation and operation. The following are some tree and tree component load considerations:

- Riser and BOP loads;
- Flowline connection loads;
- Snagged tree frame, umbilicals or flowlines;
- Thermal stresses (trapped fluids, component expansion, pipeline growth);
- Lifting loads;
- Dropped objects;
- Pressure-induced loads, both external and internal.

Non-pressure-containing structural components should be designed in accordance with AWS D1.3 [11].

The tree framework is usually designed around standard API post centers (API RP 17A [12]). This is typically, but not always true, even if the tree is designed to be guideline-less. API defines the position of four guideposts evenly spaced around the well centerline at a 6-ft radius. This equates to 101.82 in. between the posts on any side of the square corners that they form.

22.4.8.5. Thermal Analysis

Subsea Xmas tree design requires to determine the insulation coating type and thickness to achieve the overall heat transfer coefficient (U value). This insulation system should work in the service life at specific water depth and internal temperature. It should also be able to withstand the loadings expected during installation and production.

Thermal analysis and insulation coating calculations may include following issues:

- Thermal expansion coefficient;
- Insulating material compositions by layer;
- Thickness of each layer;
- Maximum temperature rating for each material type (layer);
- Graph of thermal conductivity vs. temperature for each layer;
- Insulation system U-value in air;
- Insulation system U-value at the beginning and the end of design life;
- Layout drawing showing location of insulation coating;
- Other mechanical behaviors of the coating.

FEA is typically used to analyze the insulated components to illustrate that they meet the thermal insulation criteria. The components can be analyzed individually or together in a system model. Adjacent effects from

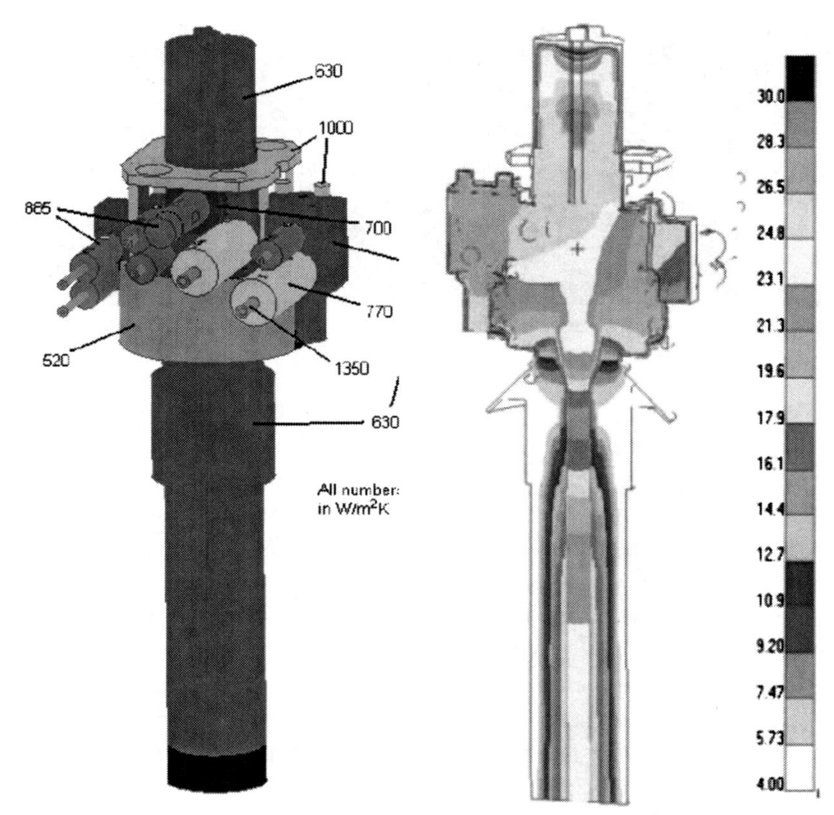

Figure 22-41 Subsea Xmas Tree Thermal Analysis using FEA [13]

neighboring components must be considered with care if two or more components are analyzed together. Figure 22-41 show the thermal analysis using FEA.

22.4.9. Subsea Xmas Tree Installation

Subsea Xmas trees can be installed either with a drill pipe or with the cable of a crane/winch, as shown in Figure 22-42. The typical size of a tree is 12 ft × 12 ft × 12 ft and typical weight is 20 to 50 tonne. This size allows trees to be installed through a moon-pool if the tree is already on the deck of a drilling vessel. Otherwise the tree will be transported by a transportation barge. The tree is lifted with the deck crane and lowered subsea. Because the cable of a crane is normally 200 to 300 m long, for deep water, the tree will be transferred to a rig winch, which has wire lengths of up to 1000 m.

Figure 22-42 Tree Installation by Drill Pipe (Left) and Rig Winch (Right)

As introduced in Chapter 5, the installation vessel for a subsea Xmas tree can be a jack-up, semisubmersible, or drill ship, based on the water depth of the system, as illustrated in Figure 22-43.

In a VXT configuration, the tubing hanger and downhole tubing are run prior to installing the tree, whereas for an HXT the tubing hanger is typically landed in the tree, and hence the tubing hanger and downhole tubing can be retrieved and replaced without requiring removal of the tree. By the same token, removal of an HXT normally requires prior removal of the tubing hanger and completion string.

VXT systems are run on a dual-bore completion riser (or a monobore riser with bore selector located above the LRP and a means to circulate the

Jack up (<100m WD)

Semisub (500~1500m WD)

Drillship (>1500 WD)

Figure 22-43 Installation Vessels

annulus, usually via a flex hose from the surface). The TH of an HXT is run on casing tubular joints, thereby saving the cost of a dual-bore completion riser; however, a complex landing string is required to run the TH. The landing string is equipped with isolation ball valves and a disconnect package made especially to suit the ram and annular BOP elevations of a particular BOP.

Guidance of trees onto the subsea wellhead is usually performed by guidelines that go from the surface to the PGB of wellhead. Guide wires are pushed into the guideposts of the tree and the tree is then lowered subsea. However, the guidelines are usually used in water depths of less than 500 m, because of the limit of wire length on the rig. For deeper water depths, a DP vessel, which uses thrusters to keep the vessel in location, may be needed to land and lock the tree onto the wellhead.

Typical procedures for installing a vertical Xmas tree via a drill pipe through a moon-pool are as follows (see Figure 22-44):

- Perform preinstallation tree tests.
- Skid tree to moon-pool.
- Push guide wires into tree guide arms.
- Install lower riser package and emergency disconnect package (EDP) on tree at moon-pool area.
- Connect the installation and workover control system (IWOCS).
- Lower the tree to the guide base with tubing risers, as shown in sequence 1 of Figure 22-44.
- Lock the tree onto the guide base. Test the seal gasket.
- Perform tree valve function tests with the IWOCS.
- Retrieve the tree running tool.
- Run the tree cap on the drill pipe with the utility running tool system.
- Lower the tree cap to the subsea tree, as shown in sequence 2.

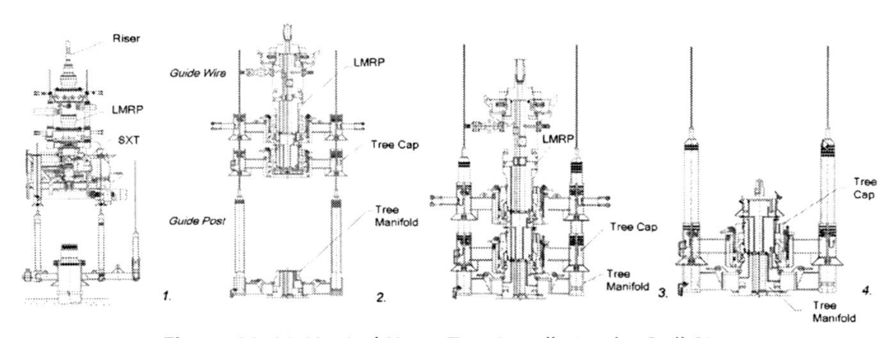

Figure 22-44 Vertical Xmas Tree Installation by Drill Pipe

- Land and lock the tree cap onto the tree mandrel, as shown in sequence 3.
- Lower the corrosion cap onto the tree cap with a drill pipe (or lifting wires). Some suppliers have developed ROV-installed corrosion caps (see sequence 4).

Typical procedures for installing a horizontal Xmas are given next. As the tubing hanger is installed in the tree, subsea completions are performed during tree installation (see Figure 22-45):

- Complete drilling.
- Retrieve the drilling riser and BOP stack; move the rig off.
- Retrieve drilling guide base.
- Run the PGB and latch onto the wellhead.
- Run the subsea HXT.
- Land the tree, lock the connector, test seal function valves with an ROV, release tree running tool (TRT).

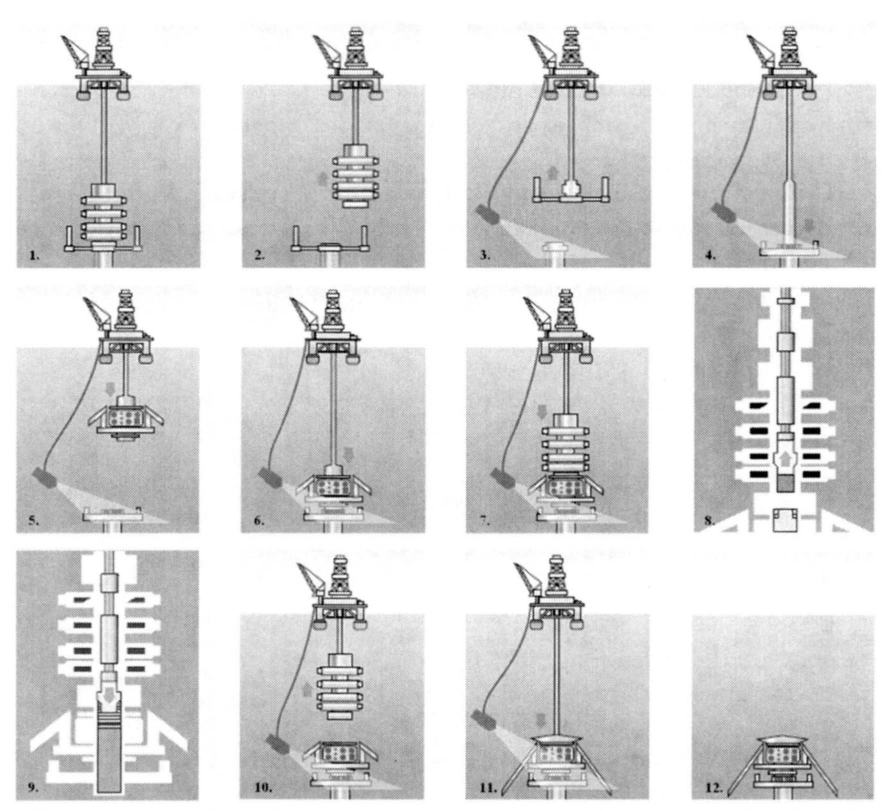

Figure 22-45 Horizontal Xmas Tree Installation Process *(Courtesy of Schlumberger)*

- Run the BOP stack onto the HXT; lock the connector.
- Run the tubing hanger; perform subsea well completion; unlatch the THRT.
- Run the internal tree cap by wireline through the riser and BOP; retrieve THRT.
- Retrieve BOP stack.
- Install debris cap.
- Prepare to start the well.

REFERENCES

[1] G.R. Samuel, G. Adolfo, Optimization of Multistring Casing Design with Wellhead Growth, Landmark Drilling & Well Services, SPE Paper 56762 (1999).
[2] American Petroleum Institute, Specification for Subsea Wellhead and Christmas Tree Equipment, first ed., API Specification 17D, 1992.
[3] A.S. Halal, R.F. Mitchell, Casing Design for Trapped Annulus Pressure Buildup, Drilling and Completion Journal (June 1994) 107.
[4] H. Matlock, L.C. Reese, Generalized Solutions for Laterally Loaded Piles, Journal of the Soil Mechanics and Foundations Division, ASCE, Vol. 86, No SM5, pp. 63–91, (1960).
[5] International Standards Organization, Design and Operation of Subsea Production System - Subsea Wellhead and Tree Equipment, ISO, 13628–4, (2007).
[6] American Petroleum Institute, Petroleum and Natural Gas Industries — Drilling and Production Equipment — Wellhead and Christmas Tree Equipment, nineteenth ed., API, 6A, (2004).
[7] National Association of Corrosion Engineers, Petroleum and Natural Gas Industries Material for Use in H_2S-Containing Environments in Oil and Gas Production, NACE MR0175 (2002).
[8] American Society of Mechanical Engineers, Process Piping, ASME, B31.3, (2008).
[9] American Petroleum Institute, Specification for Pipeline Valves, API, 6D, (2008).
[10] DNV Recommend Practice, Cathodic Protection Design, DNV, RP B401, (2005).
[11] American Welding Society, Structural Welding Code – Sheet Steel, AWS, D1.3, (2008).
[12] American Petroleum Institute, Recommended Practice for Design and Operation of Subsea Production Systems, API, 17A, (2002).
[13] K.A. Aarnes, J. Lesgent, J.C. Hubert, Thermal Design of Dalia SPS Deepwater Christmas Tree – Verified by Use of Full – Scale Testing and Numerical Simulations, OTC 17090, Offshore Technology Conference, Houston, Texas, 2005.

CHAPTER 23

ROV Intervention and Interface

Contents

Subsea Engineering Handbook
ISBN 978-0-12-397804-2, doi:10.1016/B978-0-12-397804-2.00023-0

23.1. INTRODUCTION

As field developments move into deep water, numerous subsea intervention tasks have been moved out of reach of human direct intervention. Remote-operated vehicles (ROVs) and remote-operated tools (ROTs) are required to carry out subsea tasks that divers cannot reach. An ROV is a free-swimming submersible craft used to perform subsea tasks such as valve operations, hydraulic functions, and other general tasks. An ROT system is a dedicated, unmanned subsea tool used for remote installation or module replacement tasks that require lift capacity beyond that of free-swimming ROV systems [1,2].

This chapter provides an overview of the state of the art for subsea remote intervention and vehicles, ROV technologies, and ROV capabilities and requirements for subsea operations.

23.2. ROV INTERVENTION

The term *subsea intervention* refers to all activities carried out subsea. An intervention philosophy needs to be built before designing the interface for a subsea production system (SPS). The intervention philosophy normally focuses on the following questions:

- What kind of tasks will be done subsea?
- What methods will be used to complete these jobs?
- What are the requirements to complete the intervention activities?

The first question will be discussed in this section. For the second question, two intervention methods are commonly used in the subsea engineering: (1) ROVs for inspection, cleaning, and so on, and (2) ROTs for module replacement and subsea tie-in. The answer to the third question can be found in API and NORSOK rules [1,2,3,4]. Generally, ROVs are used for these issues, which are detailed in the following sections:

- Site survey;
- Drilling assistance;
- Installation assistance;
- Operation assistance;
- Inspection;
- Maintenance and repair.

23.2.1. Site Survey

A site survey has to be carried out before offshore activities such as drilling and installation to obtain the seabed's precise bathymetry and properties.

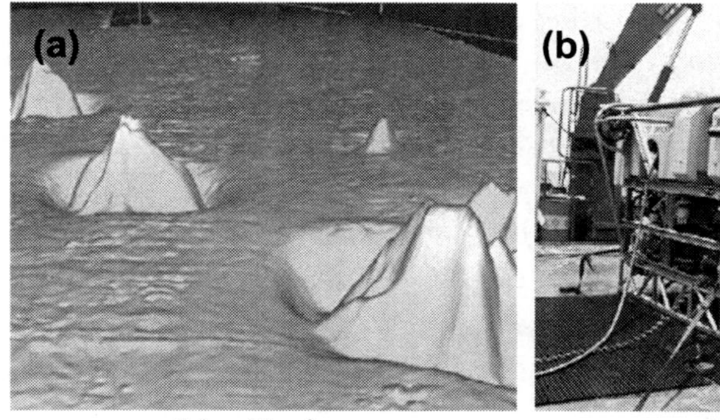

Subsea Mapping ROV Used for Mapping

Figure 23-1 Seabed Mapping with a Workclass ROV [5]

Detailed seabed mapping through precise bathymetry may be performed by a seabed reference system with differential pressure sensors and acoustic data transmission, which may be deployed and retrieved by an ROV. Seabed mapping can also be performed by an ROV carrying a multibeam echo sounder (MBE) or a side-scan sonar (SSS) as shown in Figure 23-1. A sub-bottom profiler (SBP) for sub-bottom profiling may be used to assess the quality of seabed properties for offshore installation foundation.

23.2.2. Drilling Assistance

Drilling activities for production drilling and completion normally include:

- Deployment of acoustic units such as transponders or beacons by an ROV for surface or subsea positioning;
- Bottom survey by visual observation from a ROV with video and still cameras;
- Structure setting and testing (if needed) of permanent guide base (PGB), temporary guide base (TGB), Xmas tree, BOP, etc.;
- As-built (bottom) survey by ROV visual observation with supplemental equipment.

During the entire process, the observation tasks with video cameras (often with scanning sonar as supplemental "acoustic observation") make up the majority of ROV drilling assistance. Tasks include conducting the bottom survey, monitoring the lowering of the structure and touching down, checking the structure's orientation and level with a gyrocompass and bull's-eye, respectively, and performing an as-built survey. Some necessary intervention work may have to be done with ROVs or ROTs during structure setting and testing:

- Acoustic transponder or beacon deployment and recovery;
- Debris positioning and removal from seabed and tree, including dropped objects;
- Structure position assistance with ROV pull/push;
- Guide wire deployment, recovery, and cutting during emergency conditions;
- Rigging (e.g., shackle connection and disconnection);
- Cement cleaning on guide base with brush or water jet;
- Valve operation with hydraulic torque tool or hydraulic stab-in;
- ROV-operable guide posts, replacement, and pin pull release;
- Control pod replacement if suitable for ROV (otherwise ROT);
- Anode installation by clamp and contact screw.

23.2.3. Installation Assistance

The installation of a subsea production system from the water surface to the seabed can be divided into two parts:

- Subsea equipment installation (e.g., manifold deployment, landing);
- Pipeline/umbilical installation (e.g., initiation, normal lay and laydown).

The installation methods for subsea equipment may be divided into two groups. Large subsea hardware with weights over 300 tonne (metric ton) can be installed by a heavy lift vessel where the crane wire is long enough to reach the seabed and the crane is used to both put the equipment overboard and lower it. A soft landing to the seabed may be required using an active heave compensation system with the crane. Alternatively, it may be installed with a drilling tower on a drilling rig, which can have a lifting capacity up to about 600 tonne. For smaller subsea hardware (maximum approximately 250 tonne), a normal vessel equipped with a suitable crane for overboarding the hardware may be used. The vessel normally would not have a long enough crane wire to the seabed, so the hardware is transferred from the crane wire to a winch with a high capacity and a long enough wire for lowering the equipment to the seabed once the hardware passes through the splash zone. In both installation groups, ROVs are used for observation and verification and for engagement and release of guide wires and hooks.

Subsea structures are widely positioned underwater using the long baseline (LBL) method in which transducers used for position measuring, a gyro-compass for orientation measuring, a depth sensor for depth measuring may be mounted onto structure by package(s) that will be retrieved by the ROV. The orientation control may be assisted by the ROV, and the ROV has to verify via camera that the structure is aligned and level before the structure's final set-down. ROVs may also be used to install chokes, multiphase meters, and subsea control modules. For seal pressure tests, ROVs can be used for hot stabbing.

ROVs can be used to assist in the installation of a dead anchor for pipeline/umbilical laying initiation. They can also be used to connect the pull-in line for J-tube or I-tube initiation. During normal installation and pipeline/umbilical laydown, the touchdown point is often monitored with ROVs in front and behind.

The connections between subsea production equipment and flowlines and subsea equipment and umbilicals may be completed through flying leads from the umbilical termination assembly (UTA) to the tree/manifold, well jumper from the tree to the manifold, jumper from the manifold to the PLET. The flying leads may be handled and pulled in by an ROV directly. Jumpers can be deployed from a vessel with spreader bar(s), and then positioned and connector actuated with the assistance of an ROV.

23.2.4. Operation Assistance

Main production activities normally include:

- Flow control by chokes and valves operated by hydraulic actuators through control pods and umbilicals or externally by ROV or ROT intervention;
- Monitoring of flow temperature and pressure by relevant measurement meters;
- Chemical and inhibitor injection for corrosion, waxing, and hydrate formation resistance;
- Flow separation of liquids, gases, and solids (filtering);
- Flow boosting by pumping;
- Flow heating or cooling.

During the operation phase, ROVs are normally not required except for noncritical valve actuation and possibly intermittent status checks, taking samples, etc.

23.2.5. Inspection

Inspection may be needed on a routine basis for the structures expected to deteriorate due to flowline vibration, internal erosion, corrosion, etc. Inspection includes:

- General visual inspection, including cathodic measurements and marine growth measurements;

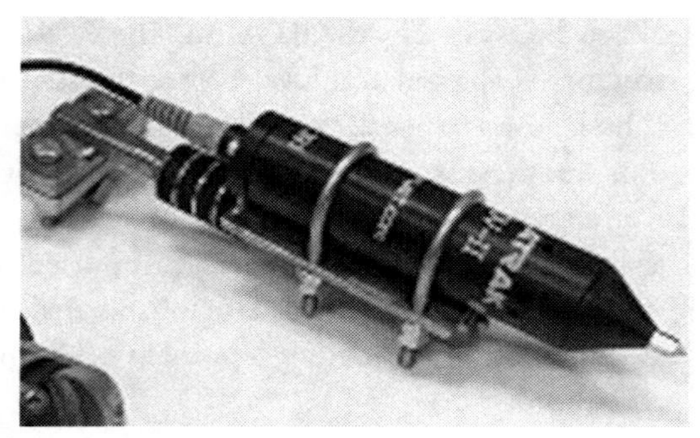

Figure 23-2 Polatrak ROVII Tip-Contact CP Probe *(Courtesy of Deepwater Corrosion Services, Inc. [6])*

- Close visual inspection additionally requiring physical cleaning for close visual inspection, CP measurements, and crack detection by means of nondestructive testing (NDT);
- Detailed inspection including close visual inspection, crack detection, wall thickness measurements, and flooded member detection;
- Routine pipeline inspection including tracking and measurement of depth of cover for buried pipelines, which is also applicable for control umbilicals and power/control cables.

Cathodic protection (CP) potential measurements may be completed by CP probe as shown in Figure 23-2. This type of measurement is normally carried out by a workclass ROV.

Cleaning may be performed by an ROV with brushing tools or high-pressure wet jets and grit entrainment as shown in Figure 23-3.

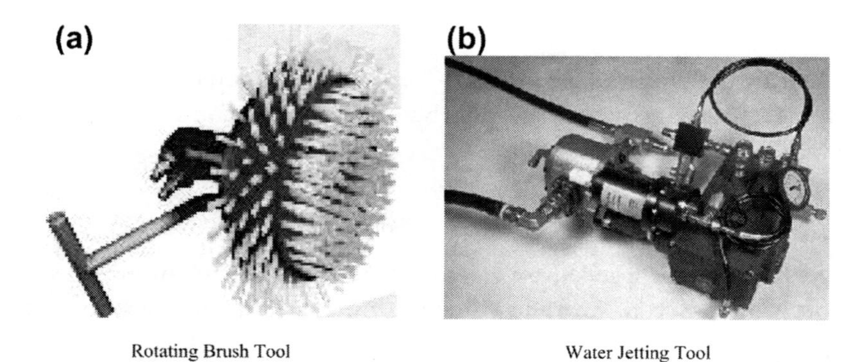

Figure 23-3 Rotating Brush and Water Jetting Tool *(Part a courtesy of Specialist ROV Tooling Services, Inc. [7]; part b courtesy of Subsea 7 Inc. [8])*

Crack detection may be performed by an ROV with magnetic particle inspection (MPI), eddy current, alternating current field measurement (ACFM) methods, etc.

23.2.6. Maintenance and Repair

Maintenance activities include repair or replacement of modules subject to wear. Maintenance is normally performed by retrieving the module to the surface and subsequently replacing it with a new or other substitute module.

Retrieval and replacement have to be anticipated during subsea equipment design. Some modules such as multimeters, chokes, and control pods are subject to removal and replacement. A completed replacement may have to be carried out due to the significant wear on or damage to non-retrievable parts of subsea equipment.

Due to the difficulty and expense of maintenance and repair, the operation may be continued with regular monitoring if the damaged module is not readily replaced and does not prevent production.

23.3. ROV SYSTEM

23.3.1. ROV Intervention System

An ROV system used in subsea engineering, as shown in Figure 23-4, can be divided into the following subsystems:

- Control room on deck for controlling the ROV subsea;
- Workover room on deck for ROV maintenance and repair;
- Deck handling and deployment equipment, such as A-frame or crane/winch;
- Umbilical to power ROV subsea and launch or recover ROV;
- Tether management system to reduce the effect of umbilical movement on the ROV;
- ROV for subsea intervention.

23.3.1.1. ROV Categories

ROV can be divided into fives classes as summarized in Table 23-1.

23.3.1.2. Topside Facilities

Suitable deck area and deck strength, external supplies, and ease of launch and recovery should be provided on deck for safe and efficient operation of ROVs.

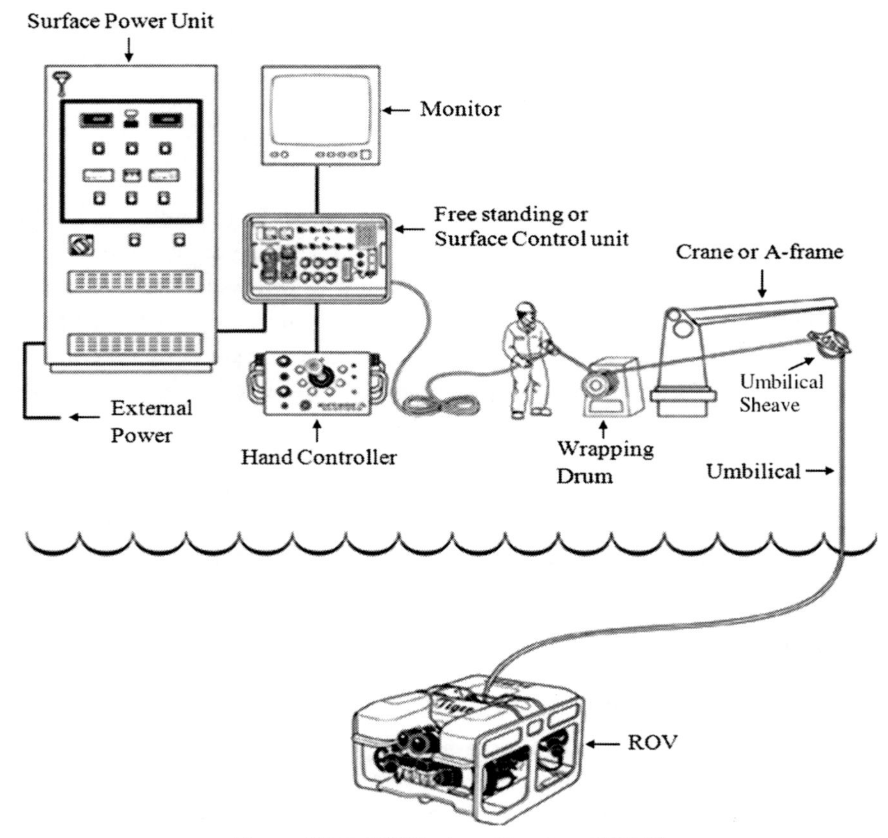

Figure 23-4 ROV System *(Courtesy SEAEYE)*

ROV control stations vary from simple PC gaming joysticks to complex and large offshore control containers/rooms on the platform or vessel. The control stations contain video displays and a set of operator/ROV interface-controlling mechanisms. A typical control container consists of operator console, lighting, electrical outlet, fire alarm and extinguishers, etc.

23.3.1.3. ROV Launch and Recovery Systems (LARS)

The LARS consists of a winch, winch power unit, crane/A-frame with fixed block, and ROV guiding system.

Generally speaking, launch and recovery activities can be achieved by a simple rope with uplift force. However, to facilitate the deployment and recovery of the rope, a reel/drum is used, and a motor is usually used to rotate the reel and provide the uplift force. The motor may be either

Table 23-1 ROV Classes [4]

ROV Classes	Horse Power	Power Sources	Water Depth	Payload	Utilization
Observation Class	<20	Electric	Limited	Minimal to None	Observation only
Observation ROVs are employed for visual inspection/monitoring and diver assistance. These systems are typically fitted with light sensors, probes and a simple grabber but carry little or no payload;					
Light Work Class	20–75	Electro-hydraulic or electric	1000–3000 m	Moderate lift and payload	Survey, Minimum drilling support
Light work ROVs are used for tasks ranging from inspection, observation and assistance during drilling operations and light subsea tasks (e.g. 150–200 kg). They can carry one or two cameras, sonar and a single manipulator arm; they are able to carry light payload tasks.					
Work Class	75–100	Electro-Hydraulic	1000–3000 m	Heavy Lift & Payload	Construction, Pipelay, Drilling and Completion
Work ROVs are employed for various tasks such as drill support, construction and repair tasks, platform cleaning, subsea tool deployment and operations (e.g. up to 500 kg). They feature higher hydraulic power and payload capacity, more sensor channels and are generally fitted with either a grabber or a seven-function manipulator arm (e.g. for drill support) or two seven-function manipulator arms (e.g. for construction tasks).					
Heavy Work Class	150+	Electro-Hydraulic	2000–5000 m	Ultra heavy Lift & Payload	Major Construction and Telecommunications
Heavy work class ROVs such as Towed or Bottom Crawling Vehicles are typically purpose built to a particular intervention task e.g. subsea trenching or pipeline repair.					

a hydraulic motor or an electromotor with/without a gear box used to reduce the rotary speed and increase the torque force. The system of motor, reel/drum, base frame, and other ancillary structures such as a brake and clutch is normally called a *winch*. A fixed block, sustained at the end of a crane boom/A-frame beam, is used to change the upward direction of the required winch force to a downward direction and position the winch on the lower structure, for example, the deck.

To restrain ROV motion while it is being lowered from the air to the water surface a LARS is used. This helps prevent, for example, damage to the umbilical by the bilge keel if side deployment is being used. The LARS may be equipped with a docking head, cursor, or guide rails, as shown in Figures 23-5, 23-6, and 23-7.

23.3.1.4. Umbilical and TMS

One characteristic difference between an ROV and an autonomous underwater vehicle (AUV), is that the ROV has an umbilical that runs between the support vessel and the ROV to transport hydraulic/electronic

Figure 23-5 Snubber-Rotator Docking Head *(Courtesy of Saab Seaeye Ltd. [9])*

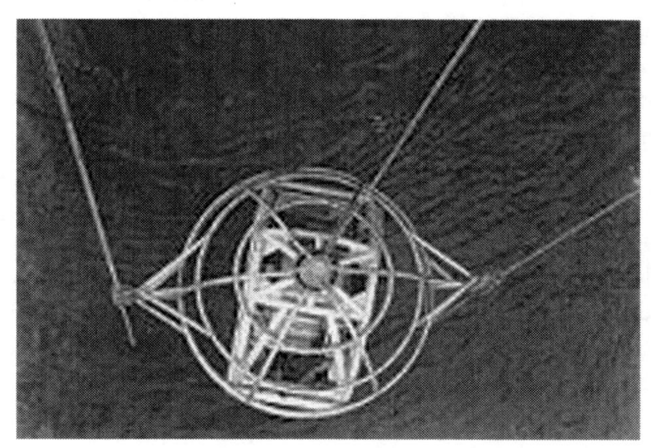

Figure 23-6 Wire-Guided Cursor [10]

Figure 23-7 Guide Rail System [10]

power from the vessel to the ROV and information gathered from the ROV to the surface. The AUV, on the other hands, is a robot that travels underwater without tethering to the surface vessel/platform. An ROV is usually armored with an external layer of steel and has torque balance capacity. The diameter and weight of the umbilical should be minimized to reduce the drag force due to waves and currents as well as lifting requirements during launch and recovery of the ROV from the water to the surface. Normally the umbilical has a negative buoyancy, and the umbilical may be attached with buoyancy, for example, every 100 m (328.1m) to avoid entanglements between the umbilical and subsea equipment or the ROV itself during shallow-water operation.

A tether management system (TMS) is used to deploy the ROV for deepwater applications where the umbilical with negative buoyancy can launch and recover the TMS and ROV. The connection cable between the ROV and TMS can be an umbilical called a *tether* that has a relatively small diameter and neutral buoyancy. The TMS is just like an underwater winch for managing the soft tether cable. A TMS has two significant advantages:

1. ROVs can be moved more easily due to deleting the force implied by the umbilical, which may be the same as the flying resistance of the ROV itself in a water depth of 200 m (656.2 ft) and increase rapidly with increasing water depth.
2. There is no need to use the ROV's own thrusters to get the ROV down to the working depth near the seabed. A powered TMS, i.e., installing some thrusters to TMS cages may be carried out to account for large drag force on TMS due to significant currents (e.g., current velocity of 1 to 1.5 knots) in some areas.

The TMS is designed to manage the tether and can be either attached to a clump weight, mainly for the observation ROV, or to a cage deployment system, as shown in Figures 23-8 and 23-9, mainly for workclass ROVs.

23.3.2. ROV Machine

23.3.2.1. Characteristics

Configuration

Most workclass ROVs have a rectangular configuration and an open Al-based frame that supports and protects the thrusters for propulsion, underwater cameras for monitoring, lights and other instruments such as closed-circuit television for observation, the gyrocompass for heading detection, depth gauges for depth detection, an echo-sounding device for altitude detection, and scanning sonars for environment inspection.

Figure 23-8 Top Hat Type TMS *(Courtesy DOF SUBSEA)*

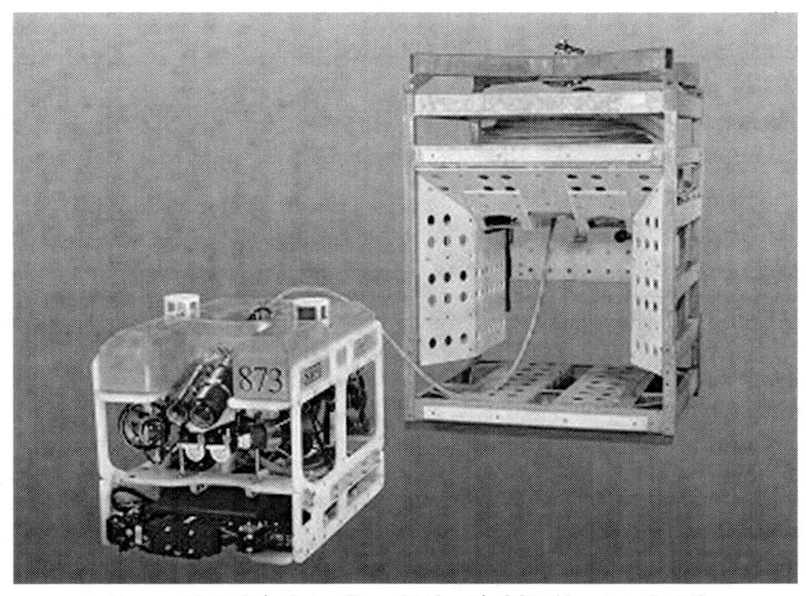

Figure 23-9 Side Entry Type TMS with ROV *(Courtesy SAAB)*

Most ROVs are near neutral buoyancy underwater. They do have a little buoyancy to make sure the ROVs can float to the water surface during emergency conditions or if they break. An ROV moves downward with a vertical thruster. Generally, the buoyancy is provided by synthetic foam material above the Al-based ROV frame. An ROV's weight is typically in the range from 1000 to 3500 kg. Examples can be seen from the table in Annex A of API RP 17H [1]. The higher buoyancy center and lower weight of gravity ensure that the ROV provides good stability performance.

Operation Depth

Traditionally, ROVs had been designed and built for operations in water depths of 100 to 1000 m (328 to 3280ft), primarily supporting drilling operations, including seabed surveys, water jetting, and seal ring installation, as well as providing light construction support and inspection work. These ROVs have payload capacities of around 250 kg with power ranging from 40 to 75 hp.

As the oil and gas industry has probed into deeper and deeper waters, demand has increased for ROVs that provide diverless solutions for such tasks as remote interventions and pipeline/umbilical tie-ins. Table 23-2 lists the operational water depths of typical ROVs.

Payload

The payload capacity of an ROV is limited by:

- ROV power;
- Structural integrity;
- Manipulator load an torque capacity;
- Current condition.

Table 23-2 Operational Water Depth of Typical ROVs

ROV Name	Challenger	Hysub 60	Pioneer HP	Super Scorpio	MRV	Diablo
Operational Water Depth (m)	1572	2000	1500	914	1500	2000
ROV Name	Triton XLS	SCV-3000	Hydra Magnum	Hydra Millennium	INNOVATOR™	
Operational Water Depth (m)	3000	3000	3144	3144	3500	

Payload capacities for typical ROVs are listed in Table 23-3.

Table 23-3 Payload Capacity of Typical ROVs

ROV Name	Challenger	Diablo	Pioneer HP	Super Scorpio	MRV
Payload Capacity (kg)	113	250	100	100	220
ROV Name	Triton XLS	SCV-3000	Hydra Magnum	Hydra Millennium	Maximum
Payload Capacity (kg)	300	100	227	318	499

23.3.2.2. Navigation System

The navigation of ROV includes general navigation and accurate navigation. The deck reckoning method and hydroacoustic method are used for general navigation. The hydroacoustic method is the most widely used today via the LBL system, in which there is a responder array on the seabed, at least one transponder set on the ROV, and one receiver set on the vessel.

Accurate navigation is used to lead the ROV to the target object. A gyrocompass provides the ROV heading and a viewing system is used that comprises imaging sonar/low light cameras and lights. Normally the power of a subsea light is between 250 and 500 W. An ROV might also have 750 to 1000 W of power for large lights and 100 to 150 W of power for small lights. There is also a cubic TV to provide 3D object configurations, that is, to obtain data about target thickness and distance between the ROV and the target.

23.3.2.3. Propulsion System

The propulsion system for an ROV consists of a power source, controller for an electric motor or a servo-valve pack for a hydraulic motor, and thrusters to adjust the vehicle condition (trim, heel, and heading) and to propel the vehicle for navigating from the TMS to the work site, and vice versa.

Being the main part of an ROV propulsion system, the underwater thrusters are arranged in several ways to allow for proper maneuverability and controllability of the vehicle through asymmetrical thrusting and varying the amount of thrust. The thrusters need to be adequately sized for countering all of the forces acting on the vehicle, including hydrodynamic and workload forces. There are a wide range of thrusters from electrically

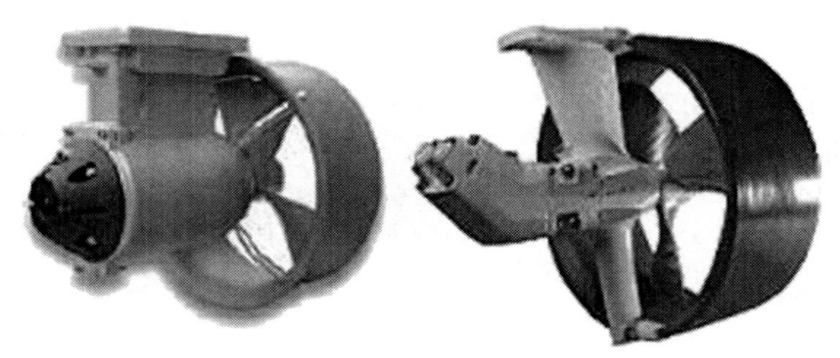

Figure 23-10 Typical Electric and Hydraulic Thrusters *(Courtesy of Sub-Atlantic [13])*

powered to hydraulically powered. In general, electrical thrusters are used for smaller vehicles, while the hydraulic ones are used for larger and workclass vehicles. Typical electric and hydraulic thruster examples are shown in Figure 23-10. When selecting thrusters for an ROV, the factors of power, efficiency, pressure, flow, weight, size, and forward /reverse characteristics should be taken into account.

23.3.2.4. Viewing System

A wide range of underwater video cameras are used in ROVs for viewing purposes, typically for navigation, inspection, and monitoring.

Camera image sensors include low-light silicon intensified targets (SITs), charge-coupled devices (CCDs), and HADs for high-definition images. Some cameras are equipped with LED lights providing illumination for close-up inspection and eliminating the need for separate lighting. Images captured by a camera are transmitted as video signals through the tether and umbilical to a video capture device on the water surface.

23.3.2.5. Manipulators

An ROV is normally equipped with two manipulators, one for ROV position stabilization, normally with a five -function arm, and the other for intervention tasks, normally with a seven-function arm. Manipulator systems vary considerably in size, load rating, reach, functionality, and controllability. They may be simple solenoid-controlled units or servo-valve-controlled position feedback units. The end of the arm is fitted with a gripper, usually consisting of two or three fingers that grasp handles, objects, and structural members for carrying out an activity or stabilizing the ROV.

23.4. ROV INTERFACE REQUIREMENTS

To facilitate subsea interventions such as drilling assistance, installation assistance, and inspection, maintenance, and repair (IMR) by ROVs, interfaces are introduced and produced for ROVs:

- Hydraulic work package and docking frame for the ROV itself;
- Hydraulic connector, valve override tool, and adapter for the ROV manipulator;

and also for subsea equipment to facilitate ROV application:

- X-mas trees fitted with a valve panel;
- Manifold valves fitted with a guidance system;
- Modules and tools equipped with a local control panel.

This section describes the requirements for the typical interfaces and is based primarily on API 17H [1].

23.4.1. Stabilization Tool

The stabilization of ROVs can be achieved in the following ways:

- Working platform, formed by utilizing part of the subsea structure;
- Suction cups, consisting of an arm attached to the ROV and a suction cup on the end of the arm which may be used when the ROV carries out manipulator operations such as cleaning, inspection, and individual valve operation;
- Grasping, which can be widely used for all items of subsea hardware;
- Docking, used with a single- or twin-docking tool deployment unit (TDU) together where the loading of a subsea equipment interface is not desirable. The receptacle is incorporated into the structure of the subsea equipment, whether as a separate bolted or welded-in unit, or incorporated as part of the subsea equipment.

Figure 23-11 shows grasping handles and Figure 23-12 shows a suction cup used for an ROV.

The interfaces should satisfy following requirements:

- Working platforms for ROVs should be flush and free from obstruction.
- The subsea structure should be a flat surface broadly adjacent to the task area for the use of a suction cup for stabilization purposes.
- Grasping intervention interfaces should be designed to withstand a minimum force of 2.2 kN (500 lbf) and a gripping force of 2.2 kN (500 lbf) applied from any direction.
- The docking probe should have a fail-safe release as shown in Figure 23-13 and overload limitation features.

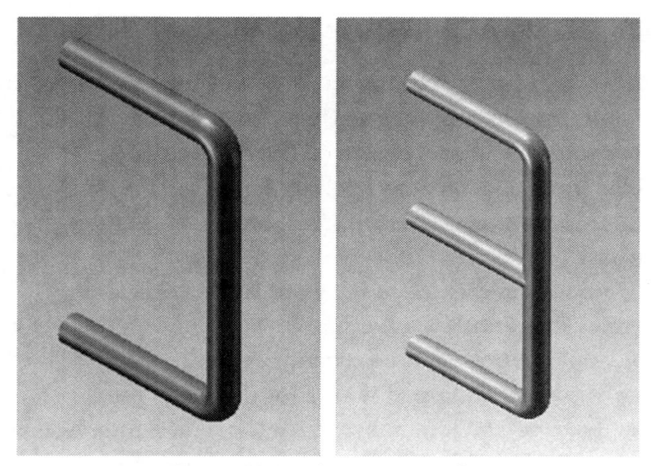

Figure 23-11 Grasping Handles

Figure 23-12 Suction Cup *(Courtesy of Canyon Offshore Inc. [14])*

23.4.2. Handles

Handles are the interface between tools such as torque tools or hot stabs and the ROV manipulators or ROV-mounted tool TDU. The interface configuration is illustrated in Figure 23-14.

The interface should satisfy following requirements:

- The stem of handles should be capable of resisting the maximum operational forces regardless of whether they are linear forces or rotary forces.

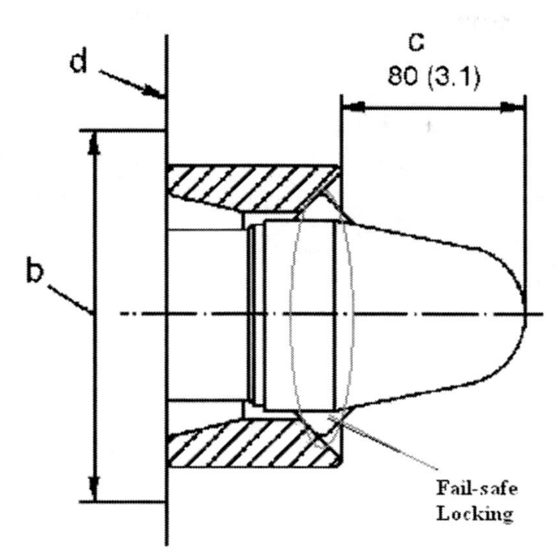

Figure 23-13 Docking with Fail-Safe Locking [1]

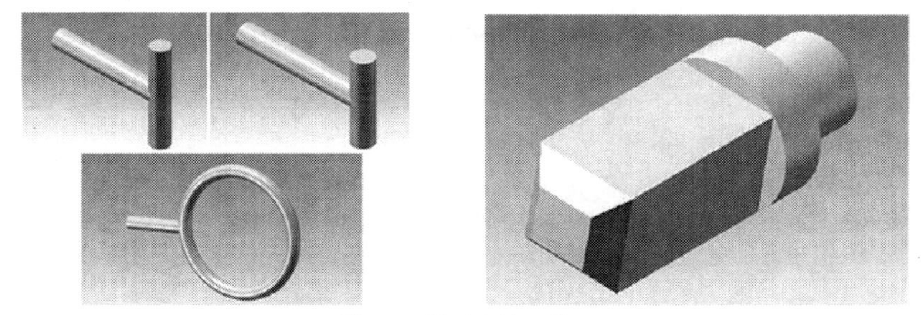

Figure 23-14 Handles

- Any out-of-line forces generated by the operator in linear applications have to be taken into account by the compliancy between the handle and the attachment to the tool.
- The handle should be marked to display the direction of movement of the handles to reduce the probability of damage.

The interface can be operated by an ROV-mounted tool TDU or manipulator.

23.4.3. Torque Tool

23.4.3.1. Low Torque

This interface provides for ROV operation of ball and needle valves, clamps, etc. The interface on the subsea equipment comprises a paddle or T-bar enclosed in a tubular housing, as shown in the following figures, which may

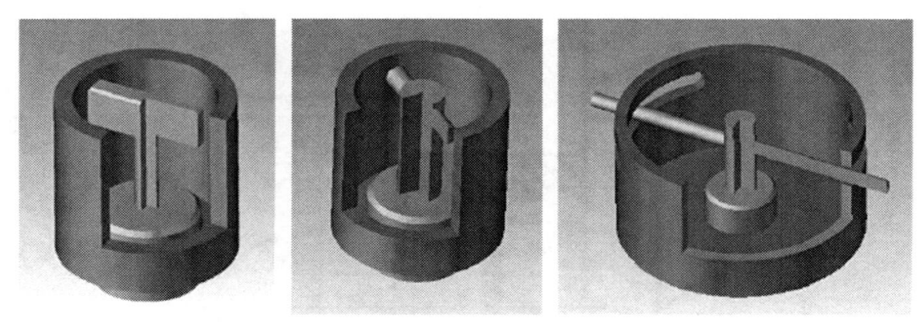

Figure 23-15 Types A, B, and C Low-Torque Interface Receptacles

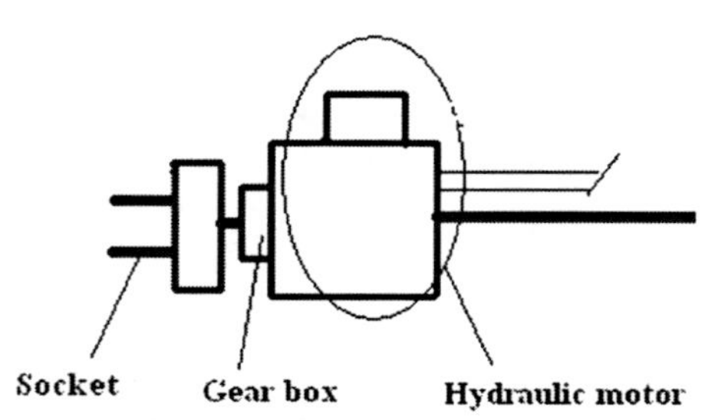

Figure 23-16 Sketch of Torque Tool Operated by an ROV

be incorporated into a panel by bolting or welding, be free standing, or be made as part of the subsea equipment.

Figure 23-15 shows types A, B, and C low-torque interface receptacles. The interface should satisfy the following requirements:

- If the interface receptacle is incorporated into a panel, the panel should be flush with the docking face.
- The interface should be manufactured from material with a minimum tensile strength of 450 MPa (65,300 psi) so that it can operate at the specified torques, but the engineer is free to specify other materials where different load conditions exist.
- Protection from marine growth and corrosion will be necessary in most environments, so consideration should be given to the use of corrosion-resistant materials or appropriate coatings.
- Access is required in the area along the drive stem axis.

The interface tool of a low-torque tool can be operated by an ROV-mounted tool TDU or manipulator as shown in Figure 23-16.

23.4.3.2. High Torque

High-torque interfaces provide for the ROV operation of tree valves, clamps, satellite control module (SCM) lockdown, shackle release, etc.

Figure 23-17 High-Torque Interface Receptacle

The interface comprises a square driving stem enclosed in a tubular housing, as shown in Figure 23-17, which can be incorporated into a panel by bolting or welding, be free standing, or be made as part of the subsea equipment.

The interface should satisfy the following requirements:

- If the interface receptacle is incorporated into a panel, the panel should be flush with the docking face.
- The interface should be manufactured from material with a minimum tensile strength of 450 MPa so that it can be operated at the specified torques.
- Protection from marine growth and corrosion are necessary in most environments, so the use of corrosion-resistant materials or appropriate coatings should be considered.
- Access is required in the area along the drive stem axis.

The interface tool for a high-torque tool, as shown in Figure 23-18, can be operated by an ROV-mounted tool TDU or manipulator.

23.4.4. Hydraulic Connection Tool

The hot stab/hydraulic connection tool interface provides for ROV typical operation of valve actuation, seal and connection testing, chemical injection, planned or emergency release, and fluid collection. The interface comprises male and female mating halves, as show in Figure 23-19. The connection allows pressurization between two isolated sections separated by seals.

The interface should satisfy the following requirements:

- The female receptacle, permanently mounted to the subsea system, should be manufactured with corrosion-resistant materials making it suitable for long-term immersion in seawater.

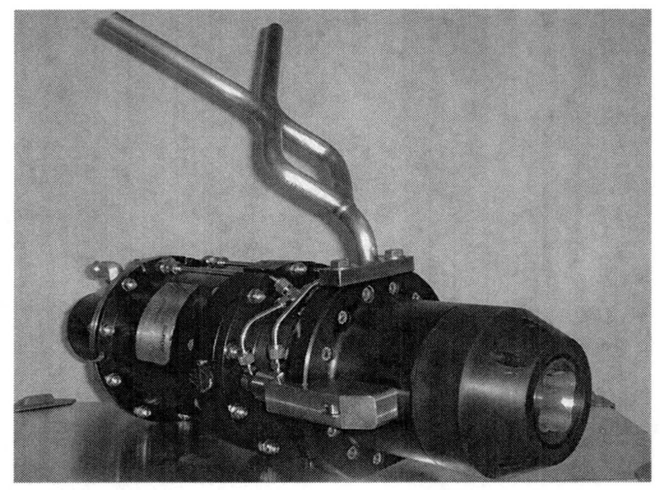

Figure 23-18 High-Torque Tool per API 17D *(Courtesy of Fugro-imp ROV Ltd. [15])*

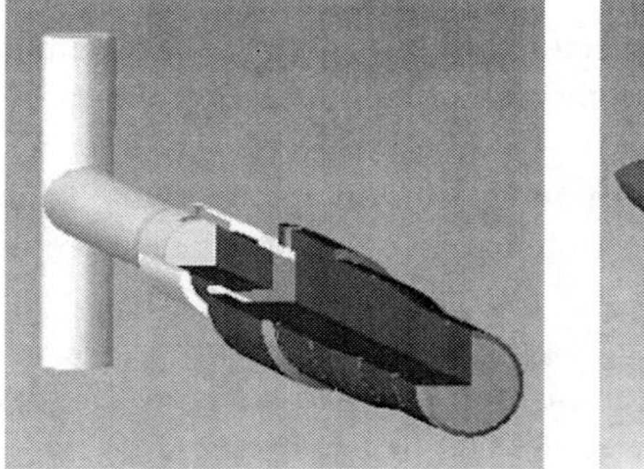

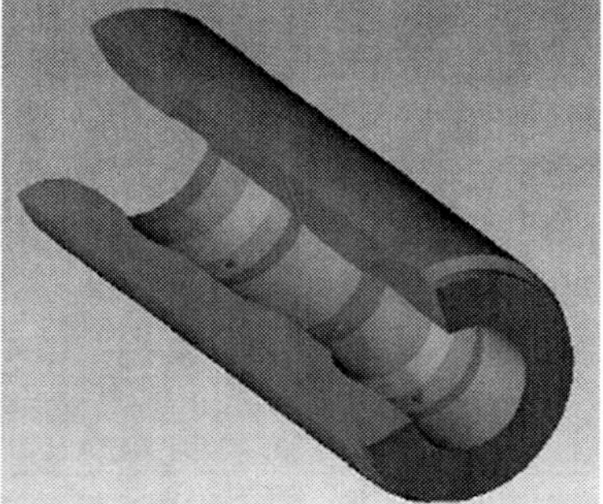

Figure 23-19 Male and Female Mating Halves of a Hot Stab Hydraulic Connection

- The hot stab receptacle should be designed so that it does not fill with debris while not in use.
- The hot stab should have a flexible joint in the handle or stab to ease connection and disconnection activities.
- The receptacle should be self-sealing and watertight when uncoupled.

Figures 23-20 and 23-21 show two different types of hot stabs from different manufacturers. The interface tool can be operated by an ROV-mounted tool TDU or manipulator.

Figure 23-20 Hot Stab *(Courtesy of Nemo Offshore Pty Ltd. [16])*

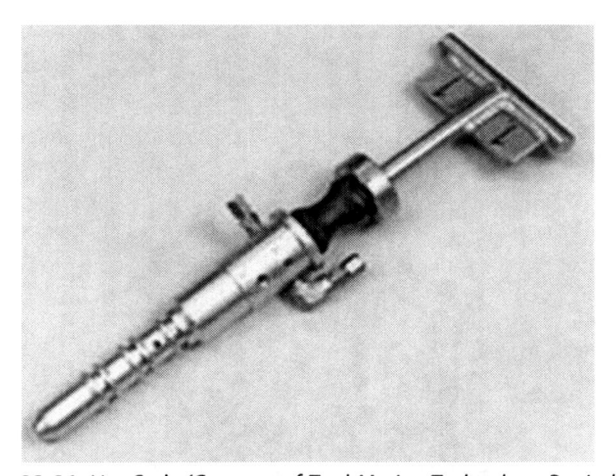

Figure 23-21 Hot Stab *(Courtesy of Tool Marine Technology Pty Ltd. [17])*

23.4.5. Linear Override Tool

The ROV interface is primarily/typically for the ROV operation of hydraulic gate valves open after fail-safe closure, where it may be incorporated as part of the valve actuator. The linear override tool (LOT) interface tool is used to transmit the axial force to the stem of hydraulically actuated gate valves.

The interface on subsea equipment comprises a flange around a central stem, and it can be operated by the LOT handled by an ROV-mounted tool (e.g., ROV + TDU/manipulator).

Figure 23-22 shows a LOT made by Subsea 7. The linear actuator override tool (LAOT) has two discrete units: the tool itself and a tool carrier, which is used to deploy, lock (actuate the valve on subsea equipment at the same time), and retrieve the tool.

Figure 23-23 shows an example of a standard linear override/push interface, which belongs to a type A interface. The interface has to satisfy the following requirements:

- The interface flange should endure the load induced by the maximum push force based on that required to open gate valves at full differential pressure and should be manufactured from material with a minimum tensile strength of 450 MPa to operate at the above loads.
- Protection from marine growth and corrosion is necessary in most environments, so the use of corrosion-resistant materials or appropriate coatings should be considered.

Figure 23-22 Linear Actuator Override Tool *(Courtesy of Subsea 7 Inc. [8])*

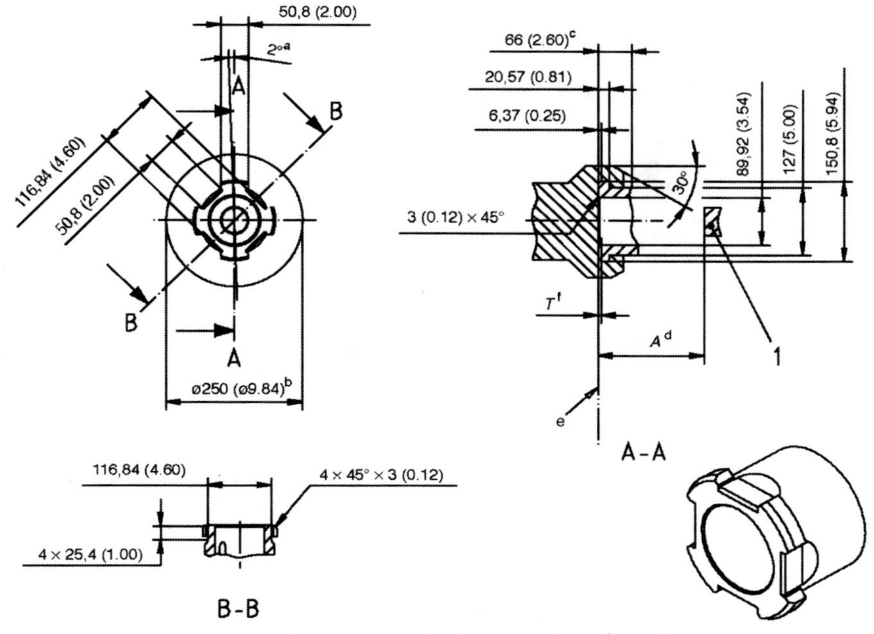

Figure 23-23 Linear Push Type A Interface [1]

- It is important to check the stroke on the TDU to ensure sufficient clearance to fully make up the linear push device and to subsequently remove it.
- Access is required in the area along the drive stem axis.

23.4.6. Component Change-Out Tool (CCO)

The ROV interface provides for ROV operation for the replacement of control pods, chokes, multiphase meters, etc. The ROV interface comprises two identical landing units, with one central lockdown receptacle and two weight receptacles including soft landing dampers or cover plates, as shown in Figure 23-24.

The ROV interface should satisfy the following requirements:

- The top face of the landing units should be positioned flush with or above the component lifting mandrel.
- Design loads for the landing units are particular to the component and should be evaluated on a case-by-case basis.
- The landing units may be structurally supported from either the structural framework or from the component base unit. The clearance

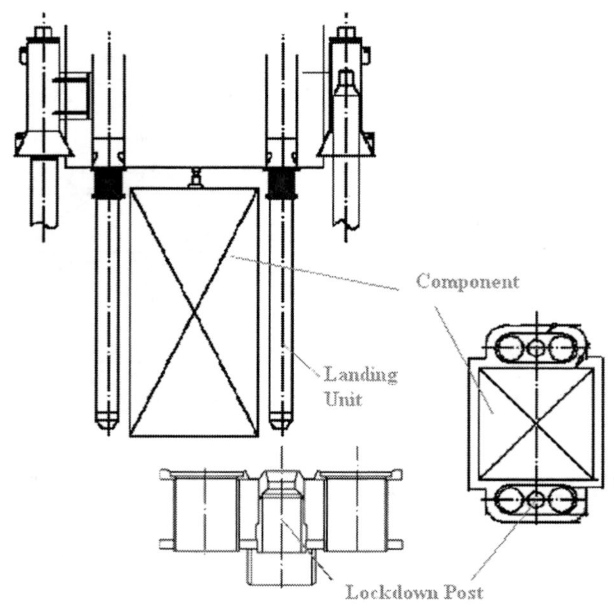

Figure 23-24 Component Change-Out Interface [1]

required for weight transfer units should be taken into account the support arrangements.

The interface can be operated by an ROV-mounted tool or an ROT with ROV positioning, as shown in the Figure 23-25.

23.4.7. Electrical and Hydraulic Jumper Handling Tool

This ROV interface provides for the ROV operation of installing the electrical flying lead, steel flying lead for transporting hydraulic fluid, and stab plate of the assembly of electric connectors or hydraulic couplings or both.

A single-connection interface handling tool can be operated by either a manipulator or TDU, whereas a multiple quick connect (MQC) stab plate can be operated by a TDU or a combination of manipulator and tool elevator. An oil-filled hose conduit connection interface can be used for a single connection.

The MQC interface should satisfy the following requirements:

- The material specification, plate shape, and thickness should be able to resist the maximum operational forces.
- A release mechanism should be incorporated for the inboard locking assembly in the event of stab plate jamming.

Figure 23-25 Component Change-Out Tool *(Courtesy of Subsea 7 Inc. [8])*

23.5. REMOTE-OPERATED TOOL (ROT)

23.5.1. ROT Configuration

An ROT system can be divided into the following subsystems, as shown in Figure 23-26:

- Deck handling and deployment equipment, such as A-frame or crane/winch;
- Controls room on deck to control the ROT subsea operation;
- Workover room on deck for ROT maintenance and repair;
- Umbilical/liftwire to power the ROT subsea and deploy or recover an ROV;
- The ROT.

A ROT is mainly used for module replacement/change-out and flowline tie-in, both of which require a handling force larger than that of an ROV. ROTs are usually deployed on liftwires or a combined liftwire/umbilical, and the lateral guidance is powered by an umbilical with dedicated thrusters, ROV assistance, or guidewires.

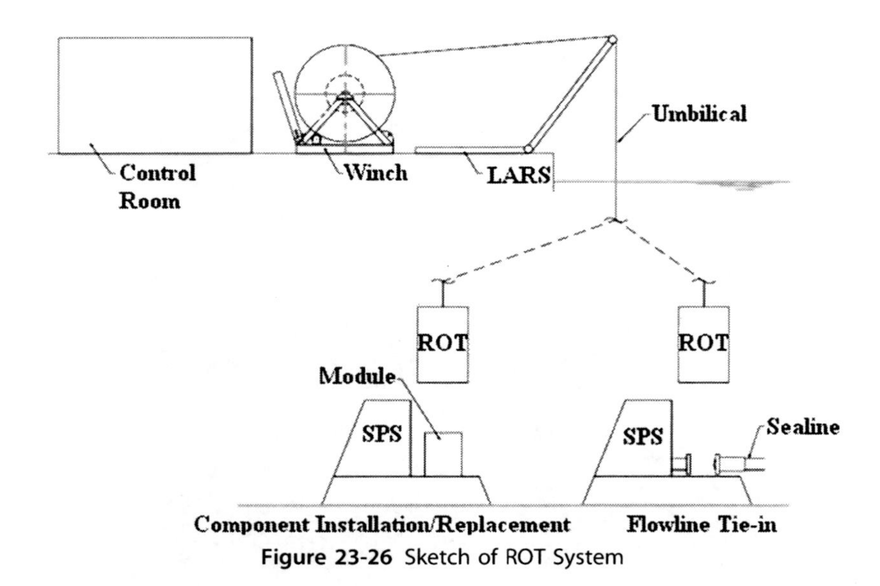

Figure 23-26 Sketch of ROT System

The ROT system should provide for safe locking of the replacement module during handling, deployment, normal operation, and emergency conditions such as power failure.

There are three generations of ROT for tie-in tools:

- A first-generation ROT is two separate tools that pull in and connect individually.
- A second-generation ROT is a combined tool that pull ins and connects together.
- A third-generation ROT is an ROV-mounted pull-in and connection tool.

23.5.2. Pull-In and Connection Tool

In this section, basic information about a pull-in and connection tool is provided. The pull-in and connection system includes the following main equipment:

- Connectors with seal assemblies. Connector type determines the required function of hydraulic supply, torque tools, integrated seal plate handling, and replacement tools;
- Pull-in and connection tools;
- Hubs, caps, and terminations;
- Pull-in porches/alignment structures for interfacing with subsea structures.

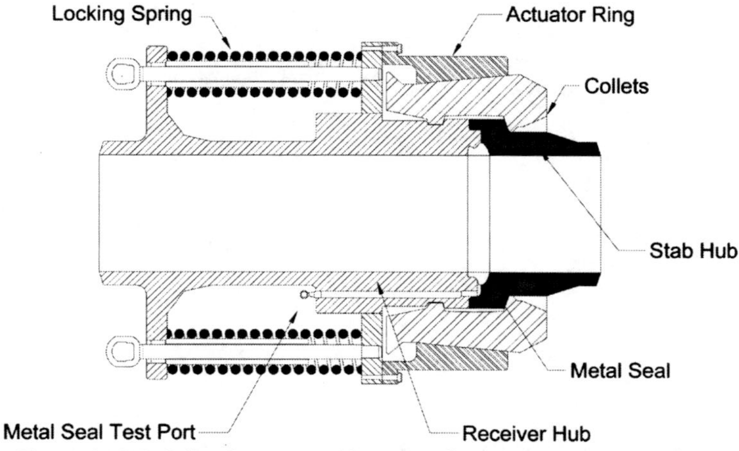

Figure 23-27 Collet Connector *(Courtesy of Cameron international [18])*

The collet connectors and clamp connectors are widely used connectors in the subsea industry. They are shown in the Figures 23-27 and 23-28. The stab hub (i.e., inboard hub) may be fixed on the subsea equipment. Figure 23-29 shows an example of a pull-in and connection tool.

The following are the main procedures required to make a connection:

- Rough alignment between inboard and outboard hubs, approached by pull-in tool or guiding funnel;
- Precise alignment after two hubs mate, completed by the connection tool;
- Connection by connecting tools (e.g., jacking screw operation by torque tools, and seal pressure test by ROV).

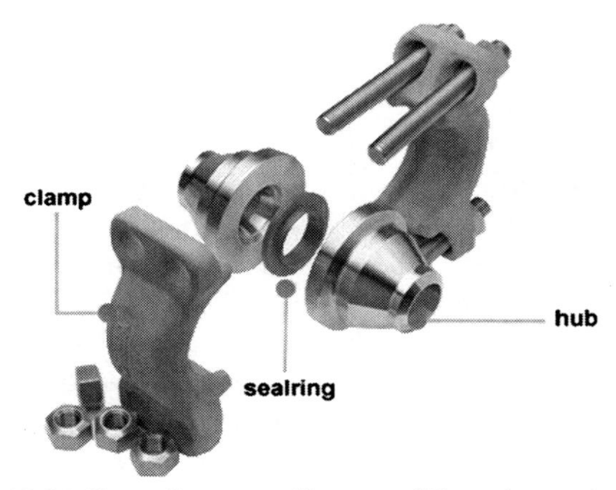

Figure 23-28 Clamp Connector *(Courtesy of Vector International [19])*

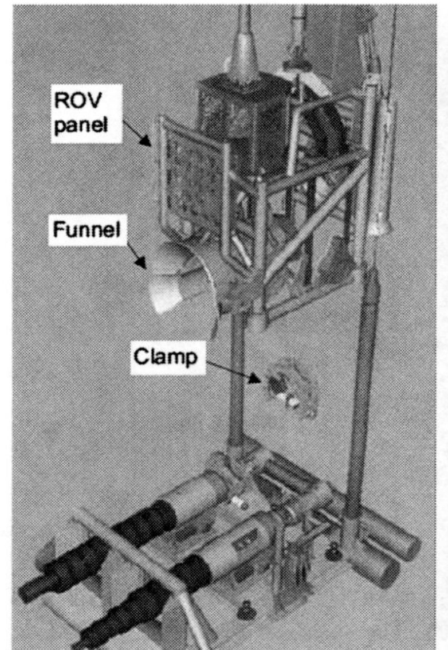

Figure 23-29 Pull-In and Connection Tool *(Courtesy of Alker Solutions [20], Formerly Kvaerner)*

To prevent intrusion of saltwater, dirt, etc., into the hub sealing area, caps should be placed on the umbilical termination head before the connecting operation.

23.5.3. Component Change-Out Tool

A component change-out (CCO) tool is used to recover and reinstall subsea modules such as:

- Chokes;
- Valves;
- Multiphase meters;
- Subsea control modules;
- Chemical injection modules;
- Pig launchers.

A CCO is often deployed by the guideline method or as a fly-to-place tool. It can be deployed to the seabed in an individual basket using the fly-to-place method, and the ROV with interface skid can dock onto the CCO and establish control of the CCO with, for example, the electrohydraulic stab plate connector. The ROV will then unlock the ROT from the basket, fly it to place, and dock onto the subsea structure. Figure 23-30 shows ROV-mounted tool deployment.

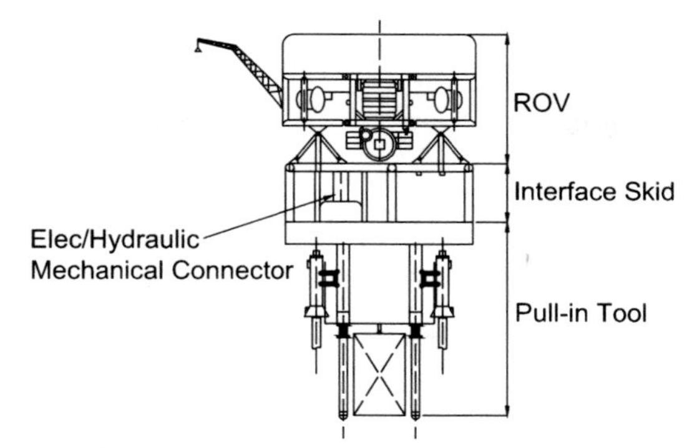

Figure 23-30 ROV Deployed ROT Sketch

This is the landing sequence for module replacement:

- Soft land the ROT to the subsea structure.
- Precisely land and align the module onto the subsea interface.

REFERENCES

[1] American Petroleum Institute, Remotely Operated Vehicle (ROV) Interfaces on Subsea Production Systems, first ed., API-RP-17H, 2004.

[2] American Petroleum Institute, Remotely Operated Tool (ROT) Intervention System, first ed., API- RP-17M, 2004.

[3] Norwegian Technology Centre, Remotely Operated Vehicle Services, Rev 1, NORSOK standard U-102, (2003).

[4] Norwegian Technology Centre, Subsea Intervention Systems, Rev 2, NORSOK standard U-007, (1998).

[5] J.R. Gert, Challenges and Experience in ROV-based Deepwater Seabed Mapping, OTC 13158, Offshore Technology Conference, Houston, Texas, 2001.

[6] Deepwater Corrosion Services Inc, http://www.stoprust.com.

[7] Specialist ROV Tooling Services Ltd., http://www.specialistrov.co.uk.

[8] Subsea 7, http://www.subsea7.com.

[9] Saab Seaeye Limited, http://www.seaeye.com.

[10] C.I. Davis, E.B. Lallier, C.L. Ross, Protective Deployment of Subsea Equipment, OTC 15089, Offshore Technology Conference, Houston, Texas, 2009.

[11] Schilling Robotics, LLC. http://www.schilling.com.

[12] SMD Robotics Ltd. http://www.smd.co.uk.

[13] Sub-Atlantic. http://www.subatlantic.co.uk.

[14] Canyon Offshore Inc. http://www.helixesg.com

[15] Fugro-ImpROV Ltd. http://www.improvltd.co.uk

[16] Nemo Offshore Pty Ltd. http://nemo-offshore.com.au

[17] Tool Marine Technology Pty Ltd. http://www.tmtrov.com.au/index.asp

[18] Cameron International Corporation. http://www2.c-a-m.com/index.cfm

[19] Vector International. http://www.vectorint.com/

[20] Alker Solutions. http://www.akersolutions.com/Internet/default.htm

PART *Four*

Subsea Umbilical, Risers & Flowlines

CHAPTER 24

Subsea Umbilical Systems

Contents

Subsea Engineering Handbook
ISBN 978-0-12-397804-2, doi:10.1016/B978-0-12-397804-2.00024-2

24.1. INTRODUCTION

Subsea umbilicals are installed between the host facility and the subsea facility. In general, the umbilical includes a catenary riser (dynamic segment) transitioning into a static segment along the seabed to the umbilical termination assembly (UTA) at the subsea facility. For shorter lengths, the segments may be identical. The umbilical pull-head will include a split flange (or other) assembly for hang-off of the umbilical at the host facility. A bend stiffener or limiter may be installed at the top of the catenary riser and at the UTA.

General requirements for the umbilical system include:

- Electrical power, control, and data signals should, as a base case, be contained on the same pair of conductors.
- Super duplex steel tubes should be used. (Other materials can be considered but substantial documentation should be required to guarantee their applicability.)

- The umbilical is fabricated in one continuous length.
- The umbilical system is designed without any planned change-out over the design life.

A subsea umbilical is a combination of electrical cables, fiber optic cables, steel tubes, and thermoplastic hoses, or two or three of these four components that execute specific functions. These components are assembled to form a circular cross section. The functions and characteristics of the four umbilical components are described and specified in the following sections of this chapter.

Umbilicals are used in various ways by the offshore industry today. The main functions are listed below and described in the following sections:

- Subsea production and water injection well control;
- Well workover control;
- Subsea manifold or isolation valve control;
- Chemical injection;
- Subsea electrical power cable.

Figure 24-1 shows a typical subsea control umbilical and its cross section.

The umbilical delivery procedure typically includes the following steps and schedule:

- Feasibility study;
- Umbilical specifications and request for quotation;
- Qualification tests for fatigue and other tests (specifications and execution);
- Long-lead item procurement;

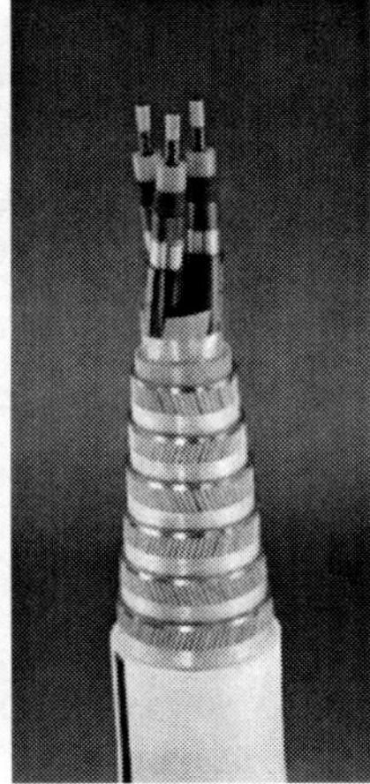
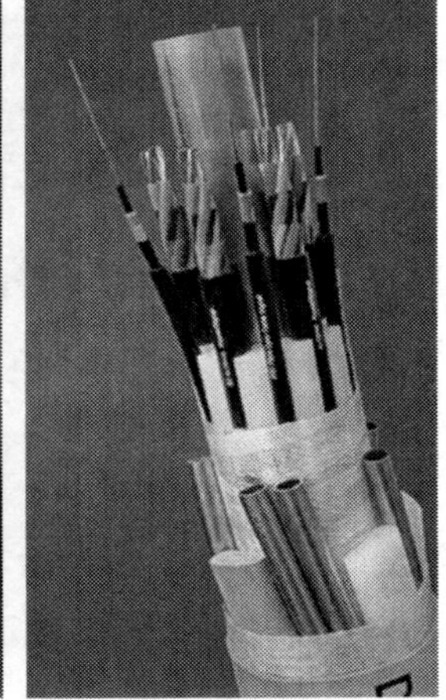

Figure 24-1 Subsea Umbilicals

- Bid evaluation;
- Supplier selection;
- Project sanction and umbilical procurement;
- Detailed umbilical design and analysis by the supplier;
- Third-party design verification by an analysis specialist;
- Prototype qualification tests;
- Umbilical manufacturing (normally requires a period of 1 year);
- System integration test;
- Umbilical delivery to host vessel;
- Commissioning;
- System start-up;
- Project management, QA/QC.

One of the earliest papers dealing with steel tube umbilical design was "Metal Tube Umbilicals—Deepwater and Dynamic Considerations" [1]. Another useful publication for further information is ISO 13628-5 [2], which is used as the standard for umbilical design and operation.

24.2. UMBILICAL COMPONENTS

24.2.1. General

A subsea umbilical consists of electrical cables, fiber optic cables, steel tubes, and thermoplastic hoses. It may also include two or three of these four components for executing specific functions. The umbilical components are designed and manufactured to meet the umbilical functional and technical requirements. Proper materials are chosen to manufacture the components, and verification and acceptance tests are done to demonstrate the conformance to the component functional and technical requirements.

The functions and characteristics of the four umbilical components are described and specified in the following sections.

24.2.2. Electrical Cable

The electrical cables are divided into two types: power cables and signal/ communication cables. Usually the power cable and signal cable are combined into one cable, which is called a power and control umbilical.

24.2.2.1. Power Cables

Power cables are used for the power supply of offshore platforms and subsea production equipment, such as control pod, pilot control valve, and electric pumps.

According to Section 7.2.2.1 of ISO 13628-5 [2], power cables voltage ratings are selected from a range of 0 V up to the standard rated voltages $U_0/U(U_m) = 3.6/6\ (7.2)$ kV RMS, where U_0, U, and U_m are as defined in IEC 60502-1 and IEC 60502-2.

24.2.2.2. Signal/Communication Cables

Signal/communication cables are usually used for the remote control/monitoring of subsea production equipment, such as operation of a pilot control valve, feedback of wellhead status, and operating parameters.

According to Section 7.2.2.2 of ISO 13628-5 [2], signal/communication cables are selected from a range of 0 V RMS up to $U_0/U\ (U_m) = 0.6/1.0$ (1.2) kV RMS, where U_0, U, and U_m are as defined in IEC 60502-1 and IEC 60502-2.

24.2.3. Fiber Optic Cable

Fiber optic cables are capable of continuous operation when immersed in a seawater environment. The fiber type is of either single-mode or multimode design. The design is as given in the manufacturer's/supplier's specifications. Individual fiber identification is by means of fiber coloring.

The fibers are contained within a package that prevents water and minimizes hydrogen contact with each fiber. The carrier package for mechanical protection and its contents are designed to block water ingress in the event that the fiber optic cable in the umbilical is severed.

24.2.4. Steel Tube

Umbilical steel tube referred to as *super duplex* steel tube is capable of continuous operation when immersed in a seawater environment and when it meets the requirements of ASTM A240 for either UNS S32750 or S39274 chemistries and the additional requirements specified herein and listed below:

- The tube is made by the pilger or cold-drawn process from tube hollows that should be 100% visually inspected prior to processing. Hollows should be demonstrated to meet the product chemistry requirements.
- The tube is in-line batch or continuously furnace or induction annealed in a nonoxidizing annealing atmosphere at a temperature and quench rate to be determined by the manufacturer.
- The tube may be built up on reels to the specified length by automatic orbital welding in accordance with preprogrammed and approved procedures.
- The tube meets all of the applicable material and process requirements of NACE MR-01-75.

- All tube lengths are cleaned to NAS 1638 Class 6 and measures taken to ensure contamination do not occur during transport and storage prior to incorporation into umbilical lengths.

24.2.5. Thermoplastic Hose

Thermoplastic hose is capable of continuous operation when immersed in a seawater environment.

24.3. UMBILICAL DESIGN

24.3.1. Static and Dynamic Umbilicals

24.3.1.1. Static Umbilicals

The design of umbilicals incorporates mechanical strength to withstand crushing and tensile loads during handling, installation, and service. The umbilical is also of sufficient weight to ensure satisfactory seabed stability. The umbilical and its pulling head/termination design allow for installation into the facility approach.

Wall thicknesses of steel tube elements are sized to meet requirements for allowable stresses under all installation and operational conditions. Other design analyses calculate:

- Maximum allowable tension and minimum breaking strength;
- Recommended back tension during lay;
- Strength of terminations;
- The effect of radial loads (collapse pressure);
- The effect of dropped objects and snagging (e.g., ship's anchor);
- The effect of installation tensioning devices;
- Maximum allowable impact loads;
- Bend radius and bending stiffness;
- Torsional balance;
- Hydrodynamic stability on seabed;
- Environmental loads and hydrodynamic stability of beach approach;
- Material and outer sheathing suitability for onshore applications.

24.3.1.2. Dynamic Umbilicals

The umbilical system is expected to operate in a static mode after installation. However, the umbilical system will be subject to dynamic loading during installation and to environmental loads in the facility approach. Further, potential unsupported spans along the seabed may be subjected to fatigue owing to vortex-induced vibrations (VIVs). Dynamic and fatigue analysis should be carried out to evaluate fatigue properties of the umbilical system given the anticipated installation and environmental loads and to establish the maximum allowable span lengths. Minimum required fatigue life may be 10 times design life.

24.3.2. Design

24.3.2.1. Conceptual Design

Preliminary Cross-Section Sizing

- Tube sizing;
- Possible interaction with vendors.

Preliminary Configuration Design

- Strength, interference, etc.;
- Preliminary component design;
- Early confirmation of feasibility.

Early Identification of Manufacturing Issues

- Issues affecting bid or spec requirements.

Service Environment

The umbilical is designed for immersion in seawater for the specified design life. Consideration should also be given to

- Storage prior to installation;
- Exposure to service fluids;
- The seabed and topsides environment in terms of radiation, ozone, temperature, and chemicals;
- Imposed dynamic conditions within the free-hanging regions
- Protection against dropped objects.

24.3.2.2. Detailed Design

Parameters

- Temperature range;
- Maximum working load;
- Minimum breaking load;
- Minimum bend radius;
- Dynamic service life.

On-Bottom Stability Study

The umbilical is designed to be sufficiently stable, when laid on the seabed, for the seabed condition and seabed current values.

The behavior of the umbilical on the seabed can be characterized by friction coefficients in the axial and lateral directions.

Cross-Sectional Design

One of the initial stages in the design of an umbilical is the placing of the components of the umbilical in the cross-section design. The cross section of an umbilical could include various items such as steel tubes for transporting hydraulic and other fluids, electrical cables, fiber optic cables, steel rods or wires for strength capacity, polymer layers for insulation and protection, and polymer fillers to fill in the spaces between the components and keep them in place.

Manufacture Design

- Lay-up;
- Sub-bundles;
- Inner sheath;
- Armoring;
- Outer sheath.

An outer sheath is applied as a continuously extruded thermoplastic sheath or as a covering of helically applied textile rovings.

24.3.3. Manufacture

The procedures for umbilical manufacturing should be in accordance with ISO 13628-5 [2].

24.3.3.1. Lay-Up

Lay-up operations are carried out in a clean, dedicated, controlled area, which is subject to a regular cleaning schedule.

Optimum fiber optic cable and hose or steel tube lay-up configurations, fillers, etc., are provided to minimize the overall diameter and weight while meeting the general performance and construction specifications and to ensure good flexibility.

Lay-Up Configuration

The minimum lay angle of the umbilical components should be confirmed. Umbilical components, steel tube, and optical fiber will be laid up in a continuous helix or planetary configuration. If hoses are used an oscillatory cabling technique will be used.

Damaging Pull

The cabling or lay-up is designed so that the individual components will not be strained, deformed, or otherwise affected when the components are subjected to a tensile pull. A table of maximum allowable tension on umbilical components throughout the manufacture process should be summarized and submitted.

Acceptance Testing

After cabling of the umbilical hoses or steel tubes, the following checks are conducted prior to extrusion of the inner/outer sheath on the umbilical. A hydraulic proof test of hoses or steel tubes is conducted after each cabling layer (1.5 times working pressure).

24.3.3.2. Inner Sheath

The operation is carried out in a clean dedicated, controlled area, which is subject to a regular cleaning routine.

The total cabled assembly is protected with an inner sheath extruded tightly over a taped assembly. The extrusion of the inner sheath OD should be monitored in two planes 90° apart.

24.3.3.3. Outer Sheath

The total cabled assembly is protected with an overall jacket extruded tightly over the assembly. The extrusion of the outer jacket OD is monitored in two planes 90° apart.

24.3.3.4. Marking

The marking of subsea umbilicals should usually be done according to Section 9.14 of ISO 13628-5 [2].

24.3.3.5. Main Manufacturers

The following is the introduction of main subsea umbilical manufactures in the world and their productions:

1. DUCO

   DUCO is a member of Technip Group. The work scope and products of DUCO include:
   - Design and verification of umbilicals and hardware;
   - Manufacture of thermoplastic hoses in sizes between 1/4" -1.5" ID with working pressure between 3000-12,500 psi;
   - Manufacture of steel tubes in sizes between 3/8" -3" ID, materials covers carbon steel, stainless steel, Duplex and Super Duplex;
   - Manufacture of opto-electric cables, low voltage cables ranging in cross sections between 2.5-25 mm^2; medium voltage cables ranging in size between 35 mm^2 to 400 mm^2 with voltage ratings up to 36 kV.

2. Kvaerner Oilfield Products
   With manufacturing sites located in Norway an USA, Aker Kvaerner provides subsea umbilicals for more than 15 years and its products include:
   - Steel tube umbilicals: from basic direct-hydraulic and electro-hydraulic umbilicals to large, integrated service umbilicals; Long continuous manufacturing without transition joints/offshore umbilical connectors;
   - Integrated production umbilicals (IPU): the IPU is designed based on the proven technology and qualified for both static and dynamic applications in shallow and deep water;
   - Power distribution umbilicals: dynamic and electrical analysis completed, PVC profile and construction technique for increased impact and crush resistance; carbon fiber rod technology used to enhance axial mechanical characteristics; Improved stability compared to conventional armoured power umbilicals.
3. Nexans
   Since its foundation in 1915, Nexans Norway has been the main supplier of subsea umbilicals in Norway. Nexans has supplied more than 2400 km of umbilical. Some typical products of umbilical include,
   - Umbilicals for Shell NaKika project;
   - Umbilicals for Statoil Snohvit field, includes 45 km long umbilical from land to subsea and represents a record-breaking length;
   - Umbilicals for BP's King, Atlantis and Thunder Horse projects in GoM.

24.3.3.6. Sample Manufacturing Plant Layout

Figure 24-2 illustrates the layout of a manufacturing plant for subsea umbilicals.

24.3.4. Verification Tests

24.3.4.1. Tensile Test

A representative length of the completed umbilical, which takes into account the end effects and pitch lengths of the umbilical components, should be subjected to a two-stage tensile loading program.

24.3.4.2. Bend Stiffness Test

A representative length of the completed umbilical is subjected to a bend stiffness test procedure.

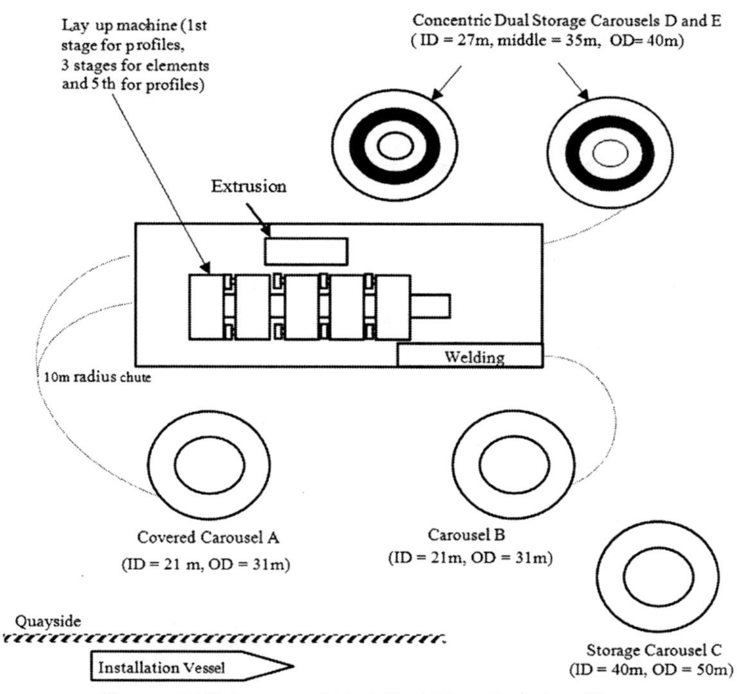

Figure 24-2 Layout of Umbilical Manufacturing Plant

24.3.4.3. Crush Test

A sample of the completed umbilical is subjected to lateral loading to allow determination of its resistance to deformation.

24.3.4.4. Fatigue Test

Mechanical testing is undertaken to determine the fatigue resistance of an umbilical. The test regime is chosen to demonstrate that a particular design or design feature is suitable to withstand the repeated flexures sustained by an umbilical during manufacture, transfer spooling, load-out, I- or J-tube pull-in, burial, and, for a dynamic installation, operational service throughout the service life.

24.3.5. Factory Acceptance Tests

The following final testing is conducted on the completed umbilical:

- Visual Inspection;
- Electric cable;

- Hoses;
- Tubes.

All test results are recorded and certified by the umbilical manufacturer.

24.3.6. Power and Control Umbilicals

As specified above, the power umbilical is used for supplying electrical power from shore to platforms, between platforms, or from platform to subsea equipment. Supplying power to offshore installations from energy sources onshore makes for smaller and lighter offshore structures, reduced personnel requirements, and lower CO_2 emission levels.

With this solution any number of installations can be linked and provided with power from an onshore power source. The power supply cable system can be expanded to form a network between offshore fields, providing flexible and safe power utilization for the oil and gas industry. Two types of submarine power cables are to be distinguished: "dry" design or "wet" design, with the former being more reliable but at a higher cost.

Figure 24-3 shows a typical power supply umbilical that has fiber optic cores for a control system. The remote control of unmanned installations is another application for submarine composite cables. Most subsea power cables installed offshore have a fiber optic element containing 8 to 32 optical fibers for signal transmission. The advantages of combining signal and power capabilities into one cable are as follows:

- Communication will not be influenced by weather or surface traffic.
- A greater bandwidth is available compared to radio frequencies.
- Optical fibers provide higher data transmission rates.

24.3.7. IPU Umbilicals

The IPU umbilical is designed to combine the function of an umbilical with that of a production or an injection flowline, and also for supplying high

Figure 24-3 Typical Power Supply Umbilical Including Fiber Optic Cores for a Control System

voltage power to potential subsea users. The temperature in the flowline is maintained through a combination of thermal insulation and active heating.

Typical IPU components and configurations include,

- A large bore central pipe, 6"-12", to transport well fluid, water, gas or other fluid that is required in large quantities;
- Around the central pipe, an annular shaped PVC matrix to keep the position of the spirally winded umbilical tubes and cables, and provide thermal insulation to the central pipe;
- Embedded in the PVC matrix, but sliding freely within it, the various metallic tubes for heating, hydraulic and service fluids, and the electrical/fiber-optic cables for power and signal;
- A protective sheath in polyethylene or similar material.

24.4. ANCILLARY EQUIPMENT

24.4.1. General

The subsea end of an electrical cable, optical fiber, thermoplastic hose, or metallic tube may be terminated in half a connector assembly, which can then be mated underwater. Alternatively, the umbilical components may be terminated directly into a subsea control pod or junction box. The design of umbilical terminations and ancillary equipment is invariably specific to a particular umbilical system and, as such, detailed specification data are the scope of ISO 13628-5 [2].

24.4.2. Umbilical Termination Assembly

A topside umbilical termination assembly (TUTA) is designed for dynamic umbilicals. A TUTA provides a termination point for the tubes, wires, and optical fibers from the bull nose assembly with stainless steel tube-to-tube fitting connections.

The TUTA assembly includes an electrical junction box for interfacing with the electrical wire and an optical junction box for interfacing with the fiber optic filaments from the bull nose assembly.

The metal tubes route to hydraulic couplers via super duplex steel tube pigtails and tube sockets where they are welded. Each set of electrical wires and fiber optic filaments is terminated with the female half of an electrical or optical connector that can be mated underwater.

24.4.3. Bend Restrictor/Limiter

The bend restrictor, shown in Figure 24-4, is also called a bend limiter and is used for preventing overstressing when the umbilical is unsupported over

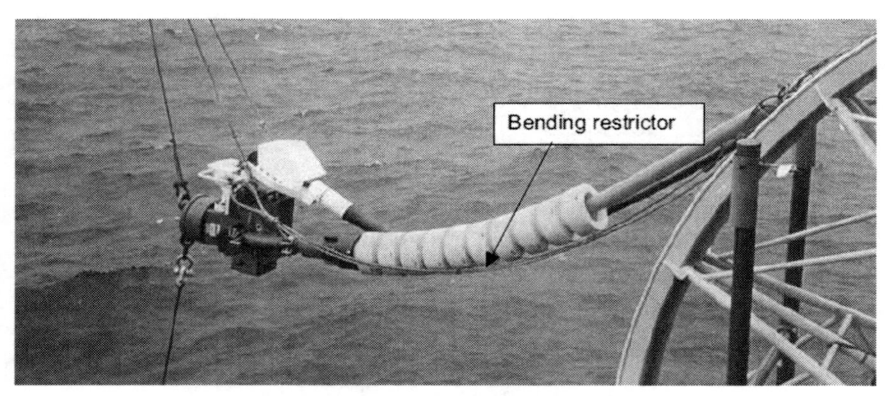

Figure 24-4 Bend Restrictor

a large free span. It is usually used where the umbilical is attached to the umbilical subsea termination point.

24.4.4. Pull-In Head

A pull-in head is used to pull the umbilical along the seabed or through an I- or J-tube. The pull-in head is designed to withstand installation loads without damage to the umbilical or its functional components. The pull-in head is designed, if possible, to allow uninterrupted travel over rollers/ sheaves and through I- or J-tube risers without damaging or snagging.

24.4.5. Hang-Off Device

The hang-off device is used for supporting the umbilical to the top of the I-tube or J-tube at the host suspension point. The hang-off point will be on a deck or outboard of the columns in the tubes. Figure 24-5 shows examples of hang-off assemblies.

24.4.6. Bend Stiffer

The bend stiffer is a device for limiting the bending radius of the umbilical by providing a localized increase in bending stiffness. It is usually a molded device. Figure 24-6 shows a bend stiffer on a working deck.

24.4.7. Electrical Distribution Unit (EDU)

The EDU shown in Figure 24-7 provides electrical distribution to a number of end devices, such as individual subsea trees on a template. The EDU is an oil-filled and pressure-compensated enclosure, within which the incoming

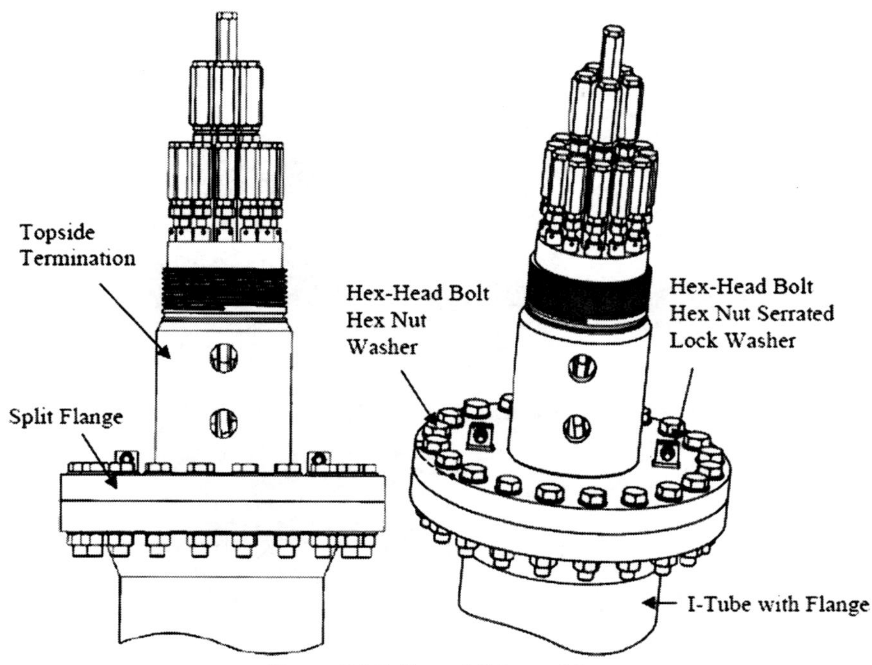

Figure 24-5 Hang-Off Assemblies

electrical power and electrical signals are distributed to two or more satellite SCMs. More than one EDU may be chained together, with each EDU serving a number of satellite SCMs.

24.4.8. Weak Link

A weak link is a device used to protect equipment that is permanently installed on a manifold or template, so that in the event of an umbilical being snagged, the umbilical will break away, activating the link and shearing jumpers connecting to the fixed subsea equipment.

24.4.9. Splice/Repair Kit

A splice/repair kit should be provided to contain the necessary materials and parts to perform repairs on both the main dynamic umbilical and its components and the in-field static umbilical and its components in the event that it should become damaged during the installation process.

24.4.10. Carousel and Reel

Any reel used is not allowed to violate the MBR of the stored umbilical. Figure 24-8 shows a powered umbilical reel.

Figure 24-6 Bend Stiffer

24.4.11. Joint Box

A joint box is used to join umbilical sublengths to achieve overall length requirements or to repair a damaged umbilical. Each umbilical end to be joined has an armored termination, if applied. The joint box is of a streamlined design, with a bend stiffer at each end if required, and of compact size to facilitate reeling storage and installation requirements.

24.4.12. Buoyancy Attachments

Depending on the installed configuration, a dynamic umbilical can necessitate buoyancy attachments in the form of collars, tanks, etc., to achieve the necessary configuration and dynamic motions. The method of attachment does not induce stress cracking in the umbilical sheath, nor allow excessive stress relaxation within the compressive zone of the attachment if clamped, nor allow excessive strain of the umbilical and its component.

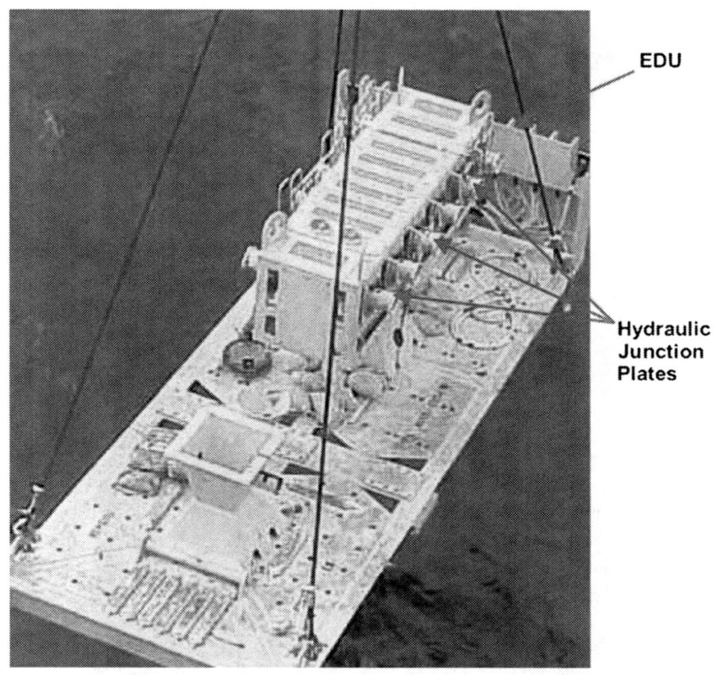

Figure 24-7 Example of a Large Umbilical Termination Assembly with a Large EDU Included

24.5. SYSTEM INTEGRATION TEST

The FAT test is always performed to ensure that the individual components and items of equipment meet the specified requirements and function correctly.

System integration testing, commonly referred to as SIT, is another process that verifies that the integrated equipment is suitable for use. In another words, it is performed to verify that equipment from various suppliers, which must interface with each other, fits and works together acceptably. These procedures are normally very project specific and relate to various equipment interfaces within the project. Figure 24-9 shows system integration testing of a control umbilical.

24.6. INSTALLATION

Subsea trees are monitored and controlled via umbilicals suspended in a catenary shape and protected at the splash zones by I- and J-tubes fixed to

Figure 24-8 Umbilical Reel *(Courtesy JDR)*

Figure 24-9 System Integration Testing of Control Umbilical

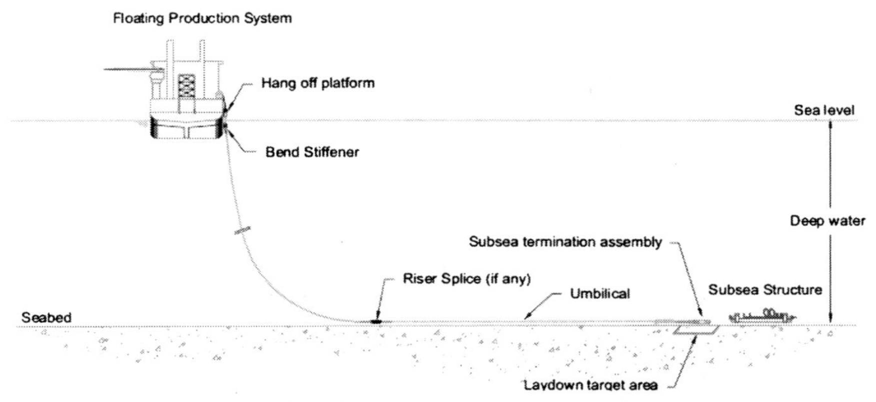

Figure 24-10 Umbilical Connecting an FPS to a Subsea Structure

the structures. Figure 24-10 illustrates an umbilical connecting an FPS to a subsea structure in deep water; Figure 24-11 shows an umbilical connecting a platform to a subsea structure in shallow water.

24.6.1. Requirements for Installation Interface

The installation vessel and its installation equipment should be in good condition and working order, and be verified according to relevant regulations and safety plans prior to vessel mobilization. In addition to the requirement of API Recommended Practice 17I, *Installation Guideline for Subsea Umbilical*, the interfaces relating to the installation of the subsea umbilical should be carefully managed including these items:

- The design and fabrication of the UTAs and their support frames;
- Determination of the design requirements for all cable crossings;
- The design of umbilical supports for crossing of pipelines;
- Protection requirements.

24.6.2. Installation Procedures

Umbilicals are laid using one of the following typical methods:

- The umbilical is initiated at the manifold with a stab and hinge-over connection or a pull-in/connection method and terminated near the subsea well with a second-end lay-down sled (i.e., infield umbilical connection from manifold to satellite well). The connection between the umbilical and the subsea well is later made using a combination of

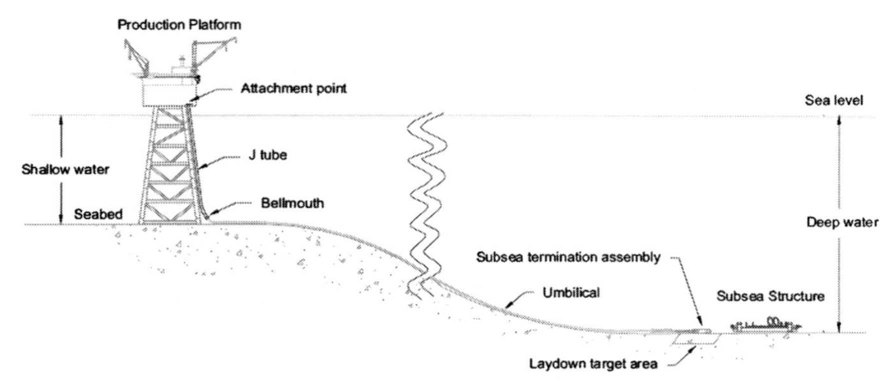

Figure 24-11 Umbilical Connecting a "Host" Platform to a Subsea Structure

the following tie-in methods: (1) rigid or flexible jumper, (2) junction plates, and (3) flying leads.

- The umbilical is initiated at the manifold with a stab and hinge-over connection or a pull-in/connection method. It is laid in the direction to the fixed or floating production system and pulled through an I- or J-tube or cross hauled from the laying vessel to the floating production vessel.
- The umbilical can also be initiated at the fixed or floating production system and terminated near the subsea structure with a second-end umbilical termination assembly (termination head, lay-down sled, umbilical termination unit, etc.). A pull-in and connection tool operated by an ROV may be used to connect the umbilical to the subsea structure.

Umbilical installation can be carried out in the following cases:

- Umbilical installation between subsea manifold and tree;
- Umbilical installation with first-end initiation at subsea structure;
- Umbilical installation with first-end initiation at floater.

24.6.3. Fatigue Damage during Installation

The issues that need to be considered when dealing with fatigue damage during installation of steel tube umbilicals are as follows:

- The contribution to accumulated plastic strain during reeling and potential retrieval;

- Low cycle fatigue during reeling and potential retrieval;
- Dynamic wave frequency fatigue contributions during the critical stages of installation, that is, midlay and handover/pull-in.

The methodology for accounting for accumulated plastic strain and low cycle fatigue will be considered in Section 24.7.2. The calculations for accumulated plastic strain and low cycle fatigue are carried out for both fabrication and installation together.

The methodology for the calculation of wave-induced fatigue damage during the critical stages of installation is similar to the in-place fatigue assessment described in Section 24.7.3. However, some aspects of installation fatigue analysis do not apply to in-place fatigue analysis:

- Since the umbilical changes configuration and is subject to different loads during various stages of installation, different umbilical models are needed to model the various stages of installation that require analysis.
- For installation fatigue analysis it is appropriate to use a time-domain approach. A frequency-domain analysis would not adequately predict the fatigue damage suffered during installation due to the highly irregular loading that the umbilical experiences during this stage of its life.

24.7. TECHNOLOGICAL CHALLENGES AND ANALYSIS

24.7.1. Umbilical Technological Challenges and Solutions

Some of the technological challenges are discussed in the following subsections.

24.7.1.1. Deep Water

The deepest umbilical installation to date (Apr, 2010) is the Perdido 1TOB umbilical in 2,946m (9,666ft) in Gulf of Mexico[3]. Some other deepwater umbilicals in Gulf of Mexico are Shell's Na Kika project in 2,316m (7,598ft), the Thunder Horse umbilical in 1,880m (6,168ft) of water, and the Atlantis umbilical in 2,134m (7,000ft) of water [4]. A challenge in design is that steel tubes are under high external pressure as well as high tensile loads. At the same time, the increased weight may also cause installation problems. This is particularly true for copper cables because the yield strength of copper is low. In ultra-deep water, a heavy dynamic umbilical may present a problem to installation and operation because its hang-off load is high.

For design and analysis of an ultra-deepwater umbilical, it is important to correctly model the effect of stress and strain on an umbilical and the friction effect. Sometimes, bottom compression may be observed for an umbilical under a 100-year hurricane scenario. In this scenario, the design solution may be to use a lazy-wave buoyancy module or to use carbon fiber rods. The use of carbon fiber rods allows umbilicals to have a simple catenary configuration, without the need for expensive, inspection/maintenance demanding buoyancy modules. The carbon fiber rods enhance axial stiffness because they have a Young's modulus close to the value of steel but with only a fraction of the weight.

One of the concerns surrounding the use carbon fiber rods is their capacity for compressive loads. Hence, it is beneficial to conduct some tests that document the minimum bending radius and compressive strength of the umbilical.

If the currents an ultra-deepwater umbilical will be subjected to are severe, it might be necessary to use strakes for VIV protection, although the use of strakes has so far not been required. The strakes may, for instance, be a 16D triple start helix with a strake height of 0.25D.

24.7.1.2. Long Distances

The length for the Na Kika, Thunder Horse, and Atlantis umbilicals is 130, 65, and 45 km, respectively. The longest yet developed is 165 km in a single length, for Statoil's Snohvit development off northern Norway. One of the constraints on umbilical length is the capacity of the installation equipment. The Nexans-operated installation vessel *Bourbon Skagerrak* can carry up to 6500 tonnes of cable, which equals a length of 260 km, assuming the umbilical unit weight is 25 kg/m.

24.7.1.3. High-Voltage Power Cables

The design constraints are the low yield strength of copper, which requires an increasing amount of protection as depths increase, and the weight of steel armoring employed to provide that protection as depths increase. Fatigue of copper cables in dynamic umbilicals is another technical challenge.

24.7.1.4. Integrated Production Umbilical (IPU®)

Heggadal [5] presented an integrated production umbilical (IPU) in which the flowline and the umbilical are combined in one single line (see Figure 24-12). The IPU cross section consists of the following elements:

- A $10^3/_4$-in. flowline with a three-layer PP coating (its thickness is 4 and 14 mm for the static portion and dynamic portion, respectively).

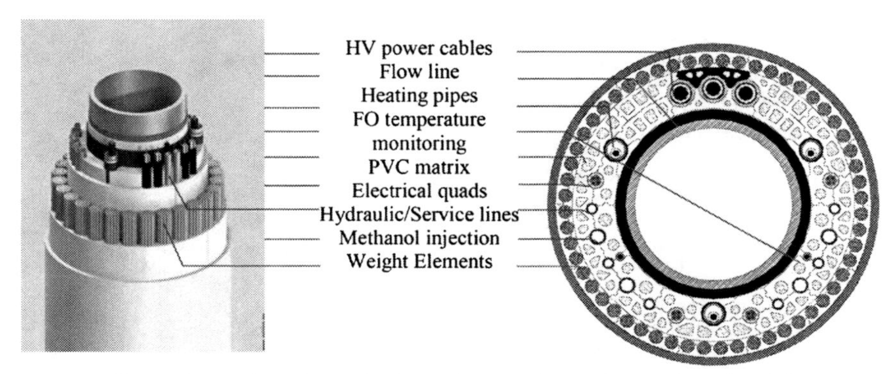

Figure 24-12 IPU Dynamic Cross Section, Super Duplex Flowline [5]

- Around the flowline, there is an annular-shaped PVC matrix that keeps in place the spirally wound umbilical tubes and cables and provides thermal insulation to the flowline.
- Embedded in the PVC matrix, but sliding freely with it, are the various metallic tubes for heating, hydraulic, and service fluids, the electrical/ fiber optic cables for power and signals, and the high-voltage cables for powering the subsea injection pump.
- It has an outer protective sheath of polyethylene 12 mm thick.

To qualify a new design concept like this, a series of analysis and qualification tests were conducted [5]:

1. Analysis
   - Global riser analysis and fatigue analysis;
   - Corrosion and hydrogen-induced cracking assessment;
   - Thermal analysis;
   - Structural analysis (production pipe, topside and subsea termination);
   - Reeling analysis;
   - Electrical analysis;
   - Reel/trawler interaction and on-bottom studies.
2. Basic Tests
   - Mechanical material tests, fatigue, corrosion, etc.
3. Fabrication Tests
   - Fabricationand closing test;
   - STS injection test;
   - QC tests and FAT;
   - Pre/postinstallation tests.

4. Prototype Tests
   - External hydrostatic test;
   - Impact test;
   - Model tensioner test;
   - Reeling and straightening trials;
   - Stinger roller trial;
   - Repair trial;
   - Vessel trial;
   - System test;
   - Dynamic riser full-scale testing.

24.7.2. Extreme Wave Analysis

An important aspect of the umbilical design process is an analysis of extreme wave/environmental conditions. A finite element model of the umbilical is analyzed with vessel offsets, currents and wave data expected to be prevalent at the site where the umbilical is to be installed. For example, in the Gulf of Mexico, this would include an analysis for a 100-year hurricane, 100-year loop current, and submerged current. The current and wave directions are applied in a far, near, and cross condition. This analysis is used to determine the top tension and angles that the hang-off location of the umbilical is likely to experience. These values are then used to design an adequate bend stiffener that will limit the umbilical movements and provide adequate fatigue life for the umbilical. Design analysis based on extreme wave analysis includes:

1. The touchdown zone of the umbilical is analyzed to ensure an adequate bending radius that is larger than the minimum allowable bending radius. It is also important to check that the umbilical does not suffer compression and buckling at the touchdown zone.
2. A polyurethane bend stiffener has been designed to have a base diameter of x inches, and cone length of y ft. This design is based on the maximum angle and its associated tension, and maximum tension and its associated angle from dynamic analysis results using the pinned finite element model.
3. The maximum analyzed tension in the umbilical was found to occur at the hang-off point for the 100-year hurricane wind load case when the vessel is in the far position.
4. The minimum tension in the umbilical may be found to occur in the TDP region for the 100-year hurricane wind load case when the vessel is in the near position.

5. The minimum bend radius (MBR) is estimated over the entire umbilical, over the TDP region and the bend stiffener region, respectively. They are to be larger than the allowable dynamic MBR.
6. The minimum required umbilical on-seabed length is estimated assuming it is subject to the maximum value of the extreme bottom tensions.

24.7.3. Manufacturing Fatigue Analysis

A certain amount of fatigue damage is experienced by a steel tube umbilical during manufacturing, and this needs to be evaluated during fatigue analysis. The two main aspects of umbilical manufacturing fatigue analysis that require attention are accumulated plastic strain and low cycle fatigue. These are explained next.

24.7.3.1. Accumulated Plastic Strain

Accumulated plastic strain is defined as "the sum of plastic strain increments, irrespective of sign and direction" in DNV-OS-F101 and DNV-RP-C203 [6,7]. Accumulated plastic strain can occur in the steel tubes of an umbilical during fabrication and installation. The accumulated plastic strain needs to be maintained within certain limits to avoid unstable fracture or plastic collapse for a given tube material and weld procedure. Accumulated plastic strain is the general criteria used by umbilical suppliers to determine whether the amount of plastic loading on the steel tubes is acceptable. An allowable accumulated plastic strain level of 2% is recommended for umbilical design.

Figure 24-13 shows a schematic of deformations that are likely to take place during the fabrication and installation of a steel tube umbilical. All of the processes shown in this diagram are likely to induce plastic strain in the umbilical.

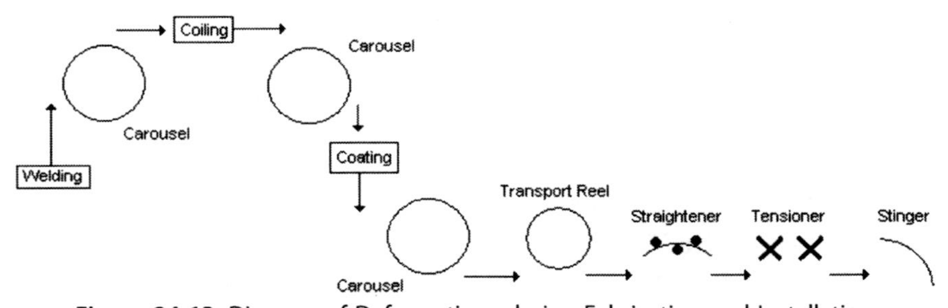

Figure 24-13 Diagram of Deformations during Fabrication and Installation

24.7.3.2. Low Cycle Fatigue

The umbilical steel tubes are subject to large stress/strain reversals during fabrication and installation. Fatigue damage in this low cycle regime is calculated using a strain-based approach.

For each stage of fabrication and installation, the fatigue damage is calculated by considering the contributions from both the elastic and plastic strain cycles. The damage calculated from low-frequency fatigue is added to that from in-service wave and VIV conditions to evaluate the total fatigue life of each tube of the umbilical.

24.7.4. In-Place Fatigue Analysis

The methodology used to assess wave–induced, in-place fatigue damage of umbilical tubes can be summarized as follows:

- Selection of sea-state data from a wave scatter diagram;
- Analysis of finite element static model;
- Umbilical fatigue analysis calculations;
- Simplified or enhanced approach;
- Generation of combined stress history;
- Rain flow cycle counting procedure or spectral fatigue damage;
- Incorporation of mean stress effects in histogram.

The first three items of fatigue analysis mentioned above are described in the following three subsections. The main difference between fatigue analysis for an umbilical and a SCR is the effect of friction when the tubes in the umbilical slide against their conduits and each other due to bending of the umbilical. The methodology discussed here for umbilical in-place fatigue analysis is based on two OTC papers [8,9]. In-place fatigue analysis is required to prove that the fatigue life of the umbilical is 10 times the design life.

24.7.4.1. Selection of Sea-State Data

The wave scatter diagram describes the sea-state environment for the umbilical in service. It is not practical to run a fatigue analysis with all of the sea states described in a wave scatter diagram. Hence, the usual methodology is to group a number of sea states together and represent these "joint sea states" with one significant wave height and wave period. The values of the wave height and wave period are chosen to be conservative.

This methodology results in the reduction of the wave scatter diagram to a "manageable" number of sea states (say, about 20 to 50). This enables the analysis to be carried out in a reasonable amount of time. It is also very important to accurately consider the percentage of time that the umbilical is expected to be affected by these different sea states.

24.7.4.2. Analysis of Finite Element Static Model

A finite element static analysis is carried out for a model representing the steel tube umbilical. The static solution is used as a starting point for a time-domain or frequency-domain dynamic FEA.

24.7.4.3. Umbilical Fatigue Analysis Calculations

Fatigue damage in an umbilical is the product of three types of stress. These are axial (σ_A), bending (σ_B), and friction stress (σ_F). The equations defining these stress terms are as follows:

$$\sigma_A = 2\sqrt{2}SD_T/A \quad (24\text{-}1)$$

$$\sigma_B = 2\sqrt{2}E\ R\ SD_k \quad (24\text{-}2)$$

where

SD_T: standard deviation of tension;

A: steel cross-sectional area of the umbilical;

E: Young's modulus;

R: outer radius of the critical steel tube;

SD_k: standard deviation of curvature.

The critical steel tube is the tube in the umbilical that experiences the greatest stress. This is usually the tube with the largest cross-sectional area and furthest from the centerline of the umbilical.

The friction stress experienced by the critical tube is the minimum of the sliding friction stress (σ_{FS}) and the bending friction stress (σ_{FB}). This is based on the theory that the tube experiences bending friction stress until a point is reached when the tube slips in relation to its conduit. At this point the tube experiences sliding friction stress. This is represented by Figure 24-14. Therefore, according to Kavanagh et al. [9]:

$$\sigma_F = \min\left(\sigma_{FS}, \sigma_{FB}\right) \quad (24\text{-}3)$$

$$\sigma_{FS} = \frac{\mu F_C}{A_t} \quad (24\text{-}4)$$

$$\sigma_{FB} = E\ R_L\ \sqrt{2}\ SD_k \quad (24\text{-}5)$$

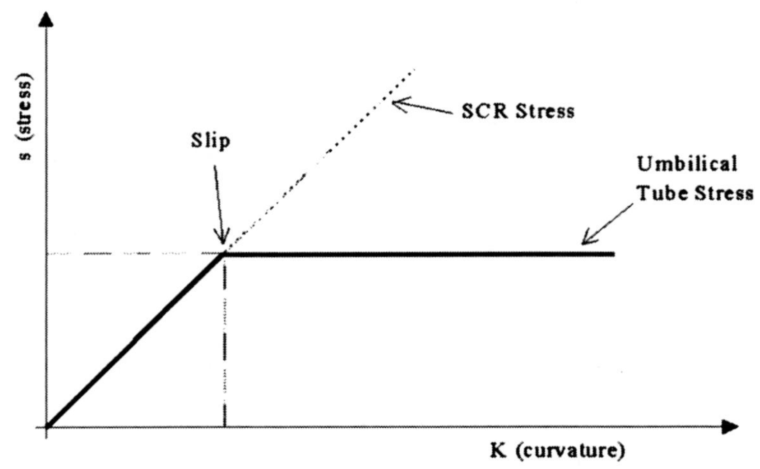

Figure 24-14 Representation of Umbilical Friction Stress [9]

where
μ: friction coefficient;
F_C: contact force between the helical steel tube (defined below);
A_t: cross-sectional area of the critical steel tube within the umbilical;
R_L: layer radius of the tube (this being the distance from the center of the umbilical to the center of the critical steel tube).

$$F_C = \left[\frac{T\sin^2\phi}{R_L} + \frac{EI_{\text{tube}}\sin^4\phi}{R_L^3}\right] L_P \tag{24-6}$$

where
T: mean tension;
ϕ: tube lay angle (this being the angle at which the tubes lie relative to the umbilical neutral axis);
EI_{tube}: individual tube bending stiffness;
L_P: tube pitch length/2.

24.8. UMBILICAL INDUSTRY EXPERIENCE AND TRENDS

Many offshore projects require the consideration of greater depths, longer umbilicals and more control functions than those common in the past. Further, most oil companies are striving to reduce the time required to complete such developments. As a result, the selection of umbilicals has become more critical as many factors such as performance, fluid compatibility, impact on interfaces, ease of deployment, etc. must be considered during the selection process.

There are two continuing trends in the offshore umbilical industry: (1) Oil production in deeper and deeper waters; (2) The desire to perform subsea Processing; Subsea field development moves to deeper and deeper water depths. This requires higher and higher demands on the umbilical systems to ensure operational integrity throughout the lifetime of the fields. The need for feeding of electrical power to subsea equipment, such as subsea pumps, is also increasing because subsea processing technology is considered more and more these years. Both these trends affect the umbilical design that the weaker cross-section components must be able to withstand the larger forces and elongations. Besides, the cross-section must include high voltage power elements.

REFERENCES

[1] R.C. Swanson, V.S. Rao, C.G. Langner, G. Venkataraman, Metal Tube Umbilicals-Deepwater and Dynamic Considerations, OTC 7713, Offshore Technology Conference, Houston, Texas, 1995.

[2] International Standards Organization, Petroleum and Natural Gas Industries, Design and Operation of Subsea Production Systems, Part 5: Subsea Umbilicals, ISO 13628-5, (2009).

[3] Technip Technology and Teamwork Achieve World Class Success for Shell Perdido, Oil & Gas Journal on line, Volume 108 (Issue 31) (April 1, 2010). http://www.ogfj.com/index.

[4] N. Terdre, Nexans Looking beyond Na Kika to Next Generation of Ultra-deep Umbilicals, Offshore, Volume 64, Issue 3, Mar 1, 2004, http://www.offshore-mag.com/index.

[5] O. Heggdal, Integrated Production Umbilical (IPU® for the Fram Ost (20 km Tie-Back) Qualification and Testing, Deep Offshore Technology Conference and Exhibition (DOT), New Orleans, Louisiana, 2004, December.

[6] Det Norske Veritas, Submarine Pipeline Systems, DNV-OS-F101, (2007).

[7] Det Norske Veritas, Fatigue Strength Analysis of Offshore Steel Structures, DNV-RP-C203 (2010).

[8] J. Hoffman, W. Dupont, B. Reynolds, A Fatigue-Life Prediction Model for Metallic Tube Umbilicals, OTC 13203 (2001).

[9] W.K. Kavanagh, K. Doynov, D. Gallagher, Y. Bai, The Effect of Tube Friction on the Fatigue Life of Steel Tube Umbilical RisersdNew Approaches to Evaluating Fatigue Life using Enhanced Nonlinear Time Domain Methods, OTC 16631, Offshore Technology Conference, Houston, Texas, 2004.

Drilling Risers

Contents

25.1. INTRODUCTION

Floating drilling risers are used on drilling semisubmersibles and drilling ships as shown in Figure 25-1. As the water depth increases, integrity of drilling risers is a critical issue. The design and analysis of drilling risers are particularly important for dual operation, dynamically positioned (DP)

Subsea Engineering Handbook
ISBN 978-0-12-397804-2, doi:10.1016/B978-0-12-397804-2.00025-4

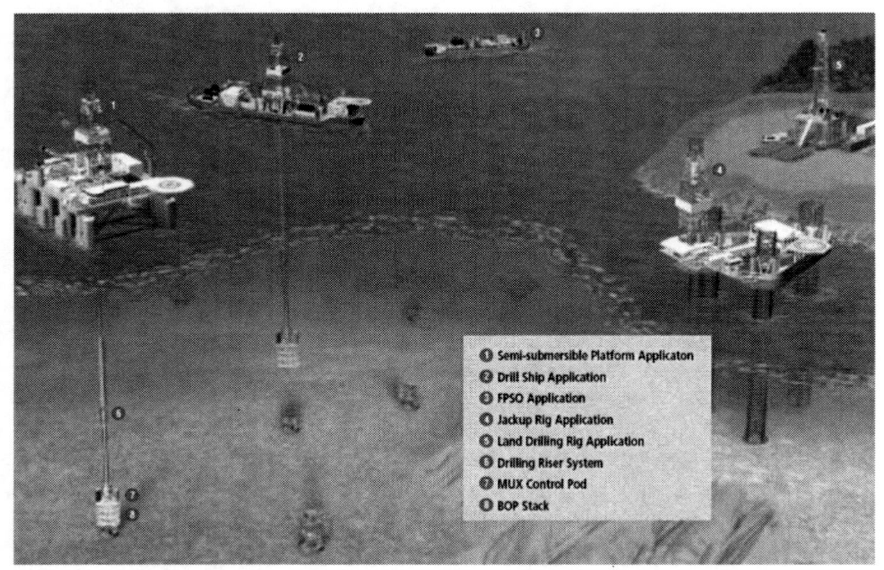

Figure 25-1 Drilling Semisubmersibles, Drilling Ships, and Drilling Risers (*Courtesy of World Oil [1]*)

semisubmersible rigs. For the integrity assurance purpose, a series of dynamic analysis needs to be carried out. The objective of the dynamic analysis is to determine vessel excursion limits and limits for running/retrieval and deployment. In recent years, qualification tests are also required to demonstrate fitness for purpose for welded joints, riser coupling and sealing systems. For risers installed in the Gulf of Mexico, vortex-induced vibrations are a critical issue. Some oil companies encourage use of monitoring systems to measure real-time vessel motions and riser fatigue damage. The monitoring results may also be used to verify the VIV analysis tools that are being applied in the design and analysis.

In this chapter, after a brief outline of the floating drilling equipment and subsea systems, the riser components and vessel data are outlined. Various methods of riser analysis are presented.

25.2. FLOATING DRILLING EQUIPMENT

25.2.1. Completion and Workover (C/WO) Risers

Two different types of risers are used for installation and intervention in and on a well: completion risers and workover risers.

A completion riser is generally used to run the tubing hanger and tubing through the drilling riser and BOP into the wellbore. The completion riser may also be used to run the subsea tree. The completion riser is exposed to external loading such as curvature of the drilling riser, especially at the upper and lower joints.

A workover riser is typically used in place of a drilling riser to reenter the well through the subsea tree, and may also be used to install the subsea tree. The workover riser is exposed to ocean environmental loads such as hydrodynamic loads from waves and currents in addition to vessel motions.

Figure 25-2 is a typical figure for a C/WO riser, taken from ISO code for *Completion and Work Riser Systems [2]*. The C/WO riser can be a common system with items added or removed to suit the task being performed. Either type of risers provides communication between the wellbore and the surface equipment. Both resist external loads and pressure loads and accommodate wireline tools for necessary operations.

Riser connectors are one of the most important riser components. As drilling depths have increased, riser connectors have evolved to address issues concerning high internal and external pressures, increasing applied bending moment and tension loads, and extreme operating conditions such as sweet and sour services. For connector design, the material selection and fabrication of bolts are critical issues.

Figure 25-3 shows key components in a typical drilling riser system. Figure 25-4 shows bolt-based riser connectors.

The *spider* is a device with retractable jaws or dogs used to hold and support the riser on the uppermost connector support shoulder during running of the riser. The spider usually sits in the rotary table on the drill floor.

The *gimbal* is installed between the spider and the rotary table. It is used to reduce shock and to evenly distribute loads caused by a rig's roll/pitch motions, on the spider as well as the riser sections.

A *slick joint*, also known as a telescope joint, consists of two concentric pipes that telescope together. It is a special riser joint designed to prevent damage to the riser and control umbilicals where they pass through the rotary table. Furthermore, it protects the riser from damage due to rig heave.

Riser joints are the main members that make up the riser. The joints consist of a tubular midsection with riser connectors in the ends. Riser joints are typically provided in 9.14- to 15.24-m (30- to 50-ft) lengths. For the sake of operating efficiency, riser joints may be 75 ft in length. Shorter

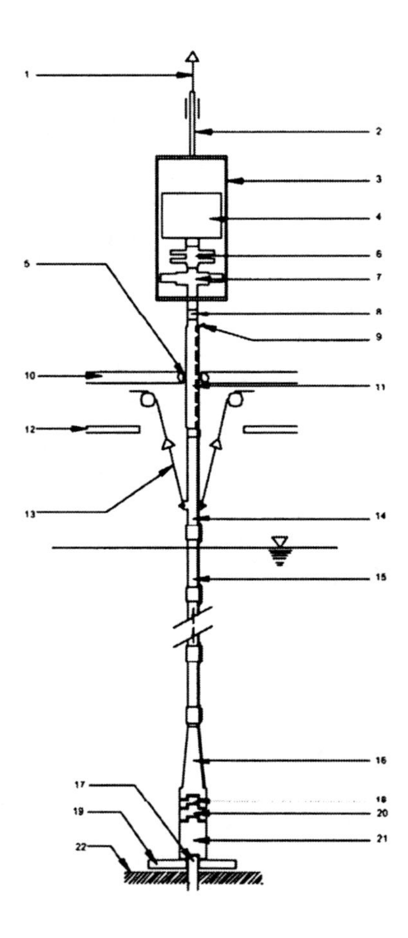

Key

1	Top drive	9	Umbilical (to hose reel)	17	Wellhead
2	Drill sub	10	Drill floor	18	Emergency disconnect package
3	Surface tree tension frame	11	Slick (cased wear) joint	19	Guide base
4	Coiled tubing injector	12	Moon pool area	20	Lower riser package
5	Roller bushing	13	Riser tension wires	21	Subsea tree
6	Surface BOP	14	Tension joint	22	Seabed
7	Surface tree	15	Standard riser joints		
8	Surface tree adapter joint	16	Stress joint		

Figure 25-2 Stack-Up Model for a C/WO Riser [2]

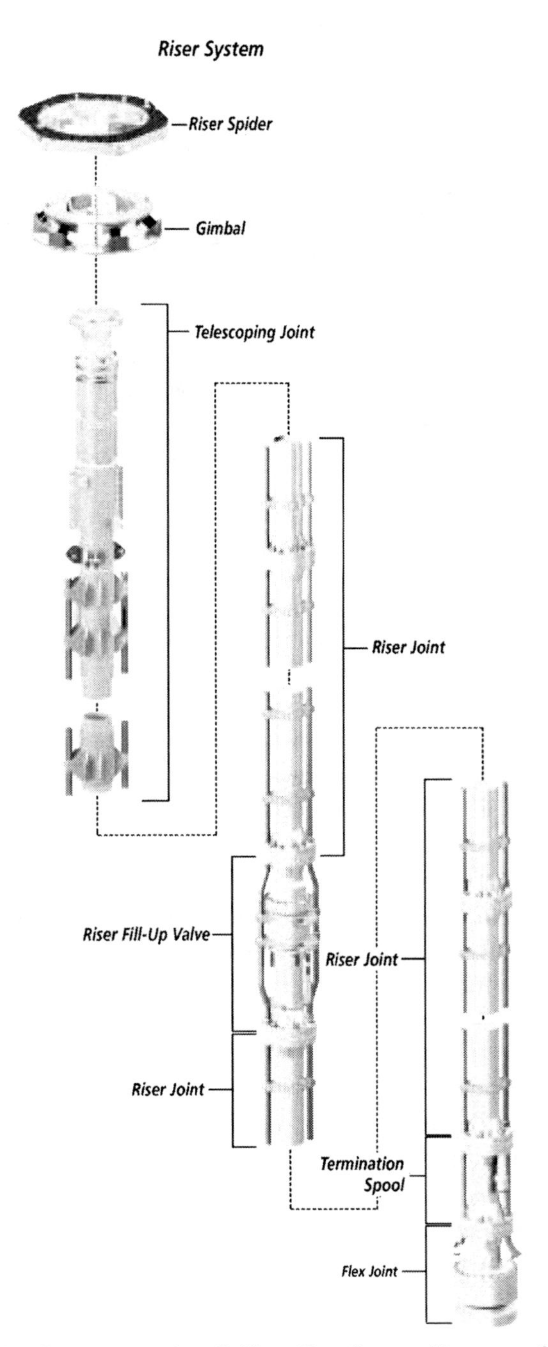

Figure 25-3 Key Components in a Drilling Riser System (*Courtesy of World Oil [1]*)

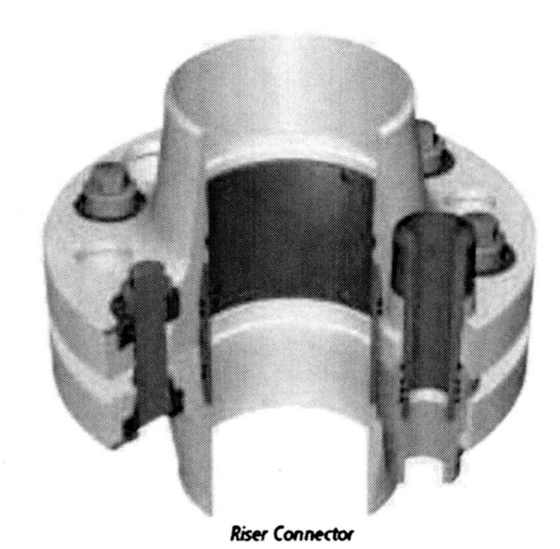

Figure 25-4 Riser Connector (*Courtesy of World Oil [1]*)

joints, called *pup joints*, may also be provided to ensure proper space-out while running the subsea tree, tubing hanger, or during workover operations.

Depending on configuration and design, a drilling riser system also consists of the following components:

1. The *BOP adapter joint* is a specialized C/WO riser joint used when the C/WO riser is deployed inside a drilling riser and subsea BOP to install and retrieve a subsea tubing hanger.
2. The *lower workover riser package (LWRP)* is the lowermost equipment package in the riser string when configured for subsea tree installation/ workover. It includes any equipment between the riser stress joint and the subsea tree. The LWRP permits well control and ensures a safe operating status while performing coiled tubing/wireline and well servicing operations.
3. An *emergency disconnect package (EDP)* is an equipment package that typically forms part of the LWRP and provides a disconnection point between the riser and subsea equipment. The EDP is used when the riser must be disconnected from the well. It is typically used in case of a rig drift-off or other emergency that could move the rig from the well location.
4. The *stress joint* is the lowermost riser joint in the riser string when the riser is configured for workover. The joint is a specialized riser joint

designed with a tapered cross section, in order to control curvature and reduce local bending stresses.

5. The *tension joint* is a special riser joint, which provides means for tensioning the C/WO riser with the floating vessel's tensioning system when in open-sea workover mode. The tension joint is often integrated in the lower slick joint.
6. The *surface tree adapter joint* is a crossover joint from the standard riser joint connector to the connection at the bottom of the surface tree.
7. The *surface tree* provides flow control of the production and/or annulus bores during both tubing hanger installation and subsea tree installation/workover operations.

25.2.2. Diverter and Motion-Compensating Equipment

A diverter is similar to a low-pressure BOP. When either gas or other fluids from shallow gas zones enter the hole under pressure, the diverter is closed around the drill pipe or kelly and the flow is diverted away from the rig.

All floating drilling units have motion-compensating equipment (as shown in Figure 25-5) installed to compensate for the heave of the rig. Compensators function as the flexible link between the force of the ocean and the rig. The equipment consists of the drill string compensator, riser

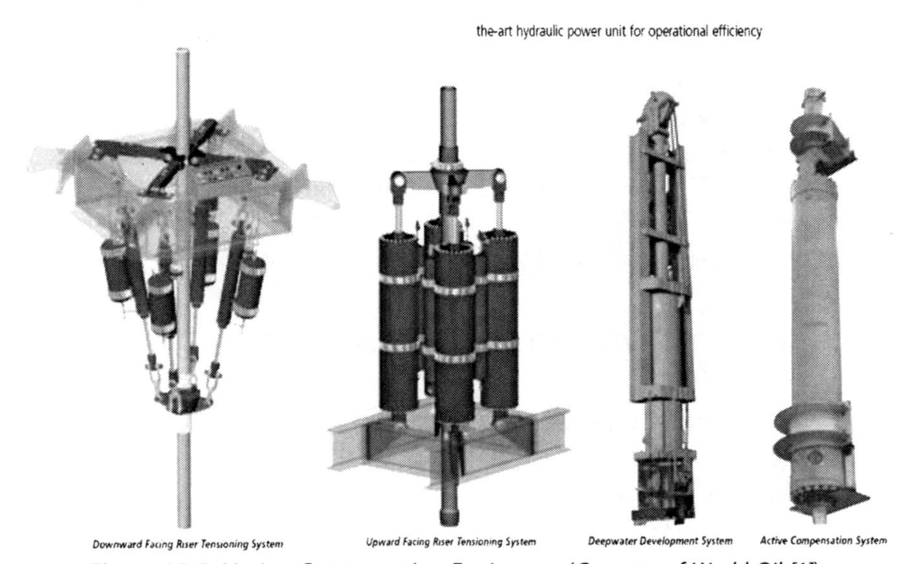

Figure 25-5 Motion-Compensating Equipment (*Courtesy of World Oil [1]*)

Figure 25-6 Complete Riser Joint (*Courtesy of* Offshore *Magazine, May 2001 [3]*)

tensioners, and guideline and podline tensioners. The drilling string compensator, located between the traveling block and swivel and kelly, permits the driller to maintain constant weight on the bit as the rig heaves. Riser tensioners are connected to the outer barrel of the slip joint with wire ropes. These tensioners support the riser, and the mud within it, with a constant tension as the rig heaves. The guideline and podline maintain constant tension on guideline wire ropes, and wire ropes that support the BOP control podlines as the rig heaves.

25.2.3. Choke and Kill Lines and Drill String

Choke and kill lines are attached to the outside of the main riser pipe (see Figure 25-6). They are used to control high-pressure events. Both lines are usually rated for 15 ksi. High pressure is circulated out of the wellbore through the choke and killed lines by pumping heavier mud into the hole. Once the pressure is normal, the BOP is opened and drilling resumes. If the pressure cannot be controlled with the heavier mud, cement is pumped down the kill line and the well is killed.

The drill string permits the circulation of drilling fluid or liquid mud. Some functions of this mud are to:

- Cool the bit and lubricate the drill string
- Keep the hole free of cuttings by forced circulation to the top.
- Prevent wall cave-ins or intrusions of the formations through which it passes.
- Provide a hydrostatic head to contain pressures that may be present.

25.3. KEY COMPONENTS OF SUBSEA PRODUCTION SYSTEMS

25.3.1. Subsea Wellhead Systems

The foundation of any subsea well is the subsea wellhead. The function of the subsea wellhead system is to support and seal the casing string in addition to supporting a BOP stack during drilling and also the subsea tree under normal operation.

Installation of equipment to the seabed is generally done by one of two methods:

1. With the use of tensioned guidelines attached to guide sleeves on the subsea structure orienting and guiding equipment into position;
2. With a guideline-less method that uses a dynamic positioning reference system to move the surface vessel until the equipment is positioned over the landing point, after which the equipment is lowered into place.

Regardless of the guidance system, the procedure in which the wellhead system is installed is as follows:

1. The first component installed is the *temporary guide base*. The temporary guide base serves as a reference point for the installation of subsequent well components compensating for any irregularities of the seabed. For guideline systems the temporary guide base also acts as the anchor point for guidelines.
2. The *conductor housing* is essentially the top of the casing conductor. The casing conductor and housing are installed through the temporary guide base, either by piling or drilling, and provides an installation point for the permanent guide base and a landing area for the wellhead housing.
3. The *permanent guide base*, which is installed on the conductor housing, establishes structural support and final alignment for the wellhead system. The permanent guide base provides guidance and support for running the BOP stack or the subsea tree.
4. The *wellhead housing* (or high-pressure housing), which is installed into the conductor housing, provides pressure integrity for the well, supports subsequent casing hangers, and serves as an attachment point for either the BOP stack or subsea tree by using a wellhead connector.
5. To carry each casing string, a *casing hanger* is installed on top of each string and the casing hangers are supported in the wellhead housing, which thus supports the loads deriving from the casing. To seal the inside annuli, an *annulus seal assembly* is mounted between each casing hanger and the well housing.

25.3.2. BOP

The marine riser is used as a conduit to return the drilling fluids and cuttings back to the rig and to guide the drill and casing strings and other tools into the hole. Geiger and Norton [4] have provided a short and relevant description on floating drilling.

The well is begun by setting the first casing string known as the conductor or structural casing—a large-diameter, heavy-walled pipe—to a depth that is dependent on soil conditions and strength/fatigue design requirements. Its primary functions are to:

- Prevent the soft soil near the surface from caving in;
- Conduct the drilling fluid to the surface when drilling ahead;
- Support the BOP stack and subsequent casing strings;
- Support the Christmas tree after the well is completed.

The depth and size of each drilling string is determined by the geologist and drilling engineer before drilling begins. When drilling from a semi-submersible or drill ship, the wellhead and BOP must be located on the seabed.

The BOP stack is used to contain abnormal pressures in the wellbore while drilling the well. The primary function of the BOP stack is to preserve the fluid column or to confine well fluids or gas to the borehole until an effective fluid column can be restored.

At the lower end of the riser is the lower flexjoint. After the hole has been drilled to its final depth, electric logs are run to determine the probable producing zones. Once it has been determined that sufficient quantities of oil and gas exist, the production tubing is then run to the zone determined to contain that oil or gas. Only after this takes place is the well "completed" by removing the BOP stack and installing the fittings used to control the flow of oil and gas from the wellhead to the processing facility.

25.3.3. Tree and Tubing Hanger System

To complete the well for production a tubing string is installed in and supported by a tubing hanger. The tubing hanger system carries the tubing and seals of the annulus between casing and tubing. To regulate flow through the tubing and annulus, a subsea tree is installed on the wellhead. The subsea tree is an arrangement of remotely operated valves, which controls the flow direction, amount, and interruption.

25.4. RISER DESIGN CRITERIA

25.4.1. Operability Limits

Table 25-1 presents the typical criteria used for determination of the operability limits for the drilling riser, defined mainly based on API 16Q [5].

In general, a DNV F2 curve is used for the weld joints and a DNV B curve for the riser connectors (coupling). Two stress concentration factors (SCFs) are used in fatigue analysis, one is 1.2 for the girth welds, and the other is roughly 2.0 for riser connectors, depending on the type of risers. In recent years, fatigue qualification testing has been performed to determine

Table 25-1 Criteria for Drilling Riser Operability Limits

Design Parameter	Definition	Drilling Conditions	Nondrilling Conditions
Low flex joint angle	Mean	1°	n/a
	Max	4°	90% of capacity (9°)
Upper flex joint angle	Mean	2°	n/a
	Max	4°	90% of capacity (9°)
Von Mises stress	Max	67% of σ_y	80% of σ_y
Casing bending moment	Max	80% of σ_y	80% of σ_y

the actual *S-N* curve data. An engineering criticality assessment (ECA) analysis is conducted to derive defect acceptance criteria for inspection.

For the drilling riser, the safety factor on fatigue life is 3 because the drilling joints can be inspected. The fatigue calculations are to account for all relevant load effects, including wave, VIV, and installation-induced fatigue. In some parts, such as the first joints nearest the lower flexjoint (LFJ), the fatigue life could be less, in which case the fatigue life will determine the inspection interval.

25.4.2. Component Capacities

For strength checks, various component capacities need to be defined such as:

- Wellhead connector;
- LMRP connector;
- LFJ;
- Riser coupling and main pipe;
- Peripheral lines;
- Telescopic joint;
- Tensioner/ring;
- Active heave draw works;
- Hard hang-off joint;
- Soft hang-off joint;
- Spider-gimbal;
- Riser running tool.

25.5. DRILLING RISER ANALYSIS MODEL

25.5.1. Drilling Riser Stack-Up Model

A schematic of a typical drilling riser stack-up was shown earlier in Figure 25-2. The weight in air and seawater for the telescopic joint, flex-joints, LMRP, and BOP need to be defined for riser analysis.

The submerged weight and dimensions (length × width × height) for the production trees, manifolds, and jumpers are required for the dual activity interference analysis. The properties of the auxiliary rig drill pipe and wire rope will also be used in the interference analysis.

The recommended tensioner forces used in the analysis are calculated based on the mud weight. In the analysis, the drilling mud densities are typically assumed to be 8.0, 12.0, 16.0 ppg, etc. The maximum allowable bending moment in the casing may be determined assuming the allowable stress is 80% of yield strength.

The hydrodynamic coefficients to be used in the analysis include the normal drag coefficient and the associated drag diameters for the bare and buoyancy joints. The tangential drag coefficient may be taken from API RP 2RD [6], Section 6.3.4.1, Equation 31. For the LMRP and the BOP, the vertical and horizontal drag areas and coefficients may be provided by suppliers.

The red alarm is typically 60 sec before disconnect point, and the yellow alarm is roughly 90 sec before the red alarm.

25.5.2. Vessel Motion Data

The required vessel motion data include the following:

- The principal dimensions of the vessel;
- The mass and inertia properties at maximum operation draft;
- Reference point locations for RAOs (Response Amplitude Operator);
- The survival draft RAOs for various wave directions;
- The maximum operating draft RAOs for various wave directions;
- The transit draft RAOs for various wave directions.

In addition, wave drift force quadratic transfer functions for surge, sway, and yaw are required to conduct irregular wave force calculations in a drift-off analysis. Wind and current drag coefficients for the vessel are also required.

25.5.3. Environmental Conditions

Generally angles denote the direction "from which" the element is coming, and they are specified as clockwise from true north.

Tidal variations will have a negligible effect on the loads acting on deepwater risers and may be negligible in the design.

Environmental conditions include:

- Omnidirectional hurricane criteria for the 10-year significant wave height and associated parameters;

- Omnidirectional winter storm criteria for 10- and 1-year return periods;
- The condensed wave scatter diagram for the full population of waves (operational, winter storm, and hurricane);
- Loop/eddy normalized profiles;
- The 10- and 1-year loop/eddy current profiles along with the associated wind and wave parameters;
- The bottom current percent exceedance and the normalized bottom current profile;
- The combined loop/eddy and bottom current normalized current profile as a fraction of the maximum;
- The combined loop/eddy and bottom current profiles for a 10-year eddy + 1-year bottom, and 1-year eddy + 1-year bottom;
- A 100-year submerged current probability of exceedance and profile duration.

Background current is the current that exists in the upper portion of the water column when there is no eddy present. Mean values of the soil undrained shear strength data, submerged unit weight profile, and ε_{50} profiles are used along the soil column to calculate the equivalent stiffness of the soil springs, for analysis of the connected riser.

25.5.4. Cyclic *p-y* Curves for Soil

The methodology for deriving a *p-y* curve for soft clay for cyclic loading was developed by Matlock [7]. A family of *p-y* curves will be required to model the conductor casing/soil interaction at various depths below the mudline.

25.6. DRILLING RISER ANALYSIS METHODOLOGY

Some key terms for drilling riser design and analysis are shown in Figure 25-7.

From a structural analysis point of view, a drilling riser is a vertical cable, under the action of currents. The upper boundary condition for the drilling riser cable is rig motions that are influenced by rig design, wave and wind loads. One of the key technical challenges for deepwater drilling riser design is fatigue of VIV due to (surface) loop currents and bottom currents.

Figure 25-8 shows a typical finite element analysis model for C/WO risers. It illustrates the process of running riser and landing, the riser being connected or disconnected or in hang-off mode.

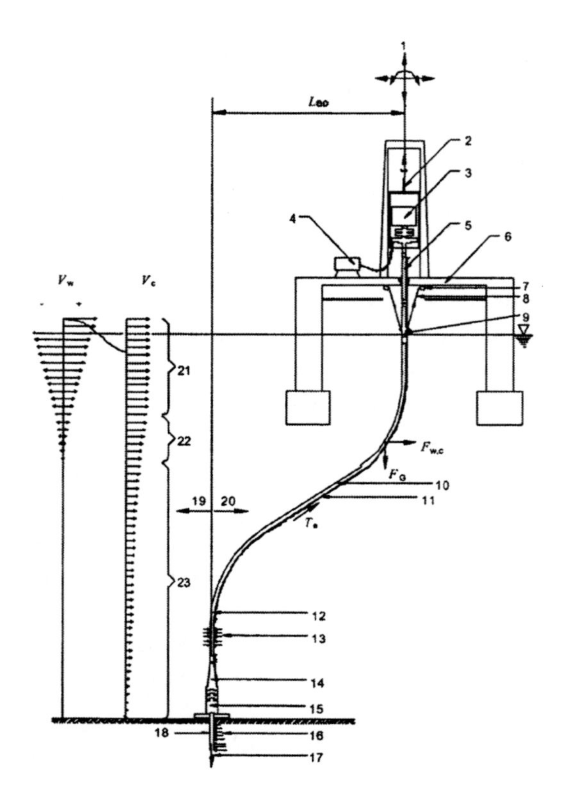

Key

1 Wave motions due to first order wave motions
2 Draw works tension and stroke
3 Surface equipment
4 Surface pressure (Choke or mud-pump)
5 Slick joint
6 Drill floor
7 Tensioner sheaves
8 Tensioner tension and stroke
9 Tensioner joint
10 Outside diameter
11 Riser joints
12 Bending stiffness
13 External pressure
14 Stress joint
15 Subsea equipment
16 Soil restraint
17 Tool
18 Conductor bending stiffness
19 Upstream
20 Downstream
21 Excitation zone
22 Shear zone
23 Damping zone
$F_{w,c}$ Wave and current forces
F_G Gravity forces
T_e Effective tension
V_w Wave velocity
V_c Current velocity
L_{so} Vessel offset (+)

Figure 25-7 Principal Parameters Involved in C/WO Riser Design and Analysis (*ISO 13628-7, 2003(E) [2]*)

25.6.1. Running and Retrieve Analysis

The goal of running/retrieval analysis is to identify the limiting current environment that permits this operation. During this operation, the riser could be supported by the hook at 75 ft above the RKB (Rotary Kelly Bushing, a datum for measuring depth in an oil well), or it could be hanging in the spider. The critical configuration is the hook support because of its

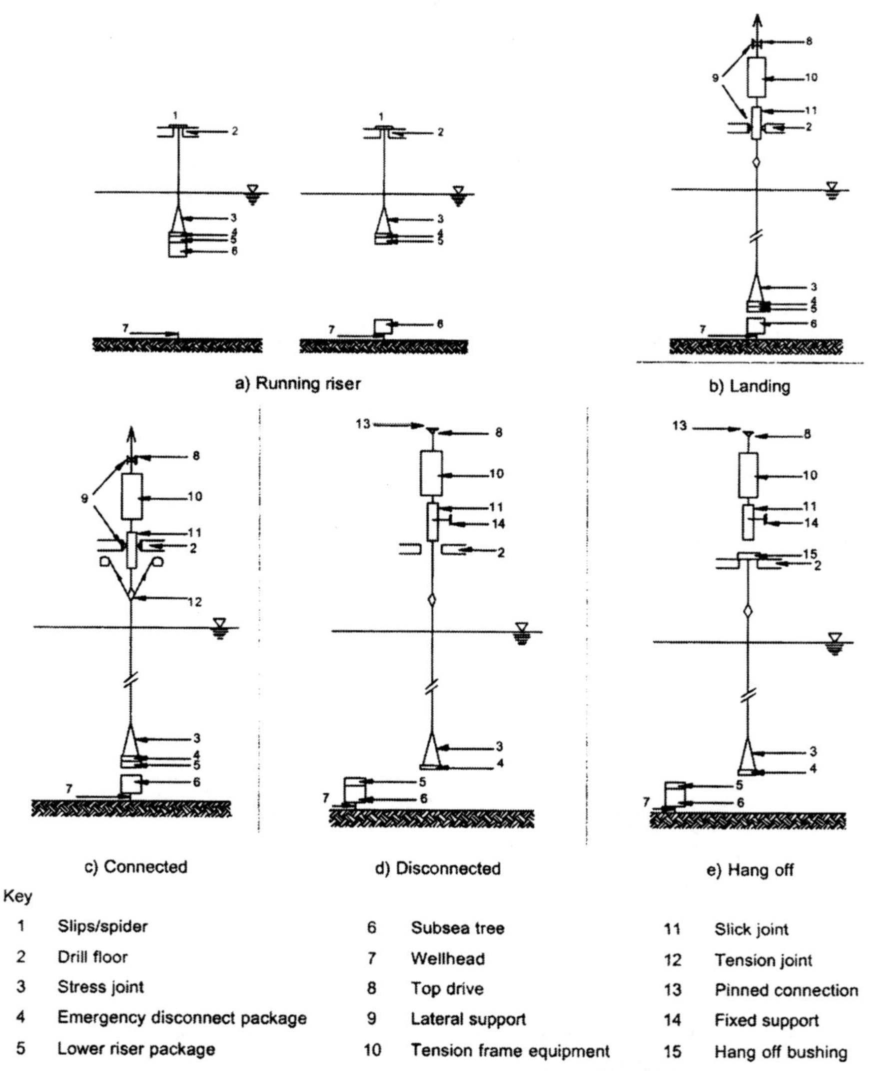

Figure 25-8 Typical Finite Element Analysis Model for C/WO Risers [2]

greater potential for contact between the joint and the diverter housing. The BOP is on the riser for deployment, and may not be on the riser if the riser is disconnected at the LMRP, in which case only the LMRP is on the riser.

The hook is considered to be a pinned support with only vertical and horizontal displacement restraints. The riser may rotate about the hook under current loading. The limiting criterion is contact between a riser joint and the diverter housing.

Static analysis is used to evaluate the effects of the current drag force. Wave dynamic action on the riser's lateral motions is not considered.

25.6.2. Operability Analysis

The objective of the operability analysis is to define the operability envelope, for various mud weights and top tensions, per the recommendations of API RP 16Q [5].

The operability envelope for the limiting criteria is computed using both static and dynamic wave analysis. The static analysis involves offsetting the rig upstream and downstream under the action of the current profile to find the limiting up and down offsets at which one limiting criterion is reached. Two current combinations are typically considered: background + bottom and eddy + bottom. Typically three mud weights are modeled with their respective top tensions.

The procedure for the dynamic analysis is the same as that for the static analysis except that wave loading is added, and the analysis is carried in the time domain, using regular waves based on H_{max}, and for at least five wave periods. The dynamic analysis predicts the maximum LFJ and upper flexjoint (UFJ) angles, which should be checked against their limiting values.

The limiting conditions for the flexjoint angles are typically as follows:

- Connected drilling for dynamic analysis:
  - Upper flexjoint mean angle < 2 degrees; 4 degrees maximum;
  - Lower flexjoint mean angle < 1 degree; 4 degrees maximum.
- Connected nondrilling:
  - Upper flexjoint max angle < 9 degrees;
  - Lower flexjoint max angle < 9 degree.

Note that the allowed limit for the upper and lower flexjoint angles is 1 degree for static analysis of connected drilling risers.

Other limitations on the dynamic riser response are as follows:

- Riser von Mises stresses < 0.67 yield stress for extreme conditions;
- Riser connector strength;
- Tensioner and TJ stroke limit.

Limitations may also arise from loading on the wellhead and conductor system:

- LMRP connector capability;
- BOP flanges or clamps;
- Wellhead connector capacity;
- Conductor bending moment (0.8 yield stress).

For drilling, usually it is the mean angles of the LFJ (1 degree) and the UFJ (2 degrees) that determine the envelope. For nondrilling conditions, usually it is the maximum dynamic bending moment in the casing that controls the envelope.

25.6.3. Weak Point Analysis

Weak point analysis forms a part of the design process of a drilling riser system. The objective of a weak point analysis is to design and identify the breaking point of the system under extreme vessel drift-off conditions should the LMRP fail to disconnect. The riser system should be designed so that the weak point will be above the BOP.

The basic assumption here is that all equipment in the load path is properly designed per the manufacturers' specifications. The areas of potential weakness in a drilling riser system are typically:

- Overloading of the drilling riser;
- Overloading of connectors or flanges;
- Stroke-out of the tensioner;
- Exceedance of top and bottom flexjoint limits;
- Overloading of the wellhead.

Evaluation criteria for weak point analysis are derived for each potential weak point of the drilling riser system. The weak point criteria determine failure of the system. The evaluation criterion for stroke-out of the tensioners is typically the tensile strength (rupture) of the tensioner lines for each line. The maximum capacity of a padeye will be based on the load that causes yield in each padeye.

The failure load capacity for a flexjoint typically corresponds to the maximum bending moment and tension combination that the flexjoint can withstand. This typically relates to additional loading following angular lock-out.

The failure load capacity for standard riser joints and conductor joints is typically taken as the maximum combined tension and bending stress that the joint can withstand before exceeding the yield stress of the riser material.

To eliminate uncertainty, a full time-domain weak point computer analysis may be conducted, as follows:

- Perform dynamic regular wave analyses for a selected combination of wind, waves, and currents and establish the dynamic amplification of the loads generated at potential weak points, especially at the wellhead connector and at the LMRP connector.

- Sensitivity analyses are typically performed to determine the effect on the weak point of varying the critical parameters such as mud weight and soil properties.
- Vessel offsets should range from the drilling vessel in the mean position to extreme vessel offsets downstream, as determined by a coupled mooring analysis.
- Following the offset analysis of the drilling riser system, the results will be processed to extract the forces and moments generated by the offset position. These are then compared with the corresponding evaluation criteria at potential weak points along the drilling riser system.

If the weak point is below the BOP, failure would have severe consequences in terms of well integrity, riser integrity, or cost. Then further analysis should be conducted to relocate the weak point to a position with less onerous consequences in the event of failure. One option that might be considered in this context is to redesign the capacity of the hydraulic connectors or bolted flanges/bindings in the drilling riser system such that the weak point occurs at one of those locations.

In a mild environment, slow drift-off generates low static and dynamic moments on the wellhead because of the mild current and the low wave height. In a fast drift-off environment, the lower riser straightens out quickly before wave action magnifies the wellhead connector static moment when the tensioner strokes out. The suggested critical environment would be a combination of high current to generate a high static moment at the wellhead connector, high waves to cause high dynamic moments, and slow wind to generate slow drift.

25.6.4. Drift-Off Analysis

Drift-off analysis is a part of the design process for a drilling riser system on a DP rig. The objective of a drift-off analysis is to determine when to initiate disconnect procedures under extreme environmental conditions or drift-off/drive-off conditions. The analysis is performed for the drilling and the nondrilling operating modes. In each mode, the analysis will identify the maximum downstream location of the vessel under various wind and current speeds and wave height/period.

The first task in a drift-off analysis is to determine the evaluation criteria by which the disconnect point will be identified. These criteria are based on the rated capacities of the equipment in the load path:

- Conductor casing, based on 80% of yield;
- Stroke-out of the tensioner/telescopic joint;

- Top and bottom flexjoints limits;
- Overloading of the wellhead connector;
- Overloading of the LMRP connector;
- Stress in the riser joint (0.67 of yield).

Coupled system analysis is used where the soil and casing, wellhead and BOP stack, riser, tensioner, and the vessel are all included in one model. Combinations of environmental actions (wind, current, and waves) are applied to the system, and the dynamic time-domain response is then computed. In this coupled vessel approach, the vessel drift-off (or vessel offset) is an output of the analysis. This approach, which accounts for soil/casing/riser/vessel interactions, is more accurate than the uncoupled approach where the vessel offset is computed separately and then applied to the vessel drift curve to the riser model in a second analysis.

Following the static and dynamic analyses of the drilling riser system, the disconnect point of the system is identified as follows:

- The vessel offset, for the specified environmental load conditions, that generates a stress or load equal to the disconnect criteria of the component is the allowable disconnect offset for that particular component.
- The allowable disconnect offset should be determined for each of the key components along the drilling riser system.
- Then the point of disconnect (POD) corresponds to the smallest allowable disconnect offset for all critical components along the drilling riser system.
- Once the vessel offset at which the riser must be disconnected has been determined, the offset at which the disconnect procedure must be initiated (red limit) will typically be based on 60 sec. This is the EDS time.
- For nondrilling, the disconnect initiation offset is adjusted by 50 ft before the EDS time. This is the modified red limit for nondrilling.
- For drilling, the disconnect initiation offset is typically 90 sec before EDS.

25.6.5. VIV Analysis

The objectives of performing VIV analysis of the drilling risers are as follows:

- Predict VIV fatigue damage.
- Identify fatigue critical components.
- Determine the required tensions and the allowable current velocity.

Following the modal solution, the results are prepared for input to Shear7 [8]. SHEAR7 is one of the leading modeling tools for the prediction of vortex-induced vibration (VIV) developed by MIT. Parameters that remain user defined are as follows:

- Mode cut-off value;
- Structural damping coefficient;
- Strouhal number;
- Single and multimode reduced velocity double bandwidth;
- Modeling of the straked riser section with VIV suppression devices.

In a VIV analysis of the drilling riser, the vessel is assumed to be in its mean position. The analysis includes these tasks:

- Generate mode shapes and mode curvatures for input to VIV analysis using a finite element modal analysis program.
- Model the riser using Shear7 based on the tension distribution determined from initial static analyses.
- Analyze the VIV response of the riser for each current profile using Shear7.
- Evaluate the damage due to each current profile.
- Plot the results in terms of VIV fatigue damage along the riser length for each current profile.

25.6.6. Wave Fatigue Analysis

A time-domain approach is adopted for motion-induced fatigue assessment of the drilling riser. No mean vessel offsets or low-frequency motions are considered for motion fatigue analysis of the drilling riser.

The procedure for performing a fatigue analysis is as follows:

- Perform an initial mean static analysis.
- Apply relevant fatigue currents statically as a restart analysis.
- Perform dynamic time-domain analyses for the full set of load cases, applying the relevant wave data for each analysis.
- Postprocess the results from the time-domain analyses to estimate fatigue damage of the drilling riser at the critical locations.

25.6.7. Hang-Off Analysis

Two hang-off configurations are assumed as follows: a *hard hang-off* in which the telescopic joint is collapsed and locked, thereby forcing the top of the riser to move up and down with the vessel; and a *soft hang-off* in which the riser is supported by the riser tensioners with all air pressure vessels (APVs) open and a crown-mounted compensator (CMC), providing a soft vertical spring connection to the vessel.

Time-domain analysis is conducted using random wave analysis and a simulation time of at least 3 hr. The hard hang-off cases are the 1-year winter storm (WS), 10-year WS, and 10-year hurricane. The soft hang-off cases are the 10-year WS and the 10-year hurricane. The goal of the time-domain dynamic analysis is to investigate the feasibility of each mode.

In a hang-off configuration model, the riser is disconnected from the BOP, and only the LMRP is on the riser. For the hard hang-off method, only the displacements are fixed. The rotations are determined by the stiffness of the gimbal-spider. The trip saver is at the main deck.

For the soft hang-off method, the riser weight is shared equally by the tensioner and the draw works. The draw works have zero stiffness. The tensioner stiffness may be estimated based on the weight of the riser supported by the tensioners and the riser stroking from wave action.

The evaluation criteria for soft and hard hang-off analyses are as follows:

- For soft hang-off, use the stroking limit for the tensioner and slip joint;
- Minimum top tension to remain positive to avoid uplift on the spider;
- Maximum top tension: rating of substructure and the hang-off tool;
- Riser stress limited to 0.67 Fy;
- Gimbal angle to prevent stroke-out;
- Maximum riser angle between gimbal and keel to avoid clashing with the vessel.

25.6.8. Dual Operation Interference Analysis

Dual operation interference analysis evaluates the different scenarios proposed for having the drilling riser in place and connected on the main rig while performing deployment activities on the auxiliary rig. The goal of this analysis is to identify limiting currents and offsets where these activities can take place without causing any clashing between the drilling riser, the suspended equipment on the auxiliary rig, or the winch. The distance between the main riser and the auxiliary rig and between the main rig and winch is an important design parameter. Note that clashing of the main riser with the moon-pool, vessel hull, or bracing will need to be assessed separately prior to finalizing the stack-up model.

A static offset will be applied according to supplied information on the dual operation activity and subsequently another static offset of the vessel due to current loading. Finally, the current loading will be added and then the system will be evaluated for minimum distance between the drilling riser, the dual operation equipment, and the vessel.

A drag amplification factor will be applied to the completion riser (off of the auxiliary rig) to account for VIV drag. No drag amplification will be added to the drilling riser (off of the main rig) to conservatively estimate its downstream offset due to the current. The auxiliary rig equipment will be considered deployed at 10, 30, 60, and 90% of water depth and upstream, whereas the main drilling riser will always be considered to be downstream and connected.

25.6.9. Contact Wear Analysis

The contact between the drill string and the bore of the subsea equipment may result in wearing of both surfaces due to the rotation and running/pulling of the drill string. The softer bore of the subsea equipment will experience more wear than the drill string and, therefore, it is the subject of this section. The wear volume estimation is based on the work of Archard [9] and others. The expression for wear is given by:

$$V_{wt} = (K/H)\, N\, S \tag{25-1}$$

where

V_{wt}: total wear volume from both surfaces, in.^{3};

K: material constant;

H: material hardness in BHN;

N: contact force normal to the surfaces, lbf;

S: sliding distance, in..

This result is based on several hundred experiments that included a wide range of material combinations. The experimental result demonstrates that the wear rate, V_w/S, is independent of the contact area and the rate of rotation or sliding speed, as long as the surface conditions do not change. Such a change can be caused by an appreciable rise in surface temperature. The H value for 80-ksi material is 197 BHN. For the flexjoint wear ring and wear sleeve, H is 176 BHN.

The normal force, N, is obtained from the contact analysis of the drill string for the load cases.

The sliding distance, S, is related to the string RPM as follows:

$$S = \pi\, d(RPM)\, t \tag{25-2}$$

where d is the diameter of the drill pipe/tool joint, and t is the time in minutes.

Substitution of Equation (25-1) into Equation (25-2) and solving for t as a function of V_w gives:

$$t = (H/K)\,(V_w/N)\,(\pi\, dRPM) \tag{25-3}$$

The drilling fluid provides lubrication with reduction in wear by comparison to the dry contact conditions. Therefore, the results of this study, which are based on unlubricated wear, will be conservative. The wear volume, V_w, can be further related to the wear thickness, t_w, by the wear geometry as discussed in the next section.

Because the goal is to find the wear thickness, the wear geometry should be considered. The wear area is the crescent bounded by the bore and the OD of the tool joint or the drill pipe. The following are the possible contact cases:

- Tool joint contact with casing;
- Tool joint contact with BOP-LMRP;
- Tool joint contact with riser joint;
- Tool joint contact with flexjoint;
- Drill pipe contact with riser joint;
- Drill pipe contact with flexjoint.

For each tension, and each position of the tool joint, the flexjoint angle is increased between 0 and 4 degrees at increments of 0.1 degree. The reaction forces at each increment are reported. As long as the drill string tension is maintained at a given angle that is greater than zero, the wear process will continue under the reaction forces. Because of the large scale of the problem, these reaction forces remain unchanged for wear thicknesses of up to 1 in. So the question becomes this: How much time does it take to wear out a certain thickness?

The first step in calculating wear is to estimate the drill string tension near the mudline since the contact reaction forces depend on this tension. To simplify the wear calculations, a conservative approach is implemented where the reaction forces for each contact location are normalized with respect to the drill string tension for the five positions of the tool joint.

A typical wear calculation procedure could be as follows:

1. Determine as input the following:
   - Angle of drilling;
   - Material Brinell hardness;
   - Tension range;
   - RPM.

2. Calculate the normal force from the reaction envelopes from the tension.
3. Obtain the sliding distance, *S*, from the $t_w - V_w$ values.
4. Obtain the time, *t*, in minutes for each t_w.

25.6.10. Recoil Analysis

The objectives of conducting a recoil analysis are to determine recoil system settings and vessel position requirements, which ensure that during disconnect the following are achieved:

1. The LMRP connector does not snag.
2. The LMRP risers clear the BOP.
3. The riser rises in a controlled manner.

Recoil analysis is not required for every specific application if the vessel has an automatic recoil system. The criteria to be considered at each stage of recoil are as follows:

- *Disconnect:* The angle of the LMRP as it leaves the BOP should not exceed the allowable departure angle of the connector. This may limit the possibility of reducing tension prior to disconnect.
- *Clearance:* The LMRP should rise quickly enough to avoid clashing with the BOP as the vessel heaves downward.
- *Speed:* The riser should not rise so fast that the slip joint reaches maximum stroke at high speed.

Requirements for modeling the riser during recoil are the same as those needed for hang-off. In addition, it must be possible to account for the nonlinear and velocity-dependent characteristics of the tensioner system. A time-domain riser analysis program can be used alone or in conjunction with spreadsheet calculations from which tensioner characteristics are derived. The analysis sequence is as follows:

- Conduct analysis of the connect riser.
- Release the base of the LMRP, to reflect unlatching, and analyze the subsequent response for a short period of time.
- Change tensioner response characteristics to simulate valve opening or closure and analyze subsequent riser response for a number of wave cycles.

The analysis is repeated to determine the necessary time delay between operations. The upstroke of the riser must be monitored to detect whether top-out occurs and at what speed. If the riser is to be allowed to stroke on the slip-joint during hang-off, vertical oscillation of the riser following disconnect must also be monitored to ensure that clashing with the BOP does not occur.

REFERENCES

[1] World Oil, Composite Catalog of Oilfield Equipment & Services, forty fifth ed., Gulf Publishing Company, Houston, 2002/03.

[2] International Standards Organization, Petroleum and Natural Gas Industries – Design and Operation of Subsea Production Systems – Part 7: Completion/Workover/Riser System, ISO 13628-7, 2005.

[3] T. Clausen, R. D'Souza, Dynamic Risers Key Component for Deepwater Drilling, Floating Production, Offshore Magazine, vol. 61, (2001), May.

[4] P.R. Geiger, C.V. Norton, Offshore Vessels, Their Unique and Applications for the Systems Designer, Marine Technology, vol. 32 (No. 1) (1995) 43–76.

[5] American Petroleum Institute, Recommended Practice for Design, Selection, Operation and Maintenance of Marine Drilling Riser Systems, API-RP- 16Q (1993).

[6] American Petroleum Institute, Design of Risers for Floating Production Systems (FPSs) and Tension-Leg Platform (TLPs), API-RP- 2RD (1998).

[7] H. Matlock, Correlations for Design of Laterally Load Piles in Soft Clay, OTC, 2312, Offshore Technology Conference, Houston, Texas, 1975.

[8] K. Vandiver, L. Lee, User Guide for Shear7 Version 4.1, Massachusetts Institute of Technology, Cambridge, 2001, March 25.

[9] J.F. Archard, Contact and Rubbing of Flat Surfaces, Journal of Applied Physics, vol. 24 (No. 8) (1953) 981.

CHAPTER 26

Subsea Production Risers

Contents

Subsea Engineering Handbook
ISBN 978-0-12-397804-2, doi:10.1016/B978-0-12-397804-2.00026-6

26.1. INTRODUCTION

A production riser system consists of conductor pipes connected to floaters on the surface and the wellheads at the seabed. Also, it is the primary device of the floating production system to convey fluids to and from the vessel. It is one of the most complex aspects of a deepwater production system [1,2].

Figure 26-1 illustrates various types of subsea riser systems used in deepwater field developments. There are essentially two types of subsea risers: rigid risers and flexible risers. A hybrid riser is achieved through a combination of these two types of risers. Four types of production risers are compared in this chapter: steel catenary risers (SCRs), top tensioned risers (TTRs), flexible risers, and hybrid risers. Then, detailed descriptions of all four categories of riser are given. Unless otherwise specified, the term

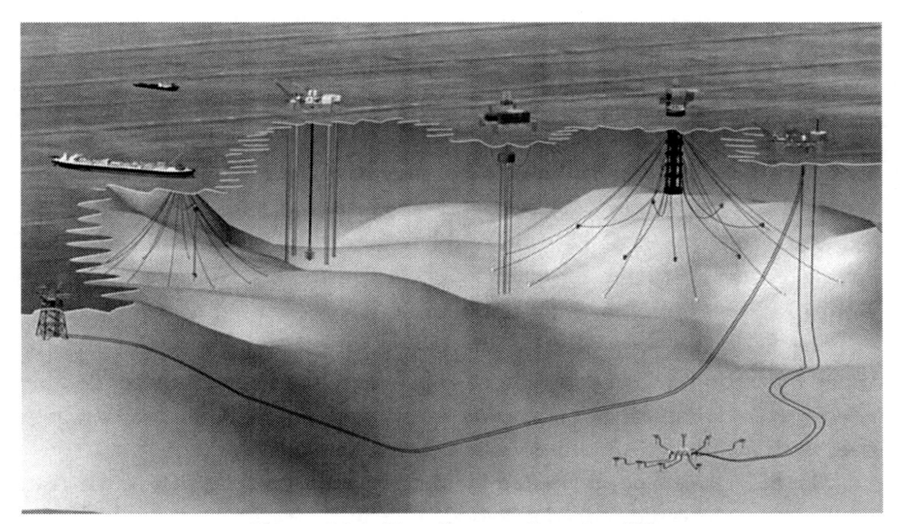

Figure 26-1 Riser Systems Overview [3]

deep water refers to water depths of more than 500 m, and ultra-deep water is defined as water depths of more than 2000 m.

Flexible risers are one of the common types of production risers. They may be deployed in a variety of configurations, depending on the water depth and environment. Flexible pipes have traditionally been limited by diameter and water depth; however, many deepwater projects in the Gulf of Mexico and Brazil are now employing SCRs for both export and import risers.

The choice between a flexible riser and an SCR is not clear cut. The purchase cost of flexible risers for a given diameter is higher per unit length, but they are often less expensive to install and are more tolerant to dynamic loads. Also, where flow assurance is an issue, the flexible risers can be designed with better insulation properties than a single steel riser.

Flexible risers and import SCRs are associated with wet trees. Top tensioned risers are almost exclusively associated with dry trees and hence do not usually compete with flexible risers and SCRs except at a very high level: the choice between wet and dry trees [4,5].

26.1.1. Steel Catenary Risers (SCRs)

SCRs were initially used as export lines on fixed platforms. SCRs clearly have similarities with free-hanging flexible risers, being horizontal at the lower end and generally within about 20° of the vertical at the top end. In this arrangement, the riser forms an extension of the flowline that is hung from the platform in a simple catenary. Relative rotational movement between riser and platform can use a flexjoint, stress joint, and pull tube to offset the movement.

The Auger platform is the first floating production facility to implement an SCR (2860 ft, Gulf of Mexico, 1993), whereby two 12-in. lines were used for oil and gas export. Since then, they have been progressively used in more severe application. SCRs were used in the late 1990s, again for exporting aboard the P-18 submersible offshore in Brazil. Recently, they have been used in large numbers as production risers associated with the Bonga FPSO. Figure 26-2 illustrates an SCR system used in deep water.

The SCR is a cost-effective alternative for oil and gas export and for water injection lines on deepwater fields, where the large-diameter flexible risers present technical and economic limitations. An SCR is a free-hanging riser with no intermediate buoys or floating devices.

The SCR is self-compensated for the heave movement; that is, the riser is lifted off or lowered onto the seabed. The SCR riser still needs a ball joint to allow rotation of the risers to be induced by waves, currents, and vessel motion at the upper end connection.

The SCR is sensitive to waves and current due to the normally low level of effective tension on the riser. The fatigue damage induced by vortex-induced vibrations (VIVs) can be fatal to the riser. Use of VIV suppression devices such as helical stakes and fairing can help reduce the vibrations to a reasonable level.

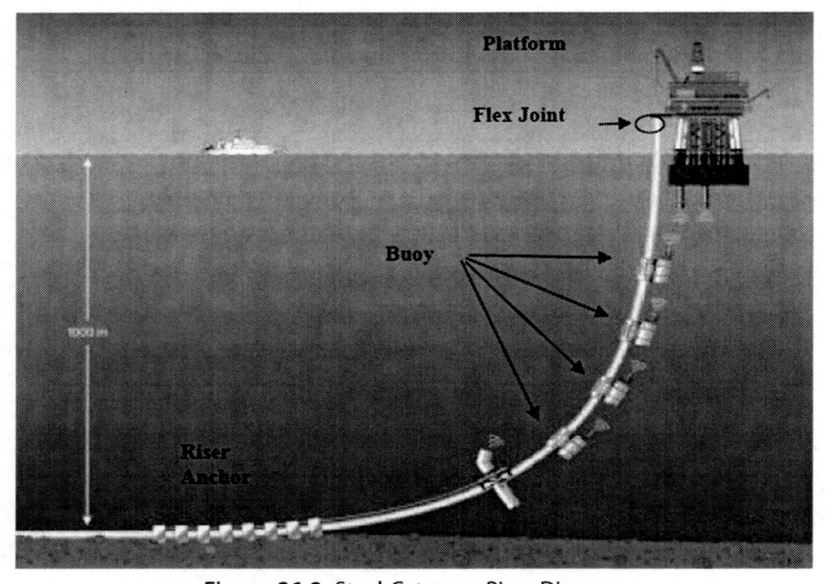

Figure 26-2 Steel Catenary Riser Diagram

26.1.2. Top Tensioned Risers (TTRs)

TTRs [6] are long circular cylinders used to link the seabed to a floating platform. These risers are subject to steady currents with varying intensities and oscillatory wave flows. The risers are provided with tensioners at the top to maintain the angles at the top and bottom under the environmental loading. The tension requirements for production risers are generally lower than those for drilling risers. The risers often appear in a group arranged in a rectangular or circular array.

TTRs rely on a top tensioner in excess of their apparent weight for stability. TTRs are commonly used on a tension leg platform (TLP) [7] or spar dry tree production platform (spar), as shown in Figure 26-3.

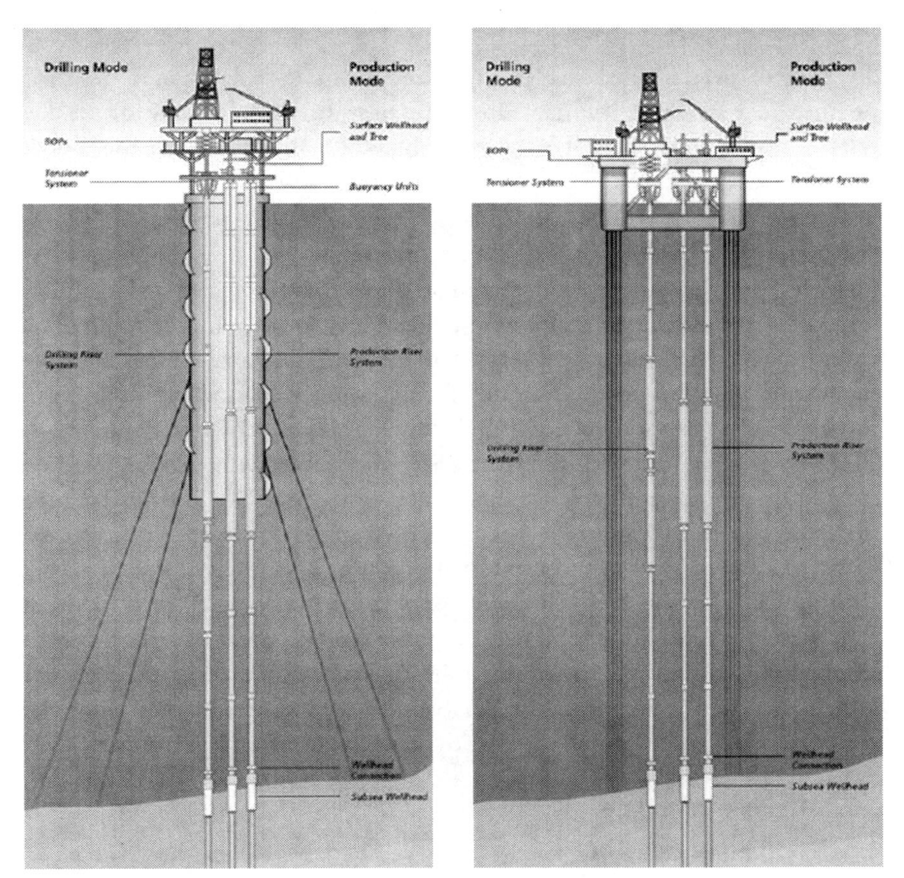

Figure 26-3 Top Tensioned Risers Used on Spar and TLP [8]

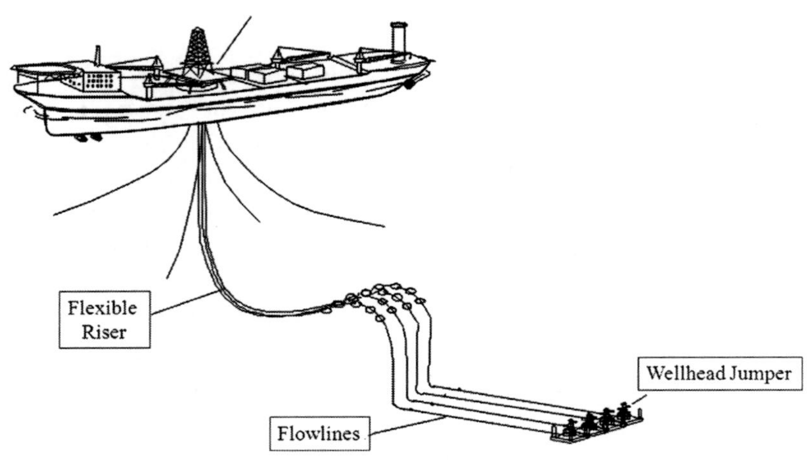

Figure 26-4 Flexible Riser Diagram

At the surface, the riser is supported from the platform by hydro-pneumatic tensioners, which allow the riser to move axially or stroke, relative to the platform. TTRs were designed for shallow-water use, but as the water depths grew, so did the need for new designs.

26.1.3. Flexible Risers

Flexible risers have been a successful solution for deepwater and shallow-water riser and flowline systems worldwide. Flexible risers are the result of an extraordinary development program that was based on flexible pipes. Flexible pipes were found to be ideally suited for offshore applications in the form of production and export risers, as well as flowlines [1], as shown in Figure 26-4.

Flexible risers are multiple-layer composite pipes with relative bending stiffness, to provide performance that is more compliant [9]. The flexible pipe structure shown in Figure 26-5 is composed of several layers (e.g., carcass) made of stainless steel to resist external pressure. The internal sheath acts as a kind of internal fluid containment barrier, the pressure armor is made of carbon steel to resist hoop pressure, the tensile armor is made of carbon steel to resist tensile loading, and the external sheath is a kind of external fluid barrier [4,5,10]. In such applications, the flexible pipe section may be used along the entire riser length or limited to short dynamic sections such as jumpers.

26.1.4. Hybrid Riser

The concept of a hybrid riser was developed based on the TTRs. Its principal feature is that it accommodates relative motion between a floating

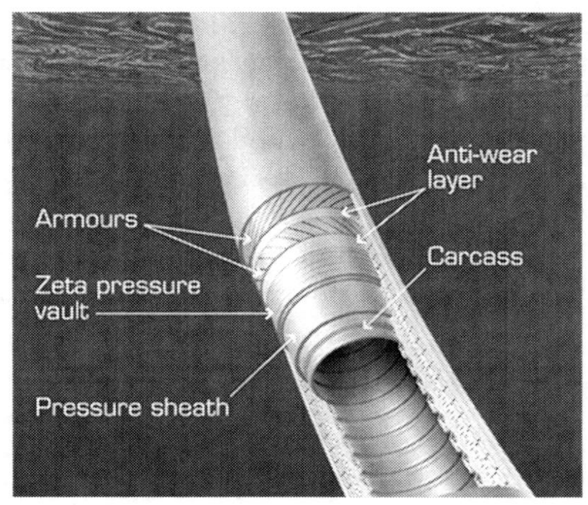

Figure 26-5 Typical Cross Section of Flexible Pipe

structure and a rigid metal riser, by connecting them with flexible jumpers [11]. Figure 26-6 illustrates a bundled hybrid riser.

The main section of the bundled hybrid riser consists of a central structural tubular section, around which synthetic foam buoyancy modules are attached. Peripheral production and export lines run through the buoyancy modules and are free to move axially in order to accommodate

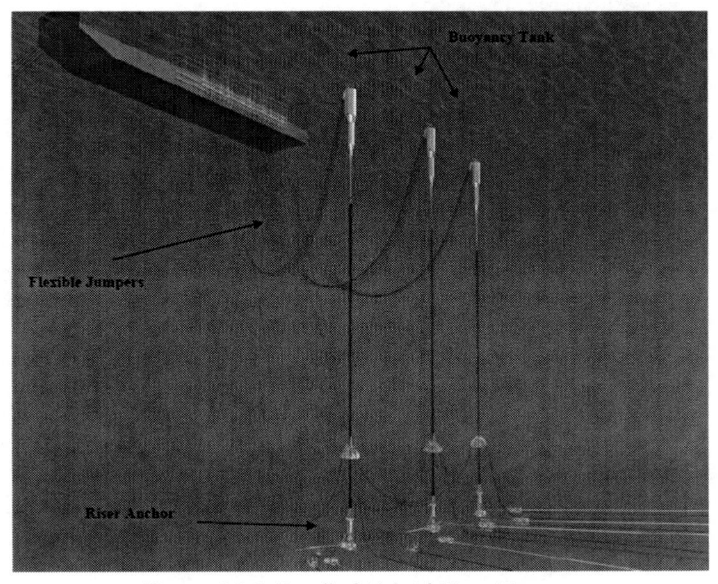

Figure 26-6 Bundled Hybrid Riser Diagram

thermal and pressure-induced extension. The central structural member is connected to the riser base by way of a hydraulic connector and stress joint. The peripheral lines are attached to hard piping on the base, which provides a connection to the subsea flowlines, and terminates in goosenecks some 30 to 50 m below the water surface. Flexible piping is attached between the goosenecks and porches on the pontoons of the semisubmersible production vessel, providing the flow path to the vessel while accommodating relative movements between the rigid riser section and the platform [12].

26.2. STEEL CATENARY RISER SYSTEMS

SCRs (see Figure 26-7) are a preferred solution for deepwater wet tree production, water/gas injection, and oil/gas export. At the end of 2008, the total number of installed deepwater SCRs was more than 100 worldwide, with the majority installed in the Gulf of Mexico (GoM). Meanwhile, about 30 SCRs are used in the detailed engineering design phase. In addition, the SCR concept has been selected for most of the tie-back projects [13].

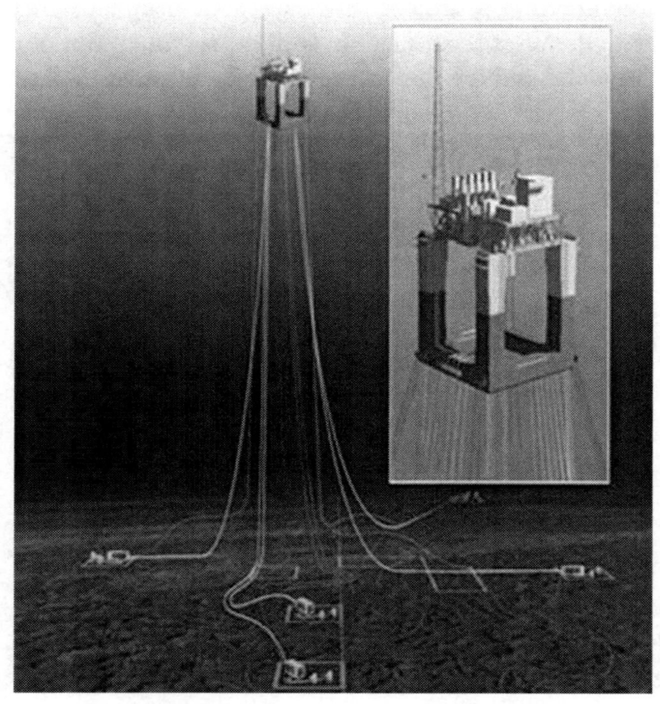

Figure 26-7 Steel Catenary Riser Configurations

The design, welding, and installation challenges associated with SCRs in ultra-deepwater floating production are primarily related to the higher hang-off tensions caused by the integration of their weight over the water depth, in combination with additional challenges from high-pressure, high-temperature and sour service [14].

26.2.1. Design Data

26.2.1.1. General Sizing

In the preliminary stage, the diameter and wall thickness of the riser and pipe must be determined to minimize the cost of the pipes. Factors that influence riser diameter and wall thickness sizing include:

- Operating philosophy: transportation strategy, pigging, corrosion, inspection;
- Well characteristics: pressure, temperature, flow rate, heat loss, slugging, well fluids and associated chemistry;
- Structural limitations: burst, collapse, buckling, postbuckling;
- Installation issues: tensioning capacity of available vessels;
- Construction issues: manufacturability, tolerances, weld procedures, inspection;
- Vessel offsets and motions;
- Metocean conditions;
- Deepwater environments.

26.2.1.2. Materials Selection

Factors to be considered in material selection include strength requirements, adequate material toughness for fracture and fatigue performance, weld defect acceptance criteria, and sweet/sour service requirements [15].

26.2.1.3. Deepwater Environmental Conditions

For deepwater environmental conditions, four parameters must be analyzed for the design phase of SCRs [16]:

- Hydrodynamic loads;
- Material properties;
- Soil interaction;
- Extreme storm situations.

26.2.1.4. Metocean Data

The location of a riser may dictate critical design conditions; for example, consider the loop currents in the GoM and the highly directional

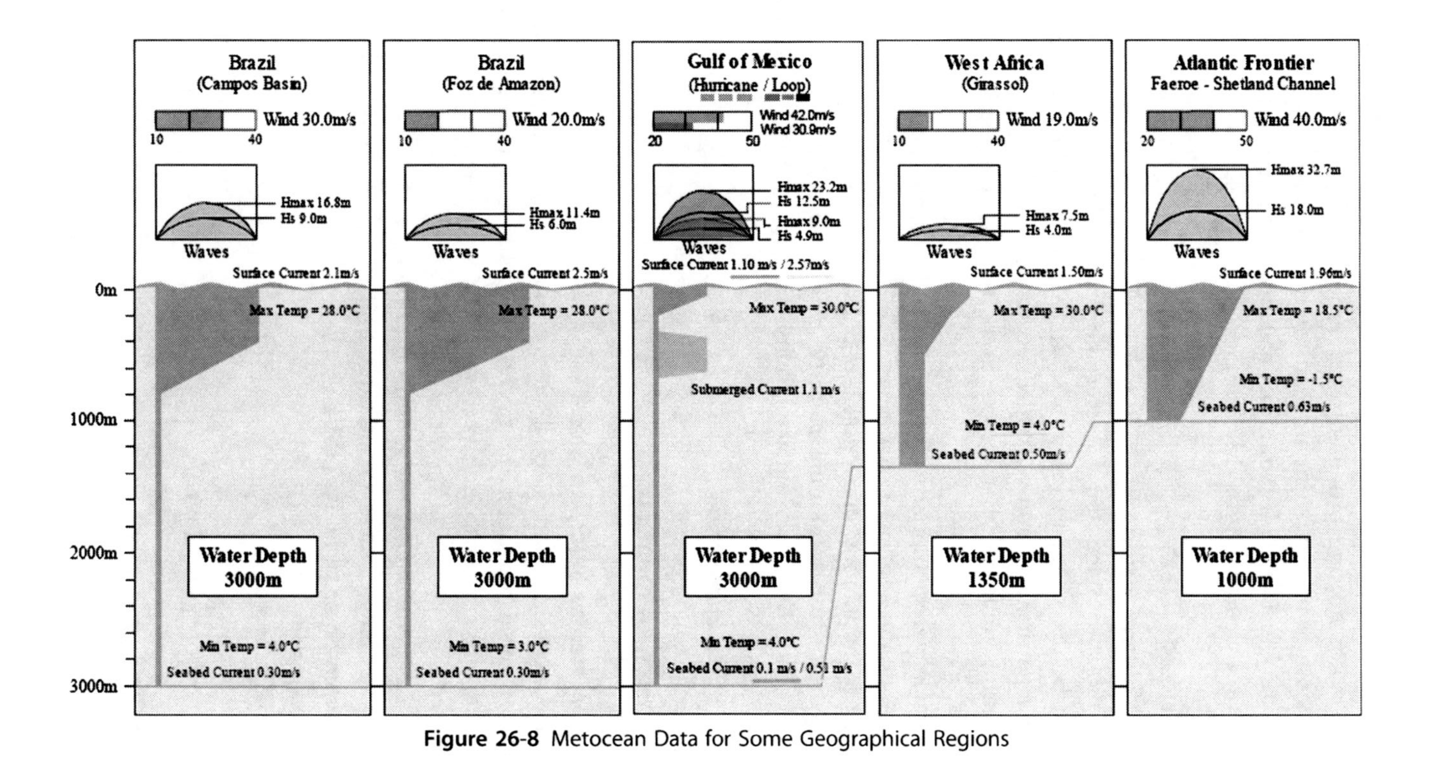

Figure 26-8 Metocean Data for Some Geographical Regions

environments of West of Africa. Vessel motions and offsets have a major influence on riser design. Figure 26-8 shows different metocean data for some geographical regions. The following metocean data are used in riser analyses:

- Water depth;
- Waves;
- Currents;
- Tide and surge variations;
- Marine growth.

26.2.1.5. Geotechnical Data

ROV footage from deepwater operating installations indicates significant and complex riser interaction with the seabed in the SCR touchdown zone (TDZ). Such behavior is largely influenced by the geotechnical properties of the seabed the riser is interacting with. The nonlinear stress/strain behaviors of soil, consolidation and remolding of soil (and associated changes in shear strength), trenching and backfilling, hysteresis, strain rate, and suction effects affect the loads imparted on the riser. It is not possible or desirable to reproduce complex interactions perfectly, but it is important to model those characteristics that have the greatest effects on riser stresses and fatigue lives.

Deepwater Soils

SCRs are most commonly used as a part of deepwater floating systems. In many deepwater locations (GoM, West Africa, Brazil), soil types found at or close to the seabed are generally underconsolidated, normally consolidated, or lightly overconsolidated clays. Much of the research to date has focused on interaction with such seabeds. However, other soil conditions are also possible. For instance, soils are generally very variable in glacial settings such as in northwestern Europe and Canada, and gravelly and bouldery stiffer clays and sands are often encountered at the seabed.

26.2.1.6. Vessel Motion Characteristics

The host vessel's motions are defined by the global performance analysis accounting for wave, wind, and current loads using either time-domain analysis or frequency-domain analysis. The motion data are expressed as time traces of vessel motions or RAOs defined at the center of gravity (CoG) for the floater for predefined loading conditions. The motions at the riser hang-off location will be transferred from the CoG via rigid body assumptions. The riser system is considered to be a cable under current loads with a boundary condition that is defined as the motions at the hang-off location.

Vessel RAOs are used throughout the whole design process and it is important for them to be well defined. The following definitions should always be provided with RAOs:

- What the motion phase angle is relative to, whether it is a lag or a lead;
- Vessel coordinate system;
- Location of point for which RAOs are given;
- Units;
- Direction of wave propagation relative to vessel.

26.2.1.7. Wave Theories

As a general rule, Airy (linear) wave theory is suitable for most applications and is applicable in both regular and random wave analysis. Other theories such as Stokes V have advantages where only regular wave analysis is performed, especially with regard to fluid particle kinematics. This theory resembles those of "real" large-amplitude regular waves.

26.2.2. Steel Catenary Riser Design Analysis

In the pre-FEED phase, an initial design is carried out to define the following:

- Riser host layout (for interface with other disciplines);
- Riser hang-off system (flexjoint, stress joint, pull-tube, etc.);
- Riser hang-off location, spacing, and azimuth angle (hull layout, subsea layout, total risers and interference consideration);
- Riser hang-off angle for each riser;
- Riser location elevation at hull (hull type, installation, and fatigue consideration);
- Global static configuration.

A static configuration may be determined based on catenary theory accounting for hang-off angle, water depth, and riser unit weight. The SCR design should meet basic functional requirements such as SCR internal (and/or external) diameters, submerged tension on host vessel, design pressure/temperature, and fluid contents.

Due considerations should be given to future tie-back porches to accommodate the variations for a hang-off system, riser diameter, azimuth angles, and the required extreme response and fatigue characteristics.

26.2.3. Strength and Fatigue Analysis

For the preparation of FEED documentation, a preliminary analysis is carried out to confirm [17]:

- Extreme response that meets the stress criteria per API 2RD [6] and extreme rotation for flexjoints;
- VIV fatigue life and the required length of strakes (or fairing);
- Wave fatigue life;
- Interference between risers and with floater hull.

In the detailed design phase, installation analyses and special analyses, such as a VIM (Vortex Induced Motion)-induced fatigue analysis, semi-submersible heave VIV(Vortex Induced Vibration) fatigue analysis, and coupled system analysis, are conducted.

26.2.4. Construction, Installation, and Hook-Up Considerations

The effects of construction and installation operations may impose a permanent deformation and residual loads/torques on the riser system while consuming a proportion of the fatigue life [18]. In-service requirements determine weld quality, acceptable levels of mismatch between pipe ends, and out-of-roundness, whereas nondestructive testing (NDT) requirements are determined from fatigue life and fracture analysis assessments.

26.2.4.1. Construction Considerations

Risers are to be constructed in accordance with related guides that are consistent with specifications such as ABS Guide for Building and Classing Subsea Pipeline Systems and Riser [2]. The pipe-laying methods and other construction techniques are acceptable provided the riser meets all of the criteria defined in these guides [19]. Guides and specifications are to be prepared to describe alignment of the riser, its design water depth and trenching depth, and other parameters, such as [20]:

- Water depth during normal pipe-laying operations and contingency situations;
- Pipe tension;
- Pipe departure angle;
- Retrieval;
- Termination activities.

26.2.4.2. Installation Considerations

The actual SCR installation philosophy may not be decided at the start of a project. Therefore, the pipeline specifications may have to be drafted with contingencies that consider all installation methods (S-lay, J-lay, and reel lay) [21]. In the case of a J-lay, the boring internal diameter is likely to be changed. Therefore, 1 mm must be added to the pipe wall thickness to allow for the change.

For smaller diameter SCRs, installation with the reel-lay procedure may be possible. Substantial engineering effort should qualify the SCRs for reeling:

- Full-scale bending trials and testing of centralizers to maintain a constant annulus for a pipe-in-pipe situation;
- Full-scale fatigue testing of the reeling riser welds. It is important to accurately calculate the bending strain in the welds during the reeling operation, to assess crack propagation;
- Careful monitoring of the allowable bending strain.

26.2.4.3. Hook-Up Consideration

The term *hook-up* is another expression for the hang-off system of a riser. The hang-off system shown in Figure 26-9 is a system that connects SCRs to the offshore platform via flexjoints or stress joints. To consider a hook-up system, certain parameters have to be determined such as the hang-off location, hang-off spacing, hang-off angle, azimuth angle, and the location of elevation at the hull structure.

26.2.5. Pipe-in-Pipe (PIP) System

Thermal insulation is required for some production risers to avoid problems with hydrate, wax, or paraffin accumulation. The use of external insulation, in some cases, might impair the riser's dynamic performance by increasing drag and reducing weight in water. However, PIP thermal insulation technology can often be used to satisfy the insulation requirements of lower

Figure 26-9 Riser Hang-Off Location

U-values, while maintaining an acceptable global dynamic response with the penalty of a heavier and perhaps more costly structure.

26.2.5.1. Structural Details

The inner and outer pipes of a PIP system, as shown in Figure 26-10, may be connected via bulkheads at regular intervals. Bulkheads limit relative expansion and can separate the annulus into individual compartments, if required. Use of bulkheads can be a good solution for pipelines, but for dynamic SCRs, one must consider the effects of high stress concentrations, local fatigue damage, and a local increase in heat loss. Alternatively, regular spacers (centralizers) can be used, which allow the inner and outer pipes to slide relative to each other while maintaining concentricity.

The items listed below are common effects for single-PIP SCRs:

- Residual curvature, which may change along the SCR during installation;
- Residual stresses due to large curvature history;
- Residual axial forces between the two pipes;
- Connection between the inner and outer pipes, including length and play of centralizers;
- Boundary conditions and initial conditions at riser terminations;
- Fatigue life consumed during installation;
- Preloading of inner and outer pipes;
- Axial forces and relative motions during operation, due to thermal expansion and internal pressure;
- Poisson's ratio effect on axial strains;
- Local stresses in inner and outer pipes due to centralizer contact, including chattering effects;

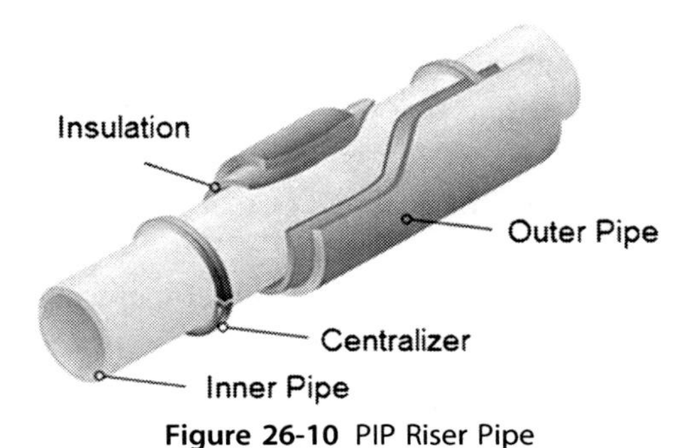

Figure 26-10 PIP Riser Pipe

- Frictional effects between inner and outer pipes;
- Thermal stresses and thermal cycling effects;
- Buckling checks (including helical buckling) due to thermal and general dynamic loading;
- Soil forces on outer pipe;
- Internal and external pressures having different effects on stress in inner and outer pipes;
- Effect of packing material in reversal of lay direction on a reel should be assessed and cross-section distortion minimized; the pipe yields as it is reeled and it is very soft at the reel contact point;
- Effects of PIP centralizers on pipe geometry during reeling;
- Wear of centralizers;
- Validity of VIV calculations (e.g., with regard to damping);
- Possible effect of any electrical heating on corrosion rates;
- Effect of damage (e.g., due to dropped objects striking outer pipe) on thermal and structural performance.

26.2.6. Line-End Attachments

SCRs can be attached to the floating vessel using a flexjoint or a stress joint.

26.2.6.1. Flexjoints

A flexjoint allows the riser system to rotate with minimum bending moment. The flexjoint normally exhibits strong nonlinear behavior at a small rotation angle and hence should ideally model as a nonlinear rotation spring or a short beam with nonlinear rotational stiffness. A correct understanding of the flexjoint stiffness is important in determining maximum stresses and fatigue in the flexjoint region [22]. Flexjoint stiffness for the large rotations that typically occur in severe storms is much less than that for the small amplitudes that occur in fatigue analysis. Temperature variations can also result in significant changes in flexjoint stiffness. In addition, it should be verified that the flexjoint can withstand any residual torque that may be in the riser following installation or released gradually from the seabed section of the line. Steps may be taken to relieve torque prior to attachment.

A flexjoint bellows system is illustrated in Figure 26-11. The bellows protect the elastomeric flex element from the explosive decompression caused by large internal pressure fluctuations in a gas-saturated environment. The cavity between the body/flex element and the bellows is sealed and

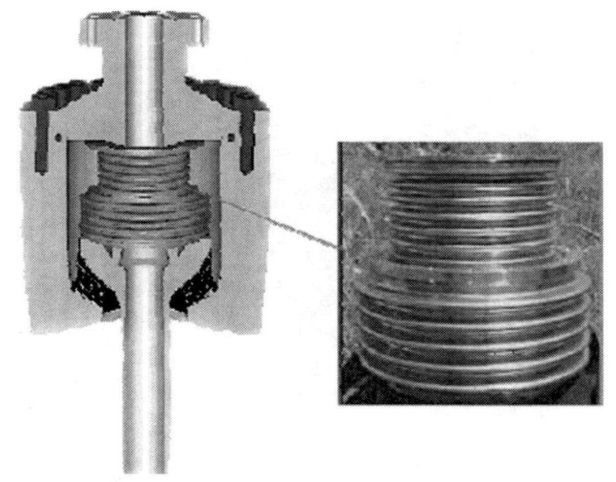

Figure 26-11 Flexjoint Bellow System for Sour Services [23]

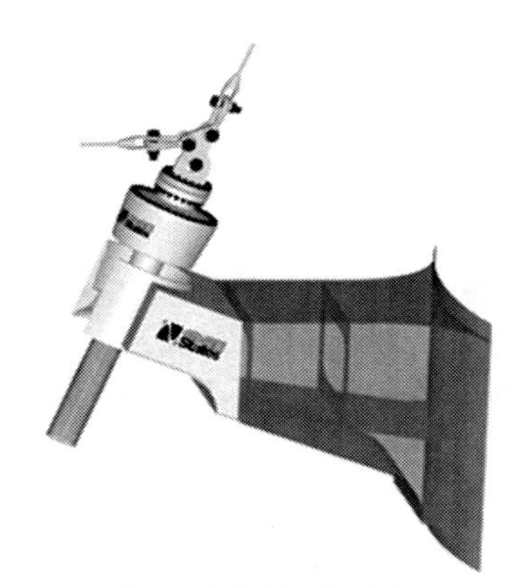

Figure 26-12 Flexjoint

filled with a water/propylene glycol-based corrosion inhibiting fluid. Figure 26-12 shows a flexjoint used with an SCR.

26.2.6.2. Stress Joints

Stress joints may be used in place of flexjoints, but they usually impart larger bending loads to the vessel. They are simple, inspectable, solid metal structures, and particularly able to cope with high pressures and temperatures. Figure 26-13 shows a stress joint, which may be made of either steel or

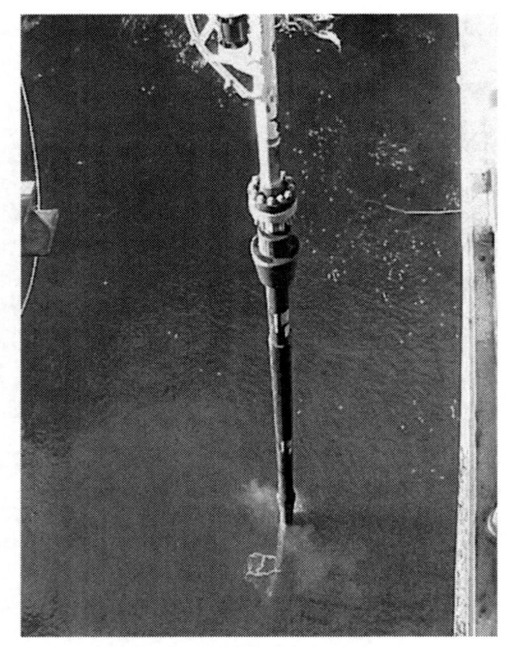

Figure 26-13 Stress Joint

titanium. The latter material has the advantage of good resistance to attack from sour and acidic well flows and, of course, gas permeation. Titanium gives lower vessel loads than steel and typically has better fatigue performance than steel.

26.3. TOP TENSIONED RISER SYSTEMS

TTRs are used as conduits between dynamic floating production units (FPUs) and subsea systems on the seafloor, for dry tree production facilities such as spars and TLPs.

TTRs are individual risers that rely on a top tension in excess of their apparent weight for stability. TTRs are commonly used on TLPs and spar dry tree production platforms. TTRs are generally designed to give direct access to the well, with the wellhead on the platform. This type of riser has to be capable of resisting the tubing pressure in case of a tubing leak or failure. The four types of TTRs are drilling risers, completion/workover risers, production/injection risers, and export risers. Table 26-1 shows the types of TTRs and their reference specifications.

Table 26-1 Types of Top Tensioned Risers

Types of Top Tensioned Riser	Applications	Specifications
Drilling	• MODU drilling risers • Surface wellhead platform drilling risers	• API RP16Q • API RP 2RD
Completion/ workover [24]	• MODU completion/workover risers • Surface wellhead platform completion/workover risers	• API RP 17G • API RP 2RD
Production/ injection	• Surface wellhead platform production risers • Subsea tie-backs	• API RP 2RD • API RP 1111
Export	• Surface wellhead export risers	• API RP 2RD • API RP 1111 • ASME B31.4 • ASME B31.8 • 30 CFR • 49 CFR

26.3.1. Top Tensioned Riser Configurations

Figure 26-14 presents a typical TTR system configuration. TTRs consist of long, flexible circular cylinders used to link the seabed where the wellheads are located to a floating platform. TTRs are provided with tension at the top to maintain the angles at the top and bottom under environmental loading

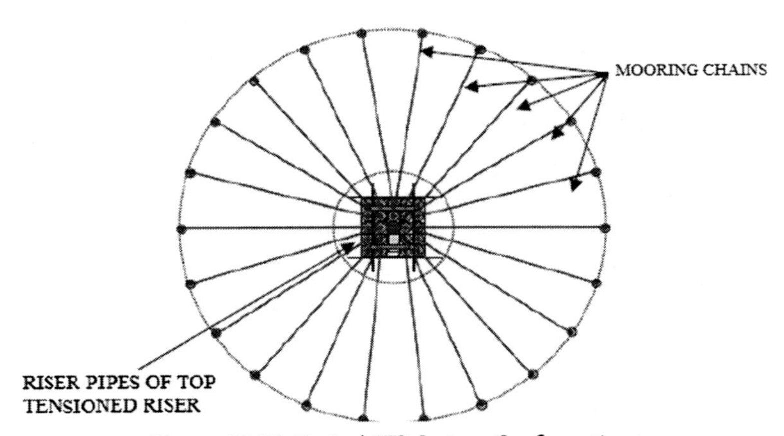

Figure 26-14 Typical TTR System Configurations

conditions. TTRs often appear in a group arranged in a rectangular or circular array.

The following major design considerations are important when designing a TTR to prevent failure:

- Allowable floater motions;
- Allowable stroke of tensioning system;
- Maximum riser top tension;
- Size of stress joints and flexjoints;
- Keel joints;
- Increasing length of riser joints;
- Design criteria for the safety philosophy of liquid barriers, valve, and seals;
- Current, interface with array, and VIV;
- Impact between buoyancy cans and hull guides.

26.3.2. Top Tensioned Riser Components

26.3.2.1. Riser System

Figure 26-15 presents a typical TTR setup for use with a TLP. The TTR configuration depends on the riser function and the number of barriers selected (single or dual). In general, the riser configuration comprises the following components:

- The main body is made up of rigid segments known as joints. These joints may be made of steel, titanium, aluminum, or composites, although steel is predominantly used.
- Successive joints are linked by connectors such as threaded, flanged, dogged, clip type, box, and pin connectors.
- The riser is supported by a tensioning system, such as traditional hydraulic tensioners, air cans, RAM tensioners, tensioner decks, and counterweights.

26.3.2.2. Buoyancy System

Figure 26-16 presents a buoyancy-based TTR for use with a spar.

For water depths exceeding 2000 ft, buoyancy systems are required to provide lift, which reduces the top tension requirements, prevents excessive stresses in the riser, and reduces the hook load during deployment/retrieval of the BOP. Both synthetic foam and air-can buoyancy systems have been used for deepwater riser systems, either individually or in combination.

Buoyancy cans (see Figure 2-17) decouple the vertical riser movement from the vessel, and can be built by the fabrication yard. However, either a heavy lift vessel or a specially designed rig is required for offshore

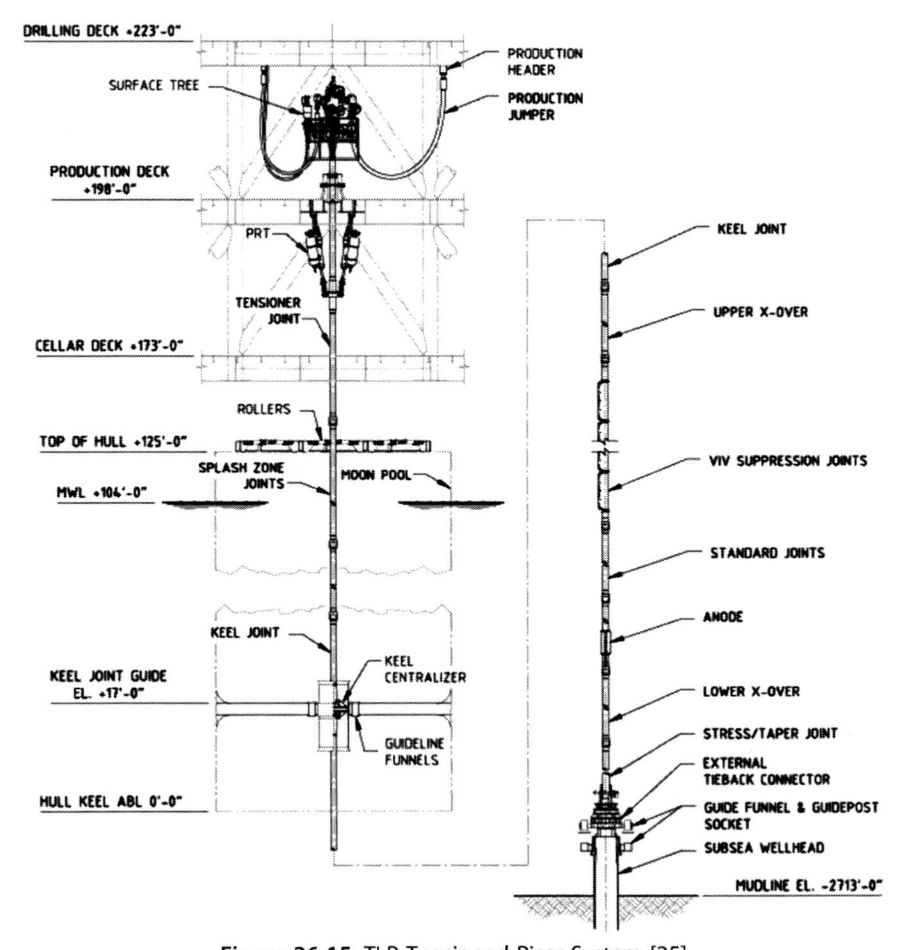

Figure 26-15 TLP Tensioned Riser System [25]

installation. It has large relative vertical motion in storm conditions and may generate lateral loads between the buoyancy cans and the spar center wall.

26.3.3. Design Phase Analysis

Before designing the TTR, several analyses need to be conducted to ensure that the riser design is up to specifications. To design a TTR system, these analyses must be performed:

- Top tension factor analysis;
- Pipe sizing analysis;
- Tensioning system sizing analysis;

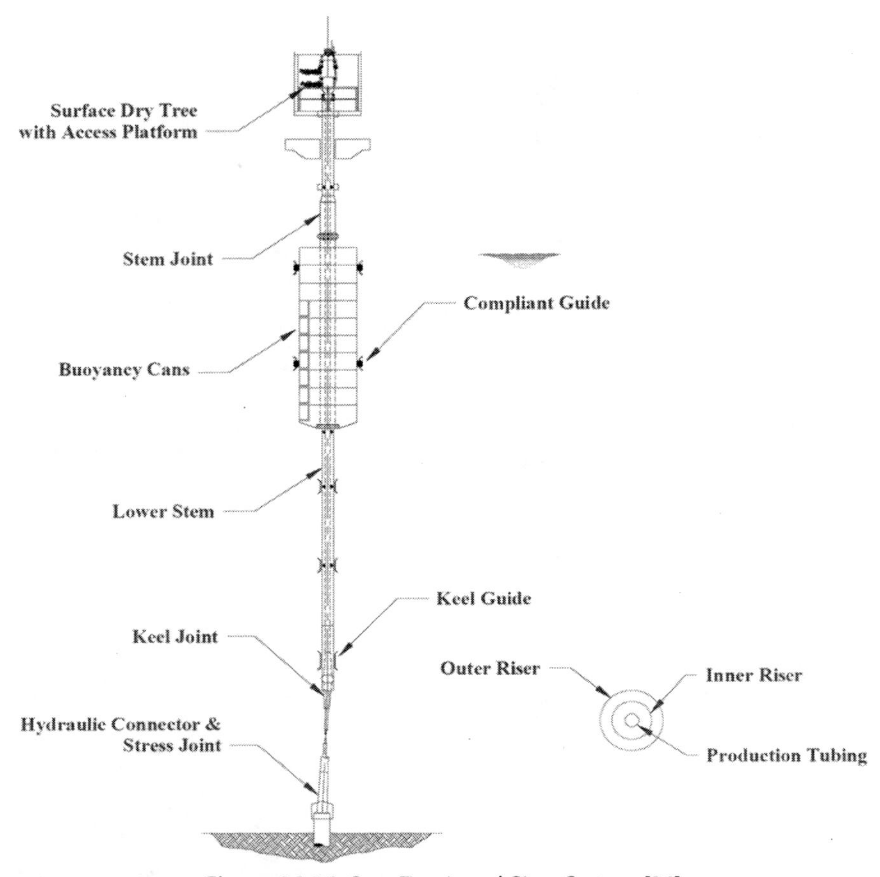

Figure 26-16 Spar Tensioned Riser System [26]

- Stroke analysis;
- Riser VIV fatigue analysis [27,28];
- Interference analysis;
- Strength analysis;
- Fatigue analysis.

26.4. FLEXIBLE RISERS

Flexible risers trace their origins to pioneering work carried out in the late 1970s. Initially flexible pipes were used in relatively benign weather environments such as offshore Brazil, the Far East, and the Mediterranean. Since then, however, flexible pipe technology has advanced rapidly and today

Figure 26-17 Typical Buoyancy Modules

flexible pipes are used in various fields in the North Sea and are also gaining popularity in the Gulf of Mexico. Flexible pipe applications include water depths down to 8000 ft, high pressures up to 10,000 psi, and high temperatures above 150°F, as well as the ability to withstand large vessel motions in adverse weather conditions.

The main characteristic of a flexible pipe is its low relative bending to axial stiffness. This characteristic is achieved through the use of a number of layers of different materials in the pipe wall fabrication. These layers are able to slip past each other when under the influence of external and internal loads; hence, this characteristic gives a flexible pipe its property of low bending stiffness. The flexible pipe composite structure combines steel armor layers with high stiffness to provide strength and polymer sealing layers with low stiffness to provide fluid integrity. This construction gives flexible pipes a number of advantages over other types of pipelines and risers such as steel catenary risers. These advantages include prefabrication, storage in long lengths on reels, reduced transport and installation costs, and suitability for use with compliant structures.

26.4.1. Flexible Pipe Cross Section

Two types of flexible pipes are in use: bonded and unbonded flexible pipes. In bonded pipes, different layers of fabric, elastomer, and steel are bonded

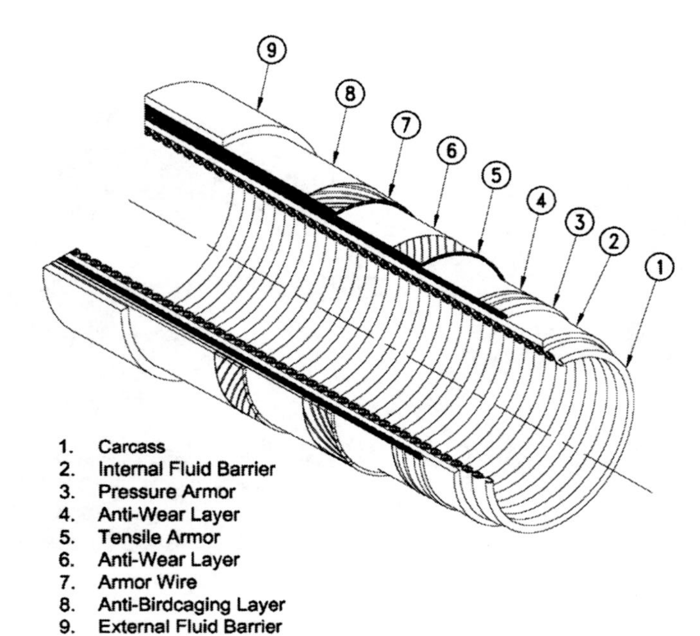

Figure 26-18 Typical Cross Section of an Unbonded Flexible Pipe [29]

together through a vulcanization process. Bonded pipes are only used in short sections such as jumpers. However, unbonded flexible pipes can be manufactured for dynamic applications in lengths of several hundred meters. Unless otherwise stated, the rest of this chapter deals with unbonded flexible pipes.

Figure 26-18 shows a typical cross section of an unbonded flexible pipe. This figure clearly identifies the five main components of the flexible pipe cross section. The space between the internal polymer sheath and the external polymer sheath is known as the pipe annulus. The five main components of the flexible pipe wall are discussed in the following sections.

26.4.1.1. Carcass

The carcass forms the innermost layer of the flexible pipe cross section. It is commonly made of a stainless steel flat strip that is formed into an interlocking profile. The main function of the carcass is to prevent pipe collapse due to hydrostatic pressure or buildup of gases in the annulus.

26.4.1.2. Internal Polymer Sheath

The internal polymer sheath provides a barrier to maintain the bore fluid integrity. Exposure concentrations and fluid temperature are key design drivers for the internal sheath. Common materials used for the internal sheath include Polyamide-11 (commercially known as Rilsan®), high-density polyethylene (HDPE), cross-linked polyethylene (XLPE), and PVDF.

26.4.1.3. Pressure Armor

The role of the pressure armor is to withstand the hoop stress in the pipe wall that is caused by the inner bore fluid pressure. The pressure armor is wound around the internal polymer sheath and is made of interlocking wires.

26.4.1.4. Tensile Armor

The tensile armor layers are always cross-wound in pairs. As their name implies, these armor layers are used to resist the tensile load on the flexible pipe. The tensile armor layers are used to support the weight of all of the pipe layers and to transfer the load through the end fitting to the vessel structure. High tension in a deepwater riser may require the use of four tensile armor layers, rather than just two.

26.4.1.5. External Polymer Sheath

The external polymer sheath can be made of the same materials as the internal polymer sheath. The main function of the external sheath is as a barrier against seawater. It also provides a level of protection for the armor wires against clashing with other objects during installation.

26.4.1.6. Other Layers and Configurations

Besides the five main layers of a flexible pipe just discussed, there are other minor layers, which make up the pipe cross section. These layers include antifriction tapes wound around the armor layers. These tapes reduce friction and wear of the wire layers when they rub past each other as the pipe flexes due to external loads. Antiwear tapes can also be used to ensure that the armor layers maintain their wound shape. These tapes also prevent the wires from twisting out of their preset configuration, a phenomenon called *birdcaging* that is a result of hydrostatic pressure causing axial compression in the pipe.

In some flexible pipe applications, the requirement for the use of high-tensile wires for the tensile armor layers arises because of high-tensile loads, and yet the presence of a "sour" environment means that these wires would suffer an unacceptable rate of HIC/SSC. A solution to this situation is to fabricate a pipe cross section with two distinct annuli rather than one.

26.4.2. Flexible Riser Design Analysis

26.4.2.1. Design Basis Analysis Document

The design document should include the following, as minimum requirements [30]:

- Host layout and subsea layout;
- Wind, wave, and current data, as well as any vessel motion that is applicable to riser analysis;
- Applicable design codes and company specifications;
- Applicable design criteria;
- Porch and I-tube design data;
- Load case matrices for static strength, fatigue, and interference analyses;
- Applicable analysis methodology.

Several types of analyses have to be carried out when conducting the design analysis for a flexible riser, including these:

- Finite element modeling and static analysis;
- Global dynamic analysis;
- Interference analysis;
- Cross-sectional model analysis;
- Extreme and fatigue analysis.

26.4.3. End Fitting and Annulus Venting Design

26.4.3.1. End Fitting Design and Top Stiffener (or Bellmouth)

The end fitting design is a critical component of the global flexible pipe design process. The main functions of the end fitting are to transfer the load sustained by the flexible pipe armor layers onto the vessel structure, and to complement the sealing of the polymer fluid barrier layers. Figure 26-19 illustrates a typical end fitting system.

The most severe location for fatigue damage in the risers is usually at the top hang-off region. The riser is protected from overbending in this area by either a bend stiffener or a bellmouth. Detailed local analyses for the curvature or bellmouth are carried out using a 2D finite element model.

26.4.3.2. Annulus Venting System

To prevent the buildup of gases in the annulus due to diffusion, a venting system is incorporated into the pipe structure to enable the annulus gases to be vented to the atmosphere. Three vent valves are incorporated into both end fitting arrangements of a pipe. The vent valves are directly connected to the annulus and are designed to operate at a preset pressure of about 30 to 45

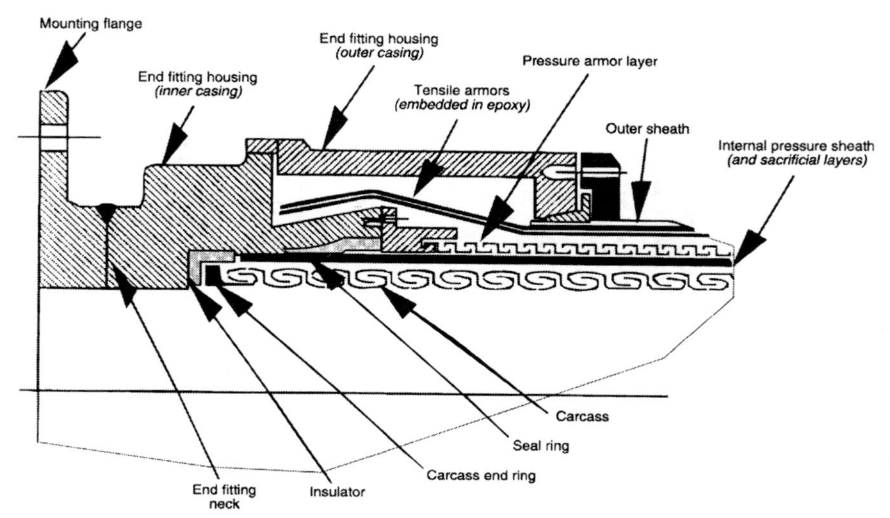

Figure 26-19 Typical End Fitting System

psi. The vent valves in the end fitting arrangement located subsea are sealed to prevent any ingress of seawater into the annulus.

26.4.4. Integrity Management

This section deals mainly with risk assessment and integrity management of flexible pipes [31]. A recognized methodology for formulating an integrity management plan involves carrying out a risk assessment and determining the risks inherent to the use of flexible pipe. Once risks have been determined, specific integrity management measures can be identified to mitigate these risks.

26.4.4.1. Failure Statistics

It is important to determine actual failure mode statistics from operational statistics in order to overcome the failure or damage efficiently.

Figure 26-20 illustrates damage and failure mechanisms. From a total of 106 flexible pipe failure/damage incidents reported by UKCS operators (not including flooded annuli), it was found that 20% of flexible pipes have experienced some form of damage or failure. Figure 26-21 shows a comparison between rigid steel pipe and flexible pipe statistics.

26.4.4.2. Risk Management Methodology

Risk is often quantified as an integer that is the product of two other values, which are known as a *probability of failure* and a *consequence rating*. A risk

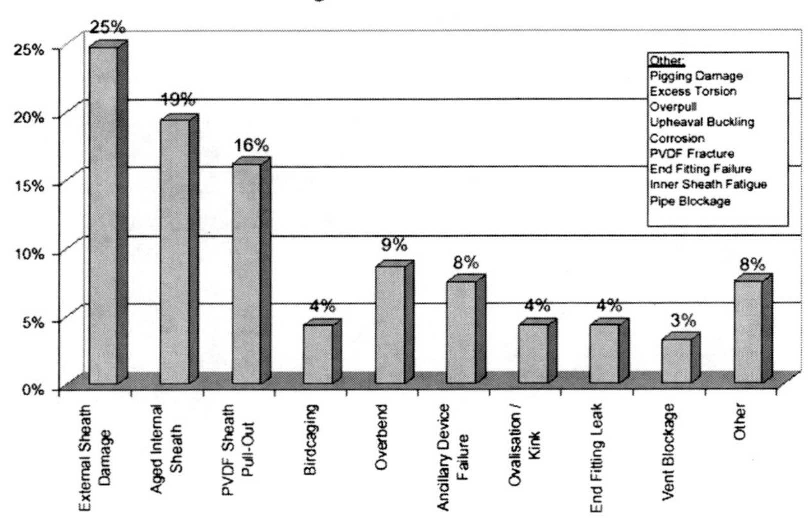

Figure 26-20 System Failure Mechanisms

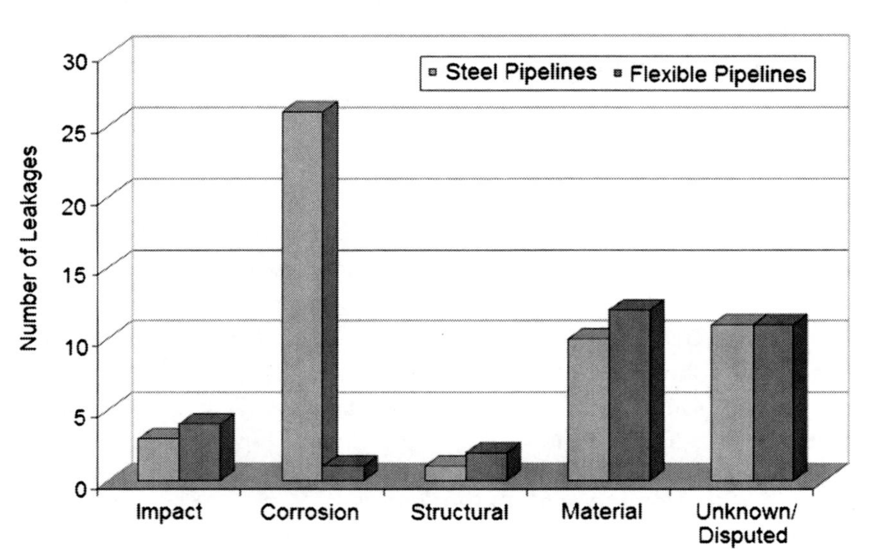

Figure 26-21 Comparison of Steel and Flexible Pipe Failure Statistics

management should take all possible failure modes into consideration, through an analysis of failure drivers (such as temperature, pressure, product fluid composition, service loads, and pipe blockage or flow restriction) and general failure modes (such as fatigue, corrosion, erosion, accidental damage, and ancillary equipment).

26.4.4.3. Failure Drivers

Five failure drivers contribute to the failure of a flexible riser system. The identified five failure drivers are [32]:

- Temperature;
- Pressure;
- Product fluid composition;
- Service loads;
- Ancillary components.

26.4.4.4. Failure Modes

Failure modes for flexible riser systems include the following:

- Fatigue;
- Corrosion;
- Erosion;
- Pipe blockage or flow restriction;
- Accidental damage.

26.4.4.5. Integrity Management Strategy

The integrity management system establishes and maintains a database for design data and field operation data, including:

- Design basis and main design findings;
- Manufacturing data relevant for reassessment of risers;
- Operating temperature and pressure with focus on temperature fluctuations;
- Fluid compositions, injected chemicals, and sand production;
- Riser annulus monitoring and polymer coupon test results;
- Sea-state conditions, vessel motions, and riser response.

26.4.4.6. Inspection Measures

Inspection measures include:

- General visual inspection (GVI)/close visual inspection (CVI);
- Cathodic protection survey.

26.4.4.7. Inspection and Monitoring Systems

A typical inspection and monitoring system includes [33]:

- Polymer monitoring: online, offline, topside, and subsea;
- Annulus monitoring: vent gas rate, annulus integrity;
- Riser dynamics: tension, angle, and curvature;
- Steel armor: inspection method, magnetic or radiograph;
- Use of existing sensors and pressure and temperature sensors.

26.4.4.8. Testing and Analysis Measures

Testing and analysis measures include:

- Coupon sampling and analysis;
- Vacuum testing of riser annulus;
- Radiography.

26.5. HYBRID RISERS

26.5.1. General Description

Hybrid risers (HRs) consist of a vertical bundle of steel pipes supported by external buoyancy. Flexible jumpers connecting the top of the riser and the vessel are used to accommodate relative motion between the vessel and riser bundle.

The first use of hybrid risers was a deepwater production riser installed in 1988 in the Green Canyon Block 29 (GC29) by Cameron (Fisher and Berner, 1988) [13]. The system was reused in 1995 for deployment in Garden Banks 388 (GB388) in the Gulf of Mexico by Enserch Exploration Inc. and Cooper Cameron Co. This first-generation hybrid riser was installed through the moon-pool of a drilling vessel.

In recent years, HRs have become popular for use offshore of West Africa, for example, for the development of Total's Girassol field, as shown in Figure 26-22. This second-generation hybrid riser was bundled, and it was fabricated at an onshore site with installation by tow-out and upending.

In this subsection, we present the current hybrid riser concept including the following main components [9,34]:

- Riser base foundation;
- Riser base spools;
- Top and bottom transition fogging;
- Riser cross section;
- Buoyancy tank;

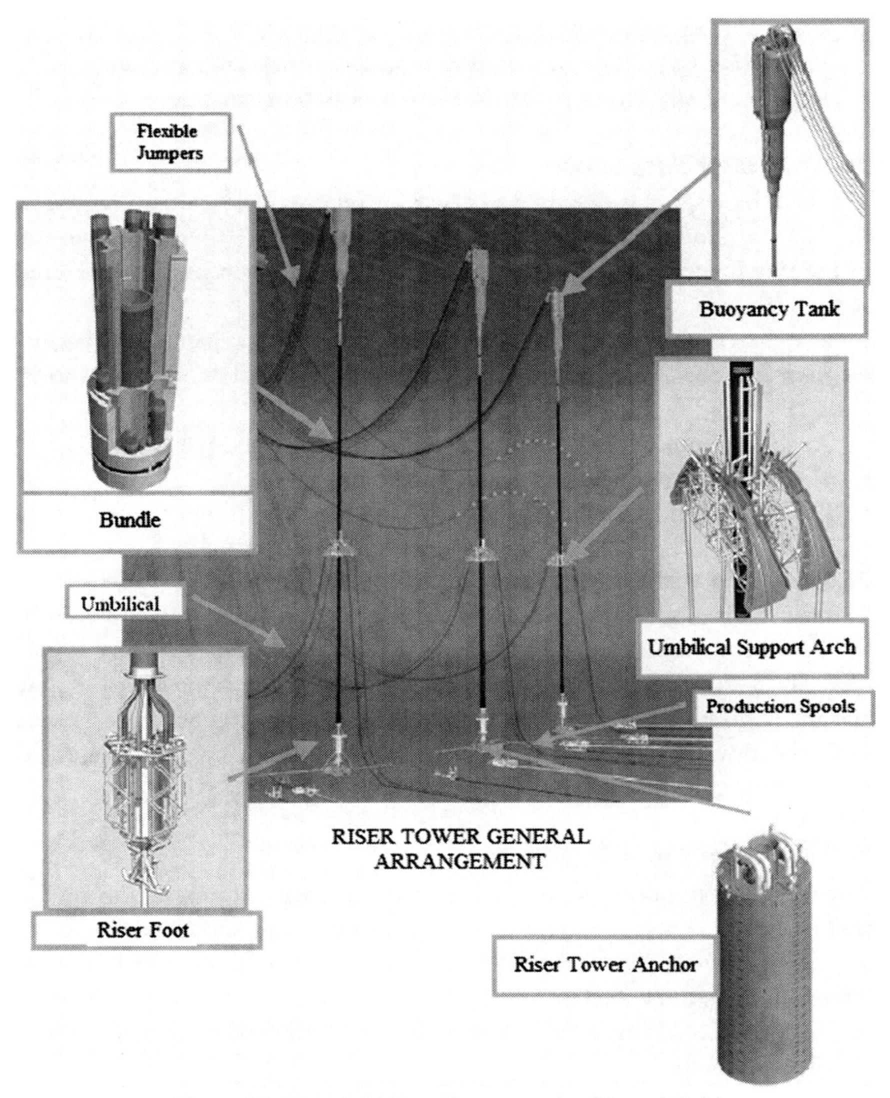

Figure 26-22 Hybrid Riser Towers at the Girassol Field

- Flexible jumper between the top of the riser and the hang-off point in the FPS.

26.5.1.1. Riser Foundation

The foundation structure for the HR consists of a steel suction anchor (and/or a gravity base) that will support the HR installation aids (such as pulling winches) and the flex element (a roto-latch connector is proposed).

Two riser base foundation approaches are possible:

- *Pinned riser base:* This type of riser base allows free riser rotation.
- *Fixed riser base:* This type of riser base resists riser rotation.

26.5.1.2. Riser Base Spools

The riser base spool is designed to allow for relative displacements between the HR base and the flowline that may be caused by the thermal expansion of the flowline or/and angular movements (due to environmental loading) at the HR base.

For the common use of a pinned riser base, the receptacle for the flex element is inclined to optimize the design of the bundle connection to its base.

At the bottom of the hybrid riser, a forged Y-shaped tee allows:

- Vertical sections to be connected to the flex element;
- Lateral branch to be ended by a CVC to the rigid jumper spool.

26.5.1.3. Top and Bottom Transition Forging

Transition forging on the bottom of the riser connects the standard riser to the flowline spool. A short thickened-wall section of pipe (adapter joint), typically 3 m long, provides a transition between the standard riser pipe and the large stiffness of the riser top assembly. A flanged connection mates with the flow spool that connects the flexible jumper and the standard riser.

26.5.1.4. Riser Cross Section

The two most important factors for describing the riser cross section are the wall thickness of the riser pipe and corrosion coatings and anodes.

26.5.1.5. Buoyancy Tank

A steel structured buoyancy tank is designed based on the following functional requirements:

- To thrust the HR dead weight.
- To provide sufficient pulling tension for the dynamic equilibrium of the HR.
- To limit maximum angular deflection of the HR with regards to its static position.

The riser is tensioned by means of an air or nitrogen-filled buoyancy tank. The tank contains a number of individual compartments separated by bulkheads. The bulkheads contain stiffeners arranged on the underside of each bulkhead plate to provide additional reinforcement. The buoyancy tank is designed to be pressure balanced with the external pressure of the water; this allows the thickness of the buoyancy tank skin to be limited to a minimal wall thickness.

26.5.1.6. Flexible Jumpers and Hook Up to Vessel

Flexible jumpers connecting the top of the riser and the vessel are used to accommodate relative motion between the vessel and riser bundle. Flexible pipe jumpers are used to connect goosenecks, located immediately below the air can and the vessel. Bend stiffeners, as shown in Figure 26-23, are used at the gooseneck and vessel termination points in order to restrict the bend radius of the jumper. The jumper's properties are very much dependent on the riser service, pigging requirements, and insulation requirements.

26.5.2. Sizing of Hybrid Risers

The key design issues for the arrangement/sizing of the hybrid riser include:

- Overall arrangement: footprints at bases, hang-off locations/spacing;
- General arrangement:
  - Riser thickness determination;
  - Riser top: jumper with the characteristics and geometry of end fittings, geometry for the buoyancy tank;
  - Riser base: forging and foundation;
- Sizing of buoyancy tank and suction anchor;

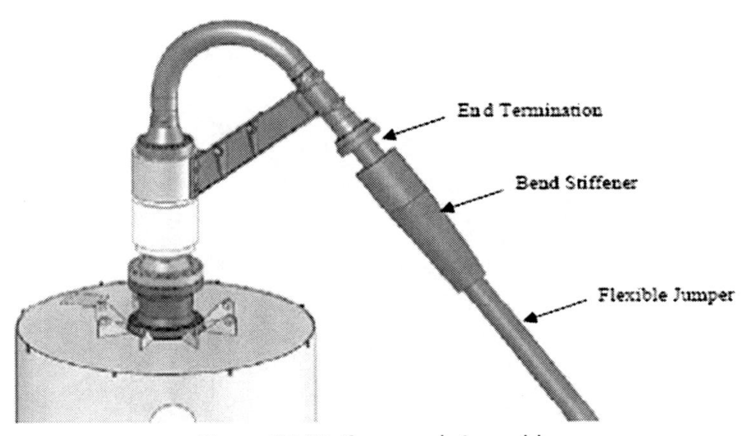

Figure 26-23 Gooseneck Assembly

- Sizing of rigid jumper spool (at hybrid riser base) connected to the gas export pipeline via collet connectors.

The riser system is sized so that:

- Tethering tension is equivalent to what can be obtained from available equipment.
- Distributed buoyancy is equivalent to what is needed for neutral buoyancy in production mode.
- The upper air can provides enough buoyancy to give nominal overpull at the riser base.
- Flexible jumper hoses have sufficient length to accommodate relative vertical movement between the riser and platform at maximum drift-off position.

26.5.3. Sizing of Flexible Jumpers

Flowline jumpers [35] are used to connect the riser base with the flowline base. The flowline jumpers are conventional subsea technology requiring five diameter bends for pigging. Figure 26-24 shows a flexible jumper being lifted from a barge during installation.

Figure 26-24 Lifting of a Flexible Jumper from a Barge

26.5.4. Preliminary Analysis

The key objective of the preliminary analysis is to get an indication of the requirements for:

- Buoyancy and tension distribution along the riser;
- VIV suppression devices.

Preliminary analysis is performed in two steps:

1. Static analysis to determine response to currents and vessel displacement conditions;
2. Time-domain regular wave analysis to determine response to time-varying loading.

26.5.5. Strength Analysis

The strength analysis should be conducted to optimize the following items for a range of possible loading conditions:

- Distributed buoyancy requirements;
- Top tension;
- Tethering tension (if any);
- Foundation loading.

The loading conditions include:

- Extreme waves;
- Extreme currents;
- Extreme vessel drift;
- Winds and currents from opposing directions.

26.5.6. Fatigue Analysis

Like any other riser analysis, a fatigue analysis for a hybrid riser should include vessel drift motions, first-order wave action, and VIV and installation-induced fatigue. VIV analysis of hybrid risers may consider current profiles of varying levels, up to and including 100-year return currents [36,37]. The total VIV fatigue damage can then be calculated using the damage from each profile and the associated percentage occurrence.

26.5.7. Riser Hydrostatic Pressure Test

A pressure test must be carried out before any riser commences operation. The filling, cleaning, gauging, batching, logging, dewatering, and hydrostatic pressure testing operations of the riser system should be performed in accordance with the requirements of DNV OS F101 and other relevant codes.

The sequence of pressure testing operations for a riser system is as follows:

- Filling;
- Cleaning and gauging;
- Hydrotesting, including temperature stabilization, pressurization, air contents check, and hydrostatic test/holding period;
- Post-testing, including depressurization and documentation;
- Rectification activities (if required), including leak location during test, dewatering for rectification, and rectification of defects;
- Final/repeat hydrotesting;
- Testing, certificates, and witnessed signature.

REFERENCES

[1] Y. Bai, Q. Bai, Subsea Pipelines and Risers, Elsevier, Oxford, 2005.
[2] American Bureau of Shipping, Guide for Building and Classing: Subsea Pipeline Systems and Risers, ABS (March, 2001)
[3] T. McCardle, Subsea Systems and Field Development Considerations, SUT Subsea Awareness Course, 2008.
[4] American Petroleum Institute, Recommended Practice for Flexible Pipe, API RP 17B, (2002).
[5] American Petroleum Institute, Specification for Unbonded Flexible Pipe, API Specification 17J, (1999).
[6] American Petroleum Institute, Design of Risers for Floating Production Systems (FPSs) and Tension-Leg Platforms (TLPs), API RP 2RD, (June, 1998).
[7] R. Burke, TTR Design and Analysis Methods, Deepwater Riser Engineering Course, Clarion Technical Conferences, (2004).
[8] World Oil, Composite Catalog of Oilfield Equipment & Services, Fourty Fifth ed., 2002, March.
[9] G. Wald, Hybrid Riser Systems, Deepwater Riser Engineering Course, Clarion Technical Conferences, 2004.
[10] S. Chakrabarti, Handbook of Offshore Engineering, Ocean Engineering Book Series, Elsevier, Oxford, 2005.
[11] V. Alliot, J.L. Legras, D. Perinet, A Comparison between Steel Catenary Riser and Hybrid Riser Towers for Deepwater Field Developments, Deep Oil Technology Conference. (2004).
[12] L. Deserts, Hybrid Riser for Deepwater Offshore Africa, OTC 11875, Offshore Technology Conference, Houston, Texas (2000)
[13] E.A. Fisher, P. Holley, S. Brashier, Development and Deployment of a Freestanding Production Riser in the Gulf of Mexico, OTC 7770, Proc. 27th Offshore Technology Conference, Houston, (1995).
[14] C.T. Gore, B.B. Mekha, Common Sense Requirements (CSRs) for Steel Catenary Risers (SCR), OTC 14153, Offshore Technology Conference, Houston, Texas, (2002).
[15] American Petroleum Institute, Specification for Line Pipe, API Specification 5L, fourty second ed., (2000).

[16] F. Kopp, B.D. Light, T.S. Preli, V.S. Rao, K.H. Stingl, Design and Installation of the Na Kika Export Pipelines, Flowlines and Risers, OTC 16703, Offshore Technology Conference, Houston, Texas, (2004).
[17] R. Franciss, Vortex Induced Vibration Monitoring System in Steel Catenary Riser of P-18 Semi-Submersible Platform, OMAE2001/OFT-1164, Proceedings of OMAE'01 (2001).
[18] E.H. Phifer, F. Kopp, R.C. Swanson, D.W. Allen, C.G. Langner, Design and Installation of Auger Steel Catenary Risers, OTC 7620, Offshore Technology Conference, Houston, Texas, (1994).
[19] G. Chaudhury, J. Kennefick, Design, Testing, and Installation of Steel Catenary Risers, OTC 10980, Offshore Technology Conference, Houston, Texas, (1999).
[20] H.M. Thompson, F.W. Grealish, R.D. Young, H.K. Wang, Typhoon Steel Catenary Risers: As-Built Design and Verification, OTC 14126, Offshore Technology Conference, Houston, Texas, (2002).
[21] K. Huang, X. Chen, C.T. Kwan, The Impact of Vortex-Induced Motions on Mooring System Design for Spar-Based Installations, OTC 15245, Offshore Technology Conference, Houston, Texas, (2003).
[22] M. Hogan, Flexjoints, ASME ETCE SCR Workshop, Houston, Texas, 2002, February.
[23] J. Buitrago, M.S. Weir, Experimental Fatigue Evaluation of Deepwater Risers in Mild Sour Service, Deep Offshore Technology Conference, Louisiana, New Orleans, (2002).
[24] International Standards Organization, Petroleum and Natural Gas Industries Design and Operation of Subsea Production Systems, Part 7: Completion/Workover Riser Systems, ISO 13628-7: 2005, (2005).
[25] R. Jordan, J. Otten, D. Trent, P. Cao, Matterhorn TLP Dry-Tree Production Risers, OTC 16608, Offshore Technology Conference, Houston, Texas, (2004).
[26] A. Yu, T. Allen, M. Leung, An Alternative Dry Tree System for Deepwater Spar Applications, Deep Oil Technology Conference, New Orleans, (2004).
[27] Massachusetts Institute of Technology, User Guide for SHEAR7, Version 2.0, Department of Ocean Engineering, 1996.
[28] Massachusetts Institute of Technology, SHEAR7 Program Theoretical Manual, Department of Ocean Engineering, 1995.
[29] Y. Zhang, B. Chen, L. Qiu, T. Hill, M. Case, State of the Art Analytical Tools Improve Optimization of Unbonded Flexible Pipes for Deepwater Environments, OTC 15169, Offshore Technology Conference, Houston, Texas, (2003).
[30] J. Remery, R. Gallard, B. Balague, Design and Qualification Testing of a Flexible Riser for 10,000 psi and 6300 ft WD for the Gulf of Mexico, Deep Oil Technology Conference, Louisiana, New Orleans, (2004).
[31] B. Seymour, H. Zhang, C. Wibner, Integrated Riser and Mooring Design for the P-43 and P-48 FPSOs, OTC 15140, Offshore Technology Conference, Houston, Texas, (2003).
[32] P. Elman, R. Alvim, Development of a Failure Detection System for Flexible Risers, 18th International Offshore and Polar Engineering Conference, (2008).
[33] D.E. Thrall, R.L. Poklandnik, Garden Banks 388 Deepwater Production Riser Structural and Environmental Monitoring System, OTC 7751, Proc. of the 27th Offshore Technology Conference, Houston, Texas, (1995).
[34] L. Deserts, Hybrid Riser for Deepwater Offshore Africa, OTC 11875, Offshore Technology Conference, Houston, Texas, (2000).
[35] M. Wu, P. Jacob, J.F.S. Marcoux, V. Birch, The Dynamics of Flexible Jumpers Connecting A Turret Moored FPSO to A Hybrid Riser Tower, Proceedings of D.O.T XVIII Conference, (2006).

[36] J.K. Vandiver, L Li, User Guide for SHEAR7, Version 2.1 & 2.2, for Vortex-Induced Vibration Response Prediction of Beams or Cables with Slowly Varying Tension In Sheared or Uniform Flow, Massachusetts Institute of Technology, 1998.
[37] J.K. Vandiver, Research Challenges in the Vortex-Induced Vibration Prediction of Marine Risers, Offshore Technology Conference, Houston, Texas, (1998).

CHAPTER 27

Subsea Pipelines

Contents

Subsea Engineering Handbook
ISBN 978-0-12-397804-2, doi:10.1016/B978-0-12-397804-2.00027-8

27.1. INTRODUCTION

Subsea pipelines are used for a number of purposes in the development of subsea hydrocarbon resources, as shown in Figure 27-1. A flowline system can be a single-pipe pipeline system, a pipe-in-pipe system, or a bundled system. Normally, the term *subsea flowlines* is used to describe the subsea pipelines carrying oil and gas products from the wellhead to the riser foot; the riser is connected to the processing facilities (e.g., a platform or a FPSO). The subsea pipelines from the processing facilities to shore are called export pipelines. The subsea pipelines include:

- Export (transportation) pipelines;
- Flowlines to transfer product between platforms, subsea manifolds, and satellite wells;
- Flowlines to transfer product from a platform to export lines;
- Water injection or chemical injection flowlines;
- Pipeline bundles.

The design process for each type of line in general terms is the same. It is this general design approach that will be discussed in this chapter. For a more

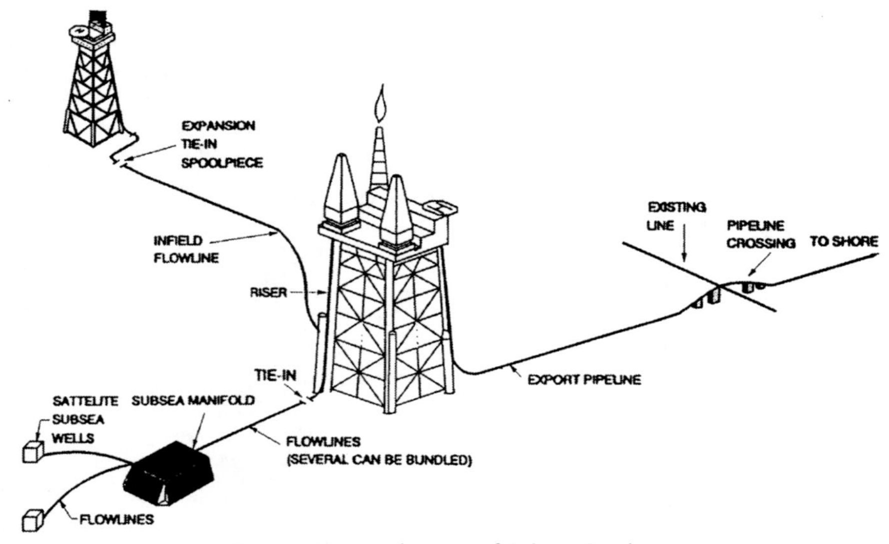

Figure 27-1 Application of Subsea Pipelines

detailed design and analysis methodology refer to the book *Subsea Pipelines and Risers* [1].

27.2. DESIGN STAGES AND PROCESS

27.2.1. Design Stages

The design of subsea pipelines during subsea field development is usually performed in the following three stages:

- Conceptual engineering;
- Preliminary engineering or pre-engineering;
- Detail engineering.

The continuity of engineering design from conceptual engineering and preliminary engineering to the detailed design is essential to a successful project. The objective and scope of each design stage vary depending on the operator and the size of the project. However, the primary aims are discussed next.

27.2.1.1. Conceptual Engineering

The primary objectives of the conceptual engineering stage are normally:

- To establish technical feasibility and constraints on the system design and construction;
- To eliminate nonviable options;
- To identify the required information for the forthcoming design and construction;
- To allow basic cost and scheduling exercises to be performed;
- To identify interfaces with other systems planned or currently in existence.

The value of the early engineering work is that it reveals potential difficulties and areas where more effort may be required in the data collection and design areas.

27.2.1.2. Preliminary Engineering or Basic Engineering

The primary objectives of the preliminary engineering are normally:

- Perform pipeline design so that the system concept is fixed. This will include:
  - Verifying the sizing of the pipeline;
  - Determining the pipeline grade and wall thickness;
  - Verifying the pipeline's design and code requirements for installation, commissioning, and operation;
- Prepare authority applications;

- Perform a material take-off (MTO) sufficient to order the line pipe.

The level of engineering is sometimes specified as being sufficient to detail the design for inclusion into an *engineering, procurement, construction and installation* (EPCI) tender. The EPCI contractor should then be able to perform the detailed design with a minimum number of variations as detailed in its bid.

27.2.1.3. Detailed Engineering

The detailed engineering phase is when the design is developed to a point where the technical input for all procurement and construction tendering can be defined in sufficient detail.

The primary objectives of the detailed engineering stage can be summarized as follows:

- Optimize the route.
- Select the wall thickness and coating.
- Confirm code requirements on strength, VIVs, on-bottom stability, global buckling, and installation.
- Confirm the design and/or perform additional design as defined in the preliminary engineering stage.
- Development of the design and drawings in sufficient detail for the subsea scope. This may include pipelines, tie-ins, crossings, span corrections, risers, shore approaches, and subsea structures.
- Prepare detailed alignment sheets based on the most recent survey data.
- Prepare specifications, typically covering materials, cost applications, construction activities (i.e., pipe lay, survey, welding, riser installations, spool piece installation, subsea tie-ins, subsea structure installation), and commissioning (i.e., flooding, pigging, hydrotesting, cleaning, drying);
- Prepare MTO and compile necessary requisition information for the procurement of materials.
- Prepare design data and other information required for the certification authorities.

27.2.2. Design Process

The object of the design process for a subsea pipeline is to determine, based on given operating parameters, the optimum pipeline size parameters. These parameters include:

- Pipeline internal diameter;
- Pipeline wall thickness;

- Grade of pipeline material;
- Type of corrosion coating and weight (if any);
- Thicknesses of coatings

The design process required to optimize the pipeline size parameters is an iterative one and is summarized in Figure 27-2. The design analysis is illustrated in Figure 27-3.

Each stage in the design should be addressed whether it is a conceptual, preliminary, or detailed design. However, the level of analysis will vary

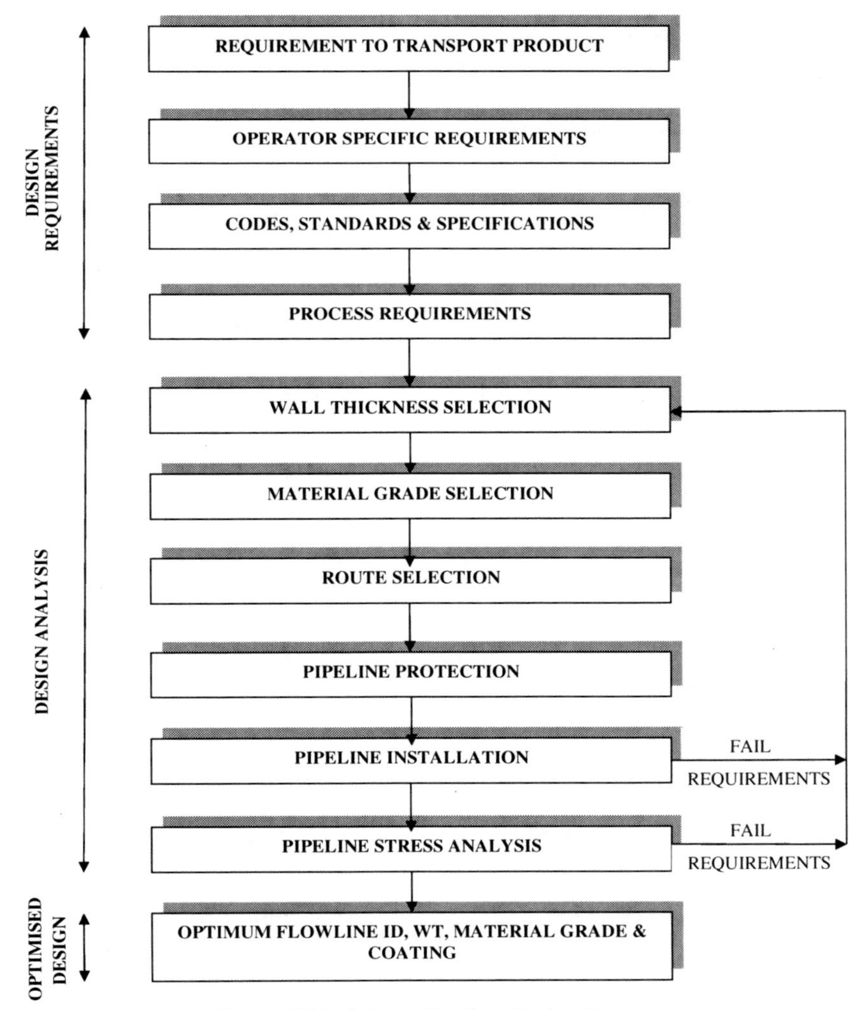

Figure 27-2 Subsea Pipeline Design Process

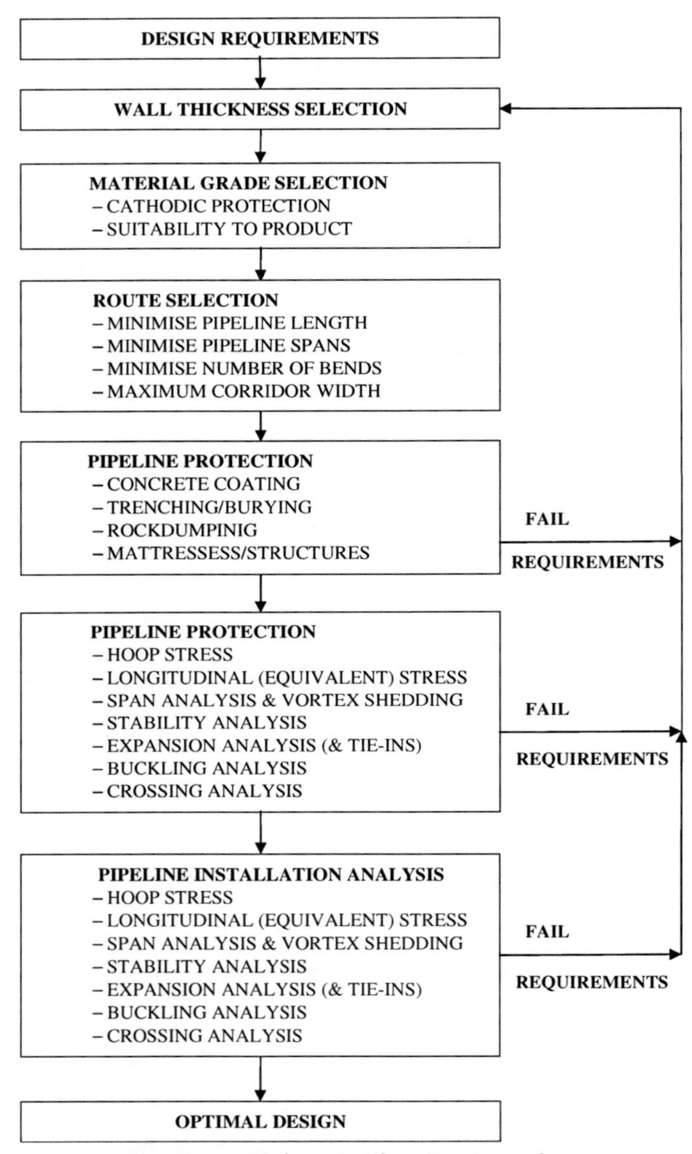

Figure 27-3 Subsea Pipeline Design Analysis

depending on the required output. For instance, reviewing the objectives of the detailed design (Section 27.2.1.3), the design should be developed such that:

- Pipeline wall thickness, grade, coating, and length are specified so that the pipeline can be fabricated.

- Route is determined such that alignment sheets can be compiled.
- Pipeline stress analysis is performed to verify that the pipeline is within allowable stresses at all stages of installation, testing, and operation. The results will also include pipeline allowable spans, tie-in details (including expansion spool pieces), allowable testing pressures, and other inputs into the design drawings and specifications.
- Pipeline installation analysis is performed to verify that stresses in the pipeline at all stages of installation are within allowable values. This analysis should specifically confirm if the proposed method of pipeline installation will not result in pipeline damage. The analysis will have input into the installation specifications.
- Analysis of global response:
  - Expansion, effective force, and global buckling;
  - Hydrodynamic response;
  - Impact.
- Analysis of local strength:
  - Bursting, local buckling, ratcheting;
  - Corrosion defects dent.

27.3. SUBSEA PIPELINE FEED DESIGN

27.3.1. Subsea Pipeline Design Basis Development

A design basis should be prepared to cover all necessary data, parameters, codes, and standards. All relevant data should be reviewed and evaluated and the criteria established for the detailed design. These should cover the following as a minimum:

- Application of design codes and standards. The pipeline system should be designed in accordance with relevant sections of the referenced codes and standards;
- System design and operating data;
- Environmental data;
- Pipeline design approach.

27.3.2. Subsea Pipeline Route Selection

Develop a pipeline route selection based on route survey information and environmental data provided by the owner. The scope of work should include at a minimum the following tasks:

- Select the pipeline route and prepare platform approach drawings.
- Design the spacing between two parallel pipelines.

- Optimize the pipeline route to minimize costs for the new pipelines including any crossings of existing subsea pipelines and submarine cables.
- Prepare the proposed pipeline route drawing.
- Determine the minimum horizontal curvature radius for the pipeline routes.
- Prepare arrangement drawings and layouts for all tie-in/interface locations (i.e., start and termination of line) clearly showing break points between facilities that will eventually be installed by a pipeline construction contractor and those that will be installed by a facilities and substructure construction contractor.
- Pipeline routing should take into account the environmental conditions, design, installation, operation, and maintenance of the pipelines.
- Pipeline routing should be selected such that the pipeline follows a smooth seabed profile avoiding wherever possible huge and multiple spans, coral growths, rock outcrops, soft or liquefiable soils, and other seafloor obstructions.
- Prepare route selection and optimization reports for the proposed pipeline route including other available options, indicating their respective pros and cons.

27.3.3. Steady-State Hydraulic Analysis

Prepare a steady-state hydraulic analysis study to confirm the selected pipeline sizes and define the operating envelope based on the production forecast given, selected pipeline route and overall pipeline configuration. Possible changes in flow rates and operational modes over the entire operational life of the pipeline should be included. The study should include:

- Pressure drop calculation and pipeline capacity curve for various flow rate scenarios and selection of optimum pipeline size;
- A sensitivity study on the pipeline size optimum selection;
- Liquid hold-up volume for various flow rate scenarios and turndown limitations;
- Liquid arrival rate for various flow rate scenarios;
- Fluid velocity and flow pattern inside the flowline for various flow rate;
- Pressure and temperature profile along the flowline;
- Occurrence of severe slugging in the flowline/riser and slug size;
- Liquid catching facilities and slug control requirements;
- Flow assurance addressing after well shutdown and restart and insulation requirements;
- Thermal insulation requirements (if any) especially for waxy crude services.

27.3.4. Pipeline Strength Analysis

Prepare a mechanical design of the pipeline:

- Select the material for the pipeline.
- Perform mechanical strength analyses of the pipeline along the proposed route to determine the optimum wall thickness.
- Perform analyses for pipeline collapse, buckling due to bending and external pressure, buckle initiation and propagation for installation and operating conditions, and the "worst case" load condition.
- Perform design of appropriate measures against possible pipeline collapse and buckling if required.

27.3.5. Pipeline Vertical and Lateral On-Bottom Stability Assessment

Perform pipeline on-bottom stability analysis in accordance with DNV RP E305 [2] (or DNV RP F109 [3]) design code or using AGA software (Subsea pipeline on-bottom stability analysis) [4]. The pipeline should be stable both laterally and vertically, incorporating "liquefaction" and self-burial consideration against the environmental loadings and soil condition. The ballast weight may need to be determined during the installation and operational phases.

An on-bottom stability analysis for the pipelines should include the following conditions:

- Installation condition;
- Operating condition;
- Hydrotest condition.

27.3.6. Installation Method Selection and Feasibility Demonstration

The following studies should be performed when designing subsea pipelines:

- Installation feasibility study for available installation methods;
- Economic comparison for all methods of pipeline installation;
- Cost comparison between preinstalled riser and conventional method;
- Tie-in methodology study that includes above-water welding (stalk on), hyperbaric welding, and subsea flange tie-in for the pipeline and riser portion;
- Preliminary analysis for the selected pipeline installation method.

Designers should consider pipeline installation via a conventional pipe-laying barge (S-lay, J-lay, and reel-lay), bottom tow, off-bottom tow, and surface tow or any other installation method.

27.3.7. Material Take-Off (MTO)

Prepare the MTO for the pipeline for the following items:

- Line pipe;
- Pipeline end fitting;
- Pipeline pull-in and hang-off assembly;
- Field joint connection;
- Any other required material identified during the design.

27.3.8. Cost Estimation

Prepare cost estimates for the flowlines, which include, but are not limited to:

- All pipeline materials and their associated fittings;
- Pipeline installation, precommissioning, and commissioning.

27.4. SUBSEA PIPELINE DETAILED DESIGN

Many design works between FEED and detailed engineering phases are similar, the detailed engineering phase extends all issues of FEED phase with more detailed designs to give the technical inputs for all procurement and construction of the pipeline project. Following design issues should be included in the detailed engineering phase based on the FEED phase.

27.4.1. Pipeline Spanning Assessment

Prepare a spanning analysis for the pipeline to include the following conditions:

- Installation condition;
- Operating condition;
- Hydrotest condition;

and the following VIV limiting criteria for dynamic analysis:

- In-line vibrations;
- Cross-flow vibrations.

27.4.2. Pipeline Global Buckling Analysis

Prepare a global buckling analysis for the pipeline for the following operating conditions:

- Lateral buckling analysis;
- Upheaval buckling analysis.

27.4.3. Installation Methods Selection and Feasibility Demonstration

Perform for the pipelines:

- An installation feasibility study for available installation methods;
- Economic comparison for the all methods of pipeline installation;
- Perform cost comparison between pre-installed riser and conventional method;
- Perform tie-in methodology study that subsea pipeline connected with onshore pipeline;
- Perform preliminary analysis for the selected pipeline installation methods.

Consultant should consider pipeline installation via a conventional pipe-laying barge, bottom tow, off-bottom tow and surface tow or any other installation methods.

27.4.4. Pipeline Quantitative Risk Assessment

The assessments carried out in this work are as follows:

- Trawl impact frequency of the pipeline;
- Pipeline response to trawl pullover load;
- Frequency/probability of anchor drop from commercial cargo vessel impacting the pipeline;
- Consequence of anchor drop to pipeline.

27.4.5. Pipeline Engineering Drawings

Prepare a set of pipeline engineering drawings as follows:

- Pipeline overall field layout;
- Pipeline route drawing;
- Pipeline approach at respective platform;
- Pipeline end fitting details;
- Pipeline field joint details;
- Pipeline or subsea cable crossing details (if any);
- Pipeline shore approach drawings (if any);
- Typical pipeline installation drawings;
- Any other drawings required.

27.5. PIPELINE DESIGN ANALYSIS

27.5.1. Wall-Thickness Sizing

The wall-thickness level for pipelines should be able to withstand pressure and pressure effect (hoop and burst strength) requirements in accordance with API-RP-1111 [5] and API 5L [6] (or CFR [7], ASME B31 [8,9], DNV-OS-F101 [10] depending on the project's requirement). A stress analysis should be carried out for the wall-thickness check.

27.5.1.1. Stress Definitions

A difference of internal pressure and external pressure in a pipeline produces stresses in the wall of a pipeline. These stresses are the longitudinal stress σ_L, the hoop stress σ_H, and the radial stress σ_R. The directions of each stress component are illustrated in Figure 27-4.

27.5.1.2. Stress Components for Thick-Wall Pipe

Figure 27-5 illustrates a thick-wall pipe with a capped end. The stress components for the thick-wall pipe are expressed as follows:

$$\begin{aligned}
\sigma_R &= \frac{p_i r_i^2 - p_e r_e^2}{r_e^2 - r_i^2} - \frac{r_i^2 r_e^2}{r^2(r_e^2 - r_i^2)}(p_i - p_e) \\
\sigma_H &= \frac{p_i r_i^2 - p_e r_e^2}{r_e^2 - r_i^2} + \frac{r_i^2 r_e^2}{r^2(r_e^2 - r_i^2)}(p_i - p_e) \\
\sigma_L &= \frac{p_i r_i^2 - p_e r_e^2}{r_e^2 - r_i^2} + \frac{F_{ext}}{\pi(r_e^2 - r_i^2)} = \text{constant} \\
\sigma_R + \sigma_H &= 2\frac{p_i r_i^2 - p_e r_e^2}{r_e^2 - r_i^2} = \text{constant}
\end{aligned} \tag{27-1}$$

27.5.1.3. Hoop Stress

Hoop stress σ_H varies across the pipe wall from a maximum value on the inner surface to a minimum value on the outer surface of the pipe, as

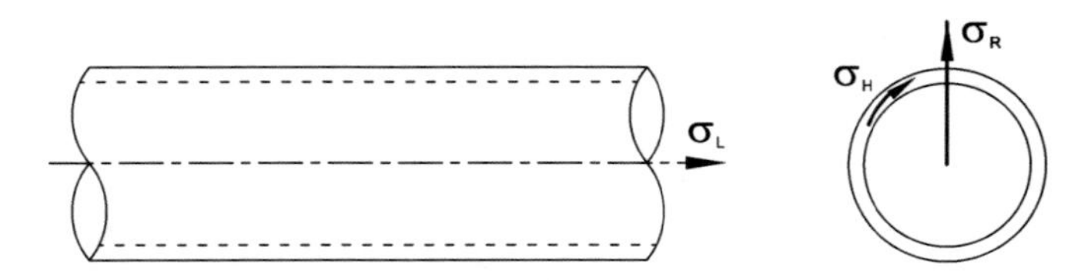

Longitudinal stress, σ_L

Hoop stress, σ_H

Radial stress, σ_R

Figure 27-4 Definitions of Pipe Wall Stress

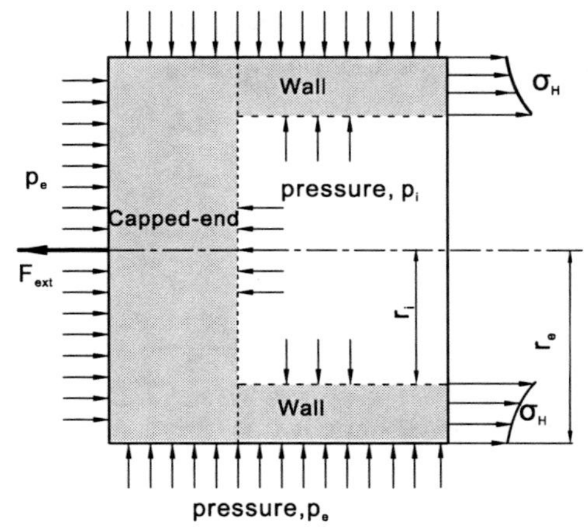

Figure 27-5 Stresses for Thick-Wall Pipe with Capped End

expressed in the hoop stress of Equation (27-1). The equation for the hoop stress is also called the Lame equation and is rewritten as follows:

$$\sigma_H = \frac{p_i D_i^2 - p_e D_e^2}{D_e^2 - D_i^2} + \frac{(p_i - p_e) D_i^2 D_e^2}{(D_e^2 - D_i^2) D^2} \tag{27-2}$$

where

σ_H: Lame hoop stress;

D: diameter at which hoop stress is calculated;

D_e: external pipe diameter;

D_i: internal pipe diameter;

p_e: external pressure of pipe;

p_i: internal pressure of pipe.

Hoop stress for a thin-wall pipe can be obtained from the force balance below, assuming the hoop stress to be constant in the radial direction:

$$\sigma_H = \frac{p_i D_i - p_e D_e}{2t} \tag{27-3}$$

where t is the minimum wall thickness of the pipeline.

The radial stress σ_R varies across the pipe wall from a value equal to the internal pressure, p_i, on the inside of the pipe wall, to a value equal to the external pressure, p_e on the outside of the pipe. The magnitude of the radial stress is usually small when compared with the longitudinal and hoop stresses; consequently it is not specifically limited by the design codes.

According to ASME B31.8 (2010) [8], Hoop stress for a thin wall ($D/t > 30$) can also be expressed in the following equation:

$$\sigma_H = (p_i - p_e)\frac{D_e}{2t} \tag{27-4}$$

Hoop stress for a thick wall ($D/t < 30$) can be calculated using the following equation:

$$\sigma_H = (p_i - p_e)\frac{D_e - t}{2t} \tag{27-5}$$

Depending on which code/standard is used, the hoop stress should not exceed a certain fraction of the specified minimum yield stress (SMYS).

27.5.1.4. Longitudinal Stress

The longitudinal stress σ_L is the axial stress experienced by the pipe wall. It consists of stresses due to:

- Bending stress (σ_{lb});
- Hoop stress (σ_{lh});
- Thermal stress (σ_{lt});
- End cap force induced stress (σ_{lc}).

The components of each are illustrated in Figure 27-6. Poisson's ratio is assumed to be 0.3 for steel. The longitudinal stress can be determined using the equation:

$$\sigma_L = \nu\sigma_{lh} + \sigma_{lb} + \sigma_{lt} + \sigma_{lc}$$

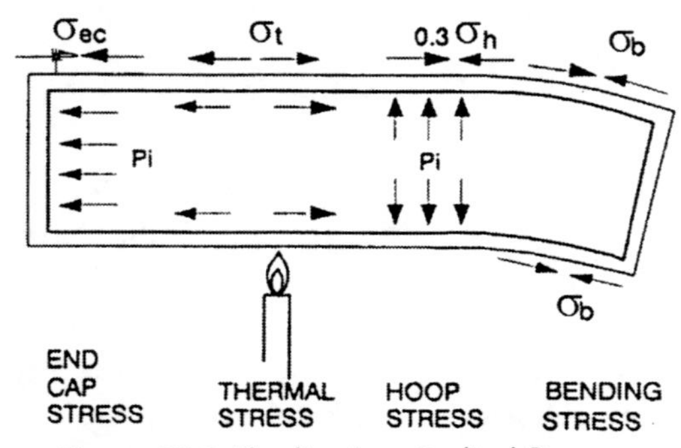

Figure 27-6 Flowline Longitudinal Stresses

The longitudinal stress in a restrained pipeline wall, $\sigma_{L,R}$, due to the temperature and Poisson's ratio pressure stress components is expressed as follows:

$$\sigma_{L,R} = \frac{2\nu(p_i A_i - p_e A_e)}{A_s} - E\alpha(T_P - T_a) \tag{27-6}$$

where A_i is the internal bore area of pipe, A_e the external area of pipe, A_s the pipe cross section area of steel, E the Young's modulus, α the thermal expansion coefficient of pipe material, T_p the temperature of pipe wall, and T_a the ambient temperature.

The longitudinal stress of an unrestrained pipeline is expressed as follows:

$$\sigma_{L,U} = \frac{(1-2\nu)\cdot(p_i A_i - p_e A_e)}{A_s} \tag{27-7}$$

Note that sign conventions should be utilized when employing this equation; that is, tensile stress is positive.

27.5.1.5. Equivalent Stress

The combined stress is determined differently depending on the code/standards utilized. However, the equivalent stress (σ_e) for a thin wall pipe can usually be expressed as:

$$\sigma_e = \sqrt{\sigma_H^2 + \sigma_L^2 - \sigma_H\sigma_L + 3\tau_{LH}^2} \tag{27-8}$$

where

σ_H: hoop stress;

σ_L: longitudinal stress;

τ_{LH}: tangential shear stress.

For high-pressure pipes with D/t ratios of less than 20 and where the shear stresses can be ignored, the form of equivalent stress may be calculated as:

$$\sigma_e = \sqrt{\frac{1}{2}[(\sigma_H - \sigma_L)^2 + (\sigma_L - \sigma_R)^2 + (\sigma_H - \sigma_R)^2]} \tag{27-9}$$

27.5.2. On-Bottom Stability Analysis

Subsea pipelines resting on the seabed are subject to fluid loading from both waves and steady currents. For regions of the seabed where damage may result from vertical or lateral movement of the pipeline, it is a design requirement for the pipe weight to be sufficient to ensure stability under the worst possible environmental conditions. In most cases this weight is provided by a concrete coating on the pipeline. In some circumstances the pipeline may be allowed to move laterally provided stress (or strain) limits are not exceeded. The first case is discussed briefly in this section since it is applied in the majority of design situations.

The analysis of on-bottom stability is based on the simple force balance or detailed finite element analysis. The loads acting on the pipeline due to wave and current action are the fluctuating drag and lift and inertia forces. The friction resulting from effective weighting of the pipeline on the seabed to ensure stability must resist these forces. If the weight of the pipe steel and contents alone or the use of rock berms is insufficient, then the stability design must establish the amount of concrete coating required. In a design situation, a factor of safety is required by most pipeline codes.

The hydrodynamic forces are derived using traditional fluid mechanics with a suitable coefficient of drag, lift and diameter, roughness, and local current velocities and accelerations.

The effective flow to be used in the analysis consists of two components:

- The steady current, which is calculated at the position of the pipeline using boundary layer theory;
- The wave-induced flow, which is calculated at the seabed using a suitable wave theory. The selection of the flow depends on the local wave characteristics and the water depth.

The wave and current data must be based on extreme conditions. For example, the wave with a probability of occurring only once in 100 years is often used for the operational lifetime of a pipeline. A less severe wave, say, of 1 year or 5 years, is applied for for pipeline installation, in which the pipeline is placed on the seabed in an empty condition and has less submerged weight.

Friction, which depends on the seabed soils and the submerged weight of the pipe, provides the equilibrium of the pipeline. Remember that this weight is reduced by the fluid lift force. The coefficient of lateral friction can vary from 0.1 to 1.0 depending on the surface of the pipeline and on the soil. Soft clays and silts provide the least friction, whereas coarse sands offer greater resistance to movement.

Figure 27-7 illustrates the component forces for a subsea pipeline stability analysis. For the pipeline to be stable on the seabed, the following relationship must be satisfied:

$$\gamma(F_D - F_I) \leq \mu(W_{\text{sub}} - F_L) \tag{27-10}$$

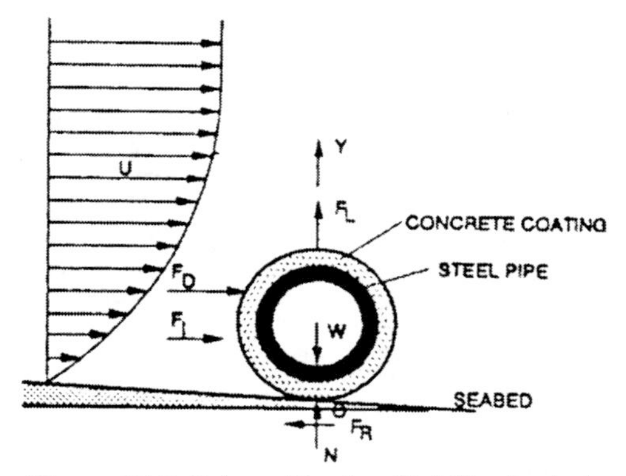

Figure 27-7 Subsea Pipeline Stability Analysis

where

γ: factor of safety, normally not to be taken as less than 1.1;

F_D: hydrodynamic drag force per unit length (vector);

F_I: hydrodynamic inertia force per unit length (vector);

μ: lateral soil friction coefficient;

W_{sub}: submerged pipe weight per unit length (vector);

F_L: hydrodynamic lift force per unit length (vector).

We can see that stability design is a complex procedure that relies heavily on empirical factors such as force coefficient and soil friction factors. The appropriate selection of values is strongly dependent on the experience of the engineer and the specific design conditions.

The goal of the subsea pipeline stability analysis described is to determine how much additional weight should be added via the coating. If the weight of the concrete required for stability makes the pipe too heavy to be installed safely, then additional means of stabilization will be necessary. The two main techniques are:

- To remove the pipeline from the current forces by trenching;
- To provide additional resistance to forces by use of anchors (rock berms) or additional weights on the pipeline.

27.5.3. Free-Span Analysis

Over a rough seabed or on a seabed subject to scour, pipeline spanning can occur when contact between the pipeline and seabed is lost over an appreciable distance (see Figure 27-8). In such circumstances, normal code requirements are that the pipeline be investigated for:

- Excessive yielding under different loads;

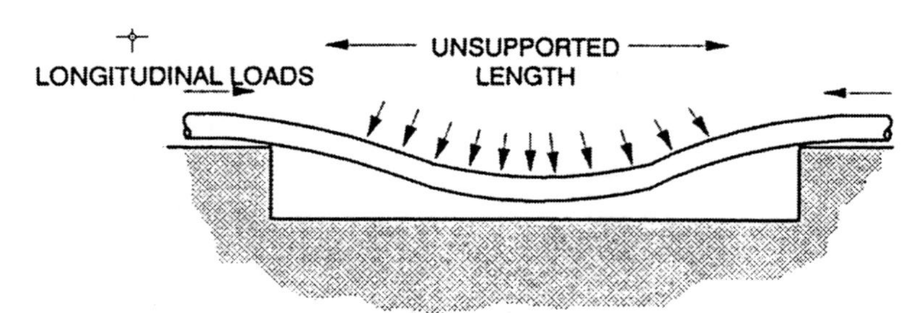

Figure 27-8 Pipeline Span Analysis

- Fatigue due to VIV;
- Interference with human activities (fishing).

Due consideration to these requirements will result in the evaluation of an allowable free-span length. Should actual span lengths exceed the allowable length, then correction is necessary to reduce the span for some idealized situations. This can be a very expensive exercise and, consequently, it is important that any span evaluations conducted be as accurate as possible. In many cases, multiple span analyses have to be conducted that account for a real seabed and *in situ* structural behavior.

The flow of waves and current around a pipeline span, or any cylindrical shape, will result in the generation of sheet vortices in the wake (for turbulent flow). These vortices are shed alternately from the top and bottom of the pipe, resulting in an oscillatory force being exerted on the span as shown in Figure 27-9. Vortex shedding means that cross currents will generate alternating loads on a pipe, resulting in vibration of the pipe.

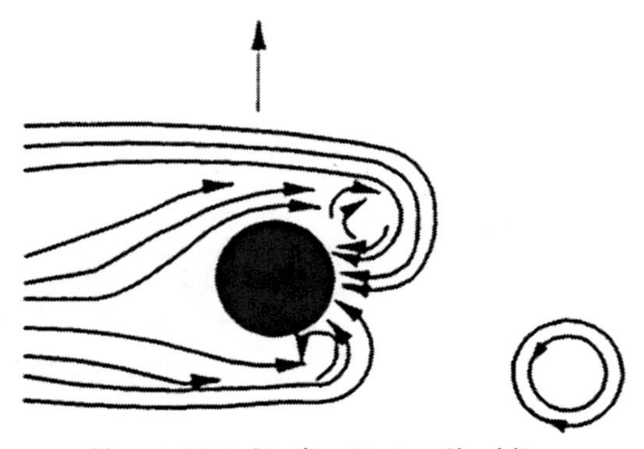

Figure 27-9 Pipeline Vortex Shedding

If the frequency of shedding approaches the natural frequency of the spanning pipeline, severe resonance can occur. This resonance can induce fatigue failure of the pipe and cause the concrete coating to crack and possibly be lost. The evaluation of the potential of a span to undergo resonance is based on the comparison of the shedding frequency and the natural frequency of the spanning pipeline. The calculation of shedding frequency is achieved using traditional mechanics although some consideration must be given to the effect of the closeness of the seabed. Simple models have traditionally been used to calculate the natural frequency of the spanning pipeline, but recent theories have shown these to be oversimplified and multiple span analyses need to be conducted. Finite element analysis for a pipeline on a seabed is widely used for the natural frequency and VIV fatigue damage analyses.

27.5.4. Global Buckling Analysis

Global buckling of a pipeline occurs when the effective compressible force within the line becomes so great that the line has to deflect and, hence, reduces these axial loads (i.e., takes a lower energy state). As more pipelines operate at higher temperatures, the likelihood of buckling becomes more pertinent.

Global buckling analysis will be performed to identify whether the global buckling is likely to occur (see Figure 27-10). If it is, then further analysis is performed to either prevent buckling or accommodate it.

A method of preventing buckling is to dump rocks on the pipeline. This induces even higher loads in the line but prevents it from buckling. However, if the rock dump does not provide enough restraint, then localized buckling may occur (i.e., upheaval buckling), which can cause failure of the pipeline.

Another method is to accommodate the buckling problem by permitting the pipeline to deflect on the seabed, by using a snake lay or buckle mitigation methods such as sleepers or distributed buoyancies. This method is obviously cheaper than rock dumping, and results in the pipeline experiencing lower loads. However, the analysis will probably have to be based on the limit-state design, because the pipe will have plastically deformed. This method is becoming more popular. This method can also be used with intermittent rock dumping; by permitting the pipeline to snake and then to rock dump, the likelihood of upheaval buckling is reduced.

The methods employed in calculating upheaval and lateral buckling, as well as the pullover response, are detailed in Ref. [12].

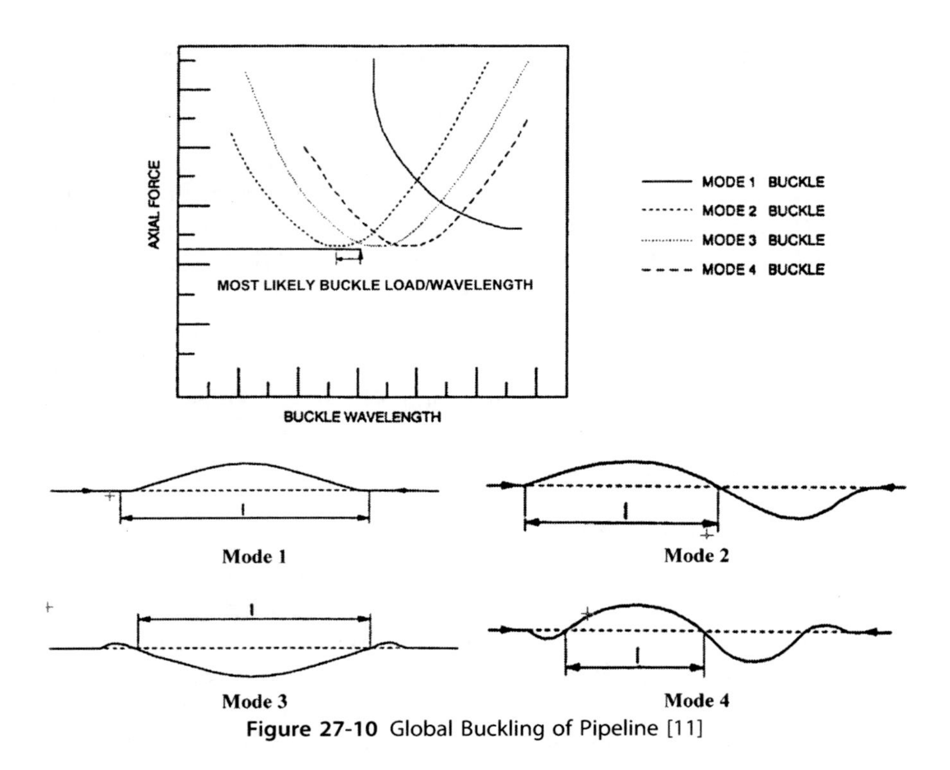

Figure 27-10 Global Buckling of Pipeline [11]

27.5.5. Pipeline Installation

Various methods are used to install subsea pipelines. The method of installation determines the type of analyses that must be performed, as discussed next.

27.5.5.1. Pipe Laying by Lay Vessel

Pipe laying (including the S-lay and J-lay) by a lay vessel involves joining pipe joints on the lay vessel, where welding, inspection, and field joint coating take place at a number of different workstations. Figure 27-11 illustrates an the S-lay method. Pipe laying progresses with the lay vessel moving forward on its anchors. The pipe is placed on the seabed in a controlled S-bend shape. The curvature in the upper section, or the overbend, is controlled by a supporting structure, called a stinger, that is fitted with rollers to minimize damage to the pipe. The curvature in the lower portion is controlled by application of tension on the vessel using special machines.

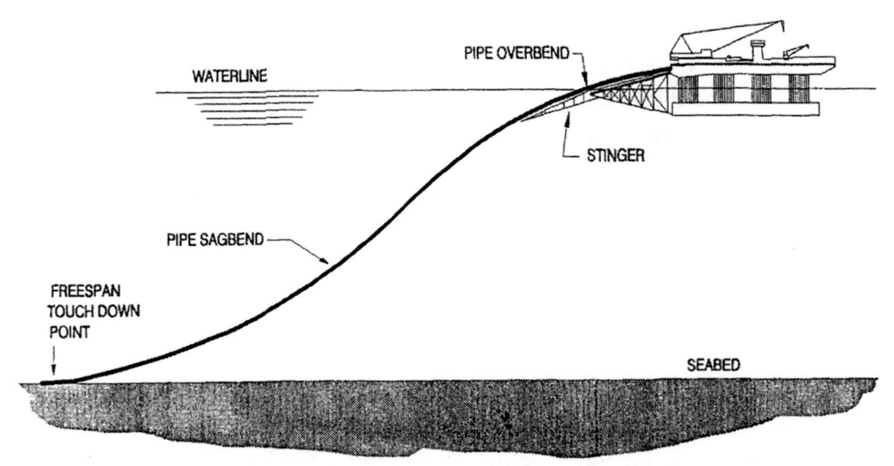

Figure 27-11 Typical Pipe Configuration during S-Lay Pipeline Installation

The pipeline designer must analyze the pipe-lay configuration to establish that the correct tension capacity and barge geometry have been used and that the pipe will not be damaged or overstressed during the laying process.

An appropriate analysis can be performed by a range of methods, from a simple catenary analysis that provides approximate solutions, to a precise analysis that uses finite element analysis. The main objective of the analysis is to identify stress levels in two main areas. The first is on the stinger where the pipe can undergo high bending, especially at the last support. Because the curvature can now be controlled, the pipeline codes generally allow a small safety factor. The second high-stress area is in the sag bend where the pipe is subject to bending under its own weight. The curvature at the sag bend varies with the pipeline's lay tension and, consequently, is less controllable than the overbend area.

In all cases the barge geometry and tension are optimized to produce stress levels in the pipe wall that stay within specified limits.

27.5.5.2. Pipe Laying by Reel Ship

The pipe reeling method has been applied widely in the North Sea and Gulf of Mexico for pipeline sizes up to 18 in. The pipeline is made up onshore and is reeled onto a large drum on a purpose-built vessel. During the reeling process the pipe undergoes plastic deformation on the drum. During installation the pipe is unreeled and straightened using a special straightening ramp. The pipe is then placed on the seabed in a configuration similar to the

J-lay configuration that is used by the laying barge, although in most cases a steeper ramp is used and overbend curvature is eliminated.

The analysis of the reeled pipe-laying method can be carried out using the same techniques as for a laying vessel. Special attention must be given to the compatibility of the reeling process with the pipeline steel grade since the welding process can cause unacceptable work hardening in higher grade steels.

A major consideration in pipeline reeling is that the plastic deformation of the pipe must be kept within limits specified by the relevant codes, such as DNV-RP-F108 [13]. Existing reel ships reflect such code requirements.

27.5.5.3. Pipeline Installation by Tow or Pull

In certain circumstances a pipeline may be installed by a towing technique where long sections of pipeline are made up onshore and towed either on the seabed or off the bottom by means of an appropriate vessel (tug or pull barge). The technique has its advantages for short lines and for bundled lines where several pipelines are collected together in a carrier. In this case difficult fabrication procedures can be carried out onshore. The design procedures for towed or pulled lines are very dependent on the type of installation required. For example, it is important to control the bottom weight of a bottom towed line to minimize towing forces and at the same time give sufficient weight for stability. Thus, a high degree of weight optimization may be needed, which can involve tighter control on pipeline wall thickness tolerances than for pipe lay methods, for example.

Detailed subsea pipeline installation is described in Chapter 5.

27.6. CHALLENGES OF HP/HT PIPELINES IN DEEP WATER

High-pressure, high-temperature (HP/HT) pipelines are increasing in deepwater areas such as West Africa and the Gulf of Mexico. In deep water, severe conditions lead to challenges in installation, flow assurance, thermal buckling, and management, all of which are critical issues for HP/HT pipeline design [14].

27.6.1. Flow Assurance

Flow assurance is one crucial factor for HP/HT pipeline design, which includes (1) steady-state performance of the flowline as governed by the overall heat transfer coefficient or U-value, and (2) transient performance of heatup and cooldown requirements. During the cooldown period after shutdown of the fluid flow, the product temperature must remain above a specified value to avoid hydrate formation. In deep water, pipeline design must seek a balance between the steady-state and transient cooldown requirements to meet the specific field conditions, along with issues of weight, installation needs, and other commercial and economic factors.

For the thermal insulation system, conventional single pipe and pipe-in-pipe (PIP) systems are two options and the final selection depends on a range of factors, such as cost, installation and thermal management. PIP systems may offer the best balance of thermal efficiency and steady-state performance for HP/HT pipelines; however, the increased weight is a factor of considerably greater significance for installation in deep water than in shallow.

The wet insulation materials used with HP/HT pipelines will result in an extremely thick coating for good steady-state performance, which introduces increased cost and seabed stability issues due to the low submerged weight. For the transient cooldown procedure for a flowline, low thermal conductivity and high thermal inertia are very important to determine the "no-touch" time during a shutdown and also the thickness of the insulation layers. The higher density also results in an increased submerged weight, which, in turn, eases seabed stability issues. Flow assurance for HP/HT pipelines is detailed in Part 2 of this book.

27.6.2. Global Buckling

HP/HT pipelines laid on the seabed are susceptible to global lateral buckling, resulting in deflections that can lead to the pipe cross section yielding. This is caused by compressive axial force building up as the pipeline tries to expand thermally but is restrained due to soil resistance between the pipe and the seabed.

The key to an effective solution is the pipe's ability to extend axially. Control of the axial and lateral movements of a pipeline can be obtained by using either a buckle initiation device or expansion spools and it will result in economic relief of the axial force. Various approaches have been tried and applied to manage pipeline global buckling.

Finite element (FE) modeling of the pipeline installation and operation has been widely used to identify the issues related to the management of thermal buckling, which includes pipeline walking under cyclic thermal loading, global buckling interactions, etc. The detailed analysis and modeling of thermal buckling will enable cost-effective mitigation options.

27.6.3. Installation in Deep Water

The installation of HP/HT pipelines is more challenging and costly in deep water than in shallow water. As the reservoir and field go deeper, the weights involved with these pipeline systems may become too heavy for conventional installation methods. PIP systems are more installation-vessel dependent than conventional pipe, and this dependency is even more pronounced with increasing water depth.

Lay rates and reliability are key factors for the pipeline installation in deep water. The high tensions associated with the J-lay method and the high bending in the steep S-lay method may cause high stress in the pipeline. The strain-based design approach has been used in many installation analyses. Pipeline installation is detailed in Chapter 5.

REFERENCES

[1] Y. Bai, Q. Bai, Subsea Pipelines and Risers, Elsevier, Oxford, 2005.
[2] Det Norske Veritas, On-Bottom Stability Design of Submarine Pipelines, DNV-RP-E305, (1998).
[3] Det Norske Veritas, On-Bottom Stability Design of Submarine Pipelines, DNV-RP-F109, (2007).
[4] Kellogg Brown & Root Inc., Submarine Pipeline On-Bottom Stability - vol. I & II, PR-178-01132, (2002).
[5] American Petroleum Institute, Design, Construction, Operation, and Maintenance of Offshore Hydrocarbon Pipelines, (Limit State Design), API-RP-1111, (2009).
[6] American Petroleum Institute, Specification for Line Pipe, API Specification 5L, fourty second ed., (2000).
[7] US Department of the Interior, Minerals Management Service, 30 CFR 250, DOI-MMS Regulations, Washington D.C., (2007).
[8] American Society of Mechanical Engineers, Gas Transmission and Distribution Piping Systems, ASME B31.8, (2010).
[9] American Society of Mechanical Engineers, Pipeline Transportation Systems for Liquid Hydrocarbons and Other Liquids, ASME B31.4, (2002).
[10] Det Norske Veritas, Submarine Pipeline Systems, DNV-OS-F101, (2000).
[11] R.E. Hobbs, In-Service Buckling of Heated Pipelines, Journal of Trans. Engineering, 110 (2) (March, 1984).
[12] Det Norske Veritas, Global Buckling of Submarine Pipelines, DNV-RP-F110, (2007).
[13] Det Norske Veritas, Fracture Control for Pipeline Installation Methods-Introducing Cyclic Plastic Strain, DNV-RP-F108, (2006).
[14] M. Dixon, HP/HT Design Issues in Depth, E & P, Oct. 2005.

INDEX

T

Lightning Source UK Ltd.
Milton Keynes UK
UKOW03n0033200813

215625UK00008B/208/P